Y0-BPT-015

ROTARY-WING AERODYNAMICS

Two Volumes Bound as One

VOLUME I:

BASIC THEORIES OF ROTOR AERODYNAMICS

(With Application to Helicopters)

by W. Z. STEPNIEWSKI

VOLUME II:

PERFORMANCE PREDICTION OF HELICOPTERS

by C. N. KEYS

Edited by W. Z. Stepniewski

Dover Publications, Inc., New York

Published in Canada by General Publishing Company, Ltd., 30 Lesmill Road, Don Mills, Toronto, Ontario.
Published in the United Kingdom by Constable and Company, Ltd.

This Dover edition, first published in 1984, is an unabridged, slightly corrected republication in one volume of the work originally published in two volumes by the Science and Technical Information Office of the National Aeronautics and Space Administration for the U.S. Army Air Mobility Research & Development Laboratory of the Aviation Systems Command. Volume I, "Basic Theories of Rotor Aerodynamics (With Application to Helicopters)," by W. Z. Stepniewski, was originally published in 1979. Volume II, "Performance Prediction of Helicopters," by C. N. Keys, originally published in 1979, is being reprinted here from the 1981 edition revised and edited by W. Z. Stepniewski. The Volume II revision was prepared by Boeing Vertol Company. Indexes have been added to the Dover edition at the end of each volume. Wanda L. Metz was Associate Editor for this work.

Manufactured in the United States of America
Dover Publications, Inc., 31 East 2nd Street, Mineola, N.Y. 11501

Library of Congress Cataloging in Publication Data

Stepniewski, W. Z. (Wieslaw Zenon), 1909-
Rotary-wing aerodynamics.

"An unabridged, slightly corrected republication in one volume of the work originally published in two volumes by the Science and Technical Information Office of the National Aeronautics and Space Administration . . . in 1979. Volume II . . . is being reprinted here from the 1981 edition revised and edited by W. Z. Stepniewski [for the U.S. Army Air Mobility Research and Development Laboratory]"—T.p. verso.
Includes bibliographical references and indexes.
Contents: Basic theories of rotor aerodynamics (with application to helicopters)—Performance prediction of helicopters.
1. Helicopters—Aerodynamics. I. Keys, C. N. II. Title.
TL716.S76 1984 629.132'3 83-20528
ISBN 0-486-64647-5

FOREWORD

In recent years, there has been an increasing volume of reports, articles, papers, and lectures dealing with various aspects of rotary-wing aircraft aerodynamics. To those who enter this domain, either as graduate students with some background in general aerodynamics, or those transferring from other fields of aeronautical or nonaeronautical engineering activities, this vast amount of literature becomes a proverbial haystack of information; often with the result of looking for a needle that isn't there. But even those who are professionally engaged in some aspects of rotary-wing technology may experience a need for a reference text on basic rotor aerodynamics.

Through my experience both as an educator and practicing engineer directly involved in various aspects of industrial aeronautics, it became doubly apparent that there was a need for a textbook that would fulfill, if not all, at least some of the above requirements.

With this goal in mind, the text entitled *Rotary-Wing Aerodynamics* was written under contract from USAAMRDL/NASA Ames. On one hand, the objective is to provide an understanding of the aerodynamic phenomena of the rotor and on the other, to furnish tools for a quantitative evaluation of both rotor performance and the helicopter as a whole.

Although the material deals primarily with the conventional helicopter and its typical regimes of flight, it should also provide a comprehensive insight into other fields of rotary-wing aircraft analysis as well.

In order to achieve this dual aim of understanding and quantitative evaluation, various conceptual models will be developed. The models will reflect physical aspects of the considered phenomena and, at the same time, permit establishment of mathematical treatment. To more strongly emphasize this duality of purpose, the adjective *physicomathematical* will often be used in referring to these models.

It should be realized at this point that similar to other fields of engineering analysis, conceptual models—no matter how complicated in detail—still represent a simplified picture of physical reality. It is obvious, hence, that the degree of sophistication of the physicomathematical models should be geared to the purpose for which they are intended. When faced with the task of developing such a model, one may be advised to first ask the following two questions: (1) whether the introduction of new complexities truly contributes to a better understanding of the physics of the considered phenomena and their qualitative and quantitative evaluation, and (2) whether the possible accuracy of the data inputs is sufficiently high to justify these additional complexities.

In this respect, one should determine whether a more complex model would truly lead to a more accurate analysis of the investigated phenomena or just, perhaps, that the procedure only looks more impressive while mathematical manipulation would consume more time and money. Furthermore, it should be realized that often in the more complex approach, neither intermediate steps nor final results can be easily scrutinized.

With respect to rotary-wing aerodynamics in general, and performance predictions in particular, one should realize that aerodynamic phenomena associated with the various regimes of flight of an even idealized, completely rigid rotor are very complicated. Furthermore, the level of complexity increases due to the fact that in reality, every rotor is non-rigid because of the elasticity of its components and/or built-in articulations. As a result, a continuous interaction exists between aerodynamics and dynamics, thus introducing new potential complexities to the task of predicting aerodynamic characteristics of the rotor. Fortunately, even conceptually simple models often enable one to get either accurate trends or acceptable approximate answers to many rotary-wing performance problems.

By following the development of basic rotor theories—from the simple momentum approach through the combined momentum and blade-element theory, vortex theory, and finally, potential theory—the reader will be able to observe the evolution of the physicomathematical model of the rotor from its simplest form to more complex ones. It will also be shown that an understanding or explanantion of the newly encountered phenomena may require modifications and additions and sometimes, a completely new approach to the representation of the actual rotor by its conceptual model. By the same token, a better feel is developed with respect to the circumstances under which a simple approach may still suffice. Finally, these simpler and more easily scrutinized methods may serve as a means of checking the validity of the results obtained by potentially accurate, but also more complicated ways which may be prone to computational errors.

Presentation of the above-outlined theories, plus considerations of airfoils suitable for rotary-wing aircraft constitutes the contents of Vol I. "Reduction to practice" of the material presented in Vol I is demonstrated in Vol II, where complete performance predictions are carried out for classical, winged, and tandem configurations including such aspects as performance guarantees and aircraft growth.

The existing need for a text conforming to the above outlined philosophy was recognized by representatives of USAAMRDL and in particular, by Mr. Paul Yaggy, then Director of USAAMRDL, and Dr. I. Statler, Director of Ames Directorate whose support made possible the contract for the preparation of the first two volumes.

To perform this task, a team was formed at Boeing Vertol consisting of the undersigned as Editor-in-Chief and author of Vol I; Mr. C.N. Keys as the principal author of Vol II; and Mrs. W. L. Metz as Associate Editor. The course of the work was monitored by Mr. A. Morse and Dr. F.H. Schmitz of Ames Directorate; while Mr. Tex Jones from USAAMRDL-Langley Directorate provided his technical assistance and expertise by reviewing the material contained in both volumes. Thanks are extended to the above-mentioned as well as the other representatives of USAAMRDL. Finally, my associates and I wish to thank the management of Boeing Vertol; especially, Messrs. K. Grina, J. Mallen, W. Walls, and E. Ratz for their support, understanding and patience.

There are, of course many more people from this country and abroad who significantly contributed to the technical contents. Their individual contributions are more specifically acknowledged in the prefaces of the individual volumes.

W. Z. Stepniewski
Ridley Park, Pa.
March 31, 1978

VOLUME I:

BASIC THEORIES OF ROTOR AERODYNAMICS

(With Application to Helicopters)

PREFACE

Volume I of the text entitled *Rotary-Wing Aerodynamics* is devoted in principle to *Basic Theories of Rotor Aerodynamics.* However, the exposition of the material is preceded by an introductory chapter wherein the concept of rotary-wing aircraft in general is defined. This is followed by comparisons of the energy effectiveness of helicopters with that of other static-thrust generators in hover; as well as with various air and ground vehicles in forward translation. While the most important aspects of rotor-blade dynamics and rotor control are only briefly reviewed, they should still provide a sufficient understanding and appreciation of the rotor dynamic phenomena related to aerodynamic considerations.

The reader is introduced to the subject of rotary-wing aerodynamics in Ch II by first examining the very simple physicomathematical model of the rotor offered by the *momentum theory.* Here, it is shown that even this simple conceptual model may prove quite useful in charting basic approaches to helicopter performance predictions; thus providing some guidance to the designer. However, the limitations of the momentum theory; i.e., its inability to account for such phenomena as profile drag and lift characteristics of blade profiles and geometry, necessitated the development of a more sophisticated conceptual rotor model.

The *combined blade-element and momentum theory* presented in Ch III represents a new approach which demonstrates that indeed, greater accuracy in performance predictions is achieved, and this would also become a source of more-detailed guidelines for helicopter design. Even with this improvement, many questions regarding flow fields (both instantaneous and time averaged) around the rotor still remain unanswered.

In the *vortex theory* discussed in Ch IV, a rotor blade is modeled by means of a vortex filament(s) or vorticity surface; thus opening almost unlimited possibilities for studying the time-average and instantaneous flow fields generated by the rotor. Unfortunately, the price of this increased freedom was computational complexity usually requiring the use of high-capacity computers.

It appears that some of the rotor aerodynamic problems amenable to the treatment of the vortex theory may be attacked with a somewhat reduced computational effort by using the approaches offered by the *velocity and acceleration potential theory.* This subject is presented in Ch V which also contains a brief outline of the application of potential methods to the determination of flow fields around three-dimensional, nonrotating bodies.

Considerations of *airfoil sections* suitable for rotors, as presented in Ch VI, completes the sequence on fundamentals of rotary-wing aerodynamics. This material provides a basis for development of the methods for helicopter performance predictions used in Vol II.

In order to create a complete series on Rotary-Wing Aerodynamics the author anticipates a third volume devoted to the application of the basic theories established in Vol I. This volume would include (1) selected problems of helicopter flight mechanics (e.g., ground effect, flight maneuvers, performance limitations, and autorotation); (2) establishment of a link between aerodynamics and design optimization; and (3) development of techniques leading to performance maximization of existing helicopters. In fairness to the aeronautical engineers and designers who have been anxiously awaiting for the publication of this series, the first two volumes are being released prior to the writing of the proposed third volume.

Returning to the present volume, the reader's attention is called to the fact that both SI metric and English unit systems are used in parallel; thus expediting an acquaintance with the metric approach for those who are not yet completely familiar with this subject.

In conclusion, I wish to express my indebtedness to the following persons who generously contributed to this volume: Professor A. Azuma of the University of Tokyo, Japan for his review of the appendix to Ch IV; and to Drs. R. Dat and J.J. Costes of ONERA, France for their valuable inputs and review of Ch V.

W. Z. Stepniewski

NOTES ON METRIC SYSTEM

In order to assist the reader in making the transition from English units to equivalent SI metric units of measure, some important aspects of the SI system encountered in applied subsonic aerodynamics and rotary-wing mechanics of flight are listed below and briefly reviewed.

BASIC METRIC SYSTEM (SI) UNITS

QUANTITY	UNIT	SYMBOL	ENGLISH EQUIVALENT
mass	kilogram	*kg*	0.0685 slug
length	meter	*m*	3.281 ft
time	second	*s*	1.0 s
temperature	Kelvin	*K*	1.8° Rankine

DERIVED UNITS

QUANTITY	DERIVATION	UNIT	SYMBOL	ENG. EQ.
force	kg m/s^2	newton	*N*	0.2248 lb
force	9.807*N*	kilogram force	*kG*	2.2046 lb
pressure	N/m^2	newton/m^2	N/m^2	0.0209 psf
pressure	kG/m^2	kilogram force/m^2	kG/m^2	4.8825 psf
density	kg/m^3	kilogram/meter^3	kg/m^3	0.00194 slug/ft^3
velocity	m/s	meter/second	m/s	3.281 fps
acceleration	m/s^2	meter/second^2	m/s^2	3.281 fps^2
acceleration of gravity	9.807 m/s^2	g	g	32.2 ft/s^2
work; energy	Nm	joule	*Nm*	0.7376 ft-lb
work; energy	kG-m	kilogram meter	*kG-m*	7.233 ft-lb
power	N m/s	watt	*W*	0.7376 ft-lb/s
power	kG-m/s	kilogram meter/s	*kG-m/s*	7.233 ft-lb/s
power	75kG-m/s	metric horsepower	*hp*	0.9863 hp*
viscosity	kg/ms	μ (coefficient)	*kg/ms*	0.0288 slug/ft-s
kinematic viscosity	μ/ρ	ν (stoke)	m^2/s	10.764 ft^2/s

English horsepower ≡ 550 ft-lb/s.

The reader's attention is called to the fact that as long as metric units are input into relationships designated as SI, and no special conversion factors are incorporated into the formulae, the obtained forces will be in newtons, pressures in newtons per meter squared, work or energy in newton meters, and power in newton meters per second. It should be noted however, that in many countries, the kilogram of force (symbolized in this text as *kG*) is widely used for the determination of weight (including that of aircraft), while such quantities as disc and/or wing loadings are measured in kG/m^2.

The popularity of the kilogram as a unit of force stems from the fact that prior to the establishment of the newton as a unit of force, the kilogram was generally accepted in engineering practice as well as in everday life. It was defined as a force resulting from the acceleration of earth's gravity acting on one kilogram of mass. However, the earth's g value varies with altitude over sea level and geographic latitude. Consequently, the definition of the kilogram of force as the weight of a kilogram of mass on the earth surface required additional specifications of earth coordinates as to where the weight is measured. The newton ($kg\ m/s^2$) as a unit of force is more universal in the context of its not being directly related to the gravity conditions encountered on this particular planet.

It should also be mentioned that in addition to the direct use of the kilogram of force in engineering practice, its indirect influence can be consistently found when dealing with the metric horsepower defined as $hp = 75\ kG\ m/s$. It should also be noted that the so-defined metric horsepower amounts to 0.9863 of its English counterpart defined as $hp = 550\ ft\text{-}lb/s$.

One final note – The most important characteristics of air according to the International Standard Atmosphere are given in the following table.

THE INTERNATIONAL STANDARD ATMOSPHERE

h $(m \times 10^{-3})$	T (K)	$\theta \equiv T/T_o$	s (m/s)	M (s/s_o)	p $(N/m^2 \times 10^{-4})$	$\delta \equiv p/p_o$	ρ (kg/m^3)	$\sigma_\rho = \rho/\rho_o$	$\sqrt{\sigma_\rho}$	μ $(kg/ms \times 10^5)$	ν $(m^2/s \times 10^5)$
0	288·2	1.000	340·3	1·000	10·132	1·000	1·225	1·000	1·000	1·789	1·461
1·0	281·7	0·977	336·4	0·989	8·989	0·887	1·112	0·908	0·953	1·758	1·581
2·0	275·2	0·955	332·5	0·977	7·950	0·785	1·007	0·822	0·907	1·726	1·715
3·0	268·7	0·932	328·6	0·966	7·012	0·692	0·909	0·742	0·862	1·694	1·863
4·0	262·2	0·910	324·6	0·954	6·166	0·609	0·819	0·669	0·818	1·661	2·028
5·0	255·7	0·887	320·5	0·942	5·405	0·533	0·736	0·601	0·775	1·628	2·211
6·0	249·2	0·865	316·5	0·930	4·722	0·466	0·660	0·539	0·734	1·595	2·416
7·0	242·7	0·842	312·3	0·918	4·111	0·406	0·590	0·482	0·694	1·561	2·646
8·0	236·2	0·820	308·1	0·905	3·565	0·352	0·526	0·429	0·655	1·527	2·904
9·0	229·7	0·797	303·8	0·893	3·080	0·304	0·467	0·381	0·618	1·493	3·196
10·0	223·2	0·775	299·5	0·880	2·650	0·262	0·414	0·338	0·581	1·458	3·525
11·0	216·7	0·752	295·1	0·867	2·270	0·224	0·365	0·298	0·546	1·422	3·898
12·0	216·7	0·752	295·1	0·867	1·940	0·192	0·312	0·255	0·505	1·422	4·558
13·0	216·7	0·752	295·1	0·867	1·658	0·164	0·267	0·218	0·467	1·422	5·333
14·0	216·7	0·752	295·1	0·867	1·417	0·140	0·228	0·186	0·431	1·422	6·239
15·0	216·7	0·752	295·1	0·867	1·211	0·120	0·195	0·159	0·399	1·422	7·300
16·0	216·7	0·752	295·1	0·867	1·035	0·102	0·167	0·136	0·369	1·422	8·540
17·0	216·7	0·752	295·1	0·867	0·885	0·087	0·142	0·116	0·341	1·422	9·990
18·0	216·7	0·752	295·1	0·867	0·757	0·075	0·122	0·099	0·315	1·422	11·686
19·0	216·7	0·752	295·1	0·867	0·647	0·064	0·104	0·085	0·291	1·422	13·670
20·0	216·7	0·752	295·1	0·867	0·553	0·055	0·089	0·073	0·269	1·422	15·989

TABLE OF CONTENTS

INTRODUCTION

This chapter introduces the reader to the concept of rotary-wing aircraft in general. The energy consumption of helicopters and tilt-rotors is compared with that of other aircraft and automobiles. A brief discussion of dynamic problems of rotors indicates the necessity for freedom of the flapping and lagging motion of rotor blades through discrete hinges or blade flexibility, or for some alternate means of aerodynamic control of blade lift around the azimuth. Then, stability of the flapping motion is examined—leading to an explanation of the harmonic presentation of blade motion and rotor control through the first harmonic inputs. Finally, the most common representatives of practical rotary-wing configurations are briefly described.

Principal notation for Chapter I

A	area	m^2 or ft^2
A	amplitude	m or ft
A_n $(n = 0,1,2,...)$	coefficient in Fourier expansion of feathering	rad or deg
a	acceleration	m/s^2, fps^2, or g's
a_n $(n = 0,1,2,...)$	coefficient in Fourier expansion of flapping	rad or deg
B_n $(n = 1,2,...)$	coefficient in Fourier expansion of feathering	rad or deg
b_n $(n = 1,2,...)$	coefficient in Fourier expansion of flapping	rad or deg
CF	centrifugal force	N or lb
c	blade chord	m or ft
D_s	specific distance	km or n.mi
d	diameter	m, or ft
g	acceleration of gravity	$9.80m/s^2$ or $32.2fps^2$
I	moment of inertia	$kg\text{-}m^2$ or $slug\text{-}ft^2$
I_s	specific impulse	s
i	angle of incidence, or imaginary unit: $\sqrt{-1}$	rad or deg
k	spring constant	N–m/rad or lb-ft/rad
L	lift	N or lb
M	moment	N–m or ft-lb
m	mass	kg or slugs
R	rotor radius	m or ft
R_s	specific range	km/kG or n.mi/lb
r	radial distance	m or ft
T	thrust	N or lb
t	time	s or hr
$tsfc$	thrust specific fuel consumption	(kG/s)/kG or (lb/s)/lb
U	velocity of flow approaching the blade	m/s or fps
V	velocity of flow in general	m/s or fps
v	downwash velocity	m/s or fps

Theory

W	weight or gross weight	N, kG, or lb
w	force/unit area (incl blade & disc loading)	N/m^2, kG/m^2, or lb/ft^2
a	rotor disc angle-of-attack	rad or deg
β	blade flapping angle	rad or deg
β_o	coning angle	rad or deg
$\eta \equiv r_{fh}/R$	relative flapping hinge offset	
θ	blade pitch angle	rad or deg
κ	damping coefficient	N- m/s or lb-ft/s
λ	root of a characteristic equation	
$\mu \equiv V_{\|}/R\Omega$	rotor advance ratio	
$\nu = 1/\tau$	frequency	hertz, Hz
ρ	air density	kg/m^3 or $slugs/ft^3$
τ	period of oscillations	s
ψ	blade azimuth angle	rad or deg
Ω	rotational velocity	rad/s
$\omega \equiv 2\pi\nu$	natural frequency of harmonic motion	rad/s

Subscripts

b	blade
c	considered, or Coriolis force
CF	centrifugal force
$crit$	critical
d	disc
$damp$	damping
F	fuel
f	flapping
fd	forward
h	hinge
L	lift
o	representative, or initial
p	phase
r	reverse
res	restoring
s	per second, specific, or shaft
T	thrust
t	tip
w	weight
ψ	azimuth
$\|$	parallel
$\perp$	perpendicular

Superscript

$-$	quantities referred to R or ρV_t^2

1. DEFINITION OF ROTARY-WING AIRCRAFT

1.1 General

From a strictly aerodynamic point-of-view, rotary-wing aircraft may be defined as configurations which, at least during takeoff and landing maneuvers, derive their lifting force directly from an open airscrew, or airscrews. These maneuvers may be performed either vertically or with ground run.

The lifting airscrew of vertical takeoff and landing (VTOL) aircraft must be directly powered. Some rotary-wing aircraft taking off and landing with a ground run—such as helicopters and tilt-rotors operating at gross weights in excess of their hovering ability—may also belong to the directly-powered group. However, there are other rotary-wing configurations; for example, the autogiro where energy to the lifting airscrew is derived indirectly through the motion of the vehicle as a whole with respect to the air mass.

In principle, all VTOL configurations, ranging from helicopters to rockets, possess the ability to hover, but until now, VTOL aircraft extensively using hovering capabilities in actual operations have been almost exclusively represented by helicopters because of their (1) low energy consumption per unit of generated static thrust, and (2) relatively low downwash in the fully developed slipstream in hover.

The first of these characteristics enables the aircraft to operate for extended periods of time in hovering and near-hovering conditions. The second contributes to the reduction of ground erosion, and also permits activities of ground personnel within the downwash covered areas.

The low energy consumption and relatively low downwash associated with static thrust result from a low loading of the lift generators (thrust-per-unit-area of the lift generating surface). It is interesting to note that there has always been a strong intuitive association of rotary-wing aircraft with low disc loading which is reflected in the commonly accepted name of *rotor* given to their lifting airscrews. In contrast, the word *propeller* is considered more applicable for the higher loaded lifting and propelling airscrews used in tilt-wing configurations.

One might argue that the airscrews of the tilt-rotor should also be called propellers since, in a completely converted forward flight, they truly propel the aircraft, while the lift is provided solely by the fixed wing. Nevertheless, the tilt-rotor airscrews are still classified as rotors. It appears hence that low disc loading remains the main characteristic separating the so-called rotary-wing aircraft from other possible configurations depending on open airscrews for vertical lift generation during takeoffs and landings.

1.2 Disc Loading

In order to provide a more quantitative definition of rotary-wing rotor loading limits, past and current trends on the level of disc loading are examined. A strict definition of this parameter would be $w = T/A$, where T is the thrust-per-rotor and A is the rotor disc area. However, in the so-called helicopter type steady-flight condition, all lift is provided by the rotor(s). Thus the thrust is usually approximately equal to the gross weight of the aircraft: $T \approx W$ and, unless stated otherwise, the disc loading is defined as $w \equiv W/A$ where A is now the total disc area of all lifting rotors of the aircraft.

Although historically there is a continuous trend to increase the disc loading of helicopters, it can be seen from Fig 1.1 that its current value appears to level off at $w \approx 50\ kG/m^2$ ($w \approx 10\ psf$) for medium and heavy gross-weight machines, while the value of the lighter aircraft appears to be much lower.

There are only a few inputs from tilt-rotor aircraft actually flown or being developed, but they seem to indicate $w \approx 70\ kG/m^2$ ($w \approx 14\ psf$) as the upper limit of the disc loading. However, this trend reflects only relatively small aircraft, while for larger machines, as in the case of helicopters, w may increase with gross weight. In view of this and from additional design studies of large aircraft, it appears that $w \approx 100\ kG/m^2$ ($w \approx 20\ psf$) can be assumed as the upper limit for the tilt-rotor concept.

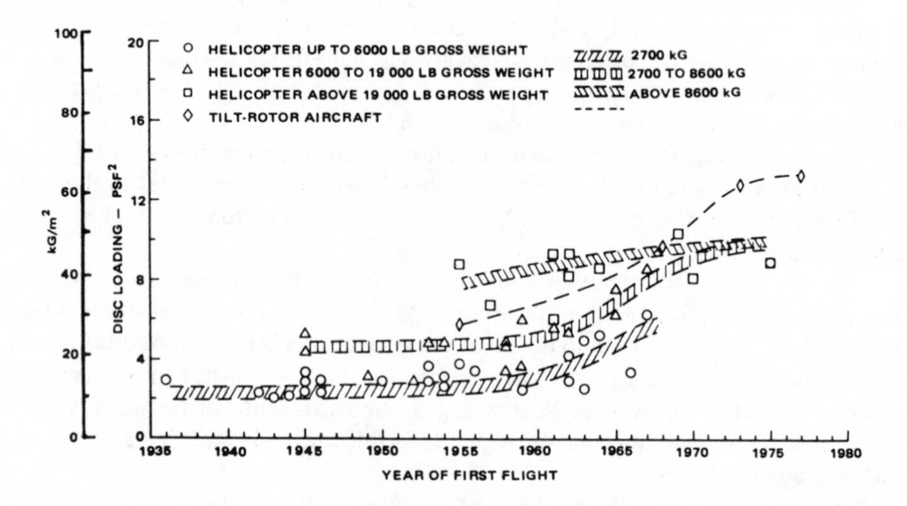

Figure 1.1 Trends in disc loading of rotary-wing aircraft

2. ENERGY CONSUMPTION OF ROTARY-WING AIRCRAFT

2.1 Hover

Of all static-thrust generators, whether air dependent or air independent, such as rockets, the rotors of rotary-wing aircraft during hover operate at the lowest loading of the thrust-generating area. Consequently, of the whole family of actual and potential VTOL aircraft, rotary wings represent the lowest level of energy consumption required for static thrust.

Specific impulse (I_s) is used to provide a comprehensive comparative scale for this energy consumption:

$$I_s = T/\dot{W}_F \tag{1.1}$$

where T is the thrust in kilograms (lb), and $\dot{W}_F$ is the rate of fuel consumption (kG/s, or lb/s). Specific impulse hence can be interpreted as the hypothetical time in seconds that a given thrust generator could operate by consuming the amount of fuel having a weight equal to the generated thrust. For VTOL configurations, this thrust can be assumed as equal to the gross weight for which $\dot{W}_F$ is determined. Denoting the thrust specific fuel consumption per unit of force in one second as $(tsfc)_s$, the expression for specific impulse can be rewritten as follows:

$$I_s = 1/(tsfc)_s. \tag{1.1a}$$

In Fig 1.2, specific impulse for air-dependent generators is shown for the static condition while the I_s of rockets, of course, is independent of the state of motion of the vehicle. The various concepts extend along the abscissa axis in the order of their increasing thrust-generating area loading. It should be noted that this ranking is synonymous with the increasing fully-developed downwash velocity (v_∞).

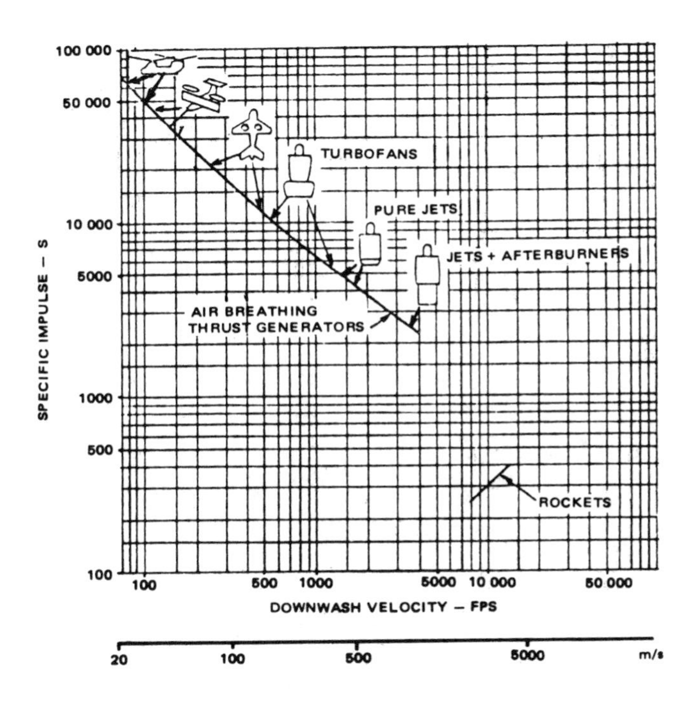

Figure 1.2 Specific impulse of various thrust generators

This figure also shows that rotary-wing aircraft with a static specific impulse of over 70 000 seconds vs a few hundred seconds for chemical rockets represent the concepts most suitable for operations where long times in hover and near-hovering conditions are required. The additional operational advantage of a low fully-developed downwash velocity v_∞ is also quite apparent; e.g., $v_\infty < 30\ m/s\ (< 100\ fps)$ for helicopters, versus $v_\infty > 2400\ m/s\ (> 8000\ fps)$ for chemical rockets.

2.2 Cruise

In order to provide a yardstick for a quantitative comparison of various modes of transportation regarding energy consumption in horizontal translation, a concept similar to that of the specific impulse is proposed. It will be called the specific distance (D_s) representing a hypothetical distance in km (n.mi) that a vehicle could travel if the weight of fuel consumed were equal to the gross weight (W). When the so-called specific range ($R_s \equiv$ *distance traveled using one unit of the fuel weight; say, km/kG)* is known, D_s can be expressed as follows:

$$D_s = R_s W. \tag{1.2}$$

It is obvious that the specific range and hence, the specific distance, depends on the speed of motion. For bouyant water vessels, as well as airships and wheel-supported ground vehicles, the R_s and D_s increase as motion speed decreases; while for aircraft (both rotary and fixed-wing), there are combinations of flight speed and altitude which maximize R_s and D_s.

Specific distances for helicopters, tilt-rotors in the airplane mode of flight, automobiles, and dirigibles are shown in Fig 1.3 as a function of speed while for other fixed-wing aircraft, D_s values are indicated at their optimum cruise speed and flight-altitude combinations.

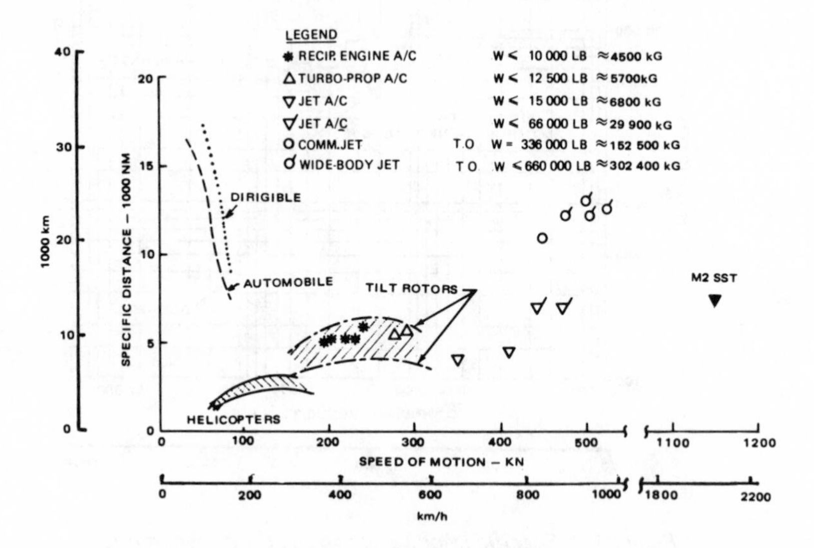

Figure 1.3 Specific distance of various vehicles

From Fig 1.3 and other studies[1], it can be seen that in contrast to hovering, the helicopter in cruise shows much higher energy consumption levels per unit of gross weight and unit of distance traveled than other means of air and ground transportation. However, this does not preclude the possibility that under actual operating conditions (due to less wasted time in terminal operations or more direct routes), even the helicopters now in operation (1960 technology) may become competititive with other vehicles as far as energy per passenger kilometer is concerned[2]. Furthermore, large strides toward improvement in helicopter energy consumption in cruise appear possible[2,3].

Another representative of rotary-wing aircraft; namely, the tilt-rotor, is in a much better position as far as cruise energy is concerned.

3. FUNDAMENTAL DYNAMIC PROBLEMS OF THE ROTOR

3.1 Asymmetry of Flow

Most of the dynamic and many of the aerodynamic problems of rotary-wing aircraft stem from the fact that an inplane velocity component ($V\|$) appears in all non-axial translatory motions of a rotor. Let us assume that the velocity vector (Fig 1.4) representing the distant incoming flow ($-\vec{V}$ ≡ *velocity of flight with an opposite sign*) forms an angle $-a_d$ with the rotor plane (positive when the incoming flow has a component in the thrust direction). Then the component—either inplane or parallel to the disc—can be expressed as

$$V\| = V \cos a_d. \tag{1.3}$$

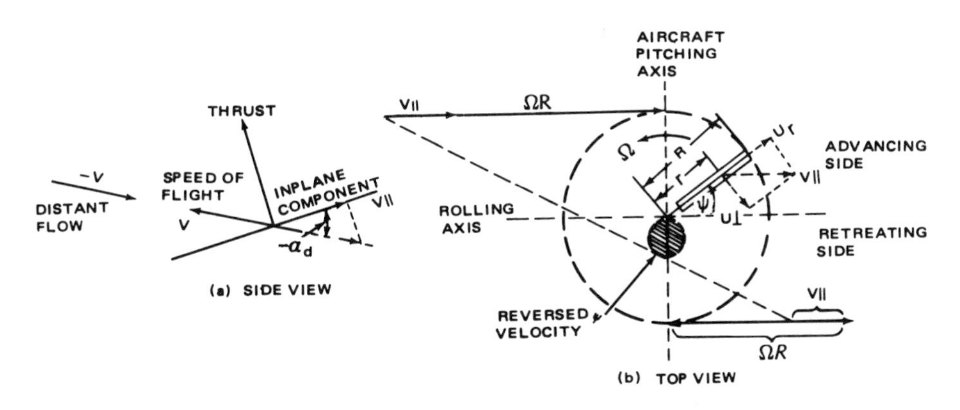

Figure 1.4 Inplane velocities of a rotor in nonaxial translation

In hovering and axial translation, every blade element experiences the velocity of the incoming flow which is solely a function of the radial location given by radius r of that element. However, the presence of the inplane component $V\|$ destroys the axial symmetry and now the air velocity encountered by the blade element is not only a function of r, but also of the blade azimuth position (ψ) measured in the direction of rotor rotation from the downwind blade position (Fig 1.4b).

In order to have a quantitative measure of the amount of asymmetry caused by the inplane component, the concept of advance ratio (μ) was introduced and defined as

$$\mu = V\|/R\Omega \tag{1.4}$$

where $R\Omega \equiv V_t$ is the tip speed of the rotor.

3.2 Asymmetry of Blade Loads

Relying on an analogy with the fixed wings, one may anticipate that from the point-of-view of lift generation by a blade element located at station $\bar{r} \equiv r/R$, the most important air velocity component of the inplane velocity $V\|$ would be that which is perpendicular to the blade axis $(U\perp_{\bar{r}})$:

$$U\perp_{\bar{r}} = V\| \sin\psi + V_t\bar{r}$$

or (1.5)

$$U\perp_{\bar{r}} = V_t(\mu \sin\psi + \bar{r}).$$

It is clear from Eq (1.5) that every element of the blade experiences a sinusoidal perturbation of its air velocity component perpendicular to the axis. As a result of this, the $U\perp_{\bar{r}}$ values on the advancing side will be higher than on the retreating side. Furthermore, the level of these differences would depend on the magnitude of the advance ratio μ. It should also be noted that on the retreating side there is a circle of diameter $d_r = \mu R$ where the blade encounters air flow from the trailing edge. This region is called the *reversed velocity area.*

The above somewhat cursory analysis should indicate that with $\mu > 0$, the advancing side of the rotor may produce a higher lift than the retreating one. For a blade rigidly attached to the hub, this would produce a rolling moment which, in principle, can be neutralized for the aircraft as a whole by pairing equally sized rotors rotating in opposite directions. Such pairing can be visualized through coaxial, side-by-side, tandems, or some other configuration.

However, as long as an unbalanced rolling moment exists within the rotor itself, a corresponding bending moment (usually very high) at the blade root and rotor shaft would also be present.

Proper aerodynamic countermeasures should be applied in order to diminish or completely eliminate the moment unbalance resulting from the presence of the rotor inplane component of the distant flow. In this respect, two solutions come to mind: (1) variation of the blade angle-of-attack in such a manner that the influence of the $\mu \sin\psi$ term in Eq (1.5) is nullified; and (2) application of other aerodynamic means of lift management such as circulation control through blowing and/or suction, flaps, and spoilers. All of these potential means of lift control should be activated in such a way that the effect sought in (1) is achieved.

The $\mu \sin\psi$ term is of the first-harmonic type with respect to the rotor rotation about its axis; hence, the countermeasures aimed at the elimination (or decrease) of the influence of this term should also be of the first-harmonic character.

3.3 Flapping Hinge

A practical solution to problems stemming from the asymmetry of flow was achieved by de la Cierva through the introduction of the flapping hinge in his autogiro in 1923.

This was done after an accident resulting from an uncontrolled roll in an earlier model having "rigid" blades.

Incorporation of the flapping hinge eliminated any possibility of transferring the blade bending moment to the hub. At the same time, it gave the blade the freedom to flap about the hinge. It will be shown later that this flapping motion has an aerodynamic effect equivalent to a reduction of the blade angle-of-attack on the advancing side, and an increase on the retreating one. It is understandable that the additional degree of freedom given to the blade led to a strong interaction between dynamic and aerodynamic effects.

More recently, schemes of controlling the rolling moment through circulation control have been proposed by Yuan[4], Cheesemen[5], Dorand[6], and Williams[7]. However, our attention will be focused on the de la Cierva *blade articulation approach*, as it represents the most widely used scheme of dealing with the azimuthal blade lift perturbations resulting from the inplane velocity. It should be emphasized that the introduction of blade articulation was probably the most important contribution to the development of practical rotary-wing aircraft.

For the sake of simplicity, it is assumed that the flapping hinge is located on the rotor axis, and that it is perpendicular to that axis (Fig 1.5).

Neglecting the possible effects of rotor generated vortices in the proximity of the blades, it may be stated that in hovering or axial translation, velocities experienced by a rotor blade and hence, the aerodynamic loadings, remain constant with the azimuth angle. In addition to the aerodynamic lift per blade (L_b) which is approximately equal to thrust per blade (T_b), there are two other forces acting in the plane passing through the rotor and blade axes: the centrifugal force, CF; and the weight of the blade, W_b. Because of the freedom of motion around the flapping hinge, the blade may deflect from the rotor plane (plane perpendicular to the shaft and passing through the hub center) and start to move along a conical surface. The angle that the generatrice–in this case, the blade axis–forms with the rotor plane is called the coning angle (β_o; up positive), and its value can be determined from the moment equilibrium conditions around the flapping hinge (Fig 1.5):

$$L_b r_L - CF \sin\beta_o r_{CF} - W_b r_W \cos\beta_o = 0 \tag{1.6}$$

or, under the small-angle assumption,

$$L_b r_L - CF\beta_o r_{CF} - W_b r_W = 0. \tag{1.6a}$$

Once the lift and mass distribution along the blade and the rotor tip speed (V_t) are known, it becomes easy to find the corresponding coning angle. The following simplifying assumptions are made to illustrate this process: (a) the mass per unit of blade span ($\bar{m}_b$) is constant; ie, $\bar{m}_b(r) = const$ and hence, $r_{CF} = (2/3)R$; (b) the weight of the blade is small in comparison with the aerodynamic lift (thrust) per blade, $W_b \ll L_b$; and (c) the lift distribution along the blade is parabolic, $r_L = (3/4)R$. Under these assumptions, Eq (16a) would yield the following solution:

$$\beta_o = (9/8)(L_b/CF) \equiv (9/4)(RL_b/m_b V_t^2) \tag{1.7}$$

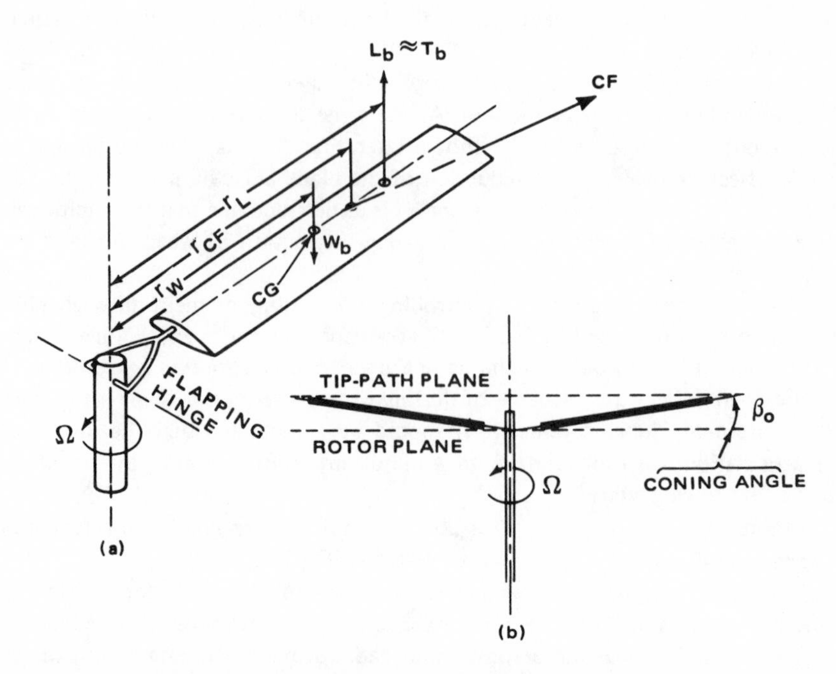

Figure 1.5 Articulated blade with flapping hinge only

where $m_b = R\bar{m}_b$ is the blade mass.

The physical and design significance of the above equations is quite obvious: even if the L_b/m_b ratio and tip speed remain the same for both large and small radii blades, the coning angle value would be higher for the blade of a larger radius. Large coning angles ($\beta_o > 9°$) are not desirable because of aerodynamic interference in forward flight. They may also introduce control errors. For this reason, relative shifting of the blade mass center toward the tip is required for large diameter rotors to remedy the tendency of β_o to grow with radius (R).

4. BLADE FLAPPING MOTION

Introduction of the freedom-to-flap leads to an important question—Will the motion of the blade about the flapping hinge be stable? Both static and dynamic stability should be considered in answering this question.

4.1 Static Stability

In order to examine the static stability of the blade, we will assume that the blade is displaced up from its position of equilibrium at an angle β_o through an angle $d\beta$ (Fig 1.6). This displacement will affect only the magnitude of the component of centrifugal force perpendicular to the blade. Under the small-angle assumption, a new value of the moment about the flapping hinge, due to the CF component perpendicular to the blade ($CF\perp$) will generally be

$$M_{CF\perp} + dM_{CF\perp} = -(CF)r_{CF}(\beta_o + d\beta) \tag{1.8}$$

but

$$-(CF)r_{CF}\beta_o \equiv M_{CF\perp}$$

and consequently,

$$dM_{CF\perp} = -(CF)r_{CF}d\beta$$

or (1.9)

$$dM_{CF\perp}/d\beta = -(CF)r_{CF}.$$

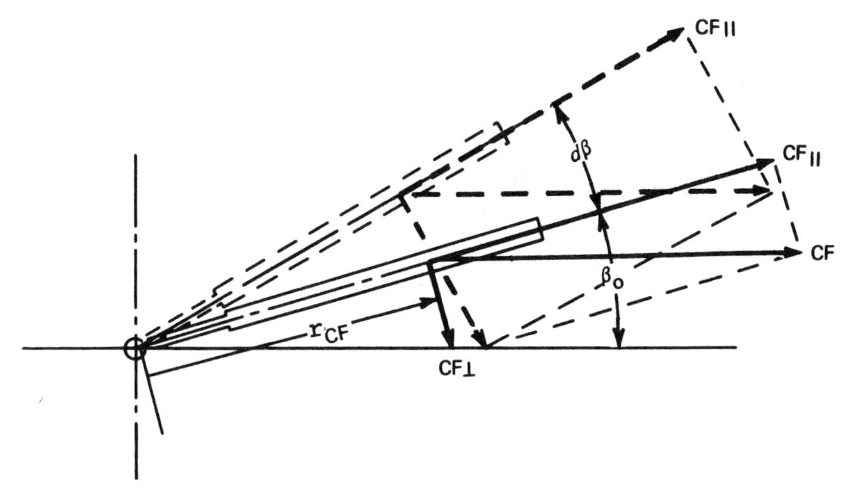

Figure 1.6 Displacement of a blade from its position of equilibrium

From Eq (1.9) it can be seen that the blade is statically stable since $dM_{CF\perp}/d\beta$ is negative. This means that when the blade is displaced from its position of equilibrium, a restoring moment is generated—tending to return the blade to its original location.

4.2 Dynamic Stability

Knowing that the blade possesses static stability, the remaining question is whether the blade flapping motion is dynamically stable; i.e., whether the displaced blade will converge in time to its original position of equilibrium.

Prior to answering this question, some new geometric and aerodynamic aspects must be considered. The geometry of a blade element can usually be described by the following three characteristics: (1) relative blade station, $\bar{r} \equiv r/R$, giving the position of the considered element along the blade span, (2) shape of the airfoil and position of the zero-lift chordline, and (3) orientation of the airfoil with respect to the rotor plane as determined by the local pitch angle $\theta_{\bar{r}}$ which represents the angle between the zero-lift blade chord and the rotor plane (Fig 1.7b).

Depending on the distribution of $\theta_{\bar{r}}$ along the radius, the blade may be either flat, $\theta_{\bar{r}}(\bar{r}) = const$; or it may be twisted, $\theta_{\bar{r}}(\bar{r}) \neq const.$

It should be noted that the pitch angles of all stations can be varied simultaneously by rotating the blade as a whole about the so-called *pitch axis* through an increment $\Delta\theta_o$.

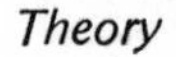

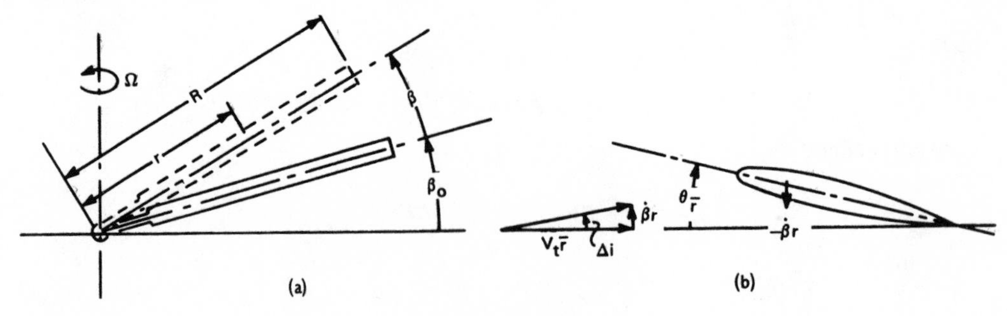

Figure 1.7 Notations

This action is called *feathering.*

For the sake of simplicity, we are considering hovering conditions; therefore, the only "distant" flow experienced by the blade element is that due to the rotor angular velocity (Ω) about its axis. The velocity vector of this flow is either in, or parallel to, the rotor plane and for an element located at $\bar{r}$, would amount to $U_{\bar{r}} = -r\Omega = -\bar{r}V_t$. Under steady-state conditions, the angle of incidence ($i_{\bar{r}}$) that a blade element makes with $U_{\bar{r}}$ is equal to its geometric pitch angle $i_{\bar{r}} = \theta_{\bar{r}}$.

Returning to the subject of dynamic stability, we will assume—as in the case of static—that the blade has been displaced up from its position of equilibrium at the coning angle β_o to a new position at an angle β (Fig 1.7a).

When released from its forced position at β, the statically stable blade will begin to move down (toward β_o) at an angular velocity; $-(d\beta/dt) \equiv -\dot{\beta}$.

It should also be noted (Fig 1.7b) that with the appearance of $-\dot{\beta}$, the angles of incidence at all blade stations will increase. At any relative station $\bar{r} \equiv r/R$, the increment $\Delta i_{\bar{r}}$, under small-angle assumptions, will be:

$$\Delta i_{\bar{r}} = -\frac{R\bar{r}(d\beta/dt)}{V_t\bar{r}} = -\frac{d\beta/dt}{\Omega} = -\dot{\beta}/\Omega. \tag{1.10}$$

Since Δi is independent of $\bar{r}$ in Eq (1.10), all blade stations will experience the same change in their angle-of-incidence as they would by feathering through an angle $\Delta\theta$. In other words, from an aerodynamic point-of-view, it has been shown that flapping is equivalent to feathering. ($\Delta\theta_o$ *due to blade rotation about the pitching axis* = Δi *due to flapping.*) This means that Δi as given by Eq (1.10) would cause exactly the same change in aerodynamic forces and moments acting on the blade as blade pitching (feathering) through an angle $\Delta\theta_o$.

As to dynamic stability, it is important to recognize that down-flapping increases the blade incidence, since Δi will be positive just as $d\beta/dt$ is negative. Consequently, as long as the blade is operating below its stall limits, down-flapping would result in an increase in the lift (thrust) per blade. Furthermore, these forces and the flapping velocity are of opposite signs; hence, the resulting additional aerodynamic moment about the flapping hinges will oppose the blade flapping motion; i.e., damping will be provided.

Knowing the slope $\partial T_b/\partial\theta_o|\theta_{oc}$ (in N per radian) in the vicinity of the considered pitch angle (θ_{o_c}), assuming that $\partial T_b/\partial\theta_o|\theta_{oc} = const$, and taking advantage of Eq (1.10), the aerodynamic damping moment encountered by a flapping blade can be expressed as follows:

$$M_{damp} = r_R(\partial T_b/\partial\theta_o)|_{\theta_{o_c}}\dot{\beta}/\Omega \tag{1.11}$$

or calling $r_R(\partial T_b/\partial\theta_o)|_{\theta_{o_c}}(1/\Omega) \equiv \kappa$ the damping coefficient, Eq (1.11) becomes

$$M_{damp} = -\kappa\dot{\beta}.$$

Similar to the case of static stability (Eq (1.8)), the restoring moment (tending to bring the blade to its original position) can be expressed as follows:

$$M_{res} = -(CF)r_{CF}\beta \tag{1.12}$$

or recognizing that $(CF)r_{CF} \equiv k$ represents the "spring" constant, Eq (1.12) becomes

$$M_{res} = -k\beta. \tag{1.12a}$$

Now, the equation of motion of the flapping blade can be written as

$$I_f\ddot{\beta} + \kappa\dot{\beta} + k = 0 \tag{1.13}$$

where I_f is the blade moment of inertia about the flapping hinge.

Eq (1.13) has the same form as the well-known equation of linear motion of a mass point with elastic restraint and damping. Its general solution (p. 130 of Ref 8) is:

$$\beta = Ae^{\lambda_1 t} + Be^{\lambda_2 t} \tag{1.14}$$

where λ_1 and λ_2 are the roots of the characteristic equation

$$I_f\lambda^2 + \kappa\lambda + k = 0. \tag{1.15}$$

These roots are:

$$\lambda_1 = -(\kappa/2I_f) + \sqrt{(\kappa/2I_f)^2 - (k/I_f)}$$

and (1.16)

$$\lambda_2 = -(\kappa/2I_f) - \sqrt{(\kappa/2I_f)^2 - (k/I_f)}.$$

The sign of the under-the-root expression govern the character of motion which, in turn, depends on the sign of the following quantity:

$$\kappa^2 - 4kI_f$$

or, rewriting the above in explicit form, it becomes

$$r_R[(\partial T_b/\partial\theta_o)|_{\theta_o}(1/\Omega)]^2 - 4(CF)r_{CF}I_f. \tag{1.17}$$

In practical designs, the first term in this expression is usually much smaller than the second one: $\kappa^2 \approx 0.032(4kI_f)$.

This means that in general, both square roots in Eq (1.16) are imaginary and thus, blade oscillations about the flapping hinge will be periodic. The solution of Eq (1.13) containing real quantities only is written in order to find the period of oscillations (τ in seconds):

$$\beta = Ce^{-(\kappa/2I_f)t} \cos\left[\sqrt{(k/I_f) - (\kappa^2/4I_f^2)}\,t\right] + De^{-(\kappa/2I_f)t} \sin\left[\sqrt{(k/I_f) - (\kappa^2/4I_f^2)}\,t\right] \tag{1.18}$$

where C and D are two constants to be determined from the boundary conditions.

The period, in seconds, of the damped oscillations expressed by Eq (1.18) will be:

$$\tau = 2\pi \frac{1}{\sqrt{(k/I_f) - (\kappa^2/4I_f^2)}} \tag{1.19}$$

and the frequency (in hertz) is $\nu = 1/\tau$.

It can be seen from Eq (1.19) that when the damping factor becomes $\kappa = 2\sqrt{I_f k}$, $\tau \to \infty$; in other words, the motion becomes non-oscillatory, of the pure subsidence type.

The above κ value is called critical damping (κ_{crit}). For $\kappa < \kappa_{crit}$, the general character of an oscillatory motion will be of the type shown in Fig 1.8.

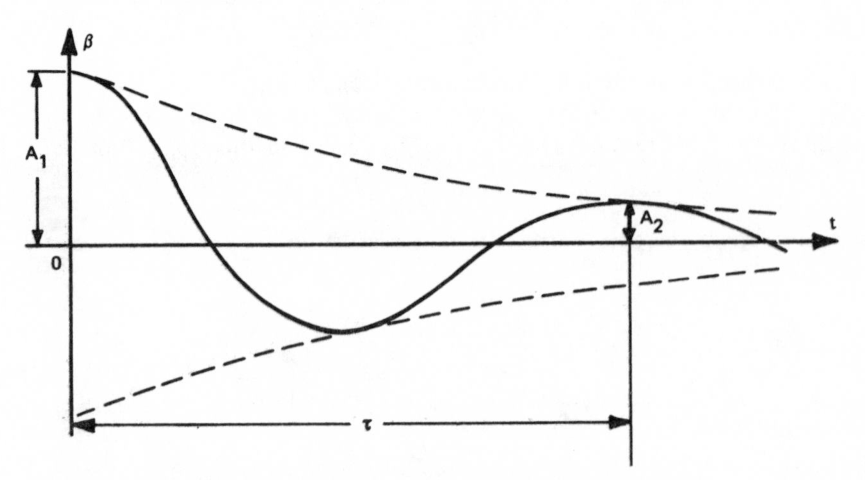

Figure 1.8 Character of the periodic damped motion

The rapidity of convergence can be measured by the ratio of two consecutive amplitudes (say, A_2/A_1):

$$A_2/A_1 = e^{-(\kappa/2I_f)(t+\tau)}/e^{-(\kappa/2I_f)t} = e^{-(\kappa/2I_f)\tau}. \tag{1.20}$$

In order to get some idea regarding the period and decay of the amplitude of an oscillating blade, the ratios appearing under the square root in Eq (1.19) will be examined.

$$k/I_f = [(1/2)MR\Omega^2][(2/3)R]/(1/3)MR^2 = \Omega^2. \tag{1.21}$$

It can be seen from Eqs (1.19) and (1.21) that when there is no damping ($\kappa = 0$), the period of oscillation is exactly equal to the time of one revolution ($\tau_o = 2\pi/\Omega$; or $\tau_o\Omega = 2\pi$). This means that the frequency of oscillations is equal to one per revolution.

The ratio of the period of oscillation with damping to that without damping (τ/τ_o) can be obtained from Eq (1.19):

$$\tau/\tau_o = 1/\sqrt{1 - (\kappa^2/4kI_f)}. \qquad (1.22)$$

But the $\kappa^2/4kI_f$ has already been examined and assuming typical values, was found to be $\kappa^2/4kI_f \approx 0.032$. It appears, hence, that for practical design rotors, it is permissible to neglect the influence of damping and to assume that

$$\tau \approx \tau_o = 2\pi/\Omega.$$

Some insight regarding consecutive amplitude ratios can be obtained by examining the expression $\kappa/2I_f$ appearing in the exponent of **e** in Eq (1.20). Noting that $T_b = w_b Rc$ where w_b is the blade loading, the derivative expressing blade-thrust versus collective-pitch variations can be expressed as

$$\partial T_b/\partial\theta_o|\theta_{o_c} = (\partial w_b/\partial\theta_o|\theta_{o_c})Rc.$$

Assuming that the blade mass is uniformly distributed along its axis, the blade moment of inertia about the flapping hinge becomes $(1/3)(W_b/g)R^2$, and the expression for $\kappa/2I_f$ can be written as follows:

$$\kappa/2I_f = Rr_T Rc(\partial w_b/\partial\theta_o)|\theta_{o_c}(1/\Omega)/(2/3)(W_b/g)R^2. \qquad (1.23)$$

In order to make the present investigation of dynamic stability more universal, the blade loading is nondimensionalized by dividing w_b by ρV_t^2: $\overline{w}_b \equiv w_b/\rho V_t^2$. The $\partial w_b/\partial\theta_o$ derivative can now be written as $\partial w_b/\partial\theta_o = (\partial\overline{w}_b/\partial\theta_o)\rho V_t^2$. Remembering that $W_b = w_{sb}Rc$; where w_{sb} is the blade structural weight-per-unit area, and $\Omega R \equiv V_t$; and further, assuming that $\overline{r}_T = 2/3$, Eq (1.23) can now be expressed as

$$\kappa/2I_f = [\partial(w_b/\rho V_t^2)/\partial\theta_o|\theta_{o_c}]g\rho V_t^2/w_{sb}. \qquad (1.24)$$

The character of the $\partial\overline{w}_b/\partial\theta_o = f(\theta_o)$ variation is shown in Fig 1.9. It can be seen that when $\theta_o \to 0$; $\partial\overline{w}_b/\partial\theta_o \to 0$, also. This obviously means that for either completely or almost completely unloaded rotors ($\theta_o \approx 0$), the damping coefficient may become very small or may even drop to zero. However, for typical operations, $\theta_o \geqslant 10°$; and it may be assumed that $\partial\overline{w}_b/\partial\theta_o \approx 0.65/rad$. This quantity, when combined with typical values encountered in practice: namely, $V_t = 200\ m/s$ ($\approx 650\ fps$), $w_{sb} = 30\ kG/m^2$ ($\approx 6\ psf$), and $\rho = \rho_o = 1.23\ kg/m^3$ ($0.002378\ slugs/cu.ft$); would result in $\kappa/2I_f \approx 5.3/s$, and an amplitude ratio of

$$A_2/A_1 \approx \mathbf{e}^{-1.13} = 0.323.$$

This means that each amplitude following another would only amount to approximately one-third that of the preceding one. Consequently, a blade having a flapping hinge perpendicular to the rotor axis would rapidly converge to its original position after having been displaced from its angle of equilibrium. In other words, it may be stated that under normal operating conditions, the flapping motion of a rotor blade would exhibit not only static, but also dynamic stability. This characteristic is one of the most important factors in making rotary-wing aircraft feasible.

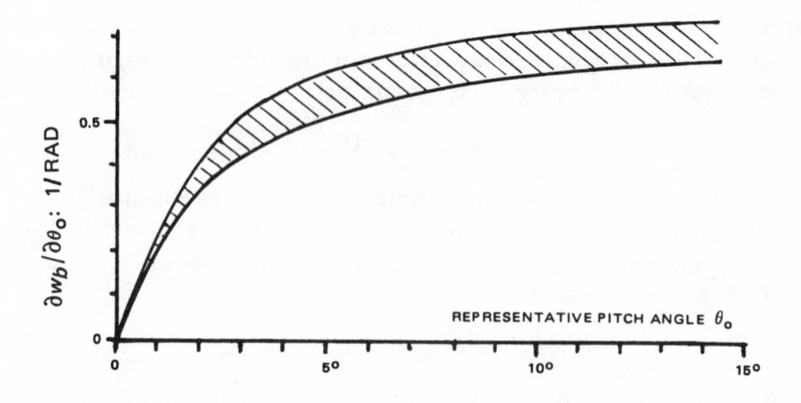

Figure 1.9 Character of the $\partial w_b/\partial \theta_o$ variation

4.3 Effect of Flapping Hinge Offset

All of the preceding considerations were performed under the assumption that the flapping hinge passes through, and is perpendicular to, the rotor axis. For simpler design solutions, as well as improved controllability due to the presence of the hub moment, the flapping hinges of practical rotary-wing aircraft are often located with some offset (usually a few percent) of the rotor radius $\eta \equiv r_f/R$, where r_f is the radius determining the location of the flapping hinge (Fig 1.9). As before, it will be assumed that (1) the flapping hinge is perpendicular to a radial plane passing through the rotor axis, (2) the blade mass (m_b) is uniformly distributed along the blade so that $m_b = R\overline{W}_b/g$ where $\overline{W}_b$ is the blade weight per unit span, and g is the acceleration of gravity, and (3) the blade is completely rigid.

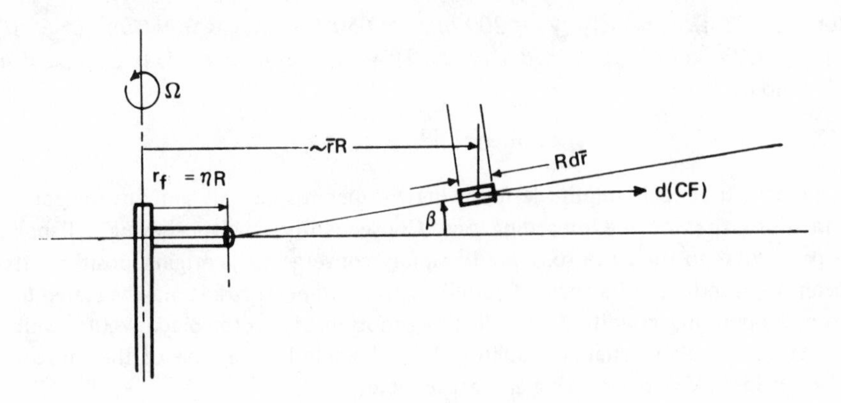

Figure 1.10 Offset flapping hinge

The question is, What effect would the hinge offset have on the flapping frequency of the blade which, for an undamped blade with the flapping hinge at the rotor axis, amounts to one-per-revolution?

To investigate this problem alone, the damping term can be neglected, and the equation of the blade motion around its flapping hinge can be written as follows:

$$I_f\ddot{\beta} + M_{res} = 0 \tag{1.25}$$

where M_{res} is the restoring moment about the flapping hinge. Keeping Eq (1.2a) in mind, Eq (1.25) can be rewritten as

$$I_f\ddot{\beta} + k\beta = 0. \tag{1.25a}$$

In Eqs (1.25) and (1.25a), I_f is the blade moment of inertia about the flapping hinge which can now be expressed as

$$I_f = \int_{\eta}^{1} (\overline{W}_b/g)RdrR^2(\overline{r}-\eta)^2 = (1/3)(W_b/g)R^2(1-\eta)^3. \tag{1.26}$$

The restoring moment M_{res} appearing in Eq (1.25) can be written as follows:

$$M_{res} = \int_{\eta}^{1} (\overline{W}_b/g)R^2\Omega^2 r(\overline{r}-\eta)R\beta dr = (1/6)W_b/g)R^2\Omega^2(1-\eta)^2(\eta+2)\beta. \tag{1.27}$$

Comparing Eqs (1.25) and (1.25a) with Eq (1.27), one would find that the spring constant k can be expressed as

$$k = (1/6)(W_b/g)R^2\Omega^2(1-\eta)^2(\eta+2). \tag{1.28}$$

For the type of equation represented by Eq (1.25a), the natural frequency of harmonic motion[8] can be expressed as

$$\omega^2 \equiv (2\pi\nu)^2 = \sqrt{k/I_f}. \tag{1.29}$$

Substituting the k value from Eq (1.28) and the I_f value from Eq (1.26) into Eq (1.29), the blade natural frequency referred to one revolution becomes

$$(\nu/n) = (\omega/\Omega) \equiv \sqrt{\omega^2/\Omega^2} = \sqrt{(\eta+2)/2(1-\eta)}. \tag{1.30}$$

In view of the fact that usually, $\eta \leqslant 1$, Eq (1.30) can be simplified:

$$\nu/n \approx \sqrt{[1 + (\eta/2)](1+\eta+\eta^2+...)} \approx \sqrt{1+(3/2)\eta} \approx 1+(3/4)\eta. \tag{1.31}$$

Values of the first natural flapping frequency for blades with an offset flapping hinge as given by the exact (Eq (1.30)), and an approximate (Eq (1.31)) formulae are shown in Fig 1.11.

Two things can be noted from the equations and figures: (1) flapping frequency increases with the flapping hinge offset, and (2) Eq (1.31) provides an acceptable approximation for the offset values of $\eta \leqslant 0.08$ usually encountered in practice for articulated rotors.

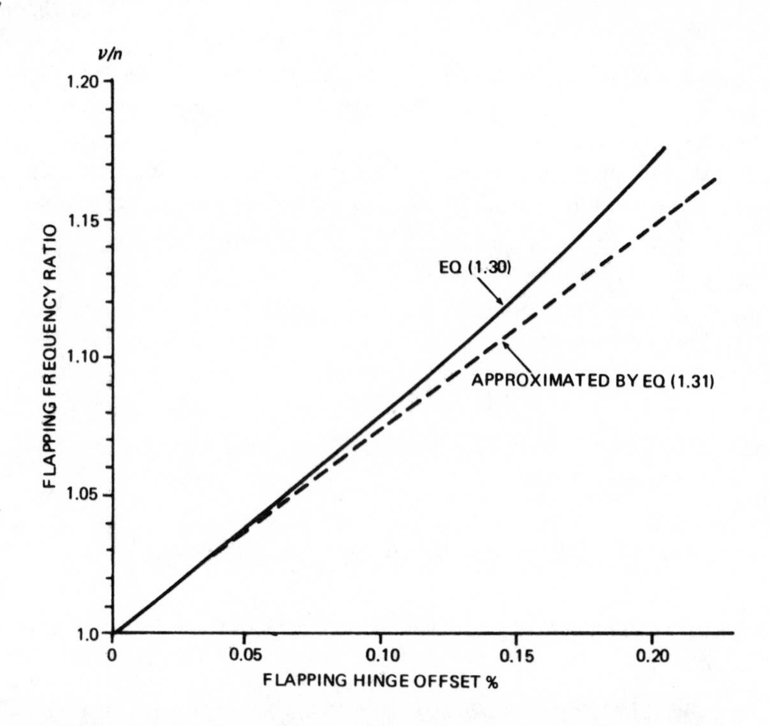

Figure 1.11 Effect of flapping hinge on blade flapping frequency

The so-called hingeless rotors exhibit various degrees of rigidity as far as blade flapping deformation is concerned. This, combined with the action of the centrifugal acceleration on the blade mass particles leads to flapping frequencies higher than one per rev, ($\nu/n > 1.0$). For most hingeless rotors, this ratio would probably not exceed the $\nu/n \approx 1.1$ value. However, in truly "rigid" blades such as the counter-rotating rotors of the ABC type, values as high as $\nu/n \approx 1.4$ may be encountered (see p. 39).

Since the value of the flapping frequency ratio of a hinged blade depends on the magnitude of hinge offset, one often encounters the flapping rigidity of hingeless blades expressed in terms of the equivalent or virtual flapping hinge offset. For example, it may be stated that a hingeless blade having a first natural frequency of $\nu/n = 1.08$ represents an equivalent flapping hinge offset of 10 percent, since an articulated blade with the actual flapping hinge located at 10 percent of radius would have the same ν/n value (Fig 1.11). This approach is convenient for such estimates as an immediate assessment of the control power through the hub moments which are proportional to the flapping hinge offset.

Although the role of flapping hinge offsets, both actual and equivalent, may be neglected in most performance tasks, it has been considered in some detail because problems related to the hinge offset are constantly encountered in such fields as control and trim analysis. For a more detailed discussion on this subject, the reader is referred to Chs 2 and 3 of Vol II.

5. ROTOR CONTROL

5.1 Rotor Thrust Inclination through Cyclic Control in Hover

In Sect 4.2 it was shown that the blade motion about the flapping hinge of a rotating rotor can be interpreted as a free oscillatory motion with damping. If, however, an external periodically varying moment is applied, the motion would then become forced oscillation. In the case of a rotor, it is evident that such an external moment can be generated through a forced variation of the blade thrust. It was also mentioned that this could be done by such means as flap deflection and circulation control through blowing or suction. At present, however, the most common way of varying the blade thrust around the azimuth is through feathering–a periodic change of the blade pitch angle θ_o as a whole. This variation of θ_o with azimuth must be periodic and consequently, can be expanded into a Fourier series containing any number of harmonics $(n \geqslant 1)$. However, in practice, control inputs in all regimes of flight and especially in hover are usually of first harmonic character:

$$\theta_{o\psi} = \theta_o - \theta_1 \cos(\psi - \psi_1) \tag{1.32}$$

where θ_1 is the maximum deviation of the blade collective pitch angle from its nominal (average) value (θ_o), and ψ_1 represents the azimuth angle at which this maximum deviation occurs (Fig 1.12a)

In practice, the blade collective pitch variation given by Eq (1.32) is accomplished through a relatively simple mechanical device called the swashplate, which was probably first proposed by Yur'iev in 1911. This device consists of a circular track which, through pilot-control inputs, can be arbitrarily inclined with respect to the plane perpendicular to the rotor axis (Fig 1.12b).

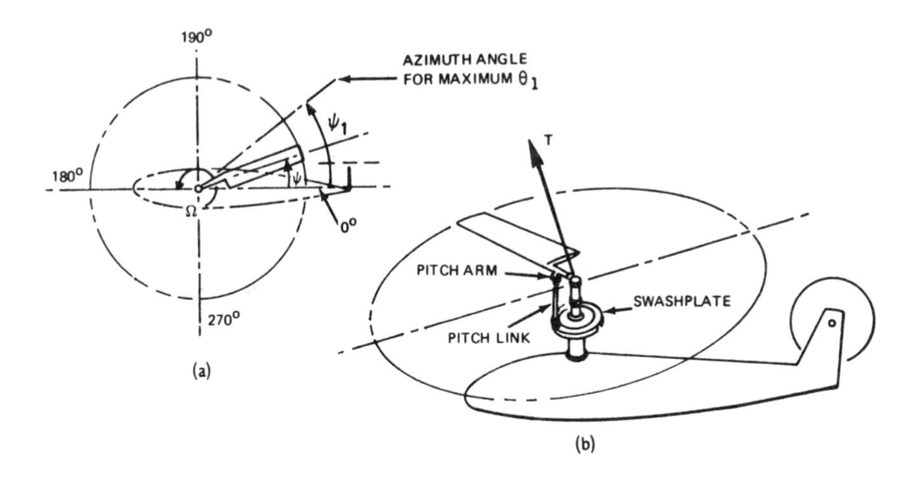

Figure 1.12 Scheme of cyclic control

The inner race of the swashplate is attached to the rotor shaft by means of a gimbal joint. This is done in such a way that it remains stationary as far as rotation about the rotor axis is concerned. In contrast, the outer race is driven at the rotational speed of the rotor. Pitch links, connecting the outer race to the blade pitch arms, transmit the inclination of the swashplate into the variation of the blade pitch angle according to Eq (1.32).

Without going into the refinements of aerodynamic theories, it may be anticipated that the first harmonic cyclic variation of the blade collective pitch would produce corresponding cyclic changes in the thrust per blade which, of course, would mean that the moment experienced by the blade about the flapping hinge would also vary cyclically. This would generate a forcing moment (M_f) which would become a function of $\Delta\theta_{o\psi} = \theta_{o\psi} - \theta_o$; $M_f = f(\Delta\theta_{o\psi})$.

Simply to illustrate the problem, it will be assumed that $\Delta\theta_{o\psi}$ reaches its maximum positive value at $\psi = 90°$; hence, $M_{f\psi} = M_{f_{max}} \sin\psi \equiv M_{f_{max}} \sin\Omega t$. Under this assumption, an analogy to the case of forced oscillation of a mass point with damping can be noticed as the equation of motion of the blade becomes

$$I_f\beta + \kappa\ddot{\beta} + k\dot{\beta} = M_{f_{max}} \sin\Omega t. \qquad (1.33)$$

The complete solution of Eq (1.33) is composed of the following terms:

$$\beta = Ae^{\lambda_1 t} + Be^{\lambda_2 t} + M_{f_{max}} \frac{(k - I_f\Omega^2)\sin\Omega t - \kappa\Omega \cos\Omega t}{(k - I_f\Omega^2)^2 + \kappa^2\Omega^2}. \qquad (1.34)$$

The first two terms represent previously discussed free, dynamically stable oscillations that quickly decrease with time (transient motion). The third term describes a simple harmonic motion with constant amplitude. It is clear, hence, that after a few revolutions, the blade will be moving according to the third term of Eq (1 .34). The e terms in this equation, can therefore be neglected and the equation can be rewritten as follows:

$$\beta = \frac{M_{f_{max}}}{\sqrt{(k - I_f\Omega^2)^2 + \kappa^2\Omega^2}} \sin(\Omega t - \psi_p) \qquad (1.35)$$

where the value of the phase lag angle ψ_p is determined by the following relationship:

$$\tan\psi_p = \kappa\Omega/(k - I_f\Omega^2). \qquad (1.36)$$

It can be seen from Eq (1.36) that when $k = I_f\Omega^2$, $\psi_p = 90°$.

For a blade with its non-offset flapping hinge located perpendicular to the rotor axis and having a constant axial mass distribution, $\overline{m}(r) = const$,

$$k = (CF)R\bar{r}_{CF} = (1/2)MR\Omega^2(2/3)R = I_f\Omega^2.$$

Consequently, the phase lag angle for this blade will be 90°. This means that when the first-harmonic pitch control is applied in such a way that its maximum value occurs at $\psi = 90°$ (θ_{max} at $\psi = 90°$) and hence $M_{f_{max}}$ also occurs at $\psi = 90°$; β_{max} will take place at $\psi = 180°$.

As to the magnitude of β_{max} (Ref 8, p. 139), it can be shown that

$$\beta_{max} \approx M_{f_{max}}/\kappa\Omega \tag{1.37}$$

but

$$M_{f_{max}} = \partial T_b/\partial\theta_o|\theta_{o_c}(\Delta\theta_{o_{max}} R\bar{r}_T).$$

Substituting the κ value appearing in Eq (1.11) into Eq (1.37), this equation is now reduced to

$$\beta_{max} \approx \Delta\theta_{o_{max}}. \tag{1.37a}$$

Eqs (1.35) and (1.37a) indicate that by applying a first-harmonic variation of the cyclic pitch to a hovering rotor with flapping hinges (zero offset hinges perpendicular to the rotor axis) a steady inclination of the tip-path plane can be obtained as long as the control input remains the same. Furthermore, the maximum value of that angle-of-inclination is equal to the maximum pitch control input ($\Delta\,\theta_{o_{max}}$). In this way, an equivalence of blade feathering and flapping has once more been demonstrated.

It was also shown that for the zero-offset flapping hinges, the phase lag angle—which is independent of damping—amounts to 90°. This means that the maximum flapping angle of the blade occurs 90° later than the maximum blade pitch angle.

In order to get some idea regarding the influence of the blade natural frequency and damping on the phase lag angle, Eq (1.36) is written as follows:

$$tan\ \psi_p = \frac{\kappa\Omega}{[(k/I_f) - \Omega^2]I_f}. \tag{1.38}$$

where k/I_f is the square of the frequency of the flapping blade with zero damping (see Eq (1.19)) times 2π: $\sqrt{k/I_f} = 2\pi\nu$. Substituting this value into Eq (1.38) and considering that $\nu = (\nu/n)n$, Eq (1.38) becomes:

$$tan\ \psi_p = \frac{\kappa\Omega}{[(\nu/n)^2(2\pi n)^2 - \Omega^2]I_f}. \tag{1.38a}$$

Since $2\pi n \equiv \Omega$, Eq (1.38a) can now be rewritten as

$$tan\ \psi_p = \frac{\kappa}{[(\nu/n)^2 - 1]\Omega I_f}. \tag{1.38b}$$

Eq (1.38b) clearly indicates that as the natural frequency of the blade about its flapping hinge becomes greater than one-per-revolution, as in the case of articulated blades with flapping hinge offsets and hingeless rotors, the phase lag angle becomes less than 90°.

In order to more clearly indicate the influence of damping, a relative damping ratio (κ/κ_{crit}) can be introduced. This is done by dividing both the numerator and denominator in Eq (1.38b) by $\kappa_{crit} \equiv 2\sqrt{I_f k} = 2I_f\Omega$. Now Eq (1.38b) becomes

$$tan\ \psi_p = \frac{2(\kappa/\kappa_{crit})}{[(\nu/n)^2 - 1]}. \tag{1.38c}$$

This relationship, illustrated in Fig 1.13, is reproduced from Gessow and Myers[9].

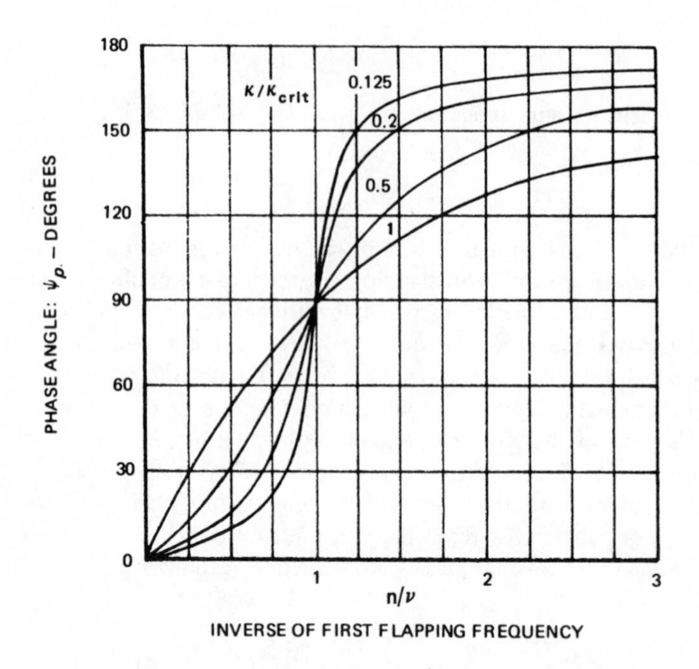

Figure 1.13 Effect of flapping frequency and damping on phase angle

5.2 Blade Flapping in Forward Flight

For simplicity, helicopter translations with an inplane velocity ($V\|$) will be called forward flight. In Sect 3.1, Eq (1.5), it was shown that the velocity component perpendicular to the blade ($U\perp$) varies with the azimuth as a *sin* ψ function. It was also pointed out that because of blade thrust, the aerodynamic moment about the flapping hinge (M_f) may also vary with the azimuth in the same manner as the $U\perp$ component. This, in turn, would lead to a variation of the aerodynamic moment with ψ: $M_f = M_f(\psi)$. Under steady-state conditions, M_f would change periodically in exactly the same way through each revolution. Similar to the previously considered case of cyclic control in hover, M_f may be regarded as a moment forcing the blade to oscillate about its flapping hinge. The following equation describes this kind of motion:

$$I_f\ddot{\beta} + \kappa\dot{\beta} + k\beta = M_f(\psi). \tag{1.39}$$

It may be expected that the steady-state solution of this equation is a harmonic function of ψ which can be expressed in terms of a Fourier series representing a sum of simple harmonic motions. Stopping at the second harmonic, the solution would be

$$\beta = a_o - a_1 \cos\psi - b_1 \sin\psi - a_2 \cos(2\psi) - b_2 \sin(2\psi) \tag{1.40}$$

where the following interpretation can be given to the coefficients a_o, a_1, a_2, b_1, and b_2:

a_o – represents the part of flapping independent of ψ. This is the same as the coning angle in hover (β_o) when all cyclic inputs are zero: $\beta_o = a_o$.

a_1 – the coefficient representing the amplitude of a pure, first-harmonic, cosine motion and, according to the previously adopted sign convention, it describes the fore and aft inclination of the rotor disc (tip-path plane) having a maximum elevation at $\psi = 180^\circ$ and minimum elevation at $\psi = 0$ (Fig 1.14).

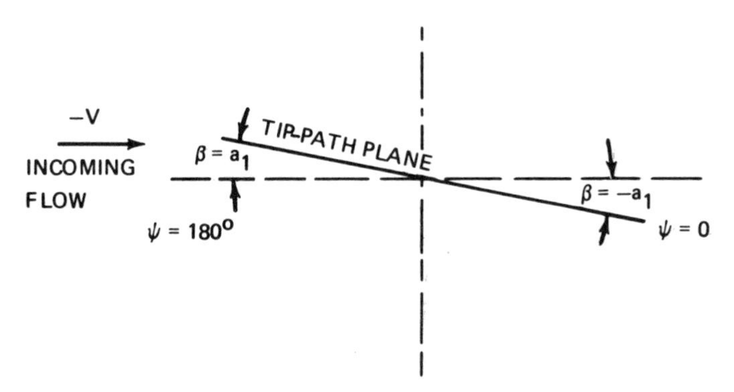

Figure 1.14 Pure cosine motion – tip-path plane seen from the left

b_1 – coefficient representing the amplitude of a pure, first-harmonic, sine motion reaching β_{min} at $\psi = 90^\circ$, and β_{max} at $\psi = 270^\circ$ (Fig 1.15).

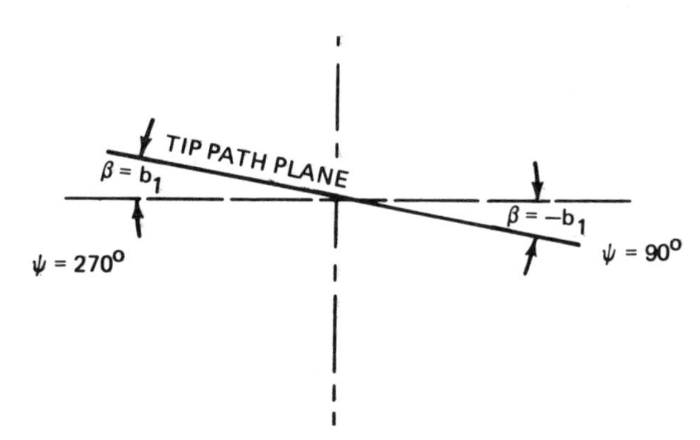

Figure 1.15 Pure sine motion – tip-path plane seen from the rear

a_2, b_2, *etc.* – coefficients representing amplitudes of the higher harmonics. For instance, when plotting the $\beta = -a_2 \cos(2\psi)$ motion vs ψ (Fig 1.16), it can be seen that two maxima appear; one at $\psi = 90^\circ$ and another at $\psi = 270^\circ$. By the same token, two minima are also present ($\psi = 0^\circ$ and $\psi = 180^\circ$), while zeroes are reached at $\psi = 45^\circ, 135^\circ, 225^\circ$, and 315°.

It is obvious that as additional harmonic terms are introduced into Eq (1.40), the determination of the path of the blade tips becomes more accurate. However, for purely performance considerations, the zero (a_o) plus the two first-harmonic terms are usually sufficient.

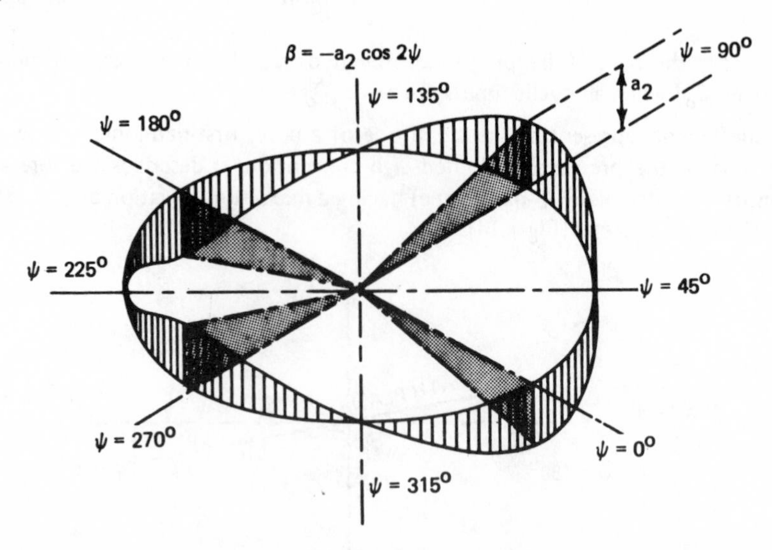

Figure 1.16 Tip-path of a blade executing second harmonic cosine motion

Some insight into the order of magnitude of the flapping harmonic coefficients encountered in practice can be obtained from the following flight measurements[9]:

$$a_o = 8.7°;\ a_1 = 6.1°;\ b_1 = 3.9°;\ a_2 = 0.5°;\ \text{and } b_2 = -0.1°.$$

Retaining only first-harmonic terms and making substitutions[10] — $a_o = \beta_o$, $a_1 = \beta_1 \cos \psi_1$, and $b_1 = \beta_1 \sin \psi_1$ — Eq (1.40) may be rewritten as follows:

$$\beta = \beta_o - \beta_1 \cos(\psi - \psi_1). \tag{1.41}$$

Eq (1.41) is identical to Eq (1.32) and clearly indicates that the tip-path plane is inclined from the plane perpendicular to the shaft through an angle β_1. It should also be noted that the tip-path plane is perpendicular to an axis called the *virtual axis of rotation*. This axis extends through the hub and lies in a plane which passes through the rotor axis and makes an angle ψ_1 with the $\psi = 0$ plane.

In general, $\psi_1 \neq 0$; therefore, the lowest point in flapping is not necessarily at the down-wind, and the highest at the up-wind, position. Should $M_f(\psi)$ be of the character $M_f = M_{f_{max}} \sin \psi$ however, then the equation of motion (1.39) would be identical to Eq (1.33) and thus, for the case of $(\nu/n) = 1$, a 90° phase lag shift may be expected. In the expression for β as given by Eq (1.41), this would mean that $\psi_1 = 0$ and the blade would experience maximum flapping angle at $\psi = 180°$ and minimum at $\psi = 0°$.

Determination of the flapping coefficient requires an extensive knowledge of rotary-wing aerodynamics. Consequently, only the details necessary for a better understanding of the physical aspects is provided in the following discussion.

Assuming, for simplicity, that the resultant force representing blade thrust (T_b) in both hover and forward flight remains at the same relative blade station $\bar{r}_T$, the thrust moment about a non-offset flapping hinge would be

$$M_f = T_b R \bar{r}_T. \tag{1.42}$$

If the blades are completely rigid, it may be anticipated that as a first approximation, T_b, as any other lift force, would vary proportionally to the square of the $U\perp$ component experienced by the blade at station r_T. This means that the thrust per blade in hover (T_{b_h}) would be proportional to $U\perp_h{}^2$:

$$T_{b_h} \sim U\perp_h{}^2 = (V_t \bar{r}_T)^2 \tag{1.43}$$

while in forward flight,

$$T_{b_{fd}} \sim U\perp_{fd}{}^2 = V_t^2 (\bar{r}_T + \mu \sin \psi)^2. \tag{1.44}$$

Combining Eqs (1.42), (1.43), and (1.44), imagining that the flapping hinges are locked perpendicularly through the rotor axis, and assuming that the blades are rigid, the ratio of blade thrust moment about the flapping hinge in forward flight ($M_{f_{fd}}$) to that in hovering (M_{f_h}) would exhibit the following proportionality:

$$(M_{f_{fd}}/M_{f_h}) \sim [1 + 2(\mu \sin \psi/\bar{r}_T) + (\mu^2 \sin^2 \psi)/r_T{}^2]. \tag{1.45}$$

Omitting the last term in Eq (1.45) as being small (at least for $\mu < 0.3$) in comparison with the two preceding ones, Eq (1.45) can be rewritten as

$$(M_{f_{fd}}/M_{f_h}) \sim [1 + 2(\mu \sin \psi/\bar{r}_T)]. \tag{1.45a}$$

From this equation and Fig 1.17, it may be assumed as a first approximation that in translatory flight with an inplane velocity component ($\mu > 0$), an aerodynamic forcing moment proportional to $\sin \psi$ will be present. Furthermore, it can be seen from Eq (1.45a) that its magnitude will also be proportional to μ. Proportionality of the forcing moment to $\sin \psi$ makes the presently considered case analogous to that of the thrust vector tilt in hover as discussed in Sect 5.1. Consequently, many of the conclusions

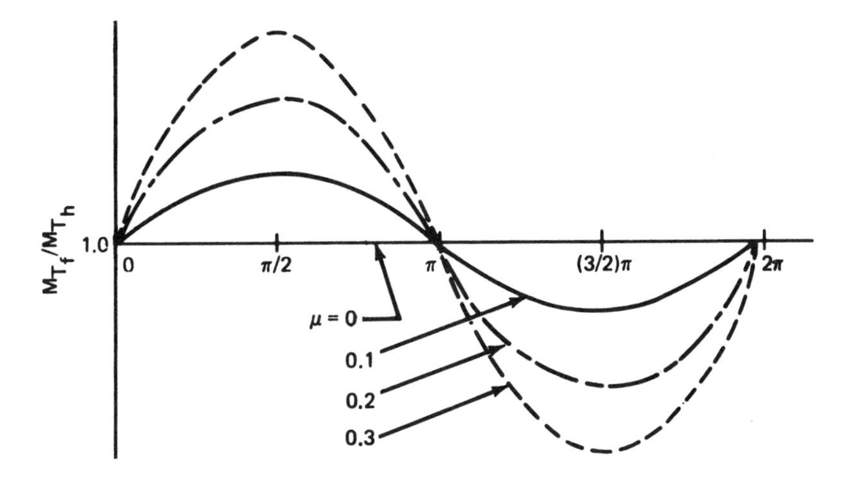

Figure 1.17 Character of the aerodynamic forcing moment due to the inplane velocity component

reached at that time can readily be applied here. For instance, it becomes evident that for flapping blades with zero offset $[(\nu/n) = 1.0]$, the phase angle in forward flight will also be $\psi_p = 90°$. In other words, maximum flapping will occur 90° after $M_{f_{fd}}$ assumes its maximum value. In the considered case, it will obviously mean that maximum elevation will take place at $\psi = 180°$ and its minimum (the largest negative), at $\psi = 0$. As to the magnitude of the flapping motion determined by the value of the a_1 coefficient in Eq (1.40), it may be anticipated that this quantity will be strongly influenced by the μ level (see Eq (1.45)).

The sideward tilt of the tip-path plane as reflected by the b_1 values in Eq (1.40) may be explained as a consequence of coning: $b_1 = f(a_o)$. Using an approach suggested by Gessow and Myers[9], the above dependence can be explained as follows: Without coning, the influence of forward velocity on the flapping blade at $\psi = 180°$ will be the same as at $\psi = 0^0$ as shown in Fig 1.18a).

However, for the coned rotor, a difference in the angle-of-attack of the blade within the leading $(90° \leqslant \psi \leqslant 270°)$ and the trailing $(270° \leqslant \psi \leqslant 450°)$ sectors may occur.

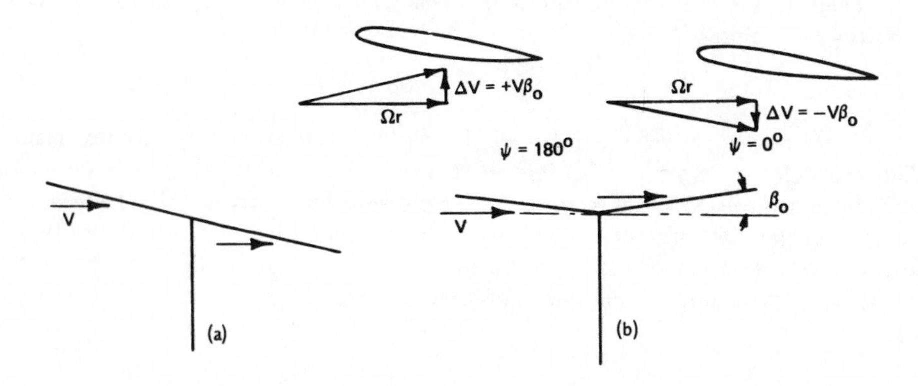

Figure 1.18 Velocity component due to coning

If we imagine that the blades are somehow restricted against flapping but retain their coning angle $a_o \equiv \beta_o$, additional forward velocity components at $\psi = 0°$ $(-\Delta V)$ and $\psi = 180°$ (ΔV) would be as shown in the upper part of Fig 1.18b.

A general expression for ΔV at all azimuth angles will be as follows:

$$\Delta V = -V\beta_o \cos\psi. \tag{1.46}$$

It is clear from this equation that the influence of the angle-of-attack variation due to coning will be first harmonic in character. This, in turn, would generate variation of the aerodynamic moment of the same character. It may be expected hence, that a new periodic motion of the blade will result. For the frequency ratio $\nu/n \approx 1.0$, it may be expected that the phase lag angle will be about 90° or, in other words, the $b_1 \sin\psi$ should approximately describe the character of the flapping angle variation due to coning angle effects.

It should be noted, however, that in addition to coning which is usually the strongest factor, there may be other causes for the *cos* ψ variation of the blade aerodynamic moment leading to lateral flapping.

There are many causes for the blade tip motions described by the higher flapping harmonics represented in Eq (1.40) by the terms containing coefficients a_2, b_2, ... For instance, existence of the reversed flow region and the presence of the $sin^2\ \psi$ and $cos^2\ \psi$ terms may be one cause of the second harmonic excitations.

5.3 Control of the Thrust Vector Inclination

Control of the thrust vector inclination, both in hovering and forward flight, can be accomplished through the following means:

1. Inclination of the rotor shaft whose axis in this case is identical to that of the control axis (Fig 1.19a).
2. Inclination of the hub only, which would be equivalent to the displacement of the control axis (Fig 1.19b).
3. Inclination of the swashplate which, again, is equivalent to the displacement of the control axis (Fig 1.19c).

Although only the tilt of the thrust vector through cyclic control inputs was discussed in Sect 5.1, it can be shown that in the absence of the inplane translatory velocity component (i.e., when $\mu = 0$ as in hovering), any of the means of thrust tilt shown in Fig 1.19 would, after a brief transient period, lead to an alignment of the thrust vector with the control axis. This means that the tip-path plane would then assume a position perpendicular to the control axis.

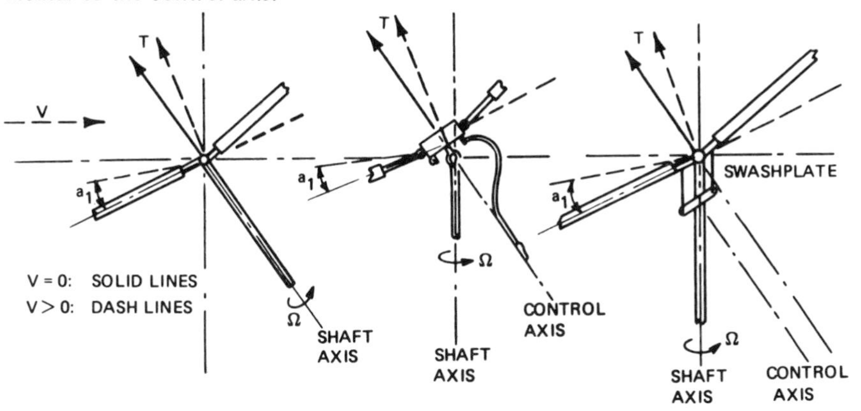

Figure 1.19 Various means of tilting the thrust vector

However, when an inplane component of the distant flow appears ($\mu > 0$), flapping would develop and the thrust vector will tilt (from its original position of alignment with the control axis) in the direction of the inplane velocity component (dash lines in Fig 1.19). This means that the tip-path plane will no longer be perpendicular to the control axis. This phenomenon of deviation of the thrust vector from the control axis is important when making a trim analysis of the whole aircraft.

Feathering which is presently achieved through mechanical inputs from a swashplate—either directly to the blade pitch arm or through a special flap, twisting the blade—represents practically the only method used for controlling the rotor thrust tilt. It is obvious that this type of control would permit one to tilt the rotor thrust vector to any position in the ground frame of reference as needed for a desired flight condition.

In order to gain more insight and understanding of the physical relationship between feathering and flapping, the following elementary example is considered:

> A simple flapping rotor with infinitely heavy blades ($a_o = 0$) is operating in forward flight. The control axis is vertical and the rotor disc is tilted upward by a_1 as depicted in Fig 1.20a. As interpreted by Gessow and Myers[9]—"An observer riding on the control axis and rotating with the blades observes that the blades flap up and down with each revolution, but they are fixed in pitch. At the same time, an observer who sits in the plane of the tips—rotating with the blades—observes that the blades do not flap at all, but do change their pitch—high, then low—once each revolution. The pitch is low on the advancing side of the rotor and high on the retreating side. Examining Fig 1.20b, it is seen that the amount of blade feathering with respect to the plane of the tips is equal in degrees to the amount of blade flapping with respect to the control axis. Fore and aft (a_1) flapping with respect to the control axis is therefore equal to lateral (B_1) feathering with respect to the axis perpendicular to the plane of the tips. The control axis is the axis of no feathering; the axis perpendicular to the plane of the tips is the plane of no flapping (except for higher harmonics)."

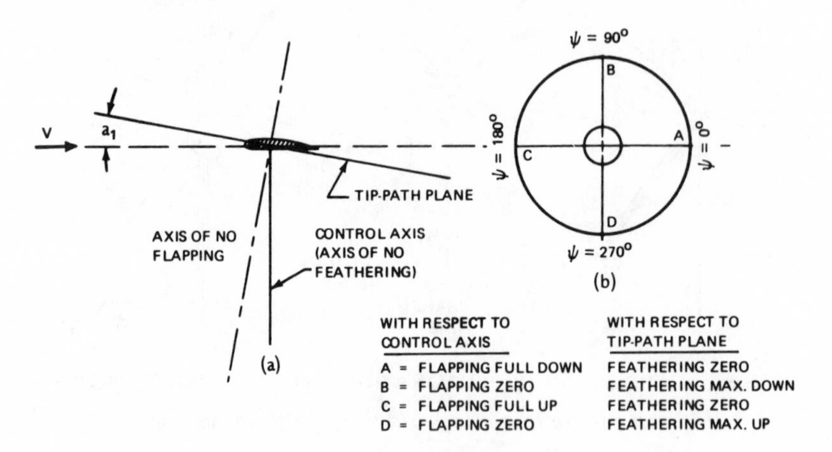

Figure 1.20 Flapping and feathering

From a practical point of view, one usually tries to use some clearly identifiable line as a reference because of its physical or design significance. In helicopters, this line is represented by the rotor-shaft axis. For this reason, it is important to know how to calculate the position of the tip-path plane (to which thrust is assumed to be perpendicular) with respect to the rotor-shaft axis.

Before establishing the necessary relationships, it should be recalled that there is no feathering with respect to the control axis; i.e., the axis perpendicular to the swashplate plane. This becomes obvious when one realizes that the attachment points of the pitch links and the blade pitch axis move in a plane parallel to the swashplate (Fig 1.20c).

As previously discussed, however, flapping may exist with respect to the control axis, and its value was given by Eq (1.40) which, when limited to the first-harmonic flapping terms, becomes

$$\beta = a_o - a_1 \cos\psi - b_1 \sin\psi. \tag{1.47}$$

Feathering motion with respect to the tip-path plane (also up to the first harmonics only) is given by

$$\theta = A_o - A_1 \cos\psi - B_1 \sin\psi. \tag{1.48}$$

Denoting flapping and feathering motions with respect to the shaft by the subscript s, the following relationships as shown in Fig (1.21)[9] are obtained:

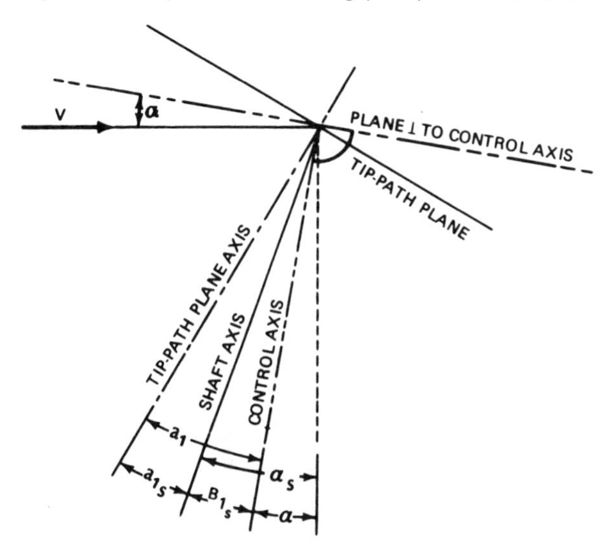

Figure 1.21 Flapping and feathering with respect to the shaft axis

$$\alpha = \alpha_s - B_{1_s} \tag{1.49}$$

$$A_o = A_{os} \tag{1.50}$$

$$a_o = a_{os} \tag{1.51}$$

$$a_1 = a_{1_s} + B_{1_s} \tag{1.52}$$

$$b_1 = b_{1_s} + A_{1_s} \tag{1.53}$$

where a represents the angle-of-attack (with respect to the distant flow V) of the plane perpendicular to the control axis.

6. BLADE LAGGING MOTION

When looking at rotor configurations incorporating flapping hinges, one would usually notice additional discrete hinges located approximately perpendicular to the rotor plane. These hinges are called lag or drag hinges, and they permit the blade to execute some motion in the rotor plane (Fig 1.22).

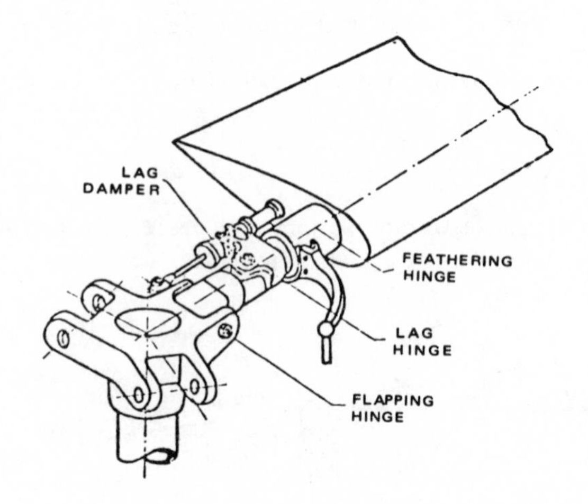

Figure 1.22 Hub with flapping and lag hinges

In hingeless rotors, one can imagine that, similar to the flapping case, the inplane flexibility would provide a virtual hinge for this type of motion. The need for a lag hinge, or provision of inplane blade flexibility can be better appreciated in light of the following consideration:

It can be shown (for instance on p.85, Ref 10) that the Coriolis (inplane) acceleration (a_c) experienced by a mass element located at a distance r from the shaft axis of a blade rotating about the rotor axis with an angular velocity Ω, and having a flapping angular velocity $\dot{\beta}$ when the blade is at an angle β, will be

$$a_c = 2r\Omega\beta\dot{\beta}. \tag{1.54}$$

Assuming a simple harmonic flapping motion, β may be expressed as

$$\beta = \beta_o - \beta_1 \cos(\psi - \psi_1) \tag{1.55}$$

where β_o is the coning angle. In the present consideration, it is assumed that $\beta_o = 0$; β_1 is the maximum (absolute) value of the flapping angle; and ψ_1 is the azimuth angle corresponding to the maximum (down) value of β. Expressing ψ in terms of rotor rotational velocity and time ($\psi = \Omega t$), Eq (1.55) can be written as

$$\beta = \beta_o - \beta_1 \cos(\Omega t - \psi_1). \tag{1.55a}$$

Consequently,

$$\mathring{\beta} = \beta_1 \Omega \sin(\Omega t - \psi_1); \qquad (1.55b)$$

and Eq (1.54) becomes

$$a_c = -2r\Omega^2 \beta_1^{\,2} \sin(\Omega t - \psi_1) \cos(\Omega t - \psi_1). \qquad (1.56)$$

However,

$$\sin(\Omega t - \psi_1) \cos(\Omega t - \psi_1) = \tfrac{1}{2} \sin(2\Omega t - 2\psi_1),$$

and

$$a_c = -r\Omega^2 \beta_1^{\,2} \sin 2(\Omega t - \psi_1). \qquad (1.57)$$

It is clear from Eq (1.57) that a_c varies as the second harmonic and thus reaches its highest positive as well as its maximum negative value twice per each revolution. It is obvious that without a lag hinge, the blade root could be subjected to a fatigue bending moment–changing from + to – twice per revolution.

For a typical medium-size helicopter (such as the hypothetical aircraft considered in Vol II) with $R = 7.6m$; $\Omega = 30\ rad/s$; and $\beta_1 = 0.1\ rad$; the inplane Coriolis acceleration experienced by the blade mass elements at $r = 0.8R$ would amount to $a_{c_{max}} \approx 5.5g$, where g is the acceleration of gravity.

Similar to the previous investigation of blade flapping stability, it can be shown that a blade having the freedom to move around an offset lag hinge would exhibit static stability. However, as far as dynamic stability is concerned, hinges perpendicular or almost perpendicular to the plane of rotation would encounter much lower aerodynamic damping. This is because the blade resistance to lagging would now be provided by drag forces which, in general, are much smaller than those associated with lift. Consequently, special hydraulic or friction dampers or other means of damping are usually installed (Fig 1.22).

Although the lag hinge and the associated inplane blade motion has a considerable effect on the dynamic problems of rotary-wing aircraft (for example, ground and air resonance), its importance from a purely performance point of view is very minor and except for hub drag effects, its presence may be ignored.

7. CONFIGURATIONS

With general background information regarding rotary-wing aircraft as a group as well as some exposure to dynamic problems particular to the rotor, the reader should be in a position to recognize the significance of various approaches to design concepts of rotors, types of control, and rotorcraft configurations.

7.1 Rotor Types

Main and tail rotors usually belong to one of the following types:

1. Fully-articulated–incorporating both flapping and lagging hinges (Figs 1.22 and 1.23a).

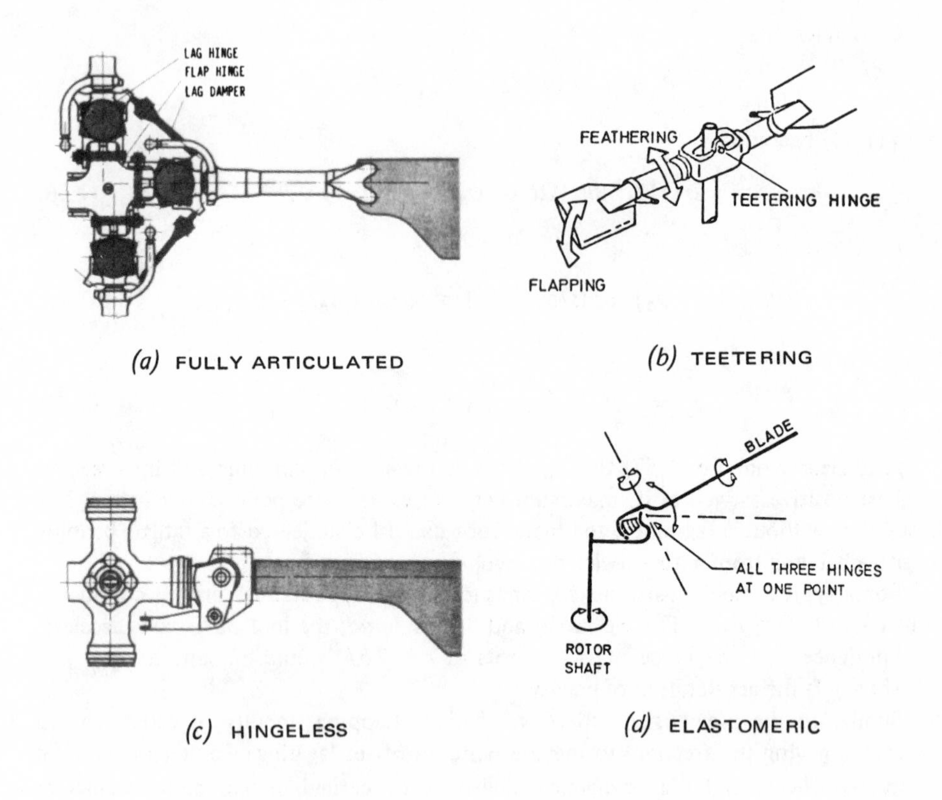

(a) FULLY ARTICULATED (b) TEETERING

(c) HINGELESS (d) ELASTOMERIC

Figure 1.23 Principal types of rotor hubs[11]

2. Teetering (Fig 1.23b)
3. Hingeless (Fig 1.23c).

A fourth type of blade suspension, represented by the elastomeric hub, is now being developed (Fig 1.23d). In this scheme, freedom of flap, lag, and feathering motions are concentrated into a single joint, thus assuring a limited spherical freedom of blade movement. Of course, many other systems of blade attachments have either been, or are being developed (for instance, bearingless and flex types).

From the point-of-view of flying qualities, aeroelastic, and vibration aspects, all of the aforementioned hub configurations may exhibit different characteristics. Nevertheless, except for the parasite drag level, the aerodynamic treatment of their performance problems may be identical for practical engineering purposes.

7.2 Types of Helicopter Control

Similar to fixed-wing aircraft, longitudinal, lateral, and directional controls are also required in helicopters. However, because of the VTOL-type operations of helicopters, a means of vertical control directly governing the height of rotary-wing aircraft within the ground reference frame is also needed. Furthermore, the large reaction torques generated by the mechanically-driven lifting rotors must be counterbalanced in order to prevent an undesirable rotation of the airframe. In multi-rotor configurations (e.g., coaxial, tandem, and side-by-side), this task is accomplished by pairing lifting rotors of equal size, but having opposite directions of rotation. With few exceptions, the tail rotor—whether open or enclosed—currently serves as a main-rotor torque-compensating device for pure helicopter single-rotor designs. The control forces and moments in pure helicopters are usually generated entirely by the lifting and tail rotors, while the nonrotating empennage surfaces serve principally as a means of trim.

Vertical control of helicopter configurations is almost always obtained through a direct variation of thrust of the lifting rotor or rotors. The most common means of accomplishing this goal is through a geometric simultaneous increase or decrease of the pitch angle of all the blades in one rotor. This is called *collective or zero-harmonic pitch control*. Other means of rotor thrust variation include circulation, or boundary-layer control of the flow around the blade airfoils. Variation of rotor rotational speed may, in principle, also be used as a means of direct thrust control, but the slow response resulting from the high rotational inertia of lifting rotors renders this approach impractical.

Longitudinal and lateral control of single-rotor and coaxial helicopters is usually accomplished by tilting the thrust vector through a cyclic application of the first-harmonic blade feathering from the swashplate. The resulting inclination of the thrust vector provides a horizontal component which pulls the rotorcraft in the desired direction while the thrust moment about the aircraft c.g. rotates the helicopter in a pitching, rolling, or coupled motion.

At this point, it should be noted that application of *cyclic control* to rotors having offset flapping hinges—either discrete or virtual—generates a hub moment in addition to that resulting from thrust-vector tilt.

In the scheme shown in Fig 1.24, it is assumed that the flapping hinge is located at a distance $r_{fh} = \eta R$ and the blades execute a flapping motion as given in Eq (1.41).

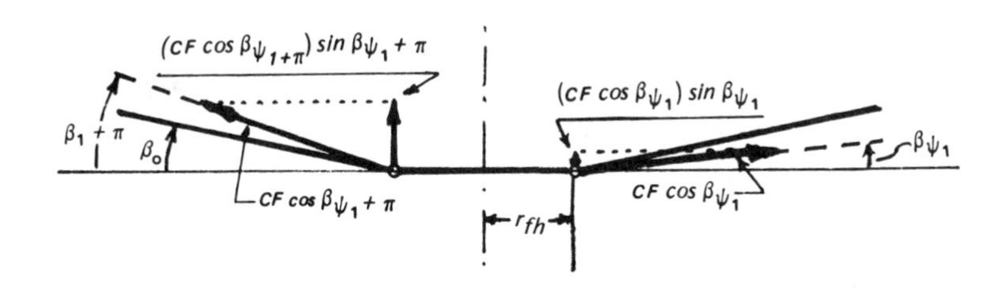

Figure 1.24 Scheme of hub-moment generation

The centrifugal force component ($CF \cos \beta$) acting along the blade can, in turn, be resolved into two forces at the flapping hinge; one, perpendicular ($CF\perp = CF \cos \beta \sin \beta$) and the other parallel ($CF\| = CF \cos^2 \beta$), to the hub plane. The $CF\perp$ component will reach its maximum value when the blade attains $\psi = \psi_1 + \pi$, and its lowest level when the blade is at $\psi = \psi_1$.

Taking Eq (1.41) into consideration, and accepting small angle assumptions, the moment generated at azimuth angle ψ by the $CF\perp$ of a single blade having an offset (η) flapping hinge will be

$$M'_{\eta\psi} = \eta R(CF)[\beta_o \cos(\psi - \psi_1) - \beta_1 \cos^2(\psi - \psi_1)]. \tag{1.58}$$

Eq (1.58) can be averaged over one revolution as follows

$$M'_{\eta} = [\eta R(CF)/2\pi] \int_{\psi-\psi_1=0}^{\psi-\psi_1=2\pi} [\beta_o \cos(\psi - \psi_1) - \beta_1 \cos^2(\psi - \psi_1)]\, d\psi$$

which becomes

$$M'_{\eta} = \tfrac{1}{2}\beta_1 \eta R(CF) \tag{1.59}$$

and for all b blades

$$M_{\eta} = \tfrac{1}{2} b\beta_1 \eta R(CF). \tag{1.60}$$

With respect to the $CF\parallel$ components, it can be seen that under the small angle assumption ($\cos \beta = 1$), their contributions will mutually cancel out within one revolution. It may be stated hence that on the strength of Eq (1.60), the hub moment generated by offset flapping hinges is equal to a product of the half amplitude of the flapping angle, offset of the flapping hinge, number of blades, and magnitude of the blade centrifugal force.

Longitudinal control of the side-by-side configuration would be obtained in the same way as for the single-rotor helicopter.

Longitudinal control of tandems can be best acheived through a differential variation of the magnitude of the rotor thrust. This may be accompanied by some tilt of the thrust vectors with respect to the rotor shaft. The first of these objectives is reached through the application of the differential collective pitch control while for the second, cyclic control inputs are used as in the single-rotor case.

Lateral control of the side-by-side configuration is usually obtained in exactly the same way as longitudinal control of the tandem; i.e., through differential collective pitch which again, may be combined with cyclic control inputs.

Lateral control of tandems is usually accomplished in the same way as for the single-rotor types; i.e., through cyclic control.

Directional control of the single-rotor helicopter is most often obtained from the tail-rotor thrust variation resulting from collective pitch inputs to the tail-rotor blades.

Directional control of coaxial types is commonly based on changing the torque distribution between the rotors. This is usually done through differential variation of the collective pitch of the upper and lower rotors. Other means of disturbing the torque equilibrium between the rotors can also be used; for instance, by actuating the spoilers which would cause an increase in the blade drag while the lift remains constant.

Directional control of the side-by-side configurations and tandems is usually achieved through differential inclination of the lifting-rotor thrust vectors. This is accomplished by proper cyclic control inputs.

It should be realized that—similar to the case of the previously discussed vertical control—inclination of the rotor thrust can be achieved by means other than directly governing the periodic variation of the blade pitch angle through swashplate inputs. First-harmonic airfoil circulation control through blowing or boundary layer control through blowing or suction can lead to the same results. There are other examples of cyclic variation of the blade thrust. Use of a servo-flap, located some distance behind the blade trailing edge, and twisting the torsionally soft blade is one of them (Kaman).

The most common types of helicopter controls are summarized in Table I-1.

Higher harmonics and other controls. In addition to the zero and first-harmonic blade-lift variation generally used for control of the helicopter as a whole, the blade lift may be varied according to higher harmonics. Such controls are primarily intended for such special purposes as vibration suppression, alleviation of structural loads, and improvements in rotor aerodynamic characteristics[12].

An extreme in rotor control freedom is represented by active feedback control. Here, as indicated by Kretz[13], an uncoupled blade-lift variation replaces the mutually dependent monocyclic control generated by the swashplate. Both theoretical and experimental studies indicate that this concept, called the *feedback rotor*, should contribute to the extension of the rotor flight envelope to high advance ratios ($\mu = 1.0$) by eliminating instabilities occurring in the high μ regime of flight. To accomplish this goal, information about the blade-flapping angle and blade stall conditions is fed into a closed-loop system (incorporating on-board computers) which commands the blade pitch angle through fast-response, hydroelectric actuators.

At this point it should be noted that because of the elevated frequencies of blade feathering oscillations associated with higher harmonics, as well as the short response time required in feedback rotors, a high angular acceleration about the pitch axis may be encountered. This in turn would necessitate fast-reacting actuators capable of developing high forces. To avoid the weight and power penalties which would result, designers must look for other means of varying blade lift (e.g., circulation control and flaps) which would not require high angular acceleration of components (blades) having large polar moments of inertia.

7.3 Conventional Helicopter

Basic rotor theories and their application to performance are limited in this volume to conventional helicopters. They may be defined as rotorcraft having a main rotor disc loading of $w \leqslant 70\ kG/m^2$, while propulsion, as well as most of the lift in all regimes of flight is provided by a subsonic rotor, or rotors. Configurations incorporating auxiliary forms of propulsive thrust are not considered here. However, rotor thrust augmentation through auxiliary wings is discussed in Vol II.

The most commonly known configurations conforming to the above definition of conventional helicopters are described below.

The *mechanically-driven, single-rotor helicopter,* at present, represents the most widely used configuration and thus, may be considered as the classical representative of conventional helicopters. This type covers a considerable range of gross weights—from small single-seaters of less than 450 kG (≈1000 lb) gross weight (Fig 1.25), to the crane type having a gross weight approaching 45 000 kG (≈100 000 lb) (Fig 1.26).

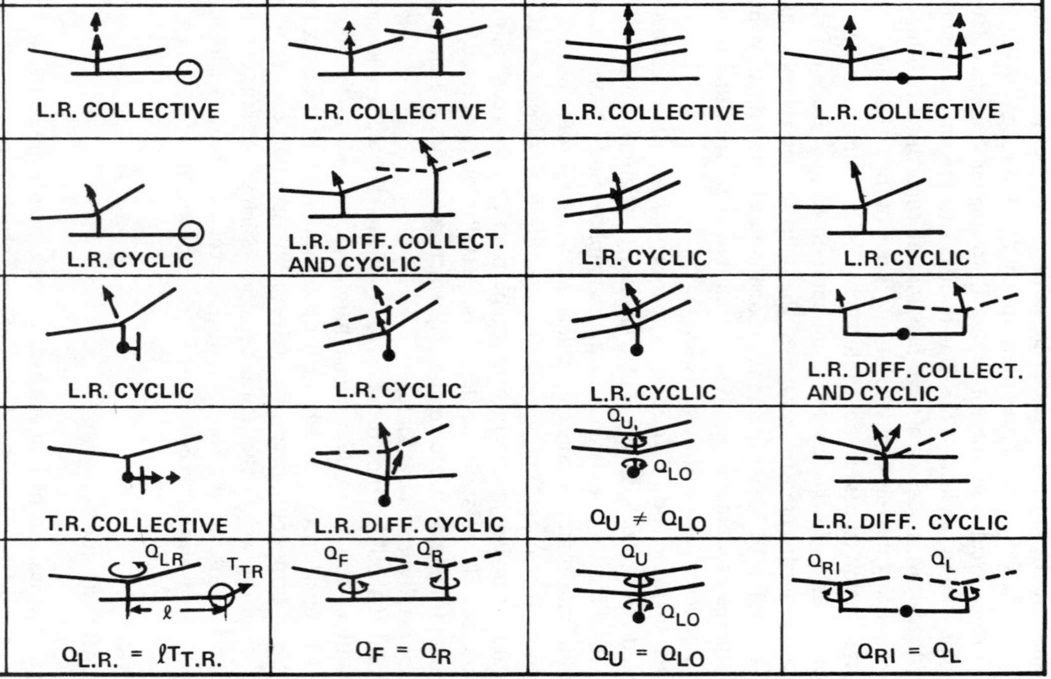

CONTROL	CONFIGURATION			
	SINGLE ROTOR	TANDEM	COAXIAL	SIDE-BY-SIDE
VERTICAL	L.R. COLLECTIVE	L.R. COLLECTIVE	L.R. COLLECTIVE	L.R. COLLECTIVE
LONGITUDINAL	L.R. CYCLIC	L.R. DIFF. COLLECT. AND CYCLIC	L.R. CYCLIC	L.R. CYCLIC
LATERAL	L.R. CYCLIC	L.R. CYCLIC	L.R. CYCLIC	L.R. DIFF. COLLECT. AND CYCLIC
DIRECTIONAL	T.R. COLLECTIVE	L.R. DIFF. CYCLIC	Q_{U_1}, Q_{LO} $Q_U \neq Q_{LO}$	L.R. DIFF. CYCLIC
MAIN ROTOR(S) TORQUE BALANCE	Q_{LR}, T_{TR}, ℓ $Q_{L.R.} = \ell T_{T.R.}$	Q_F, Q_R $Q_F = Q_R$	Q_U, Q_{LO} $Q_U = Q_{LO}$	Q_{RI}, Q_L $Q_{RI} = Q_L$

NOTATIONS: L.R. – LIFTING ROTOR; T.R. – TAIL ROTOR; F – FRONT; R – REAR; L – LEFT; RI – RIGHT; LO – LOWER; U – UPPER; T – THRUST; Q – TORQUE

TABLE I-1 SCHEMES OF HELICOPTER CONTROLS

Figure 1.25 Small mechanically-driven, single-rotor helicopter in ground trainer configuration (Del-Mar Engineering DH-2A — W ≈ 295 kG)

Figure 1.26 Large mechanically-driven, single-rotor helicopter (MIL MI-10 — W ≈ 41 000 kG)

All of these configurations incorporate tail rotors with special aerodynamic problems of their own.

Main rotors usually consist of one of the following types: *fully-articulated*, incorporating both flapping and lagging hinges (Fig 1.23a); *teetering* (Fig 1.23b); *hingeless* (Fig 1.23c), or *elastomeric* (Fig 1.23d).

Tandem. The tandem is the second most widely used configuration of the conventional helicopters. At present, it appears that this application is more and more directed toward helicopters of higher gross weights ranging from $W \approx 8600$ *kG* (19 000 lb) for the light transports (Fig 1.27) to $W \approx 60\ 000$ *kG* (130 000 lb) as represented by the new class of projected heavy-lift helicopters (Fig 1.28).

Figure 1.27 Light transport helicopter (Boeing Vertol 107-II – $W \approx 8600$ kG)

Figure 1.28 Artist's concept of XCH-62 heavy-lift tandem helicopter ($W \approx 60\,000$ kG)

As indicated in Table I-1, mutual cancellation of the rotor torques eliminates the need for a tail rotor, and yaw control is usually obtained by differential lateral inclination of the fore and aft rotor thrust vectors through the application of proper cyclic control. Although the horizontal thrust components required for forward translation of the helicopter are chiefly obtained by rotating the aircraft about its c.g. through differential thrust of the front and rear rotors, an additional horizontal component is generated in the same way as for single-rotor configurations; i.e., through cyclic control using swash-plate inputs.

This need for thrust-vector inclinations; especially, in the differential direction (yaw control), favors the application of articulated rotor hubs with both flapping and lagging hinges, or with elastomeric suspension of the blades.

As far as rotor arrangements are concerned, there appears to be a tendency toward the overlapping type. This leads to aerodynamic interferences occurring in both vertical translation and hover. In an oblique translation with a velocity component in the aircraft plane of symmetry, an aerodynamic interference between the rotors appears whether they are of the overlapping or non-overlapping types.

Coaxial. The coaxial configuration, although practically nonexistent in the U.S., represents a rather considerable class numerically, chiefly due to its popularity in the USSR (Fig 1.29).

Figure 1.29 Coaxial configuration
(Kamov – $W \approx 7300$ kG)

The coaxial helicopter may be considered as an extreme of the tandem with both rotors completely overlapping. Consequently, aerodynamic problems of this configuration may be treated as an extension of those of the tandem. Because of the mutual torque compensation by the main rotors, no antitorque rotor is required for this concept. In order to avoid structural complexity by the provision of a special tail rotor purely for yaw control, the latter is usually achieved by various means of alternating torque distribution between the two main rotors.

A new concept in coaxial configurations is represented by the ABC (advanced blade concept) flight research helicopter (Fig 1.30).

Because of truly rigid blades ($\nu/n \approx 1.4$) and a design in which large root and shaft bending moments can be tolerated, higher lift can be developed on the advancing and then, on the retreating side of each rotor. Therefore, some stall problems resulting from the necessity of operating the blades at high lift coefficients on the retreating side can be eliminated. The unbalanced rolling moments within each individual rotor will cancel each other for the aircraft as a whole.

Figure 1.30 Sikorsky 8-69 (XH-59A) prototype for evaluation of the ABC rotor system

Side-by-side. As far as pure helicopters are concerned, it appears that the side-by-side configurations (similar to the tandem) are most widely applied in the heavier gross-weight classes up to $W \approx 105\,000\ kG$ (232 000 lb). One such helicopter is depicted in Fig 1.31.

Figure 1.31 Heavy-lift type helicopter of the side-by-side configuration (MIL Mi-12 – $W \approx 105\,000\ kG$)

Similar to the tandem, yaw moments are usually obtained through differential inclination (this time, in the fore and aft direction) of the rotor thrust vectors. This requirement in turn leads to the preference for either completely articulated rotors, or rotors equipped with relatively soft blades in the flapping plane.

From an aerodynamic point of view, the side-by-side configuration can usually be treated in the first approximation as being composed of two independent rotors. However, aerodynamic interaction encountered between the rotor and the supporting structure, or the wing, should normally be considered. These interferences can easily be understood on the basis of consideration of the vertical drag of the classical and winged single-rotor helicopters in hover and axial translation. A detailed discussion on this subject can be found in Chs II and IV of Vol II.

Helicopters with Reaction-Driven Rotors. Helicopters equipped with reaction-driven rotors usually belong to the single-rotor category. There are several possible ways of obtaining reaction drive; however, it appears that prevailing concepts incorporate rotors with driving devices located in the tip region of the blades. Blade driving force is obtained by discharging either air or a mixture of air and combustion products. Depending on the temperature of the exhaust gases, these systems can be categorized as representing (a) cold, (b) warm, or (c) hot cycles. In the fifties, small cold-cycle helicopters such as the Djinn achieved operational applications. However, it is now believed that helicopters with reaction-driven rotors might be potentially competitive with the mechanically-driven concepts within the class of heavy, 50 000-kG (110 000-lb), and very heavy-lift, 100 000-kG (220 000-lb) and higher, helicopters (Fig 1.32).

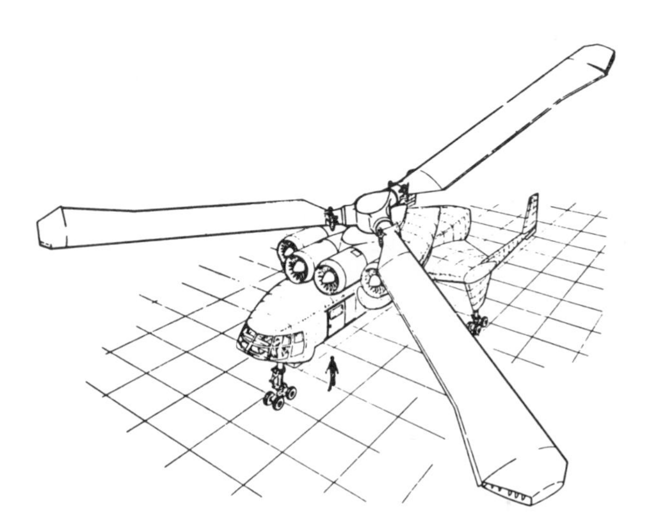

Figure 1.32 Artist's concept of a heavy-lift helicopter with a warm-cycle, jet-driven rotor

For this heavy-lift category, it appears that the warm-cycle would probably be the most suitable, although the hot-cycle, or even concepts based on special turbofans or turbojet engines mounted at the blade tips, cannot be completely excluded.

In principle, the reaction drive of the rotor eliminates the need for torque compensation. Nevertheless, small tail rotors are sometimes envisioned in order to provide yaw control in hover and low forward speeds and take care of the friction moments transmitted through the hub.

Thermodynamics of the cycle and its overall efficiency represent special problems of these configurations. However, basic aerodynamics of the classic single rotor should also be applicable to the reaction types.

In addition to the above discussed configurations, there may be many others.

However, as long as they can be classified in an aerodynamic sense as belonging to the family of conventional helicopters defined at the beginning of this section, the reader should have no difficulty in dealing with either performance predictions or other aerodynamic problems.

7.4 Tilt Rotor

Aerodynamic problems and performance predictions of tilt-rotors (Fig 1.33) in the helicopter regime of flight can be treated exactly as those of the side-by-side type. Special problems associated with the role of the wing, especially under forward-flight conditions, are somewhat similar to those of the winged helicopter discussed in Ch IV of Vol II.

Figure 1.33 Bell XV-15 tilt-rotor in the helicopter regime of flight

References for Chapter I

1. Stepniewski, W.Z. *Civilian Vertical-Lift Systems and Aircraft.* The Aeronautical Journal, Vol 78, No 762, June 1974.

2. Davis, S.J. and Stepniewski, W.Z. *Documenting Helicopter Operations from an Energy Standpoint.* NASA CR-132878, Nov. 1974.

3. Davis, S.J. and Rosenstein, H.J. *Identifying and Analyzing Methods for Reducing Energy Consumption of Helicopters.* NASA CR-144953, Nov. 1975.

4. Yuan, S.W. *Preliminary Investigation of the Karman-Yuan Helicopter Rotor System.* 11th Annual AHS Forum, Apr. 1955.

5. Cheeseman, I.G. and Seed, A.R. *The Application of Blowing to Helicopter Rotors.* Journal of the Royal Aeronautical Society, Vol 71, July 1967.

6. Dorand, R. *The Application of the Jet Flap to Helicopter Rotor Control.* Journal of the Helicopter Assn. of Great Britain, Vol 13, Dec. 1959

7 Williams, R.M. *Recent Development in Circulation Control Rotor Technology.* AGARD–CPP–111, Sept. 1972.

8. v. Karman, T. and Biot, M.A. *Mathematical Methods in Engineering.* McGraw Hill, 1940.

9. Gessow, A. and Meyers, G.C., Jr. *Aerodynamics of Helicopters.* MacMillan, 1952.

10. Nikolsky, A.A. *Helicopter Analysis.* John Wiley & Sons.

11. Reichert, G. *Basic Dynamics of Rotors.* AGARD–LS–63, 1973.

12. Kelly, M.W. and Rabbott, J.P. *A Review of Advanced Rotor Research*. Paper 77-33-17, AHS 33rd Annual Forum, May 1977.

13. Kretz, M. *Relaxation of Rotor Limitations by Feedback Control.* Paper 77-33-36, AHS 33rd Annual Forum, May 1977.

MOMENTUM THEORY

Basic momentum relationships are used in the development of physicomathematical models of lifting airscrews. The actuator disc concept is presented as a model of the ideal rotor, and performance predictions of helicopters incorporating such rotors are outlined. Induced power penalties associated with nonuniform downwash distributions and tip losses are considered. Momentum theory is applied to estimates of induced power losses of a tandem in forward flight. Finally, limitations of the simple momentum theory in modeling actual rotors is discussed.

Principal notation for Chapter II

A	area	m^2, or ft^2
D	drag	N, or lb
E	energy	N- m, or ft-lb
H	total head	N/m^2, or lb/ft^2
HP	horsepower	735N-m/s, or 550 ft-lb/s
h	altitude or height	m, or ft
k	ratio of actual power to ideal power	
k_v	download factor	
m	mass	kg, or slugs
ℓ	distance	m, or ft
P	power	N- m/s, or ft-lb/s
p	pressure	N/m^2, or psf
R	rotor or slipstream radius	m, or ft
r	radial distance	m, or ft
T	thrust	N, or lb
V	velocity	m/s, or fps
W	weight, or gross weight	N, or lb
w	area, disc loading	N/m^2, or psf
α	angle of thrust inclination	rad, or deg
γ	angle	rad, or deg
Δ	increment	
κ	ratio of ideal power available to P_{id_h}	
ρ	air density	kg/m^3, or $slugs/ft^3$

Subscripts

av	available
ax	axial
c	climb
d	descent
ds	downstream

e	effective, or economic
eq	equivalent
f	forward
fp	flight path
fr	front
free	free
fs	freestream
h	hover
ho	horizontal
id	ideal
ind	induced
mix	mixed
o	initial
opt	corresponding to maximum range
re	rear
s	slipstream
u	ultimate
up	upstream
v	vertical
w	wind

Superscripts

$\rightarrow$	vector
$-$	dimensionless relative value
$\cdot$	differential with respect to time
$'$	resultant flow

1. INTRODUCTION

The idea of using an airscrew as a direct lift-producing device is not new. Sketches and models made by Leonardo da Vinci indicate that he worked along these lines at the end of the 15th century. Nevertheless, with the emergence of fixed-wing aircraft and dirigibles, the airscrew found its principal application as a device producing the thrust required to overcome the drag in forward flight. Therefore, the airscrew theory was originally developed chiefly for its application as a propeller. Later, when the helicopter concept began to receive more attention and practical thought, the already developed propeller theories served as a guide for analysis of the helicopter rotor.

Obviously, propeller theories were primarily concerned with the movement of the airscrew along its axis and generation of static thrust. But the most attractive feature of helicopters and VTOL aircraft is their ability to climb vertically and to hover without any motion of the aircraft as a whole. Consequently, for both of these regimes of flight, the study of propellers provided theories which could be applied directly to the helicopter and rotor/propeller VTOL's.

Through the 19th century, theories were developed in conjunction with the steadily growing application of propellers in naval systems and thus, became another source of information and inspiration in rotor analysis. The works of Rankine (1865) and Froude (1878-1889) which served as a foundation for the theory of ship propellers, also served as a basis for the momentum theory considered in this chapter. As their approach represents one of the simplest explanations of the problem of thrust generation by a propeller or helicopter rotor, it is also extremely suitable for the physical interpretation of numerous flight phenomena. It seems advisable, therefore, to begin the study of rotary-wing aerodynamics by becoming acquainted with the principles of the simple momentum theory.

2. SIMPLEST MODEL OF THRUST GENERATION

The basic relationships of Newtonian mechanics can lead to the development of a simple, but at the same time quite universal, physicomathematical model of a thrust generator. Without going into any details regarding either geometric characteristics or the modus operandi of the thrust-generator itself, it is simply assumed that under static conditions as well as in translation at velocity $\vec{V}$ with respect to the ideal (frictionless and incompressible) fluid of density ρ and pressure p_o, the as-yet-undefined device is somehow capable of imparting linear momentum to the medium.

For the sake of convenience, it is usually postulated that the thrust generator remains stationary while a very large mass of fluid moves past it at a uniform velocity $(-\vec{V})$, (Fig 2.1). It will also be assumed that the thrust coincides with the positive axis of a coordinate system having its origin at the "center" of the thrust generator.

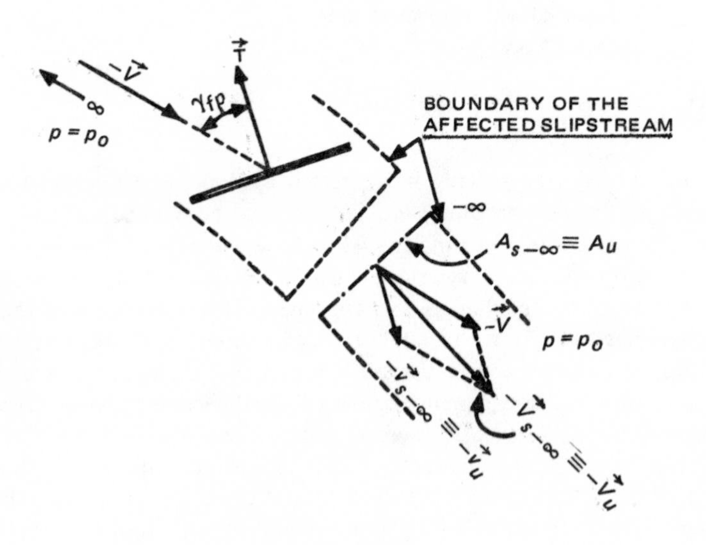

Figure 2.1 Simplest physicomathematical model of a thrust generator

The thrust generator interacts with the fluid by imparting uniformly distributed linear momentum to a distinct streamtube bound by a surface through which the mass cannot be exchanged. This means that by the law of continuity, the mass flow within the tube is the same at every section, while both velocity and pressure of the fluid alter. However, at some point far downstream, the pressure returns to p_o, and the incremental velocity variation reaches its ultimate value of $-\vec{v}_u$ uniformly distributed over the final tube cross-section area, A_u.

Knowledge of $\vec{v}_u$ and A_u in addition to the already known $\vec{V}$ and ρ represents all the necessary information for determining the thrust $\vec{T}$ generated by this very simple physicomathematical model, as well as for computing the power required in that process.

According to the laws of classical mechanics, the direction of the generated thrust ($\vec{T}$) will be opposite to that of $\vec{v}_u$, while its magnitude will be equal to the rate of momentum change within the streamtube between its final and initial values. Denoting the rate of mass flow by $\dot{m}$, force $\vec{T}$ becomes

$$\vec{T} = -\dot{m}(\vec{V}_u - \vec{V}) \tag{2.1}$$

where the resultant velocity of flow far downstream ($\vec{V}_u$) is $\vec{V}_u = \vec{V} + \vec{v}_u$. Consequently, Eq (2.1) becomes

$$\vec{T} = -\dot{m}\vec{v}_u. \tag{2.1a}$$

Furthermore, since $\dot{m} = V_u A_u \rho$, the above equation can be rewritten as

$$\vec{T} = -|\vec{V} + \vec{v}_u| A_u \rho \vec{v}_u \tag{2.1b}$$

where $|\ |$ denotes the absolute (scalar) value of the resultant vector, $\vec{V}_u$; while $\vec{v}_u$ from now on will be known as the *fully-developed induced velocity*.*

Inspection of Eqs (2.1a) and (2.1b) can teach us that in dynamic generation of a given thrust ($\vec{T}$), a tradeoff can be made between the magnitude of a fully-developed induced velocity ($\vec{v}_u$), and the mass flow which, in turn, depends on (a) cross-section of the streamtube far downstream, A_u; (b) absolute value of the resultant velocity of flow within the tube itself far downstream, V_u; and (c) density of fluid, ρ. This of course would permit further tradeoffs between the above parameters within a constant $\dot{m}$.

The above general conclusions already contribute to some understanding of dynamic thrust generation, but in order to get a still deeper insight into this matter, it is also necessary to consider the power (P) required in the process. This can be done by examining the difference in the rate of flow of kinetic energy through a cross-section of the streamtube far downstream in the ultimate wake ($\dot{E}_u$) and far upstream ($\dot{E}_{up}$).

---

*It should be noted that when SI units (kg/s for mass flow and m/s for velocity) are used in the above equations, thrust T is obtained in newtons which can be converted to kilogram force by multiplying the answer by $1/9.807 \approx 0.102$. In English units; i.e., when mass flow is computed in slugs/s and velocity in fps, T is given in pounds.

$$P = \dot{E}_u - \dot{E}_{up} = \tfrac{1}{2}\dot{m}(\vec{V}_u^2 - \vec{V}^2) \qquad (2.2)$$

Remembering that $\vec{V}_u = \vec{V} + \vec{v}_u$, and performing subtraction as indicated in Eq (2.2), one obtains

$$P = \tfrac{1}{2}\dot{m}v_u(2V + v_u), \qquad (2.2a)$$

but $\dot{m}\vec{v}_u = -\vec{T}$; hence, Eq (2.2a) can be rewritten as follows:

$$P = -(\vec{T}\cdot\vec{V} + \tfrac{1}{2}\vec{T}\cdot\vec{v}_u). \qquad (2.2b)$$

The above equation shows that power required by our simple model of a thrust generator is equal to minus the sum of a dot (scalar) product of the velocity of the incoming flow and the developed thrust, and one-half of another dot product of thrust and fully-developed induced velocity.

Since $\vec{T}$ and $\vec{v}_u$ are in opposite directions (forming a 180° angle), their dot product $(\vec{T}\cdot\vec{v}_u)$ is negative. Hence, the second term in the parentheses in Eq (2.2b) is always negative and in view of the minus sign in front of the parentheses, the contribution of that product to the power required is always positive. In other words, a power input is always required to cover energy losses associated with the induced velocity needed in the process of dynamic thrust generation.

The first term $(\vec{T}\cdot\vec{V})$ in the parentheses of Eq (2.2b) can either be negative (angle between thrust $\vec{T}$ and the flight path $\vec{V}$ is $0 \leqslant \gamma_{fp} < 90°$), or positive ($90° < \gamma_{fp} \leqslant 180°$). In the latter case, the total amount of power required to be inputted into the thrust generator may decrease, reduce to zero, or even become negative (windmill).

For the case of actual flight, Eq (2.2b) can be rewritten in nonvectorial notations:

$$P = TV\cos\gamma_{fp} + \tfrac{1}{2}Tv_u. \qquad (2.2c)$$

It should be noted at this point that in developing our simplest mathematical model, an ideal fluid was assumed, with no dissipation of energy through friction or energy transfer to the fluid under the form of turbulent wakes. Furthermore, a uniform distribution of the fully-developed induced velocity has been assumed.

It will be shown in Sect 5 that uniformity of the fully-developed downwash velocity is synonymous with minimization of the power required to generate a given thrust under assumed conditions of flight velocity, air density, and the cross-section area of the fully-developed slipstream. Thus, the power expressed by Eqs (2.2b) or (2.2c) may be called the *ideal power* (P_{id}) required, while its part related to the thrust-generating process and represented by the second term in Eqs (2.2b) or (2.2c) is called *ideal induced power* ($P_{id_{ind}}$). It should also be noted that the assumption of a steady-state motion signifies that the thrust T is in balance with all of the other forces (aerodynamic, gravitational, etc.) acting on the thrust generator.

It may be of interest to find out that even this simplest physicomathematical model of a thrust generator may prove helpful in understanding some important trends in the VTOL field. For instance, (Eq (2.2c) indicates that for the case of static thrust generation which, for VTOL, is synonymous with hovering out-of-ground effect, the ideal power required would be:

$$P_{id_h} = \tfrac{1}{2} T v_u \tag{2.3}$$

where, in SI units, the answer will be in N m/s and in English units in ft-lb/s. Thrust in kilograms of force ($kG = 9.807N$) per horsepower will be

$$T/P_{id_h} = 150/v_u \quad \text{kG/hp}$$

or in pounds, (2.3a)

$$T/P_{id_h} = 1100/v_u \quad \text{lb/hp.}$$

The above expression clearly shows that in order to obtain the highest possible power economy (maximum thrust per horsepower) in the generation of static thrust, one should strive for the lowest possible induced velocity in the fully-developed slipstream. The relationship given in Eq (2.3a) is plotted in Fig 2.2 (reproduced from Ref 1) and it remarkably well indicates the actual trend in static power loading from helicopters to rockets.

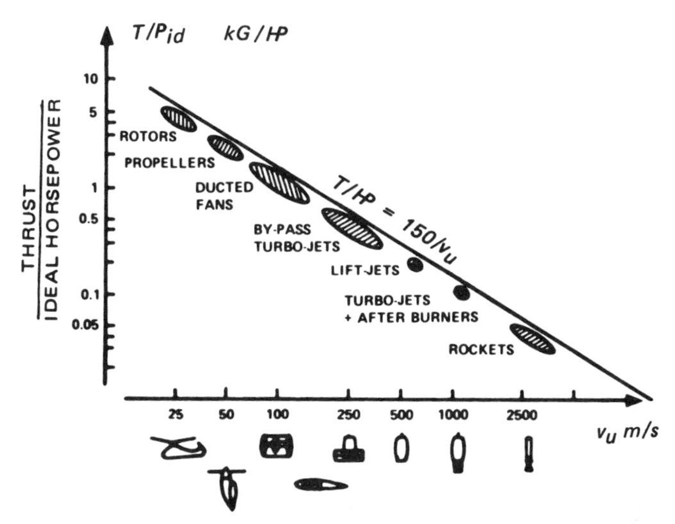

Figure 2.2 Trend in static thrust to ideal power ratios for various VTOL concepts

In the previously considered model of the thrust generator, no attempt was made to describe its geometry nor to explain the actual process of imparting linear momentum to the fluid. For this reason, our simplest model, although helpful in predicting some general trends, would be of little help in dealing with practical problems of design and performance predictions of rotary-wing aircraft. Thus, another model reflecting at least some geometric characteristics of open airscrews (rotors and propellers) is required.

3. ACTUATOR DISC

The *actuator disc*, while still representing a very simple physicomathematical model, is better suited for the simulation of an open airscrew than the model described above. In this case, a disc—perpendicular to the generated thrust and capable of imparting axial momentum to the fluid as well as sustaining pressure differential between its upper and lower surface—is substituted for an actual rotor-propeller. This concept may be considered as a sublimation of the idea of an airscrew having a large number of blades.

As in the previous section, it is assumed that the disc remains stationary while a large mass of fluid of density ρ and pressure p_o flows around it at a velocity $-\vec{V}$. Under these conditions, the mechanism of thrust generation is explained as follows: Fluid passing either through the disc or in its vicinity acquires induced velocity v which is uniform over the entire disc and is directed opposite to the thrust. It is again assumed that the fluid is ideal. Consequently, rotation of the disc does not encounter any friction or form drag as it imparts purely linear momentum (in the $-T$ direction) to the passing fluid. This of course means that there is no rotation of the slipstream.

3.1 Induced Velocity and Thrust in Axial Translation

Axial translation in climb can obviously be either in the thrust direction as in vertical climb of a rotorcraft, or in the opposite direction as in vertical descent. In the first case let us consider a rotor or propeller moving in the thrust direction (climbing) at a constant velocity V_c, while developing thrust T. Here, an equivalent motion is substituted when the thrust generator remains stationary while the air flows past it in the axial direction (far from the rotor) at a speed of $-V_c$ (Fig 2.3).

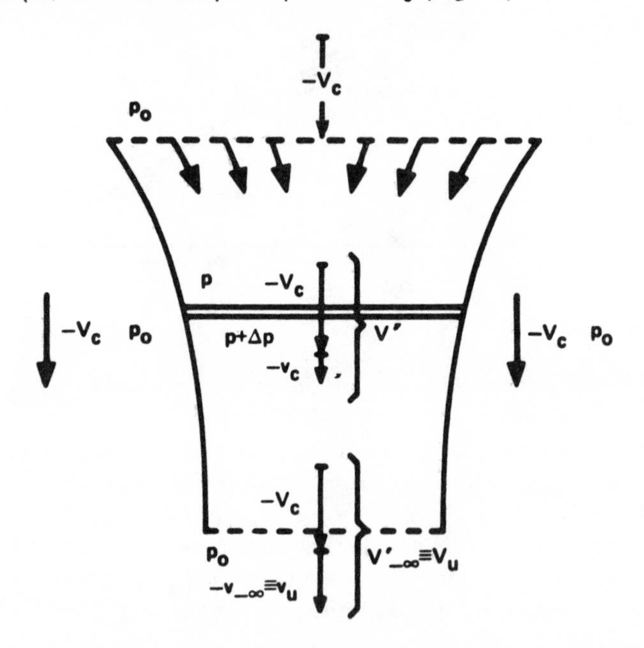

Figure 2.3 Scheme of flow corresponding to vertical climb

Similar to the simple model, a single axis coordinate system is selected with its positive direction coinciding with that of the thrust T. Air particles approaching the actuator disc acquire some additional velocity that reaches a $-v_c$ value at the disc itself.

After passing through the disc, the speed of flow increases still further until, far downstream, the induced velocity reaches its ultimate value of $-v_u$ while the resultant velocity of flow becomes $V_u = -V_c - v_u$, and pressure returns to that of the surrounding air; i.e., it becomes p_o. Remembering that the mass flow within the streamtube is constant, its shape will probably resemble that shown in Fig 2.3.

In order to physically explain the thrust-generating mechanism of the actuator disc, it may be assumed that pressure above the disc is lower than that below it. Because of this discontinuity in pressure, Bernoulli's equation can only be applied separately to the upstream and downstream parts of the flow tube. For the upstream part of the tube, the total head (H_o) should be the same.

$$H_o = p_o + \tfrac{1}{2}\rho V_c^2 = p + \tfrac{1}{2}\rho(V_c + v_c)^2 \tag{2.4}$$

where v_c is the induced velocity at the disc and p is the pressure just above the disc surface.

The same is true for the total head (H_{ds}) of the downstream part:

$$H_{ds} = p_o + \tfrac{1}{2}\rho(V_c + v_u)^2 = p + \Delta p + \tfrac{1}{2}\rho(V_c + v_c)^2 \tag{2.5}$$

where $p + \Delta p$ represents the pressure just below the disc, with Δp being the pressure differential at the disc.

Subtracting H_o from H_{ds}, one obtains:

$$\Delta p = \rho(V_c + \tfrac{1}{2}v_u)v_u. \tag{2.6}$$

Consequently, thrust developed by a disc of radius R can be expressed as $T = \pi R^2 \Delta p$ or, in terms of Eq (2.6):

$$T = \pi R^2 \rho(V_c + \tfrac{1}{2}v_u)v_u. \tag{2.7}$$

On the other hand, according to Eq (2.1a), the total thrust T can be expressed in this case as

$$T = \pi R^2 \rho(V_c + v_c)v_u. \tag{2.8}$$

Equating the right-hand sides of Eqs (2.7) and (2.8), one finds that

$$v_c = \tfrac{1}{2}v_u \tag{2.9}$$

or

$$v_u = 2v_c. \tag{2.9a}$$

Substitution of this new value of v_u into Eq (2.8) results in

$$T = 2\pi R^2 \rho(V_c + v_c)v_c \tag{2.10}$$

or denoting the total thrust generating area of any actuator disc-like device as A, Eq (2.10) may be rewritten more generally as

$$T = 2A\rho(V_c + v_c)v_c. \tag{2.10a}$$

Eq (2.10a) can be solved for v_c, thus obtaining an expression for induced velocity at the disc

$$v_c = -\tfrac{1}{2}V_c + \sqrt{\tfrac{1}{4}V_c^2 + (T/2A\rho)}. \qquad (2.11)$$

Remembering that $T/A \equiv w$ is the disc loading or, in more general terms, the thrust area loading, it can be rewritten as follows:

$$v_c = -\tfrac{1}{2}V_c + \sqrt{\tfrac{1}{4}V_c^2 + (w/2\rho)}. \qquad (2.11a)$$

In hover or under any static conditions at a speed $V_c = 0$;

$$T = 2A\rho v_h^2, \qquad (2.12)$$

while the induced velocity (v_h) becomes

$$v_h = \sqrt{w/2\rho}. \qquad (2.13)$$

As to the direction of the induced velocity vector ($\vec{v}_c$), it was assumed at the beginning of the development of relationships given by Eqs (2.11) to (2.13) that $\vec{v}_c$ is opposite to the thrust $\vec{T}$, and in the same direction as the relative flow of air ($-\vec{V}_c$) resulting from climb. Under these assumptions, positive values of v_c simply indicate that the initial assumption of the $\vec{v}_c$ direction is correct.

It can be seen that the last term under the square root sign in Eq (2.11a) is the square of induced velocity in hovering. Eq (2.11a) can then be rewritten as follows:

$$v_c = -\tfrac{1}{2}V_c + \sqrt{\tfrac{1}{4}V_c^2 + v_h^2}. \qquad (2.14)$$

The above expression can be nondimensionalized by dividing both sides by v_h and defining two nondimensional velocities: (1) the nondimensional induced velocity, $\overline{v}_c \equiv v_c/v_h$, and (2) the nondimensional rate of climb, $\overline{V}_c \equiv V_c/v_h$ or, more generally, the nondimensional rate of axial translation in the direction of thrust, $\overline{V}_{ax} \equiv V_{ax}/v_h$. For the case of climb, Eq (2.14) can now be rewritten as

$$\overline{v}_c = -\tfrac{1}{2}\overline{V}_c + \sqrt{\tfrac{1}{4}\overline{V}_c^2 + 1}. \qquad (2.14a)$$

This equation shows how the nondimensional induced velocity varies (decreases) from its maximum hover value of 1.0 as the rotor starts to axially translate in the direction of thrust at a rate of $\overline{V}_c$ or $\overline{V}_{ax}$. The relationship expressed by Eq (2.14a) is shown graphically in the right side of Fig 2.4.

Descent in General. Eqs (2.11) to (2.14a) which establish relationships between thrust and induced velocity in vertical climb can be modified for vertical descent. This can be simply done by changing the sign of V_c and $\overline{V}_c$ in the first term on the right side of Eqs (2.11) to (2.14a) from "–" to "+". However, in order to clearly indicate that the considered case now refers to vertical descent, the symbol $V_d \equiv -V_c$ is used for dimensional, and $\overline{V}_d \equiv -\overline{V}_c$ is used for nondimensional rates of descent, while the corresponding induced velocities will be symbolized as v_d and $\overline{v}_d$, respectively. Thus,

$$v_d = \tfrac{1}{2}V_d + \sqrt{\tfrac{1}{4}V_d^2 + (T/2A\rho)}, \qquad (2.15)$$

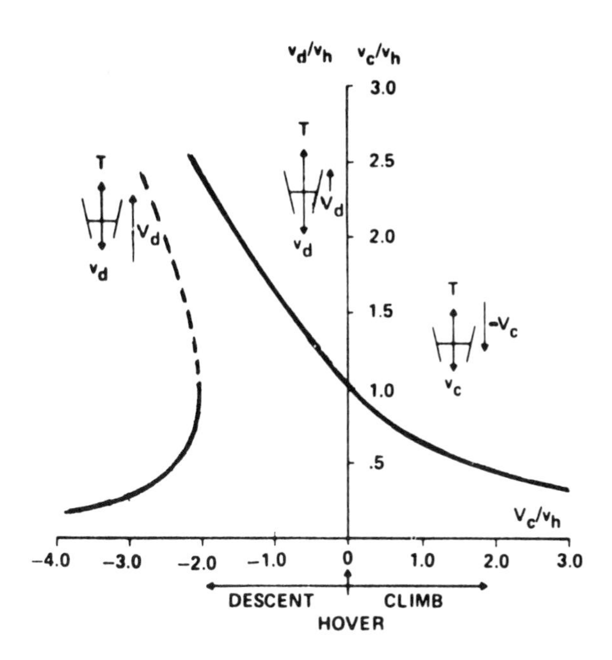

Figure 2.4 Nondimensional induced velocity vs nondimensional rate of climb or descent

$$v_d = \tfrac{1}{2}V_d + \sqrt{\tfrac{1}{4}V_d^2 + (w/2\rho)}\,, \tag{2.15a}$$

or

$$v_d = \tfrac{1}{2}V_d + \sqrt{\tfrac{1}{4}V_d^2 + v_h^2}\,. \tag{2.15b}$$

In nondimensional form where $\overline{v}_d \equiv v_d/v_h$ and $\overline{V}_d \equiv V_d/v_h$,

$$\overline{v}_d = \tfrac{1}{2}\overline{V}_d + \sqrt{\tfrac{1}{4}\overline{V}_d^2 + 1}\,. \tag{2.15c}$$

A plot representing Eq (2.15c) was added to Fig 2.4 as an extension of the $\overline{v}_c = f(V_c)$ curve to the descent region.

It can be seen from Eq (2.15a) that in vertical descent, $\overline{v}_d$ increases with the increasing $\overline{V}_d$ and when $\overline{V}_d \gg 1.0$, $\overline{v}_d \rightarrow \overline{V}_d$.

It should be realized at this point that the above-presented results are based on the conceptual model, assuming that regardless of the high level of the rate of vertical descent, downwash velocity through the rotor is still so high that the resultant flow through the rotor is always in a direction opposite to the thrust (i.e., downward). This obviously forces an increase of $\overline{v}_d$ with that of $\overline{V}_d$ until at high levels of $\overline{V}_d$, the $\overline{v}_d$ values start to approach those of $\overline{V}_d$. Consequently, the rate of flow through the disc approaches zero $[\pi R^2 \rho(v_d - V_d) \rightarrow 0]$ and the whole concept of the actuator disc acting on the air passing through it becomes meaningless. It can be seen hence that the validity of the physicomathematical model based on the general type of flow shown in Fig 2.3, although apparently satisfactory for the case of vertical climb, hovering, and moderate rates of vertical descent, should be reexamined for fast vertical descent.

Fast Descent. Fast vertical descents can be defined as those cases where the rate of descent is sufficiently high to sustain an unbroken flow in the thrust direction (i.e., up) within the whole streamtube. The necessary condition would obviously be $|V_d/v_d| > 2.0$. When this condition is fulfilled, one can imagine a special shape of a streamtube effected by a stationary rotor submerged in a large air mass flowing in the positive direction at speed V_d (Fig 2.5).

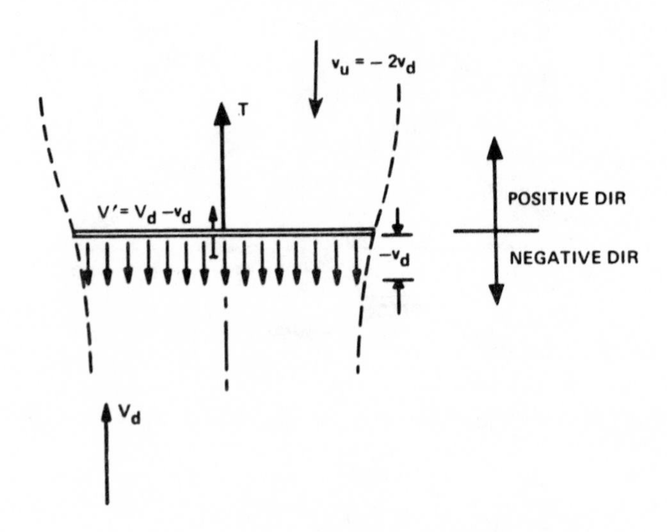

Figure 2.5 Flow patterns in vertical descent

Far below the rotor, the airstream velocity is $V_d \equiv -(-V_c)$. However, as the flow approaches the disc, the rate of flow is reduced under the influence of the rotor downwash and the airstream widens.

If the downwash at the rotor itself is $-v_d$, then the rate of flow through the rotor disc will obviously be $V' = V_d - v_d$. It can be shown, as for the rate of climb, that the downwash far downstream from the rotor (in this case, above the rotor) will reach a maximum value equal to twice the induced velocity at the disc: $v_u = 2v_d$.

Under the assumption of $|V_d/v_d| > 2.0$ (Fig 2.5), the relationship for thrust can be written as follows:

$$T = 2\pi R^2 \rho (V_d - v_d) v_d.$$

Substituting $v_h{}^2$ for $T/2\pi R^2 \rho$ and solving the above equation for v_d, one obtains:

$$v_d = \tfrac{1}{2} V_d - \sqrt{\tfrac{1}{4} V_d^2 - v_h^2}\,, \tag{2.16}$$

or, in nondimensional form,

$$\bar{v}_d = \tfrac{1}{2}\bar{V}_d - \sqrt{\tfrac{1}{4}\bar{V}_d^2 - 1}\,. \tag{2.16a}$$

A plot representing Eq (2.16a) is also added to Fig 2.4 (solid-line portion of left-hand side curve).

It should be noted that the selection of the "–" sign before the square root in Eqs (2.16) and (2.16a) assures that $|V_d/v_d| \geqslant 2.0$. This means that the condition of a continuous unidirectional (up) flow within the streamtube is fulfilled. By contrast if the sign before the square root is "+", then in general, $|V_d/v_d| \leqslant 2.0$. At high rates of vertical descent where $\bar{V}_d \gg 1.0$, the ratio $|V_d/v_d|$ would approach 1.0 (broken line portion of the left-hand side curve in Fig 2.4).

As in the case of the previously considered model (Fig 2.3) in the above limiting case, no flow would go through the disc. However, even for the intermediate values of $1.0 < |V_d/v_d| < 2.0$, a situation would exist in which the concept of a continuous streamtube is incompatible with the unidirectional flow. It can be seen that for $|V_d/v_d|$ within the above limits, the flow in the streamtube above the disc would come to a stop and then reverse its direction. It may be expected hence that in reality, a new different flow pattern would establish itself before the postulated complete reversal of the direction of flow occurs.

For the limiting case of $|V_d/v_d| = 2.0$ which corresponds to $|\bar{V}_d| = 2.0$, the air far above the disc will come to rest.

3.2 Contraction and Expansion of the Slipstream

Knowledge of the fully-developed induced velocity ($v_u = 2v_c$ or $v_u = 2v_d$) would permit determination of the slipstream contraction or expansion (ratio of the radius of the fully-developed slipstream, R_u, to that of the actuator disc, R). For the case of climb or more generally, axial translation in the thrust direction,

$$R_u/R = \sqrt{[1 + (V_c/v_c)]/[2 + (V_c/v_c)]}\,. \tag{2.17}$$

Dividing V_c and v_c by v_h, Eq (2.17) becomes:

$$R_u/R = \sqrt{[1 + (\bar{V}_c/\bar{v}_c)]/[2 + (\bar{V}_c/\bar{v}_c)]}\,. \tag{2.17a}$$

Substituting the $\bar{v}_c$ values from Eq (2.14a) into the above equation, the following relationship expressing R_u/R in terms of $\bar{V}_c$ is obtained:

$$R_u/R = \sqrt{\left(\tfrac{1}{2}\bar{V}_c + \sqrt{\tfrac{1}{4}\bar{V}_c^2 + 1}\right)\Big/2\sqrt{\tfrac{1}{4}\bar{V}_c^2 + 1}}\,. \tag{2.17b}$$

It can be seen from Eq (2.17b) that for $\bar{V}_c = 0$ (i.e., in hover), $R_{u_h}/R = 0.707$. As $\bar{V}_c$ increases, so does the R_u/R ratio; and for $\bar{V}_c \gg 1.0$, contraction of the slipstream tends to disappear (Fig 2.6).

When vertical climb is replaced by vertical descent, $\bar{V}_c$ becomes negative, $\bar{V}_c < 0$; and the far-away flow is now directed upward. However, if it is postulated that the resulting flow in the streamtube remains continuously directed in the negative direction (opposite to the thrust vector) as in the case of climb (Fig 2.3), then the R_u/R ratio would tend to decrease as the absolute value of $-\bar{V}_c$ becomes higher. Finally, for $\bar{V}_c \to -\infty$, $R_u/R \to 0$.

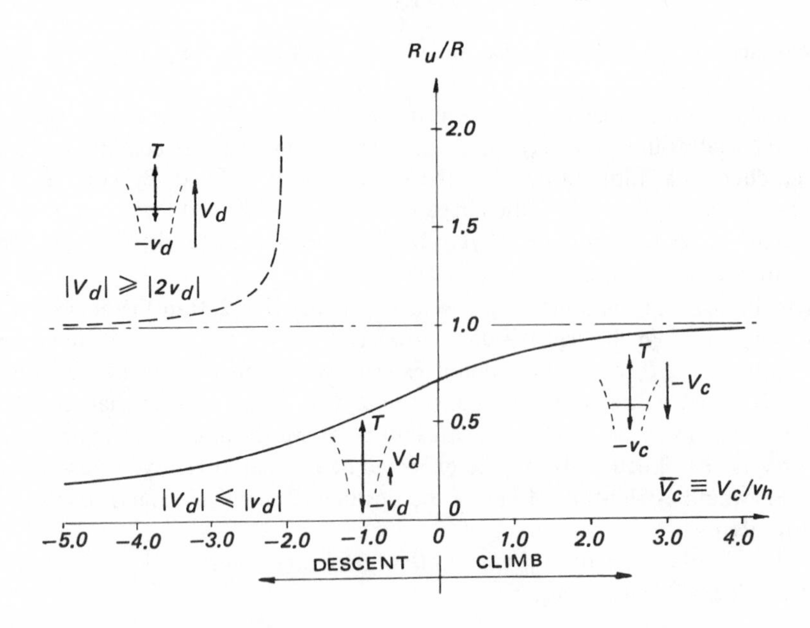

Figure 2.6 Ratio of fully-developed wake radius to that of the actuator disc vs V_c

It is obvious that maintaining the above postulated flow pattern would require that $|v_d|/|V_d| > 2.0$. This requirement, in practice, can probably be fulfilled in partial-power descent at low descent speeds, but outside of this regime of flight, the very high induced velocities needed in this flow scheme would not qualify it as a model properly reflecting physical reality.

For the flow schemes in vertical descent corresponding to the so-called fast descent depicted in Fig 2.5, the R_u/R ratio is as follows:

$$(R_u/R)_d = \sqrt{[(\overline{V}_d/\overline{v}_d) - 1]/[(\overline{V}_d/\overline{v}_d) - 2]}\,. \tag{2.18}$$

Substituting the $\overline{v}_d$ values given in Eq (2.16a) into the above equation, the following is obtained:

$$(R_u/R)_d = \sqrt{\left(\tfrac{1}{2}\overline{V}_d + \sqrt{\tfrac{1}{4}\overline{V}_d^2 - 1}\right)\Big/ 2\sqrt{\tfrac{1}{4}\overline{V}_d^2 - 1}}\,. \tag{2.18a}$$

The above equation is also plotted in Fig 2.6. From this figure as well as an inspection of Eq (2.18a), it can be seen that for $\overline{V}_d \gg 1.0$, $(R_u/R)_d \to 1.0$. However, as $|\overline{V}_d| \to 2.0$; $R_u/R \to \infty$, while $V_d - 2v_d \to 0$. This would mean that the air flow in the streamtube (Fig 2.5) would come to rest with respect to the rotor disc, while the whole mass of air outside of the tube would still flow upward at a speed equal to V_d. Of course, this situation is not acceptable from a physical point of view, and it is more reasonable to assume that before this ultimate state of air coming to rest is reached, a new flow pattern would be created.

3.3 Ideal Power in Climb and Hovering

As in the case of the simplest thrust generator model, power required by the actuator disc either for climb or hover may again be called the ideal power. This is justified by the previously made assumptions that (a) there are no friction or form drag losses, (b) the whole disc up to the limit of its geometric dimensions is participating in thrust generation, and (c) the downwash velocity is uniform at any slipstream cross-section.

Expressions for the ideal power in vertical climb (P_{id_c}) can readily be obtained from the relationship previously established in Sect 2. When a proper value of the fully-developed downwash velocity ($v_u = 2v_c$) and $\cos \gamma_{fp} = 1.0$ are introduced into Eq (2-2c), then HP_{id_c} becomes

$$\text{(SI)} \qquad HP_{id_c} = T(V_c + v_c)/735.$$
$$\text{(ENG)} \qquad HP_{id_c} = T(V_c + v_c)/550. \qquad (2.19)$$

Eq (2.19) shows that for an idealized rotor as modeled by the actuator disc developing thrust T, the power required in steady climb at a rate V_c is equal to the product of the thrust and the sum of the rate of climb plus the induced velocity at the disc (v_c).

Thus, in the case of a steady vertical ascent with no download when $T = W$ (W being the weight of the aircraft in newtons or pounds), the total ideal power needed to climb is equal to the power required to overcome gravity (WV_c) plus the ideal induced power (Wv_c).

By substituting the expression for v_c from Eq (2.11a) into Eq (2.19), the following explicit relationship between HP required and rate of climb V_c (for $T = W$) is obtained:

$$\text{(SI)} \qquad HP_{id_c} = W[\tfrac{1}{2}V_c + \sqrt{\tfrac{1}{4}V_c^2 + (W/2A\rho)}]/735.$$
$$\text{(ENG)} \qquad HP_{id_c} = W[\tfrac{1}{2}V_c + \sqrt{\tfrac{1}{4}V_c^2 + (W/2A\rho)}]/550. \qquad (2.20)$$

Eq (2.20) is also valid for an axial motion in the direction of thrust when V_{ax} is substituted for V_c, and T for W.

Remembering that $W/2\rho A \equiv w/2\rho = v_h^2$, Eq (2.20) can be rewritten in either N-m/s or ft-lb/s as follows:

$$P_{id_c} = W(\tfrac{1}{2}V_c + \sqrt{\tfrac{1}{4}V_c^2 + v_h^2}), \qquad (2.20a)$$

or in terms of nondimensional rate of climb $\overline{V}_c \equiv V_c/v_h$, it becomes

$$P_{id_c} = Wv_h(\tfrac{1}{2}\overline{V}_c + \sqrt{\tfrac{1}{4}\overline{V}_c^2 + 1}). \qquad (2.20b)$$

P_{id_c} can be expressed as the ideal power required in hover P_{id_h} times a factor κ: $P_{id_c} \equiv \kappa P_{id_h}$ where, in turn, $P_{id_h} = Wv_h$. Substituting the above expressions into Eq (2.20b), one obtains

$$\kappa = \tfrac{1}{2}\overline{V}_c + \sqrt{\tfrac{1}{4}\overline{V}_c^2 + 1}. \qquad (2.20c)$$

In the case of hovering or under static thrust conditions in general, $V_c = 0$, and Eq (2.19) would reduce to the following expression for HP required:

$$\text{(SI)} \qquad P_{id_h} = Tv/735 .$$
$$\text{(ENG)} \qquad P_{id_h} = Tv/550 . \qquad (2.21)$$

The above result could have been obtained directly from Eq (2.3), remembering that $v_u = 2v_h$ for the actuator disc.

Substituting the value for v_h from Eq (2.13) and rewriting Eq (2.21) in terms of the reciprocal of horsepower loading; i.e., $(P/T)_h$ and disc loading w, one obtains

$$\text{(SI)} \qquad (HP/T)_h = \sqrt{w/2\rho}/735$$
$$\text{(ENG)} \qquad (HP/T)_h = \sqrt{w/2\rho}/550 \qquad (2.21a)$$

where, for the case of hovering with no download, $T = W$ and thus, $w \equiv T/A = W/A$.

Eq (2.21) indicates that if a physicomathematical model based on the actuator disc concept could truly represent actual rotor/propellers, there would be no lower limit for the power required to produce a given static thrust. It would only be necessary to make the disc loading (w) as low as possible. It should be noted that in spite of the fact that the profile drag is neglected in the actuator disc concept, the disc loading is still the most important parameter as far as achieving a set goal of minimizing the HP/T ratio of actual rotary-wing aircraft is concerned. However, high structural weight and operational difficulties encountered by aircraft having low disc-loading rotors force the designers to optimize their designs around higher w values as indicated in Fig 1.1.

3.4 Vertical Climb Rates

Nondimensional rates in vertical climb can be simply obtained by solving Eq (2.20c) for $\overline{V}_c$:

$$\overline{V}_c = \kappa - (1/\kappa) . \qquad (2.22)$$

Remembering that the κ factor represents a ratio of the ideal power available to the ideal power required in hovering ($\kappa \equiv P_{id_{av}}/P_{id_h}$), Eq (2.22) can be plotted as shown in Fig 2.7.

Problems of predicting the rate of vertical ascent as well as absolute and/or operational ceiling (corresponding to a prescribed rate-of-climb value) of actual rotary-wing aircraft can be reduced to the case of the ideal actuator disc considerations; hence, a way of determining the dimensional rate of climb vs altitude becomes of interest.

Dimensional rates of climb for $T = W$ can easily be obtained from Eq (2.22) by multiplying both sides of this equation by $v_{id} = \sqrt{W/2\pi R^2 \rho}$ and expressing the κ ratio in terms of ideal available power ($P_{id_{av}}$) and ideal hovering power, $P_{id_h} = W\sqrt{W/2\pi R^2 \rho}$. After performing these operations and expressing powers in units of HP, the following is obtained.

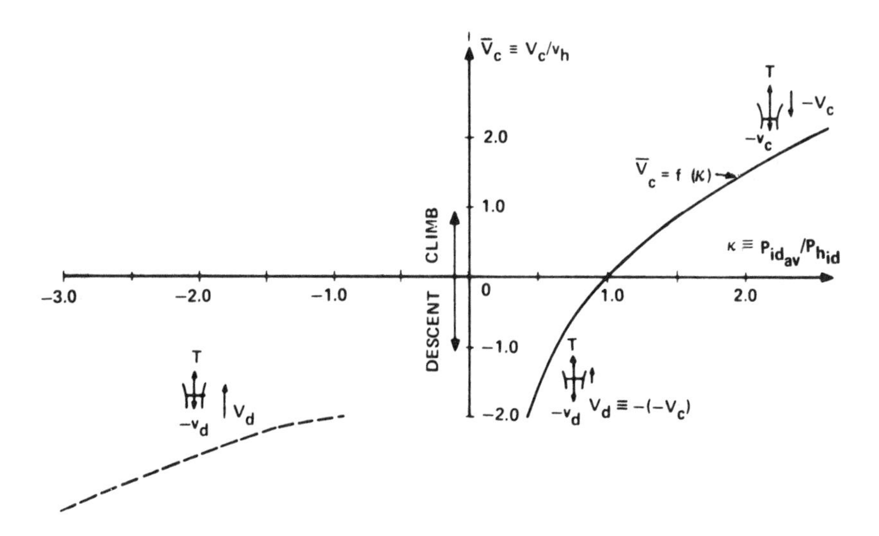

Figure 2.7 Nondimensional rate of climb and descent vs power setting

$$\text{(SI)} \qquad V_c = (735\,HP_{id_{av}}/W) - (W^2/1470\,\pi R^2 \rho HP_{id_{av}}).$$

$$\text{(ENG)} \qquad V_c = (550\,HP_{id_{av}}/W) - (W^2/1100\,\pi R^2 \rho HP_{id_{av}}). \qquad (2.23)$$

As the variation of $HP_{id_{av}}$ vs altitude should be known, then $HP_{id_{av}}/W$ can easily be computed for any altitude h. The same applies to $W^2/\pi R^2 \rho\, HP_{id_{av}}$, and the vertical rate of climb at any altitude can readily be obtained from Eq (2.23).

If the relationship between the power available from the rotor ($HP_{id_{av}}$) and air density, ρ, can be expressed as a simple algebraic function, then by setting $V_c = 0$, Eq (2.23) can be solved for ρ_H; i.e., the density corresponding to the absolute ceiling. From this value of ρ_H, the absolute ceiling can readily be found from tables of standard atmosphere.

When there is a defined requirement for rate of vertical climb at the so-called operational hovering altitude (say, $V_c = 120\ m/min = 2.0\ m/s$) the hovering ceiling can be found by substituting the desired value of V_c into Eq (2.23) and solving for ρ. However, the relationship between engine power and density cannot usually be expressed simply, and a graphical method such as that shown in Fig 2.8 can be used,or a suitable computer program for solving this equation through an iteration process must be established.

3.5 Vertical Descent Rates

Rates in partial-power descent will be considered first for the $|V_d/v_d| < 1.0$ case. Under these conditions, the general flow is still down and, according to the previously developed rules, the ideal power (in HP) required for this process—according to Eq (2.2b) with $\vec{T}\cdot\vec{V}_d$ being negative—will be as follows:

$$P_{id_d} = T(v_d - V_d) \qquad (2.24)$$

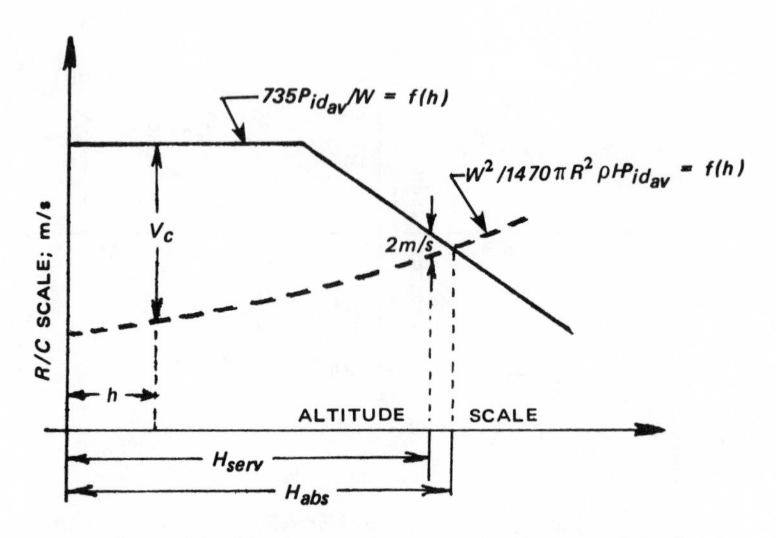

Figure 2.8 Absolute and service ceiling in vertical climb

or, assuming $T = W$,

$$P_{id_d} = W(v_d - V_d). \tag{2.24a}$$

Substituting the v_d value from Eq (2.15a) into Eq (2.24a), relationships similar to those given by Eq (2.20) can be obtained:

$$P_{id_d} = W[-\tfrac{1}{2}V_d + \sqrt{\tfrac{1}{4}V_d^2 + (W/2A\rho)}] \tag{2.25}$$

or

$$P_{id_d} = W(-\tfrac{1}{2}V_d + \sqrt{\tfrac{1}{4}V_d^2 + v_h^2}) \tag{2.25a}$$

In substituting $\overline{V}_d \equiv |V_d/v_h|$ and defining, as in the case of climb, $P_{id_d} \equiv \kappa_d P_{id_h}$ where $0 \leqslant \kappa_d \leqslant 1.0$, one obtains:

$$\kappa_d = -\tfrac{1}{2}\overline{V}_d + \sqrt{\tfrac{1}{4}\overline{V}_d^2 + 1}. \tag{2.25b}$$

The solution of Eq (2.25b) for $\overline{V}_d$ in terms of κ_d gives

$$\overline{V}_d = -\kappa_d + (1/\kappa_d). \tag{2.26}$$

A plot of the above relationship is added to Fig 2.7. It can be seen from this figure that as the actuator disc begins to descend at some nondimensional rate $\overline{V}_d$ or, in other words, when $\overline{V}_c$ changes its sign from the positive to the negative, power required to produce a given amount of thrust becomes lower than that required in hovering ($\kappa < 1.0$). Conversely, when power supplied to the rotor is reduced below the hovering level, it starts to descend in the so-called partial-power descent. It should be noted, however, that both Fig 2.7 and Eq (2.25b), indicating no power required ($\kappa = 0$), can only be approached at very high $\overline{V}_d$ values ($\kappa \to 0$ when $\overline{V}_d \gg 1.0$). It may be recalled at this point that the relationships given in Eqs (2.25) through (2.26) were based on a physical concept; assuming that the resultant flow everywhere within the slipstream tube is in the direction of the downwash at the disc (down) as dictated by the condition $|2v_d| > |V_d|$, while the flow outside the tube moves in the direction of thrust (up).

Let us now look at the other concept of flow; namely, when $|V_d| \geqslant |2v_d|$, the flow within the streamtube is always in the direction of thrust (up), just the same as the whole mass of air. In this case, since $|V_d| \geqslant |2v_d|$, P_{id_d} as given by Eqs (2.24) and (2.24a) will be negative.

$$P_{id_d} = -T(V_d - v_d) \tag{2.27}$$

or

$$P_{id_d} = -W(V_d - v_d)\,. \tag{2.27a}$$

Eqs (2.27) and (2.27a) indicate that power is delivered by the actuator disc and thus, this particular descent or more generally, exposure to the airflow with velocity $|V_d| \geqslant |2v_d|$ in the thrust direction, is called the *windmill* state. Substituting the induced velocity value (v_d) as given by Eq (2.16) into Eq (2.27a), and making the same rearrangements as in the previous case, one obtains

$$P_{id_d} = -W(\tfrac{1}{2}V_d - \sqrt{\tfrac{1}{4}V_d^2 - v_h^2})\,. \tag{2.28}$$

Rates in partial-power descent in nondimensional form can be obtained from Eq (2.28):

$$\kappa_d = -\tfrac{1}{2}\overline{V}_d + \sqrt{\tfrac{1}{4}\overline{V}_d^2 - 1} \tag{2.28a}$$

and finally,

$$\overline{V}_d = -[\kappa + (1/\kappa)]\,. \tag{2.29}$$

It can be seen from Eq (2.28a) that for $\overline{V}_d > 2$, the κ_d is negative; i.e., power is delivered by the rotor. When $\overline{V}_d < 2$, there is no real solution to Eq (2.28a), which means that the assumed physical concept of the model pictured in Fig 2.5 is no longer applicable. For the limiting case of $\overline{V}_d = 2$, the power ratio reaches its lowest value for the windmill state; namely, $\kappa = -1.0$, and the corresponding rate of descent also attains its lowest value, $\overline{V}_d = 2$. In order to maintain a steady-state operation, energy delivered by the rotor should be consumed or dissipated at the rate of its generation. Actual rotors dissipate energy because of the existence of profile power. However, the rotor profile power of such rotary-wing aircraft as helicopters and tilt-rotors does not usually exceed 30 percent of the ideal hovering power. This obviously means that even for the limiting case of the lowest power delivered in the windmill stage ($\kappa = -1$), all of that power cannot be dissipated as profile power losses. It may be expected hence, that the corresponding rate of descent of $\overline{V}_d = 2.0$ or, in other words, $V_d = 2\sqrt{w/2\rho}$ would be too conservative. Indeed, the above velocity is higher than those usually observed in actual flight tests. Furthermore, it is again emphasized that the requirement of air coming to rest in the slipstream above the rotor while the remaining mass moves at steady velocity V_d is physically doubtful.

It appears that the actuator disc concept, when applied to cases of vertical ascent and hovering, does not encounter any logical difficulties. By contrast, in vertical descent, inadequacies of the assumed model become quite obvious. It may be expected, hence, that the actuator disc approach may provide reasonably good guidance for both understanding and even approximate performance predictions in vertical climb and in hover and, perhaps, at partial power descents with power levels only slightly lower than that required in hover. However, even in the latter case, wind-tunnel tests performed by Yaggy[2] with a tilt-wing propeller, and discussed by this author[3], leave some doubts regarding the validity of Eq (2.29). As to the whole spectrum of vertical descent; i.e., from partial power to pure autorotation ($\kappa = 0$), the search for a completely satisfactory physicomathematical model still continues. In the meantime, analytical gaps are being plugged through experimental results.

3.6 Induced Velocity and Thrust in Nonaxial Translation

Through application of the actuator disc concept to the case of axial translation, a basic relationship for thrust in this type of motion was established which may be expressed as follows:

> The thrust developed by a rotor moving along its axis at a speed $\vec{V}_{ax}$ is equal to the rate of mass flow through the disc times the doubled induced velocity at the disc. In this case, the rate of mass flow is clearly defined as the product of the disc area ($A \equiv \pi R^2$) times the air density ρ, times the resultant flow through the disc: $\vec{V}' = \vec{V}_{ax} + \vec{v}$ which, in axial translation, is identical to an algebraic sum.

The accuracy of the above relationship has been proven within the limits of validity of the simple momentum theory. Unfortunately, as far as exposure of an actuator disc to velocity $-\vec{V}$ (opposite to the flight speed) with an inplane component is concerned (Fig 2.9), no rigorous development of the formula for thrust based on the simple momentum approach can be offered. Consequently, the relationship proposed by Glauert[4,5], although unproven for the time being, will be accepted. This relationship, when expressed in words, sounds exactly the same as those given for axial translation. However, in nonaxial translation, the resultant speed of flow through the disc should always be interpreted as the vectorial sum of the distant flow velocity ($-\vec{V}$) and induced velocity in forward flight ($\vec{v}_f$). Later in Chs IV and V it will be shown that Glauert's basic formula can be rigorously proven with the help of vortex and potential theories. In the meantime, the vectorial definition of this relationship can be translated into analytical expressions as follows.

Denoting the scalar value of the resultant speed of flow at the disc by V', Eq (2.10), giving thrust in axial translation, can now be generalized into the following expression by substituting V' for $|\vec{V}_c + \vec{v}_c|$:

$$T = 2\pi R^2 \rho V' v_f. \tag{2.30}$$

Analogous with Eq (2.10), it is postulated that far downstream, the induced velocity v_f is doubled; i.e., $v_u = 2v_f$. In making this assumption, $\pi R^2 \rho V'$ becomes the mass

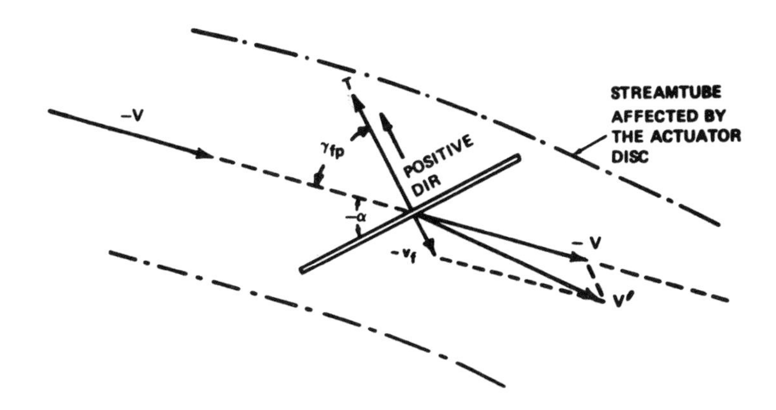

Figure 2.9 Actuator disc in an oblique flow

flow through the streamtube (affected by action of the rotor), whose cross-section is πR^2.

An indirect support for the validity of Eq (2.30) can be found by comparing it with a formula giving lift developed by a wing having $2R$ span and downwash v distributed uniformly along the span. In this case, the lift developed in horizontal flight is expressed by exactly the same formula as Eq (2.30)[6].

Accepting the validity of Eq (2.30), the induced velocity in forward flight can readily be expressed as

$$v_f = T/2\pi R^2 \rho V'. \tag{2.31}$$

It should be noted, however, that the above expression in its present form does not permit determination of the v_f value since V' is also dependent on v_f $(\vec{V}' = \vec{V} + \vec{v}_f)$. In order to solve Eq (2.30) for v_f, V' must be first expressed in terms of V and v_f.

With the same notations as those in Fig 2.9, V' becomes:

$$V' = \sqrt{(v_f - V\sin a)^2 + (V\cos a)^2}.$$

Substituting the above value into Eq (2.30) and performing the necessary manipulation—remembering that $T/\pi R^2 \equiv w$ is the disc loading—the following fourth-degree equation for v_f is obtained:

$$v_f^4 - 2V v_f^3 \sin a + V^2 v_f^2 - (w/2\rho)^2 = 0. \tag{2.32}$$

However, $(w/2\rho)^2 = v_h^4$ where v_h is the induced velocity in hovering or under static thrust conditions in general, and Eq (2.32) can be presented in nondimensional form:

$$\overline{v}_f^4 - 2\overline{V}\,\overline{v}_f^3 \sin a + \overline{V}^2 \overline{v}_f^2 - 1 = 0 \tag{2.32a}$$

where $\overline{V} \equiv V/v_h$ and $\overline{v}_f \equiv v_f/v_h$. Eqs (2.32) and (2.32a) can be solved by Newton's method and its more modern derivatives adapted to computer techniques. Also graphical solutions may be quite useful in that respect.

It should be noted that for the axial translation in the direction of thrust; i.e., when $a = -90°$, and $V = V_{ax}$ or $V = V_c$, Eq (2.32a) can be reduced to a quadratic form and the solution is identical to that of Eq (2.14a). Also of interest may be another limiting condition; namely, when $a = 0$. In the latter case, Eq (2.32a) is reduced to a biquadratic form and the solution for $\overline{v}_f$ can also be easily obtained:

$$(\overline{v}_f)_{a=0} = \sqrt{-\tfrac{1}{2}\overline{V}^2 + \sqrt{\tfrac{1}{4}\overline{V}^4 + 1}}. \qquad (2.33)$$

Relationships of $\overline{v} = f(\overline{V})$ for $a = -90°$ and $a = 0°$ are plotted in Fig 2.10. Thus, these two curves represent limiting cases of nondimensional induced velocity $\overline{v}_f$ vs nondimensional speed of flow $\overline{V}$ (speed of flight with the opposite sign). All other cases corresponding to intermediate angle-of-attack (a) values will be included within these two curves. Of course, for horizontal flight when absolute values of a are small, the trend indicated by the $a = 0$ curve should be quite representative. By examining Fig 2.10, it should also be noted that for $\overline{V} \geqslant 3.0$, $\overline{v}$ values tend to converge to a common limit regardless of the magnitude of a.

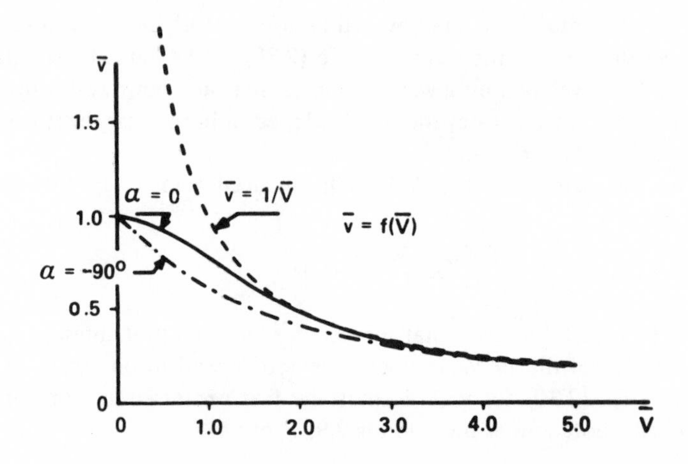

Figure 2.10 Nondimensional induced velocity vs nondimensional speed of flow

Furthermore, starting from $\overline{V} \geqslant 3.0$, the nondimensional induced velocity can be approximated by the simple relationship of

$$\overline{v}_f = 1/\overline{V}. \qquad (2.34)$$

Eq (2.34) could have been directly derived from Eq (2.30) by assuming that $V' \approx V$. Eq (2.30) would then become

$$T = 2\pi R^2 \rho V v_f \qquad (2.35)$$

and consequently,

$$v_f = T/2\pi R^2 \rho V \qquad (2.36)$$

which can be easily transformed into the form of Eq (2.34).

3.7 Power Required in Nonaxial Translation

Using the notations in Fig 2.11, and substituting the proper quantities for $\vec{T}\cdot\vec{V}$ and $\frac{1}{2}\vec{T}\cdot\vec{v}_u$ into Eq (2.2b), ideal power required (P_{id_f} in N-m/s or ft-lb/s) in nonaxial translation (forward flight) can be obtained:

$$P_{id_f} = -T(V_f \sin a - v_f) \qquad (2.37)$$

where the expression in the parentheses represents the axial component (V'_{ax}) of flow at the disc: $V'_{ax} = V_f \sin a - v_f$. It can be seen that when $a \leqslant 0$, Eq(2.37) is positive; i.e., in this type of flow, power must be delivered to the rotor modeled by the actuator disc.

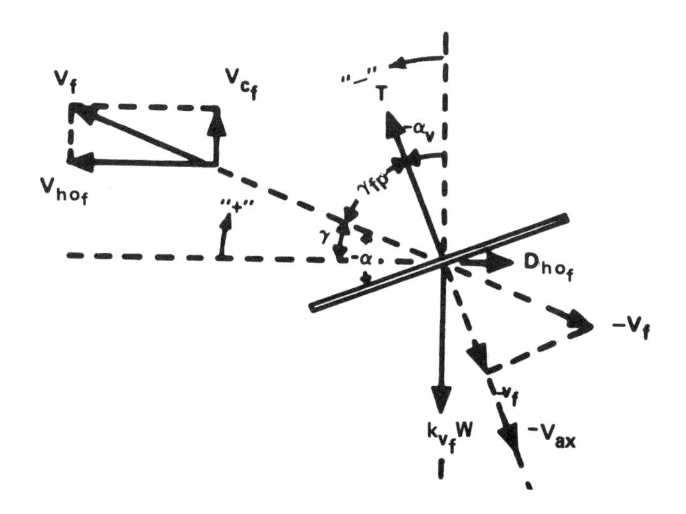

Figure 2.11 Rotary-wing aircraft (modeled by the actuator disc) in forward flight

For a helicopter moving in the gravitational coordinate system at velocity of flight $\vec{V}_f$, P_{id_f} can be obtained by rewriting Eq (2.37):

$$P_{id_f} = -T(V_f \cos \gamma_{fp} - v_f)\,. \qquad (2.37a)$$

However, $\gamma_{fp} = 90° - (\gamma + a_v)$, and Eq (2.37a) can be presented in the following form:

$$P_{id_f} = (V_f \sin \gamma)(T \cos a_v) + (V_f \cos \gamma)(T \sin a_v) + Tv_f\,. \qquad (2.38)$$

It should be realized that

$V_f \sin\gamma \equiv V_{cf}$ — rate of climb in forward flight

$V_f \cos\gamma \equiv V_{ho}$ — horizontal component of the speed of flight

and in a steady-state flight:

$T \cos a_v = k_{vf} W$ — vertical thrust component, balancing aircraft gross weight times the k_{vf} coefficient, accounting for the vertical drag in forward flight

$T \sin a_v = D_{ho}$ — horizontal thrust component required to overcome the horizontal component of the total drag.

Taking the above relationships into consideration, the ideal power required in forward flight can be expressed as follows:

$$P_{id_f} = V_{ho} D_{ho} + V_{cf} k_{vf} W + T v_f \tag{2.39}$$

where $T = \sqrt{(k_{vf} W)^2 + D_{ho}^2}$, or $T = W\sqrt{k_{vf}^2 + (D_{ho}/W)^2}$, and D_{ho}/W is the horizontal drag-to-weight ratio of the aircraft as a whole at the considered speed. For those cases when $k_{vf} \approx 1.0$ and D_{ho}/W are considered small, it may be assumed that $T \approx W$, and Eq (2.39) may be written as follows:

$$P_{id_f} = W[(D_{ho}/W) V_{ho} + V_{cf} + v_{ho}]. \tag{2.39a}$$

Thus, ideal power required by a helicopter modeled by the actuator disc mounted on a realistic body; i.e., generating drag forces when moving through the air, would consist of a sum of three distinct terms: (a) power required to overcome the horizontal component of the total drag, (b) power required to overcome gravity in climb, and (c) the induced power associated with the process of thrust generation.

Horizontal flight represents a particular case when the ideal power required is composed of drag and induced terms only. Induced power reaches its maximum in hovering and then decreases with increasing speed of flight. In contrast, starting with zero in hover, the drag power increase is roughly proportional to the cube of forward speed, and the resulting total ideal power-required curve should exhibit the character shown in Fig 2.12.

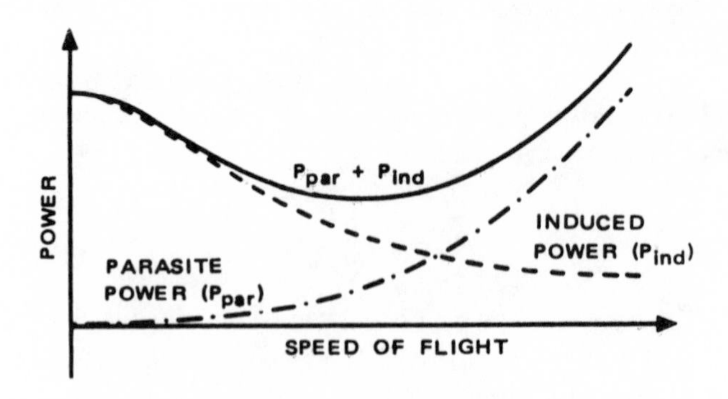

Figure 2.12 Parasite and induced power vs speed of flight

3.8 Thrust Tilt in Forward Flight

In the pure helicopter, the rotor performs a dual function of lifting and propelling in all regimes of powered flight. Using the notations in Fig 2.11, the horizontal force equilibrium condition can be expressed as

$$T \tan\text{-}a_v = D_{h\,o}$$

or

$$-a_v = \tan^{-1}(D_{h\,o}/T)\,. \tag{2.40}$$

In many practical considerations, the small angle assumption as well as the $W \approx T$ condition can be justified. Thus, Eq (2.40) can be simplified to the following form:

$$-a_v \approx (D_{h\,o}/W)\,. \tag{2.40a}$$

3.9 Induced Power in Horizontal Flight

In principle, Eqs (2.32) and (2.32a) allow one to calculate induced velocity at any speed in horizontal flight ($V_{h\,o}$) once the tilt of the thrust vector (a_v) and hence, the a angle ($a = a_v$) corresponding to that speed, is computed from either Eq (2.40) or (2.40a). However, the iteration method required to solve Eqs (2.32) and (2.32a) may be tedious unless a suitable computer program is available. For this reason, simpler approaches may be of some value.

At low flying speeds, the assumption that $V_{h\,o} = V'$ is no longer acceptable and thus, Eqs (2.34) and (2.36) cannot be used. However, the rotor tilt, $a_v = a$, required in steady flight at low velocities will be so small that $a_v \approx 0$. This implies that the induced velocity $v_{h\,o}$ is perpendicular to the flying speed $V_{h\,o}$ (Fig 2.13).

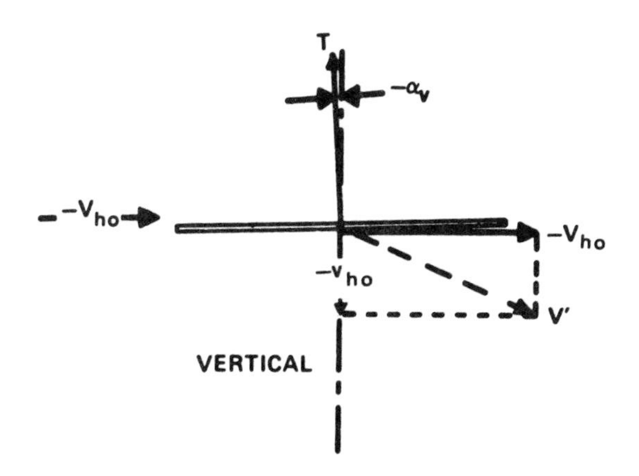

Figure 2.13 Velocity at the disc at a low horizontal speed

Under the foregoing assumptions, Eq (2.33) can be used. Also, approximate values of $\bar{v}_{ho}$ can be obtained from the nondimensional graph of Fig 2.10. However, for those cases when the $a_v \approx 0$ assumption is not acceptable, the following graphical method, which permits consideration of the existence of the thrust tilt angle $a_v \neq 0$, can be used.

A curve giving $v = f(V')$ is drawn from Eq (2.30) using the same scale for both ordinate and abscissa axes. This will obviously be a hyperbola whose point $v = V'$ will correspond to the hover condition.

The value of the induced velocity must satisfy Eq (2.30) as given by the graph in Fig 2.14, as well as the other relationship of

$$\vec{V}' = \vec{V}_{ho} + \vec{v}_{ho}.$$

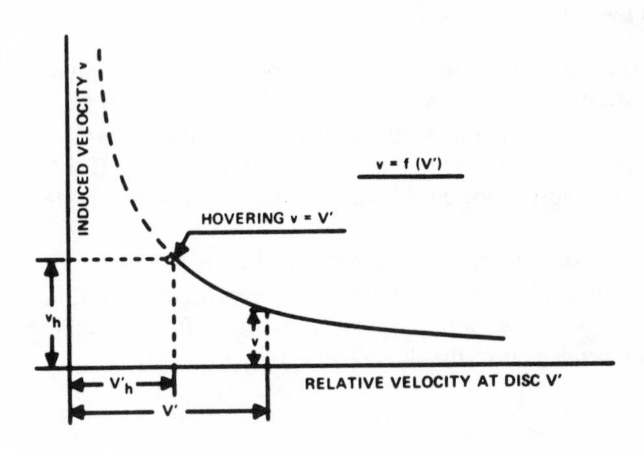

Figure 2.14 Induced velocity vs relative velocity at the disc

By referring to Figs 2.13 and 2.14, one can see that a simple graphical method can be employed in finding the downwash velocity in horizontal flight.

The value of V_{ho} and the direction of v_{ho} (tilt of the rotor a_v) are known. Assuming a value of V', the corresponding value of v_{ho} is found from Fig 2.14. These values of V' and v_{ho} must also satisfy the vectorial relationship shown in Fig 2.15. This means that the head of the vector V' must lie on line *a-b* parallel to the rotor axis, while the length of *a-b* must be equal to the value of v_{ho} corresponding to the assumed V'. If the V' and v_{ho} chosen the first time do not fulfill these conditions, a new value of V' should be assumed and the whole procedure repeated. By cutting and trying, the correct pair of values of V' and v_{ho} satisfying both Eq (2.30) and the vectorial sum condition can easily be found.

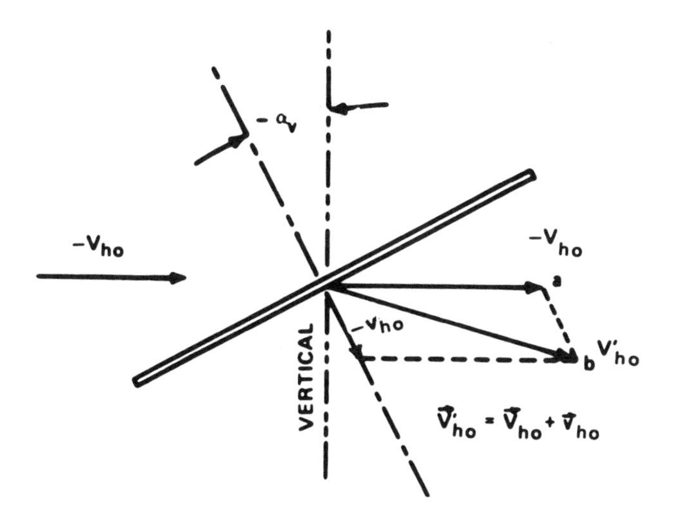

Figure 2.15 Velocities at the disc in horizontal flight at high-thrust inclination

3.10 Rate of Climb in Forward Flight

When the total power available at the actuator disc modeling a rotor ($P_{id_{av}}$) is known, the rate of climb can be found in the following manner.

It is assumed that the rotor thrust inclination-a_v remains the same in ascending flight as it would be in horizontal flight at a speed V_{ho} equal to the horizontal component of the actual flying speed V_f. Hence, if the drag as a function of forward speed of the helicopter is known, then the rotor thrust tilt-a_v can readily be computed. Also, since the ideal power available at the rotor $P_{id_{av}}$ (N m/s or ft-lb/s) is known, the total rate of axial flow through the disc U (m/s or fps) can be found from Eq (2.37), where U is substituted for the axial component V'_{ax}:

$$U = P_{id_{av}}/k_{v_f}W. \tag{2.41}$$

On the other hand, for small a_v's, the rate of climb can be expressed (Fig 2.16) as:

$$V_c = U - (V_f a_v + v_f). \tag{2.42}$$

and, substituting Eq (2.41) for U,

$$V_c = (P_{id_{av}}/k_{v_f}W) - (V_f a_v + v_f). \tag{2.43}$$

The value of v_f in Eq (2.43) is yet unknown, but it may readily be found with the help of a simple graph such as that shown in Fig 2.16. From the head of vector V_{ho}, a line is drawn parallel to the disc axis. Again, from the head of vector U, a line normal to the disc axis and intersecting the first line, is drawn. By approximation, point A may be considered as the head of a vector representing the relative velocity of the slipstream V'.

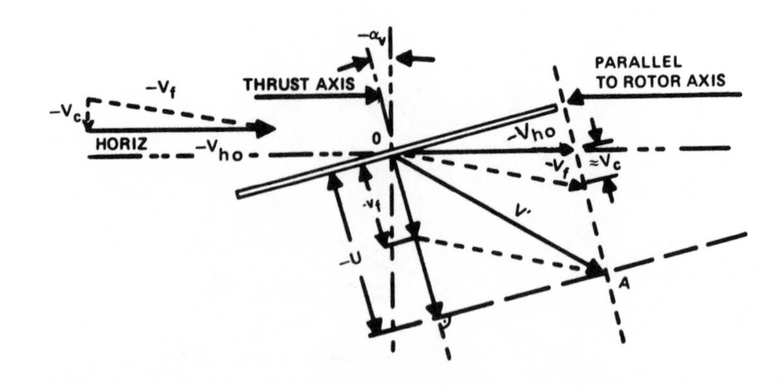

Figure 2.16 Velocity scheme in forward flight climb

0-A indicates the magnitude of V', and the corresponding v_f can be found easily from the graph in Fig 2.14, $v_f = f(V')$. When the v_f value is introduced into Eq (2.43), the rate of climb will be obtained.

In finding the induced velocity v_f at any altitude, it must be remembered that v_f is inversely proportional to the air density (Eq 2.30). A graph of $v_f = f$ (altitude) with the scale of V' remaining constant could be helpful for altitude calculations. However, a procedure based on the principle of excess power is usually accurate enough for all practical purposes in determining the rate of climb. It can be seen from Eq (2.39) that

$$V_{cf} = [P_{id_{av}} - (V_{ho}D_{ho} + Tv_f)]/k_{v_f}W.$$

In the above equation, the expression in the parentheses represents the power required in horizontal flight $P_{idh_{o\,req}}$ at a speed V_{ho} while $P_{id_{av}}$ should be interpreted as the ideal power available at the actuator disc (rotor). V_{cf} can now be expressed in m/s or fpm , while both powers are in horsepower, and weight in N or lb:

$$\text{(SI)} \qquad V_{cf} = 735\,(HP_{id_{av}} - HP_{idho_{req}})/k_{v_f}W.$$
$$\text{(ENG)} \qquad V_{cf} = 550(HP_{id_{av}} - HP_{idho_{req}})/k_{v_f}W. \qquad (2.44)$$

In many cases, it may be assumed that the vertical load factor $k_{v_f} = 1.0$.

Service and absolute ceilings in forward flight can be obtained from Eq (2.44) by finding the maximum rate of climb $(V_{cf})_{max}$ at several altitudes and plotting $(V_{cf})_{max}$ values versus altitude. It is obvious that the altitude at which $(V_{cf})_{max}$ reaches some prescribed value; say, $(V_{cf})_{max} = 70\ m/min$, will represent the service ceiling (Fig 2.17).

The method of finding the rate of climb from the excess power can be accepted for higher flying speeds (V_e and higher) when climbing would not appreciably change the rate of flow through the disc. For low forward speeds, the graphical method previously outlined is more suitable.

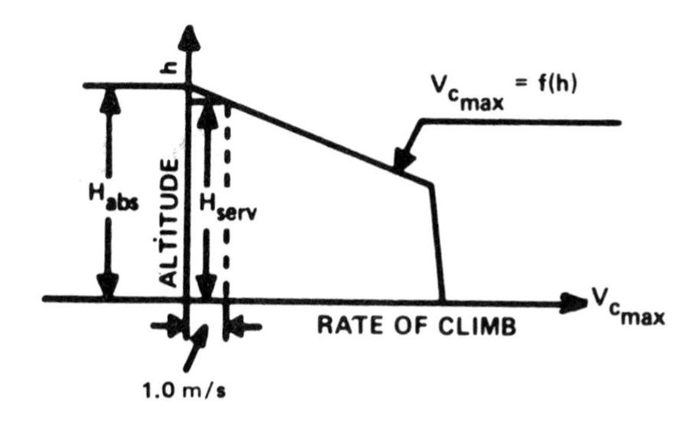

Figure 2.17 Maximum forward climb diagram and determination of absolute and service ceilings

3.11 Partial and zero-power descent in forward flight

In forward flight with a horizontal velocity component V_{ho}, it can be seen from Eq (2.44) that V_{cf} becomes negative as the ideal rotor power available becomes lower than that required in horizontal flight. In other words, the aircraft begins to descend at a rate $V_{df} \equiv -V_{cf}$. In a particular case where $HP_{id_{av}} = 0$, this rate of descent becomes

$$\text{(SI)} \qquad V_{df} = 735\, HP_{id\, ho_{req}} / k_{v_f} W.$$

$$\text{(ENG)} \qquad V_{df} = 550 HP_{id\, ho_{req}} / k_{v_f} W. \tag{2.45}$$

In this equation, V_d is in m/s or fps, and W in N or lb.

If power is taken from the rotor (say, to drive some accessories), then the first term in Eq (2.44) also becomes negative and obviously, the absolute value of $-V_{cf}$ (i.e., at a rate of descent V_d) would become higher than that given by Eq (2.45).

At this point it should be noted that the validity of Eqs (2.44) and (2.45) is based on the assumption that contrary to the case of axial and near-axial translation, neither climb nor descent would noticeably alter the value of induced velocity corresponding to V_{ho}. Consequently, in climb and descent, the induced power would remain the same as for the case of horizontal flight. It can be seen from Fig 2.10 that for speeds of forward translation in excess of $\overline{V}_f \approx 2$, the variation in the inflow angles that may be encountered in either climb or descent of practical rotary-wing aircraft would not noticeably alter the induced velocity values. Furthermore, for a rate of climb (or descent) on the order of $|\pm \overline{V}_c / \overline{V}_f| \leqslant 1/3$ (not likely to be exceeded in practice) the difference between the total rate of flow (V') through the slipstream at the disc and that corresponding to the horizontal component may be ignored. Consequently, it may be assumed that for $\overline{V}_f \geqslant 2$, the induced velocity and hence, the induced power in forward flight (or descent) with a horizontal component V_{ho} would be the same as in a purely horizontal flight at the same speed. As a result, the method of predicting $\pm V_c$ on the basis of either an excess or shortage of power available with respect to power required in horizontal flight is justified.

For some exceptional cases of extremely high rates of climb or descent at $\overline{V}_f \geqslant 2$, the associated variations of induced velocities and powers from those corresponding to horizontal flight should be considered.

4. FLIGHT ENVELOPE OF AN IDEAL HELICOPTER

Much may be learned and many performance problems more simply solved by substituting the flight envelope of an idealized helicopter for that of actual rotary-wing aircraft.

A complete flight envelope for a constant flight altitude would be contained within the limits of the maximum rate of climb at the highest possible $HP_{id_{av}}$, and a rate of descent corresponding to $HP_{id_{av}} = 0$. These rates of climb and descent can be shown as a function of the horizontal component of the speed of flight given in either an absolute dimensional $[\pm V_{cf} = f(V_{ho})]$ or nondimensional $[\pm \overline{V}_{cf} = f(\overline{V}_{ho})]$ form. In both cases, the general character of the graph will obviously remain the same as shown in Fig 2.18.

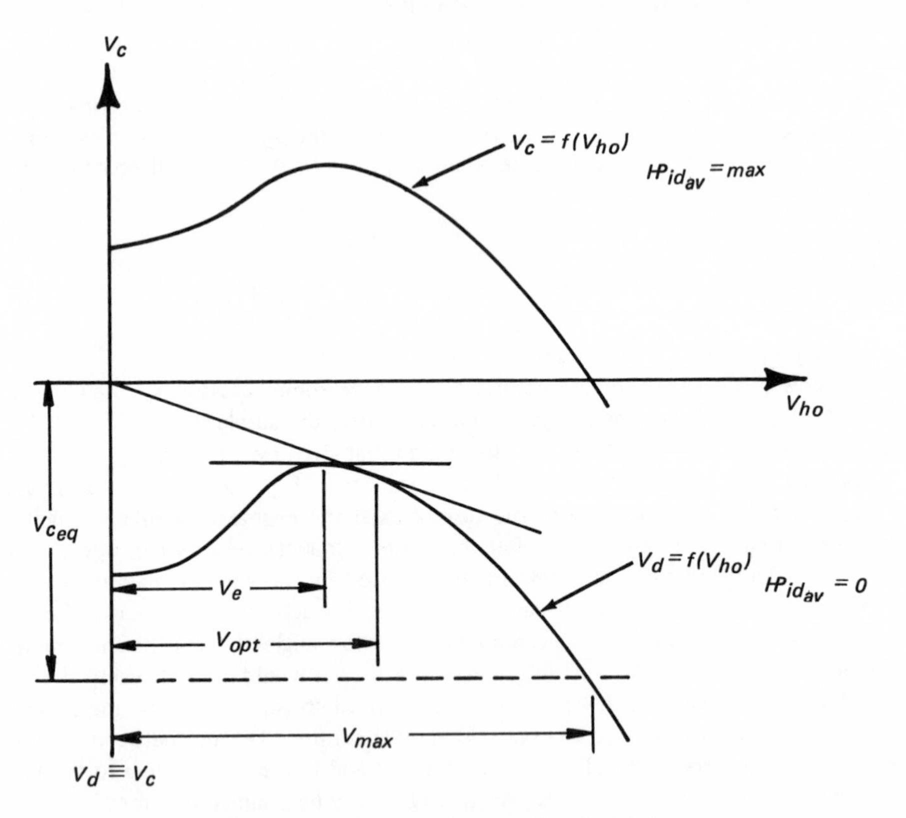

Figure 2.18 Flight envelope of the ideal helicopter

In establishing the flight envelope in vertical ascent and descent at low forward speeds ($\overline{V}_{ho} < 2.0$), the V_{cf} and V_{df} values should be calculated by the procedures outlined in Sects 3.2.2 and 3.2.3. However, for higher forward speeds ($\overline{V}_{ho} \geqslant 2$) where the principle of excess ideal power available can be directly used, a new interpretation for the relationships between $P_{id_{av}}$, $V_d = f(V_{ho})$, and $V_{cf} = f(V_{ho})$ is possible. This approach, outlined below, may be of interest in dealing with some of the problems of performance optimization.

Equivalent rate-of-climb. Assuming that $k_{vf} = 1.0$, the second term in the parentheses of Eq (2.44) becomes the rate of descent at $HP_{id_{av}} = 0$ as given by Eq (2.45), while the first term may be called the equivalent rate of climb in m/s or fps: (SI) $V_{c_{eq}} \equiv 735\ HP_{id_{av}}/W$; or (ENG) $V_{c_{eq}} \equiv 550\ HP_{id_{av}}/W$.

Using the above interpretation of Eq (2.44), the rate of climb at any value of the horizontal component (V_{ho}) of flight velocity (V_f) can be expressed as the difference between $V_{c_{eq}}$ and rate of descent at that particular speed, when $HP_{id_{av}} = 0$:

$$V_{cf} = V_{c_{eq}} - V_{df}. \tag{2.46}$$

The above equation can also be presented in nondimensional form by dividing both sides by the induced velocity in hovering:

$$\overline{V}_{cf} = \overline{V}_{c_{eq}} - \overline{V}_{df}. \tag{2.46a}$$

Consequently, once $V_{df} = f(V_{ho})$ or $\overline{V}_{df} = f(\overline{V}_{ho})$ relationships are established for $HP_{id_{av}} = 0$, then it becomes easy to obtain rates of climb in forward flight ($\overline{V}_{ho} \geqslant 2$) for all values of $HP_{id_{av}}$. This is done by calculating the equivalent rates of climb ($V_{c_{eq}}$ or $\overline{V}_{c_{eq}}$) corresponding to $HP_{id_{av}}$ and performing the subtraction indicated by Eqs (2.46) and (2.46a), either arithmetically or graphically (see Fig 2.18).

Examination of the $V_{df} = f(V_{ho})$ for $HP = 0$ will help one to single out several, operationally important, speeds of flight, V_f. Note that it may usually be assumed that $V_f \approx V_{ho}$).

Economic speed, V_{ho_e} or simply, V_e, corresponds to the lowest rate of descent for unpowered flight (Fig 2.18). In powered horizontal flight, it corresponds to the lowest value of power required. In flight with excess power, it assures the highest rate of climb ($V_{cf_{max}}$).

Optimum speed $V_{ho_{opt}}$ or simply, V_{opt}, assures the best gliding ratio in unpowered flight. In the presence of a tailwind (V_w) or headwind ($-V_w$), this gliding ratio showing the ratio of distance traveled (ℓ) to altitude lost (h) will be

$$(\ell/h) = (V_{ho} \pm V_w)/V_d.$$

For a zero-wind condition, this ratio becomes maximum at $V_{ho} = V_e$, representing the abscissa of the point of tangency of a straight line drawn from the origin of coordinates to the $V_d = f(V_{ho})$ curve. In the presence of a headwind $-V_w$ or tailwind V_w, the tangents will be drawn from the $V_{ho} = V_w$ point in the first case, and $V_{ho} = -V_w$ in the second (Fig 2.18).

In horizontal powered flight, the thus-determined speeds of flight would assure maximum range under zero wind and $\pm V_w$ conditions, as this would minimize the ideal amount of energy (E_{id}) required per unit of distance traveled ($\ell = 1.0$) and unit of gross weight (W):

$$E_{id}/W|_{\ell=1} = P_{id_{req}}/(V_h \pm V_w)W$$

where $P_{id_{req}}$ is in Nm/s or ft-lb/s, and W in N or lb.

However, $P_{id_{req}}/W = V_{idd_f}$ and consequently, the condition of minimizing the above expression becomes the same as for the optimum glide angle in unpowered flight.

Maximum speed of horizontal flight $V_{ho_{max}}$ or simply V_{max}, is reached when $V_{ceq} = V_{df}$ or, in other words, ideal power available becomes equal to the ideal power required.

5. EFFECTS OF DOWNWASH CHARACTERISTICS ON INDUCED POWER

5.1 Uniform Downwash – No Tip Losses

In the physicomathematical models based on the actuator disc concept, it was assumed that the induced velocity in axial flow is uniform both at the disc and in the fully-developed wake. For oblique flows, especially those characterized by high speeds ($\bar{V}_f \gg \bar{v}_f$), no explicit scheme was postulated regarding induced velocity distribution over the disc itself; the only assumption being that the downwash is uniform in any cross-section of the fully-developed slipstream and is equal to twice the average induced velocity at the disc. Furthermore, it was also stated that in either case, the disc is equally effective in thrust generation up to the limit of its geometric dimension; i.e., up to its radius R. The induced velocity associated with this type of given thrust generation is called the ideal induced velocity and the corresponding induced power, the ideal power.

It can be proven that the above flow conditions are synonymous with the minimization of induced power required for generation of a given thrust; i.e., for making the P_{ind}/T ratio a minimum. For the sake of simplicity, the problem of the P_{ind}/T optimization will be considered for cases of hovering and high-speed horizontal translation only ($\bar{V}_{ho} \gg \bar{v}_{ho}$).

Hovering. In steady-state hovering–no controls application–it may be assumed that because of the axial symmetry of flow, the variation of downwash velocity at the disc (v) is a function only of the radial position at the considered point on the disc. This means that for any annulus of radius $r = R\bar{r}$, and width $dr = Rd\bar{r}$ (where $\bar{r} \equiv r/R$), the induced velocity $v_{\bar{r}} = const$ (Fig 2.19).

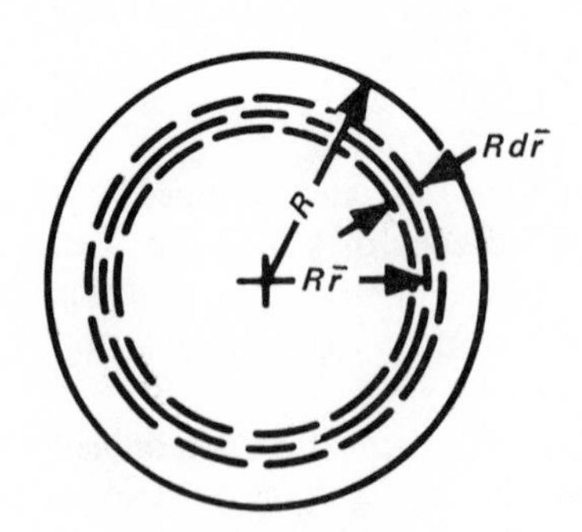

Figure 2.19 Elementary annulus of the actuator disc

With respect to the variation of $v_{\bar{r}}$ from one station $\bar{r}$ to another, this relationship can be expressed as a product of the ideal induced velocity v_{id} corresponding to a given thrust T and some function of $\bar{r}$:

$$v_{\bar{r}} = v_{id} f(\bar{r}) \tag{2.48}$$

where $f(\bar{r})$ is assumed to be a continuous, differentiable function of r, at least within the $0 \leqslant \bar{r} \leqslant 1.0$ interval.

Elementary thrust produced by an annulus $Rd\bar{r}$ wide and having radius $R\bar{r}$ can be expressed as follows:

$$dT_{\bar{r}} = 4\pi R^2 \rho v_{\bar{r}}^2 \bar{r} d\bar{r}, \tag{2.49}$$

while the total thrust will be

$$T = 4\pi R^2 \rho \int_0^{1.0} v_{\bar{r}}^2 \bar{r} d\bar{r}. \tag{2.50}$$

The corresponding elementary and total induced power will be

$$dP_{ind} = 4\pi R^2 \rho v_{\bar{r}}^3 \bar{r} d\bar{r}, \tag{2.51}$$

and

$$P_{ind} = 4\pi R^2 \rho \int_0^{1.0} v_{\bar{r}}^3 \bar{r} d\bar{r}. \tag{2.52}$$

Expressing $v_{\bar{r}}$ in Eqs (2.50) and (2.52) according to Eq (2.48), the P_{ind}/T ratio can be written as follows:

$$P_{ind}/T = v_{id} \int_0^{1.0} [f(\bar{r})]^3 \bar{r} d\bar{r} \Big/ \int_0^{1.0} [f(\bar{r})]^2 \bar{r} d\bar{r}. \tag{2.53}$$

It can be seen from Eq (2.53) that for a given v_{id}, the P_{ind}/T ratio is a function of $\bar{r}$ only. Consequently, conditions for an extremum for the P_{ind}/T value can be sought by differentiating Eq (2.53) with respect to $\bar{r}$ and equating the numerator of the so-obtained fraction to zero. This would result in the following equation:

$$\int_0^{1.0} [f(\bar{r})]^2 \bar{r} d\bar{r} \int_0^{1.0} \Big\{ [f(\bar{r})]^3 + 3[f(\bar{r})]^2 f'(\bar{r}) r \Big\} d\bar{r}$$

$$- \int_0^{1.0} [f(\bar{r})]^3 \bar{r} d\bar{r} \int_0^{1.0} \Big\{ [f(\bar{r})]^2 + 2[f(\bar{r})] f'(\bar{r}) \bar{r} \Big\} d\bar{r} = 0, \tag{2.54}$$

where $f'(\bar{r})$ is the first derivative of $f(\bar{r})$.

An inspection of Eq (2.54) indicates that when $f(\bar{r}) = const$ within the $0 \leqslant \bar{r} \leqslant 1.0$ interval and hence, $f'(\bar{r}) = 0$, the left side of the equation reduces to zero. Physical considerations indicate that the above condition of $f(\bar{r}) = const$ would make the P_{ind}/T ratio a minimum.

To enable one to visualize the character of the disc area loading associated with various downwash distributions, $dT_{\bar{r}}$ as given by Eq (2.49) is divided by the elementary annulus area $(2\pi R^2 \bar{r} d\bar{r})$ over which $dT_{\bar{r}}$ is generated. This leads to the following relationship:

$$\Delta p_{\bar{r}} \equiv dT_{\bar{r}}/2\pi R^2 \bar{r} d\bar{r} = 2\rho v_{\bar{r}}^2 . \tag{2.55}$$

This expression can be nondimensionalized by first substituting Eq (2.48) for $v_{\bar{r}}$, and then dividing both sides of Eq (2.55) by the dynamic pressure corresponding to the ideal induced velocity required to generate a given thrust or, more generally, produce a desired disc loading (w). When so-modified, Eq (2.55) becomes

$$\Delta \bar{p}_{\bar{r}} = 4[f(\bar{r})]^2 \tag{2.55a}$$

where $\Delta \bar{p}_{\bar{r}} \equiv \Delta p_{\bar{r}} / \tfrac{1}{2} \rho v_{id}^2$.

It can be seen from Eq (2.55a) that for the particular case of $f(r) = const = 1.0$, the pressure differential over the disc becomes constant, as was assumed in Sect 3.1, and is equal to 4 times the dynamic pressure of induced velocity.

Horizontal Flight. To gain an insight into induced power aspects and spanwise load distribution of rotors in horizontal flight, a technique was developed which would (1) relate the shape of the downwash distribution in the fully developed wake to the induced power required to produce a given thrust, and (2) determine the type of the corresponding span loading of the disc modeling an actual rotor.

According to the momentum theory, the thrust $dT_{\bar{x}}$ developed by the strip of the disc $Rd\bar{x}$ wide located at $x = R\bar{x}$ can be expressed as the rate of flow of the vertical momentum through a corresponding strip of the slipstream cross-section located far downstream where the induced velocity reaches its ultimate value (Fig 2.20).

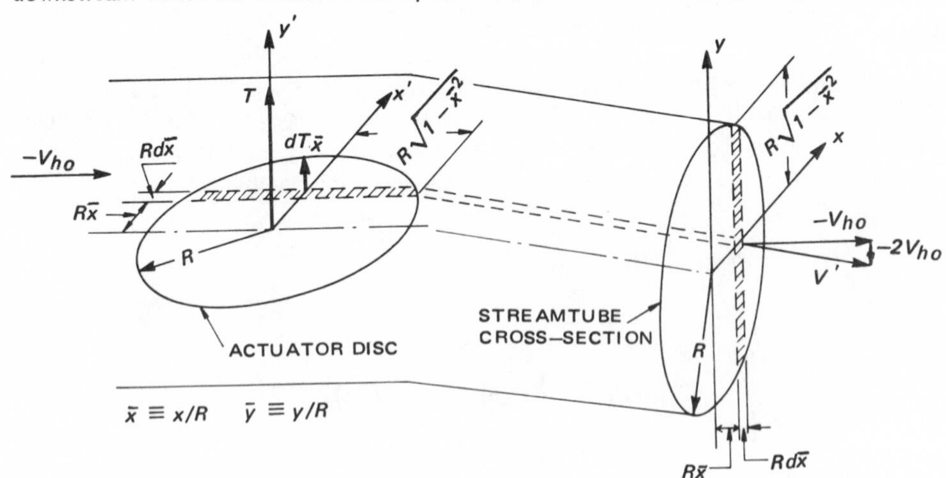

Figure 2.20 Scheme of horizontal flight

Assuming that $|V_{ho}| \gg |v_{ho}|$ and hence, $V' \approx V_{ho}$, this elementary thrust becomes

$$dT_{\bar{x}} = 4R^2 \rho V_{ho} v_{\bar{x}} \bar{y} d\bar{x}\,, \tag{2.56}$$

while the corresponding elementary induced power will be

$$dP_{ind\bar{x}} = 4R^2 \rho V_{ho} v_{\bar{x}}^2 \bar{y} d\bar{x} \tag{2.57}$$

where $v_{\bar{x}}$, the average induced velocity at the disc at station $\bar{x}$, is assumed to double its value in the fully-developed slipstream.

In both equations, $\sqrt{1 - \bar{x}^2}$ can be substituted for $\bar{y}$ with the following results:

$$dT_{\bar{x}} = 4R^2 \rho V_{ho} v_{\bar{x}} \sqrt{1 - \bar{x}^2}\, d\bar{x} \tag{2.56a}$$

and

$$dP_{ind\bar{x}} = 4R^2 \rho V_{ho} v_{\bar{x}}^2 \sqrt{1 - \bar{x}^2}\, d\bar{x}\,. \tag{2.57a}$$

As in the case of hovering, $v_{\bar{x}}$ can be expressed in terms of the ideal induced velocity in horizontal flight as given by Eq (2.36) times some known or assumed function $f(\bar{x})$:

$$v_{\bar{x}} = v_{id_{ho}} f(\bar{x}) \tag{2.58}$$

where function $f(\bar{x})$ should be continuous and hence, differentiable within the $-1.0 \leqslant \bar{x} \leqslant 1.0$ limits.

Once $f(\bar{x})$ is known, the total thrust and the corresponding induced power can be obtained by substituting the right side of Eq (2.58) for $v_{\bar{x}}$, and then integrating Eqs (2.56a) and (2.57a) within the $\bar{x} = -1.0$ to $\bar{x} = 1.0$ limits. If a symmetry exists with respect to the vertical plane perpendicular to the x axis, and passing through the y axis, these integrals become

$$T = 2 \int_0^{1.0} dT_{\bar{x}} \tag{2.59}$$

and

$$P_{ind} = 2 \int_0^{1.0} dP_{ind\bar{x}}\,. \tag{2.60}$$

The character of the spanwise load distribution along the disc span (lateral diameter) associated with various $f(\bar{x})$ functions can best be presented in nondimensional form. This can be done by keeping in mind Eq (2.58) and remembering that the thrust corresponding to uniform downwash distribution is $T = 2\pi R^2 \rho V_{ho} v_{id_{ho}}$. Dividing both sides of Eq (2.56a) by the so-defined T,

$$d\bar{T}_{\bar{x}}/d\bar{x} = (2/\pi) f(\bar{x}) \sqrt{1 - \bar{x}^2}\,. \tag{2.61}$$

Decades ago, in a situation analogous to the momentum interpretation of the thrust generated by an actuator disc in forward flight (Figure 2.20), it was proven by the fixed-wing theory that the P_{ind}/T ratio is minimized when the downwash velocity in a fully-developed wake is uniform. While no further proof of this statement is necessary at this point, it should be noted that when $f(\bar{x}) = 1.0$ and hence $v_{\bar{x}} = v_{idh_o}$, the integration indicated by Eq (2.59) will lead to the basic Glauert relationship given by Eq (2.35), and the accompanying span-loading as given by Eq (2.61) is elliptical. Although this latter result could also be accepted solely on the basis of the proof offered by fixed-wing aerodynamics, the above analysis was developed as a tool for future investigations of the various effects of nonuniform downwash distributions in forward flight.

5.2 Nonuniform Downwash and Tip Losses – The k_{ind} Factor

Definition of the k_{ind} Factor. In the preceding chapter, it was shown that nonuniform downwash distribution (both in axial and in forward flight) causes the induced power to rise above its optimum level. It may also be expected that various aerodynamic phenomena occurring at the outer rim of the disc may reduce its thrust-generating effectiveness in that region; therefore, the effective disc radius (R_e) becomes smaller than R; i.e., $R_e/R \equiv \bar{r}_e < 1.0$. This would also contribute to an increase in the P_{ind} associated with generation of a given thrust.

The deviations of the actual induced power from its ideal level resulting from the above-discussed, and other phenomena such as mutual rotor interference can be conveniently gauged through the k_{ind} factor,

$$k_{ind} = P_{ind}/P_{id}\,. \tag{2.62}$$

In hovering, the induced power of a disc with nonuniform distribution of the downwash velocity, and the effective relative radius $\bar{r}_e < 1.0$, can be obtained from Eq (2.52) by substituting $v_{id}f(\bar{r})$ for v_r and changing the limits of integration from 0 to 1.0, to 0 to $\bar{r}_e$:

$$P_{ind} = 4\pi R^2 \rho v_{idh}{}^3 \int_0^{\bar{r}_e} [f(\bar{r})]^3 \bar{r}\,d\bar{r} \tag{2.63}$$

Introducing the P_{ind} value as given by Eq (2.63) into Eq (2.62), and then dividing the result by the expression $P_{id} = 2\pi R^2\,\rho\, v_{idh}{}^3$, the following expression for the k_{ind} factor in hovering is obtained:

$$k_{ind_h} = 2\int_0^{\bar{r}_e} [f(\bar{r})]^3 \bar{r}\,d\bar{r}\,. \tag{2.64}$$

The procedure used to determine the k_{ind} factor in horizontal flight is similar to that outlined above. In Eq (2.57a), $v_{idh_o}f(\bar{x})$ is substituted for $v_{\bar{x}}$ and the 1 under the square root is replaced by $\bar{x}_e{}^2$ where $\bar{x}_e = \bar{r}_e$ is one-half of the relative effective disc span. The integration indicated by Eq (2.60) is now performed within the 0 to $\bar{x}_e$ limits instead of the *0 to 1.0* limits, and the obtained expression is divided by the ideal induced power in horizontal flight $(2\pi R^2 \rho V_{ho} v_{idho}{}^2)$:

$$k_{indho} = (4/\pi) \int_0^{\bar{x}_e} [f(\bar{x})]^2 \sqrt{\bar{x}_e{}^2 - \bar{x}^2}\, d\bar{x} \,. \tag{2.65}$$

It should be emphasized at this point that in horizontal flight as the downwash velocity in the slipstream becomes nonuniform, the streamtube would no longer retain its circular cross-section as shown in Fig 2.20. Nevertheless, it is assumed that in spite of the cross-section deformation, the area dA_x of a section element associated with a location x will remain the same as in uniform downwash considerations; i.e., $dA_{\bar{x}} = 2R^2\sqrt{\bar{x}_e{}^2 - \bar{x}^2}\, dx$. This would obviously mean that the corresponding elementary thrust and induced power can still be expressed by Eqs (2.56a) and (2.57a) respectively, and Eq (2.65) remains valid.

5.3 Examples of k_{ind} Values and Types of Loading in Hover (Figure 2.21)

Uniform Downwash with Tip Losses. To produce the same thrust as for the ideal case *(a)*, the induced velocity *(b)* should be $v_{ind} = v_{id}/\bar{r}_e$ and consequently,

$$f(\bar{r}) = 1/\bar{r}_e \,, \tag{2.66}$$

which, substituted into Eq (2.64), yields

$$k_{ind_h} = 1/\bar{r}_e \,. \tag{2.67}$$

The nondimensional disc area loading remains uniform *(b)* as in the case of no tip losses *(a)*, but its value as given by Eq (2.55a) will now be

$$\Delta\bar{p} = 4/\bar{r}_e{}^2 \,. \tag{2.68}$$

Triangular Downwash Distribution with $v_{\bar{r}} = 0$ *at* $\bar{r} = 0$. This type of downwash distribution can be described by the following expression for the $f(\bar{r})$ function:

$$f(\bar{r}) = \nu\bar{r} \tag{2.69}$$

where the coefficient ν should be determined on the condition that the total thrust corresponding to the nonuniform downwash distribution and tip losses must be equal to the ideal conditions $(T_{id} = 2\pi R^2 \rho v_{id}{}^2)$. Substituting $\nu v_{id}\bar{r}$ for $v_{\bar{r}}$ in Eq (2.50), integrating within the limits of *0* to $\bar{r}_e$, and equating the so-modified expression to the ideal thrust; one would find that $\nu = 1.414/\bar{r}_e{}^2$ which, substituted into Eq (2.64) gives $k_{ind_h} = 1.13/\bar{r}_e$ *(c)*.

Substituting $1.414\bar{r}/\bar{r}_e$ for $f(\bar{r})$ in Eq (2.55a), one finds that

$$\Delta\bar{p}_{\bar{r}} = (8/\bar{r}_e{}^4)\bar{r}^2 \,. \tag{2.70}$$

This means that in order to produce a triangular downwash distribution, the local disc area loading should vary as a square of $\bar{r}$ *(c)*.

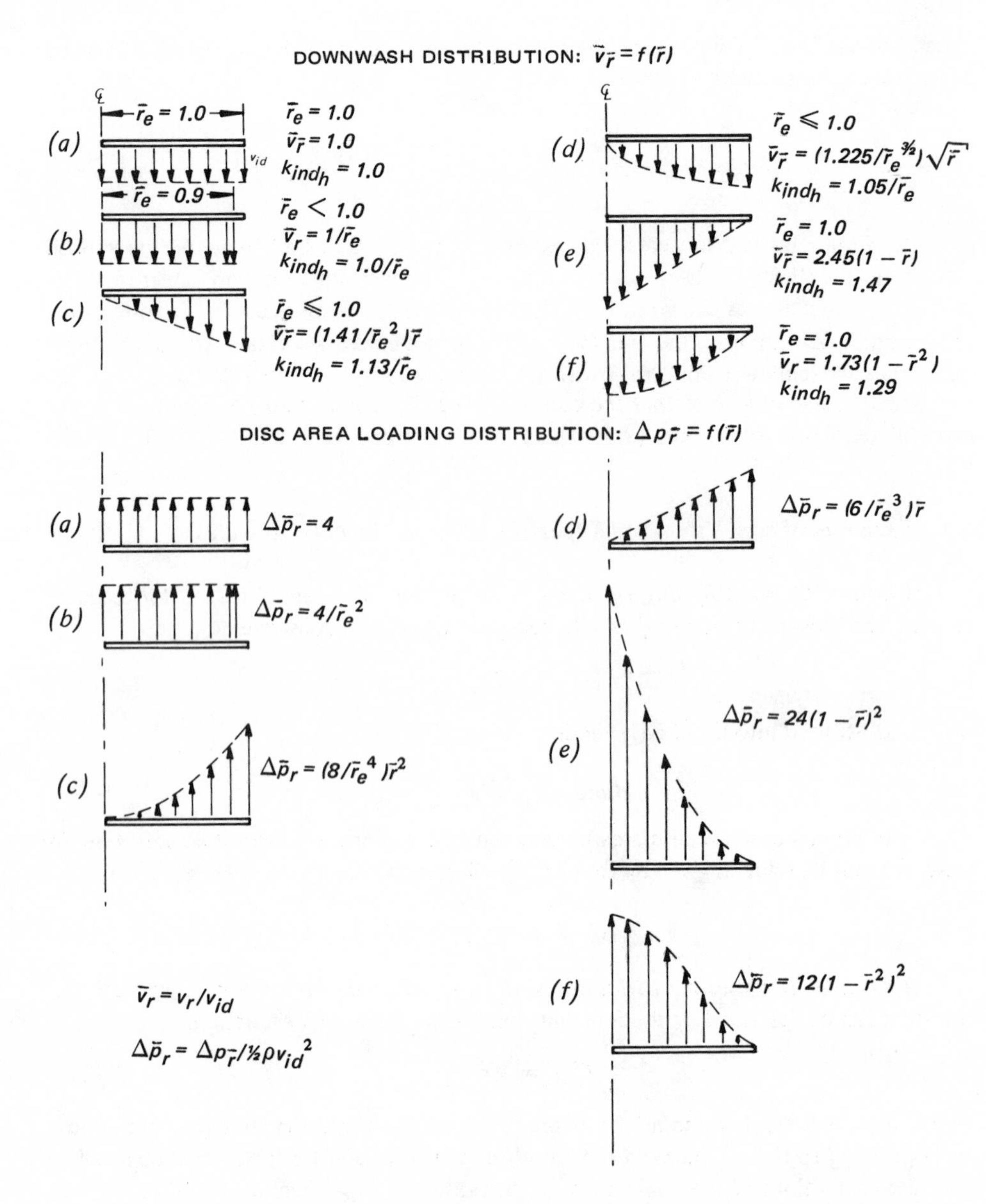

Figure 2.21 Hover k_{ind_h} values and $\Delta\bar{p}_{\bar{r}}$ variations for the following induced velocity distributions:

(a) uniform with no tip losses
(b) uniform with tip losses
(c) triangular with or without tip losses
(d) proportional to $\sqrt{\bar{r}}$ with or without tip losses
(e) reversed triangular with no tip losses
(f) proportional to $(1 - \bar{r}^2)$ with no tip losses

Other Patterns of Downwash Distribution. k_{indh} values and $\Delta\bar{p}_{\bar{r}}$-variations corresponding to the patterns of induced velocity distribution given by $f(\bar{r}) = \nu\sqrt{\bar{r}}$, $f(\bar{r}) = \nu(1-\bar{r})$, and $f(\bar{r}) = \nu(1-\bar{r}^2)$ are also shown in *(d)*, *(e)*, and *(f)*.

It can be seen that the nonuniformity of the downwash distribution and excessive tip losses may push the induced power required to generate a given thrust in hover way over its ideal value. However, some nonuniform downwash patterns are more detrimental than others. For instance, $f(\bar{r}) = \nu\sqrt{\bar{r}}$ *(d)* results in $k_{indh} = 1.05/\bar{r}_e$ only, while those with maximum downwash at the disc center and zero at the disc edge appear especially unfavorable, showing the k_{indh} factor to be as high as 1.47 and 1.29 in schemes *(e)* and *(f)*. In those cases where some induced velocities are directed opposite to the general direction of flow in the slipstream, the k_{ind} values may be even higher.

It should also be noted that the $\bar{r}_e$ in *(e)* and *(f)* does does not appear in either of the expressions for $f(\bar{r})$ or k_{indh}. This omission is the result of the low induced velocities at the outer disc rim; therefore, the variation of $\bar{r}_e$ values within practical limits $(0.9 \leqslant \bar{r}_e \leqslant 1.0)$ has very little influence on the level of the ν coefficient and k_{indh}.

With respect to the $\Delta\bar{p}_{\bar{r}}$-distribution, it can be easily seen that in addition to the previously discussed cases of uniform and triangular downwash distribution, the parabolic distribution of *(d)* would require a triangular variation of $\Delta\bar{p}_r$; for the reversed triangular distribution of *(e)*, the $\Delta\bar{p}_r$ variation should be proportional to $(1-\bar{r})^2$; and for the $f(\bar{r}) = \nu(1-\bar{r}^2)$ distribution of *(f)*, $\Delta\bar{p}_r$ would be proportional to $(1-\bar{r}^2)^2$.

5.4 k_{ind} Values and Types of Span-Loading in Horizontal Flight (Figure 2.22)

An analogy to the hovering case can be drawn here — once the type of downwash distribution $f_1(\bar{x})$ is known (e.g., uniform, parabolic, or triangular), then $f(\bar{x}) = \nu f_1(\bar{x})$. The value of ν can now be determined from the condition that thrust produced by the $f_1(\bar{x})$-type downwash distribution should be equal to that corresponding to the ideal case. This approach will be applied to a few selected examples.

Uniform Downwash Without and With Tip Losses. It can be seen from Eq (2.65) that for the ideal case when $f_1(\bar{x}) = 1.0$ and $\bar{x}_e = 1.0$, $\nu = 1.0$ and as expected, $k_{indh_o} = 1.0$ also *(a)*. If there are tip losses $(\bar{x}_e < 1.0)$ while the downwash still remains uniform, then, contrary to hover, the thrust equality condition would lead to $\nu = 1/\bar{x}_e^2$ and $v_{\bar{x}} = v_{idh_o}/\bar{x}_e^2$. The corresponding induced power factor would be $k_{indh_o} = 1/\bar{x}_e^2$ as shown in *(b)*, and the span load distribution will retain its elliptical shape.

Triangular Downwash — $f(\bar{x}) = \nu\bar{x}$. The condition of equality of thrust expressed in terms of ideal induced velocity and that given by Eq (2.59) with the integration limits $0 - \bar{x}_e$, and $\bar{x}_e^2$ substituted for 1 in Eq (2.56a) leads to:

$$\nu = 4/\pi \int_0^{\bar{x}_e} \bar{x}\sqrt{\bar{x}_e^2 - \bar{x}^2}\, d\bar{x} \tag{2.71}$$

which, integrated within the indicated limits, gives $\nu = \pi/(4/3)\bar{x}_e^3 = 2.356/\bar{x}_e^3$. Substituting $f(\bar{x}) = \nu\bar{x}$ into Eq (2.65) and integrating from $\bar{x} = 0$ to $\bar{x} = \bar{x}_e$ gives

$$k_{indh_o} = 1.338/x_e^2. \tag{2.72}$$

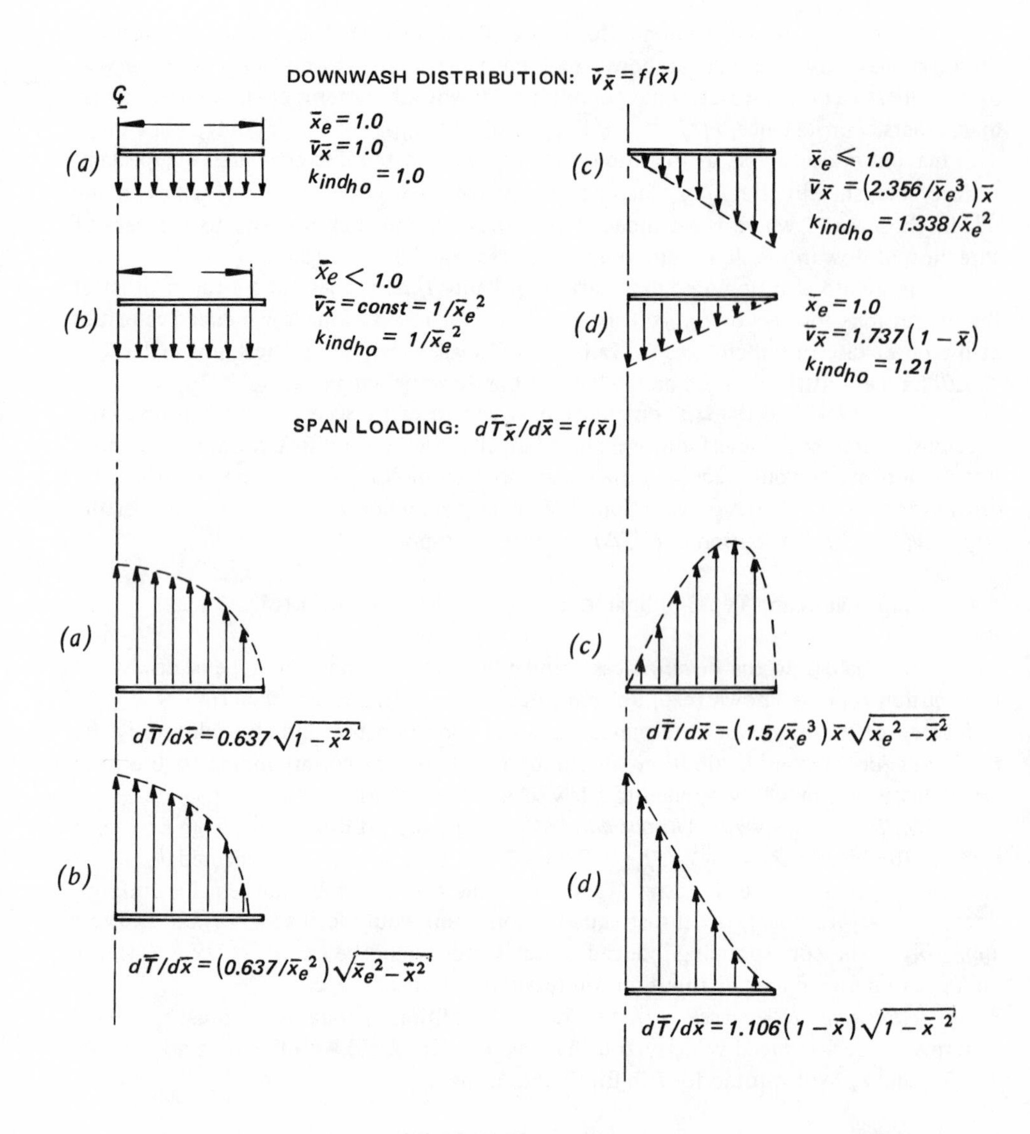

Figure 2.22 Examples of $k_{ind_{ho}}$ values and $d\,T_{\bar{x}}/d\bar{x}$ variation in horizontal flight for the following induced velocity distributions: (a) uniform with no tip losses; (b) uniform with tip losses; (c) triangular with tip losses; and (d) reversed triangular with no tip losses

while the corresponding relative span loading *(c)*, as obtained from Eq (2.61), will be

$$dT_x/dx = (1.5/x_e^3)x\sqrt{1 - x^2}. \tag{2.73}$$

Reversed Triangular Distribution – $f(\bar{x}) = \nu(1 - \bar{x})$. An example of the type of downwash distribution where induced velocities decrease to zero at the tip is given in *(d)*. Similar to the previously considered case of hovering, the influence of span losses can be ignored here, and it may be assumed that $\bar{x}_e = 1.0$. Under this assumption, the condition of thrust equality leads to $\nu = 1.737$, while the induced power factor becomes $k_{ind_{ho}} = 1.21$. The associated relative span loading will be

$$dT_x/dx = 1.106(1 - x)\sqrt{1 - x^2}. \tag{2.74}$$

As indicated in Fig 2.22, horizontal or more generally, forward flight, deviations in the average downwash from its optimum uniform value may also lead to considerable losses in induced power. It should also be noted that some types of downwash distribution, such as that exemplified by the triangular type, are especially damaging.

Although tip losses are always detrimental, their significance, like that of the hovering case, is related to the basic shape of downwash distribution. They appear least detrimental in downwash types where induced velocity approaches zero toward the tips of the disc.

Similar to the case of hover, the presence of an upwash within the generally downward directed induced velocity field would contribute to a considerable increase of the $k_{ind_{ho}}$ values.

As to span loading associated with various types of downwash distribution, the reader is referred to the lower part of Fig 2.22.

6. TANDEM ROTOR INTERFERENCE IN HORIZONTAL FLIGHT

6.1 The Model

An investigation of the induced power of non-overlapping or slightly overlapping tandems in forward flight may serve as an additional example of the application of the momentum theory to the basic problems of mutual rotor interference in forward flight. To achieve the double goal of a better understanding as well as a quantitative evaluation of these problems, the tandem is modeled by two actuator discs, of the same radius R, representing the rotors (Fig 2.23). It is again assumed that the aircraft is stationary, while a large mass of air moves past it at velocity $-\vec{V}$ (inverse of the speed of flight $\vec{V}$).

Since the actuator disc does not affect the approaching fluid, it may be postulated that–in analogy to the tandem biplane–the influence of the rear rotor on the front rotor may be neglected. By contrast, the effects of the flow tube extending downstream from the front rotor and entering the "sphere of influence" of the rear rotor represent the physical concept of mutual rotor interference. It may be anticipated that the geometric position of the rear rotor with respect to the streamtube affected by the front rotor, as determined by h_{re}, would represent one of the most important parameters in the determination of induced power.

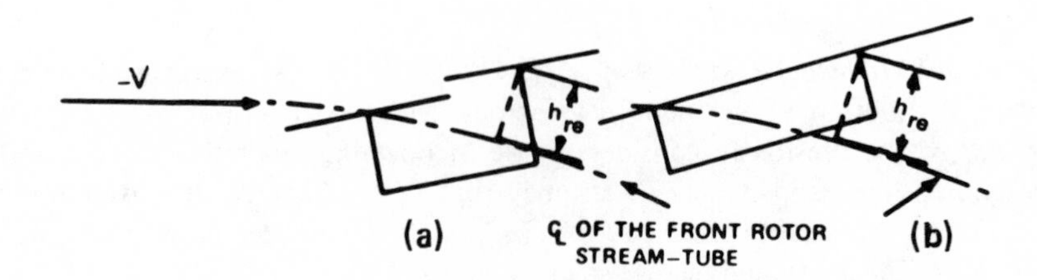

Figure 2.23 Representation of (a) slightly overlapping, and (b) nonoverlapping tandems in horizontal flight

To facilitate the task, it will be assumed that the forward speed is high enough to justify the small-angle assumption in determining deflection of the flow due to the induced velocity; therefore, the rate of flow through the streamtube affected by the rotor becomes almost equal to the speed of flight ($V' \approx V$).

6.2 Axial Flow Velocities and Induced Power

The average induced velocity of the front rotor (v_{fr}), developing thrust (T_{fr}), can be expressed in the same way as for the isolated rotor:

$$v_{fr} = T_{fr}/2\pi R^2 \rho V . \tag{2.75}$$

In the following section, it will be shown that the downwash velocity at the trailing edge of the rotor attains its full far–downstream value equal to twice the average induced velocity. Hence, it is logical to assume that the air approaching the rear rotor already has a downward component equal to $2v_{fr}$.

The induced velocity associated with thrust T_{re} of the rear rotor will be

$$v_{re} = T_{re}/2\pi R^2 \rho V . \tag{2.76}$$

Should the rear rotor be completely submerged in the slipstream of the front rotor, the total axial component (V_{ax}) of the rate of flow through the disc associated only with lift generation by both rotors will be

$$V_{ax_{re}} = 2v_{fr} + v_{re} . \tag{2.77}$$

Consequently, the induced power (in Nm/s or ft-lb/s) of the rear rotor will be

$$P_{ind_{re}} = T_{re}(2v_{fr} + v_{re}) , \tag{2.78}$$

and the total induced power of both rotors becomes

$$P_{ind} = [T_{fr}\, v_{fr} + T_{re}(2v_{fr} + v_{re})] . \tag{2.79}$$

For a particular case when the front and rear rotors are producing the same thrust, and the latter is fully submerged in the slipstream of the front one, the induced power of the tandem would be equal to twice that of the two isolated rotors producing the same thrust.

However, in numerous practical cases, the rear rotor is not fully submerged in the streamtube affected by the front rotor. This may be due to the geometry of the aircraft, its trim position in flight, and finally, the downward deflection of its front rotor slipstream. In order to deal with all of these cases, a simplified picture of the interaction between the slipstream of the front and rear rotors is conceived, imagining that the airstreams penetrate each other in the manner shown in Fig 2.24.

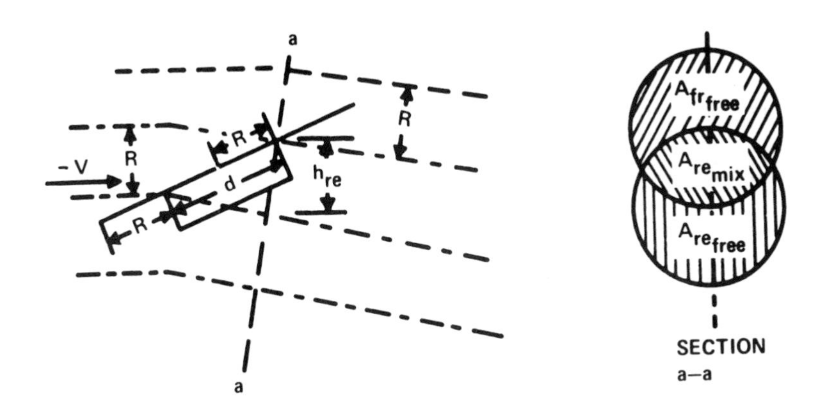

Figure 2.24 Streamtube mix of two rotors

It may be anticipated that within those regions where the streamtube affected by the front rotor does not penetrate into that influenced by the rear rotor, the downwash in the unmixed part of the rear rotor streamtube will remain as given by Eq (2.76). In those regions where the airstream influenced by the front rotor penetrates that influenced by the rear one, Eq (2.77) is assumed to be valid.

Consequently, the induced power of the whole helicopter can be broken down into three components and computed separately. The first one will be of the front rotor, and will remain the same as for the previously discussed case.

$$P_{ind_{fr}} = v_{fr}T_{fr}\,. \tag{2.80}$$

As far as the rear rotor is concerned, the part of its induced power that may be "credited" to the nonmixed part of the rear-rotor airstream can be expressed as

$$P_{ind_{re_{free}}} = T_{re}v_{re}(A_{re_{free}}/\pi R^2)\,. \tag{2.81}$$

For that part where the two streams mix together, the induced power can be determined as

$$P_{ind_{re_{mix}}} = T_{re}(2v_{fr} + v_{re})(A_{re_{mix}}/\pi R^2)\,, \tag{2.82}$$

where $A_{re_{mix}}$ represents the mixed area, and $A_{re_{free}}$ represents the freestream (Fig 2.24). The total induced power of the helicopter will be the sum of all three components:

$$P_{ind} = T_{fr}v_{fr} + (T_{re}/\pi R^2)[v_{re}A_{re_{free}} + (2v_{fr} + v_{re})A_{re_{mix}}] . \qquad (2.83)$$

For the particular case of rotors producing the same thrust equal to $T/2$, the expression for induced power is reduced to the following:

$$P_{ind} = (T^2/\pi R^2 \rho V)\ [\tfrac{1}{4} + (1/4\pi R^2)(A_{re_{free}} + 3A_{re_{mix}})] . \qquad (2,84)$$

In order to convert Eqs (2.83) and (2.84) to horsepower, the results of the calculations conducted in SI units should be divided by 735; or in English, by 550.

It should be realized that the above-considered induced power of the tandem configuration is still for an idealized case, as it assumes uniform downwash distribution and no tip losses. However, P_{ind}, as given by Eq (2.84) will be higher than the truly ideal one corresponding to two isolated rotors, each developing a thrust of one-half T.

6.3 The k_{ind} Factor

Similar to the k_{ind_h} factor discussed previously in Sect 5.2, the $k_{ind_{ho}}$ factor can be defined as a ratio of the induced power as given by Eq (2.84) to that of the ideal induced power of two isolated rotors of the same radius, each developing a thrust equal to one-half T:

$$k_{ind_{ho}} \equiv P_{ind}/2P_{id}|_{\frac{1}{2}T} .$$

Performing the necessary substitution, one obtains:

$$k_{ind_{ho}} = 2[\tfrac{1}{4} + (1/4\pi R^2)(A_{re_{free}} + 3A_{re_{mix}})] . \qquad (2.85)$$

It can be seen that for rotors located so that the centerline of the front rotor airstream passes through the hub of the rear rotor ($h_{re} = 0$), the $k_{ind_{ho}}$ would be equal to 2.0. However, when corrections resulting from tip losses are introduced, the $k_{ind_{ho}}$ factor for the case of $h_{re} = 0$ becomes higher than 2.

The graph plotted in Fig 2.25 shows the variation in the $k_{ind_{ho}}$ factor (including tip losses) vs elevation of the hub of the rear rotor over the centerline of the front rotor airstream tube as predicted by this momentum approach[7].

For comparison, the $k_{ind_{ho}}$ values are shown as computed for a non-overlapped tandem on the basis of the Mangler-Squire theory[8].

In addition, results representing an average of over 60 points obtained by Boeing-Vertol in wind-tunnel tests of a universal tandem helicopter model are also shown in this figure. The lower of the two Boeing curves represents direct total power measurements (including blade profile drag contribution). The upper curve gives the induced power ratios (true $k_{ind_{ho}}$ factor values) which were computed by assuming that profile power amounts to 25-percent of the total power. It can be seen that the $k_{ind_{ho}}$ values predicted by the simple momentum approach agree quite well with those obtained from the wind-tunnel tests.

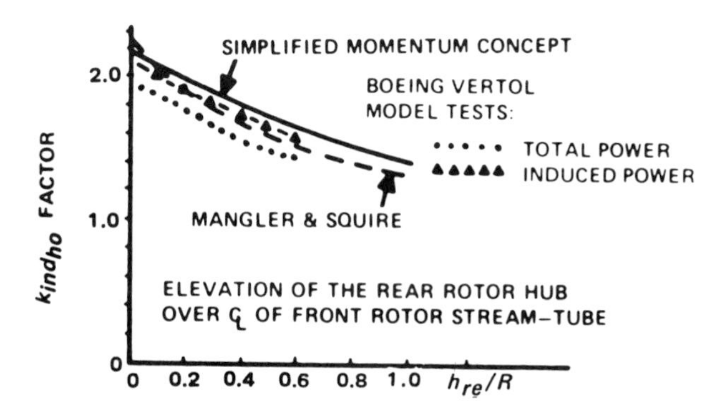

Figure 2.25 The k_{ind} factor for a tandem in horizontal flight vs rear-rotor elevation

7. INDUCED VELOCITY DISTRIBUTION ALONG DISC CHORDS

Momentum theory can also provide some insight into such problems as induced velocity distribution along the fore-and-aft rotor disc chords in horizontal flight.

To demonstrate the basic methodology of attacking this task, only a simple case of an actuator disc having a uniform surface loading ($\Delta p = w = const$) over its entire area is considered here. However, once understood, this approach can easily be extended to include other cases where Δp varies over the disc area.

The general flow pattern is assumed to be similar to that shown in Fig 2.20, including the assumption that $V_{ho} \gg v_{ho}$. However, to facilitate the present study, special coordinate systems and notations were introduced as shown in Fig 2.26.

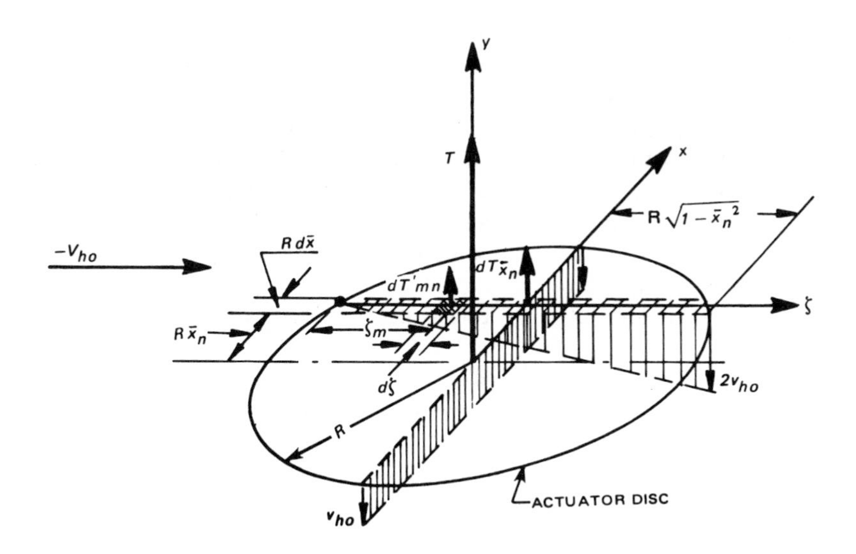

Figure 2.26 Notations and coordinate systems for determination of the chordwise induced velocity distribution

From this figure it should be noted that in addition to the x, y coordinates, an auxiliary scale ζ is introduced in such a way that its origin coincides with the leading edge of a chordwise disc element dx wide and located at a distance x_n.

In order to accomplish the task of determining induced velocity distribution along an arbitrary chord located at a distance x_n from the fore-and-aft disc diameter, attention is first concentrated on an infinitesimal element $d\zeta$ long, dx wide, and located at a distance ζ_m along the examined chord. The total length of that chord will obviously be $\ell_n = 2\sqrt{R^2 - x_n^2}$.

Thrust dT'_{mn} developed by the $d\zeta\, dx$ element will be $dT'_{mn} = w\, d\zeta\, dx$, where w is the disc loading. According to the momentum theory interpretation developed in the subsection of Sect 5.1 on horizontal flight, dT'_{mn} can be expressed as follows:

$$w\, d\zeta\, dx = 2\sqrt{R^2 - y_n^2}\, dx\, \rho V_{ho} 2 dv_{ho_{nm}} \tag{2.86}$$

where $2\sqrt{R^2 - y_n^2}$ is the chord of the slipstream cross-section associated with the disc strip located at x_n, and $dv_{ho_{nm}}$ is the induced velocity at the element $d\zeta\, dx$.

Remembering that the radius of the slipstream cross-section for $V_{ho} \gg v_{ho}$ is the same as that of the disc; i.e., $\sqrt{R^2 - y_n^2} = \sqrt{R^2 - x_n^2}$, Eq (2.86) can be rewritten as

$$dv_{ho_{nm}} = (w/4\rho V_{ho}\sqrt{R^2 - x_n^2})\, d\zeta. \tag{2.87}$$

Since the point at ζ_m was arbitrarily selected, then it may be stated that at any point of the disc chord located at x_n, Eq (2.87) is valid, and consequently,

$$(dv_{ho}/d\zeta)_n = w/4\rho V_{ho}\sqrt{R^2 - x_n^2}. \tag{2.88}$$

It may be imagined that on the scale of infinitesimal dimensions, the "far-downstream" distance from a point nm where dv_{ho} occurs can be expressed as a (not too large) number j of elementary lengths $d\zeta$. This means that the value of dv_{ho} would be doubled at a distance $j\,d\zeta$ downstream from the point where it was first generated.

Now, looking upstream from a point located on a strip n at a distance ζ_m, one would find that all the elements of that strip from the leading edge to a point located at $\zeta_m - j\,d\zeta$ developed induced velocities which are already twice as high as those given by Eq (2.87). In other words, the slope of the induced speed growth would be twice as high as that given by Eq (2.88).

It may be stated hence, that the total induced velocity occurring at point nm will be

$$v_{ho_{nm}} = 2(dv_{ho}/d\zeta)_n(\zeta_m - j\,d\zeta) + (dv_{ho}/d\zeta)_n j\,d\zeta. \tag{2.89}$$

Neglecting the infinitesimals vs the finite quantities, Eq (2.89) becomes

$$v_{ho_{nm}} = 2(dv_{ho}/d\zeta)_n \zeta_m. \tag{2.90}$$

However, ζ_m can be expressed as the half-chord length $\sqrt{R^2 - x_n^2}$ times $\bar{\zeta}_m$, where $\bar{\zeta}_m \equiv \zeta_m/\sqrt{R^2 - x_n^2}$.

Expressing the induced velocity derivative in Eq (2.90) according to Eq (2.88) and substituting $\bar{\zeta}_m\sqrt{R^2 - x_n^2}$ for ζ_m, the following is obtained:

$$v_{ho_m} = (w/2\rho V_{ho})\bar{\zeta}_m . \tag{2.91}$$

It can be seen from this equation that the influence of the spanwise location of the chord (x_n) has disappeared, which means that the expressed relationship is valid for any disc chord. At the leading edge of the disc, the induced velocity is zero and then grows linearly until it reaches a value equal to the average induced velocity of horizontal flight (Eq (2.36)) along the whole lateral diameter of the disc, where $\zeta_m = 1.0$. Along the trailing edge, the induced velocity becomes twice its average value ($\zeta_m = 2.0$) as shown in Fig 2.26.

8. CONCLUDING REMARKS RE SIMPLE MOMENTUM THEORY

The application of physicomathematical models based on the simple momentum approach to thrust generation in both axial and oblique translation has provided some understanding of the basic relationships between such important design parameters as disc loading and power required per unit of thrust. A satisfactory interpretation of power required in climb, in both vertical ascent and forward translation, and some understanding of the partial-power and no-power vertical descent phenomena have been achieved; although in the latter case, it became clear that the physical assumptions required in the structure of the model based on the simple momentum approach were not completely convincing. The influence of nonuniform downwash distribution and tip losses on induced power was shown. Qualitatively valid methods for estimating induced power losses of tandem rotors in forward flight were developed. Finally, some insight was gained regarding chordwise induced velocity distribution at the disc in horizontal flight.

It may be stated hence, that the simple momentum theory contributes to a better understanding of many basic aspects of performance of rotary-wing aircraft. Furthermore, many performance predictions (e.g., vertical flight, forward flight, and average downwash velocity at various tilt angles of the rotor) can be performed by substituting proper conceptual models based on the momentum theory for actual helicopters. In addition to obtaining quantitative results which are often sufficiently accurate, the momentum approach may provide a clarity of the overall picture that could be lacking when more complicated theories are applied.

However, the presently discussed theory encounters serious limitations in providing guidance for rotor design, as it singles out disc loading as the only important parameter. Consequenly, it does not provide any insight into such rotor characteristics as ratio of the blade area to the disc area (solidity ratio, σ), blade airfoil characteristics, tip speed values with all the associated phenomena of compressibility, etc. Even when discussing the influence of nonuniformity of induced velocity and tip losses on the k_{ind} factor in hover and forward flight, the simple momentum approach did not provide a physical concept that could explain the nonuniformities of downwash velocities or the presence of tip losses.

In order to be able to investigate the influence of such rotor design parameters as blade geometry (planform, airfoils, and twist), airfoil characteristics, solidity ratio, and tip speed on helicopter performance, a new physicomathematical model reflecting all of these quantities must be conceived. The combined blade element and momentum theory considered in the next chapter provides a more refined model.

References for Chapter II

1. Poisson-Quinton, P. *Introduction to V/STOL Aircraft Concepts and Categories.* AGARDograph 126, May 1968.

2. Yaggy, P.F. and Mort, K.W. *Wind-Tunnel Tests of Two VTOL Propellers in Descent.* NASA TN D-1766, 1963.

3. Stepniewski, W.Z. *The Subsonic VTOL and GETOL in Perspective.* Aerospace Engineering, Vol 21, No. 4, April 1962.

4. Durand, W.F. *Aerodynamic Theory.* Vol IV, Div L: Airplane Propellers by H. Glauert. Durand Reprinting Committee, California Institute of Technology, 1943, p. 319.

5. Glauert, H. *On Horizontal Flight of a Helicopter.* R&M 1157, 1928.

6. Durand, W.F. *Aerodynamic Theory.* Vol IV, Div J: Induced Power of the Tandem Configuration in Horizontal Flight by A. Betz. Durand Reprinting Committee, California Institute of Technology, 1943, p. 49,50.

7. Stepniewski, W.Z. *A Simplified Approach to the Aerodynamic Rotor Interference of Tandem Helicopters.* Proceedings of the West Coast American Helicopter Society Meeting, Sept 21 and 22, 1955.

8. Mangler, K.W. and Squire, H.B. *The Induced Velocity Field of a Rotor.* R&M No 2642, 1953.

BLADE ELEMENT THEORY

The concept of modeling rotor blades through an assembly of aerodynamically independent airfoil elements is developed, and then combined with the momentum theory in order to determine flow conditions at each element. This approach is first applied to axial translation (with hover as a limiting case) of single and overlapping tandem rotors, and then is extended to include forward regimes of flight. In this way, a more refined tool for complete performance predictions of a helicopter, and a better design guide than that offered by the momentum theory alone is provided.

Principal notation for Chapter III

A	area	m^2, or ft^2
AR	aspect ratio	
a	section lift-curve slope	rad^{-1}, or deg^{-1}
b	number of blades	
C_D	wing or body drag coefficient	
C_L	wing or body lift coefficient	
C_P	rotor power coefficient: $C_P \equiv P/\pi R^2 \rho V_t^3$	
C_Q	rotor torque coefficient: $C_Q \equiv Q/\pi R^3 \rho V_t^2$	
C_T	rotor thrust coefficient: $C_T \equiv T/\pi R^2 \rho V_t^2$	
c	chord	m, or ft
c_d	section drag coefficient	
c_ℓ	section lift coefficient	
c_m	section moment coefficient	
c_p	blade area power coefficient	
c_q	blade area torque coefficient	
c_t	blade area thrust coefficient	
D	drag	N, or lb
d	diameter	m, or ft
f	equivalent flat plate area	m^2, or ft^2
HP	horsepower	75kGm/s, or 550 ft-lb/s
h	height	m, or ft
k_{ind}	ratio of actual to induced power	
k_v	vertical download coefficient: $k_v \equiv T/W$	
L	lift	N, or lb
M	Mach number	
ov	overlap	
P	power	Nm/s, or ft-lb/s
Q	torque	Nm, or ft-lb
R	rotor radius	m, or ft
R_e	Reynolds number	

Theory

r	radial distance	m, or ft
s	speed of sound	m/s, or fps
T	thrust	N, or lb
tr	blade taper ratio: $tr \equiv c_i/c_t$	
U	velocity of flow approaching the blade	m/s, or fps
V	velocity in general	m/s, or fps
v	downwash velocity	m/s, or fps
W	weight	N, or lb
w	disc loading	N/m^2, or lb/ft^2
w_f	equivalent flat-plate area loading	N/m^2, or lb/ft^2
α	angle-of-attack	rad, or deg
β	flapping angle	rad, or deg
Δ	increment	
ϵ	small, but finite, increment of distance	m, or ft
Γ	airfoil sweep angle	rad, or deg
ζ	ordinate	
η	efficiency	
θ	blade section pitch angle	rad, or deg
λ	inflow ratio	
μ	advance ratio: $\mu \approx V/V_t$	
ρ	air density	kg/m^3, or slug/ft^3
σ	rotor solidity ratio: $\sigma \equiv b\tilde{c}R/\pi R^2$	
ϕ	inflow angle	rad, or deg
ψ	azimuth angle	rad, or deg
Ω	rotor rotational speed	rad/s
ω	angular velocity	rad/s

Subscripts

a	aerodynamics	
ax	axial	
b	blade	
c	climb	
e	effective	
eq	equivalent	
fo	forward	
h	hover	
ho	horizontal	
i	inboard, or initial	
id	ideal	
ind	induced	
inp	inplane	
L	lift	
o	zero station	station
ov	overlap	
par	parasite	

pr	profile
R	rotor
r	station r
res	resultant
rot	rotational
rr	rear rotor
t	tip
th	thermal
tot	total
tr	transmission
u	ultimate
$\perp$	perpendicular
$\parallel$	parallel

Superscripts

$-$	nondimensional
$\sim$	average

1. INTRODUCTION

The purpose of this chapter is to construct a physicomathematical model of the rotor in order to eliminate or, at least, to alleviate the previously-discussed limitations of the momentum theory.

The blade element, or strip theory, provides a model which will permit one to determine, as precisely as possible, aerodynamic forces and moments acting on various segments of the blade. This is done by imagining that the blade is composed of aerodynamically independent, chordwise-oriented, narrow strips or elements.

From purely geometric considerations, it is relatively easy to determine the total velocity of air flow approaching any blade element for any given flight condition, as well as the component of that flow perpendicular to the blade axis. If information regarding values of the lift, drag, and moment coefficients existing at each blade strip could somehow be obtained, then knowing the chord lengths of the strips and the magnitude of the flow velocity component perpendicular to the blade axis, the lift, drag, and pitching moment per unit length of the blade span can be computed. Integrating (either graphically or numerically) those unit loads over the entire blade span, the total lift, drag or, more important, torque about the rotor axis of rotation, and pitching moment experienced by the blade as a whole can be obtained.

Hence, it is clear that a precise determination of aerodynamic coefficients at various blade stations becomes the key to the successful application of the blade element concept.

Attempts to obtain this information were first made in the so-called *primitive blade element theory*, developed almost entirely by S. Drzewiwcki between 1892 and 1920 (Ref 1, p. 211).

At that time, it was commonly accepted that the angle-of-attack of a blade section was the angle between the zero-lift chord of a particular section and the normal (perpendicular to the blade axis) component of flow resulting from the translation of the rotor as a whole and rotation of the blade about the rotor axis. The role of local induced velocity was completely ignored in this approach.

Knowing the so-defined section angle-of-attack, it was further assumed that lift and drag coefficients experienced by the blade section would be the same as those of a fixed wing of "proper aspect ratio" at the same angle-of-attack. However, an uncertainty still existed as to what should be regarded as the proper aspect ratio. In this respect, various authors suggested aspect ratios ranging from $AR = 6$ to $AR = 12$, while others considered the actual blade ratio as being most representative. The uncertainty as to the true angle-of-attack at every element of the blade constituted a serious logical drawback of the rotor model based on the primitive blade element theory. Along with a better understanding of the two-dimensional (sectional) airfoil characteristic, it became clear that if the complete pattern of air flow (including induced velocities) in the immediate vicinity of the blade element were known, then the forces and moments acting on that element could be accurately predicted using sectional coefficients c_ℓ, c_d, and c_m obtained after due consideration of the existing Reynolds and Mach numbers and **steadiness** of the flow. Indeed, the entire current philosophy of predicting rotor performance and airloads can be characterized as an effort to depict, as accurately as possible and at every instant of time, the flow fields in the immediate vicinity of the blade elements.

Combining the blade element and momentum theories discussed in this chapter probably represents one of the simplest ways of finding time-average induced velocities at various points of the rotor disc. Although this represents a definite step in the right direction, it should be realized that the proposed approach can not account for instantaneous flow changes. Consequently, its usefulness is greatly limited when dealing with the aeroelastic and some airload problems where knowledge of the variation of forces and moments with time is essential. By contrast, the combined blade element–momentum theory may offer simple, but sufficiently accurate, computational methods for many practical tasks of performance prediction. This is especially true during the concept-formulation phase of preliminary design of rotary-wing aircraft. Here, the main value of this theory clearly lies in its ability to indicate the influence, on helicopter performance, of such important design parameters as tip speed, rotor solidity, blade planform, twist, and airfoil characteristics in addition to disc loading whose significance was stressed by the momentum approach. All of these relationships, should provide a valuable guide in the process of aircraft optimization for any set of mission requirements.

2. AXIAL TRANSLATION AND HOVERING

2.1 Basic Considerations of Thrust and Torque Predictions

By analogy with the momentum theory, basic concepts of the blade element theory will be initially examined using the case of a rotor in axial translation in the direction of thrust (climb).

Consider that the blade of a rotor of radius R is composed of narrow elements $dr \equiv R d\bar{r}$ wide, having a chord c, an incidence (pitch) angle with respect to the rotor plane θ, and a defined airfoil section. In general, these three quantities may vary along the blade span or, in other words, may depend on the radial position of the blade element as defined by $r \equiv R\bar{r}$ (Fig 3.1).

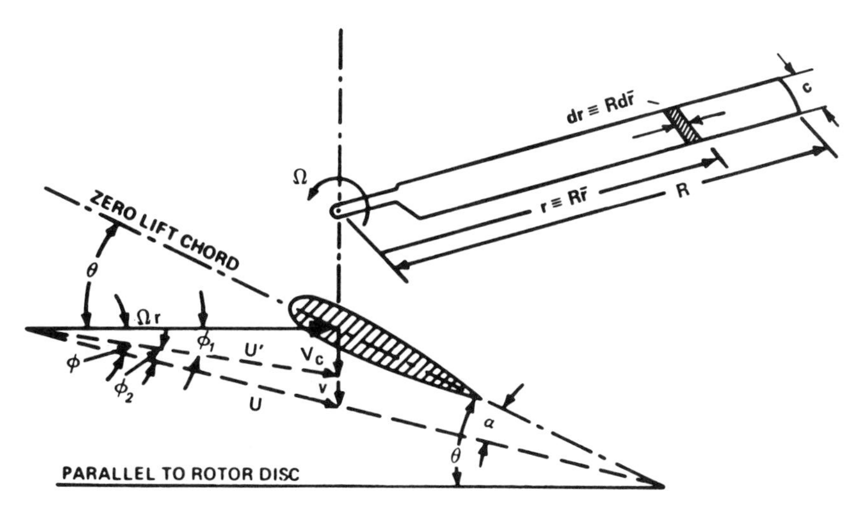

Figure 3.1 Blade element concept

The rotor is composed of *b* blades, and is assumed to be turning at a rotational velocity $\Omega \equiv V_t/R$ (where V_t is the tip speed) while moving along its axis in the direction of thrust, at a speed V_c.

If the pitch angle; i.e., the angle between the zero lift-line of the element and the rotor disc, of an element located at radius r is θ_r, its angle-of attack a_r, as shown in Fig 3.1 with r subscripts omitted, will be:

$$a_r = \theta_r - (\phi_{1_r} + \phi_{2_r})$$

where ϕ_{1_r} is the angle due to the rate of climb $\phi_{1_r} = tan^{-1}(V_c/\Omega r)$ and ϕ_{2_r} is the induced angle: $\phi_{2_r} = tan^{-1}(v_r/\Omega r)$.

The angle-of-attack of the element at station r can now be expressed as

$$a_r = \theta_r - tan^{-1}[(V_c + v_r)/\Omega r] \tag{3.1}$$

or in those cases where v_c and v_r are small in comparison with Ωr,

$$a_r = \theta_r - (V_c + v_r)/\Omega r \tag{3.1a}$$

where, of course, all angles are expressed in radians.

If the induced velocity at some radius r were known, it would be possible to estimate accurately the lift and drag of the blade element using section coefficients. The section lift coefficient $c_{\ell_r} = a_r \alpha_r$, where a_r is the slope of the lift curve for the airfoil of the considered blade element. The a_r value, of course should correspond to the operational conditions of that particular blade element; i.e., its Mach and Reynolds numbers as well as the influence of unsteady aerodynamics should be considered.

The above-mentioned aspects, as well as other phenomena affecting airfoil characteristics in the particular environment of rotary-wing operation, will be discussed in Ch VI.

The magnitude of the lift ($dL_r \perp U_r$) experienced by the blade element of width dr and chord c_r will be:

$$dL_r = \tfrac{1}{2}\rho a_r \alpha_r U_r^2 c_r dr. \qquad (3.2)$$

When V_c and v_r are small in comparison with Ωr, which is often true for the working part of the blade, $U_r \approx \Omega r$. Therefore, substituting Eq (3.1a) for α_r in Eq (3.2), it becomes

$$dL_r = \tfrac{1}{2}a_r\rho\left[\theta_r - \frac{V_c + v_r}{\Omega r}\right]c_r(\Omega r)^2\, dr. \qquad (3.3)$$

The elementary profile drag $(dD_{pr_r} \| U_r)$ experienced by the blade element at radius r will be

$$dD_{pr_r} = \tfrac{1}{2}\rho c_{d_r} U_r^2 c_r dr \qquad (3.4)$$

or, assuming that $\Omega r \approx U_r$:

$$dD_{pr_r} = \tfrac{1}{2}c_{d_r}\rho(\Omega r)^2 c_r dr. \qquad (3.5)$$

Consequently, the elementary thrust (dT_r) will be

$$dT_r = dL_r \cos\phi_r - dD_{pr_r} \sin\phi_r \qquad (3.6)$$

and the corresponding elementary torque is

$$dQ_r = (dL_r \sin\phi_r + dD_{pr_r} \cos\phi_r)r \qquad (3.7)$$

where ϕ_r is given by the relationship

$$\tan\phi_r = (V_c + v_r)/\Omega r.$$

When ϕ_r is small,

$$\tan\phi_r \approx \sin\phi_r \approx \phi_r,$$

and

$$dT_r = dL_r - dD_r[(V_c + v_r)/\Omega r]. \qquad (3.8)$$

For the working part of the blade, usually, $dD_r[(V_c + v_r)/\Omega r] \ll dL_r$, and Eq (3.8) becomes

$$dT_r \approx dL_r. \qquad (3.8a)$$

Similarly, the simplest formula for the elementary torque will be:

$$dQ_r = [dL_r(V_c + v_r)/\Omega r + dD_{pr_r}]r \tag{3.9}$$

or, in light of Eq (3.8a):

$$dQ_r = [dT_r(V_c + v_r)/\Omega r + dD_{pr_r}]r. \tag{3.9a}$$

Remembering that elementary power (dP_r in Nm/s, or ft-lb/s) required by the considered blade element is

$$dP_r = dQ_r\Omega$$

and substituting Eq (3.9a) into the above relationship, dP_r can be expressed as follows:

$$dP_r = dT_r(V_c + v_r) + dD_{pr_r}r\Omega. \tag{3.10}$$

It can be noticed from Eq (3.10) that the power required by a blade element in axial translation in the direction of thrust (climb) contains two terms previously identified in the momentum theory; namely, (1) $dT_r V_c$; i.e., power associated with an axial translation at a speed V_c, and (2) $dT_r v_r$; i.e., power associated with induced velocity (induced power). However, a third term that was not present in the momentum considerations appears in Eq (3.10). This is $dD_{pr_r}r\Omega$ which, of course, represents the profile power required by the blade element moving through the air at a speed $r\Omega$.

2.2 Combined Blade-Element and Momentum Theory

Induced velocity (v_r) for various values of r can be determined by combining blade element and momentum theories as probably originally proposed by Klemin[2]. This would provide the missing link for a more accurate estimation of dT_r, dQ_r, and hence, dP_r values.

Using the notations of Fig 3.2, the thrust (dT_r) produced by an elementary annulus of width dr and radius r can be expressed—in analogy to Eq (2.49)—according to the momentum theory as

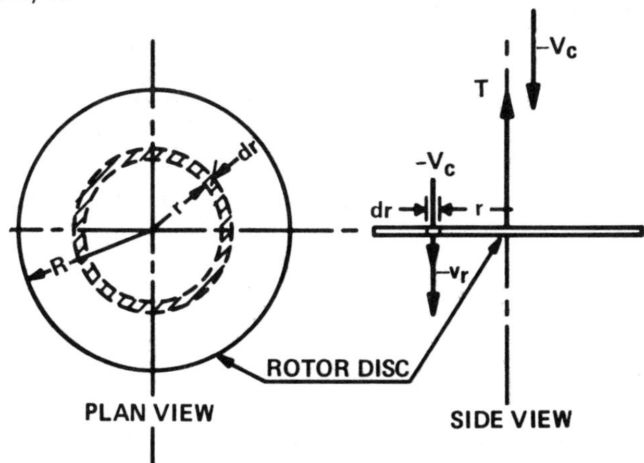

Figure 3.2 Elementary annulus of the rotor disc

$$dT_r = 4\pi\rho(V_c + v_r)v_r r\,dr \qquad (3.11)$$

where v_r is the induced velocity at the rotor disc.

On the other hand, according to the blade element theory and under the assumptions discussed in the preceding paragraphs, the elementary thrust experienced by b number of blades can be expressed as

$$dT_r = \tfrac{1}{2}c_{\ell_r}(\Omega r)^2\rho\, bc_r dr. \qquad (3.12)$$

Using the nondimensional notation ($\bar{r}$) for radial location as expressed in Ch II, the following nondimensional quantities are obtained:

$$\left.\begin{aligned} r &= R\bar{r} \\ dr &= Rd\bar{r} \\ r\Omega &= V_t\bar{r} \end{aligned}\right\} \qquad (3.13)$$

where $V_t \equiv R\Omega$ is the tip speed.

Equating the right sides of Eqs (3.11) and (3.12), introducing the notations as given in Eq (3.13), and remembering that if the pitch angle at station $\bar{r}$ is $\theta_{\bar{r}}$, then

$$c_{\ell_{\bar{r}}} = a_{\bar{r}}[\theta_{\bar{r}} - (V_c + v_{\bar{r}})/V_t\bar{r}],$$

and the following basic equation is obtained:

$$8\pi R v_{\bar{r}}^2 + (V_t a_{\bar{r}} bc_{\bar{r}} + 8\pi R V_c)v_{\bar{r}} + V_t V_c a_{\bar{r}} bc_{\bar{r}} - V_t^2 a_{\bar{r}} bc_{\bar{r}}\bar{r}\theta_{\bar{r}} = 0. \qquad (3.14)$$

The above equation can be solved for the induced velocity at station $\bar{r}$:

$$v_{\bar{r}} = V_t\left[-\left(\frac{a_{\bar{r}}bc_{\bar{r}}}{16\pi R} + \frac{V_c}{2V_t}\right) + \sqrt{\left(\frac{a_{\bar{r}}bc_{\bar{r}}}{16\pi R} + \frac{V_c}{2V_t}\right)^2 + \frac{a_{\bar{r}}bc_{\bar{r}}\bar{r}\theta_{\bar{r}}}{8\pi R} - \frac{a_{\bar{r}}bc_{\bar{r}}V_c}{8\pi R V_t}}\right] \qquad (3.15)$$

In hovering, when $V_c = 0$, Eq (3.15) is simplified to the following:

$$v_{\bar{r}} = V_t\left[-\frac{a_{\bar{r}}bc_{\bar{r}}}{16\pi R} + \sqrt{\left(\frac{a_{\bar{r}}bc_{\bar{r}}}{16\pi R}\right)^2 + \frac{a_{\bar{r}}bc_{\bar{r}}\bar{r}\theta_{\bar{r}}}{8\pi R}}\right]. \qquad (3.16)$$

If, in addition, the blade is of rectangular shape (chord c is constant), then the rotor solidity σ can be expressed as follows:

$$\sigma = bcR/\pi R^2 = bc/\pi R;$$

hence,

$$bc = \sigma\pi R.$$

Further assuming that the lift slope a may be considered the same for the whole blade span, Eq (3.16) becomes:

$$v_{\bar{r}} = V_t \left[-\frac{a\sigma}{16} + \sqrt{\left(\frac{a\sigma}{16}\right)^2 + \frac{a\sigma\bar{r}\theta_{\bar{r}}}{8}} \right] \tag{3.17}$$

Knowing the variation of the blade twist angle with its span, $\theta_{\bar{r}}(\bar{r})$, the blade pitch angle at any station $\bar{r}$ can be expressed analytically. For example, in the case of linear twist:

$$\theta_{\bar{r}} = \theta_o - \theta_{tot}\bar{r}$$

where θ_o is the pitch angle at zero station, while θ_{tot} expresses the total angle of the washout. Introducing the above expression into Eq (3.17), the formula for downwash distribution of a linearly twisted rectangular blade is obtained:

$$v_{\bar{r}} = V_t \left[-\frac{a\sigma}{16} + \sqrt{\left(\frac{a\sigma}{16}\right)^2 + \frac{a\sigma\bar{r}}{8}\left(\theta_o - \theta_{tot}\bar{r}\right)} \right] \tag{3.17a}$$

Eqs (3.15) to (3.17a) permit one to compute downwash velocity ($v_{\bar{r}}$) in vertical ascent, or in hovering for a rotor with any number of blades (b) of any planform and any pitch distribution. Knowing the true downwash value at any blade station $\bar{r}$, it is possible to find (with the help of two-dimensional airfoil characterisitics) the true values of thrust and torque experienced by every blade element (see Eqs (3.3), (3.4), (3.8), and (3.9)). Furthermore, one may acquire some feeling regarding performance advantages (magnitude of the k_{ind} factor) either in hovering or in climb resulting from particular combinations of the chord (planform) and twist distribution. It would also be possible to learn about the section lift coefficient variation along the blade span $[(c_\ell = f(\bar{r})]$.

Section lift distribution along the blade is illustrated by the following example for the hover case of a rotor with rectangular, untwisted blades. For $\theta_{tot} = 0$, Eq (3.17a) can be presented in a nondimensional form as $(v_{\bar{r}}/V_t) = f(\bar{r})$.

$$\frac{v_{\bar{r}}}{V_t} = -\frac{a\sigma}{16} + \sqrt{\left(\frac{a\sigma}{16}\right)^2 + \frac{a\sigma}{8}\,\theta_o\bar{r}}\,. \tag{3.18}$$

Assuming that $a = 5.73/rad$, several values of θ_o: $\theta_o = 4^\circ \approx 0.07\ rad$, $8^\circ \approx 0.14\ rad$, and $12^\circ \approx 0.21\ rad$; and two solidity ratios of $\sigma = 0.05$ and 0.10; v_r/V_t is computed from Eq (3.18) and shown in the lower part of Fig 3.3.

Knowing that $v_{\bar{r}}/V_t = f(\bar{r})$, the angle-of-attack at a station $\bar{r}$ can be obtained from Eq (3.1),

$$a_{\bar{r}} = \theta_o - tan^{-1}[(v_{\bar{r}}/V_t)/\bar{r}]$$

and consequently,

$$c_{\ell\bar{r}} = aa_{\bar{r}} = a\left\{\theta_o - tan^{-1}[(v_{\bar{r}}/V_t)/\bar{r}]\right\}. \tag{3.19}$$

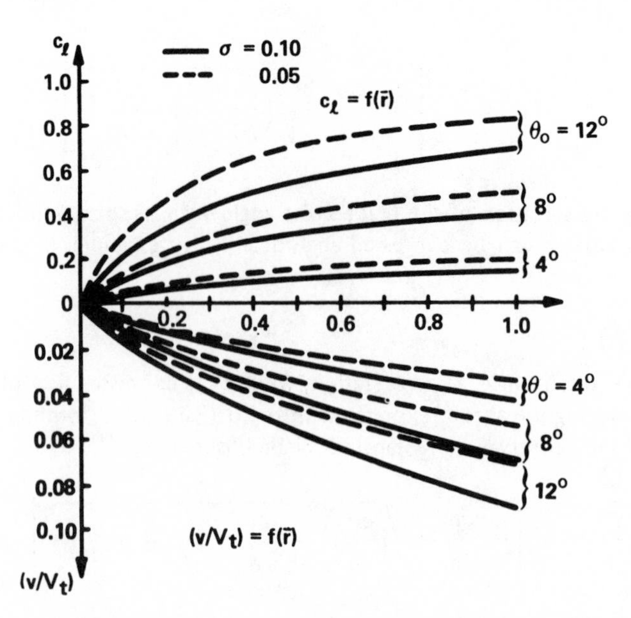

Figure 3.3 Examples of sectional lift coefficient and relative induced velocity distribution along the flat blade

The above-determined section lift coefficients are shown in the upper part of Fig 3.3.

Evaluation of the Conceptual Model. The physicomathematical model of the rotor based on the combined momentum and blade element theory provided the means of determining radial distribution of time-average induced velocities of a rotor with blades of defined geometry and known airfoil-section characteristics. This, in turn, would permit one to calculate the thrust developed by the rotor, both in hovering and climb, as well as the corresponding power required in those regimes of flight. Also, knowledge of the section lift distribution along the blade span at various θ_o values should give some idea regarding the appearance of stall at high collective pitch angles.

However, the above-discussed model has not given any indication regarding the existence of tip losses. Also, the inflow and wake structures still remain undetermined, except for the information obtained from the momentum theory that downstream, the slipstream should contract because of the increase of the induced velocity to twice that of its at-the-disc value.

Nevertheless, in spite of all the above shortcomings, the combined momentum and blade-element theory provides enough insight into the operation of real-life rotors to warrant its application to performance predictions of rotors in axial translation, and especially, in hovering. Presenting some performance aspects in nondimensional form is often more convenient than dealing with dimensional qualities; therefore, the most important dimensionless coefficients used in rotary-wing aerodynamics are discussed below.

2.3 Nondimensional Coefficients

Similar to fixed-wing practice, nondimensional thrust and torque, or power coefficients of a rotor can be defined on the basis of: (1) either disc or total blade areas, (2) air density, (3) the square of characteristic velocity (tip speed V_t), and (4) in the case of torque, rotor radius (R). Nondimensional coefficients based on the disc area (πR^2) are defined as follows:

thrust coefficient	$C_T \equiv T/\pi R^2 \rho V_t^2$
torque coefficient	$C_Q \equiv Q/\pi R^3 \rho V_t^2$
power coefficient	$C_P \equiv P/\pi R^2 \rho V_t^3$

and those referred to the total blade area (A_b):

thrust coefficient	$c_t \equiv T/A_b \rho V_t^2 = C_T/\sigma$
torque coefficient	$c_q \equiv Q/A_b R \rho V_t^2 = C_Q/\sigma$
power coefficient	$c_p \equiv P/A_b \rho V_t^3 = C_P/\sigma$.

The dimensional quantities in the above expressions are either in SI or English units, as indicated in the Principal Notations at the beginning of this chapter.

In addition to the above a priori-defined coefficients, expressions for the average lift coefficient, $\bar{c}_{\ell h}$; the average profile drag coefficient, $\bar{c}_d$; and the average total drag coefficient $\bar{C}_{D_h}$ for hover can easily be developed.*

Average Lift Coefficient. Within the validity of small angle assumptions ($dT = dL$), the thrust of a blade element $cRd\bar{r}$ located at $r = R\bar{r}$ can be expressed as

$$dT = \tfrac{1}{2}\rho bcRc_\ell V_t^2 \bar{r}^2 d\bar{r}.$$

Assuming that c_ℓ along the whole radius is constant and equal to $\bar{c}_{\ell h}$, and remembering that $bcR = \sigma\pi R^2$, the above expression can be integrated from $\bar{r} = 0$ to $\bar{r} = 1.0$; leading to

$$\bar{c}_{\ell h} = 6T/\sigma\pi R^2 \rho V_t^2 = 6C_T/\sigma = 6c_t. \tag{3.20}$$

However, if a cutout exists at the blade root so that the inboard station $\bar{r}_i \neq 0$, and tip losses reduce the effective relative rotor radius to $\bar{r}_e$, then integration of dT would be performed from $\bar{r}_i$ to $\bar{r}_e$, resulting in the following:

$$\bar{c}_{\ell h} = 6T/\sigma\pi R^2 \rho V_t^2 (\bar{r}_e^3 - \bar{r}_i^3). \tag{3.20a}$$

It can be seen from Eq (3.20c) that a customary cutout (usually $\bar{r}_i \leqslant 0.2$) would have little effect on the $\bar{c}_{\ell h}$ level. By contrast, tip losses may noticeably influence the average lift coefficient. For instance, $\bar{r}_e = 0.95$ would increase $\bar{c}_{\ell h}$ by about 13 percent.

Assuming uniform distribution of downwash velocity over the entire disc, still another useful expression for $\bar{c}_{\ell h}$ can be developed by substituting a relationship based on ideal induced velocity (v_{id}) for T in Eq (3.20); i.e., $T = 2\pi R^2 \rho v_{id}^2$.

$$\bar{c}_{\ell h} = 12(v_{id}/V_t)^2/\sigma. \tag{3.21}$$

---

*Lower-case letters are used for $\bar{c}_{\ell h}$ and $\bar{c}_d$ in order to emphasize their relationship to section coefficients, while $\bar{C}_{D_h}$ is obviously based on the total drag of the blade, including the induced drag.

Conversely, for a given value of the average lift coefficient, the corresponding (v_{id}/V_t) ratio would be:

$$(v_{id}/V_t) = 0.289\sqrt{\bar{c}_{\ell_h}\sigma}. \quad (3.21a)$$

Average Profile and Total Drag Coefficients. Similar to the average lift coefficient, the average profile $(\bar{c}_d)$ and total drag $(\bar{C}_D)$ coefficients can be obtained. The contribution of b blade elements located at $\bar{r}R$ to the rotor profile power (in N-m/s or ft-lb/s) will be

$$dP_{pr} = \tfrac{1}{2}\rho bcR\ c_d V_t^3 \bar{r}^3\, d\bar{r}. \quad (3.22)$$

Again assuming that c_d is constant along the blade and is equal to $\bar{c}_d$, the above equation is integrated within $\bar{r} = 0$ to $\bar{r} = 1.0$ limits in order to obtain profile power in hovering (P_{pr_h}):

$$P_{pr_h} = (1/8)\sigma\pi R^2 \rho V_t^3 \bar{c}_d \quad (3.22a)$$

and consequently,

$$\bar{c}_d = 8P_{pr_h}/\sigma\pi R^2 \rho V_t^3. \quad (3.23)$$

The total average drag coefficient $(\bar{C}_{D_h})$ by analogy with Eq (3.23) will be

$$\bar{C}_{D_h} = 8P_{R_h}/\sigma\pi R^2 \rho V_t^3 \quad (3.24)$$

where P_{R_h} represents the total rotor power in hovering (in N-m/s or ft-lb/s).

2.4 Rotor Profile Power in Axial Translation

Determination of $\bar{c}_{d_{ax}}$. When integrating Eq (3.22a) it was assumed that the profile drag coefficient and blade chord were constant along the blade span. Actually, one may expect that the blade chord c and usually, the profile drag coefficient c_d, vary along the blade span. This latter variation is caused by the fact that at any particular station $\bar{r}$, $c_{d_{\bar{r}}} = f(R_{e_{\bar{r}}}, M_{\bar{r}}, c_{\ell_{\bar{r}}}, etc)$. Here, the *etc* could mean airfoil section geometry, surface roughness, special boundary layer conditions as influenced by the centrifugal acceleration field of rotating blades and, in some special cases, the application of BLC. All of these parameters may be dependent on $\bar{r}$, thus making $c_{d_{\bar{r}}} = f(\bar{r})$. Consequently when integrating Eq (3.22), $c_{d_{\bar{r}}}$ and $c_{\bar{r}}$ should be retained under the sign of the integral:

$$P_{pr_h} = \tfrac{1}{2}bR\rho V_t^3 \int_0^{1.0} c_{\bar{r}}\, c_{d_{\bar{r}}}\, \bar{r}^3\, d\bar{r}. \quad (3.25)$$

Equating the right sides of Eqs (3.22a) and (3.25), the following expression for the equivalent $\bar{c}_d$ in hovering or any axial translation is obtained:

$$\bar{c}_{d_h} = (4b/\sigma\pi R)\int_0^{1.0} c_{\bar{r}}\, c_{d_{\bar{r}}}\, \bar{r}^3\, d\bar{r}. \quad (3.26)$$

When $c_{\bar{r}} = c = const$, Eq (3.26) becomes

$$\bar{c}_d = 4 \int_0^{1.0} c_{d_{\bar{r}}} \bar{r}^3 \, d\bar{r}. \tag{3.26a}$$

The integration indicated in Eqs (3.26) and (3.26a) is usually either numerically or graphically performed because of the complicated nature of the $c_{d_{\bar{r}}} = f(\bar{r})$ relationship.

It should be mentioned at this point that since blade surface roughness is one of the most important parameters influencing $c_{d_{\bar{r}}}$ values, some allowance should be made for the increase in profile drag by multiplying the $c_{d_{min}}$ values from airfoil wind-tunnel data by a roughness coefficient. Depending on the blade construction and state of the blade surface, the magnitude of the roughness coefficient may vary within rather wide limits—from 1.15 for smooth blades to 1.5 or even more for sand and/or rain-eroded ones (see Ch II, Vol II). For laminar airfoil sections which are very sensitive to surface roughness, roughness correction factors may exceed 1.5 whenever the blades are in less than perfect condition. Details for accounting for the influence of such other parameters as R_e and M on c_{d_r} can also be found in Ch II of Vol II.

Approximate Determination of $\bar{c}_{d_{ax}}$. Because of the complexity of finding $\bar{c}_{d_{ax}}$ from Eqs (3.26) and (3.26a), the following approximate method can often be used.

1. For given conditions of pitch angle, tip speed, and air density, the section lift coefficient $c_{\ell_{0.75}}$ is determined at $\bar{r} = 0.75$.

2. Assuming that the lift coefficient obtained in 1 also exists at $\bar{r} = 0.8$, the corresponding c_d value is computed, taking into account the Reynolds and Mach numbers existing at $\bar{r} = 0.8$ and $c_{\ell_{0.75}}$.

3. The c_d determined in 2 is corrected for blade surface roughness, multiplying the $c_{d_{min}}$ component of the total profile drag coefficient by a suitable correction factor. Assuming that this c_d value exists along the whole blade, the profile power is computed from Eq (3.22a).

It should be emphasized, however, that since the profile power increases due to compressibility may be quite considerable (see Ch VI of this volume, and Ch II, Vol II), the shortcut method may miss the blade area (mostly outboard) where rapid profile drag occurs. The problem of compressible drag should be thoroughly investigated if there is any possibility that an unfavorable combination of lift coefficient and Mach number may exist. This should be done throughout the whole operational range of the rotor, and especially when high thrust and elevated tip speed is combined with low air density and temperature.

While lowering the tip speed (reduction of M) may appear as the simplest way of alleviating the compressible-drag problem, in some cases this approach may prove disadvantageous as the higher blade-element lift coefficients $c_{\ell_{\bar{r}}}$ resulting from a reduced V_t —even in combination with the lower Mach number—might still produce higher drag coefficients than those associated with the original $c_{\ell_{\bar{r}}}$ and M combination.

2.5 Tip Losses

While a clearer insight into the physical and computational aspects of tip losses will be gained through the vortex theory discussed in the next chapter, the existence of this phenomenon should be acknowledged even when the rotor thrust is calculated by the combined momentum blade-element theory. This is usually done by assuming that the predicted aerodynamic lift extends up to some blade station; i.e., to the so-called effective blade radius ($r_e = R\bar{r}_e$) and ends abruptly at that station.

There are numerous theoretical or empirical formulae for predicting tip losses. For instance, Prandtl gives a simple, but only approximate formula for propellers which is based on the vortex theory (Ref 1, p. 265):

$$\bar{r}_e = 1 - \left(1.386\lambda/b\sqrt{1 + \lambda^2}\right) \tag{3.27}$$

where b is the number of blades, λ is the inflow ratio of the propeller, and $\lambda \equiv V_{ax}/V_t$; V_{ax} being the axial velocity at the disc.

Some authors simply recommend expressing the effective radius as:

$$r_e = R - 0.5\bar{c}$$

where $\bar{c}$ is the average blade chord.

Dividing the above equation by R, the following is obtained:

$$\bar{r}_e = 1 - 0.5(\pi\sigma/b). \tag{3.28}$$

Sissingh[3] proposes the following expression:

$$\bar{r}_e = 1 - c_{ro}\,[1 + 0.7(tr)]/1.5R$$

where c_{ro} is the chord length at the root end and (tr) is the blade taper ratio.

By analogy with the preceding case, the above expression for rectangular blades can be presented as follows:

$$\bar{r}_e = 1 - 3.56\sigma/b. \tag{3.29}$$

Wald[4] recommends another expression which, for a rotor with rectangular blades, can be written as follows:

$$\bar{r}_e = 1 - 1.98\sqrt{C_T/b}. \tag{3.30}$$

For comparison, tip losses as given by Eqs (3.27) through (3.30) were computed for a hovering rotor at SL, STD; out-of-ground effect where $w = 40\ kG/m^2 = 392\ N/m^2$; $V_t = 200\ m/s$; $b = 4$; and $\sigma = 0.10$. The results are shown in Table III-1.

It can be seen from this table that $\bar{r}_e$ values calculated by different formulae are somewhat different. However, their average amounts to $\bar{r}_e = 0.952$ and this, or an even slightly higher value of $\bar{r}_e = 0.96$, appears as a reasonable number for the approximate tip-loss factor in hovering.

2.6 Rotor Thrust and Power in Climb and Hovering

Knowledge of induced velocity distribution along the blades and hence, of sectional lift coefficients permits calculation of the thrust of the rotor using the blade element

EQUATION	re
(3.27)	0.978
(3.28)	0.961
(3.29)	0.912
(3.30)	0.956

TABLE III–1

approach. However, once $v = f(\bar{r})$ has been established, it may often be more convenient to use the momentum theory in estimating the total thrust T.

After substituting $R\bar{r}$ for r, and $Rd\bar{r}$ for dr, Eq (3.11) becomes

$$dT = 4\pi R^2 \rho (V_c + v) v\bar{r} d\bar{r}$$

and consequently,

$$T_c = 4\pi R^2 \rho \int_{\bar{r}_i}^{\bar{r}_e} (V_c + v)\, v\bar{r}\, d\bar{r}. \quad (3.31)$$

while, in hovering,

$$T_h = 4\pi R^2 \rho \int_{\bar{r}_i}^{\bar{r}_e} v^2\, \bar{r}\, d\bar{r}. \quad (3.32)$$

When induced velocity vs blade span is given in nondimensional form, v/V_t, Eq (3.32) becomes:

$$T_h = 4\pi R^2 \rho V_t^2 \int_{\bar{r}_i}^{\bar{r}_e} (v/V_t)^2 \bar{r}\, d\bar{r}. \quad (3.32a)$$

In all of the above three equations, $\bar{r}_i$ is the inboard station where the actual blade begins. Since the analytical relations between v and $\bar{r}$ are rather complicated, a computerized numerical or graphical integration is more suitable for practical purposes.

A scheme of graphical integration of Eq (3.32a) is shown in Fig 3.4. In this case, since the induced velocity is expressed in a nondimensional form, the area A can be used in both the SI and English measuring systems, while $R^2 \rho V_t^2$ is expressed in the proper units.

In the case of a rotor having rectangular untwisted blades the $(v/V_t) = f(\bar{r})$ is expressed by the relatively simple relationship of Eq (3.18). In addition, if the influence of M and R_e on the variation of the lift slope along the blade radius can be neglected (i.e., it may be assumed that the section lift slope $a = const$ for all stations); then Eq (3.18) is introduced into Eq (3.32a) and, for simplicity, the integration is performed within the 0 to $\bar{r}_e$ limits with the following results:

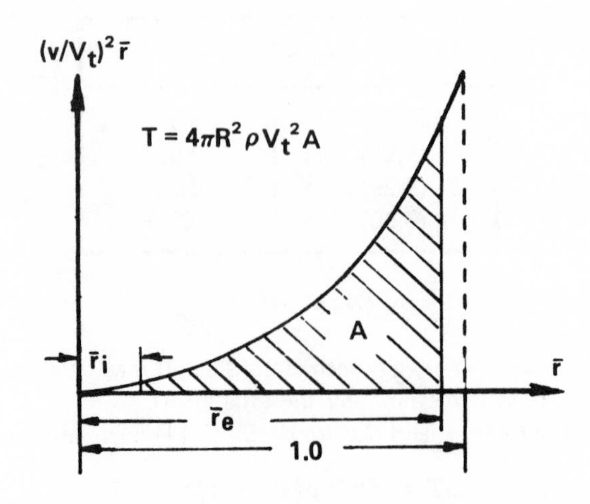

Figure 3.4 Graphical integration of thrust

$$T = 4\pi R^2 \rho V_t^2 \left[A^2 \bar{r}_e^2 + (1/3)B\bar{r}_e^3 + \frac{4A(2A^2 - 3B\bar{r}_e)\sqrt{(A^2 + B\bar{r}_e)^3}}{15B^2} \right]$$

or (3.33)

$$C_T = 4\left[A^2 \bar{r}_e^2 + (1/3)B\bar{r}_e^2 + \frac{4A(2A^2 - 3B\bar{r}_e)\sqrt{(A^2 + B\bar{r}_e)^3}}{15B^2} \right]$$

where $A = \sigma a/16$, and $B = \sigma a\theta_o/8$.

It should be noted that the terms containing A^5 were omitted in the above two equations. This is permissible as long as B is not approaching zero, which obviously means that the blade pitch angle is $\theta_o \gg 0$.

For a rotor having linearly twisted blades, expressions for T and C_T can be obtained in a way similar to the preceding case.

Maximum Thrust in Hovering. In some practical problems, it may be important to know the maximum thrust that can be obtained from a given rotor, assuming that the power available is sufficient to retain a given tip speed. To solve this problem, it would be necessary to figure out which sections of the blade would stall first and at what pitch angle. The blade pitch angle corresponding to the beginning of stall can easily be found by the following procedure.

The maximum attainable section lift coefficient ($c_{\ell_{max}}$) at various stations $\bar{r}$ should be estimated by first taking into consideration the blade airfoil sections, the effect of Reynolds and Mach numbers and finally, all other secondary effects such as unsteady aerodynamic phenomena (should this be justified by the rate of the pitch change) and boundary-layer interaction.

It should be noted that for the rectangular or moderately tapered blade, R_e increases toward the tip, while M becomes higher regardless of the blade planform. The first of these increases nearly always leads to higher $c_{\ell_{max}}$ values, while elevated Mach numbers, which may be encountered in the outer portion of the blade, tend to reduce the

$c_{\ell_{max}}$ (see Ch VI). As a result of these conflicting influences between Reynolds and Mach numbers, $c_{\ell_{max}}$ will first increase along the radius until it reaches its maximum at some blade station, and then its value will decrease toward the tip. The character of this variation is shown as a solid line in Fig 3.5.

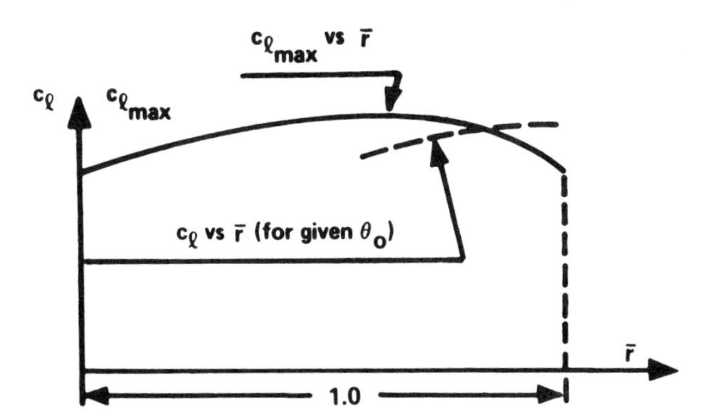

Figure 3. 5 Determination of the blade stall regions

Using the procedure leading to Eq (3.19), the section lift-coefficient values for some assumed representative collective pitch value (θ_o) can be calculated, and after plotting them vs $\bar{r}$ (broken line in Fig 3.7), one could see, at this particular θ_o level, whether there would be a possibility of stall and if so, within which segment of the blade it would occur.

Blade stall can also be studied by determining the local pitch angle at the various blade stations at which stall would take place.

Equating the right sides of Eqs (3.11) and (3.12), assuming $V_c = 0$ (for hover), and remembering that $r \equiv R\bar{r}$ and $dr \equiv Rd\bar{r}$, the relationship between induced velocity $v_{\bar{r}}$ at a station $\bar{r}$, and lift coefficient of the blade element $c_{\ell_{\bar{r}}}$ can be obtained:

$$v_{\bar{r}} = V_t\sqrt{(1/8\pi)(bc/R)c_{\ell_{\bar{r}}}\bar{r}} \tag{3.34}$$

or

$$v_{\bar{r}}/V_t = \sqrt{(1/8\pi)(bc/R)c_{\ell_{\bar{r}}}\bar{r}}. \tag{3.34a}$$

For rectangular blades ($bc/R = \pi\sigma$), Eq (3.34a) becomes

$$v_{\bar{r}}/V_t = \sqrt{(1/8)\,\sigma c_{\ell_{\bar{r}}}\bar{r}}. \tag{3.34b}$$

When $c_{\ell_{max}}$ reaches its maximum value, so does the ($v_{\bar{r}}/V_t$) ratio. Consequently, within the limits of validity of small-angle assumptions, the pitch angle of the blade element at station $\bar{r}$ corresponding to the beginning of stall will be

$$\theta_{\bar{r}_{st}} = c_{\ell_{max_{\bar{r}}}}/a_{\bar{r}} + (v_{\bar{r}}/V_t)_{max}/\bar{r}. \tag{3.35}$$

After the $\theta_{\bar{r}_{st}}$ values have been calculated for several blade stations, one will have a fairly clear picture of the pitch angle θ_o (of the blade as a whole) where stall begins.

If the maximum possible thrust from a blade of given planform and airfoil section is required, it is possible to select a twist distribution wherein the section lift coefficients simultaneously reach their maximum values over the largest possible part of the blade. When the blade is non-twisted (no washout), it is obvious that the lowest value of $\theta_{\bar{r}_{st}}$ is the blade pitch angle at which stall begins.

Using the previously established formulae, the rotor thrust value corresponding to the beginning of stall can now be calculated from the known $v_{\bar{r}}/V_t = f(\bar{r})$ relationship. The so-obtained thrust can be assumed as being close to its maximum value. However, a more thorough study of T_{max} can be made when T is calculated for several values of θ_o greater than the θ_o at the beginning of stall. In this case, at those blade stations where the local stall pitch angle $\theta_{\bar{r}_{st}}$ has been exceeded, the downwash velocity $v_{\bar{r}}$ should be computed from Eqs (3.34a) and (3.34b), using the post-stall c_ℓ value (region a–b, Fig 3.6).

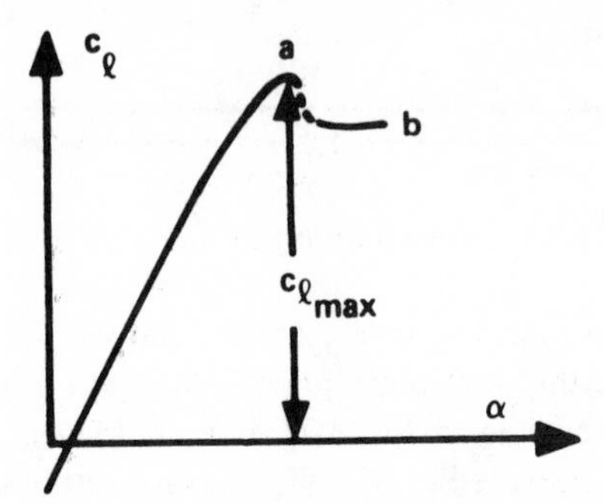

Figure 3.6 Typical c_ℓ vs α curve

The above discussion of the hovering case should enable the reader to deal with stall and maximum thrust problems in vertical flight at a given rate-of-climb.

Induced Power in Axial Translation. In vertical translation (climb or descent), as well as in hovering, the induced power can be estimated by following the outline given in Sect 5.1 of Ch II. Once the $v_{\bar{r}}$ values or the $v_{\bar{r}}/V_t$ ratios are computed as a function of $\bar{r}$, the whole procedure of finding the induced power—either in hover (P_{ind_h}) or in climb (P_{ind_c})—becomes quite simple. Eq (2.52) is rewritten using the velocity ratio approach:

$$P_{ind} = 4\pi R^2 V_t^3 \rho \int_{\bar{r}_i}^{\bar{r}_e} (v_{\bar{r}}/V_t)^3 \bar{r}\, d\bar{r} \qquad (3.36)$$

where, depending on the unit system used, P_{ind} will be in either N–m/s or ft-lb/s.

Because of the complexity of the expressions for $v_{\bar{r}} = f(\bar{r})$ or $v_{\bar{r}}/V_t = f(\bar{r})$, the most practical procedure for finding P_{ind} will be to perform the above integration, as in the case of thrust, either on a computer or graphically. The latter procedure—analogous to that illustrated in Fig 3.4—is shown in Fig 3.7.

Knowing P_{ind}, the induced power factor $k_{ind} \equiv P_{ind}/P_{id}$ can be determined; again following procedures established in Sect 5.2 of Ch II. This will be done analytically using, as an example, a rotor with rectangular untwisted blades where a closed-form formula for

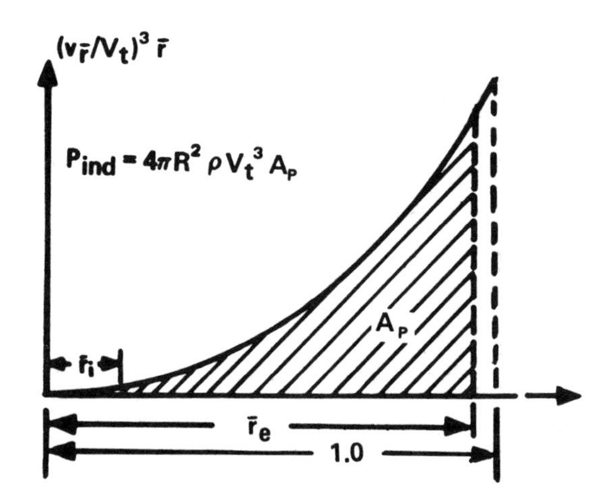

Figure 3.7 Graphical integration of induced power

P_{ind_h} can be easily obtained in the following steps: (1) substitute the $v_{\bar{r}}/V_t$ ratio given in Eq (3.18) into Eq (3.36), (2) define $a\sigma/16$ as A and $a\sigma\theta_o/8$ as B, (3) integrate within the limits of $\bar{r}_i = 0$ to $\bar{r}_e$, and (4) neglect the terms containing A^5 and A^7. Consequently,

$$P_{ind_h} = 4\pi R^2 V_t^3 \rho \left[\frac{2}{5} \frac{(3B\bar{r}_e - 2A^2)\left(\sqrt{A^2 + B\bar{r}_e}\right)^3 A^2}{B^2} + \frac{2}{35} \frac{5\left(\sqrt{A^2 + B\bar{r}_e}\right)^7 - 7A^2\left(\sqrt{A^2 + B\bar{r}_e}\right)^5}{B^2} - 2A^3\bar{r}_e^2 - AB\bar{r}_e^3 \right]. \quad (3.37)$$

Equating thrust expressed in terms of ideal induced velocity (v_{id}) to the thrust given by Eq (3.33), the following formula for the equivalent $v_{id_{eq}}$ is obtained:

$$v_{id_{eq}} = 1.41 V_t \sqrt{A^2\bar{r}_e^2 + \frac{1}{3}B\bar{r}_e^3 + \frac{4}{15}\frac{A(2A^2 - 3B\bar{r}_e)\left(\sqrt{A^2 + B\bar{r}_e}\right)^3}{B^2}}; \quad (3.38)$$

therefore, the k_{ind_h} factor becomes

$$k_{ind_h} = P_{ind}/2\pi R^2 \rho v_{id_{eq}}^3. \quad (3.39)$$

Assuming the following geometric and operational characteristics for a rotor with untwisted rectangular blades: $R = 7.6\ m$, $\sigma = 0.1$, $\bar{r}_e = 0.96$, $\theta_o = 10° = 0.17\ rad$, $V_t = 213\ m/s$, $a = 6.0/rad$, and $\rho = 1.23\ kg/m^3$, the thrust computed from Eq (3.33) would amount to $T = 70\ 100\ N \approx 7150\ kG$. The induced power (Eq (3.37)) would be $P_{ind} = 1\ 019\ 445\ N\text{-}m/s = 1387\ HP$, and the induced power factor, computed from Eqs (3.38) and (3.39) is $k_{ind_h} = 1.17$.

Blade Twist and Chord Distribution for Uniform Downwash in Hovering. The operating conditions of a rotor are specified as: (1) the effective disc loading $w_e \equiv T/\pi\bar{r}_e^2 R^2$, tip speed V_t, and air density ρ, or (2) simply as the average lift coefficient

$\bar{c}_{\ell_h}$ and V_t. The problem consists of finding a blade twist and/or chord distribution which, under the so-defined operating conditions, would produce a uniform induced velocity. In the first case, the induced velocity which we try to make uniform would be $v = \sqrt{w_e/2\rho}$ while in the second, using nominal $\bar{c}_{\ell_h}$, it can be expressed as:

$$v = \left(0.289 V_t \sqrt{\bar{c}_{\ell_h}\sigma}\right) \Big/ \bar{r}_e.$$

However, from Eq (3.34), the lift coefficient which should exist at station $\bar{r}$ in order to produce the required uniform downwash is:

$$c_{\ell_{\bar{r}}} = 8\pi R(v/V_t)^2 / bc_{\bar{r}}\bar{r}, \tag{3.40}$$

For a rectangular blade and assuming that $\bar{r}_e = 1.0$, the v/V_t ratio substituted from Eq (3.21) gives the following $c_{\ell_{\bar{r}}}$ distribution in terms of $\bar{c}_{\ell_h}$:

$$c_{\ell_{\bar{r}}} = (2/3)(\bar{c}_{\ell_h}/\bar{r}) \ . \tag{3.40a}$$

It is clear from Eqs (3.40) and (3.40a) that the required $c_{\ell_{\bar{r}}}$ increases toward the root of the blade as $\bar{r}$ decreases. For a tapered blade, this need for an increase in $c_{\ell_{\bar{r}}}$ can be at least partially offset by the chord enlargement (Eq(3.40)). Nevertheless it is obvious that the special condition of a uniform downwash can only be fulfilled down to the value of $\bar{r}$ where the required $c_{\ell_{\bar{r}}}$ does not exceed the maximum lift coefficient $c_{\ell_{max}}$.

The blade pitch angle ($\theta_{\bar{r}}$) at station $\bar{r}$ required to produce downwash v can now be readily obtained. Under small angle assumptions, it will be:

$$\theta_{\bar{r}} = (c_{\ell_{\bar{r}}}/a_{\bar{r}}) + (v/V_t\bar{r}). \tag{3.41}$$

For rectangular blades, the right side of Eq (3.40a) can be substituted for $c_{\ell_{\bar{r}}}$ in Eq (3.41) and v/V_t can be expressed in terms of $\bar{c}_{\ell_h}$; thus, Eq (3.41) becomes

$$\theta_{\bar{r}} = (2/3)(\bar{c}_{\ell_h}/a_{\bar{r}}\bar{r}) + \left(0.289\sqrt{\bar{c}_{\ell_h}\sigma}/\bar{r}_e\bar{r}\right). \tag{3.41a}$$

For $\sigma = 0.1$ and $\bar{c}_{\ell_h} = 0.3$, 0.4, and 0.5, the $\theta_{\bar{r}} = f(\bar{r})$ is computed from Eq (3.41a) and plotted in Fig 3.8, and $c_{\ell_{\bar{r}}}$ from Eq (3.40a) is also shown, thus giving some idea regarding the blade twist and the section-lift coefficient required to achieve a uniform downwash in hover. Because of the large twist angles required toward the root, and the high section lift coefficient, the goal of uniform induced velocity over most of the blade (down to say, station $\bar{r} = 0.1$) is not practical. It should be noted, however, that even a modest linear twist ($\theta_t \approx -10°$) should well approximate the ideal twist distribution for blade stations $0.4 \leqslant \bar{r} \leqslant 1.0$.

Power Losses Due to Slipstream Rotation. Slipstream rotation was not considered in the momentum theory; however, it can be seen that lift generation by the blade elements may introduce some rotation to the slipstream which, although not contributing to the lift, will create a new requirement for power as additional energy is carried away in the wake.

Slipstream rotation may even be present in a nonviscous fluid due to the fact that downwash v_L associated with the blade element lift is not parallel to the rotor axis, but is perpendicular to the resultant flow at the lifting line of that element as shown in Fig

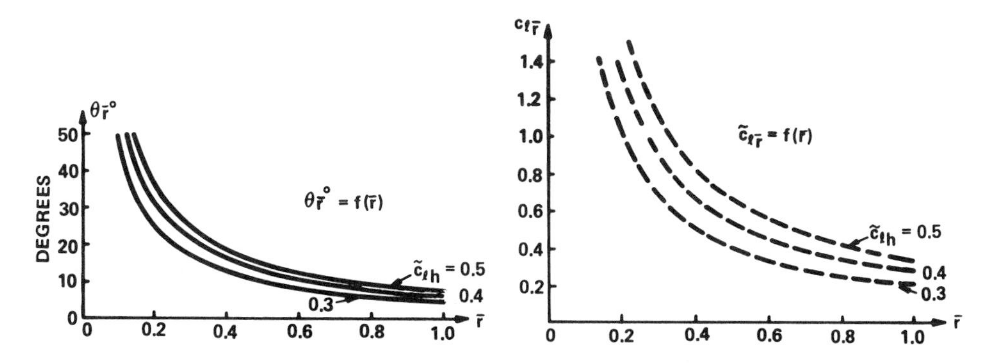

Figure 3.8 Blade twist and section lift distribution required for uniform downwash

3.9. It can be seen from this figure that $v_{rot}/v = v_L / V_{res}$. Assuming that $v_L \approx v$ and $V_{res} \approx V_t \bar{r}$, the following is obtained:

$$v_{rot} = v(v/V_t\bar{r}). \tag{3.42}$$

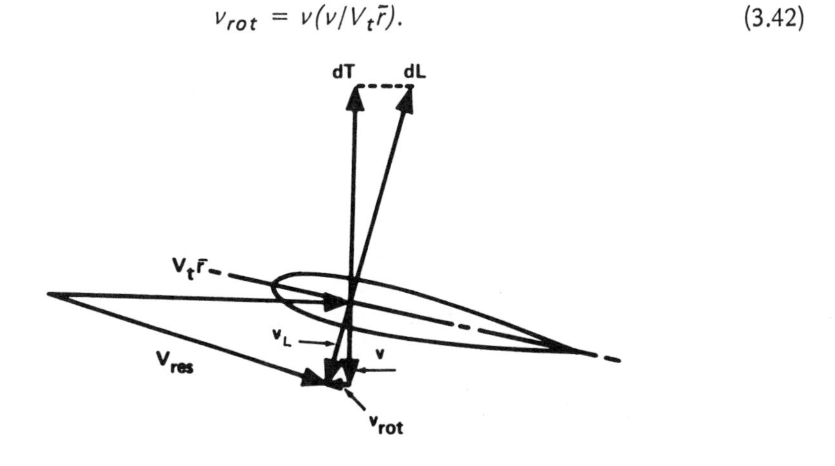

Figure 3.9 Slipstream-rotating component (v_{rot}) of lift-induced velocity (v_L)

The power loss (in N-m/s or ft-lb/s) due to the rotation of the fluid passing at a rate of $2\pi R_u{}^2\, \rho\, v\, \bar{r}\, d\bar{r}$ through an annulus located in the ultimate wake and having a radius $R_u \bar{r}$ and width $R_u\, d\bar{r}$ will be:

$$dP_{rot} = \pi R^2 R_u{}^2 \rho v \omega_u{}^2 \bar{r}^3 d\bar{r} \tag{3.43}$$

where ω_u is the angular speed of the slipstream rotation in the fully-developed wake. Defining the speed of slipstream rotation in the disc plane by ω, remembering that $R_u = R/\sqrt{2}$, and applying the principle of conservation of angular momentum, one will find that

$$\omega_u = 2\omega \tag{3.44}$$

and Eq (3.43) can be rewritten as follows:

$$dP_{rot} = 2\pi R^2 v\rho(R\bar{r}\omega)^2 \bar{r}d\bar{r}, \quad (3.45)$$

but $R\bar{r}\omega \equiv v_{rot} = v^2/V_t\bar{r}$ (Eq (3.42)); hence,

$$dP_{rot} = [2\pi R^2 \rho v^3 (v/V_t)^2/\bar{r}]\,d\bar{r}. \quad (3.45a)$$

Next assuming that the induced velocity v is constant over the disc and performing integration from $\bar{r}_i$ to $\bar{r}_e$, the following is obtained:

$$P_{rot} = 2\pi R^2 \rho v^3 (v/V_t)^2 \log_e(\bar{r}_e/\bar{r}_i). \quad (3.46)$$

However, $2\pi R^2 \rho v^3$ is the ideal induced power in hovering (P_{id}); hence,

$$P_{rot} = P_{id}(v/V_t)^2 \log_e(\bar{r}_e/\bar{r}_i). \quad (3.47)$$

For instance, assuming that $\bar{r}_i = 0.25$ and that $\bar{r}_e = 0.96$,

$$P_{rot} \approx 1.345 P_{id}(v/V_t)^2. \quad (3.48)$$

Since v/V_t is usually less than 0.07 for contemporary helicopters, power losses due to slipstream rotation should not exceed 0.7 percent.

2.7 Thrust and Induced Power of Intermeshing and overlapping rotors

Definition of Overlap. In multirotored aircraft, the rotors may be arranged in such a way that the stagger distance (d_s) between the axes of any two rotors may be smaller than the sum of their respective radii (R_1 and R_2). When $d_s < (R_1 + R_2)$, mutual interference occurs in axial translation and hovering, leading to aerodynamic characteristics different than those of isolated ones. As to the arrangement of the rotors, they may be either coplanar (pure intermeshing), or one may be elevated above the other, thus forming an overlapping configuration (Fig 3.10).

The amount of overlap (ov) for $R_1 = R_2$ can be expressed as a nondimensional number (or in percent) as follows:

$$ov = 1 - (d_s/d) \quad (3.49)$$

where d is the rotor diameter.

A physicomathematical model based on the combined momentum and blade element theory may be helpful in understanding aerodynamic interference of intermeshing and overlapping rotors in hovering and in axial translation. Consequently, the basic philosophy of this approach will be outlined and the results briefly discussed here, while details of the complete derivation of the presented relationship can be found in Ref 5.

The considered case will be limited to either truly coplanar rotors or those having a relatively small vertical displacement (say $h_{rr} \leqslant 0.10R$). This would permit treating them as being intermeshing. Thus, it will be assumed that a common rate of flow is established within the overlapped, or intermeshed, area.

Induced Velocity. The induced velocity (v_{ov}) at any point of the overlapped region can be obtained by considering an elementary area dA of diameter dr (Fig 3.11), the location of which is determined by r_1 and r_2.

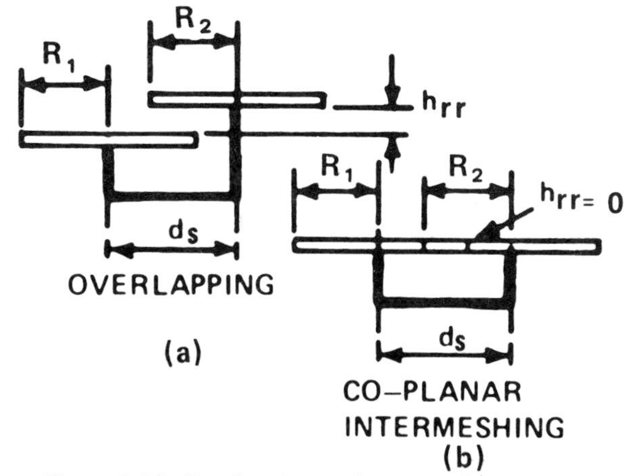

Figure 3.10 Overlapping and intermeshing rotors

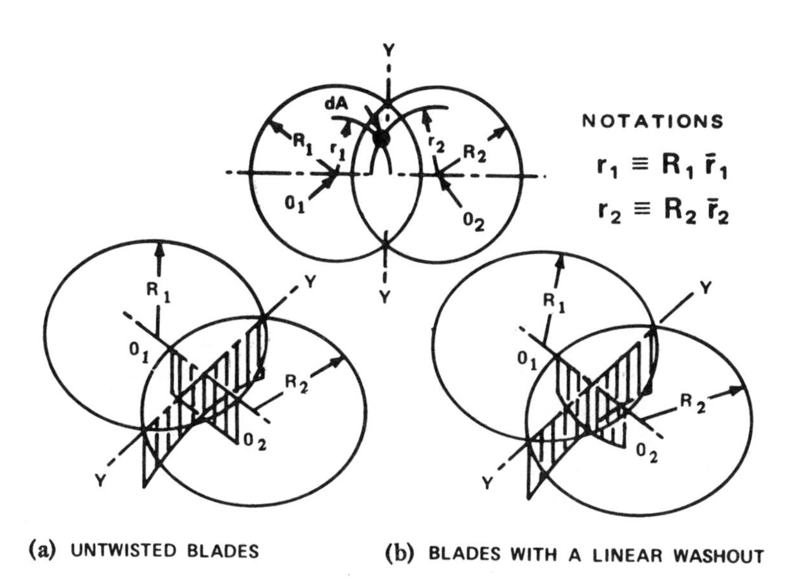

Figure 3.11 Notations and theoretical character of downwash distribution of overlapping rotors in hover

The rate of flow through the element dA will be $v_{ov}\rho \tfrac{1}{4}\pi dr^2$ and, according to the momentum theory, the thrust produced will be:

$$dT = (\pi/4)2\rho v_{ov}^2\, dr^2\,. \tag{3.50}$$

On the other hand, the elementary thrust dT—developed over the dA area due to the action of Rotor 1—can be expressed according to the blade element theory. This can be considered as the $(\pi dr^2/4)/2\pi r_1 dr$ fraction of the total thrust generated by the whole annulus of radius r_1 and the width dr as given by Eq (3.12):

$$dT_1 = (1/16)\rho(\Omega r_1)^2 c_{\ell_1} bc(dr^2/r_1). \tag{3.51}$$

Similarly, for Rotor 2,

$$dT_2 = (1/16)\rho(\Omega r_2)^2 c_{\ell_2} bc(dr^2/r_2). \tag{3.51}$$

where c_ℓ, with a suitable subscript, is the section lift coefficient at the corresponding radius $r_1 \equiv \bar{r}_1 R$ or $r_2 \equiv \bar{r}_2 R$. But the elementary thrust calculated using the momentum theory should be equal to the sum of the dT_1 and dT_2 thrust produced from the blade element theory:

$$dT = dT_1 + dT_2. \tag{3.53}$$

Substituting Eqs (3.50), (3.51), and (3.52) into Eq (3.53) and further simplifying under the assumption that the blades have a uniform chord and linear twist, the following formula for downwash at any point of the overlapped area is obtained:

$$v_{ov} = V_t\left\{-(1/8)\sigma a + \sqrt{[(1/8)a\sigma]^2 + (1/8)a\sigma[\theta_o(\bar{r}_1 + \bar{r}_2) + \theta_t(\bar{r}_1 + \bar{r}_2{}^2)]}\right\}. \tag{3.54}$$

When the blade is nontwisted, Eq (3.54) becomes

$$v_{ov} = V_t\left[-(1/8)\sigma a + \sqrt{[(1/8)a\sigma]^2 + (1/8)a\sigma\,\theta_o(\bar{r}_1 + \bar{r}_2)}\right]. \tag{3.54a}$$

It can be seen from the above equation that the sum $\bar{r}_1 + \bar{r}_2$ remains the same at all points along the $0_1 - 0_2$ axis joining the centers of the rotors. This means that for rotors with untwisted blades, the downwash velocity is constant along this axis within the overlapped area. Along the y–y axis, the sum $r_1 + r_2$ increases with distance from the 0_1–0_2 line. This implies that the downwash velocity for flat blades increases toward the sharp edges of the overlapped area (Fig 3.10a). Fig 3.10b shows the character of induced velocity distribution for blades with a linear washout.

In the limiting case when the overlap amounts to 100 percent (i.e., when the rotors are coaxial and hence $\bar{r}_1 = \bar{r}_2 = \bar{r}$), Eqs (3.54) and (3.54a) become identical with those expressing the downwash for an isolated rotor with the exception that the solidity of this single rotor is equal to the sum of solidities making a coaxial arrangement.

Experimental results[5] from direct induced velocity measurements using a bank of pitot-static anemometers in model tests of coplanar, or slightly vertically displaced overlapping rotors (with an overlap of up to 40 percent) seem to confirm the predicted trend (Fig 3.12).

Determination of Thrust and Induced Power. Knowing the induced velocity at all points of the overlapped area and keeping in mind the expressions giving this value in the nonoverlapped portion, it is easy to calculate the thrust and induced power for each part of the total projected area of the overlapping rotors.

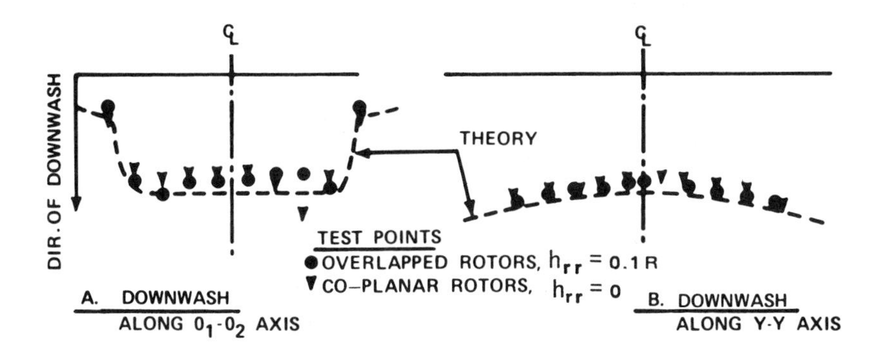

Figure 3.12 Examples of predicted and measured downwash of overlapping and intermeshing rotors incorporating 37½ percent overlap

Figure 3.13 shows a relative reduction in thrust of the overlapped configuration with respect to that of two isolated rotors. This figure also gives a comparison with test results.

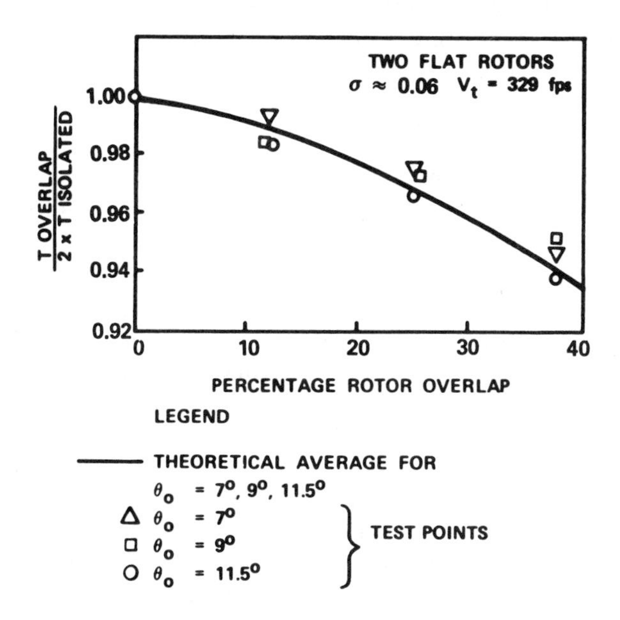

Figure 3.13 Thrust ratio vs percentage of rotor overlap

The k_{ind_h} factor of the overlapped configuration vs percentage of overlap is plotted in Fig 3.14. It can be noted that for a helicopter having 37½ percent overlap, the increase in induced power would amount to about 10 percent over that required by two isolated rotors jointly producing the same thrust.

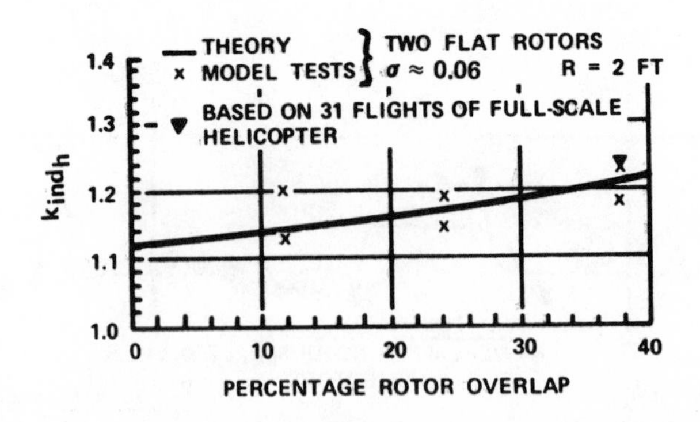

Figure 3.14 Ratio of actual-to-ideal induced power vs percentage of overlap

2.8 Rotor Power, and Aerodynamic and Overall Efficiencies in Hover

Total Rotor Power. In axially symmetric regimes of flight, including hovering, the advent of the combined blade-element and momentum theory paved the way for a more realistic prediction of rotor power required than that offered by the pure momentum approach. This was accomplished through the incorporation of blade profile power into a conceptual model and establishing a means of determining time-average induced velocity distribution along the disc radius. Consequently, the total power required by a rotor in hover can now be presented as a sum of the induced and profile powers.

$$P = P_{ind} + P_{pr} \tag{3.55}$$

where values can be computed using the methods outlined in the preceding sections of this chapter. However, for many practical engineering tasks, it may be more convenient to rewrite Eq (3.55) in terms of rotor ideal power.

$$P = P_{id}\, k_{ind_h} + P_{pr} \tag{3.55a}$$

where the k_{ind_h} factor reflects induced power losses due to both nonuniform downwash distribution and tip loss, and P_{pr} is given in terms of the average blade profile drag coefficient $\tilde{c}_d$ (Eq (3.22a)).

Aerodynamic Efficiency (Figure-of-Merit). In parallel with improved analytical methods of estimating rotor power as well as more refined testing techniques, there appeared a need for a quantitative indicator which would measure the deviation of actual rotor power obtained from its estimated ideal value.

For the static-thrust case (hovering), this is usually done through the so-called figure-of-merit *(FM)*, which is also known as aerodynamic efficiency (η_a):

$$FM \equiv \eta_a = P_{id}/P \tag{3.56}$$

where P_{id} and P are ideal and actual powers required by the rotor at a given level of the thrust generator loading as given by the total thrust T per rotor or, more generally, by the nominal disc loading $w \equiv T/\pi R^2$.

Remembering that $P_{id} = T\sqrt{w/2\rho}$; $P = P_{id}\, k_{ind} + (1/8)\,\sigma\pi R^2\, \tilde{c}_d\, V_t^{\,3}$, and $\tilde{c}_{\ell_h} = 6w/\sigma\rho V^2$; the P_{id} and P values can be substituted into Eq (3.56) and the numerator and denominator divided by $T\sqrt{w/2\rho}$, thus leading to the following:

$$\eta_{a_h} = 1/[k_{ind} + \tfrac{3}{4}(\tilde{c}_d/\tilde{c}_{\ell_h})(V_t/\sqrt{w/2\rho})] \tag{3.57}$$

or, since $\sqrt{w/2\rho} = v_{id_h}$, Eq (3.57) becomes

$$\eta_{a_h} = 1/[k_{ind} + \tfrac{3}{4}(\tilde{c}_d/\tilde{c}_{\ell_h})(V_t/v_{id_h})]\,. \tag{3.57a}$$

For contemporary helicopters, the average lift-to-profile-drag coefficient ratios would probably be included in these limits: $50 \leqslant (\tilde{c}_{\ell_h}/\tilde{c}_d) \leqslant 70$. Assuming, in addition, that $k_{ind_h} = 1.12$, the η_{a_h} values were computed from Eq (3.57a) and plotted vs (V_t/v_{id_h}) in Fig 3.15.

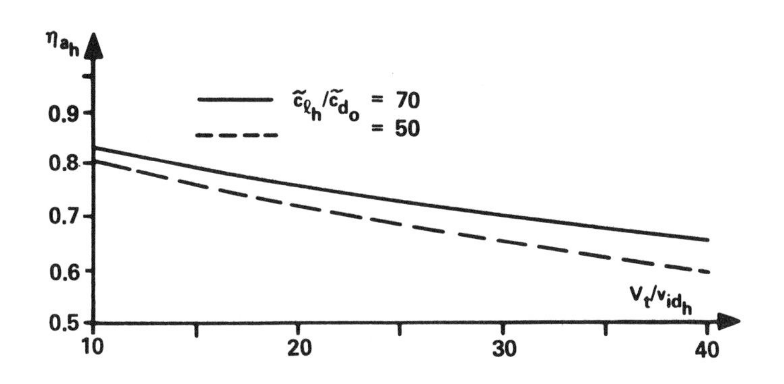

Figure 3.15 Aerodynamic efficiency in hover (η_{a_h}) vs (V_t/v_{id_h}) ratio

From this figure, it can be seen that with a disc loading of $w = 40\ kG/m^2$ (*8 psf*) and $V_t = 215\ m/s$ (*700 fps*), $(V_t/v_{id_h}) \approx 17$ and aerodynamic efficiencies as high as $\eta_a \approx 0.78$ may be expected. However, for $w = 20\ kG/m^2$, and the same V_t, then $(V_t/v_{id_h}) \approx 24$, and η_a may drop to *0.7*.

Overall Efficiency. The figure-of-merit obviously provides a measure for evaluating only the aerodynamic aspect of the excellence of design. In order to compare the overall (total) efficiency (η_{tot}) of various rotary-wing aircraft (both shaft and tip-jet driven) in hovering, a method based on a comparison of the ideal power required to the thermal energy consumed in a unit of time may be more suitable[6]. As long as jet fuels with an approximately equal heating value of about 10 300 Cal/kg (18 500 Btu/lb) are used, the rate of thermal power input (P_{th}) in N-m/s or ft-lb/s per unit (N or lb) of thrust generated can be defined as follows:

SI	$P_{th} = 1220\ tsfc$
Eng	$P_{th} = 4000\ tsfc$

and consequently, the overall total efficiency becomes

$$\text{SI} \qquad \eta_{tot} = v_{id_h}/1220\ tsfc$$
$$\text{Eng} \qquad \eta_{tot} = v_{id_h}/4000\ tsfc \tag{3.58}$$

where *tsfc* is used to denote the thrust specific fuel consumption (kG/hr-kG or lb/hr-lb) with respect to the total rotor thrust T.

Recalling Eqs (1.1) and (1.1a), the above expressions can also be written in terms of specific impulse (I_s); thus Eq (3.58) becomes

$$\text{SI} \qquad \eta_{tot} = I_s v_{id_h}/4\,267\,000$$
$$\text{Eng} \qquad \eta_{tot} = I_s v_{id_h}/14\,000\,000. \tag{3.58a}$$

For instance, for a helicopter where $w = 35\ kG/m^2$ (7.0 psf), $v_{id_h} \approx 11.8\ m/s$ (38 fps), and $I_s = 70\,000\ s$; $\eta_{tot} \approx 0.19$.

3. FORWARD FLIGHT

3.1 Velocities

As in the case of vertical ascent and hovering, the blade element approach could provide the proper means for predicting aerodynamic forces and moments acting on the blade in forward flight. Again, it is necessary to know the magnitude and direction of the relative air flow in the immediate vicinity of the investigated element of the blade. Once this information is available, actual computation of forces and moments should be based on the two-dimensional (section) airfoil characteristics, including Reynolds and Mach number effects, special aspects of unsteady aerodynamics and, if possible, proper corrections for oblique flow at various azimuth angles, effect of blade centrifugal field on the boundary layer, etc. (see Ch VI).

In the general case of a steady-state flight of a helicopter, the rotor axis is tilted from the vertical through an angle a_v; while the aircraft is moving at constant speed V_f along the inclined path where V_c and V_{ho} respectively, are the vertical (rate-of-climb) and horizontal components.

Using the concept of a stationary rotor as shown in Fig 3.16, the speed $-V_f$ can be resolved into two components; one, axial (perpendicular to the airscrew disc) and the other, parallel to the disc (inplane). Obviously, the axial components will be:

$$V_{ax} = -V_c \cos a_v + V_{ho} \sin a_v. \tag{3.59}$$

The inplane component will be

$$V_{inp} = -V_{ho} \cos a_v + V_c \sin a_v. \tag{3.60}$$

In those cases when the tilt angle a_v is small (as it usually is in all helicopter flight regimes), Eqs (3.59) and (3.60) can be simplified as follows:

$$V_{ax} = -V_c + V_{ho} a_v \tag{3.59a}$$

and

$$V_{inp} = -V_{ho} + V_c a_v. \tag{3.60a}$$

As a first approximation, only the component of V_{inp} perpendicular to the blade axis ($V_{b\perp}$) will be considered for computing forces acting on the elements. Measuring the azimuth angle from the blade downwind position (Fig 3.17), and assuming that the

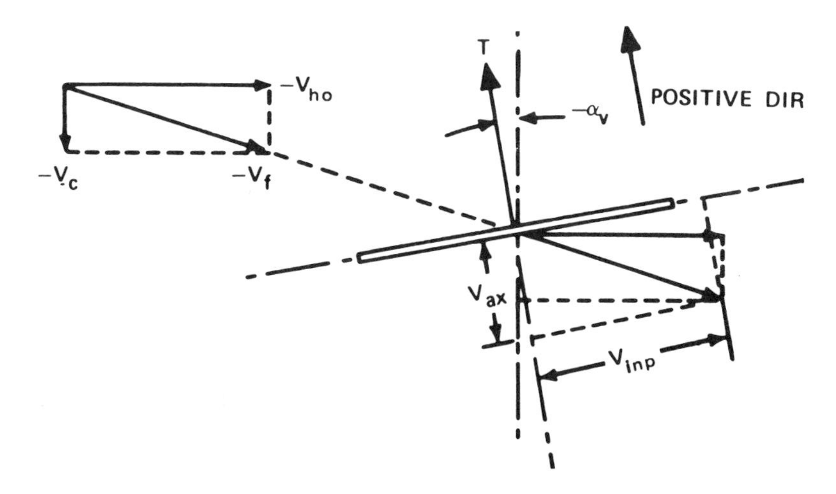

Figure 3.16 Air flows at the disc in forward translation

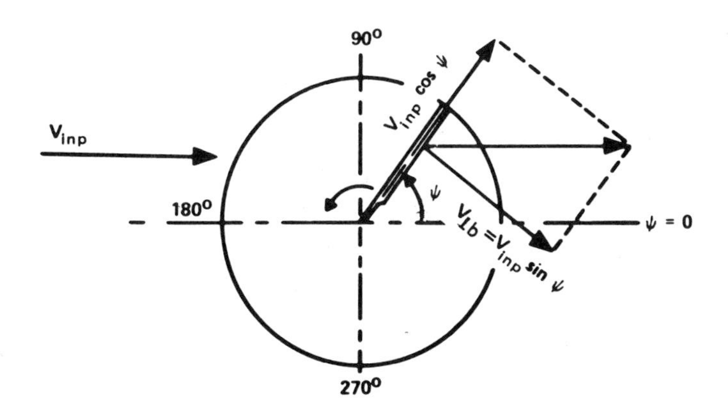

Figure 3.17 Air velocities at a blade element, due to flow in the disc plane

flow is positive when coming toward the blade leading edge, the following is obtained:

$$V_{b\perp} = V_{inp} \sin \psi. \tag{3.61}$$

In the case of horizontal flight (which is most often analyzed in detail), V_{inp} may be considered (for small a_v) as identical to the speed of flight V_{ho}, and Eq (3.61) becomes:

$$V_{b\perp} = V_{ho} \sin \psi. \tag{3.61a}$$

In addition to the $V_{b\perp}$ component, blade elements experience flow due to the rotor rotation at a tip speed of $V_t = R\Omega$. Consequently, for a blade element located at an azimuth angle ψ, and at a distance $r \equiv R\bar{r}$, the total component of the inplane velocity perpendicular to the blade ($U_\perp$) will be:

$$U_{\perp}(\bar{r},\psi) = V_t\bar{r} + V_{ho}\sin\psi. \quad (3.62)$$

Assuming that the blade tips do not change their position with respect to the tip-path plane representing the rotor disc, the air flow in the immediate neighborhood of a blade element will be as shown in Fig 3.18. Designating the pitch angle of a blade element as $\theta(\bar{r},\psi)$, the corresponding angle-of-attack $a(\bar{r},\psi)$ will be

$$a(\bar{r},\psi) = \theta(\bar{r},\psi) - \phi(\bar{r},\psi) \quad (3.63)$$

where $\phi(\bar{r}\psi)$ is the total inflow angle (positive in the clockwise direction),

$$\tan^{-1}\phi(\bar{r},\psi) = V_{ax_{tot}}(\bar{r},\psi)/U_{\perp}(\bar{r},\psi) \quad (3.64)$$

and, in turn, $V_{ax_{tot}}$ is the sum of V_{ax} as defined by Eq (3.59a) and the induced velocity $v(\bar{r},\psi)$ at the considered element.

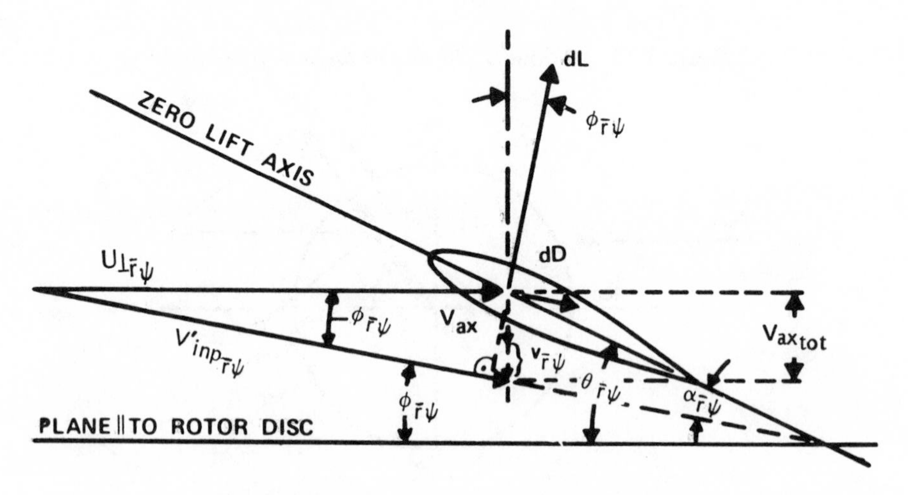

Figure 3.18 Air velocities at a blade element

3.2 Thrust and Torque (General Considerations)

If the value of the $V_{ax_{tot}}$ for a blade element at station $\bar{r}$ and azimuth angle ψ were known, it would be possible to compute the following forces experienced by this element: (1) lift $dL(\bar{r},\psi)$, (2) profile drag $dD_{pr}(\bar{r},\psi)$, and (3) total drag $dD(\bar{r},\psi)$.

Within the limits of the small-angle assumptions (the validity of which should always be checked prior to starting the actual calculations), these elementary aerodynamic forces and torques can be expressed as follows:

Thrust:

$$dT(\bar{r},\psi) \approx dL(\bar{r},\psi) = \tfrac{1}{2}R\rho\,(V_t\bar{r} + V_{ho}\sin\psi)^2\, a(\bar{r},\psi)[\theta(\bar{r},\psi) - \phi(\bar{r},\psi)]c_{\bar{r}}\, d\bar{r}$$

Total Drag:

$$dD(\bar{r},\psi) = dL(\bar{r},\psi)\phi(\bar{r},\psi) + \tfrac{1}{2}R\rho[(V_t\bar{r} + V_{ho}\sin\psi)^2\, c_{\bar{r}}\, c_d(\bar{r},\psi)]\, d\bar{r}$$

Profile Drag:

$$dD_{pr}(\bar{r},\psi) = \tfrac{1}{2}R\rho[(V_t\bar{r} + V_{ho}\sin\psi)^2\, c_{\bar{r}}\, c_d(\bar{r},\psi)]\, d\bar{r}$$

Total Torque:

$$dQ(r,\psi) = R\bar{r}dD(\bar{r},\psi)$$

Profile Torque:

$$dQ_{pr}(\bar{r},\psi) = R\bar{r}dD_{pr}(\bar{r},\psi) \tag{3.65}$$

For a rotor with rectangular blades, the following will result:

Thrust:

$$T = \tfrac{1}{4}\sigma R^2 \rho \int_{\bar{r}_i}^{\bar{r}_e} \int_0^{2\pi} \left\{(V_t\bar{r} + V_{ho}\sin\psi)^2 a(\bar{r}\psi)[\theta(\bar{r},\psi) - \phi(\bar{r},\psi)]\right\} d\bar{r}d\psi$$

Total Torque:

$$Q = (bR/2\pi) \int_{\bar{r}_i}^{\bar{r}_e} \int_0^{2\pi} \phi(\bar{r},\psi)\bar{r}dL(\bar{r},\psi)d\psi + \tfrac{1}{4}\sigma R^3 \rho \int_{\bar{r}_i}^{1.0} \int_0^{2\pi} (V_t\bar{r} + V_{ho}\sin\psi)^2 c_d(\bar{r},\psi)\bar{r}d\bar{r}\,d\psi$$

Profile Torque:

$$Q_{pr} = \tfrac{1}{4}\sigma R^3 \rho \int_{\bar{r}_i}^{1.0} \int_0^{2\pi} (V_t\bar{r} + V_{ho}\ \sin\psi)^2 c_d(\bar{r},\psi)\bar{r}d\bar{r}\,d\psi$$

Total Power:

$$P_R = Q\Omega$$

Profile Power:

$$P_{R_{pr}} = Q_{pr}\Omega, \tag{3.66}$$

where Ω is the angular velocity of the rotor.

The integrations indicated by Eq (3.66) can usually be performed either on a computer or graphically. In both cases, the thrust and/or torque for the whole blade is determined for selected azimuth angles, then the average for a complete revolution is found and the result multiplied by the number of blades. However, in order to apply the above-described procedure, the downwash velocity $v(\bar{r},\psi)$ at each point of the rotor disc must be known (Fig 3.18)

In those cases when the induced part represents only a small fraction of the total axial inflow velocity ($V_{ax_{tot}}$), the deviations of the actual induced velocity (at various points on the rotor or propeller disc) from its average value are no longer important. It may be assumed hence, that the downwash is uniform over the whole disc, and can be computed according to the simple momentum theory (Eq (2.36)). An additional small improvement can be gained by including tip losses ($\bar{r}_e < 1.0$).

If information regarding the type of disc area loading is available, a chordwise distribution of time-average induced velocities can be obtained by the methods described in Sect 7 of Ch II. A further refinement of this approach—based on combining the blade-element and momentum theories—is outlined in Sect 3.3.

However, with the current state of the art of rotary-wing aerodynamics, the most precise determination of $v(r\psi)$ values—both time-average and instantaneous—can, in principle, be obtained through the application of the vortex theory described in Ch IV, and to a lesser extent, by the potential theory discussed in Ch V.

3.3 Downwash Distribution Along the Rotor Disc Chord

Taking into account such rotor characteristics as number of blades, blade planform, twist, pitch angle, and section lift-curve slope, the blade element theory allows one to determine the variation of the area loading along any of the chords of the rotor disc. Once the above information is available, the chordwise induced velocity distribution can be computed by using the method developed in Sect 7 of Ch II. The procedure described below is based on the case of downwash distribution along the fore-and-aft rotor disc diameter; expecting that the reader now has a grasp of the principles of this approach and should be able extend it to other disc chords as well.

Downwash Distribution Along the Fore-and-Aft Rotor Diameter. Similar to Sect 7 of Ch II, the case of horizontal flight at speed V_{ho} is considered, assuming that $V_{ho} \gg v_{ho}$ and consequently, the resultant flow through the rotor $V' \approx V_{ho}$, while the basic notations of Fig 2.26 as adapted to the present case are shown in Fig 3.19.

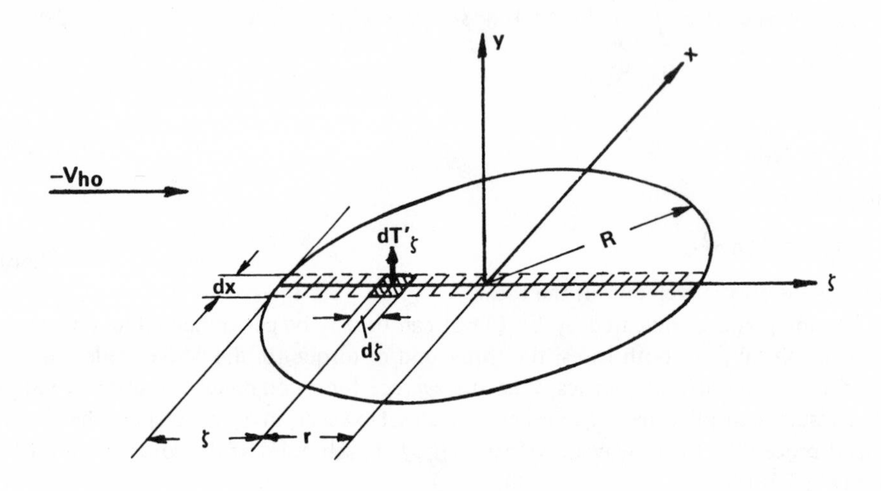

Figure 3.19 Basic Notations

In order to determine the elementary thrust dT' using the blade-element approach, it should be noted that the considered element corresponding to the coordinate ζ is also located at radius $r \equiv R\bar{r} = |R - \zeta|$.

If all b blade elements through one complete revolution (2π) maintained the same angle-of-attack and experienced the same velocity component perpendicular to the blade axis $(U_{\perp\bar{r}})$ that they had at point ζ, their collective thrust would be $dT_{2\pi}$.

However, only part of this thrust may be credited to the thrust generation at the considered element $dx\, d\zeta$; and therefore, would represent only a $dx/2\pi R\bar{r}$ fraction of $dT_{2\pi}$:

$$dT'_{\zeta} = dT_{2\pi}(dx/2\pi R\bar{r}). \tag{3.67}$$

In order to determine $dT_{2\pi}$, it is necessary to know: (a) the geometric pitch angle (θ_{ζ}) when the blade element is at a point defined by the coordinate ζ, (b) all the components of the axial inflow immediately ahead of the element, and (c) axial velocity components due to the flapping motion (if present). At this point, it should be recalled that this flapping motion is geometrically equivalent to the variation of the pitch angle of the considered blade element by the amount,

$$\Delta\theta_{\zeta} = R\bar{r}\dot{\beta}/U_{\perp\zeta}$$

where $\dot{\beta}$ is the flapping velocity at the appropriate azimuth angle (in this case, $\psi = 180$ and 0 *degrees*), and $U_{\perp}$ is the velocity component perpendicular to the rotor blade axis at station $\bar{r}R$ when it passes over the point defined by the ordinate ζ.

Summing up, it may be stated that knowing the blade twist, and either the variation of the pitch angle with azimuth as required to eliminate flapping or the rate of flapping at each azimuth angle, it will be possible to determine both the geometric and the equivalent pitch angle θ_{ζ} of each blade element along the fore-and-aft disc axis.

The actual angle-of-attack of the blade element (α_{ζ}) will be equal to the difference between θ_{ζ} and the total inflow angle resulting from (a) rotor disc inclination α_v with respect to the direction of the incoming velocity $(-V_{h\,o})$ and (b) that due to downwash (v_{ζ}) at the considered element. Therefore,

$$\alpha_{\zeta} = \theta_{\zeta} - (V_{ho}\alpha_v + v_{\zeta})/U_{\perp\zeta} \tag{3.68}$$

where, for the considered case of the fore-and-aft diameter, $U_{\perp\zeta} = V_t\bar{r}$.

Now, the elementary thrust $dT_{2\pi}$ of all b blades corresponding to blade strips $dr = d\zeta$ wide and having a chord $c_{\bar{r}}$ will be

$$dT_{2\pi} = \tfrac{1}{2}\rho(V_t\bar{r})^2\, bc_{\bar{r}}a_{\zeta}\left\{\theta_{\zeta} - [(V_{ho}\,\alpha_v + v_{\zeta})/V_t\bar{r}]\right\}d\zeta \tag{3.69}$$

where a_{ζ} is the sectional lift-curve slope at station ζ.

Substituting Eq (3.69) into Eq (3.67) leads to the following:

$$dT'_{\zeta} = (1/4\pi)\rho V_t^2\,\bar{r}(bc_{\bar{r}}/R)a_{\zeta}\left\{\theta_{\zeta} - [(V_{ho}\alpha_v + v_{\zeta})/V_t\bar{r}]\right\}dx\,d\zeta \tag{3.70}$$

The expression in front of $dx\,d\zeta$ is the local disc loading, $w_{\zeta} = dT'_{\zeta}/dx\,d_{\zeta}$. It was shown in Sect 7 of Ch II that knowledge of the local disc loading at any point along a chord of the disc would permit one to determine the slope of the chordwise variation of induced velocity at that particular point (see Eq 2.88)).

In Ch II, the disc area loading was assumed uniform, while various rotor disc chords were considered. By contrast, the present task consists of finding the induced velocity distribution along the fore-and-aft diameter only, while the disc loading varies from one point of the chord to another. Taking these differences into account, and using the notations shown in Fig 3.19, Eq (2.88) can be rewritten as follows:

$$(dv/d\zeta)_\zeta = w_\zeta/4\rho V_{ho}R. \tag{3.71}$$

It was also shown in Sect 7 of Ch II that at any small but finite, distance ϵ downstream from the point of induced velocity generation, the induced velocity slope becomes twice as high as that given by Eq (3.71):

$$(dv/d\zeta)_{\zeta+\epsilon} = w_\zeta/2\rho V_{ho}R. \tag{3.72}$$

By substituting the expression in front of $dx\,d\zeta$ in Eq (3.70) for w_ζ, a formula for the induced velocity slope just downstream of point ζ could be obtained. One should note, however, that v_ζ whose value is unknown, appears in the expression for w_ζ. This value can be defined as:

$$v_\zeta = \int_0^\zeta (dv/d\zeta)_{\zeta+\epsilon}d\zeta,$$

and Eq (3.72) can be written as

$$(dv/d\zeta)_{\zeta+\epsilon} = (a_\zeta bc_{\bar{r}}V_t\bar{r}/8\pi R^2\mu)\left\{\theta_\zeta - (a_v\mu/\bar{r}) - \left[\int_0^\zeta (dv/d\zeta)_{\zeta+\epsilon}d\zeta/V_t\bar{r}\right]\right\} \tag{3.73}$$

where $\mu \equiv V/V_t$.

Where the blades have a constant chord, and it is assumed that $a_\zeta = a = const$, then remembering that $V_t = R\Omega$, Eq (3.73) becomes

$$(dv/d\zeta)_{\zeta+\epsilon} = (\sigma a\Omega\bar{r}/8\mu)\left\{\theta_\zeta - (a_v\mu/\bar{r}) - \left[\int_0^\zeta (dv/d\zeta)_{\zeta+\epsilon}d\zeta/V_t\bar{r}\right]\right\}. \tag{3.73a}$$

Eqs (3.73) and (3.73a) can be evaluated by the finite differences step-by-step procedure. This will yield not only the desired $(dv/d\zeta)_{\zeta+\epsilon}$ slope at every point $\zeta+\epsilon$, but wlll also determine the sought relationship $v = f(\zeta)$. In order to get these results, the whole fore-and-aft diameter is first divided into a number of segments of length $\Delta\zeta$ small enough so that within each of them the induced velocity slope may be considered constant. Next, at the mid-point of each segment, values of θ_ζ and, in the case of a variable blade planform of the chord $(c_{\bar{r}})$–perhaps, a_ζ as well–should be determined.

Starting with the first segment and gradually progressing to the following ones, a general expression for the downwash slope at the nth segment can be developed which, for a rotor with rectangular blades and $a = const$, becomes

$$(dv/d\zeta)_n = (V_t\sigma a\bar{r}_n/8R\mu)\left\{\theta - (a_v/\bar{r}_n)\mu - \left[\sum_1^{n-1}(dv/d\zeta)_k\Delta\zeta/V_t\bar{r}_n\right]\right\} \tag{3.74}$$

or introducing the nondimensional $\Delta\bar{\zeta} \equiv \Delta\zeta/R$, substituting $\Delta\bar{\zeta}R$ for $\Delta\zeta$ in Eq (3.74), and remembering that $V_t = R\Omega$, the above equation becomes

$$(dv/d\zeta)_n = (a\sigma\Omega\bar{r}_n/8\mu)\left\{\theta_n - (a_v/\bar{r}_n)\mu - \left[\sum_1^{n-1}(dv/d\zeta)_k\Delta\bar{\zeta}/\Omega\bar{r}_n\right]\right\}. \qquad (3.74a)$$

The increment of induced velocity corresponding to the nth segment will be $\Delta v_n = (dv/d\zeta)_n\,\Delta\zeta$. Adding these increments, a complete $v = f(\zeta)$ relationship can be obtained.

The method described above was used to calculate the induced velocity distribution along the fore-and-aft rotor diameter for the case represented in Fig 3.20, for which experimental downwash measurements were available (Fig 6, Ref 7).

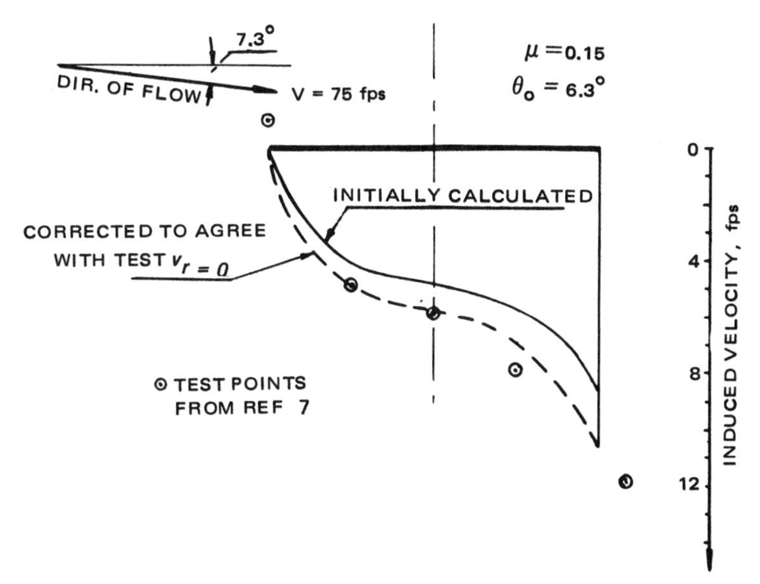

Figure 3.20 Comparison between calculated and measured induced velocity along the fore-and-aft disc axis

It can be seen from this figure that a good agreement has been obtained between the predicted (broken line) and the measured values (points).

An approach similar to that presented above can be applied to any fore-and-aft disc chord. However, in expressing the induced velocity slope according to the momentum theory, Eq (2.88) with the proper values of x_n should now be used.

3.4 Blade Profile Drag Contribution to Rotor Power and Drag (Simple Approach)

The blade element theory may be quite helpful for a better insight into problems of the blade profile drag contributions to the rotor power required in forward flight as well as rotor overall drag.

In the simple approach, only the component of the resultant air velocity perpendicular to the blade at the $\bar{r}$, ψ coordinates $(U_{\perp\bar{r}\,\psi})$ is considered, while the influence of the velocity component parallel to the blade is neglected.

Under these assumptions, the elementary profile drag experienced by a blade strip $dr = Rd\bar{r}$ wide, and located at a distance $r = R\bar{r}$ from the rotor axis when the blade is at an azimuth angle ψ, will be

$$dD_{\bar{r}\psi} = \tfrac{1}{2} R \rho c_{\bar{r}} c_{d_{\bar{r}\psi}} U_{\perp_{\bar{r}\psi}}^2 \, d\bar{r}. \tag{3.75}$$

and the corresponding elementary torque is

$$dQ_{\bar{r}\psi} = R\bar{r} dD_{\bar{r}\psi}. \tag{3.76}$$

The drag component in the direction opposite to that of flight $(dD_{fo_{\bar{r}\psi}})$ becomes:

$$dD_{fo_{\bar{r}\psi}} = dD_{\bar{r}\psi} \sin\psi. \tag{3.77}$$

Defining the flow approaching the blade from the leading edge direction as positive, and using the notations from Fig 3.21, $U_{\perp_{\bar{r}\psi}}$ becomes:

$$U_{\perp_{\bar{r}\psi}} = V_t\bar{r} + V\sin\psi \tag{3.78}$$

or, with $\mu \equiv V/V_t$,

$$U_{\perp_{\bar{r}\psi}} = V_t(\bar{r} + \mu \sin\psi). \tag{3.78a}$$

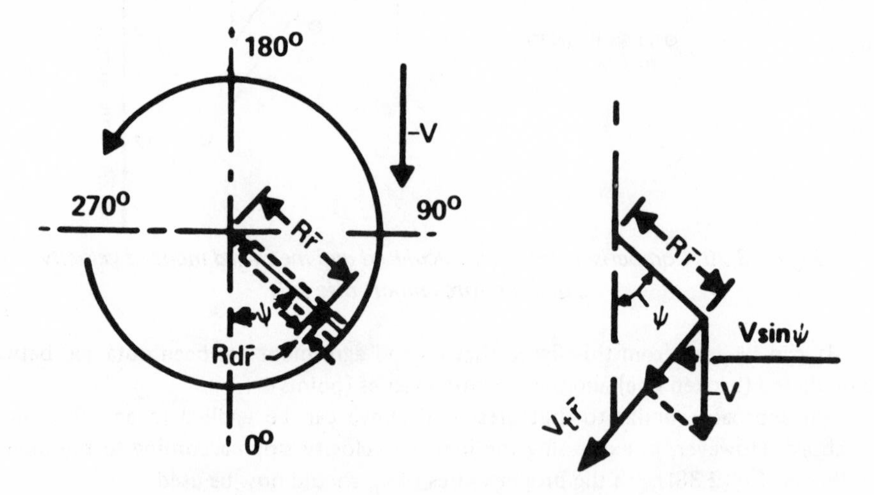

Figure 3.21 Air velocity perpendicular to the blade

Since, in general, $dD_{\bar{r}\psi}$ varies with the azimuth angle ψ as well as the blade element position $\bar{r}$, the values of the torque (Q_{pr}) and drag $(D_{fo_{pr}})$ contributions of all b blades averaged over one full rotor revolution must be obtained through integration with respect to $\bar{r}$ as well as ψ.

Substituting the value for $U_{\perp_{\bar{r}\psi}}$ from Eq (3.78a) into Eqs (3.76) and (3.77), the following is obtained:

$$Q_{pr} = (bR^2 \rho V_t^2/4\pi) \int_0^{1.0} \int_0^{2\pi} c_{\bar{r}} c_{d_{\bar{r}\psi}} \bar{r} (\bar{r} + \mu \sin\psi)^2 \, d\bar{r} \, d\psi. \tag{3.79}$$

When the $c(r,\psi)$ and $c_d(r,\psi)$ variations indicated in Eq (3.79) cannot be neglected, then the Q_{pr} values are usually evaluated through numerical procedures on the computer or graphically. However, Eq (3.79) can be easily integrated under the following simplifying assumptions:

1. The drag coefficient is equal to $\bar{c}_d$ and is constant along the blade.
2. The drag coefficient does not vary as the blade changes its azimuth.
3. The blade chord is constant.

Under these conditions, Eq (3.79) is reduced to (N-m or ft-lb):

$$Q_{pr} = (1/8)\,\sigma\pi R^3 \rho V_t^2 \bar{c}_d(1+\mu^2) \tag{3.80}$$

and the corresponding profile power in N-m/s or ft-lb/s becomes

$$P_{pr} = (1/8)\sigma\pi R^2 \rho V_t^3 \bar{c}_d(1+\mu^2) \tag{3.80a}$$

or, in horsepower:

$$\text{SI} \qquad HP_{pr} = \sigma\pi R^2 \rho V_t^3 \bar{c}_d(1+\mu^2)/5880$$

$$\text{Eng} \qquad HP_{pr} = \sigma\pi R^2 \rho V_t^3 \bar{c}_d(1+\mu^2)/4400. \tag{3.80b}$$

But $(1/8)\sigma\pi R^2 \rho V_t^3 \bar{c}_d = P_{pr_h}$; i.e.,the profile drag in hovering, therefore

$$P_{pr} = P_{pr_h}(1+\mu^2). \tag{3.80c}$$

Similar to Q_{pr}, the blade profile drag contribution to the rotor drag, in N or lb, can be expressed as

$$D_{pr} = (bR\rho V_t^2/4\pi)\int_0^{1.0}\int_0^{2\pi} c_{\bar{r}} c_{d_{\bar{r}\psi}}(\bar{r} + \mu\sin\psi)^2 \sin\psi \, d\bar{r}\, d\psi \tag{3.81}$$

and again, when $c(\bar{r},\psi) \neq const$ and $c_d(\bar{r},\psi) \neq const$, the above integration must be performed by numerical or graphical means. However, under the same simplifying assumptions, Eq (3.81) becomes

$$D_{pr} = \tfrac{1}{4}\sigma\pi R^2 \rho V_t^2 \mu\bar{c}_d \tag{3.82}$$

and the corresponding power in N-m/s or ft-lb/s is

$$P_{D_{pr}} = \tfrac{1}{4}\sigma\pi R^2 \rho V_t^3 \mu^2 \bar{c}_d \tag{3.83}$$

or, in HP:

$$\text{SI} \qquad HP_{D_{pr}} = \sigma\pi R^2 \rho V_t^3 \mu^2 \bar{c}_d/2940$$

$$\text{Eng} \qquad HP_{D_{pr}} = \sigma\pi R^2 \rho V_t^3 \mu^2 \bar{c}_d/2200. \tag{3.83a}$$

Remembering the previously quoted expression for P_{pr_h}, Eq (3.83) can be rewritten as follows:

$$P_{D_{pr}} = 2P_{pr_h}\mu^2. \tag{3.83b}$$

3.5 Further Study of Blade Profile Drag Contribution to Rotor Power and Drag

It may be recalled that equations expressing the contribution of the blade profile drag to the rotor power, Eq (3.80c) and drag, Eq (3.83b) were developed under the following assumptions:

1. Influence of oblique flow on the profile drag coefficient of the blade airfoil was neglected.
2. Only the flow components perpendicular to the blade were considered in determining the drag of any blade element.
3. Influence of the reversed flow region on the retreating side, and differences in profile drag coefficient associated with this type of flow were neglected.

For a more realistic evaluation of the profile drag and associated power, a new study is performed without the above limiting constraints, but with $c = const.$

Using the notations in Fig 3.22, it is possible to express the total resultant speed $U_{\bar{r}\psi_{tot}}$ at any blade element located at station $\bar{r}$ and at an azimuth ψ by the following:

$$U_{\bar{r}\psi_{tot}} = V_t\sqrt{(\bar{r}+\mu^2 \sin\psi)^2 + \mu^2 \cos^2\psi}. \tag{3.84}$$

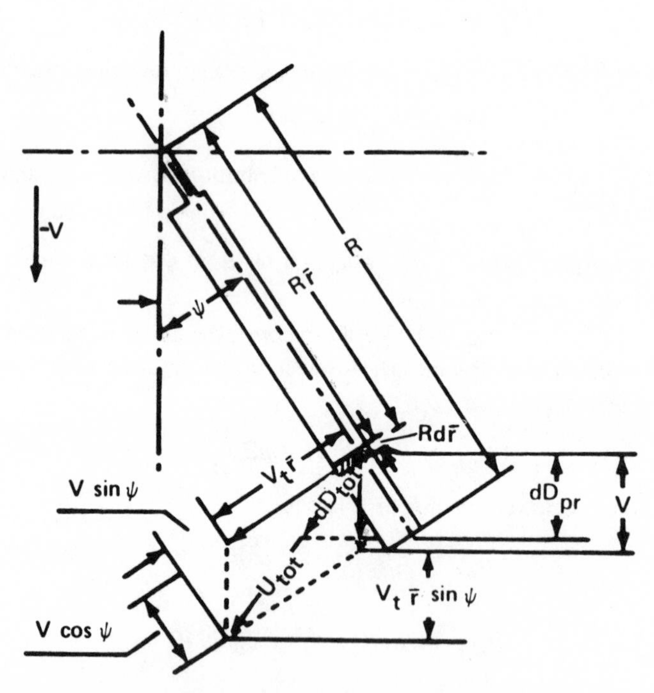

Figure 3.22 Flow directions and drag components at a blade element

Consequently, the total profile drag $(dD_{\bar{r}\psi_{tot}})$ of the blade element in the direction of the resultant local air flow becomes:

$$dD_{\bar{r}\psi_{tot}} = \tfrac{1}{2}\rho V_t^2 [(\bar{r}+\mu \sin\psi)^2 + \mu^2 \cos^2\psi] c_{d_{\bar{r}\psi}} c R d\bar{r} \tag{3.85}$$

where $c_{d_{\bar{r}\psi}}$ is the profile drag coefficient at station $\bar{r}$ and azimuth ψ, with due consideration of the actual flow conditions; i.e., direction of flow (oblique, reversed) in addition to the usually considered influences of Reynolds and Mach numbers.

The contribution of $dD_{\bar{r}\psi_{tot}}$ to the rotor drag component $dD_{pr_{r\psi}}$ can be expressed as follows (see Fig 3.22):

$$dD_{pr_{\bar{r}\psi}} = [(V_t\bar{r}\sin\psi + V)/U_{\bar{r}\psi})]dD_{\bar{r}\psi_{tot}}. \tag{3.86}$$

Making the proper substitutions from Eqs (3.84) and (3.85), Eq (3.86) becomes:

$$dD_{pr_{\bar{r}\psi}} = \tfrac{1}{2}\rho cRV_t^2 c_{d_{\bar{r}\psi}}(\bar{r}\sin\psi + \mu)\sqrt{(\bar{r} + \mu\sin\psi)^2 + \mu^2\cos^2\psi}\,d\bar{r}. \tag{3.87}$$

The total contribution to the parasite drag of a rotor equipped with b blades of constant chord c will be (in N, or lb):

$$D_{pr} = (1/4\pi)\rho bcRV_t^2 \int_0^1\int_0^{2\pi} c_{d_{\bar{r}\psi}}(\bar{r}\sin\psi + \mu)\sqrt{(\bar{r} + \mu\sin\psi)^2 + \mu^2\cos^2\psi}\,d\bar{r}\,d\psi. \tag{3.88}$$

After $\sigma\pi R^2$ is substituted for bcR, Eq (3.88) is rewritten in a form similar to Eq (3.82):

$$D_{pr} = \tfrac{1}{4}\rho\sigma\pi R^2\mu V_t^2(1/\pi\mu)\int_0^1\int_0^{2\pi} c_{d_{\bar{r}\psi}}(\bar{r}\sin\psi + \mu)\sqrt{(\bar{r} + \mu\sin\psi)^2 + \mu^2\cos^2\psi}\,d\bar{r}\,d\psi. \tag{3.89}$$

By multiplying Eq (3.89) by $V_t\,\mu = V$, an expression for the resulting power (in N-M/s or ft-lb/s, is obtained, while in horsepower (SI units) it becomes:

$$HP_{D_{pr}} = (\mu^2/2940)\rho\sigma\pi R^2 V_t^3(1/\pi\mu)\int_0^1\int_0^{2\pi} c_{d_{\bar{r}\psi_{tot}}}(\bar{r}\sin\psi + \mu)\sqrt{(\bar{r} + \mu\sin\psi)^2 + \mu^2\cos^2\psi}\,d\bar{r}\,d\psi. \tag{3.90}$$

In English units, a *2200* numerical coefficient would replace the *2940* appearing in Eq (3.90).

By examining Eqs (3.89) and (3.90), it can readily be seen that the expression "$(1/\pi\mu)$ times the double integral, ..." replaces the $\bar{c}_d$ terms in Eqs (3.82) and (3.83). Thus, it may be considered as a true average drag coefficient ($\bar{c}_{d_{pr}}$) as far as rotor contribution to the power and parasite drag is concerned:

$$\bar{c}_{d_{pr}} = (1/\pi\mu)\int_0^1\int_0^{2\pi} c_{d_{\bar{r}\psi}}(\bar{r}\sin\psi + \mu)\sqrt{(\bar{r} + \mu\sin\psi)^2 + \mu^2\cos^2\psi}\,d\bar{r}\,d\psi. \tag{3.91}$$

Because of the difficulties which may be encountered in expressing the variation of c_d with $\bar{r}$ and ψ under an analytical form, it will probably be more practical to perform the integration in Eq (3.91) either numerically or graphically.

Comparing the values of $\bar{c}_{d_{pr}}$ determined from Eq (3.91) with those of $\bar{c}_d$, some feeling may develop as to the magnitude of error resulting from neglecting skin friction and all other effects of skewed air flow with respect to the blade axis.

As an illustrative example, the variation of the $\tilde{c}_{d_{pr}}$ coefficient from Eq (3.91) was computed vs μ. In this simplified example, it was assumed that no compressibility effects were encountered and variations in the blade lift coefficient were such that $c_{d_{\Gamma\psi}}$ values were influenced only by the direction of flow with respect to the blade radial axis.

The following numerical values were used in the present example: (a) airfoil thickness, t/c = *15 percent*, (b) the profile drag at the 'sweep-back' angle $\Gamma = 0°$, is $c_d = 0.009$, (c) the friction drag at $\Gamma = 90°$ is $c_f = 0.0067$, and (d) the drag coefficient in the completely reversed flow: i.e., at $\Gamma = 180°$ is $c_{d_{180}} = 0.035$. Furthermore, it was assumed that for the oblique flow, the profile drag coefficient varies according to the following formula from p. 211 of Ref 8:

$$c_d(\Gamma) = c_f[1 + 2(t/c)\cos\Gamma + 60(t/c)^4\cos\Gamma].$$

Results of the computation of $\tilde{c}_{d_{pr}}$ are shown in Fig 3.23. A glance at this figure will indicate that the blade profile drag contribution to the parasite drag at first increases rather rapidly with μ and then very slowly as it approaches asymptomatically the average drag coefficient of the stopped rotor ($\mu = \infty$). For μ values encountered in high-speed cruise ($\mu \approx 0.35$), it may be assumed that on the average,

$$\tilde{c}_{d_{pr}} = 1.5\tilde{c}_d.$$

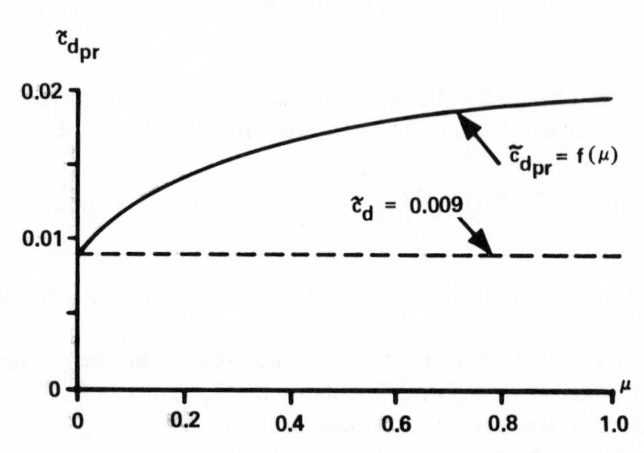

Figure 3.23 Total equivalent profile drag coefficient versus μ

Substituting the above value instead of "$(1/\pi\,\mu)$ times the double integral, ..." into Eqs (3.89) and (3.90), the following D_{pr} is obtained (in N, or lb):

$$D_{pr} = (3/8)\rho\sigma\pi R^2 V_t^2 \mu\tilde{c}_d \qquad (3.92)$$

and for horsepower:

$$\text{SI} \qquad HP_{D_{pr}} = (3/5880)\,\sigma\pi R^2 \rho\, V_t^3 \mu^2 \tilde{c}_d$$

$$\text{Eng} \qquad HP_{D_{pr}} = (3/4400)\sigma\pi R^2 \rho\, V_t^3 \mu^2 \tilde{c}_d \qquad (3.93)$$

It should be realized that one-third of the values indicated by Eq (3.93) represents the profile power in hover (HP_{pr_h}) and consequently, Eq (3.93) can be rewritten as follows:

$$HP_{D_{pr}} = 3\mu^2 HP_{pr_h}. \tag{3.94}$$

Considerations similar to the preceding ones would indicate that the profile power in forward flight increases more rapidly than indicated by Eq (3.80c), and the following expression probably describes this power rise more accurately:

$$HP_{pr} = HP_{pr_h}(1 + 1.7\mu^2). \tag{3.95}$$

Therefore, total power required to overcome the blade profile drag will be the sum of Eqs (3.94) and (3.95):

$$HP_{pr_{rot}} = HP_{pr_o}(1 + 4.7\mu^2). \tag{3.96}$$

3.6 Contribution of Blade Element Induced Drag to Rotor Torque and Power

In performance predictions of helicopters in forward flight, it is usually sufficient to estimate the induced power from the momentum theory and then correct those estimates through proper factors ($k_{ind_{fo}}$). There are also cases where more detailed studies of the induced power in forward flight of helicopters are required. To provide the necessary insight into these problems, a simplified approach to the estimation of the rotor torque (Q_{ind}) due to the blade element induced drag is briefly outlined.

Let $dL_{\bar{r}\psi}$ be the lift generated by a blade element located at a station $\bar{r}$, while the blade itself is at an azimuth ψ; then for those flight conditions and $\bar{r}$ values where the small angle assumptions are valid, the elementary induced drag due to lift $dL_{\bar{r}\psi}$ may be expressed as follows (Fig 3.24):

$$dD_{L_{\bar{r}\psi}} = [(v_{\bar{r}\psi}/V_t)/(\bar{r} + \mu \sin\psi)]\, dL_{\bar{r}\psi}. \tag{3.97}$$

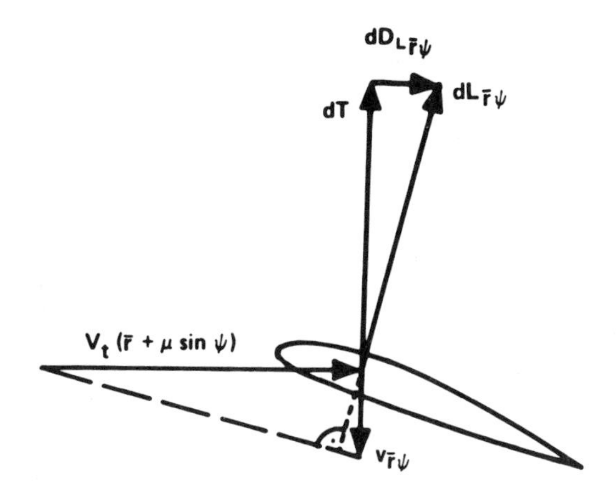

Figure 3.24 Forces and velocities at a blade element

The corresponding elementary torque for all b blades $(dQ_{ind_{\bar{r}\psi}})$ will be

$$dQ_{ind_{\bar{r}\psi}} = bR\bar{r}[v_{\bar{r}\psi}/V_t)/(\bar{r} + \mu \sin \psi)]\, dL_{\bar{r}\psi} \tag{3.98}$$

and the elementary power, in N-m/s or ft-lb/s is $dP_{ind\,\bar{r}\psi} = \Omega dQ_{ind_{\bar{r}\psi}}$ which, in view of the fact that $\Omega R = V_t$, can be written as follows:

$$dP_{ind_{\bar{r}\psi}} = b\bar{r}[v_{\bar{r}\psi}/(\bar{r} + \mu \sin \psi)]\, dL_{\bar{r}\psi}. \tag{3.99}$$

However, once the induced velocity, $v_{\bar{r}\psi} = f(\bar{r},\psi)$, as well as those of the actual or equivalent blade pitch angle $\theta_{\bar{r}\psi} = f(\bar{r},\psi)$ are known, then $dL_{\bar{r}\psi}$ can be computed from Eq (3.1). Now, both Q_{ind} and P_{ind} can be obtained by evaluating the following integral, either by computer or by graphical means:

$$Q_{ind} = (b/2\pi\Omega) \int_{\bar{r}_i}^{\bar{r}_e} \int_0^{2\pi} [v_{\bar{r}\psi}/ (\bar{r} + \mu \sin \psi)]\, r dL_{\bar{r}\psi}\, d\psi \tag{3.100}$$

while the induced power in N-M/s or ft-lb/s becomes

$$P_{ind} = \Omega Q_{ind}. \tag{3.101}$$

and in horsepower,

$$\text{SI} \qquad P_{ind} = (1/735)\Omega Q_{ind}$$

$$\text{Eng} \qquad P_{ind} = (1/550)\Omega Q_{ind}. \tag{3.101a}$$

3.7 Contribution of Blade Element Induced Drag to Rotor Drag

Under small-angle assumptions, the elementary induced drag due to the (dL) produced by a blade strip located at station r when the blade is at an azimuth ψ is

$$dD'_{ind_{\bar{r}\psi}} = dL_{\bar{r}\psi}(v_{\bar{r}\psi}/U_{\perp\,\bar{r}\psi})$$

and the contribution of $dD'_{ind_{\bar{r}\psi}}$ to the total drag of the rotor can be expressed as follows:

$$dD_{ind_{r\psi}} = [(v_{\bar{r}\psi}/V_t(\bar{r} + \mu \sin \psi)]\, \sin \psi\, dL_{\bar{r}\psi}. \tag{3.102}$$

Again, the induced part of the rotor drag (D_{ind}) could be obtained as a double integral of Eq (3.102) within the limits of $\bar{r} = \bar{r}_i$ to $\bar{r} = \bar{r}_e$, and $\psi = 0$ to $\psi = 2\pi$. However, in order to give the reader some idea as to the magnitude of induced drag, it will be assumed that at some representative blade station $\bar{r}$, both the induced velocity and elementary lift remain constant throughout a complete rotor revolution; i.e., $v_{\bar{r}\psi} = v_{\bar{r}} = const$, and $dL_{\bar{r}\psi} = dL_{\bar{r}} = const$. Under these assumptions, the integration of Eq (3.102) must be performed within the $\psi = 0$ to $\psi = 2\pi$ limits only, thus making the following contribution to the rotor drag:

$$dD_{ind_{\bar{r}}} = (dL_{\bar{r}}/2\pi)(v_{\bar{r}}/V_t)\int_0^{2\pi}[\sin\psi/(\bar{r} + \mu\sin\psi)]\,d\psi. \tag{3.103}$$

As an example, the $(dD_{ind_{\bar{r}}}/dL_{\bar{r}})$ ratio has been evaluated from Eq (3.103) for the following conditions: $\bar{r} = 0.7$, $\mu = 0.3$, $V_t = 200\ m/s$, and $w = 40\ kG/m^2$; resulting in $v_{ind} \approx 12.6\ m/s$ and $v_{ind}/V_t \approx 0.063$. This leads to $dD_{ind_{0.7}}/dL_{0.7} \approx -0.02$.

It should be noted that the sign of the above ratio is negative, which means that instead of drag, a propulsive force will be obtained. This somewhat surprising result is due to the assumption of the constancy of both elementary lift and induced velocity throughout the revolution, which makes the absolute values of the negative drag component on the retreating side higher than the positive value on the advancing side.

The magnitude of the $dD_{ind_{0.7}}/dL_{0.7}$ ratio is small, and under actual operating conditions, would probably be even smaller than in the present example. Consequently, the contribution of the blade element induced drag to the total drag of the rotor is usually neglected.

3.8 Propulsive Thrust and Power Required in Horizontal Flight

Let it be assumed that f stands for the equivalent flat-plate area representing parasite drag, D_{par}, of the helicopter as a whole, except for that contributed by the actual blades. This means that in addition to airframe drag, f would also represent the drag of the lifting rotor hub(s), including blade shanks and the entire tail rotor (if present).

In Sect 3.7, it was shown that contributions of the blade element induced drag to the rotor drag may usually be neglected. Hence, the total drag of a helicopter in forward flight will be:

$$D = D_{par} + D_{pr}$$

where $D_{par} = \frac{1}{2}\rho V^2 f$. For a helicopter with n identical rotors, the total drag becomes:

$$D = (1/2)\rho V^2 f + (3/8)n\rho\sigma\pi R^2 V_t^2 \mu\bar{c}_d. \tag{3.104}$$

Thrust Inclination. Thrust inclination in forward flight (within the validity of small angle assumptions) can now be expressed:

$$-a_v = D/W = [(1/2)\rho V^2 f + (3/8)n\rho\sigma\pi R^2 V_t^2 \mu\,\bar{c}_d]/W. \tag{3.105}$$

Defining $W/f \equiv w_f$ as the equivalent flat plate area loading and remembering that $W/\sigma\pi R^2 \equiv w_b$ is the blade loading, Eq (3.105) can be rewritten as follows:

$$-a_v = (1/2)\rho V_t^2\mu[(\mu/w_f) + (3/4)n(\sigma/w)\bar{c}_d]. \tag{3.105a}$$

or

$$-a_v = (1/2)\rho V_t^2\mu[(\mu/w_f) + (3/4)(\bar{c}_d/w_b)]. \tag{3.105b}$$

It is apparent from Eq (3.105b) that reduction of the blade area (high w_b) without a corresponding increase in the $\bar{c}_d$ values would be beneficial for reducing a_v. Reduction of the parasite drag (increase of w_f) is obviously always beneficial.

Total Rotor Power Required in Horizontal Translation. The total horsepower required by rotorcraft in the horizontal helicopter regime of flight will be the sum of the induced, profile, and parasite power. Thus, assuming that the total thrust of all n lifting rotors is $T = k_{v_{ho}}W$ and the actual disc loading is $w_{ho} = k_{v_{ho}}w$, where w is the nominal disc loading, $w \equiv W/A$, the total rotor power in horsepower (in SI units) becomes:

$$HP_R = (1/735)[(Wwk_{v_{ho}}^2 k_{ind_{ho}}/2\rho V_{ho}) + (1 + 4.7\mu^2)(n/8)\sigma\pi R^2 \rho V_t^3 \tilde{c}_d + \tfrac{1}{2}\rho V_{ho}^3 f] \quad (3.106)$$

In English units, the numerical coefficient preceding the brackets would be *550* instead of *735*.

It should be noted that it is often assumed—especially in preliminary-design calculations—that the vertical download coefficient $k_{v_{ho}} = 1.0$. For more details regarding the methods of determining $k_{v_{ho}}$ values, the reader is directed to Ch III of Vol II.

Assuming that transmission losses, power required for the tail rotor (if present), and various auxiliary equipment represent a fixed fraction of the required HP_R, the shaft horsepower, HP_S can be simply expressed as:

$$HP_S = HP_R/\eta_{tr} \quad (3.107)$$

where η_{tr}—the so-called transmission efficiency—reflects all of the above-mentioned losses, while the HP_R is given by Eq (3.106). For more details regarding transmission, accessory, and other losses, see Chs I–III of Vol II.

Eq (3.106) indicates that real rotorcraft would require additional power resulting from the blade profile drag, in contrast to the ideal helicopter considered in Ch II. Consequently, a graph of the power required components of real helicopters (Fig 3.25) would differ from that of the ideal rotor mounted on an airframe encountering parasite drag (Fig 2.12).

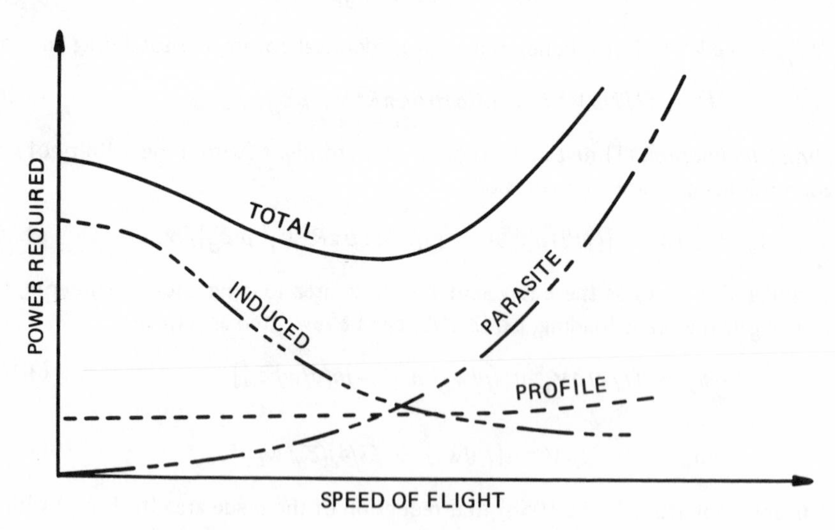

Figure 3.15 Power components of a real helicopter

3.9 Rotor and Helicopter Efficiency in Horizontal Flight

All of the significant components of power required in the horizontal helicopter-type regime of flight have been accounted for. Having this information, it may be desirable to develop a way of evaluating the degree of success in achieving a purely aerodynamic efficiency for the rotor system or, an overall excellence of the rotorcraft as a whole. The usefulness of the 'yardstick' used in such an evaluation would increase if it permitted one to grade the rotor systems and rotary-wing aircraft not only within their own groups, but also to compare them with other lifting systems, other aircraft, and even with various land and water vehicles as well. Some possible means for such a comparison are discussed in the following sections.

Weight- (or Lift)to-the-Equivalent-Drag Ratio. Determination of the gross weight of any vehicle to the equivalent-drag ratio (W/D_e) represents one of the possible ways of evaluating the overall efficiency of that vehicle in horizontal translation. For aircraft, it may usually be assumed that the lift is equal to the gross weight; hence, the term lift-to-the-equivalent-drag ratio (L/D_e) is often used when referring to the overall efficiency of the rotorcraft, or aerodynamic efficiency of the rotor alone.

The equivalent drag of vehicles using shaft-type engines can be based on the SHP required and defined as follows:

$$\left.\begin{array}{lll} \text{SI} & D_e = 735\,SHP/V_{m/s} & \text{N} \\ \text{SI} & D_e = 75\,SHP/V_{m/s} & \text{kG} \\ \text{Eng} & D_e = 550\,SHP/V_{fps} & \text{lb} \end{array}\right\} \quad (3.108)$$

where the speed of horizontal flight is in either m/s or fps. When the speed of flight is in km/h or in kn, it becomes

$$\left.\begin{array}{lll} \text{SI} & D_e = 204.2\,SHP/V_{km/h} & \text{N} \\ \text{SI} & D_e = 20.83\,SHP/V_{km/h} & \text{kG} \\ \text{Eng} & D_e = 325\,SHP/V_{kn} & \text{lb} \end{array}\right\} \quad (3.108a)$$

The weight-to-equivalent-drag ratio becomes

$$\left.\begin{array}{lll} \text{SI} & W/D_e = W\,V_{m/s}/735\,SHP & W \text{ in N} \\ \text{SI} & W/D_e = W\,V_{m/s}/75\,SHP & W \text{ in kG} \\ \text{Eng} & W/D_e = W\,V_{fps}/550\,SHP & W \text{ in lb} \end{array}\right\} \quad (3.109)$$

or

$$\left.\begin{array}{lll} \text{SI} & W/D_e = W\,V_{km/h}/204.2\,SHP & W \text{ in N} \\ \text{SI} & W/D_e = W\,V_{km/h}/20.83\,SHP & W \text{ in kG} \\ \text{Eng} & W/D_e = W\,V_{kn}/325\,SHP. & W \text{ in lb} \end{array}\right\} \quad (3.109a)$$

The typical shape of $W/D_e = f(V)$ for a helicopter is shown in Fig 3.26.

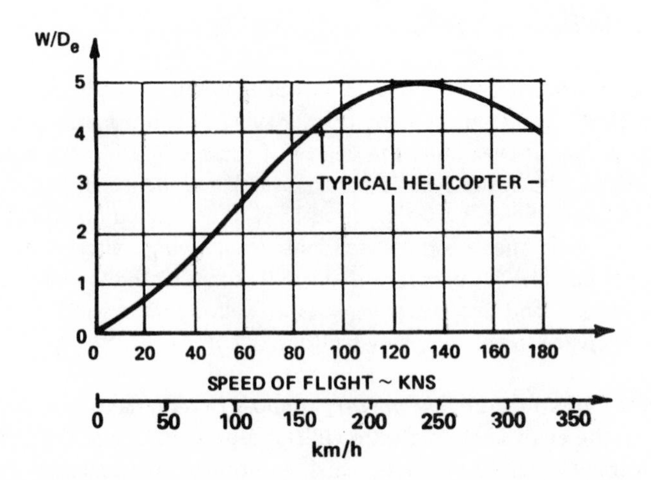

Figure 3.26 Typical character of W/D_e vs speed of flight

Eqs (3.109) and (3.109a) permit one to grade shaft-driven rotary-wing aircraft within their own group as well as to make a comparison with other types of aircraft and vehicles as long as they are powered by shaft-type engines. However, the definitions given by these equations are not suitable for jet-driven helicopters, jet-propelled aircraft, and vehicles in general which use other than shaft-type engines for propulsion. Therefore, as in the case of hover (Sect 2.8), expenditure of thermal energy instead of SHP may be used to provide a more general basis for comparison. In this approach, the ratio of weight-to-equivalent-drag based on the thermal power of vehicles using jet fuel is $(W/D_e)_{th}$. When V is in m/s or fps:

$$\text{SI} \qquad (W/D_e)_{th} = 1220\, tsfc / V_{m/s}$$

$$\text{Eng} \qquad (W/D_e)_{th} = 4000\, tsfc / V_{fps} \tag{3.110}$$

where *tsfc* is the hourly fuel consumption per unit of the gross weight of an aircraft or vehicle in general.

Specific Power. Horsepower required per unit of gross weight, when shown vs horizontal speed $(HP/W) = f(V_{ho})$ may also serve as one of the possible means of assessing the overall efficiency of rotorcraft, and can be used as a guideline in comparing them with other shaft-powered aircraft and vehicles in general. This HP/W ratio obviously has the dimension of velocity: (N-m/s)/N or (ft-lb/s)/lb. However, it may be more convenient to present this quantity in a nondimensional (coefficient) form.

The specific power coefficient $C_{P/T}$ (P/T in the subscript symbolizing the power-to-thrust, or gross-weight, ratio) can be obtained from the specific power (either P/T or P/W) by dividing that quantity by the ideal resultant velocity of flow through the disc (V'_{id} in m/s or fps) corresponding to the case of the actuator disc (ideal rotor) in horizontal translation in the absence of parasite drag. Under these assumptions, V'_{id} can be expressed according to the momentum theory as follows:

$$V'_{id} = \sqrt{v_{id}^2 + V_{ho}^2}$$

and consequently, with P in N m/s or ft-lb/s; T (or W) in N or lb; and V' in m/s or fps, the specific power coefficient becomes

$$C_{P/T} = P/T\sqrt{v_{id}^2 + V_{ho}^2}\,. \tag{3.111}$$

It is obvious that the so-defined $C_{P/T}$ can be based on (a) rotor power of an isolated rotor, (b) rotor power of a rotorcraft as a whole, or (c) shaft power of the aircraft.

A comparison of the actual $C_{P/T}$ values with ideal values can provide a measure of 'goodness' of an isolated rotor, or a rotary-wing aircraft as a whole, in forward flight. This approach is similar to the figure-of-merit, aerodynamic, and overall efficiency discussed for the case of hover in Sect 2.8.

For the ideal power required in horizontal flight (with hover as a limiting case) $P = Tv_{id}$, while v_{id} is expressed as in Eq (2.33) and then the rate of flow through the rotor becomes:

$$V'_{id} = v_{id_h}\sqrt{\tfrac{1}{2}\bar{V}_{ho}^2 + \sqrt{\tfrac{1}{4}\bar{V}_{ho}^4 + 1}} \tag{3.112}$$

and the ideal specific power coefficient, $(C_{P/T})_{id}$ expressed in terms of the nondimensional velocity in forward flight, $\bar{V}_{ho} \equiv V_{ho}/v_{id_h}$, is

$$(C_{P/T})_{id} = \sqrt{-\tfrac{1}{2}\bar{V}_{ho}^2 + \sqrt{\tfrac{1}{4}\bar{V}_{ho}^4 + 1}}\Big/\sqrt{\tfrac{1}{2}\bar{V}_{ho}^2 + \sqrt{\tfrac{1}{4}\bar{V}_{ho}^4 + 1}}\,. \tag{3.113}$$

$(C_{P/T})_{id} = f(\bar{V})$, computed from Eq (3.113), is shown in Fig 3.27.

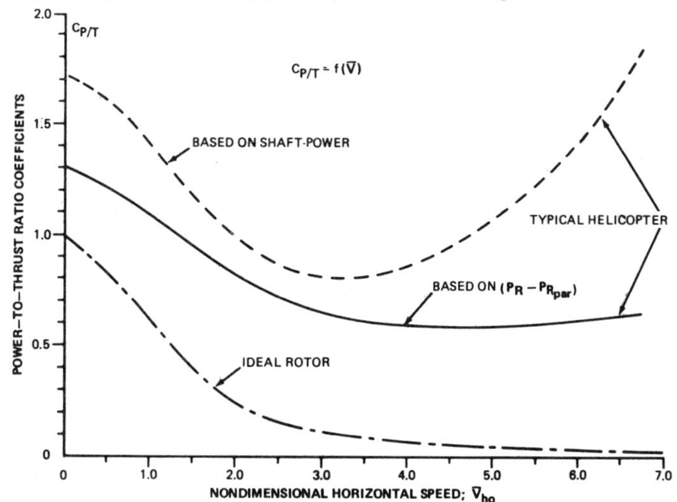

Figure 3.27 Examples of $C_{P/T}$ vs nondimensional velocity of forward flight

For comparison, the $C_{P/T}$ values for a typical helicopter based on both shaft and rotor power minus the parasite part, are also shown in this figure.

Figure-of-Merit or Efficiency Concepts in Horizontal Flight. Similar to the case of hover, the figure-of-merit (FM_{ho}) concept may be adapted to horizontal flight. It appears, however, that an alternate term of efficiency—with various qualifying adjectives—may be more descriptive than the figure-of-merit. This will become evident from the following examples.

Rotor figure-of-merit in its usual sense, also called aerodynamic efficiency $(\eta_a)_{ho}$, can be defined as:

$$\eta_{a_{ho}} \equiv FM_{ho} = (C_{P/T})_{id}/(C_{P/T})_R \tag{3.114}$$

where $(C_{P/T})_R$ is based on rotor power required, excluding all power expenditure for overcoming the parasite drag. A plot of $\eta_{a_{ho}}$ for a typical rotor system is shown in Fig 3.28. In hovering, Eq (3.114) becomes identical with Eq (3.55) previously developed in Sect 2.8.

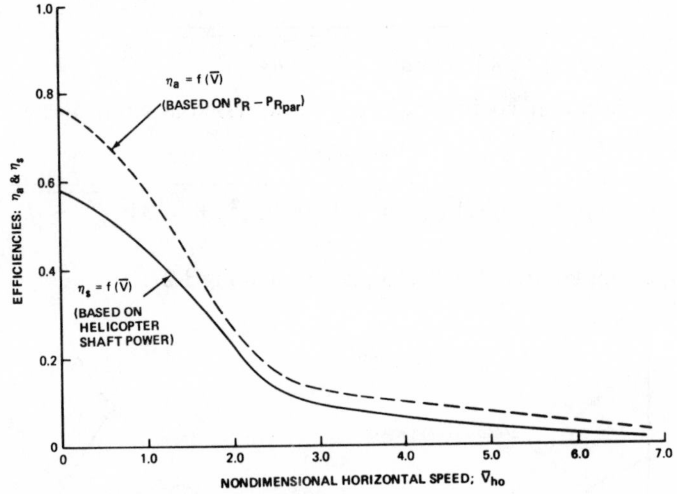

Figure 3.28 Examples of the aerodynamic and shaft power-based efficiencies

Shaft power-based efficiency, $\eta_{s_{ho}}$, may be defined in the following way:

$$\eta_{s_{ho}} \equiv FM_{s_{ho}} = (C_{P/T})_{id}/(C_{P/T})_S \tag{3.114a}$$

where $(C_{P/T})_S$ is computed on the basis of shaft power required by rotary-wing aircraft throughout the whole regime of horizontal flight and hover (Fig 3.28).

Total or overall efficiency, η_{th}, may represent the final generalization of the concept by basing $C_{P/T}$ values on the rate of expenditure of thermal energy:

$$\eta_{th} \equiv FM_{th} = (C_{T/P})_{id}/(C_{P/T})_{th}. \tag{3.114b}$$

By plotting the above defined efficiencies vs nondimensional speed of horizontal flight, one would be able to judge the extent to which a particular design approaches the ideal represented by the actuator disc. Furthermore, a means of comparing various rotary-wing aircraft in the helicopter regime of flight is obtained.

4. CONCLUDING REMARKS

The basic philosophy of the blade element theory, consisting of investigating aerodynamic phenomena occuring within a narrow chordwise strip of the blade, led to the development of a powerful, but simple, computational tool for the determination of forces and moments experienced by the blade as a whole. This approach became possible through the use of two-dimensional airfoil characteristics, reflecting not only blade section geometry (chord length and airfoil shape), but also such operational parameters as Reynolds and Mach numbers, and, if necessary, special aspects of unsteady aerodynamics.

Since the proper application of airfoil-section coefficients requires a complete definition of the flow—velocity magnitude and direction—in the immediate vicinity of the blade, knowledge of induced velocities at the blade becomes essential. The combined blade-element and momentum theory represents one of the possible methods for determining the induced velocity field of a rotor. However, it should be noted that the flow picture obtained in this way, although more realistic than that provided by the momentum theory, is still somewhat idealized. This is chiefly due to the fact that air movements associated with thrust generation are represented as time-average values.

If one would investigate the rotor wake with not-too-sensitive anemometers (pitot-static tubes for instance), one would probably find that indeed, the measured flows in axial translation under static conditions, and even in horizontal flight, are in good agreement with those predicted by the combined momentum blade-element theory; especially, inboard from the blade tip regions. However, should more sensitive velocity measuring devices such as hot-wire anemometers or laser beams be used, or the actual flow in the wake visualized by smoke or other means, then one would realize that the velocity field of the rotor is time-dependent and subject to various fluctuations.

The combined blade-element and momentum theory does not explain many of the aerodynamic phenomena. It is also of no help when dealing with such problems as tip losses, and of very little assistance in investigating the influence of the number of blades per rotor, or their index angle, on induced power. With the exception of the last two problems, limitations of the combined blade-element and momentum theory are not too important from a performance point-of-view, but may become significant as far as predictions of blade air loads, vibratory excitations, and understanding noise generation are concerned.

The so-called vortex theories, which will be reviewed in the following chapter, should provide a more suitable physical representation of time-dependent aerodynamic events, and will describe in more detail mutual blade interference, and give a clue as to the blade-load variation in the tip region.

Nevertheless, there is still a place in rotary-wing aerodynamics for the combined blade-element and momentum theory, since it provides the designer with a simple means of investigating the influence of important design parameters on performance.

In the pure momentum approach, it was recognized that the disc loading was the only significant design parameter. By contrast, the blade-element theory permits one to investigate the influence of such quantities as tip speed, blade loading, blade airfoil characteristics, blade planform, and blade twist distribution. This allows the designer to examine the interplay between these parameters; which becomes especially useful in establishing general design and performance trends and philosophies, thus opening the way toward design optimization.

References for Chapter III

1. Glauert, H. *Airplane Propellers.* Durand's Aerodynamic Theory. Vol IV, Div L. Durand Reprinting Committee, CIT, 1943.

2. Klemin, A. *Principles of Rotary-Wing Aircraft.* AERO Digest, May 1 and June 1, 1945.

3. Sissingh, G. *Contribution to the Aerodynamics of Rotary-Wing Aircraft.* NACA TM No. 921, 1939.

4. Wald, Q. *The Effect of Planform on Static Thrust Performance.* Sikorsky Aircraft, SER 442, November 1944.

5. Stepniewski, W. Z. *A Simplified Approach to the Aerodynamic Rotor Interference of Tandem Helicopters.* Proceedings of the West Coast AHS Forum, September, 1955.

6. Stepniewski, W.Z. *Suggested Ways of Comparing Overall Efficiency of Helicopters.* Journal of Aeronautical Sciences, Vol 19, June 1952.

7. Heyson, H.H. *Preliminary Results from Flow-Field Measurements Around Single and Tandem Rotors in the Langley Full-Scale Tunnel.* NACA TN 3242, November 1954.

8. Hoerner, S.F. *Aerodynamic Drag.* Published by the Author, 1952.

VORTEX THEORY

The sequence of the material presented in this chapter reflects, to some extent, the chronology of vortex theory development. The basic properties of vortices in an ideal fluid—as determined by the Biot-Savart, Helmholtz, and Kelvin laws—are considered first, followed by the early application of simple vortex systems to modeling of hovering rotors having an infinite number of blades. Consideration of the horizontal translation of a lifting airscrew with a flat wake serves as an introduction to contemporary applications of the vortex theory to rotary-wing aerodynamics. Hover and vertical climb are re-examined for cases reflecting a finite number of blades modeled by single vortices or vorticity surfaces, while the wake is either rigid or is free to form its own shape. These approaches are later extended to studies of forward flight. An examination of the importance of such real-fluid properties as compressibility and viscosity concludes this presentation of the proper vortex theory. Finally, the so-called local momentum theory, although seemingly belonging to the preceding material, is added as an appendix to this chapter for the reason that some basic knowledge of vortex mechanics is needed for an understanding of this particular approach.

Principal notation for Chapter IV

AR	aspect ratio	
a	lift curve slope	rad^{-1} or deg^{-1}
b	number of blades	
C_T	rotor thrust coefficient	
c	blade, or wing chord	m or ft
c_ℓ	section lift coefficient	
d	distance	m or ft
e	vortex core radius	m or ft
$\vec{i}, \vec{j}, \vec{k}$	unit vectors in Cartesian coordinates	
K_u, K_v, K_w	iteration factors	
ℓ	lift per unit length	N/m or lb/ft
M	Mach number	
M	moment	N- m or ft-lb
N	exponent	
$\mathbf{O}$	Biot-Savart operator	
p	pressure	N/m^2 or psf

p	parameter	
R	rotor radius	m or ft
r	radial distance	m or ft
$\bar{r}$	nondimensional radial distance; $\bar{r} \equiv r/R$	
$\vec{r}$	position vector	m or ft
s	speed of sound	m/s or fps
s	distance along a curve	m or ft
T	thrust	N or lb
t	time	s
U	velocity of flow approaching the vortex	m/s or fps
U, V, W	influence functions	
u, v, w	induced velocity components along x, y, z axes	m/s or fps
V	velocity	m/s or fps
$\vec{v}$	total induced velocity	m/s or fps
x, y, z	Cartesian coordinates	m or ft
$\tilde{x}, \tilde{y}, \tilde{z}$	$\tilde{x} \equiv x/r$; $\tilde{y} \equiv y/r$; $\tilde{z} \equiv z/r$	
α	angle-of-attack	rad or deg
β	flapping angle	rad or deg
Γ	circulation	m^2/s or ft^2/s
γ	circulation per unit length	m/s or ft/s
Δ	increment	
δ	small increment	
θ	blade section pitch angle (to zero-lift chord)	rad or deg
ϑ	angle as defined	rad or deg
Λ	angle as defined	rad or deg
λ	inflow ratio; $\lambda \equiv V_{ax}/V_t$	
μ	rotor advance ratio; $\mu \equiv V_{inp}/V_t$	
ξ, η, ζ	ordinates	m or ft
ρ	air density	kg/m^3 or $slugs/ft^3$
σ	rotor solidity ratio	
σ_{ej}	induced velocity influence coefficient	
τ	time	s
ϕ	inflow angle	rad or deg
φ	velocity potential	m^2/s or ft^2/s

ψ	blade azimuth angle	rad or deg
Ω	rotor rotational velocity	rad/s
ω	solid angle	steradian

Subscripts

av	average
b	blade
c	curvature
f	final
i	initial
ℓ	lower
r or $\bar{r}$	at station r or $\bar{r}$
t	tip
tg	tangential
u	upper
β	due to flapping
$\parallel$	parallel
$\perp$	perpendicular

Superscripts

$\rightarrow$	vector
$-$	nondimensional
$\sim$	average, or as defined
$\wedge$	as defined

1. INTRODUCTION

The simple momentum and the combined blade element-momentum theories permit one to investigate aerodynamic events created by the presence of a lifting or propelling airscrew only within the confines of the slipstream. Outside of this limited sphere of influence, the fluid—whether flowing or stationary (hovering)—is assumed to be completely unaffected by the presence of the airscrew. This is a serious conceptual limitation as one would like to be able to investigate the flow fields induced by a rotor within the whole unlimited volume of fluid without apriori space restrictions—and only then decide whether some regions of the surrounding space should be eliminated from the investigation.

At the beginning of this century, Joukowsky and Kutta proved that lift generation was related to the presence of a vortex exposed to a flow having a velocity component perpendicular to the vortex filament and thus opened the way for modeling both fixed and rotating wings through vortices. This approach eliminated the slipstream-only restrictions of the momentum and blade-element approaches since, by analogy with electromagnetic induction, the influence of a vortex in an ideal fluid is unlimited, even though according to the Biot-Savart law, the strength of that influence decreases with distance.

An additional benefit resulting from the application of vortices to airscrew modeling was the possibility of examining time-dependent flows—an aspect that was missing in the theories discussed in the two preceding chapters.

The actual application of the vortex concept to modeling of airscrews started with the development of the 'vortex theory' whose foundations were laid by Joukowsky[1] in the USSR in the late teens and early twenties, and in the West by Goldstein[2] in the twenties. Joukowsky was also the first (late teens) to apply the vortex concept to the solution of the problem of a hovering rotor with an infinite number of blades. The same task was considered much later (1937) in the West by Knight and Hefner[3]. However, the ever-growing application of the vortex theory to rotary-wing aerodynamics began in the fifties. In the West, the works of Gray, Landgrebe, Loewy, Miller, Piziali and others paved the way for that growth. Probably most of the important contributions in that domain were listed and summarized by Landgrebe and Cheney[4]. In the USSR, there was also a large number of researchers who, by pursuing an independent, although somewhat parallel, course contributed to further development of the vortex theory. In that respect, the names of Baskin, Vil'dgrube, Vozhdayev and Maykapar come to one's mind. Their most significant efforts, with those of many others, have been collected in book form[1].

To gradually introduce the reader to the more complex aspects of the vortex theory, the material in this chapter is, in principle, presented along the lines of the historic development of that theory. After a brief review of the Biot-Savart law, the simplest concept of the wake as represented by a rotor with an infinite number of blades is considered first. Then more and more sophisticated physicomathematical models are examined and the following schemes in particular will be considered in an incompressible fluid: (a) those based on linearized theory (rigid wake), where it is assumed that first tip vortices and then all the fluid elements forming the vortex sheet move rectilinearly with a uniform velocity; (b) those incorporating corrections for the deformation of the wake caused by slipstream contraction (especially in hover and low-speed axial translation) which are called quasi-linearized, or semirigid; and (c) those based on the concept of nonlinear interactions among the vortices. This latter approach is also called the free-wake concept. An examination of the role of viscosity and compressibility of the fluid concludes this chapter.

The material presented in this way should enable the reader to follow the philosophy of the airload and performance prediction computer programs in Vol II (Sect 1 of Ch II, and 3.2 of Ch III) as well as to be prepared to study the more advanced original papers now appearing in ever-increasing numbers. For those who are especially interested in theoretical aspects of the application of the vortex theory to rotary-wing aerodynamics, the book by Baskin et al[1]—frequently referred to in this chapter—is highly recommended.

2. GENERAL PRINCIPLES OF ROTOR MODELING BY VORTICES

In an incompressible medium, the vector field of induced velocity can be completely determined with the help of a suitable vortex system. However, to achieve this goal in a compressible fluid, one has to rely on an additional field of sources.

Fortunately, most of the problems encountered in rotary-wing aerodynamics can be treated as being incompressible. Consequently, classical expressions of the Biot-Savart law can usually be applied when establishing the relationships between the strength and geometry of vortices $\Gamma = f(x,y,z)$ (defined by the strength of circulation and shape of the filament), and velocities $\vec{v} = f(x,y,z)$ induced by them in the surrounding fluid:

$$\vec{v} = f(\vec{\Gamma}).$$

The basic Biot-Savart law can be modified for those cases where compressibility should not be ignored; and thus, within the limits of linearized theory, one can obtain an induced velocity vector field from the known vortex system without having to resort to an additional field of sources[1].

Incorporation of the vortex theory into the creation of physicomathematical models of airscrews in various regimes of operation opened new possibilities for a more precise treatment of the time-average flow phenomena. However, it proved especially valuable for consideration of instantaneous flows which could not be handled by the airscrew model concepts based on the simple momentum and combined momentum and blade-element theories.

Many design and analytical problems can now be attacked with the help of the approaches offered by the vortex theory; for instance, the significance of the number of blades, tip losses, upwash, impulsive loading, impulsive noise generation, etc. Furthermore, when one begins to think about large rotors with a small number of blades rotating at 2 rps or even slower, the whole concept of a continuous steady flow within a well-defined streamtube which may be acceptable for a rotor configuration with a large number of blades (Fig 4.1a), does not appear to properly represent the physical reality (Fig 4.1b).

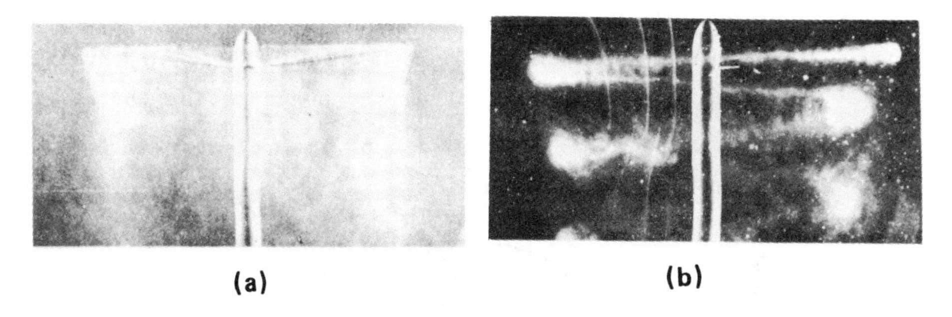

Figure 4.1 Examples of flow visualization for (a) six-bladed, and (b) one-bladed rotor in hover[5]

As to the structure of the complete vortex system of an airscrew, theoretical considerations postulate—and various visualization techniques such as smoke in air[5], dyes and bubbles in water tunnels[6], and laser[7,8] tend to confirm—that such a system can be represented by (a) the so-called bound vortices which are attached to the blade and are directed along their longitudinal axes, and (b) free vortices which actually form the wake. In the latter category one may distinguish, in turn, the so-called shed vortices which, at the moment of leaving the blade, are parallel to its axis, and trailing vortices springing along the blade span in the original direction either perpendicular or approximately perpendicular to the blade axis. Among the trailing vortices, those leaving the blade tips—the so-called tip vortices—usually dominate the flow picture in all regimes of airscrew operation (Figs 4.1 and 4.2).

Figure 4.2 Example of flow visualization of a rotor in nonaxial translation[6]

It is therefore understandable why the early physicomathematical models of the rotor had logical structures built around the tip vortices alone. However, as indicated by Gray in Fig 4.3[4], the complete vortex system of a real airscrew is more complicated since the shed and trailing vortices may form a surface of vorticity of a generally helical shape, called the vortex sheet, behind each blade. This surface has an axis that is either perpendicular (axial flow) or skewed (flow with an inplane component) with respect to the airscrew disc. In limiting cases of inflow with relatively high inplane velocity components, it may be assumed that the whole wake degenerates into a flat ribbon of vorticity.

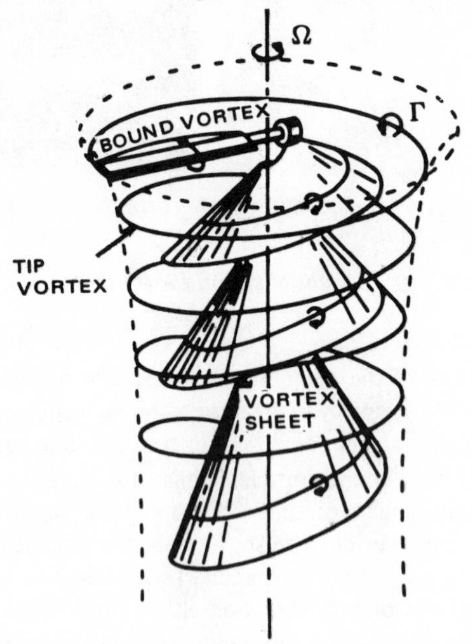

Figure 4.3 Scheme of the Wake structure for a single blade in hover[4]

All vortex surfaces represent a discontinuity in the flow field, due to the rotational motion of the fluid particles. However, outside of these surfaces, the flow can usually be considered as irrotational, or potential.

Should the wake of a lift generator remain invariable with time*, then the perturbations (velocities) induced in the fluid flowing past it could be computed with clear conscience using the Biot-Savart law which does not contain any provision for the time required to transmit a signal from vortices to the point where the induced velocity is determined. However, in reality, because of the interaction between vortices, flow fluctuations, and various maneuvers, the geometry of the wake, and strength of vortices forming that wake may vary with time. Furthermore, even when the wake stays basically invariable, the distances of various blade stations from the vortices may change with time. For instance, this would be the case in determining induced velocity at a point located in the coordinate system rotating with the blade (say at some station along its span at a given position along the chord). Consequently, noninstantaneous transmittal of aerodynamic signals might have some significance.

Examination of flow visualization pictures (e.g., Figs 4.1b and 4.2) focus one's attention on another property of real fluids; namely viscosity. It can be clearly seen from these figures that due to its influence, vortices dissipate downstream in the wake. This phenomena has been examined for both fixed-wing[9] and rotary-wing[10] aircraft. Although the process of vortex dissipation is quite complicated, its significance should be kept in mind when airscrew physicomathematical models based on the vortex theory are developed. At this point, it may be added that some first-order corrections related to the existence of viscosity can be directly incorporated by properly modifying the classic Biot-Savart relationships[1].

3. VORTICES IN IDEAL FLUID

3.1 Basic Laws

Single Filaments. All early attempts in the development of the vortex theory were based on the classic approach, treating air as an incompressible fluid. In addition, all the assumptions of classical hydro and aero mechanics regarding vortices were retained. In any textbook on this subject (e.g., Ref 11), one can find that fluid motion associated with the existence of a vortex can be broken down into (a) rotation of the infinitesimally narrow vortex core (filament), and (b) irrotational (potential) flow outside of the core itself. Two-dimensional flow induced by an infinitely long straight vortex of strength Γ occurs in concentric circles, and tangential velocity (v_{tg}) at any given point of a circle of radius r is constant and equal to

$$v_{tg} = \Gamma/2\pi r. \tag{4.1}$$

To avoid infinite v_{tg} values at the vortex center, it is assumed, even in the classic theory, that the core has a finite cross-section. Consequently, v_{tg} increases only to its finite maximum value at the border of the core, which is assumed to rotate as a rigid

*Unstalled fixed wing, or a rotor with an infinite number of blades may represent such a case.

body, and the character of the tangential velocity distribution around the vortex may be expected to be as shown in Fig 4.4.

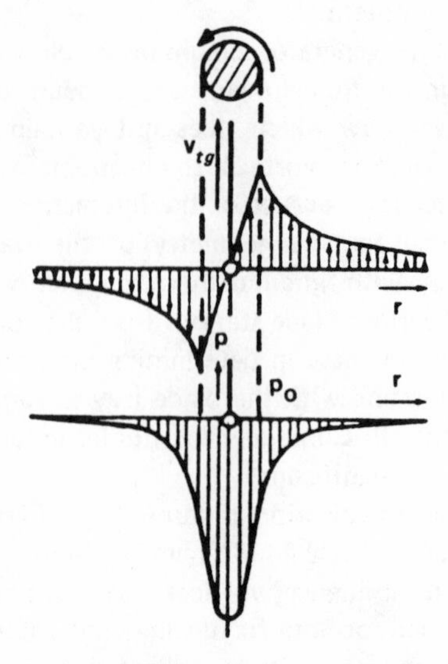

Figure 4.4 Velocity and pressure distribution in the interior and neighborhood of a rectilinear vortex[11]

The behavior of vortex filaments in an ideal fluid is governed by the following theorems of Helmholtz and Kelvin:

1. "No fluid can have a rotation if it did not originally rotate."

2. "Fluid particles which at any time are part of a vortex line (filament) always belong to the same vortex line."

3. "Vortex filaments must be either closed lines or end on the boundaries of the fluid."

As to the strength of circulation, Eq (4.1) indicates that

$$\Gamma = 2\pi r v_{tg}.$$

But the right-hand side of this equation can be interpreted as a line integral

$$\Gamma = \int^{C} v_{tg} ds$$

where v_{tg} is the tangential velocity corresponding to any radius r, and the integration is performed along the circle of that particular radius.

This relationship can be generalized; thus one can determine the circulation inside a region bordered by any closed curve (C) which contains a single vortex filament, or a system of vortex filaments:

$$\Gamma = \int^{C} v \cos \vartheta \, ds \tag{4.2}$$

where ϑ is the angle between the velocity vector and the tangent of the path along which the circulation is computed.

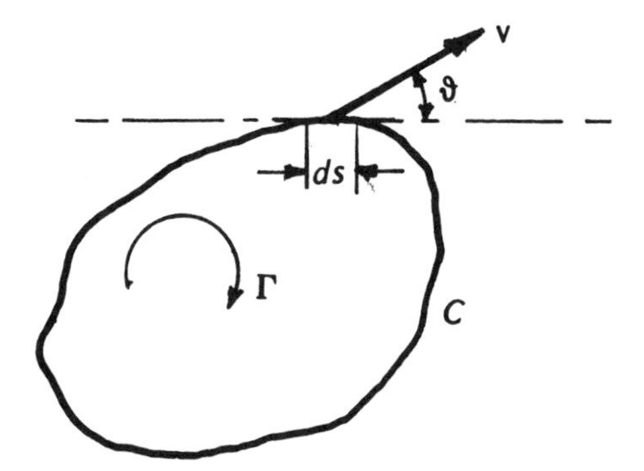

Figure 4.5 Determination of circulation

Consequently, if the velocity distribution within a particular region of a two-dimensional flow is known, the associated circulation may be found by the use of Eq (4.2).

Vortex Surface. When vortex filaments are very close to each other, they may be considered as forming a continuous surface of vorticity (vortex sheet). Denoting the circulation per unit width of the sheet as γ, the relationship between its value and that of the tangential velocity component just above v_u and below v_ℓ the vortex sheet can be found using the principle expressed by Eq (4.2).

Circulation $\delta\Gamma$ associated with a δs-wide element of the vortex sheet can be determined as a line integral of velocity taken around that element (Fig 4.6). Since δs is small, it may be assumed that the velocity components perpendicular to the sheet are the same at both ends of the element. Thus, their contributions to the line integral would cancel each other, and the resulting value would be

$$\delta\Gamma = v_u\,\delta s - v_\ell\,\delta s;$$

but $\gamma = \delta\Gamma/\delta s$, hence

$$\gamma = v_u - v_\ell. \tag{4.3}$$

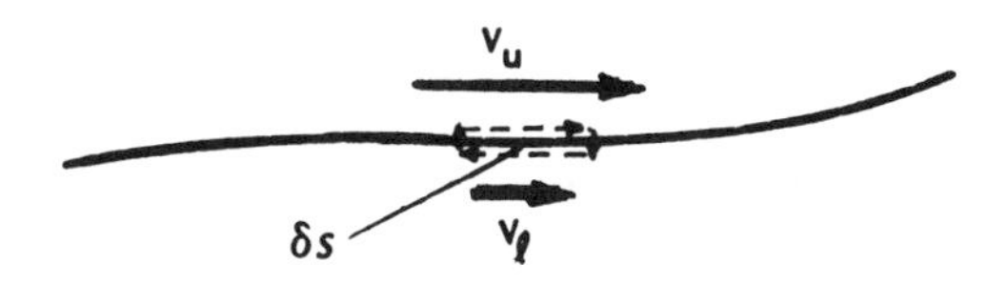

Figure 4.6 Circulation associated with a vortex sheet

Eq (4.3) shows that a vortex sheet represents a surface of discontinuity of the tangential component of the velocity of flow, and the strength of circulation per unit width of the sheet (γ) represents the amount of that discontinuity.

3.2 Biot-Savart Law

Three-Dimensional Vortex. The Biot-Savart law is one of the principal tools for determination of the flow field induced by a system of vortex filaments. In vectorial notations, the elementary velocity $d\vec{v}$ induced at a point P by an element $d\vec{s}$ belonging to vortex filament C of strength Γ is expressed as follows (Fig 4.7):

$$d\vec{v} = (\Gamma/4\pi)[(d\vec{s} \times \vec{d})/d^3], \tag{4.4}$$

while the total induced velocity (v) at that point becomes

$$\vec{v} = (\Gamma/4\pi) \oint^{C} [(d\vec{s} \times \vec{d})/d^3] \tag{4.4a}$$

where the integral sign indicates that an integration is carried along the line of filament C, $d\vec{s}$ represents an element of that filament, and $\vec{d}$ is the distance between the point in space where we want to determine velocity $\vec{v}$ and the element $d\vec{s}$. The latter is so directed on the filament C that looking along $d\vec{s}$ one should see the circulation around C in the clockwise direction. The cross-product $d\vec{s} \times \vec{d}$ has a value of $ds\, d \sin(\phi)$ and has a direction perpendicular to a plane defined by $d\vec{s}$ and $\vec{d}$, while ϕ denotes an angle between the vectors. In Prandtl's *Fundamentals*[11], the following was used to describe $\vec{v}$:

> The velocity $\vec{v}$ is obtained by adding together the contributions of the individual filament element $d\vec{s}$, and the contribution of this element is perpendicular to $d\vec{s}$ and $\vec{d}$, and is inversely proportional to the square of the distance d from the point in question. This, however, is exactly the law of Biot and Savart in electrodynamics from which the magnetic field in the neighborhood of a current-carrying wire can be calculated.

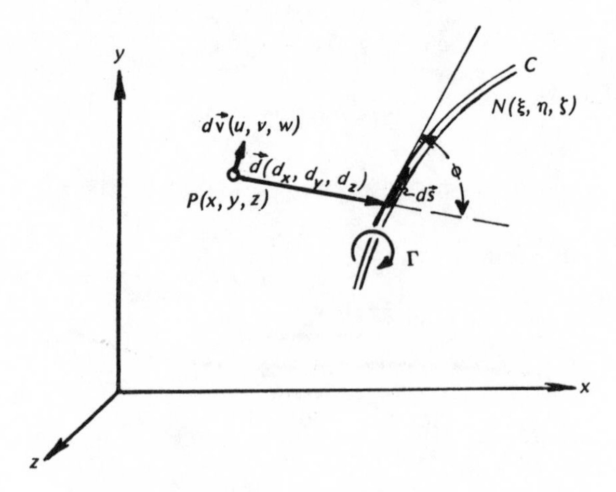

Figure 4.7 Induction of velocity $\vec{v}$ at a point P

Going from vectorial notations to Cartesian coordinates, vector $\vec{v}$ can be resolved into components $(\vec{u}, \vec{v}, \vec{w})$ along the x, y, and z axes of that coordinate system. Baskin, et al[1], outlined the following procedure: Let the equation of line C (Fig 4.7) be given in a parametric form:

$$\xi = \xi(p), \; \eta = \eta(p), \; \zeta = \zeta(p) \tag{4.5}$$

where p is the parameter (usually an angle for a curved vortex filament). As parameter p varies from its initial (p_i) to its final value (p_f), point $N(\xi, \eta, \zeta)$ describes curve C. Now, vectors $\vec{d}$, and $\vec{ds}$ (which may be considered a derivative of $\vec{d}$), can be expressed in light of Eq (4.5) as follows:

$$\begin{aligned} \vec{d} &= (\xi - x)\vec{i} + (\eta - y)\vec{j} + (\zeta - z)\vec{k} \\ ds &= \left(\frac{d\xi}{dp}dp\right)\vec{i} + \left(\frac{d\eta}{dp}dp\right)\vec{j} + \left(\frac{d\zeta}{dp}dp\right)\vec{k} \end{aligned} \tag{4.6}$$

where $\vec{i}$, $\vec{j}$, and $\vec{k}$ are unit vectors of the coordinate sytem x, y, z.

Introducing the above expressions into Eq (4.4a), the x, y, z components of the induced velocity vector are obtained:

$$u = \frac{\Gamma}{4\pi}\int_{p_i}^{p_f}\left[\frac{d\eta}{dp}(z-\zeta) - \frac{d\zeta}{dp}(y-\eta)\right]\frac{dp}{d^3} \tag{4.7}$$

$$v = \frac{\Gamma}{4\pi}\int_{p_i}^{p_f}\left[\frac{d\zeta}{dp}(x-\xi) - \frac{d\xi}{dp}(z-\zeta)\right]\frac{dp}{d^3} \tag{4.8}$$

$$w = \frac{\Gamma}{4\pi}\int_{p_i}^{p_f}\left[\frac{d\xi}{dp}(y-\eta) - \frac{d\eta}{dp}(x-\xi)\right]\frac{dp}{d^3} \tag{4.9}$$

where

$$d = \sqrt{(x - \xi)^2 + (y - \eta)^2 + (z - \zeta)^2}. \tag{4.10}$$

Vortex in a Plane. If the vortex filament lies in a plane, Eq (4.4a) can be simplified as follows:

$$v = \frac{\Gamma}{4\pi}\oint^{C} \frac{ds}{d^2} \sin \phi. \tag{4.11}$$

When the actual shape of a vortex filament can be approximated by linear segments (as for instance, AB in Fig 4.8), Eq (4.4) may be modified to a form more convenient for finding, in practice, the induced velocity increment (Δv) corresponding to this particular segment of the vortex filament.

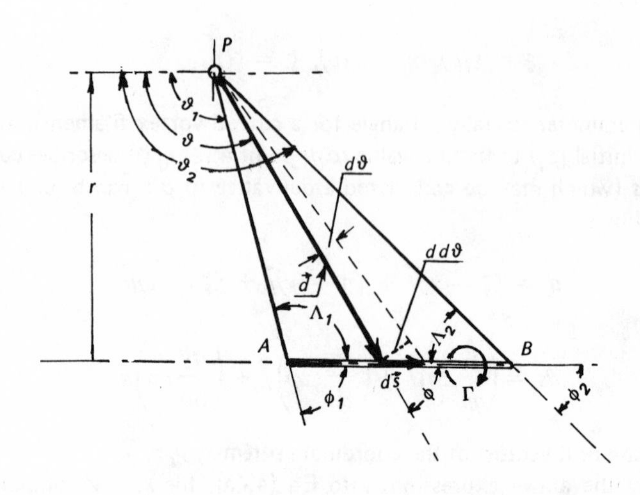

Figure 4.8 Increment of velocity Δv induced (up from the paper plane) by vortex segment A-B

It can be seen from this figure that $ds = d\,d\vartheta/\sin\phi$ and $d = r/\sin\phi$, where r is the distance of point P from the AB axis. Consequently, ds can be expressed as $ds = (r/\sin^2\phi)d\vartheta$. Substituting the above determined values for ds and d, Eq (4.11) becomes

$$\Delta v = (\Gamma/4\pi r)\int^{AB} \sin\phi\, d\vartheta,$$

but since $\vartheta = \pi - \phi$ and $\sin\phi = \sin\vartheta$, the above equation can be rewritten within limits of integration from $\vartheta_1 = \pi - \phi_1$ to $\vartheta_2 = \pi - \phi_2$:

$$\Delta v = (\Gamma/4\pi r)\int_{\vartheta_1}^{\vartheta_2} \sin\vartheta\, d\vartheta. \tag{4.12}$$

Performing the above integration and remembering the relationships between ϑ_1 and ϑ_2, and ϕ_1 and ϕ_2, the following is obtained:

$$\Delta v = (\Gamma/4\pi r)[\cos\phi_2 + \cos(\pi - \phi_1)].$$

Looking at Fig 4.8 one would note that $\pi - \phi_1 = \Lambda_1$ and $\phi_2 = \Lambda_2$, and the above equation can be rewritten as follows:

$$\Delta v = (\Gamma/4\pi r)(\cos\Lambda_1 + \cos\Lambda_2). \tag{4.13}$$

When the vortex filament AB becomes very long; i.e., when it extends from $-\infty$ to ∞, then $\Lambda_1 = \Lambda_2 = 0$, and Eq (4.13) is reduced to Eq (4.1).

Interaction between vortices. It should be noted at this point that although vortices induce velocities within the fluid they, in turn, may be subjected to the action of the fluid. According to Helmholtz's theorem, there is no exchange of either mass or momentum between the vortex core (filament) and the rest of the fluid; hence, if a vortex filament were located in the mass of moving fluid, it would move with the fluid. This obviously means that velocity fields induced by a system of vortices can produce reciprocal motion of those vortices belonging to the system. Thus, in an incompressible fluid, the classic Biot-Savart law should, in principle, provide a sufficient tool for determining from an initially given system of vortices, not only the associated field of induced velocities, but the resulting motion, or perturbation of motion, of the vortices themselves.

3.3 Kutta-Joukowsky Law

Circulation and Lift. The Kutta-Joukowsky law represents another important tool in the application of the classic vortex theory. It permits one to establish a relationship between the loads on the blades and strength (circulation Γ) of bound vortices. This law states that lift (ℓ) per unit span of a blade (or wing in general)[11] can be expressed as:

$$\ell = \rho U_{\perp} \Gamma \tag{4.14}$$

where Γ is the strength of circulation around the considered blade section, ρ is the air density, and $U_{\perp}$ is the air velocity component perpendicular to the vortex filament at that section.

If c is the chord and c_{ℓ} is the section lift coefficient at the considered station, then Eq (4.14) can be rewritten as:

$$c_{\ell} = 2\Gamma/U_{\perp}c \tag{4.15}$$

or

$$\Gamma = \tfrac{1}{2} c_{\ell} U_{\perp} c. \tag{4.16}$$

Trailing Vortices. Considering the Helmholtz and Kutta-Joukowsky laws, Eq (4.16) would indicate that whenever the value of the product of $c_{\ell} U_{\perp} c$ varies along the rotor blades, then over a length dr, a vortex filament of strength $d\Gamma = -(\partial\Gamma/\partial r)\, dr$ should leave the blade. By the same token, if the circulation around the blade remains constant along its entire span, then the vortex bound to the blade would leave it at the tip and root only. In this case, in hovering or vertical ascent, the vortex system would resemble that shown in Fig 4.9, where vortices springing out from the tip form an approximately helical line (also see Fig 4.1b).

As to the root vortex, it is apparent that since Ωr is zero at the rotor axis, the only remaining velocity (in the case of vertical flight or hovering) is axial. Hence, it is not difficult to imagine that each vortex leaving the blade at the root combines with those of the other blades to form one common vortex line along the rotor axis.

Constant Blade Circulation and C_T. In hovering or axial translation at speeds such that velocity ($U_{\perp \bar{r}}$) experienced at various blade stations $\bar{r}$ (where $\bar{r} \equiv r/R$) may be considered as $U_{\perp \bar{r}} \approx V_t \bar{r}$, it is easy to express the constant blade circulation in terms of the

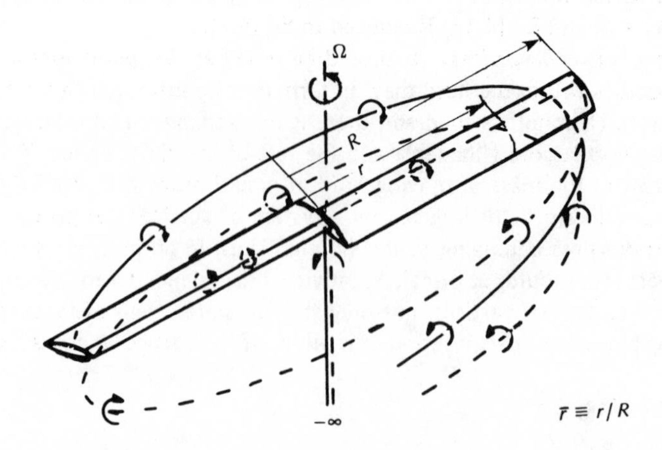

Figure 4.9 Scheme of vortex system in axial translation for constant circulation along the blade

rotor thrust coefficient (C_T), tip speed (V_t), rotor radius (R), and number of blades (b).

Eq (4.14) indicates that for $U\perp_r = V_t\bar{r}$, the load distribution along the blade is triangular, reaching its maximum value per unit of blade length of $\ell_{max} = \Gamma_b \rho V_t$ at the tip, where Γ_b is the circulation along the blade. The total load (thrust) per blade T_b will hence, be $T_b = \Gamma_b \rho V_t R/2$. However, $T_b = T/b$ where T is the total thrust. Consequently,

$$\Gamma_b = 2T/b\rho V_t R. \tag{4.17}$$

Multiplying the numerator and denominator of the above expression by $\pi R V_t$, we obtain

$$\Gamma_b = 2C_T \pi R V_t/b. \tag{4.17a}$$

4. ELEMENTARY CONSIDERATIONS OF HOVERING ROTORS

4.1 Rotor with a Single Cylindrical Wake

Assumptions. The task consists of determining induced velocities at the disc of a rotor hovering far from the ground. To solve this problem, it is assumed that the rotor has a very large (infinite) number of blades. This implies that distances between the consecutive vortex helixes are so small that instead of having either helical vortex sheets or helixes of tip vortices corresponding to the number of blades, the whole wake below the rotor may be considered as filled with vorticity. It is further assumed that (a) circulation along the blade is constant, and (b) there is no slipstream contraction. Consequently, the whole vortex wake would consist of a single cylinder of vorticity having radius R, and extend downstream from the rotor disc ($y_1 = 0$) to infinity ($y_2 = -\infty$) (physical concept somewhat similar to that of Fig 4.1a). Vortices bound to the rotor blades and the root vortices extending along the cylinder axis would complete the vortex system of the rotor (Fig 4.10).

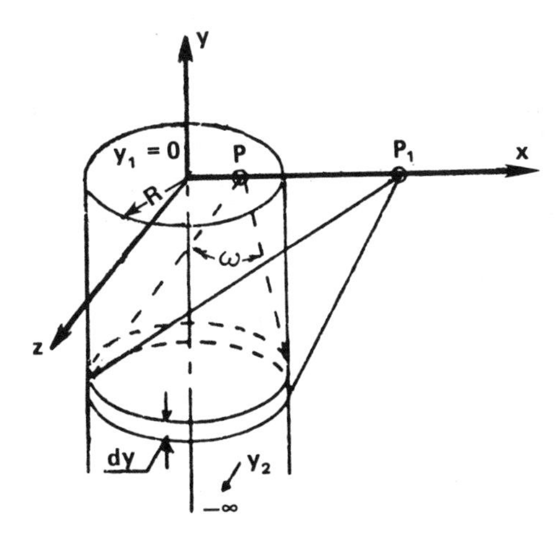

Figure 4.10 Scheme of the cylindrical wake of a stationary rotor

Coordinate System. The Cartesian *xyz* coordinate system shown in this figure will be used in most of the cases considered in this chapter. It can be seen that the y axis is directed in the thrust (vertical) direction. This is done in order to retain the traditional v symbol for the induced velocity component (downwash) directly associated with thrust generation. In forward flight regimes, the x axis will coincide with the direction of rotor displacement with respect to the still air mass.

Analysis. As to the contribution of various vortices to the downwash induced at the rotor disc, it can be seen that neither root nor bound vortices would have any influence in that respect. Root vortices (distributed along the y axis), or those representing any y component (if present) on the surface of the cylindrical wake, being perpendicular to the rotor disc cannot induce any velocities in the direction normal to the disc plane. They can only contribute to the rotation of the slipstream.

It is easy to show that the contribution of the bound vortices to the creation of downwash in the plane of the disc (under the assumption of a large number of blades) will also be zero. By selecting an arbitrary point on the disc, it can be seen that the tendency to create downwash by the sum of all the vortices located to one side of a line drawn through the chosen point and the center of the disc will be exactly equal in magnitude but opposite in sign to the sum of those located at the other side of the line. It is obvious, hence, that under these assumptions, no downwash can be created by the bound vortices. Consequently, only the vortices forming circles parallel to the rotor disc on the cylindrical wake can contribute to the generation of the induced velocity normal to the rotor plane[3].

The Biot-Savart law indicates that at a point P, the velocity potential φ_P corresponding to a closed vortex filament of strength Γ will be (Ref 11, p. 104):

$$\varphi_P = (\Gamma/4\pi)\omega$$

where ω is the solid angle subtended by that closed vortex line. The velocity (v_P) induced at that point will be:

$$\vec{v}_P = \overrightarrow{grad}\,\varphi_P = (1/4\pi)\,\overrightarrow{grad}\,\Gamma\omega. \tag{4.18}$$

In view of the above, the induced velocity (i.e., that in the direction of the y axis) will be obtained by differentiating $\Gamma\omega$ with respect to y. However, under the assumption of a cylindrical wake with no slipstream contraction, the intensity of circulation per unit of length in the y direction ($\gamma \equiv \delta\Gamma/\delta y$) remains unchanged, which also means that $d\Gamma/dy = const.$

It may be assumed that between the two coordinates y_2 and y_1, the gradient of angle ω can simply be expressed as the difference in $\omega(y_2)$ and $\omega(y_1)$ values. Now, Eq (4.18) can be rewritten as follows:

$$v_P = (1/4\pi)(d\Gamma/dy)[\omega(y_2) - \omega(y_1)]. \tag{4.19}$$

It was assumed that the rotor was far from the ground and thus, its wake extended far below: $y_2 = -\infty$ (Fig 4.10). For any point P located in the plane of the rotor disc but inside the circle of the vortex cylinder, $\omega(y_2) = 0$, since the supporting ring is infinitely far from point P. For $y_1 = 0$, $\omega(y_1) = 2\pi$, and the point P lies in the plane of the supporting ring. This means that the induced velocity in the plane of the disc becomes

$$v_P = -\tfrac{1}{2}(d\Gamma/dy) = const. \tag{4.19a}$$

In view of the fact that the above equation was established for an arbitrary point of the rotor disc, the induced velocity should be constant over the whole disc area.

Should point P be located outside of the rotor disc at P_1, both $\omega(y_2)$ and $\omega(y_1)$ equal zero, thus, according to Eq (4.19), no flow (at least in the y direction) would be induced by the assumed wake.

The downwash velocity far below the rotor ($v_{-\infty}$) can also be found with the help of Eq (4.19): If point P is moved far below the rotor, then obviously, $\omega(y_2)$ still remains equal to zero, but $\omega(y_1) = 4\pi$. Therefore,

$$v_{-\infty} = -(d\Gamma/dy). \tag{4.19b}$$

Comparing Eqs (4.19a) and (4.19b), one sees immediately that $v_{-\infty} = 2v$, as indicated previously by the momentum theory.

Taking Eq (4.17) into consideration, the total value of the circulation transferred to the wake during one revolution of the rotor becomes

$$\Delta\Gamma = 2T/\rho V_t R \tag{4.20}$$

and

$$d\Gamma/dy = -\Delta\Gamma/\Delta y \tag{4.21}$$

where $\Delta y = v\Delta t$ is the distance traveled by the wake during the time $\Delta t = 2\pi R/V_t$ corresponding to one complete rotor revolution. Making the necessary substitutions for $\Delta\Gamma$ and Δy in Eq (4.21), and then introducing the so-obtained $d\Gamma/dy$ values into Eq (4.19a) will result in the following:

$$v = \sqrt{T/2\pi R^2 \rho}.$$

The above expression is immediately recognized as the formula for the ideal induced velocity previously obtained in the momentum theory.

Although the above results are correct, at first glance it may appear that some logical contradictions exist in the physicomathematical model, as the increase of velocity in the downstream direction postulated by Eq (4.19b) should cause contraction of the wake convecting the shed vorticity; but in spite of this, a constant cross-section of the slipstream was still assumed.

However, closer examination of the consequences of slipstream contraction indicates that the relationship established for the cylindrical wake would still be valid if continuity of the vortex surface far downstream were assumed. Obviously, $\omega(y_2)$ and $\omega(y_1)$ appearing in Eq (4.19) will not change; being 0 and 2π, respectively, and the value of $d\Gamma/dy$ will also remain constant as in the cylindrical case. To visualize this result, one must imagine the vortex filaments forming the rotor wake as being of small, but finite, diameter. Since they are 'packed' so tightly that consecutive vortex rings touch each other–thus forming a continuous vortex sheet–the number of vortex filaments per unit of the wake length will be the same in the contracted slipstream region as in the non-contracted portion.

4.2 Circulation Varying Along the Blade Surface

Wake Vorticity. From the point-of-view of practical application, cases wherein circulation varies along the blade should be of greater interest than $\Gamma(r) = const.$ In the simplest physicomathematical model of a rotor with varying circulation when hovering out-of-ground effect, it may be assumed that a vortex filament leaving the blade at a particular station moves (as in the preceding case) along a circular cylinder of a radius equal to that of the blade station from which the filament originated. This would result in wakes wherein vortices are distributed along concentric cylindrical surfaces. For a large number of blades, it may be assumed that a continuous distribution is obtained along each cylindrical surface.

The strength of each vortex filament leaving the blade is equal to the corresponding variation of circulation along the radius, but is of the opposite sign.

Knowing $\Gamma = f(r)$ for the blade (Fig 4.11), the variation of circulation for a blade element dr wide and located at a distance r from the rotor axis will be $(d\Gamma/dr)dr$. Thus, a vortex of this strength should leave the blade at station $r + dr$. It was shown that vortices springing out at a given radius do not produce any downwash in the plane of the disc outside of the radius at which they separate from the blade. This means that the considered vortex of strength $-(d\Gamma/dr)dr$ leaving the blade at station $r + dr$ has no influence on the downwash velocities of blade elements outboard of this station. The vortex, however, may affect the inboard elements, and this influence can be easily estimated.

The strength of the vortex at station $r + dr$ will be

$$\Gamma_{r+dr} = \Gamma_r + (d\Gamma/dr)dr.$$

Differentiating the above expression with respect to y, and neglecting infinitesimals of higher order, the following is obtained:

$$d\Gamma_{r+dr}/dy = d\Gamma_r/dy.$$

Induced Velocities. Substituting $d\Gamma_r/dy$ instead of $d\Gamma_{r+dr}/dy$ into Eq (4.19a), the corresponding velocity induced at the rotor disc at station $r + dr \approx r$ becomes

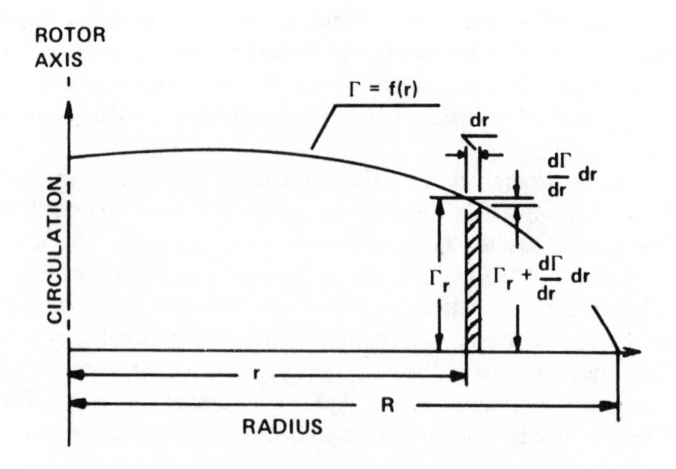

Figure 4.11 Circulation varying along the blade

$$v_r = -\tfrac{1}{2}(d\Gamma_r/dy). \tag{4.23}$$

The downwash created by the inboard elements at station r (0 to r) will be

$$v_{0-r} = -\frac{1}{2}\frac{d[\Gamma_r + (d\Gamma/dr)dr - \Gamma_r]}{dy}. \tag{4.24}$$

Neglecting the infinitesimals of higher order, Eq (4.24) reduces to zero.

These results indicate that under the assumption of a large number of blades and cylindrical wakes, the circulation existing at any blade element influences the downwash at that particular element only. Thus, the blade elements may be treated as being reciprocally independent, as was assumed in the combined momentum and blade-element theory.

It is now possible to establish relationships between the circulation, the geometry of the rotor (chord, number of blades, etc.), and the characteristics of the airfoil section at each blade station without considering any possible ramifications from any other point along the blade.

The lift coefficient of a blade section at a given radius r is:

$$c_{\ell_r} = a_r(\theta_r - \phi_r).$$

where values of a_r (section-lift curve slope), θ_r (blade pitch angle), and ϕ_r (total inflow angle) are all determined at station r. In hover, under small angle assumptions, $\phi_r = v_r/\Omega_r$, and

$$c_{\ell_r} = a_r[\theta_r - (v_r/\Omega_r)]. \tag{4.25}$$

Substituting Ωr in Eq (4.16) for the velocity of flow $U_{\perp}$, and expressing c_ℓ according to Eq (4.25), the total circulation at station r for b number of blades is readily obtained:

$$\Gamma_r = \tfrac{1}{2}a_r bc_r[\theta_r - (v_r/\Omega r)]\Omega r. \tag{4.26}$$

In the case of hover, the vortex filaments spring out from each blade station and move down with the slipstream. It is clear that the distance (d_y) traveled by the vortices springing from all b blades along the rotor axis during one revolution will be:

$$d_y = 2\pi\, v_r/\Omega. \tag{4.27}$$

Now, the average change of circulation along the rotor axis y in the negative direction, $-(d\Gamma/dy)$ can be obtained by dividing Eq (4.26) by Eq (4.27):

$$-(d\Gamma/dy)_r = (1/4\pi)a_r bc_r[\theta_r - (v_r/\Omega r)]\,\Omega^2 r/v_r. \tag{4.28}$$

In analogy to Eq (4.19a), $v_r = -\frac{1}{2}(d\Gamma/dy)_r$; therefore, expressing this derivative according to Eq (4.28) and switching to the $\bar{r}$ notations ($\bar{r} \equiv r/R$) for determining the position of the considered element on the blade, the following equation—now in $v_{\bar{r}}$—is obtained:

$$8\pi R v_{\bar{r}}^2 + V_t a_{\bar{r}} bc_{\bar{r}} v_{\bar{r}} - V_t^2 a_{\bar{r}} bc_{\bar{r}} \bar{r}\theta_{\bar{r}} = 0. \tag{4.29}$$

A glance at the above equation will indicate that it is identical to Eq (3.14) obtained from the combined momentum and blade-element theory when $V_c = 0$.

5. SIMPLE ROTOR WAKE MODELS IN FORWARD FLIGHT

5.1 Development of the Concept

In the consideration of axial rotor translations (with hovering as a limiting case), the simplest physicomathematical model of the rotor vortex system was built around the tip and root vortices only. A similar approach can also be taken for nonaxial regimes of flight. However, the basic differences between the two cases should be noted: (1) A complete axial symmetry existed in steady-state vertical climb and in hover, and the blade azimuth angle had no influence on the circulation. Consequently, constancy of the circulation along the blade radius ($\Gamma_b = \Gamma(r) = const$) represented a sufficient condition for obtaining the rotor wake models discussed in Sect 4.1 of this chapter. (2) In contrast, it should be anticipated that the azimuth angle could influence the blade circulation value in forward flight; hence, in order to obtain the desired simple wake with no shed or trailing vortices along the blade span, it must be stated explicitly that

$$\Gamma_b = \Gamma(r,\psi) = const. \tag{4.30}$$

In such a case, the vortices would leave the blade only at the tips and roots, and at low nonaxial translational speeds the wake would appear as shown in Fig 4.12a. Furthermore, assuming that there is no slipstream contraction, all of the tip vortices would form helical lines on the surface of an elliptic cylinder of constant cross-section (Fig 4.12b). In both cases, the root vortices can be imagined as forming a single filament along the wake axis.

As the inplane velocity component increases, one may also imagine that the wake assumes a more and more skewed position with respect to the rotor disc plane, while its thickness measured perpendicularly to the cylinder axis becomes progressively reduced.

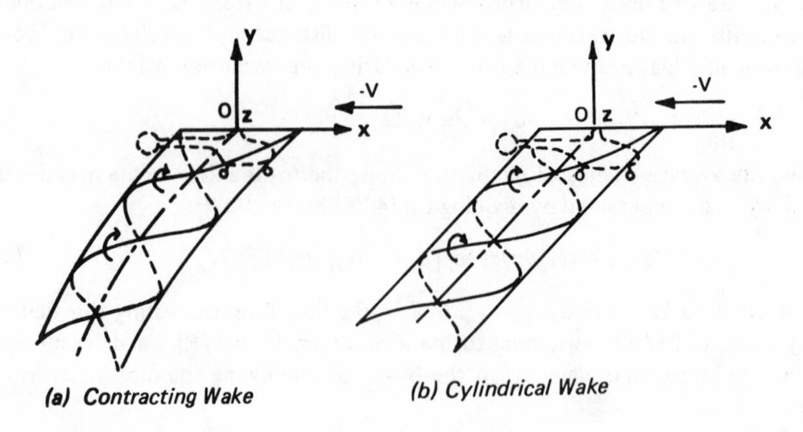

Figure 4.12 Tip and root vortex wake of a two-bladed rotor at low horizontal speeds

Finally, at the limit (as, for instance, in high-speed horizontal flight), it may be assumed that the whole wake degenerates into a constant-width flat 'ribbon' of vorticity trailing behind the rotor (Fig 4.13). This would obviously represent the simplest model for the wake of a rotor in horizontal translation.

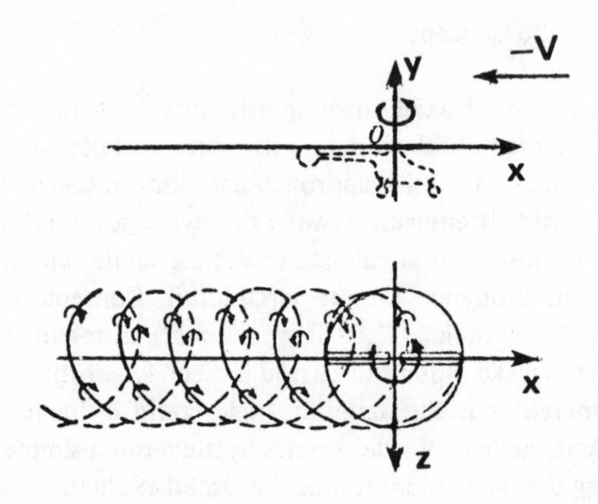

Figure 4.13 Flat wake formed by tip vortices of a two-bladed rotor at $\mu = 0.25$[1]

Having established the physical model of the wake, further procedures in finding induced velocities at points of interest (say, at the rotor disc) can be visualized as follows. First, obtain analytical expressions for the shape of vortex filaments in the *xyz* coordinate system and then using suitable Biot-Savart formula given in Sect 3.2, try to get the solution either in a closed form, or one based on the use of special functions (e.g., elliptic, Legendre, Bessel, etc.) which are available in tabulated form. If, for some reason, the above approach proves too difficult, then advantage may be taken of computer techniques. In this case, the actual shape of vortex filaments would be approximated by broken lines, with the length of these straight-line segments selected so as to reach a desirable compromise between accuracy and time of the computations.

5.2 Consequences of the $\Gamma_b(\bar{r},\psi) = const$ Assumption

At this point, it may be of interest to examine the extent to which the $\Gamma_b(\bar{r},\psi) = const$ assumption could reflect a reality. In order to do this, the corresponding variation of blade loading with radius and azimuth will be examined.

Relationship between Rotor Thrust and Circulation per Blade. Assuming that the induced velocity at the blade is small in comparison with the velocity component $U_\perp$ perpendicular to the blade axis of the resultant flow at some relative station $\bar{r} \equiv r/R$, the elementary thrust (lift) dT_b per length $d\bar{r}R$ of one blade, according to the Kutta-Joukowsky law, can be expressed as

$$dT_b = \rho\,\Gamma_b U_\perp R d\bar{r} \tag{4.31}$$

where Γ_b is the circulation around the blade and $U_\perp = (\bar{r} + \mu \sin\psi)V_t$ (Fig 4.14). Consequently,

$$dT_b = \rho\Gamma_b V_t R(\bar{r} + \mu \sin\psi)d\bar{r}. \tag{4.31a}$$

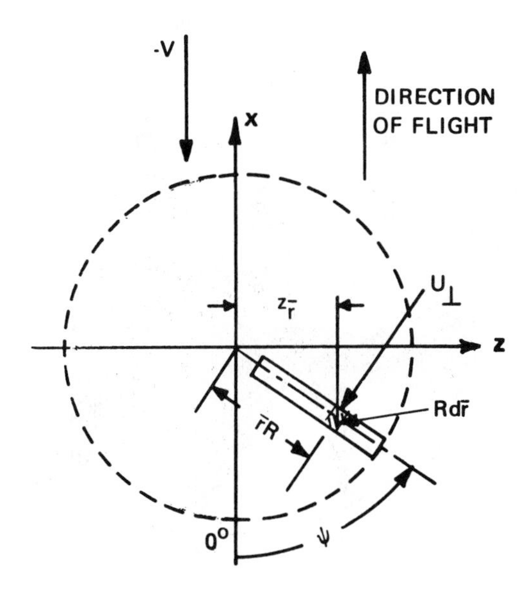

Figure 4.14 Notations

Integrating the above equations within limits of $\bar{r} = 0$ to $\bar{r} = 1.0$, and averaging the obtained lift per blade over one revolution, a relationship between the total rotor thrust T of all b blades, and the circulation per blade can easily be established:

$$T = \frac{b\rho\Gamma_b V_t R}{2\pi} \int_0^{1.0}\int_0^{2\pi} (\bar{r} + \mu \sin\psi)d\bar{r}d\psi$$

from which

$$T = \tfrac{1}{2} b\rho\Gamma_b V_t R \tag{4.32}$$

or conversely,

$$\Gamma_b = 2T/b\rho R V_t. \tag{4.33}$$

It can be seen from the above equation that the relationship (under the $\Gamma_b(\bar{r}) = const$ assumption) between T and Γ_b is identical to Eq (4.17) and independent of the advance ratio μ.

In order to further examine the practical consequences of the assumption that blade circulation remains constant along the radius and is independent of the azimuth, let us examine variations of the load-per-unit length of the blade span $(dT_b/Rd\bar{r})$ as a function of both $\bar{r}$ and ψ. From Eq (4.31a),

$$dT_b/Rd\bar{r} = \rho\Gamma_b V_t(\bar{r} + \mu \sin\psi). \tag{4.34}$$

Thrust Offset. The above equation indicates that the load over the disc would become asymmetric with respect to the xOy plane as μ begins to increase. This is illustrated in Fig 4.15 for a typical high-speed advance ratio of $\mu = 0.35$.

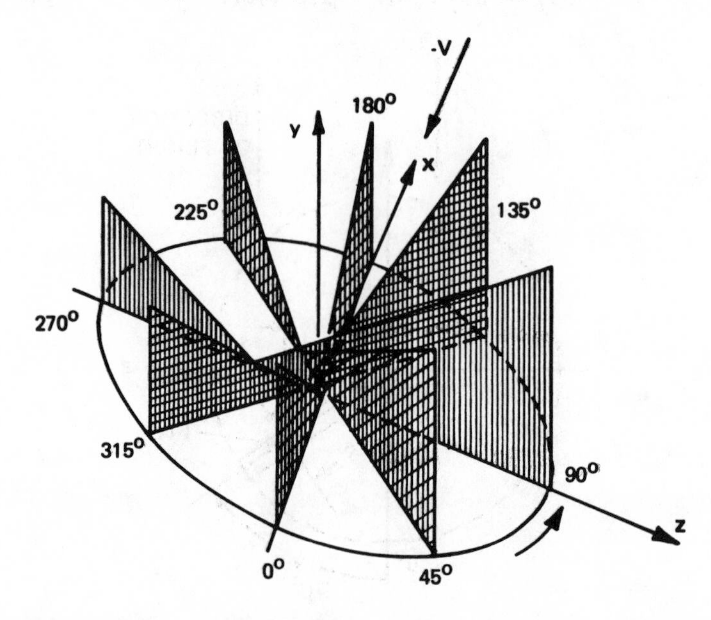

Figure 4.15 Character of load distribution at $\mu = 0.35$ for $\Gamma_b(\bar{r},\psi) = const$

It can be seen from this figure that the resultant thrust would be displaced to the advancing side of the disc, thus producing an unbalanced rolling moment M_x.

The magnitude of this moment can be easily calculated; remembering that the elementary rolling moment (dM_x) shown in Fig 4.14 is:

$$dM_x = -dT_b R\bar{r} \sin\psi.$$

Denoting dT_b in the above expression as given in Eq (4.31a) and performing the double integration for all b blades from $\bar{r} = 0$ to $\bar{r} = 1.0$, and from $\psi = 0$ to $\psi = 2\pi$, the following is obtained:

$$M_x = -\tfrac{1}{4} b\rho \Gamma_b R^2 V_t \mu \qquad \text{where } \mu \equiv |V|/V_t. \tag{4.35}$$

The relative thrust (lift) offset ($\bar{z}_{M_x} \equiv z/R$) can be found by dividing Eq (4.35) by T (as given by Eq (4.32)) and by R. This results in

$$\bar{z}_{M_x} = \tfrac{1}{2}\mu. \tag{4.36}$$

Such large thrust (lift) offsets are not realistic for single-rotor configurations, although they may be tolerated in helicopters using paired contrarotating 'rigid' rotors; for instance, the ABC type shown in Fig 1.30. Consequently, for most of the cases encountered in practice, the validity of the physicomathematical model based on the assumption of constant circulation must be critically examined before using—for the sake of simplicity—the $\Gamma(\bar{r}, \psi) = const$ approach.

Circulation Variation Required to Eliminate Thrust Offset. Elimination of the rolling moment can be achieved by postulating that the blade thrust moment, with respect to its actual (in articulated rotors) or virtual (in hingeless configurations) flapping axis, remains constant regardless of the azimuth value.

$$M_b = \int_0^{1.0} R\bar{r}\, dT_b = const. \tag{4.37}$$

Expressing dT_b according to Eq (4.31a) in Eq (4.37) and changing the notation for Γ_b to $\Gamma_b(\psi)$ in order to emphasize the dependence of circulation on the azimuth, the integral from Eq (4.37), with constant multipliers omitted, becomes

$$\Gamma_b(\psi)[1 + (3/2)\mu \sin\psi] = const \tag{4.38}$$

where $\Gamma_b(\psi)$ can now be interpreted as circulation averaged over the entire blade length when the blade is at an azimuth ψ. Consequently, the required character for the variation of the average blade circulation with ψ will be

$$\Gamma_b(\psi) = const/[1 + (3/2)\mu \sin\psi]. \tag{4.39}$$

This can also be expressed as

$$\Gamma_b(\psi) = \tilde{\Gamma}_b + \Delta\Gamma_b(\psi), \tag{4.40}$$

where $\tilde{\Gamma}_b$ is the blade circulation averaged over one revolution.

Wake of a Rotor with No Thrust Offset. Since in almost all practical rotary-wing configurations, very little or no thrust offset is permitted, the average blade circulation must vary with the azimuth angle as indicated by Eq (4.39). This in turn implies that shed vortices appear in the wake. The actual circulation also usually varies within the blade span, which leads to the presence of trailing vortices springing from the blade. It may be stated hence, that in the majority of cases encountered in forward flight, the blade circulation is a function of $\bar{r}$ and ψ.

The example in Fig 4.16 of the actual character of circulation variation for an articulated rotor in forward flight shows that indeed, circulation varies not only with the azimuth, but with blade stations ($\bar{r}$) as well.

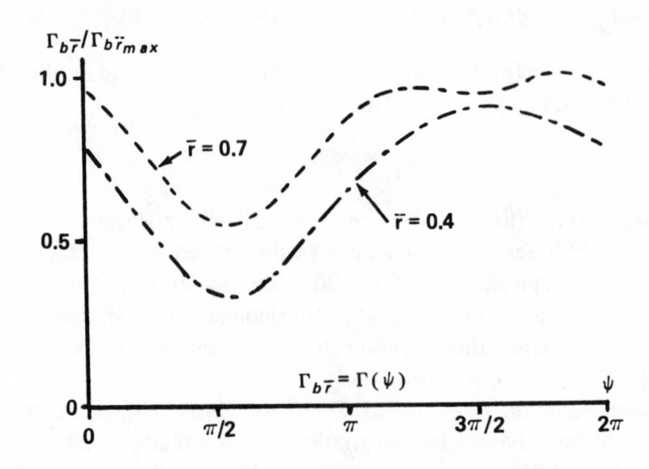

Figure 4.16 An example of circulation variation for an articulated rotor in forward flight

Because of the fact that generally, $\Gamma_b = \Gamma(\bar{r},\psi)$, the whole conceptual model of the vortex system reflecting the reality of rotor operation in forward flight may become more complicated than the simple scheme based on the existence of tip and root vortices only. A discussion of the flat wake is presented in the following section as an example of how even in a simplified model, some of the $\Gamma_b = \Gamma(\bar{r},\psi)$ aspects are taken into consideration.

6. FLAT WAKE

It was indicated in Sect 5.1 that the so-called flat wake represents a limiting case when all of the vortices transferred to the slipstream of a rotor moving at a relatively high forward speed (usually purely horizontal) are reduced to a single ribbon of vorticity. In spite of the basic simplicity of such a vortex system, it merits attention, as many practical problems of rotary-wing aerodynamics can be quantitatively examined using a mathematical model based on the flat-wake concept. Furthermore, studying a wake of this type should provide the reader with introductory material, later facilitating working with more complicated vortex systems. The basic presentation of this subject closely follows the development of the wake concept by Baskin et al[1], and then a somewhat different approach by Ormiston[14] is briefly discussed.

6.1 Variable Circulation Along the Blade

Determination of the Vortex System. It was shown that the assumption of $\Gamma_b(\psi) = const$ may lead to unbalanced rolling moments. Nevertheless, for the sake of simplicity, Baskin et al assume that the variation of circulation with azimuth may be neglected for practical calculations if, at each blade station, the circulation is averaged over a complete rotor revolution. In this way, a conceptual model is obtained where the radial change $\Gamma_b(\bar{r})$ of the ψ-averaged blade circulation becomes the only variation to be considered.

It should be recalled that acceptance of $\Gamma_b(\psi) = const$ would imply that there are no shed vortices. Bound vortices, as in the case of hover, would not generate any vertical induced velocities, but would contribute to the slipstream rotation by inducing tangential velocities above and below the rotor disc. For single rotors, these effects may be neglected since these tangential velocities are usually no higher than 0.5 percent of the tip speed. For the two-rotor configuration (coaxial and intermeshing), they may be considered in a more detailed analysis.

Because of the assumption that $\Gamma_b(r) \neq const$, the trailing vortices would spring from various blade stations and form a system which, when viewed at a particular instant in time, would consist of cycloids of various shapes, depending on the station of their origin (Fig 4.17).

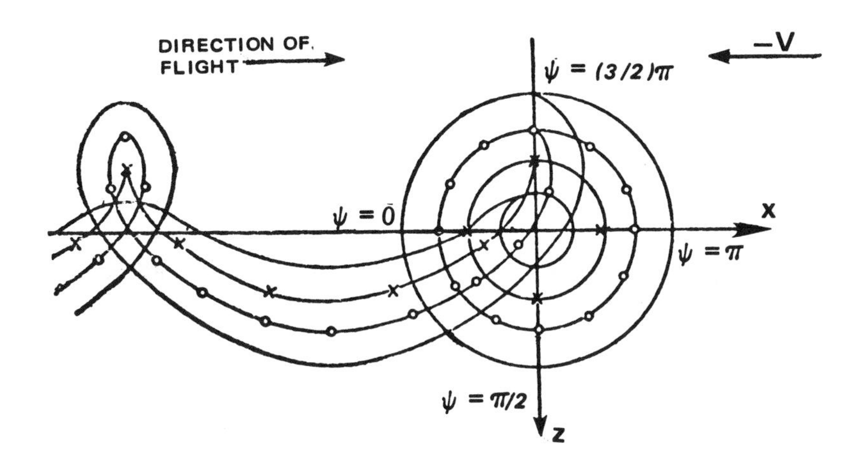

Figure 4.17 Trailing vortices of a rotor with circulation varying along the blades

Using the notations given in this figure, the equation of a single vortex filament shed from the tip can be expressed quite simply as

$$\begin{aligned} x &= Vt - R\cos\psi \\ z &= R\sin\psi \end{aligned} \tag{4.41}$$

where V is the velocity of the distant flow (with the proper sign), ψ is the blade azimuth at time t, and R is the rotor radius. Denoting ψ_o as the azimuth angle at which the tip vortex separates from the blade, t can now be expressed in terms of the azimuth angles ψ and ψ_o: $t = (\psi - \psi_o)/2\pi(rps)$ where $rps = V_t/2\pi R$. Introducing the above values into Eq (4.41), the following is obtained:

$$\begin{aligned} x &= \mu R[(\psi - \psi_o) - \cos\psi] \\ z &= R\sin\psi \end{aligned} \tag{4.41a}$$

where μ is the tip-speed ratio (which, in the assumed coordinate system, should be taken with the minus sign).

The trailing vortices leaving the blade along its span can be imagined as discrete filaments *if* the blade circulation jumps by steps. However, if the circulation varies in a continuous manner ($d\Gamma_{b\bar{r}}/d\bar{r}$, finite along the whole blade), then a continuous sheet of vorticity would also form. This sheet can be visualized as consisting of separate, Δs wide, ribbons of vorticity. As discussed in Sect 4.2 of this chapter, the circulation strength of such a ribbon springing from a blade segment $\Delta r = \Delta\bar{r}R$ long will be

$$\Delta\Gamma_{b\bar{r}} = -(d\Gamma_{b\bar{r}}/d\bar{r})\Delta\bar{r}R, \tag{4.42}$$

while its shape will be represented by a cycloid whose equation (in analogy to Eq (4.41a)) can be written in a nondimensional form as follows:

$$\begin{aligned} \bar{x} &= \mu(\psi - \psi_o) - \bar{r}\cos\psi \\ \bar{z} &= \bar{r}\sin\psi \end{aligned} \tag{4.43}$$

where

$$\bar{x} \equiv x/R, \text{ and } \bar{z} \equiv z/R.$$

In order to determine the strength of circulation γ per unit width of a ribbon of vorticity Δs wide, let us look at Fig 4.18. It can be seen from this figure that taking all b blades into account,

$$\gamma = (\Delta\psi_o/2\pi)\, b\Delta\Gamma_{b\bar{r}}/\Delta s$$

where $\Delta s = \Delta x \sin\vartheta$, while $\tan\vartheta = d\bar{z}/d\bar{x} = \bar{r}\cos\psi/(\mu + \bar{r}\sin\psi)$.

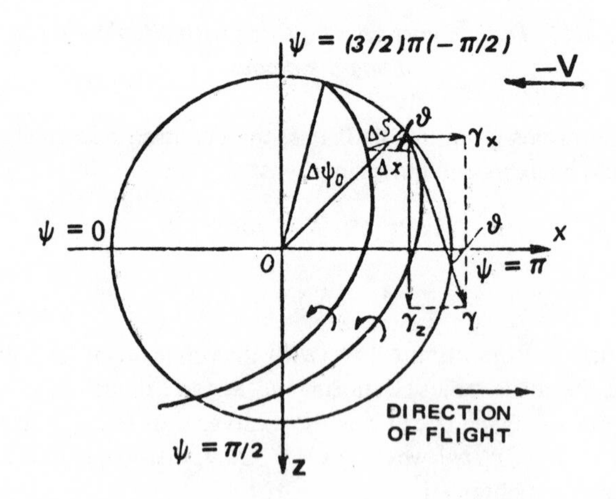

Figure 4.18 Elementary vortex strip

However, according to Eq (4.43), the nondimensional distance Δx between the two cycloids encompassing the considered vortex strip is $\Delta\bar{x} = \mu\Delta\psi_o$, and

$$\Delta\psi_o/\Delta s = 1/\mu R \sin\vartheta, \qquad \gamma = b\Delta\Gamma_{b\bar{r}}/2\pi R\mu \sin\vartheta.$$

Equivalent Rectilinear Vortex System. The cycloidal vortex strip can be resolved into two components; one, parallel to the x, and another to the z axis (Fig 4.18 and 4.19) having a circulation strength per unit width of γ_x and γ_z, respectively. The x component may be called longitudinal, and the z component lateral.

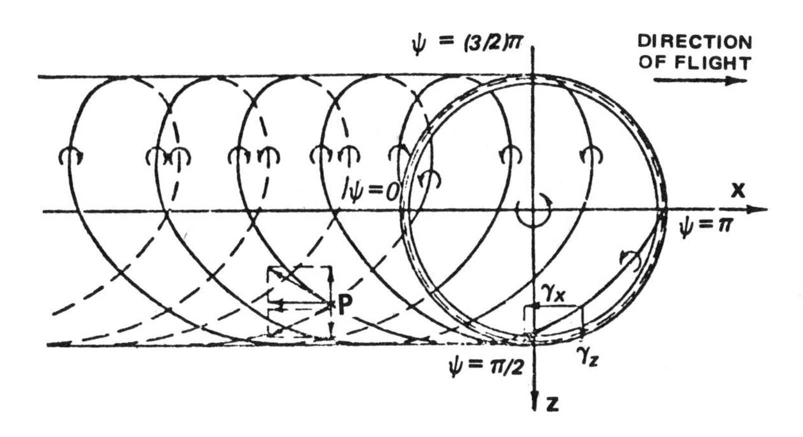

Figure 4.19 Vortex filaments from a four-bladed rotor

The values of these components within various ranges of the azimuth angle are as follows:

$$\gamma_x = \begin{cases} \gamma\cos\vartheta = \dfrac{b\Delta\Gamma_{b\bar{r}}}{2\pi R\mu}\,\dfrac{\mu + \bar{r}\sin\psi}{\bar{r}\cos\psi} & \text{for } \dfrac{\pi}{2} \leqslant \psi \leqslant \dfrac{3}{2}\pi \\[2ex] -\gamma\cos\vartheta = -\dfrac{b\Delta\Gamma_{b\bar{r}}}{2\pi R\mu}\,\dfrac{\mu + \bar{r}\sin\psi}{\bar{r}\cos\psi} & \text{for } -\dfrac{\pi}{2} \leqslant \psi \leqslant \dfrac{\pi}{2} \end{cases} \tag{4.44}$$

and

$$\gamma_z = \begin{cases} \gamma\sin\vartheta = \dfrac{b\Delta\Gamma_{\bar{r}}}{2\pi R\mu} & \text{for } \dfrac{\pi}{2} < \psi < \dfrac{3}{2}\pi \\[2ex] 0 & \text{for } \psi = \dfrac{\pi}{2} \text{ and } \psi = \dfrac{3}{2}\pi \\[2ex] -\gamma\sin\vartheta = \dfrac{b\Delta\Gamma_{\bar{r}}}{2\pi R\mu} & \text{for } -\dfrac{\pi}{2} < \psi < \dfrac{\pi}{2}. \end{cases} \tag{4.45}$$

It can be seen from Fig 4.19 that lateral components (γ_z) of vortices which spring from a circle of radius $\bar{r}$ will exist only inside of that circle. Outside, they will cancel each other (point P, Fig 4.19). By contrast, the γ_x components will add to each other and consequently, the strength per unit width of the longitudinal vortex will double outside of the circle of radius $\bar{r}$. All of the above considerations indicate that the original system of cycloidal vortices shown in Figs 4.17, 4.18, and 4.19 can be replaced by a simpler system of rectilinear vortices more suitable for mathematical treatment (Fig 4.20).

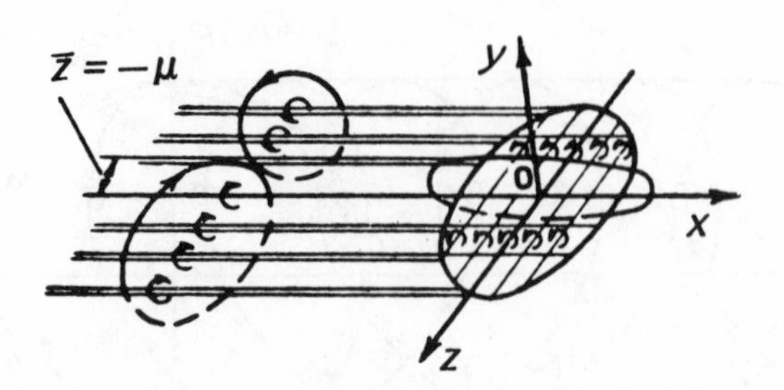

Figure 4.20 System of rectilinear vortices equivalent to that of cycloidal vortices

6.2 Determination of Induced Velocities

Linear Problem. As long as the assumption of a 'rigid' wake is maintained, no interaction exists between induced velocity $\vec{v}$ and the wake structure; thus, the problem becomes linear. This means that in computing the resultant $\vec{v}$, the components generated by various vortex subsystems forming the wake (in this case, lateral and longitudinal vortices) can be computed separately and then superimposed.

This process of summation can be best performed by determining the contributions of the lateral and longitudinal vortices to the components (v_x, v_y, v_z) of vector $\vec{v}$ along the x, y, z axes. At this point, it should be noted that rectilinear vortices can induce velocities only in planes perpendicular to them. Consequently, lateral vortices contribute only to components v_x and v_y located in the xOy plane, or in planes parallel to it, while the longitudinal ones contribute to v_y and v_z. Since both longitudinal and lateral vortices contribute to the v_y component, the first contribution will be symbolized by v_{y_0} and the second, by v_{y_1}.

Velocities Induced by Lateral Vortices. The magnitude of the elementary velocity vector $d(\Delta \vec{v})$ induced at some point $C(x_1, y_1, z_1)$ (Fig 4.21) by an element of lateral vorticity dx wide and dz long located at a point (x, z), and having strength γ_z per unit width, Eq (4.11) can be expressed as follows:

$$d(\Delta v) = \gamma_z (\sin \phi/4\pi d^2)\, dx\, dy \tag{4.46}$$

where

$$d = \sqrt{(x - x_1)^2 + y_1{}^2 + (z - z_1)^2},$$

and

$$\sin \phi = \sqrt{(x - x_1)^2 + y_1{}^2}/d.$$

It can be seen from Fig 4.21 that the y and x velocity components will be

$$d(\Delta v_{y_1}) = -d(\Delta v)\cos\theta, \text{ and } d(\Delta v_{x_1}) = -d(\Delta v)\sin\theta. \tag{4.47}$$

where

$$\sin\theta = y_1/\sqrt{(x-x_1)^2 + y_1^2}, \quad \text{and} \quad \cos\theta = (x-x_1)/\sqrt{(x_2-x_1)^2 + y_1^2}.$$

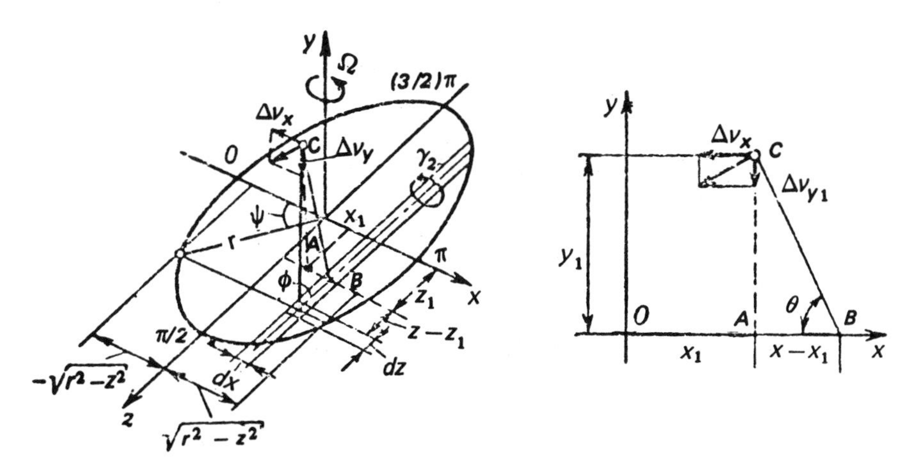

Figure 4.21 Velocities induced by lateral vortices

By substituting Eq (4.46) into Eq (4.47), and expressing (a) all the above angles in terms of r, x, and z, and (b) γ_z as shown in Eq (4.45); the Δv_{y1} and Δv_{x1} values can be obtained by performing double integration. This was done once with respect to x (from $-r$ to r), and once with respect to z (from $-\sqrt{r^2 - z^2}$ to $\sqrt{r^2 - z^2}$). A closed-form solution was obtained in Ref 1, with the following results:

$$\Delta v_{y_1} = (b\Delta\Gamma_{br}/4\pi R\mu)\Delta\hat{v}_{y_1} \tag{4.48}$$

and

$$\Delta v_x = (b\Delta\Gamma_{br}/4\pi R\mu)\,\Delta\hat{v}_x. \tag{4.49}$$

In Eq (4.48), the vertical component Δv_{y_1} is given in terms of $\Delta\hat{v}_{y_1}$ which, in turn, is expressed as follows:

$$\Delta\hat{v}_{y_1} = \tfrac{1}{2}\pi \int_{-1}^{1} \bar{Y}_y\, dz \tag{4.50}$$

where

$$\bar{Y}_y = \frac{1}{\sqrt{\left(\tilde{x}_1 + \sqrt{(1-\tilde{z}^2)}\right)^2 + \tilde{y}_1 + (\tilde{z}-\tilde{z}_1)^2}} - \frac{1}{\sqrt{\left(\tilde{x}_1 - \sqrt{(1-\tilde{z}^2)}\right)^2 + \tilde{y}_1^{\,2} + (\tilde{z}+\tilde{z}_1)^2}}$$

while

$$\tilde{x}_1 = x_1/r, \ \tilde{y}_1 = y_1/r, \ \tilde{z}_1 = z_1/r.$$

It can be seen from Eqs (4.48) and (4.50) that the character of the distribution of velocity induced by lateral vortices along the fore and aft rotor diameter is as shown in Fig 4.22.

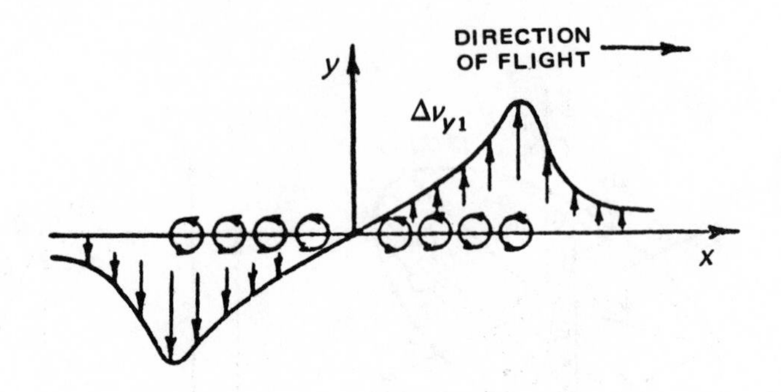

Figure 4.22 Character of Δv_{y_1} distribution along the fore-and-aft rotor disc diameter

With respect to the horizontal component (Δv_x), the following notations were used in Eq (4.49):

$$\Delta\hat{v}_x = -(1/2\pi) \int_{\tilde{z}=-1}^{\tilde{z}=1} \bar{Y}_x \, d\tilde{z} \tag{4.51}$$

where

$$\bar{Y}_x = \frac{\tilde{y}_1}{\tilde{y}_1{}^2 + (z - z_1)^2} \left\{ \frac{\tilde{x}_1 + \sqrt{1 - \tilde{z}^2}}{\sqrt{\left(\tilde{x}_1 + \sqrt{1-\tilde{z}^2}\right)^2 + \tilde{y}_1{}^2 + (\tilde{z} - \tilde{z}_1)^2}} - \frac{\tilde{x}_1 - \sqrt{1-\tilde{z}^2}}{\sqrt{\left(\tilde{x}_1 - \sqrt{1-\tilde{z}^2}\right)^2 + \tilde{y}_1{}^2 + (\tilde{z} - \tilde{z}_1)^2}} \right\}$$

The following can be deduced from Eqs (4.48) and (4.51):

1. At points symmetrical to the yOz plane, the $\Delta\hat{v}_x$ values are the same:

$$\Delta v_x(-\tilde{x}_1) = \Delta\hat{v}_x(\tilde{x}_1) .$$

2. For $\tilde{y}_1 > 0$ (above the rotor disc), $\Delta\hat{v}_x < 0$ (induced velocities opposite to the direction of flight); while for $\tilde{y}_1 < 0$ (below the disc), $\Delta\hat{v}_x > 0$.

3. For $\tilde{x}_1 \to \infty$, $\Delta\hat{v}_x \to 0$.
4. For $\tilde{y}_1 \to \infty$, $\Delta\hat{v}_x \to 0$.

Δv_{y_1} and Δv_x represent velocity components induced by lateral vortices associated with an annulus of radius r and width Δr. In order to obtain complete velocity values v_{y_1} and v_x induced by the whole lateral vortex system, Δv values associated with all annuli of the disc from the root $(\bar{r}_o)$ to the tip ($\bar{r} = 1.0$) should be summed up:

$$v_{y_1} = \sum_{\bar{r}=\bar{r}_o}^{\bar{r}=1.0} \Delta v_{y_1} \qquad v_x = \sum_{\bar{r}=\bar{r}_o}^{\bar{r}=1.0} \Delta v_x$$

To facilitate calculations, the values of $\hat{v}_x$ and $\hat{v}_{y_1}$ are presented in graphical form in Ref 1.

Velocities Induced by Longitudinal Vortices in the yOz Plane. It was shown in Fig 4.19 that outside of the circle of radius r from which the vortex springs, the strength of the circulation of a strip of longitudinal vorticity dz wide $(\Delta\Gamma_x = 2\gamma_x\, dz)$ is twice that which is inside the circle. However, inside this circle, the contribution of longitudinal vortices located ahead of the z axis is equal to that behind it. Consequently, the previously discussed vortex system which exhibits a jump in circulation strength when passing through the circle of radius r (Fig 4.23a) can be replaced by another one having a uniform circulation strength of $2\gamma_x$ per unit width, which extends from the z axis, where $x = 0$, to $x = -\infty$ (Fig 4.23b).

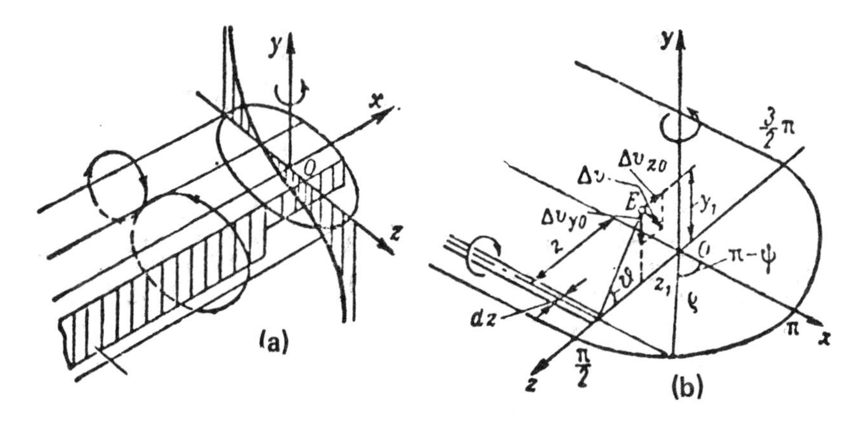

Figure 4.23 Replacement of the (a) longitudinal vortex system by system (b)

Incremental velocity $d(\Delta v)$ induced at some point $E\ (0, y_1, z_1)$ of the yOz plane by a vortex element dz wide located at a distance z can be expressed as one-half of that which would be generated by a rectilinear vortex extending from $-\infty$ to ∞ (Eq 4.1):

$$d(\Delta v) = -(2\gamma_x/4\pi d)\,dz \tag{4.52}$$

where $d = \sqrt{y_1^{\,2} + (z - z_1)^2}$.

The downwash (vertical component of this elementary induced velocity) becomes

$$d(\Delta v_{y_0}) = -d(\Delta v)\cos\vartheta, \qquad \cos\vartheta = (z - z_1)/d$$

or, in light of Eq (4.52), this changes to

$$d(\Delta v_{y_0}) = (\gamma_x/2\pi d^2)(z - z_1)dz.$$

Expressing γ_x according to Eq (4.44), and integrating from $z = -r$ to $z = r$, the following is obtained:

$$\Delta v_{y_0} = (b\Delta\Gamma_{b\bar{r}}/4\pi R\mu)\,\Delta\hat{v}_{y_0} \tag{4.52}$$

where

$$\Delta\hat{v}_{y_0} = (1/\pi)\int_{-\bar{r}}^{\bar{r}} \frac{(\bar{z} - \bar{z}_1)(\mu + \bar{r}\sin\psi)}{[\bar{y}_1^{\,2} + (\bar{z} - \bar{z}_1)^2]\,\bar{r}\cos\psi}\,d\bar{z}. \tag{4.53}$$

Evaluation of $\Delta\hat{v}_{y_0}$ along the z axis $(\tilde{y}_1 = 0)$ would indicate the following[1]:

$$\Delta\hat{v}_{y_0} = \begin{cases} -1 - \dfrac{\mu_* + \tilde{z}_1}{\sqrt{\tilde{z}_1^{\,2} - 1}} & \text{for } \tilde{z}_1 < -1 \\[2ex] -1 & \text{for } -1 \leqslant \tilde{z}_1 \leqslant 1 \\[2ex] -1 + \dfrac{\mu_* + \tilde{z}_1}{\sqrt{\tilde{z}_1^{\,2} - 1}} & \text{for } \tilde{z}_1 > 1. \end{cases} \tag{4.54}$$

where $\mu_* \equiv \mu/\bar{r}$.

Eqs (4.53) and (4.54) indicate that the longitudinal vortices belonging to the system of vorticity springing from the rotor at radius r generate uniform induced velocities along the lateral diameter of that circle. An example of the above result for $\mu = 0.15$ and $\bar{r} = 1.0$ is shown in Fig 4.24.

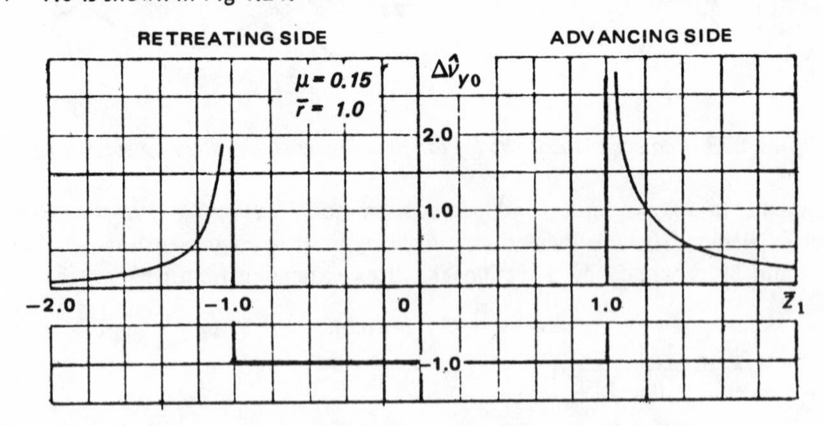

Figure 4.24 Example of the character of $\Delta\hat{v}_{y_0}$ vs $\bar{z}$[1]

Induced velocities along the z axis which are caused by vortices leaving the blades at all radii of the disc can be determined by the following method. In Eq (4.52), $\Delta\Gamma_{b\bar{r}}$ is expressed according to Eq (4.42) and then the integration is performed from the root (r_0) to the tip (R):

$$v_{y_0}(\bar{z}_1) = -\frac{b}{4\pi R\mu}\int_{r_0}^{R} \Delta\hat{v}_y(z_1)\,\frac{d\Gamma_{br}}{dr}\,dr. \tag{4.55}$$

Taking advantage of Eq (4.54), expressions can be obtained for components along the lateral rotor axis of the induced velocity resulting from longitudinal vortices. For points inside the rotor disc located on the advancing side between the tip and root stations, this becomes:

$$v_{y_0}(\bar{r}_0 \leqslant \bar{z}_1 \leqslant 1) = -\frac{b(\mu+\bar{z}_1)}{4\pi R\mu}\int_{\bar{r}_0}^{\bar{z}_1}\frac{(d\Gamma_{br}/d\bar{r})\,d\bar{r}}{\sqrt{\bar{z}_1^{\,2}-\bar{r}^2}}. \tag{4.56}$$

If $\bar{z}_1 > 1$, then

$$v_{y_0}(\bar{z}_1 > 1) = -\frac{b(\mu+\bar{z}_1)}{4\pi R\mu}\int_{\bar{r}_0}^{1}\frac{(d\Gamma_{br}/d\bar{r})\,d\bar{r}}{\sqrt{\bar{z}_1^{\,2}-\bar{r}^2}}, \tag{4.57}$$

On the retreating side inside the disc $(-\bar{r}_0 \geqslant \bar{z}_1 \geqslant -1)$,

$$v_{y_0}(-\bar{r}_0 \geqslant \bar{z}_1 \geqslant -1) = \frac{b(\mu+\bar{z}_1)}{4\pi R\mu}\int_{\bar{r}_0}^{\bar{z}_1}\frac{(d\Gamma_{br}/d\bar{r})\,d\bar{r}}{\sqrt{\bar{z}_1^{\,2}-\bar{r}^2}} \tag{4.58}$$

and outside $(\bar{z}_1 < -1)$,

$$v_{z_0}(\bar{z}_1 < -1) = \frac{b(\mu+\bar{z}_1)}{4\pi R\mu}\int_{\bar{r}_0}^{1}\frac{(d\Gamma_{br}/d\bar{r})\,d\bar{r}}{\sqrt{\bar{z}_1^{\,2}-\bar{r}^2}}. \tag{4.59}$$

With the help of Cauchy's theorem, it can be proven that in spite of the discontinuities in Eqs (4.56) and (4.58) associated with $|\bar{z}_1| = \bar{r}$; and in Eqs (4.57) and (4.59) resulting from $|\bar{z}_1| = 1$, the integrals appearing in these equations have finite values.

It can be seen from Eq (4.58) that for $\bar{z}_1 = -\mu$ on the retreating side, $v_{y_0} = 0$; while Eqs (4.54), (4.56), and (4.58) indicate that v_{y_0} also equals zero in the $-\bar{r}_0 \leqslant \bar{z}_1 \leqslant \bar{r}_0$ region. Furthermore, it can be deduced from Eqs (4.56) and (4.58) plus Eqs (4.57) and (4.59) that for the same $|\bar{z}_1|$ magnitudes, v_{y_0} values on the advancing side of the rotor are higher than those on the retreating side.

In many practical computations, circulation distribution along the blade, $\Gamma_{b\bar{r}} = \Gamma(\bar{r})$, may be given in graphical form. Then, a finite increment approach instead of integration should be taken. For instance, instead of Eq (4.58) for $-r_0 \geqslant z_1 \geqslant -1$, the following formula should be used:

$$v_{y_0}(\bar{z}_1) = \frac{b(\mu + \bar{z}_1)}{4\pi R\mu} \sum_{\bar{r}=\bar{r}_0}^{\bar{r}=\bar{z}_1} \frac{\Delta\Gamma_{b\bar{r}}}{\sqrt{\bar{z}_1^{\,2} - \bar{r}^2}} \tag{4.60}$$

Some conception of how the type of circulation variation along the blade span affects the character of the associated v_{y_0} velocities can be obtained by assuming three types of $\Gamma_{b\bar{r}} = \Gamma(\bar{r})$ distributions; namely, (a) triangular ($\Gamma_{b\bar{r}} = a\bar{r}$), (b) constant along the blade span ($\Gamma_b = const$), and (c) curvilinear where $\Gamma_{b\bar{r}} = 0$ at both ends of the blade ($\Gamma_{b\bar{r}} = a\bar{r}^2(1-\bar{r})$). The necessary derivatives of $d\Gamma_{b\bar{r}}/d\bar{r}$ appearing in Eqs (4.56) through (4.59) and the resulting v_{y_0} values can easily be computed. The results obtained for $\mu = 0.2$, and shown in Fig 4.25, were reproduced from Ref 1.

Lateral Component of Velocities Induced by the Longitudinal Vortices. The lateral component (Δv_{z_0}) of velocity induced by vortices as shown in Fig 4.23 is

$$d(\Delta v_{z_0}) = d(\Delta v) \sin\vartheta; \quad \sin\vartheta = y_1/d.$$

Similar to the development of Eq (4.52), it can be determined that

$$d(\Delta v_{z_0}) = \frac{2\gamma_x}{4\pi d^2}\, y_1\, dz$$

and

$$\Delta v_{z0} = \frac{b\Delta\Gamma_{b\bar{r}}}{4\pi R\mu} \Delta\hat{v}_{z0} \tag{4.61}$$

where

$$\Delta\hat{v}_{z0} = \frac{1}{\pi} \int_{-1}^{1} \frac{\tilde{y}_1(\mu_* + \sin\psi)d\tilde{z}}{[\tilde{y}_1^{\,2} + (\tilde{z} - \tilde{z}_1)^2]\cos\psi}.$$

Velocities Induced by Longitudinal Vortices in Planes Parallel to the yOz Plane. In order to determine velocities induced by longitudinal vortices in a plane parallel to the yOz plane and located at a distance x_1 (Fig 4.26), the following approach was used in Ref 1. The strip of longitudinal vorticity associated with radius r (or $\bar{r}$ in nondimensional notations) may be assumed to be composed of two systems whose individual contributions, using a linear approach, can be superimposed. The first system would consist of a strip of double strength ($2\gamma_x$) vorticity extending to $-\infty$ from the plane passing

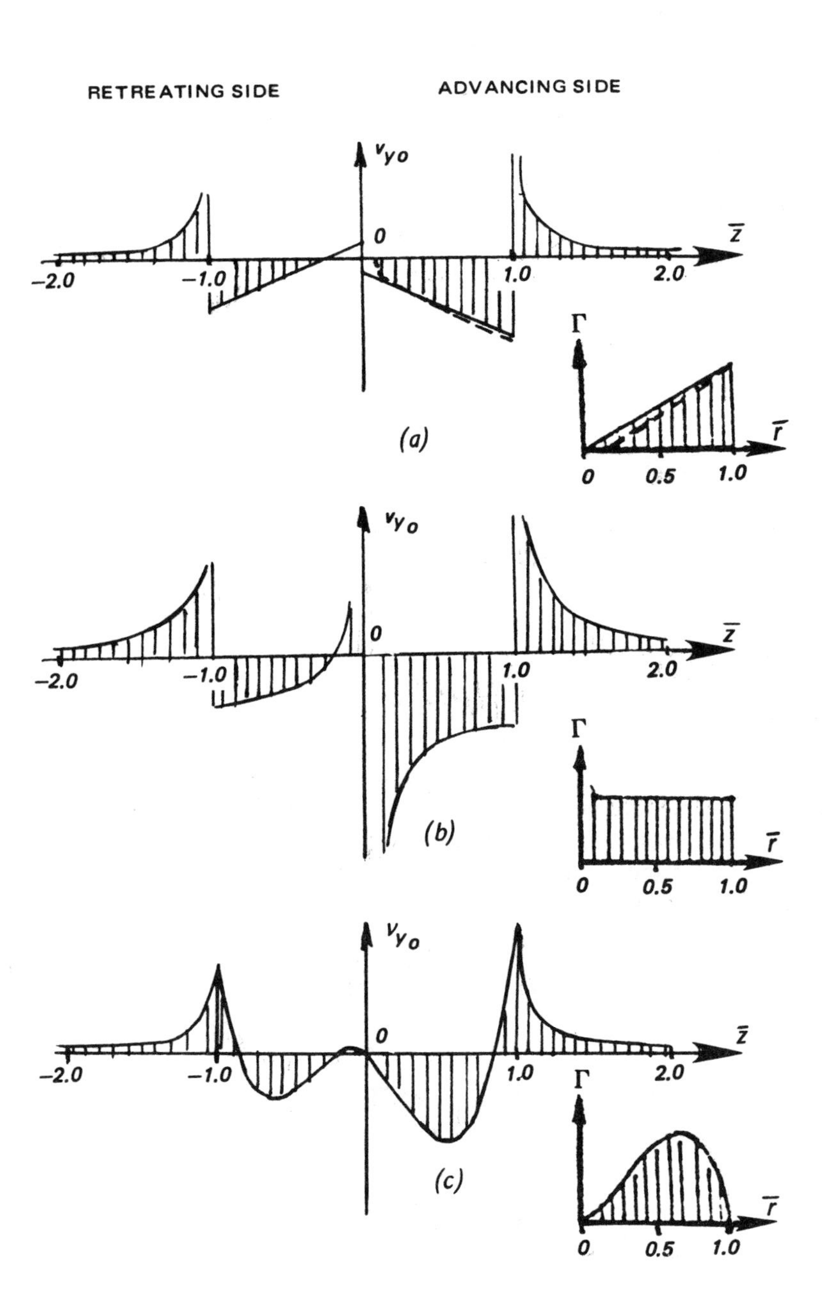

Figure 4.25 Types of circulation distribution along the blade and associated characteristics of $v_{y0}(\bar{r})$

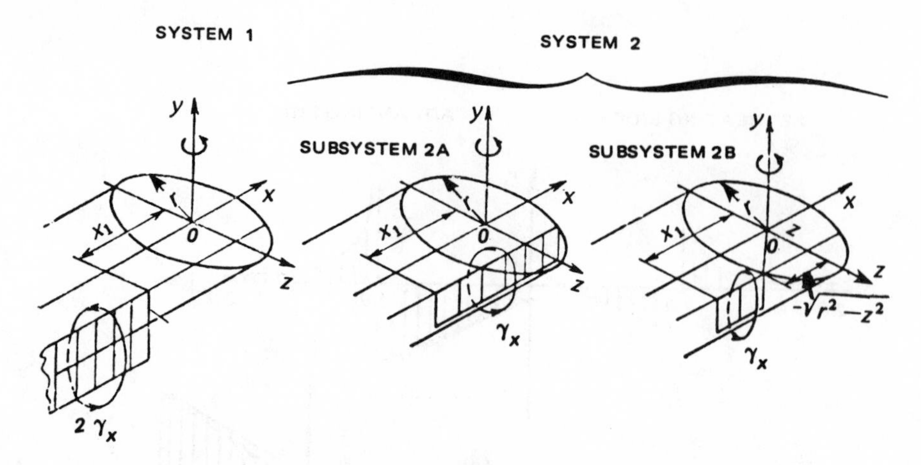

Figure 4.26 Equivalent system of longitudinal vorticity

through x_1. Induced velocity components Δv_{y_0} and Δv_{z_0} generated by this system can be readily computed from Eqs (4.52) and (4.61).

The second system would be formed by two subsystems (a) and (b) consisting of strips of vorticity of finite length, and having a strength of γ_x per unit width. Subsystem (a) would extend from the plane of interest to the leading edge of the circle of radius r, while subsystem (b) would extend only to the trailing edge of the circle. The total system will induce velocities with vertical and lateral components of Δv_{y_2} and Δv_{z_2}, respectively.

When calculating Δv_{y_2} and Δv_{z_2}, the strength of the vorticity of subsystem (b) should be considered negative for $|x_1| < \sqrt{r^2 - z^2}$, while the absolute strength of vortex filaments belonging to this system is the same (γ_x).

The vertical component due to the combined effects of subsystems (a) and (b) is expressed in Ref 1 as follows:

$$\Delta v_{y_2} = \frac{b\Delta\Gamma_{br}}{4\pi R\mu} \Delta\hat{v}_{y_2} \tag{4.62}$$

where

$$\Delta\hat{v}_{y_2} = -½(\mu_* \widetilde{M} + \widetilde{N}).$$

Similarly,

$$\Delta\hat{v}_{z_2} = \frac{b\Delta\Gamma_{br}}{4\pi R\mu} \Delta\hat{v}_{z_2} \tag{4.63}$$

where

$$\Delta\hat{v}_{z_2} = -½(\mu_* \widetilde{K} + \widetilde{L}). \tag{4.63}$$

The values of functions $\widetilde{M}$, $\widetilde{N}$, $\widetilde{K}$, and $\widetilde{L}$ are given in graphical form in Ref 1. It was also indicated that at points symmetrical with respect to that of the rotor disc *(xOz* plane*)*, velocities $\Delta\hat{v}_{y_2}$ remain the same; while $\Delta\hat{v}_{z_2}$ are of the same magnitude, but with opposite signs:

$$\Delta\hat{v}_{y_2}(\tilde{x}_1, -\tilde{y}_1, \tilde{z}_1) = \Delta\hat{v}_{y_2}(\tilde{x}_1, \tilde{y}_1, \tilde{z}_1)$$
$$\Delta\hat{v}_{z_2}(\tilde{x}_1, -\tilde{y}_1, \tilde{z}_1) = -\Delta\hat{v}_{z_2}(\tilde{x}_1, \tilde{y}_1, \tilde{z}_1). \quad (4.64)$$

For points symmetrical with respect to the vertical plane (yOz) passing through the z axis, the signs of velocities $\Delta\hat{v}_{y_2}$ and $\Delta\hat{v}_{z_2}$ are reversed, while their magnitudes remain the same:

$$\Delta\hat{v}_{y_2}(-\tilde{x}_1, \tilde{y}_1, \tilde{z}_1) = -\Delta\hat{v}_{y_2}(\tilde{x}_1, \tilde{y}_1, \tilde{z}_1)$$
$$\Delta\hat{v}_{z_2}(-\tilde{x}_1, \tilde{y}_1, \tilde{z}_1) = -\Delta\hat{v}_{z_2}(\tilde{x}_1, \tilde{y}_1, \tilde{z}_1). \quad (4.64a)$$

For $\tilde{x}_1 \to \infty$, it is shown that $\Delta\hat{v}_{y_2} \to \Delta\hat{v}_{y_0}$, and $\Delta\hat{v}_{z_2} \to \Delta\hat{v}_{z_0}$. However, in practice, for points located downstream at $\tilde{x}_1 \leqslant -10$, it can be assumed that $\Delta\hat{v}_{y_2}$ and $\Delta\hat{v}_{z_2}$ values are equal to those of $\Delta\hat{v}_{y_0}$ and $\Delta\hat{v}_{z_0}$ for the same coordinates $\tilde{y}_1$ and $\tilde{z}_1$.

Velocities Induced at an Arbitrary Point in Space. Having determined induced velocity components generated by various systems of vortices which left the rotor from a circle of radius r, the complete components along the x, y, and z axes of velocity induced by these vortices at an arbitrary point in space can be computed. For the sake of simplicity, subscript b will be omitted from the symbol of circulation referred to one blade (Γ instead of Γ_b), and equations for the induced velocity components become:

$$\Delta v_x = \frac{b\Delta\Gamma}{4\pi R\mu}\Delta\hat{v}_x, \quad (4.65)$$

$$\Delta v_y = \frac{b\Delta\Gamma}{4\pi R\mu}\Delta\hat{v}_y, \quad (4.66)$$

and

$$\Delta v_z = \frac{b\Delta\Gamma}{4\pi R\mu}\Delta\hat{v}_z, \quad (4.67)$$

where

$$\Delta\hat{v}_y = \Delta\hat{v}_{y_1} + \Delta\hat{v}_{y_0} + \Delta\hat{v}_{y_2}$$

and

$$\Delta\hat{v}_z = \Delta\hat{v}_{z_0} + \Delta\hat{v}_{z_2}.$$

For $x \to \infty$, it can be shown that $\Delta\hat{v}_{y_1} \to 0$; consequently, when $\Delta\hat{v}_{y_2} \to \Delta\hat{v}_{y_0}$,

$$\Delta\hat{v}_{y,-\infty} = 2\Delta\hat{v}_{y_0}. \quad (4.68)$$

This means that at distances sufficiently far downstream from the rotor, the vertical component of induced velocity becomes twice as large as its value in the yOz plane, as long as the y and z coordinates remain the same.

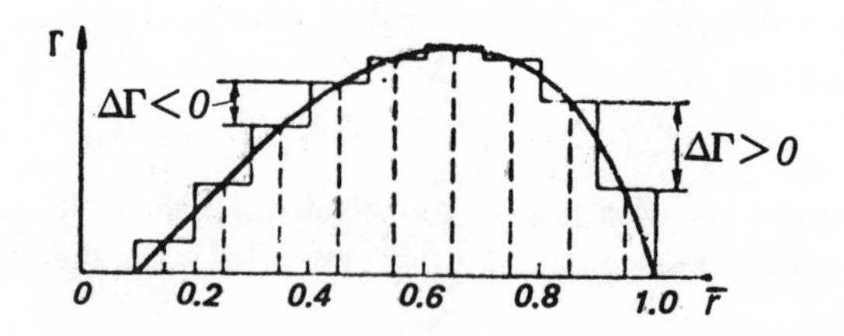

Figure 4.27 Distribution of circulation averaged along the blade

If the circulation distribution along the blade, averaged over one revolution, is known (Fig 4.27), then the procedure of determining induced velocity components (v_x, v_y, v_z) at an arbitrary point (x_1, y_1, z_1) can be visualized as follows.

The continuous curve of $\Gamma = \Gamma(\bar{r})$ is approximated by a segmented line by taking nondimensional radius increments of, say, $\Delta\bar{r} = 0.1$ which provides sufficient accuracy for practical calculations—assuming that the mean circulation for each interval is equal to the actual one at the $\bar{r} = (\bar{r}_n + \bar{r}_{n+1})/2$ station. In this way, incremental variations $(\Delta\Gamma_{\bar{r}})$ at stations $r = 1.0, 0.9, ...0.1$ will be obtained (with the proper sign) as

$$\Delta\Gamma_r = \Gamma_{r-\frac{1}{2}\Delta r} - \Gamma_{r+\frac{1}{2}\Delta r}. \tag{4.69}$$

The sought values of the induced velocity components , as well as inputs representing circulation are expressed in Ref 1 as nondimensional quantities; namely, $\bar{v}_x = v_x/\Omega R$; $\bar{v}_y = v_y/\Omega R$; $\bar{v}_z = v_z/\Omega R$; $\bar{\Gamma} = \Gamma/\Omega R^2$; and $\Delta\tilde{\Gamma}_r = \Delta\Gamma_r/\bar{\Gamma}_{av}$*, where $\bar{\Gamma}_{av}$ is the circulation for one blade averaged over a complete revolution. Expressions for $\bar{v}_x$, $\bar{v}_y$, and $\bar{v}_z$ can be written as follows:

$$\bar{v}_x(\bar{x}_1, \bar{y}_1, \bar{z}_1) = (b\bar{\Gamma}_{av}/4\pi\mu)\hat{v}_x(\bar{x}_1, \bar{y}_1, \bar{z}_1) \tag{4.70}$$

$$\bar{v}_y(\bar{x}_1, \bar{y}_1, \bar{z}_1) = (b\bar{\Gamma}_{av}/4\pi\mu)\hat{v}_y(\bar{x}_1, \bar{y}_1, \bar{z}_1) \tag{4.71}$$

$$\bar{v}_z(\bar{x}_1, \bar{y}_1, \bar{z}_1) = (b\bar{\Gamma}_{av}/4\pi\mu)\hat{v}_z(\bar{x}_1, \bar{y}_1, \bar{z}_1) \tag{4.72}$$

where

$$\hat{v}_x(\bar{x}_1, \bar{y}_1, \bar{z}_1) = \sum_{\bar{r}=\bar{r}_0}^{\bar{r}=1} \Delta\hat{v}_x^{(\bar{r})}(\bar{x}_1, \bar{y}_1, \bar{z}_1)\Delta\tilde{\Gamma}_{\bar{r}} \tag{4.73}$$

$$\hat{v}_y(\bar{x}_1, \bar{y}_1, \bar{z}_1) = \sum_{\bar{r}=\bar{r}_0}^{\bar{r}=1} \Delta\hat{v}_y^{(\bar{r})}(\bar{x}_1, \bar{y}_1, \bar{z}_1)\Delta\tilde{\Gamma}_{\bar{r}} \tag{4.74}$$

$$\hat{v}_z(\bar{x}_1, \bar{y}_1, \bar{z}_1) = \sum_{\bar{r}=\bar{r}_0}^{\bar{r}=1} \Delta\hat{v}_z^{(\bar{r})}(\bar{x}_1, \bar{y}_1, \bar{z}_1)\Delta\tilde{\Gamma}_{\bar{r}} \tag{4.75}$$

---

*It is obvious that $\Delta\tilde{\Gamma}_{\bar{r}}$ can also be expressed as a ratio of dimensional values of the appropriate circulation.

It can be seen from Eqs (4.70) through (4.72) that in order to calculate the desired velocity components at some point $(\bar{x}_1, \bar{y}_1, \bar{z}_1)$, it is necessary to know the overall average circulation around the blade $\bar{\Gamma}_{av}$ and the values of $\hat{v}_x$, $\hat{v}_y$, and $\hat{v}_z$. Some idea regarding the practical procedure of determining the v components can be acquired by considering the necessary steps for computing $\hat{v}_y(\bar{x}_1, \bar{y}_1, \bar{z}_1)$ by assuming, for example, that $\bar{x}_1 = -1.2$; $\bar{y}_1 = 0.2$; $\bar{z}_1 = 0.2$ and $\mu = 0.2$. For a blade divided into ten segments, Eq (4.74) can be written in explicit form as follows:

$$\hat{v}_y(...) = \Delta\hat{v}_y^{(1)}(...)\Delta\tilde{\Gamma}_1 + \Delta\hat{v}_y^{(2)}\Delta\tilde{\Gamma}_2 + ... + \Delta\hat{v}_y^{(10)}\Delta\tilde{\Gamma}_{10}$$

where $\Delta\tilde{\Gamma}_1, ..., \Delta\tilde{\Gamma}_{10}$ represents incremental jumps in the relative blade circulation (with respect to the average) at blade stations $\bar{r} = 0.1$ to $\bar{r} = 1.0$. Then, $\tilde{x}_1$, $\tilde{y}_1$, $\tilde{z}_1$ and μ_* are computed for each of the considered blade stations. For instance, for $\bar{r} = 0.1$, those values would be $\tilde{x}_1 \equiv \bar{x}_1/\bar{r} = -12$; $\tilde{y}_1 \equiv \bar{y}_1/\bar{r} = 2$; $\tilde{z}_1 \equiv \bar{z}_1/\bar{r} = 2$, and $\mu_* \equiv \mu/\bar{r} = 2$, respectively. Having computed these values, $\Delta\hat{v}_y^{(1)}$ can be evaluated using the graphs given in Ch II of Ref 1. Repeating the above procedure for other blade stations, the numerical values of all $\Delta\hat{v}_y$ multipliers can be determined.

The same procedure can be used for other $\Delta\hat{v}$ components and other points of interest; thus, if the values of $\bar{\Gamma}_{av}$ and $\Delta\tilde{\Gamma}_1 ... \Delta\tilde{\Gamma}_{10}$ are known, it is possible to map the induced velocity field in any region of space influenced by a rotor in forward flight.

Average Circulation at a Blade Station. According to the Kutta-Joukowsky law expressed by Eq (4.14), elementary thrust dT developed by a strip $dR \equiv Rd\bar{r}$ wide, and located at radius $r = R\bar{r}$ can be expressed using the notion of the average circulation $(\Gamma_{\bar{r}})$ at that blade station.

$$dT_{\bar{r}} = \rho\Omega\bar{r}R\Gamma_{\bar{r}}Rd\bar{r}. \tag{4.76}$$

However, circulation at station $\bar{r}$ actually varies with the azimuth angle symbolized by $\Gamma(\bar{r},\psi)$ and the elementary thrust $dT_{\bar{r}}$ should be expressed as follows:

$$dT_{\bar{r}} = (1/2\pi)\rho \int_0^{2\pi} \Gamma(\bar{r},\psi)U_{\perp}Rd\bar{r}d\psi. \tag{4.77}$$

Elementary thrust expressed by Eqs (4.76) and (4.77) should be the same, hence the right-hand sides of these equations can be equated to each other. This results in the following expression for the average circulation at station $\bar{r}$:

$$\Gamma_{\bar{r}} = (1/2\pi\bar{r}) \int_0^{2\pi} \Gamma(\bar{r},\psi)\bar{U}_{\perp}d\psi \tag{4.78}$$

where

$$\bar{U}_{\perp} = U_{\perp}/\Omega R.$$

Neglecting the thrust component caused by the profile drag, the circulation at station $\bar{r}$ and azimuth ψ given in Eq (4.16) can be written in terms of the local blade section lift coefficient (c_ℓ) only;

$$\Gamma(\bar{r},\psi) = \tfrac{1}{2}c_{\ell_r}c_r U_\perp. \tag{4.79}$$

In order to express circulation in nondimensional form $[\bar{\Gamma}(\bar{r},\psi) \equiv \Gamma(\bar{r},\psi)/R^2\Omega]$, both sides of the above equation are divided by $R^2\Omega$ which leads to

$$\bar{\Gamma}(\bar{r},\psi) = \tfrac{1}{2}c_{\ell_{\bar{r}}}\bar{c}_{\bar{r}}\bar{U}_\perp. \tag{4.80}$$

where $\bar{c}_{\bar{r}}$ is the nondimensional chord $\bar{c}_{\bar{r}} \equiv c_{\bar{r}}/R$ at station $\bar{r}$.

The section lift, as in Ch III, can be given in terms of the local angle-of-attack $[a(\bar{r},\psi)]$ of the blade station and the two-dimensional lift-curve slope a: $a(\bar{r},\psi) = \theta(\bar{r},\psi) + \phi(\bar{r},\psi)$ where $\theta(\bar{r},\psi)$ (see Fig 3.17) is the local blade pitch angle including twist and all control inputs, and $\phi(\bar{r},\psi)$ is the total inflow angle, taking into account the following: (1) the axial component of the forward velocity, $V \sin a_d$, (2) local induced velocity, $v(\bar{r},\psi) \approx v_y(\bar{r},\psi)$, and (3) the velocity component due to flapping, $v_\beta(\bar{r},\psi)$.

Under the small angle assumption, the section angle-of-attack can be written as follows:

$$a(\bar{r},\psi) = \theta(\bar{r},\psi) + \frac{Va_d + v_y(\bar{r},\psi) + v_\beta(\bar{r},\psi)}{U_\perp}$$

or, dividing the numerator and denominator of the last term of the above equation by $V_t \equiv \Omega R$, a becomes

$$a(\bar{r},\psi) = \theta(\bar{r},\psi) + [\mu a_d + \bar{v}_y(\bar{r},\psi) + \bar{v}_\beta(\bar{r},\psi)]/\bar{U}_\perp. \tag{4.81}$$

Using Eq (4.81) and remembering that $c_\ell(\bar{r},\psi) = aa(\bar{r},\psi)$, Eq (4.80) becomes

$$\bar{\Gamma}(\bar{r},\psi) = \tfrac{1}{2}a\bar{c}_{\bar{r}}[\theta(\bar{r},\psi)\bar{U}_\perp + \bar{v}_y(\bar{r},\psi) + \bar{v}_\beta(\bar{r},\psi) + \mu a_d]. \tag{4.82}$$

Substituting Eq (4.82) into Eq (4.78), switching to nondimensional circulation $(\bar{\Gamma} \equiv \Gamma/R^2\Omega)$, and performing the indicated integration, the following expression is obtained for the circulation at a particular blade station:

$$\bar{\Gamma}_{\bar{r}} = \frac{1}{2}\,a\bar{c}_{\bar{r}}\left[\theta(\bar{r},\psi)\bar{r}\left(1 + \frac{1}{2}\frac{\mu^2}{\bar{r}^2}\right) + \mu a_d + \bar{v}_w(\bar{r}) + \frac{1}{2\pi\bar{r}}\int_0^{2\pi} \bar{v}_{\beta_\psi}\bar{U}_\perp d\psi\right]. \tag{4.83}$$

In the above equation, the last integral represents the contribution of flapping motion. As long as the motion is assumed to be of the first-harmonic type, the integral is equal to zero. This means that although the circulation at various azimuth angles may vary because of blade flapping, its total contribution averaged over one complete revolution would be zero.

Eq (4.83) contains a new symbol, $\bar{v}_w(\bar{r})$ which stands for

$$\bar{v}_w(\bar{r}) = \frac{1}{2\pi\bar{r}}\int_0^{2\pi} \bar{v}_y(\bar{r},\psi)\bar{U}_\perp d\psi = \frac{1}{2\pi}\int_0^{2\pi} \bar{v}_y(\bar{r},\psi)d\psi + \frac{\mu}{2\pi\bar{r}}\int_0^{2\pi} \bar{v}_y(\bar{r},\psi)\sin\psi\, d\psi. \tag{4.84}$$

On the right-hand side of this equation, the expanded expression for $\bar{v}_w(\bar{r})$ represents the induced velocity at station $\bar{r}$ averaged over one complete rotor revolution. Hence, it will be called

$$\bar{v}_{\bar{r}} = \frac{1}{2\pi}\int_0^{2\pi} \bar{v}_y(\bar{r},\psi)\, d\psi. \tag{4.85}$$

It is shown in Ref 1 that $\bar{v}_{\bar{r}}$ is related to $\bar{\Gamma}_{\bar{r}}$ in the following way:

$$\bar{v}_{\bar{r}} = \frac{b\bar{\Gamma}_{\bar{r}}}{4\pi\bar{V}'}, \tag{4.86}$$

where b is the number of blades and $\bar{V}' \equiv V'/\Omega R$ is the nondimensional resultant velocity of flow through the disc. For high-speed horizontal flight, it may be assumed that $\bar{V}' \approx \mu$.

The physical significance of Eq (4.86) lies in the fact that it shows, as in the case of hover discussed in Sect 4.2 of this chapter, that the value of induced velocity averaged over a complete revolution at a given blade station ($\bar{r}$) depends only on the level of circulation at that station–also averaged over one complete revolution. It is not influenced by the circulation at other blade stations.

For the second integral on the right side of Eq (4.84), a new symbol of $\bar{v}_\mu(\bar{r})$ can be introduced:

$$\bar{v}_\mu(\bar{r}) = (\mu/2\bar{r})\bar{v}_{s_1} \tag{4.87}$$

where

$$\bar{v}_{s_1}(\bar{r}) = (1/\pi)\int_0^{2\pi} \bar{v}_y(\bar{r},\psi)\sin\psi\, d\psi \tag{4.88}$$

represents a coefficient at $\sin\psi$ in the development of $\bar{v}_y(\bar{r},\psi)$ into the Fourier Series. Again, it is shown in Ref 1 that Eq (4.87) can be evaluated in terms of (a) the averaged nondimensional circulation ($\bar{\Gamma}_{\bar{r}}$) at blade station $\bar{r}$, (b) number of blades b, and (c) advance ratio μ:

$$\bar{v}_\mu(\bar{r}) = -(\mu/\bar{r}^2)(b\bar{\Gamma}_{\bar{r}}/4\pi). \tag{4.89}$$

Substituting Eqs (4.86) and (4.89) for the integrals on the right side of Eq (4.84), we have:

$$\bar{v}_w(\bar{r}) = -\left(1 + \frac{\mu^2}{\bar{r}^2}\right)\frac{b\bar{\Gamma}_{\bar{r}}}{4\pi\mu}. \tag{4.90}$$

Now, substituting Eq (4.90) into Eq (4.83) and solving it for $\bar{\Gamma}_{\bar{r}}$, the following is obtained:

$$\bar{\Gamma}_{\bar{r}} = B\left[\theta_{\bar{r}}\bar{r}\left(1 + \frac{1}{2}\frac{\mu^2}{\bar{r}^2}\right) + \mu a_d\right] \tag{4.91}$$

where

$$B = \frac{4\pi a\mu\bar{c}}{8\pi\mu + ab\bar{c}[1 + (\mu^2/\bar{r}^2)]}$$

Relationships developed during the preceding discussion of the flat-wake concept should enable one to write a suitable computer program for determination of the induced velocity field. Those interested in achieving this goal without the aid of computers can find the necessary inputs in the graphical presentations of functions $\widetilde{K}$, $\widetilde{L}$, $\widetilde{M}$, and $\widetilde{N}$ in Ref 1.

6.3 Validity of the Flat Wake Concept

Figs 4.28 and 4.29, reproduced from Ref 1, give some idea as to the limits of the validity of the flat-wake concept. In Fig 4.28, the downwash velocity $\bar{v}_y = f(\psi)$ at $\bar{r} = 0.7$ was computed for a rotor with flat untapered blades; having $\sigma = 0.07$; and operating at $a_d = 0°$, $\mu = 0.15$, and $C_T = 0.006$. The $\bar{v}_y(\psi)$ relationship represented by dashed lines was first computed for circulation constant with the azimuth and varies only with ($\bar{r}$) according to Eq (4.91). Then, the variation of circulation with the azimuth (solid lines with crosses) was considered; assuming that $\Gamma = \Gamma_0 + \Gamma_1 \sin\psi + \Gamma_2 \cos 2\psi$. It can be seen from this figure that the assumption of $\Gamma(\psi) = const$ at $\bar{r} = 0.7$ has little influence on the $\bar{v}_y$ values also computed at $\bar{r} = 0.7$. In both cases, a satisfactory agreement with test measurements is shown, although the $\bar{v}_y$ values computed on the $\widetilde{\Gamma}(\psi) = const$ assumption appear to be even closer to the experimental results.

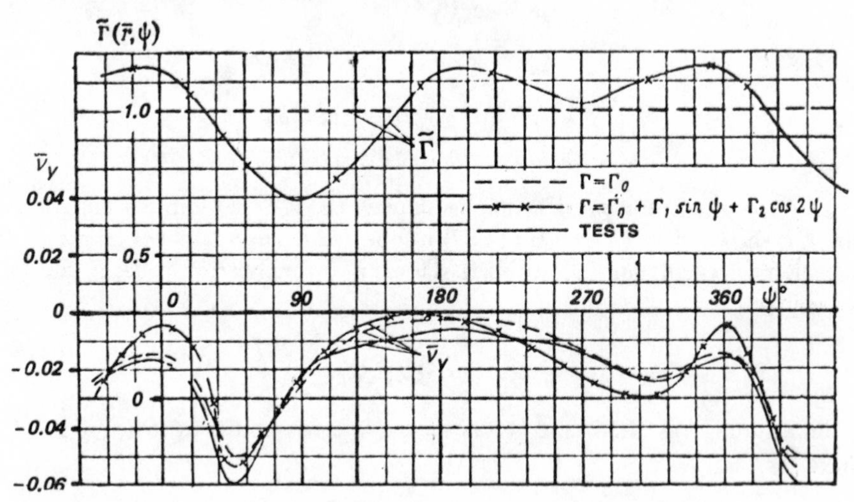

Figure 4.28 A comparison of predicted and measured $\bar{v}_y = f(\psi)$ at $\bar{r} = 0.7$[1]

Figure 4.29 provides further insight into this comparison by showing computed vs measured $\bar{v}_y = f(\bar{r})$ at various azimuth angles along the rotor diameter.

It may appear somewhat surprising that the measured downwash values at the advance ratio of $\mu = 0.15$ agreed so well with those predicted by the flat-wake approach. It should be realized that at low μ values, the amount of average deflection of the wake in the vicinity of the disc is quite noticeable and then doubles further downstream.

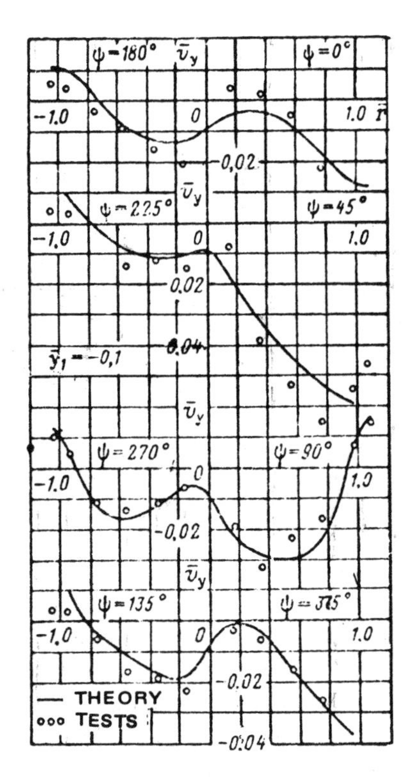

Figure 4.29 A comparison of predictions and tests of $\bar{v}_y(\bar{r})$ at several azimuth angles[1]

The order of magnitude of the wake deflection with respect to the velocity vector of the distant horizontal flow can easily be obtained from the simple momentum theory which indicates that the ratio of induced velocity at the disc (v) to V_{ho} is $v/V_{ho} = C_T/2\mu^2$. For the considered example, this ratio amounted to $v/V_{ho} = 0.13$, or about $7.5°$. Although the results of the analysis were good at this particular wake deflection, it may be expected that the flat-wake concept may lose its validity for larger slipstream inclinations.

These inherent limitations are recognized in Ref 1 where it is indicated that the flat-wake approach should be used only when the advance ratio $\mu \geqslant 1.62\sqrt{C_T}$. It should be noted that for the C_T value shown in Figs 4.28 and 4.29, this μ limit value would amount to $\mu \geqslant 0.12$, and thus the considered case was still within the validity range of the flat-wake model.

6.4 Other Approaches to the Flat-Wake Concept

In the United States, Ormiston[14] investigated the flat-wake concept and indicated that because of less computational complexity, this approach may be especially suitable for preliminary design stages. However, similar to Baskin et al, he recognized the limitations of mathematical models based on this concept. Consequently, he recommends that it be applied to advance ratios where $\mu \geqslant 0.15$.

Similar to the previously considered case, the flat vortex sheet representing the wake is broken down into the simple vortex forms shown in Fig 4.30.

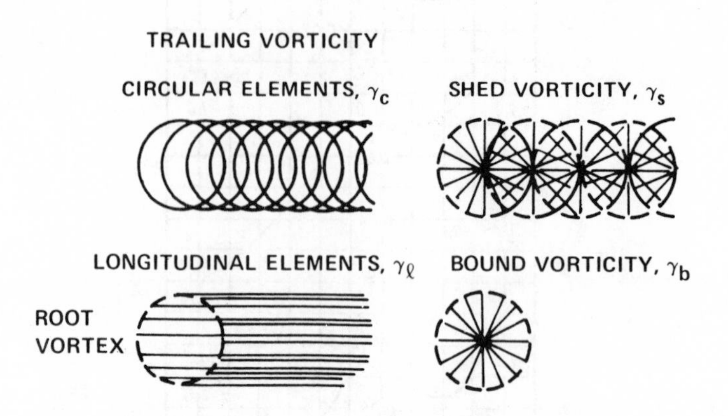

Figure 4.30 Schematic of vortex decomposition[14]

The trailing vorticity is decomposed into circular (γ_c) and longitudinal (γ_ℓ, including root vortices) components. Shed vorticity is represented by radial vortex filaments (γ_s). Finally, bound vorticity (γ_b) is assumed to be radially distributed with its center at a fixed point of the wake (rotor hub). For the special case of bound circulation constant with the azimuth, γ_s vanishes and γ_b does not contribute to the downwash generation (see Sect 4.1 of this chapter). In general, however, all four types of vorticity, plus the root vortex, should be considered.

We will now consider the wake vorticity as consisting of circular and radial elements which are resolved into orthogonal x and y components amenable to the Biot-Savart interpretations[1].

An example of the character of downwash distribution at the rotor disc obtained in this way is shown in Fig 4.31. This was done by assuming that $\Gamma(\bar{r},\psi) = const$, which postulates the existence of a strong root vortex, whose importance in the determination of downwash velocity can be appreciated by comparing Fig 4.31b with 4.31c.

7. HOVERING AND VERTICAL CLIMB OF A ROTOR HAVING A FINITE NUMBER OF BLADES

Development of the Wake Concept. The simple wake models of a rotor in hover and forward flight discussed in Sects 4 through 6 of this chapter should have exposed the reader to most of the basic techniques that may be encountered in dealing with the application of the vortex theory to various rotary-wing aerodynamic problems within the limits of incompressible and inviscid flow. In order to provide a practical example of the application of the acquired basic knowledge to more sophisticated wake concepts developed for hover OGE and vertical climb, the problem of determining aerodynamic loading of a rotor having a finite number of blades under static conditions and during axial translation in the thrust direction will be considered.

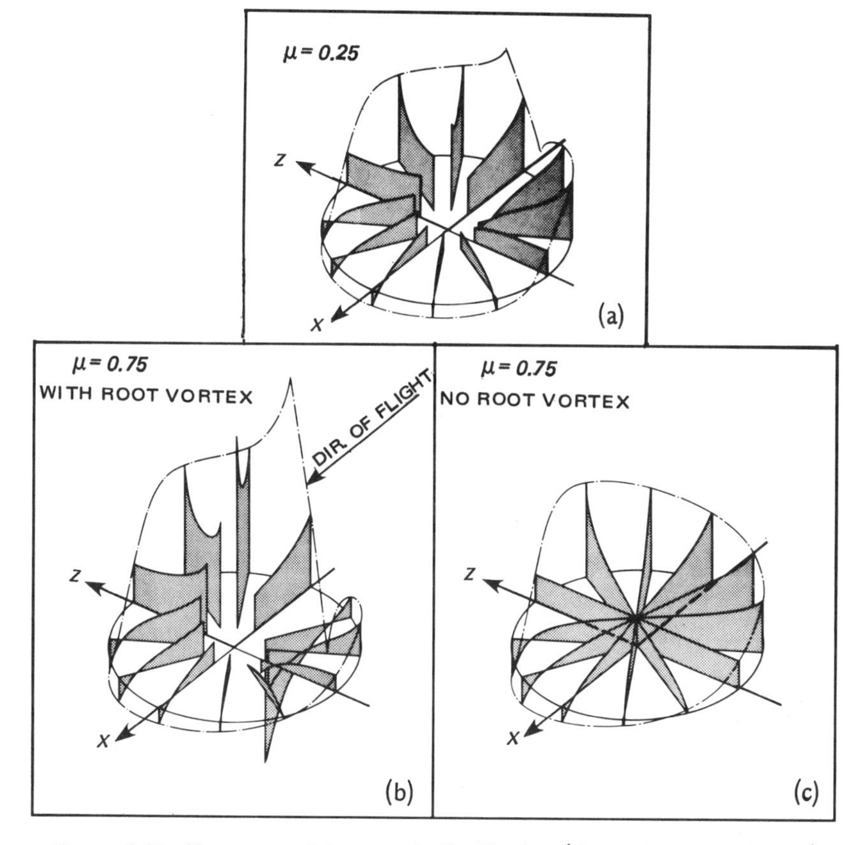

Figure 4.31 Character of downwash distribution (shown here as going up) at the rotor disc for $\Gamma(r,\psi) = const$

In Refs 4 and 5, many investigators tried to improve the blade load and hence, rotor performance methods by developing more realistic rotor-wake models than those discussed in Sect 4 of this chapter. Three of these approaches, dealing with a rotor in hover or vertical climb, are discussed below in the order of increasing complexity of the assumed vortex system.

7.1 Noncontracting Wake

A vortex system consisting of bound vortices and noncontracting wake generated by the individual blades may be considered as one of the simplest conceptual models of a rotor having a finite number of blades. An example of this approach can be found in Ch VI of Ref 1. Here, the blades are represented by segments of the lifting-line filaments with the circulation varying along their span: $\Gamma(\bar{r}) \neq const$. Furthermore, it is assumed that trailing vortices springing from the lifting line move downstream in a rectilinear motion at a constant speed equal to the sum of the distant incoming flow (V_∞) and the induced velocity averaged over the disc ($\tilde{v}$). In this way, a helical vortex sheet formed behind each blade would be inscribed into a cylinder of the same radius as that of the rotor (Fig 4.32a). Any individual vortex filament on the vortex sheet would form a helix of constant pitch (Fig 4.32b). Such helix can be described by a relatively simple parametric equation.

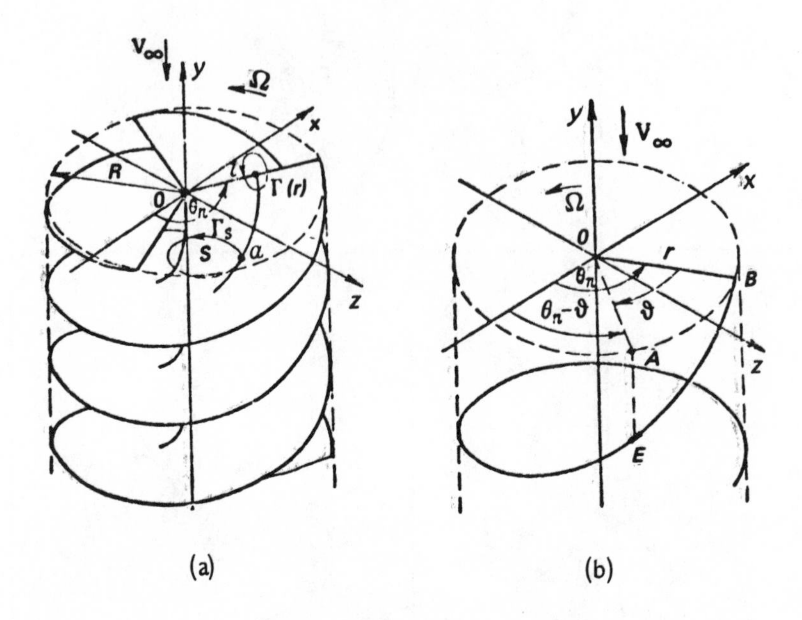

Figure 4.32 Noncontracting wake

For practical computations, the blade wake is "discretized" by dividing the blade into m segments and assuming step variations in circulation at each segment, and a discrete vortex of strength Γ_1, Γ_2, ...Γ_m leaving the blade at each step (Fig 4.33). Knowing the Γ_i values and having equations of the corresponding vortex filament shapes, the Biot-Savart relationships can be used to compute the induced velocities ($\vec{v}$) in the space around the rotor. In particular, velocities induced at control points (j) along the blade lifting line can be determined as

$$\bar{v} = \sum_{i=1}^{m} \bar{v}_{i_j} \bar{\Gamma}_i \quad (j = 1, 2, 3 ... m) \tag{4.92}$$

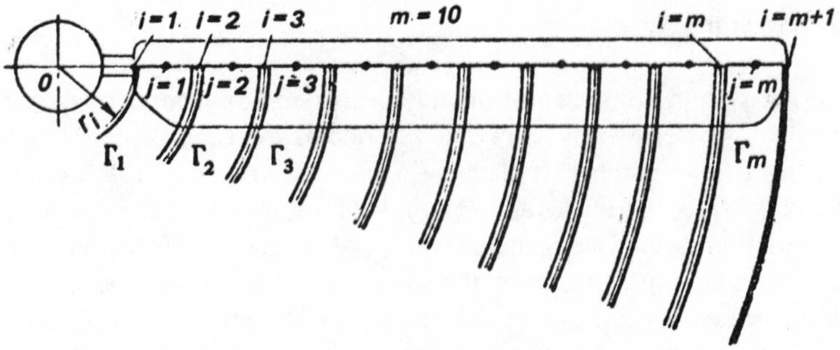

Figure 4.33 System of discrete vortices of a blade

where $\bar{v}_{i_j}$ are the proper influence coefficients, while $\bar{\Gamma}_i$ values can be related to the blade geometry and its operating conditions. Details of this operation can be found in Ref 1, where it is shown that by using this model of the vortex system of a rotor, a good agreement can be obtained between computed and measured performance values, including hover (Fig 4.34).

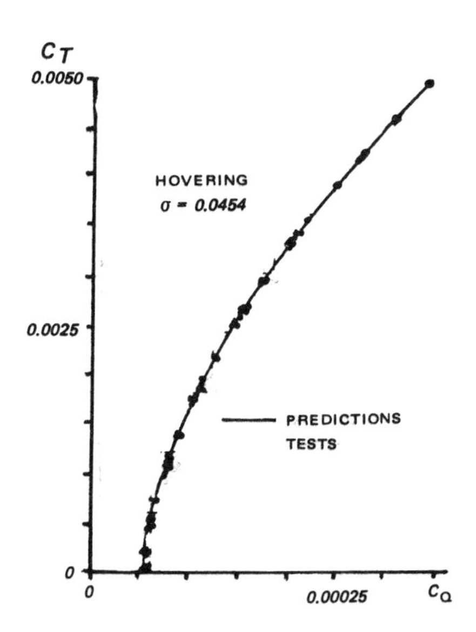

Figure 4.34 Comparison of predicted and measured $C_T = C_T(C_Q)$[1]

Reservations Regarding Noncontracting Wake. In spite of the good agreement demonstrated in rotor performance predictions, other investigators indicated that neglecting the wake contraction may lead to a seriously disturbed blade loading, especially an underestimation of loads in the tip region[15]. For this reason, different approaches aimed at the incorporation of wake contraction were developed. In some cases, this was done on a rigorous basis with a minimum of simplifying assumptions, while in others, semi-empirical corrections were added. The method developed by Erickson and Ordway[16] can be cited as an example of the rigorous approach, while that of Davenport, Magee, et al[17] represents the philosophy of semi-empirical corrections.

7.2 Free Wake

The Model. The essence of the method of Erickson and Ordway (developed in cooperation with Borst and Ladden) consisted of finding a way of determining the deformation of a blade vortex sheet as it moves downstream. These deformations would comprise an elongation of the sheet along the rotor axis, accompanied by a radial contraction and tangential distortions. In addition, the edges of the sheet are locally unstable and tend to roll up. All of these deformations may be potentially important in the case of static operation (hover), since their magnitude is larger than in vertical climb, and they occur close to the rotor disc.

Because of the type of considered flow (purely axial), a rotor-fixed polar coordinate system (Fig 4.35) appears suitable for this case. As before, the y axis coincides with the rotor axis; however, this time it is more convenient to assume that its positive direction is the same as that of the downwash flow. The rotor induced velocity $\vec{v}$ at a point $P(y, r, \vartheta)$ may therefore be considered as a vectorial sum of the following components: v (axial), u (radial), and w (tangential). All b blades are assumed to stay in the $y=0$ plane, and the $\vartheta=0$ axis coincides with that of the first indexed blade ($i=1$); thus, the blade positions may be given as

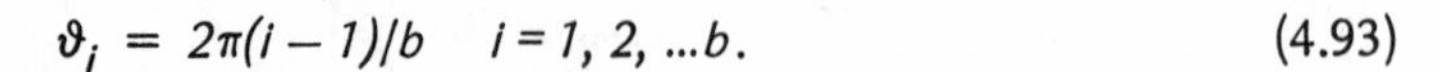

$$\vartheta_i = 2\pi(i-1)/b \quad i = 1, 2, \ldots b. \tag{4.93}$$

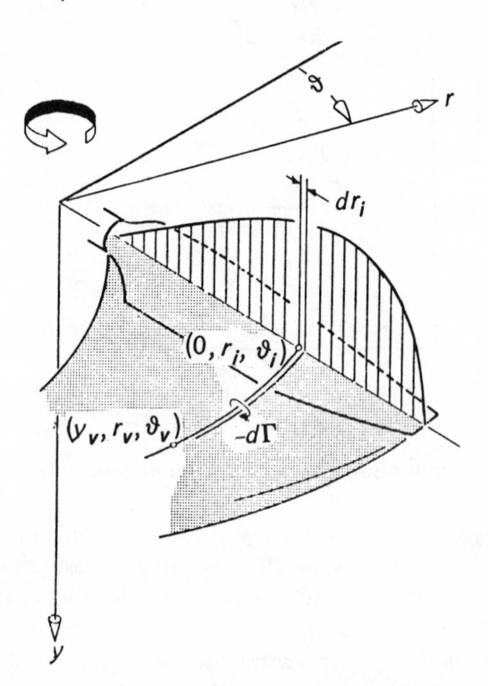

Figure 4.35 Polar coordinates

As before, the individual blades are modeled by straight-line segments of bound vortices whose circulation varies with the blade radius: $\Gamma(\bar{r}) \neq const.$

Induced Velocities. If one takes into account both bound and trailing vortices, then at any instant of time at a specified point in space, using the Biot-Savart law, the induced velocity components can be expressed as follows:

$$u = \sum_{i=1}^{b} \int_{R_o}^{R} \left[\Gamma(r_i)\, \mathbf{U}_B - \frac{d\Gamma(r_i)}{dr_i}\, \mathbf{U}_T \right] dr_i$$

$$v = \sum_{i=1}^{b} \int_{R_o}^{R} \left[\Gamma(r_i)\, \mathbf{V_B} - \frac{d\Gamma(r_i)}{dr_i}\, \mathbf{V_T} \right] dr_i \tag{4.94}$$

$$w = \sum_{i=1}^{b} \int_{R_o}^{R} \left[\Gamma(r_i)\, \mathbf{W_B} - \frac{d\Gamma(r_i)}{dr_i}\, \mathbf{W_T} \right] dr_i$$

where R_o and R, respectively, are the cutout and the blade radii, Γ is the strength of bound circulation at $(0, r_i, \vartheta_i)$, and $-d\Gamma$ is the strength of an elementary vortex ribbon leaving the blade at $(0, r_i, \vartheta_i)$. As defined in Ref 16, $\mathbf{U_B}$, $\mathbf{V_B}$, and $\mathbf{W_B}$, respectively, are influence functions for the u, v, and w velocity components at point (y, r, ϑ), induced by a bound vortex of unit strength and unit length. Similarily, $\mathbf{U_T}$, $\mathbf{V_T}$, and $\mathbf{W_T}$ represent the influence functions for the arbitrarily deformed trailing vortex ribbon of unit strength and semi-infinite length which sprang from the blade at $(0, r_i, \vartheta i)$. As in the previously considered cases, the influence of the bound vortices can often be neglected, since their contribution to the axial (and obviously, radial) induced velocity components is zero, and is only negligible to the tangential ones. Explicit expressions for influence functions are given in Ref 16.

Formulation of a Free Wake. In this approach, it can be clearly seen that no preconceived assumptions are made regarding the wake form. The wake is free to assume any shape that may result from the mutual interaction between the total vortex system of the rotor, and the velocity field induced by this system. The mechanism for achieving the ultimate wake shape (assuming that eventually, a steady-state can be reached) can be imagined as follows: The elements of vorticity forming the trailing (free) vortex sheets are carried in the rotor slipstream only under the influence of the velocity components given in Eq (4.94). During an infinitesimal increment of time $d\tau$, an element of vorticity will be carried for the following distances in the y and r directions, and through an angle ϑ:

$$dy = v\,d\tau; \qquad dr = u\,d\tau; \qquad d\vartheta = [\Omega + (w/r)]\,d\tau. \tag{4.95}$$

At time t (considered here as a parameter), the distances y, r, and the corresponding angle ϑ traveled by the considered element of the wake will be

$$r_v = \int_0^t u_v\,d\tau; \quad y_v = \int_0^t v_v\,d\tau; \quad \vartheta_v = \vartheta_i + \Omega t + \int_0^t (w/r)\,d\tau. \tag{4.96}$$

Eq (4.94) can be combined with Eq (4.96) and thus express the dependence of induced velocities on the position of the blade vortex sheets. These combined expressions are given in the following form in Ref 16:

$$\begin{aligned} u &= \mathbf{O_u}\,(\Gamma,\ -d\Gamma/dr;\ \ u, v, w) \\ v &= \mathbf{O_v}\,(\Gamma,\ -d\Gamma/dr;\ \ u, v, w) \\ w &= \mathbf{O_w}(\Gamma,\ -d\Gamma/dr;\ \ u, v, w) \end{aligned} \tag{4.97}$$

where $\mathbf{O}_u$, $\mathbf{O}_v$, and $\mathbf{O}_w$ respectively, are the Biot-Savart operators for the radial, axial, and tangential velocity components.

Blade Circulation. In Eq (4.97), circulation distribution along the blade and hence, its derivative, are still undetermined. In order to relate these quantities to rotor geometry and its operating conditions, it is recalled that the blade section lift coefficient at some station r_i can be expressed as

$$c_\ell = 2(dL/dr_i)/\rho U_{\perp_i}^2 c. \tag{4.98}$$

where c is the local blade chord and as before, $U_{\perp_i} \approx \Omega r + w_i$ is the component perpendicular to the blade axis of the resultant flow encountered by the blade at station r_i (Fig 4.36).

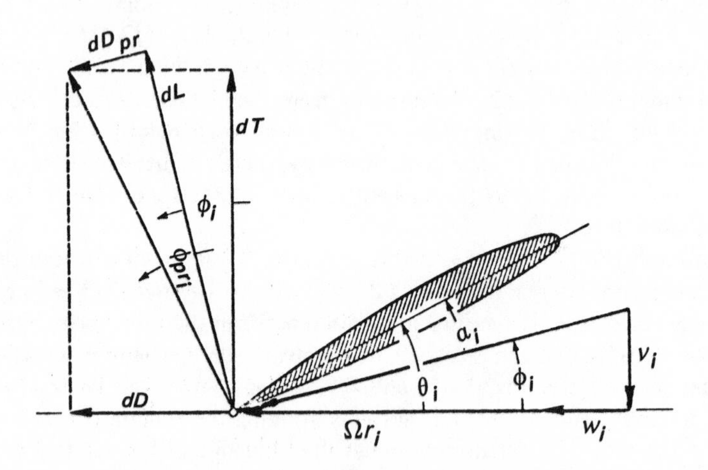

Figure 4.36 Flow scheme at a blade station

Taking Eq (4.16) into consideration, circulation Γ at blade station r_i can be written as

$$\Gamma = \tfrac{1}{2} c_\ell U_{\perp_i} c. \tag{4.99}$$

Final Determination of Wake Shape and Induced Velocity Values. Eqs (4.97) and (4.99) now form a complete set of equations which should permit one to determine the rotor wake shape and the associated field of induced velocities. The solution to this set of equations can be obtained through an iteration process using numerical techniques adapted to large capacity computers.

To facilitate this process, Eq (4.97) is rewritten under the following form:

$$\begin{aligned} u &= u - K_u[u - \mathbf{O}_u(\Gamma,...)] \\ v &= v - K_v[v - \mathbf{O}_v(\Gamma, ...)] \\ w &= w - K_w[w - \mathbf{O}_w(\Gamma, ...)] \end{aligned} \tag{4.100}$$

where K_u, K_v, and K_w are the iteration factors which are "chosen as necessary to achieve convergence and may be constant or depend on any variable of the problem."

The iteration process begins by some approximation for the inflow as well as the blade circulation and a guess at the induced velocity field in the rotor wake. In this respect, the combined blade element and momentum theory can provide a good approximation for the blade circulation $\Gamma = \Gamma(r)$ in Eq (4.99) and the induced velocity values.

Knowledge of the v_i and w_i values at the intermediate steps of the iteration process permits one to obtain the corresponding $\Gamma(r)$ and $-d\Gamma/dr$. The relationship of this quantity through Eq (4.99) to blade geometry $c(r)$ and $\theta_t(r)$, operating conditions θ_o and V_t, and blade airfoil characteristics can be easily established (Fig 4.36)

Once the fixed values of v_i and w_i are obtained, the rotor thrust and power can be computed by integrating sectional inputs in a manner similar to that previously discussed in Ch III.

In principle, the above outlined rigorous approach should provide a suitable tool for an exact prediction of blade loading and rotor performance in hover. In practice, however, correlation with tests is often not as good as that resulting from simpler methods (Fig 4.37). Furthermore, the free-wake method is somewhat limited because of the amount of computer time required to obtain acceptably accurate solutions. For this reason, semi-empirical techniques were developed, which also take into account rotor-wake interactions.

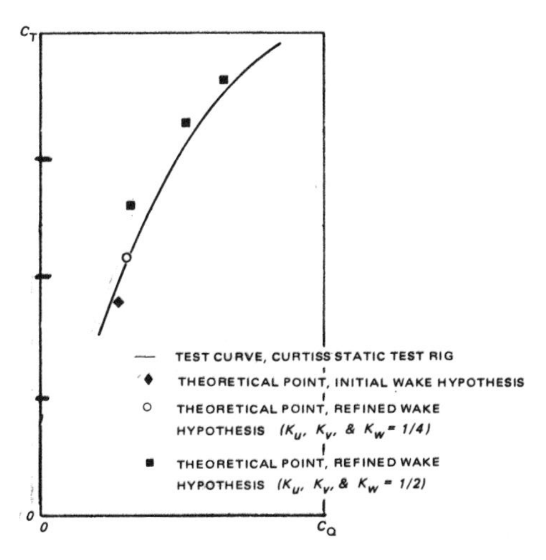

Figure 4.37 Comparison of predictions with test results[16]

7.3 Semi-Empirical Approach.

Wake Contraction. A method developed by Davenport, Magee, et al[17] is briefly reviewed as an example of an approach in which wake contraction effects are evaluated through empirically-established influence coefficients. An additional interest in this method stems from the fact that it forms a basis for the computer program used in Ch II, Vol II, of this textbook.

Similar to the two previously considered cases, each blade is modeled by a bound vortex filament with the circulation strength varying along the radius $\Gamma(r) \neq const$. It was also assumed that $\Gamma(\psi) = const$; consequently, there are no shed vortices in the wake.

In order to facilitate numerical calculations, it is postulated (as in the case of the simplified approach) that blade circulation varies in steps; thus resulting in a wake consisting of discrete trailing vortices. In contrast to the simple case considerations, the blade intervals associated with jumps in circulation are not assumed equal, but by following the so-called cosine law, they become shorter at the tip:

$$r_j = 1 - \cos\vartheta \tag{4.101}$$

where $\vartheta = \frac{1}{2}\pi/n$; n being the total number of blade segments to be considered (usually $n \geqslant 13$). This cosine law contributes to a greater accuracy of analysis, especially in the blade tip region.

It was previously emphasized that contrary to such cases as vertical climb at a high rate, the flow that carries vortex filaments in the wake during hover is solely induced by these vortices. Consequently, it becomes of utmost importance to establish, in the beginning, the so-called *nominal* wake whose shape would approximate as closely as possible the wake corresponding to the velocity field induced by the rotor. Contraction of the slipstream represents one of the most important inputs as far as the shape of the wake is concerned. The simple momentum theory as discussed in Sect 3.1, Ch II, gives only the ratio of the slipstream radius far downstream (R_∞) to that at the rotor disc (R), which is assumed to be the same as that of the rotor itself. In hover, this ratio was $R_\infty/R = 0.707$ regardless of rotor C_T values.

By modeling the wake with the help of vortex rings of finite cross-section[18], a trend between the ultimate wake contraction and rotor thrust coefficient was established. It was shown that R_∞/R is influenced by C_T values as well as by the ratio of the ring vortex core radii to the rotor radii. In the presently discussed approach, however, the influence of this parameter was ignored and a single-line relationship for $(R_\infty/R) = f(C_T)$ was assumed (Fig 4.38).

Slipstream Acceleration Parameters. Knowledge of the relative contraction of the distant wake, together with the known radius at the disc, give only two extreme cross-sections of the wake boundary. In order to get some idea as to its shape at the intermediate locations, it is assumed that the axial velocity in the rotor slipstream varies as follows:

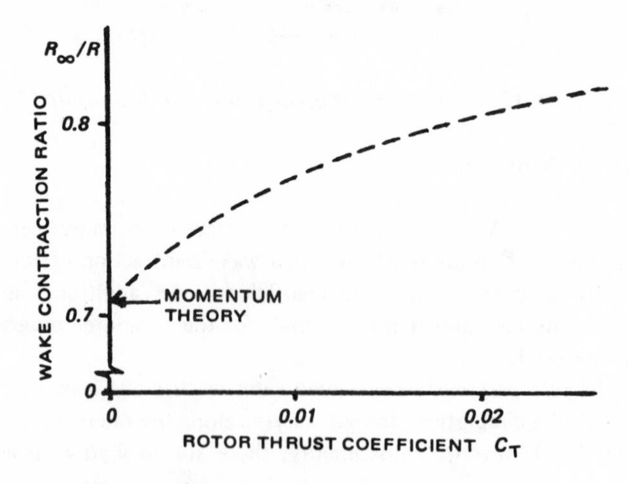

Figure 4.38 Wake contraction ratio variation

$$v(y)/v_{\infty} = \tfrac{1}{2}[2 - (1 - A)e^{-Ny/R}] \tag{4.102}$$

where A is the slipstream cross-section area ratio; and for the case of hover, v_{∞} is determined from the simple momentum theory:

$$v_{\infty} = V_t\sqrt{2C_T/[1 - (R_o/R)^2]};$$

R_o being the blade root radius. Values of A were computed from the variation of R_{∞}/R given in Fig 4.38 and the continuity equation.

As the flow moves downstream and the slipstream contracts, the flow accelerates and its rate of acceleration would depend on the magnitude of the exponent N in Eq (4.102). Assuming that the value of N is known, the nominal wake geometry of vortices carried out by the flow inside of the slipstream can be figured out. As to the tip vortices which lie on the boundary of the rotor slipstream and the outside flow—the latter being zero in the case of hovering—it is assumed in Ref 17 that these vortices are convected axially at a rate equal to the average of the speed of the slipstream and the outside flow.

Values of the exponent N were obtained empirically by computing the performance of several propellers and rotors in hover, using various magnitudes of N and noting those which best correlate with test results. This data was then plotted vs thrust coefficient, and a single envelope over the entire test range was obtained (Fig 4.39).

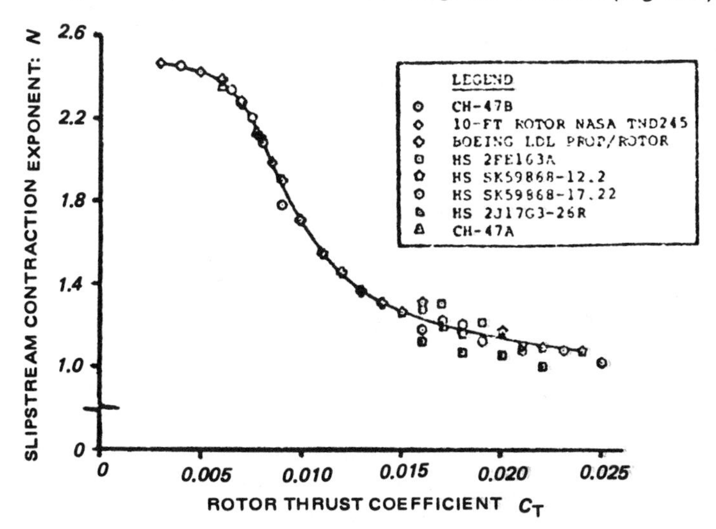

Figure 4.39 Slipstream acceleration parameter

Induced Flow and Blade Loading. Establishment of the nominal shape of the wake permits one to approximately establish positions of the discrete trailing vortex filaments in the rotor slipstream. It now becomes possible to develop a procedure leading to the achievement of the following goals:

(1) Matching the loading (via blade element angle of attack) to the induced flow at the blade which depends on vortex wake strength *and* shape, and

(2) Matching the wake shape to the induced flow at all points of the wake.

This is done in the following way: In addition to the previously mentioned blade stations $\bar{r}_j$ $(j=1,2,...n)$ where circulation undergoes a step variation in its value, another set of control points is selected for induced velocity calculations between the jumps in circulation. Further procedure is carried out in the following steps:

(1) Inputs necessary to determine the nominal wake are obtained from the data regarding the desired thrust T, tip speed V_t, air density, and the velocity of axial translation if $V_c \neq 0$. Consequently, the location in space of all the vortex filaments representing the rotor trailing vorticity can be established (Fig 4.40). The Biot-Savart law is now applied through the proper computer program to determine the velocity component induced at each control point by each vortex filament of unit circulation strength. As in previous cases, the so-determined velocities are called "influence coefficients."

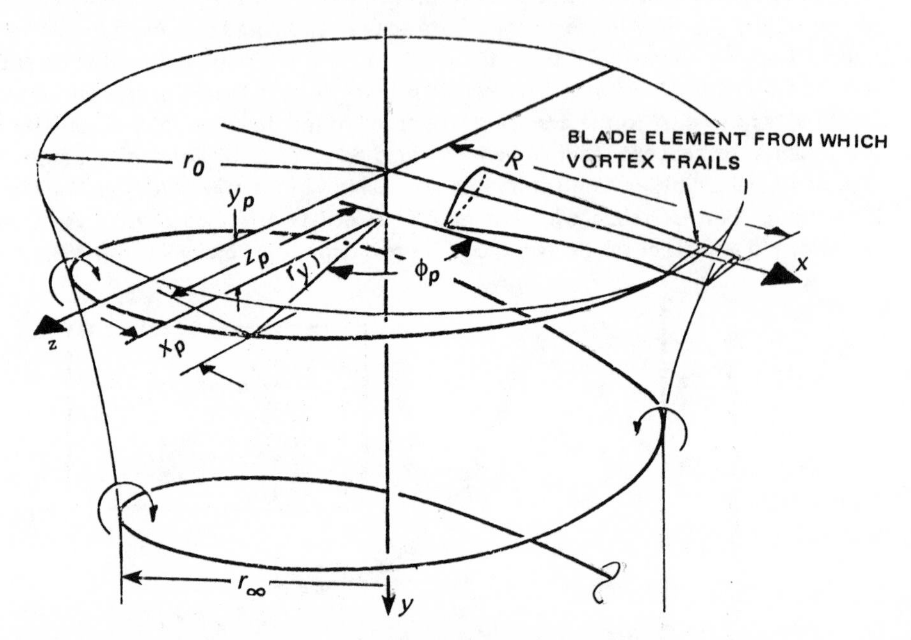

Figure 4.40 Geometry of a single vortex filament

(2) From the assumed $\Gamma(\bar{r})$ of the bound vortex, $\Delta\Gamma_j$ corresponding to each vortex filament is determined. Induced velocities at control points $i = 1, 2, ...n$ are computed from the known vortex strength $\Delta\Gamma_j$ and the corresponding influence coefficients. Appropriate adjustments are made to reflect the departure of the vortex filament geometry from the nominal (originally used to compute the influence coefficients).

(3) Using the influence coefficients obtained in step (1) and the $\Delta\Gamma_j$ values in step (2), the induced velocities at control points j are computed. Having the new induced velocity values, adjustments are made to reflect the departure of the vortex filament positions from those which were originally assumed.

(4) Taking into consideration the blade geometry (collective pitch, chord, and twist distribution) as well as the previously obtained induced velocity values, new blade loading at each control point is computed (see Fig 4.36, and the associated procedure) using tabulated two-dimensional airfoil data.

The last three steps are repeated until an acceptable agreement between the blade loading and velocity components is obtained.

Now, the thrust and torque of the rotor as a whole are computed in a way similar to the previously discussed procedures.

A final check is performed by comparing the so-obtained thrust with the originally assumed one. If there is not a satisfactory agreement, a new collective pitch value is assumed and the entire procedure repeated.

An example of the degree of correlation between the computed and measured blade loading values is given in Fig 4.41.

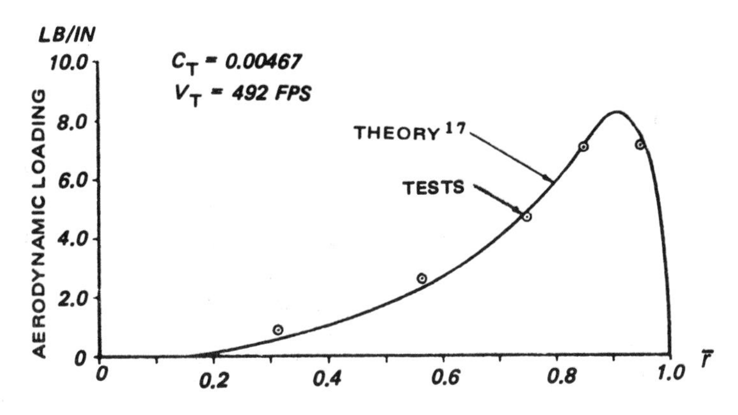

Figure 4.41 Blade loading correlation

The discussion presented above should contribute to an understanding of the method of refining the rotor wake geometry through empirical inputs. The reader interested in computational procedures associated with this approach is directed to Ref 17.

8. APPLICATION OF THE LIFTING–SURFACE THEORY (HOVER AND VERTICAL CLIMB)

8.1 Statement of the Problem

In the preceding discussion of rotors having a finite number of blades, the blades were modeled by a single bound vortex filament. However, even on a purely intuitive basis, one may expect that such a model might be accurate for slender blades, but would not be adequate for blades having a low aspect ratio. One may also expect that in the blade tip regions where three-dimensional effects are more noticeable, greater accuracy in predicting air loads should be obtained if such aspects as chordwise variation of the load is taken into consideration. For this reason, a place may be found in rotary-wing aerodynamics for a conceptual model based on the lifting-surface approach.

Boundary Conditions and Basic Assumptions. The goal of modeling a blade by a thin lifting surface may be stated as follows: Knowing the geometry of the blade and its motion through the medium, the associated field of induced velocities and aerodynamic loading of the surface must be determined. In this respect, one may visualize several boundary conditions which must be fulfilled in a meaningful physical model: (a) No flow may penetrate the lifting surface. This means that the component of the resultant flow normal to the lifting surface must be $V\perp_s = 0$. (b) The Kutta-Joukowsky condition of finite velocity of flow at the trailing edge must also be fulfilled.

It is usually assumed that the blade is not stalled and thus, the flow around the lifting surface is smooth and without separation. Under these conditions, only the lifting surface itself, and possibly its wake, represents a surface of discontinuity in the otherwise irrotational potential flow.* Consequently, it can be imagined that the lifting surface is formed by a continuous layer of properly distributed vorticity which, in turn, may be approximated by a grid of horseshoe-shaped vortices of finite strength. The central segments of these horseshoes are parallel to the blade axis, and remain bound to the blade while the free ends first move along the lifting surface and are then carried in the rotor slipstream. Should the horseshoe vortices be infinitesimal, then a continuous vortex sheet would stream from the blade. With the finite-size horseshoe vortices, discrete vortex filaments would be present in the rotor wake.

As discussed in Sect 7 of this chapter, it may be assumed that the rotor slipstream either retains a constant cross-section, or it contracts. For instance, in Ch VIII of Ref 1, it was assumed that no slipstream contraction occurs. As a result, in such regimes of flight as vertical climb, all trailing vortices would move downstream along straight lines at a uniform speed equal to the sum $(V_c + v_c)$ of the incoming flow $(-V_c)$, and the ideal induced velocity at the disc $(-v_c)$ (see Eq (2.11), etc.).

The relationship between the induced velocity field and strength of a continuous vortex sheet can be expressed with the help of the Biot-Savart law, leading to an integral equation. However, by approximating the vortex sheet system as an assembly of discrete horseshoe vortices, this equation can be replaced by a system of linear algebraic equations. This latter approach, discussed in detail in Ch VIII of Ref 1, is briefly reviewed here.

8.2 Discrete Vortices

Equations of the Vortex System. The blades are indexed by n $(n = 1, 2, \ldots b)$ and a Cartesian coordinate system $Oxyz$ is attached to the number one blade (Fig 4.42). The parametric equation for the lifting surface modeling the *n-th* blade may be written as

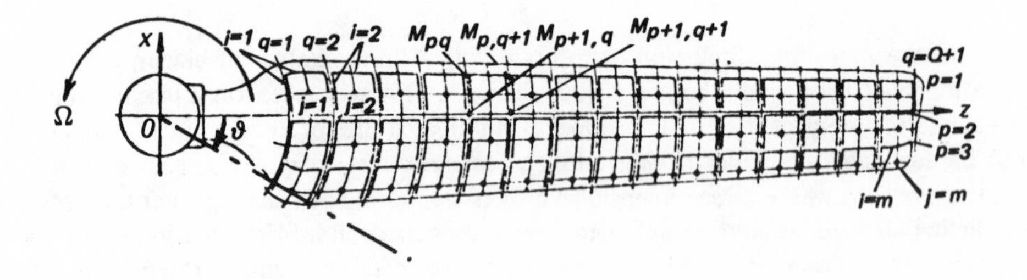

Figure 4.42 Approximation of lifting surface by discrete vortices

---

**For a more detailed discussion of potential flow, see Ch V.*

$$\bar{x} = \bar{r}\sin(\delta_n - \vartheta), \qquad \bar{y} = f(\bar{r}, \vartheta), \qquad \bar{z} = \bar{r}\cos(\delta_n - \vartheta) \tag{4.103}$$

where $\delta_n = 2\pi(n-1)/b$, $\bar{r}$ is nondimensional radius ($\bar{r} \equiv r/R$) which, together with an angle ϑ (positive when moving away from the z axis in the direction opposite to that of rotor rotation), represent the polar coordinates of point $(\bar{x}, 0, \bar{z})$ for $n = 1$.

The grid of discrete vortices replacing the lifting surface can be constructed as follows: Let us imagine a number $(Q + 1)$ of cylindrical surfaces

$$\bar{x}^2 + \bar{y}^2 = \bar{r}_q^{\,2} \qquad (q = 1, 2, 3, \dots Q + 1) \tag{4.104}$$

which cut the lifting surface (Fig 4.42). The intersecting lines of these surfaces are defined by an index "q". In turn, each of the q lines is divided into P segments of equal length.

A system of discrete horseshoe-shaped vortices (indexed $i = 1, 2, \dots m$ where $m = PQ$) is substituted for the lifting surface. Each of the i vortices consists of a bound segment extending along the line p $(p = 1, 2, \dots P)$ between nodes M_{pq} and $M_{p,q+1}$, and a pair of trailing vortices consisting of two segments; first going along the lifting surface (along the line q) and then, starting from point P, taking the shape of a half-infinite helix as a result of the vortex filament being carried in the rotor slipstream.

Induced velocities are computed at control points j $(j = 1, 2, \dots m)$ located between the bound and trailing vortices.

Under the previously mentioned assumption of a uniform flow (no contraction) in the rotor slipstream, parametric equations for the shape of trailing vortices can easily be written.

Determination of Induced Velocities. The procedure for obtaining induced velocities may be visualized as follows: Induced velocity components, generated at point j by a bound vortex of unit strength, are defined as $\bar{u}_{pj}$, $\bar{v}_{pj}$, and $\bar{w}_{pj}$, while those due to a trailing helical vortex of a strength equal to -1 are called $\bar{u}_{qj}$, $\bar{v}_{qj}$, and $\bar{w}_{qj}$; and $u_{q+1,j}$, $v_{q+1,j}$, and $w_{q+1,j}$. Detailed expressions for the above components can be found in Ch VIII of Ref 1. Here, only the relationships for velocity components induced by the i horseshoe vortex of unit strength located on the n-th blade are given:

$$\begin{aligned} \bar{u}_{ij}^{n} &= \bar{u}_{pj} + \bar{u}_{qj} - \bar{u}_{q+1,j} \\ \bar{v}_{ij}^{n} &= \bar{v}_{pj} + \bar{v}_{qj} - \bar{v}_{q+1,j} \\ \bar{w}_{ij}^{n} &= \bar{w}_{pj} + \bar{w}_{qj} - \bar{w}_{q+1,j}. \end{aligned} \tag{4.105}$$

It is also indicated in Ref 1 that programming expressions for $\bar{u}_{qj}$, $\bar{v}_{qj}$, and $\bar{w}_{qj}$ and those with index $q+1$, as well as Eq (4.105) should be done in such a way that first, the velocities of components $\bar{u}_{qj}$, etc., are computed for $p = P$. For computations of velocities induced by the helical vortices extending from nodes of the line $p = P - 1$, there is no need to again perform the integration along the semi-infinite region. It is sufficient to add the velocities generated by segments of helical vortices included between lines $p = P - 1$ and $p = P$ to the previously obtained velocities. A similar procedure is followed in determining induced velocities from helical vortices springing from the line $p = P - 2, P - 3, \dots 1$. It is possible to simplify the procedure by substituting broken lines having apexes at the nodes of the grid for the curved q lines.

In order to determine induced velocities $\bar{v}_j(\bar{u}_j, \bar{v}_j, \bar{w}_j)$ generated by the rotor at control point "j" of blade 1, components of velocity vector $\bar{v}_{ij}(\bar{u}_{ij}, \bar{v}_{ij}, \bar{w}_{ij})$ induced by the vortices with identical indices "i" and unit strength must be calculated. Summing up the velocities given for the *n-th* blade by Eq (4.105), one obtains:

$$\bar{u}_{i_j} = \sum_{n=1}^{k} \bar{u}_{i_j}^{n}, \qquad \bar{v}_{i_j} = \sum_{n=1}^{k} \bar{v}_{i_j}^{n}, \qquad \bar{w}_{i_j} = \sum_{n=1}^{k} \bar{w}_{i_j}^{n}. \tag{4.106}$$

Since vortices with identical indices "i" have the same circulation $\bar{\Gamma}_i$, the desired components of vector $\bar{v}_j$ will be

$$\bar{u}_j = \sum_{i=1}^{m} \bar{u}_{i_j}\bar{\Gamma}_i, \qquad \bar{v}_j = \sum_{i=1}^{m} \bar{v}_{i_j}\bar{\Gamma}_i, \qquad \bar{w}_j = \sum_{i=1}^{m} \bar{w}_{i_j}\bar{\Gamma}_i. \tag{4.107}$$

In this way, computation of the velocities in Eq (4.107) corresponding to given values of $\bar{\Gamma}_i$ is reduced to the determination of coefficients in Eq (4.106) on a computer. The problem now consists of determining circulation $\bar{\Gamma}_i$ on the basis of boundary conditions.

8.3 Boundary Conditions

Nonpenetration of the Lifting Surface. It was mentioned at the beginning of this section that physical considerations require that the flow around the airfoil does not go through the lifting surface modeling the blade. In order to translate this physical requirement into mathematical language, Eq (4.103) is presented in the following form:

$$\bar{y} = f(\bar{x}, \bar{z}). \tag{4.108}$$

Then, $F = \bar{y} - f(\bar{x}, \bar{z}) = 0$ and the unit vector $\vec{n}^{\circ}$—normal to the lifting surface—can be expressed as

$$\vec{n}^{\circ} = \overrightarrow{grad}\, F / |grad\, F| \tag{4.109}$$

where projections of $\overrightarrow{grad}\, F$ on the coordinate axes are

$$\partial F/\partial \bar{x} = -f_x, \qquad \partial F/\partial \bar{y} = 1, \qquad \partial F/\partial \bar{z} = -f_z \tag{4.110}$$

where $f_x = \partial f/\partial \bar{x}$, and $f_z = \partial f/\partial \bar{z}$.

If $V(V_{xj}, V_{yj}, V_{zj})$ represents the resultant velocity vector of flow at the lifting surface at point j, the condition of nonpenetration can be written as

$$\vec{V}_j \cdot \vec{n}^{\circ} = 0. \tag{4.111}$$

Due to the rotation of the rotor about its axis, the components of nondimensional velocity along the x, y, and z axes at point j are $-\bar{z}_j$, 0, and $\bar{x}_j$; hence, the nondimensional velocity components of the total flow at that point would be

$$\bar{V}_{xj} = \bar{u}_j - \bar{z}_j; \qquad \bar{V}_{yj} = \bar{v}_j - \bar{V}_{\infty}; \qquad \bar{V}_{zj} = \bar{w}_j + \bar{x}_j. \tag{4.112}$$

Taking into account Eqs (4.110) and (4.112) in the expanded product given by Eq (4.111), the following is obtained:

$$\bar{v}_j - \bar{u}_j f_{x_j} - \bar{w}_j f_{z_j} = \bar{x}_j f_{z_j} - \bar{z}_j f_{x_j} + \bar{V}_\infty. \qquad (4.113)$$

Derivatives f_{x_j} and f_{z_j} are determined at point j from the known geometry of the lifting surface (say, assumed to be composed of the mean lines of airfoil sections forming the blade) and known blade collective pitch angle.

Substituting $\bar{u}_j$, $\bar{v}_j$, and $\bar{w}_j$ values as given by Eq (4.107) into Eq (4.113), a system of linear algebraic equations for the determination of the sought relationships is obtained.

$$\sum_{i=1}^{m} \bar{\gamma}_{ij}\bar{\Gamma}_i = g_j \qquad (4.114)$$

where

$$\bar{\gamma}_{ij} = \bar{v}_{ij} - \bar{u}_{if} f_{x_j} - \bar{w}_{ij} f_{z_j}, \qquad \bar{g}_j = \bar{x}_j f_{z_j} - \bar{z}_j f_{x_j} + \bar{V}_\infty.$$

In the first order approximation for conventional airscrews, $\bar{w}_{if} f_{z_j}$ and $\bar{x}_j f_{z_j}$ can be neglected; therefore

$$\bar{\gamma}_{ij} = \bar{v}_{ij} - \bar{u}_{ij} f_{x_j}, \qquad \bar{g}_j = \bar{V}_\infty - \bar{z}_j f_{x_j}.$$

If $\vec{\bar{V}}_\infty = -\vec{\bar{V}}_c$ is low, then $|\bar{u}_{ij}| \ll |\bar{v}_{ij}|$ and correspondingly, $\bar{\gamma}_{ij} \approx \bar{v}_{ij}$.

The Kutta-Joukowsky Condition. As to the second boundary condition; namely, the Kutta-Joukowsky requirement of a finite velocity at the trailing edge, it is indicated in Ref 1 that for the assumed distribution of discrete vortices and control points, this condition is automatically fulfilled.

Comparison with Tests. A comparison of predicted and measured nondimensional circulation values vs nondimensional blade stations shows that for both relatively slender ($AR = 12.5$) and more stubby ($AR = 9.1$) blades, a better approximation of $\bar{\Gamma} = \bar{\Gamma}(\bar{r})$ in the tip region is obtained by the lifting-surface than by the lifting-line approach (Fig 4.43).

9. THREE-DIMENSIONAL WAKE MODELS IN FORWARD FLIGHT

9.1 Regimes of Forward Flight Requiring Advanced Wake Concepts

Performance predictions at medium ($\mu > 0.15$) and high-speed flight regimes of helicopters may quite often be adequately treated with the help of procedures rooted in the combined momentum and blade-element theory. For those cases where more detailed knowledge of the induced velocity field outside of the immediate rotor slipstream may be required, the previously discussed flat-wake concept may provide the necessary information.

At the other extreme of the flight spectrum; i.e., hover and vertical ascent, again the combined momentum and blade-element theory is quite often adequate for practical performance predictions. For problems requiring more detailed knowledge of blade loading, vortex theory methods based on the more sophisticated wake concepts discussed in the preceding sections can be used. It appears hence, that except for the transition region from $\mu \approx 0$ to $\mu \approx 0.15$, the various approaches previously discussed should prove adequate in dealing with aerodynamic problems of the helicopter flight spectrum.

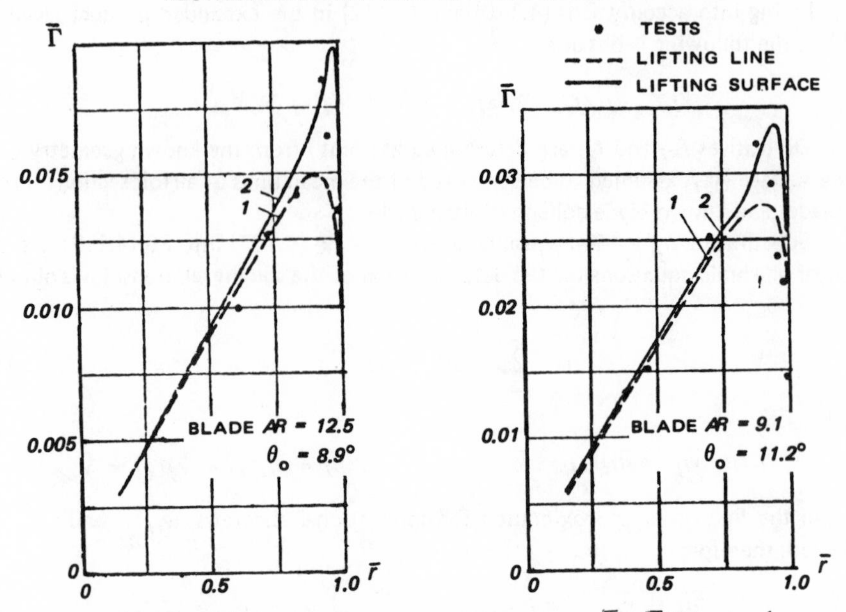

Figure 4.43 Comparison of measured with predicted $\bar{\Gamma} = \bar{\Gamma}(\bar{r})$ *values*[1]

With respect to performance predictions in this particular regime of flight, the momentum theory may provide suitable guidance for the determination of the character of induced power variations between previously established power values in hover and say, $\mu \approx 0.15$. In this region, induced power usually represents the largest fraction of the total rotor power. Consequently, the known $P_{ind}(V)$ trend may serve as a guide for the establishment of a sufficiently accurate $P_R(V)$ relationship for engineering practice. Here, accuracy may be further improved by taking into account the profile power variation $P_{pr}(V)$ indicated by the blade-element theory.

However, there are aerodynamic problems associated with transition flight where the above-mentioned approaches are inadequate, and more sophisticated physicomathematical models than those offered by the momentum, or the combined momentum and blade-element theories are required. The following tasks may be cited as examples of such requirements.

(a) Determination of time-dependent blade airloads as may be needed for helicopter vibration and rotor stress analysis.

(b) Mapping of the induced velocity field to determine aerodynamic interaction with other rotors in the multirotor, or the tail rotor in single-rotor configurations.

In the past, transition chiefly represented a transient stage between vertical takeoff and cruise, and from cruise to hover and vertical landing. As a result of this, a thorough understanding and precise analysis of this regime of flight was of lesser importance to helicopter designers and analysts than those of hover and high-speed forward flight. However, with the increasing interest of the military in nap-of-the-earth flying, low-speed horizontal translation is now commanding the growing attention of rotary-wing practitioners and thoereticians.

9.2 Rotor Vortex-System Models

Similar to the previously considered case of hover and vertical climb, one may imagine several rotor vortex-system models of various degrees of sophistication which, in principle, may be suitable for low-speed, forward-flight analysis.

First of all, the blades may be modeled either by segments of a vortex filament or by lifting surfaces. As to the structure of the rotor wake, here again, in the simplest concept, it may be assumed that the wake is rigid. This means that it is formed by free vortex elements leaving each individual blade and carried downstream along straight lines at a speed equal to the sum of the inflow velocity due to the rotor inclination with respect to the distant flow, and induced velocity averaged over the disc. In this case, obviously neither contraction nor further deflection of the wake further downstream is implied. Since such wake is not time dependent, induced velocity components generated by various rotor vortex subsystems can be computed independently and then superimposed.

At the other extreme, it may be assumed that the wake is free to take any shape that may result from the interaction of the induced velocity field and the free (both trailing and shed) vortices.

Between these extremes, similar to Sect 8 of this chapter, the so-called semi-rigid wakes may be imagined where prescribed wake deformations from the simplest form are superimposed through either experimental or analytically established corrections.

Of the three basic possibilities, two examples will be briefly reviewed; namely, one representing the rigid, and another, the free-wake approach.

9.3 Rigid Wake

Wake Structure. An example of the three-dimensional, rigid-wake concept can be found in Ch X of Ref 1, where it is recommended that it be applied to those flight regimes when $\mu < 1.41\sqrt{C_T}$. As to the structure of the wake itself, Baskin et al assume that flow in the wake downstream from the rotor is uniform, and as shown in Fig 4.44a, at a speed $\vec{V}'$,

$$\vec{V}' = \vec{V}_{\infty} + \vec{v}_{av} \tag{4.115}$$

where $\vec{V}_{\infty}$ is the velocity of the distant flow, and $\vec{v}_{av}$ is the average induced velocity at the disc.

A Cartesian coordinate system $(Oxyz)$ originates at the hub center with the y axis coinciding with the rotor axis of rotation, while $\vec{V}_{\infty}$ is in the xOy plane. The rotor rotates about the y axis at an angular velocity Ω. Each blade of the b-bladed rotor is modeled by a segment of the bound vortex filament where circulation $\Gamma_b = \Gamma(\bar{r}, \psi)$. Vector $\vec{V}'$ forms an angle $a_{V'}$ with the plane of rotation; thus

$$V'_x = -V' \cos a_{V'}, \qquad V'_y = V' \sin a_{V'}, \qquad V_z = 0. \tag{4.116}$$

The wake of the vorticity leaving each blade forms a vortex sheet. Let us look at a particular element of this vortex sheet which separated from point A of the bound vortex filament OB when it was at azimuth ψ_A (Fig 4.44b). As the considered bound vortex filament moved through angle ϑ to a new azimuth position ψ, the trailing vorticity element traveled downstream to point E. The coordinates at point A are

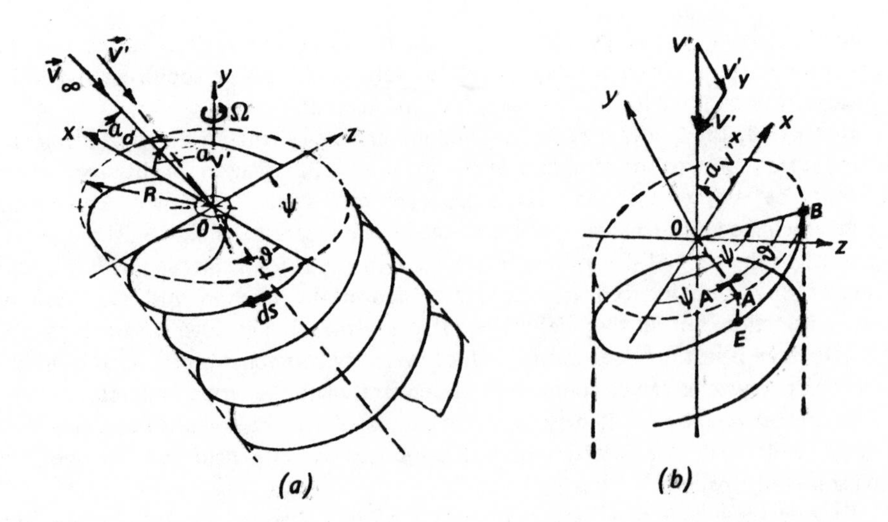

Figure 4.44 Rigid wake of (a) a four-bladed rotor and (b) a single-vortex filament

$$\xi_A = -r\cos(\psi - \vartheta), \qquad \eta_A = 0, \qquad \zeta_A = r\sin(\psi - \vartheta). \tag{4.117}$$

The time of rotation of the blade modeled by filament OB through an angle ϑ is $\Delta t = \vartheta/\Omega$, and the corresponding displacement of the trailing vortex element from its original point of separation (A) is

$$\Delta\xi_A = -V'(\vartheta/\Omega)\cos a_{V'}; \qquad \Delta\eta_A = V'(\vartheta/\Omega)\sin a_{V'}; \qquad \Delta\zeta_A = 0. \tag{4.118}$$

Coordinates of the vortex element when it reaches point E can now be presented in a nondimensional form (linear dimensions divided by R and velocity by ΩR) as follows:

$$\begin{aligned} \bar{\xi} &= \bar{V}'\vartheta\cos a_{V'} - \bar{r}\cos(\psi - \vartheta), \\ \bar{\eta} &= \bar{V}'\vartheta\sin a_{V'}, \\ \bar{\zeta} &= \bar{r}\sin(\psi - \vartheta) \end{aligned} \tag{4.119}$$

where

$$\bar{r}_o \leqslant \bar{r} \leqslant 1.0, \qquad 0 \leqslant \psi \leqslant 2\pi, \qquad 0 \leqslant \vartheta \leqslant \infty.$$

Eq (4.119) represents the desired parametric equation of the blade vortex wake. It should be noted that for a blade at azimuth angle ψ and known $\bar{V}'$ and $a_{V'}$ it is possible, from Eq (4.119), to get an equation of a single skewed helix representing the complete path of the vortex filament (Fig 4.44b) which sprang from blade point B corresponding to the relative blade station $\bar{r}$). Obviously, this can be done by assigning the proper $\bar{r} = const$ and $\psi = const$ values for those two parameters and letting $\vartheta \to \infty$. By the same token, by permitting $\bar{r}$ to vary within its limits of $\bar{r}_o \leqslant \bar{r} \leqslant 1.0$, the whole family of skewed helixes forming a sheet of trailing vorticity left by a single blade can be obtained. It can be visualized that along these skewed helical lines extend elementary trailing vortices with circulation

$$-(\partial/\partial\bar{r})[\bar{\Gamma}(\bar{r},\psi-\vartheta)]\,d\bar{r}.$$

It is also possible from Eq (4.119) to get an equation of a family of straight-line segments located on the vortex sheet, and parallel to the rotor plane-of-rotation. This can be done by assuming that both ψ and $\vartheta = const$, and letting $\bar{r}$ vary within limits of $\bar{r}_o \leqslant \bar{r} \leqslant 1.0$. In turn, by varying ϑ within limits of $0 \leqslant \vartheta \leqslant \infty$, a complete blade wake will be outlined by these segments. It may be imagined that along the so-obtained straight lines extend elementary shed (lateral) vortices of strength

$$-(\partial/\partial\vartheta)[\bar{\Gamma}(r,-\vartheta)]\,d\vartheta.$$

Induced Velocities. Having established the mathematical model of the wake and knowing the law of blade circulation variation with both radius and azimuth, it should be possible to evaluate the instantaneous as well as time-averaged velocities induced by the rotor in the surrounding space. In this process, advantage of the superposition principle can be taken by separately computing velocities induced by the various groups of the total vortex system of the rotor. For example, instantaneous induced velocities can be obtained from independently computed components due to trailing, shed, and bound vortices. Then, time-averaged components can be determined for the same vortex groups.

Expansion of induced velocities into Fourier Series and ways of determining coefficients representing induced velocities through their harmonics is also possible. In this process, Baskin et al emphasized a classic mathematical approach, where integral expressions were evaluated with the help of Legendre equations and polynomials. As a result of this philosophy, it is now possible to deal with the rigid-wake model by using graphs and tables similar to those presented in Ch X of Ref 1, without the necessity for high-capacity computers. By the same token, the above-discussed model may also serve as a basis for a suitable computer program.

9.4 Free Wake

Basic Concept. Studies by Sadler may be cited as an example of the free-wake concept and its application to the determination of blade airloads and associated induced velocity fields under steady-state conditions[19] and during maneuvers[20]. In both cases, dynamic aspects were included by taking into consideration not only blade flapping, but also blade bending and torsional harmonic deformation. Here, attention will be focused on the aerodynamic side of the problem only.

The basic philosophy of the free-wake approach can be appreciated by citing the following excerpts from the Summary of Ref 19:

> Rotor wake geometries were predicted by a process similar to the start-up of a rotor in a free stream. An array of discrete trailing and shed vortices is generated with vortex strengths corresponding to stepwise radial and azimuthal blade circulations. The array of shed and trailing vortices is limited to an arbitrary number of azimuthal steps behind each blade. The remainder of the wake model of each blade is an arbitrary number of trailed vortices. Vortex element end points were allowed to be transported by the resultant velocity of the free-stream and vortex-induced velocities. Wake geometry, wake flow, and wake-induced velocity influence coefficients were generated by this program for use in the blade loads portion of the calculations.

Blade loads computations included the effects of nonuniform inflow due to a free wake, nonlinear airfoil characteristics, and response of flexible blades to the applied loads. The resulting nonlinear equations were solved by an iterative process to determine the distribution of blade shears, bending moments, and twisting moments.

From the above, it can be expected that in contrast to the rigid wake which could be managed through classical mathematical techniques, the computational procedure required in the free-wake method could only be accomplished by the use of high-capacity computers. Fig 4.45 gives some idea regarding the most important steps in an actual computational program.

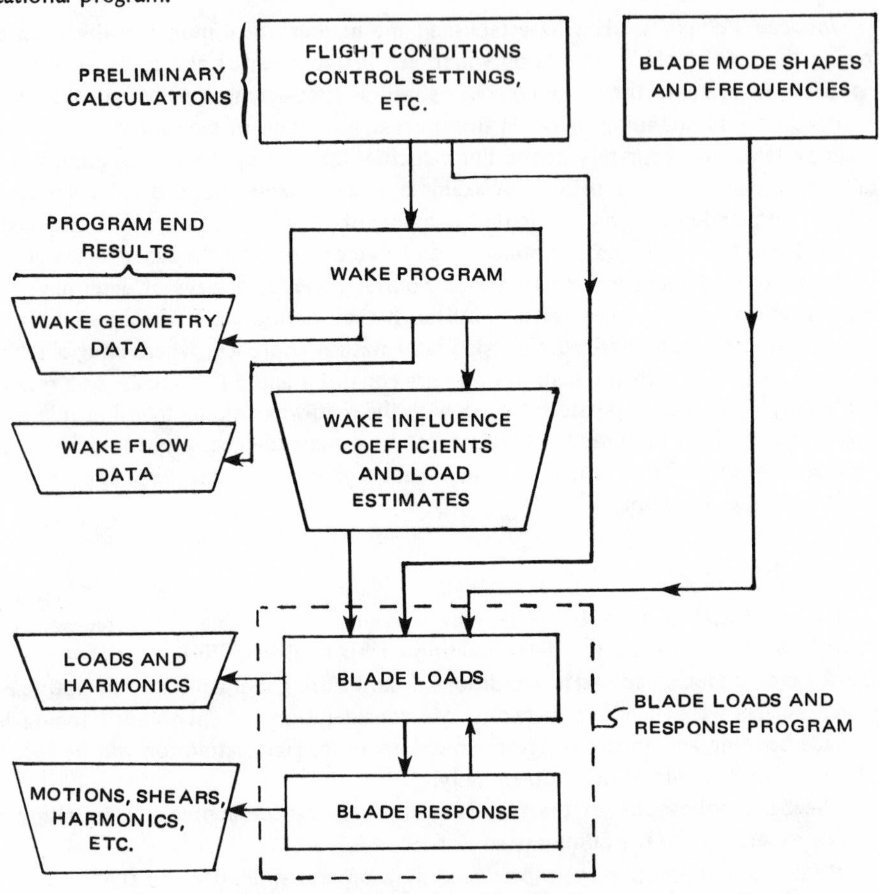

Figure 4.45 Flow diagram of program usage

Wake Geometry. Wake geometry, of course, is time-dependent, and its variation is examined in the following way: At time *0*, the blades are assumed to be at some azimuthal and flapping positions without wake vortices.

The blades then rotate through an azimuthal increment, $\Delta\psi$, and shed and trail vortex elements of unknown strength, but with known positions. The strength of the vortices that are shed immediately behind the blade are then determined, and include the effect of their own self-induced velocities. All vortex element end points not attached to the blade are then allowed to translate as the blade is stepped forward for a time Δt with velocities as determined by the free stream and induced velocities. Here, $\Delta t = \Delta\psi/\Omega$, where Ω is the rotational speed. This completes a typical first step in the wake geometry calculation.

A similar procedure is applied by further rotating the blades through increments of $\Delta\psi$. As in the first step, the vortices just behind the blade have unknown strength; however, the strength of those in the wake is already known. The resulting wake vortex system can be approximated by a mesh of discrete straight-line segments representing both shed and trailing vortices (Fig 4.46).

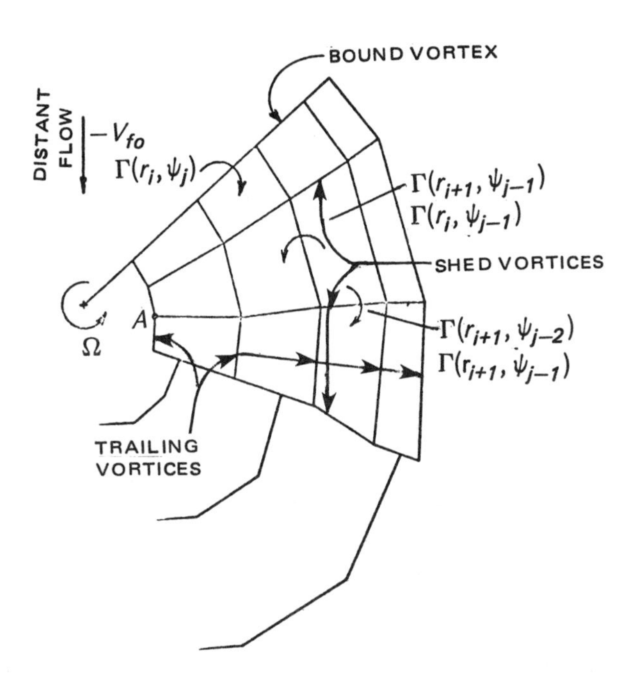

Figure 4.46 Wake model with combination of 'full-mesh' wake, and 'modified' wake of trailing vortices only

The strength of the bound vortices, as well as the induced velocities at the end (nodal) points of vortex elements and the elements themselves, are computed. The nodal points of the vortex mesh are assumed to be convected with the velocity resulting from the freestream flow and induced velocities.

It was indicated in Ref 19 that if the complete mesh of vortices is retained, but for the sake of reducing computer time, $\Delta\psi$ is made large, considerable inaccuracy in the whole procedure would result. By the same token, if the $\Delta\psi$ step values are selected sufficiently small to assure acceptable accuracy, computer time would become prohibitive. As a compromise, only the trailing vortices were retained further downstream, and the full mesh was confined to represent the wake immediately behind the blade (Fig 4.46).

Finite-Size Cores of Vortices. Contrary to the previously considered cases, here the wake and blade vortices are assumed to have finite-sized cores of rotational fluid, and the core sizes at the blade are controlled by an input parameter so as to be adjustable to improve agreement between calculated and experimental or theoretical data, as desired.

In Fig 4.46 all vortex elements are assumed to be straight for purpose of calculations, except where induced velocity determinations are performed for an end point (i.e., point A in Fig 4.46) at the inboard junction of the trailing and shed vortices. At such a point, the neighboring elements are assumed to be arcs with curvature determined by three appropriate end points.

Induced Velocities. Induced velocity components generated at any point (C) of the space surrounding the rotor can be computed for each straight-line segment of the vortex sheet using Eq (4.13) which, in Refs 19 and 20, is written (with the notations shown in Fig 4.47) under the following form:

$$\Delta v = (\Gamma/4\pi d)(\cos\theta_A - \cos\theta_B). \tag{4.120}$$

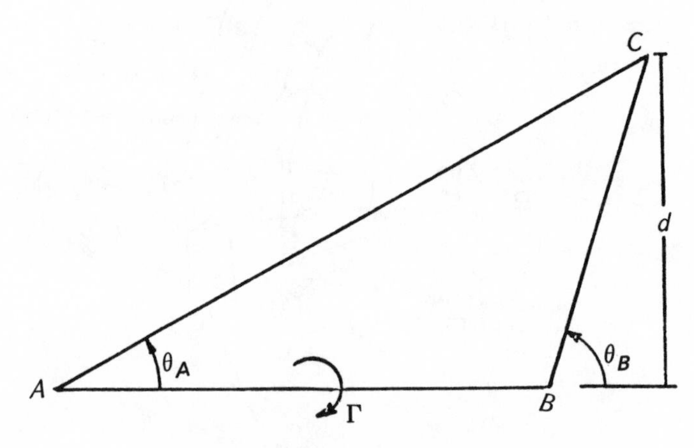

Figure 4.47 Vortex induced flow model

The total induced velocity at any point of interest is obtained by summing the contributions of all the vortex mesh segments. However, when computing the vortex induced flow at a point adjoining that vortex element, Eq (4.120) becomes indeterminable. Following the approach developed by Crimi[21], an expression for induced flow by adjoining finite-core-radii vortex elements was developed:

$$\Delta v_{s_i} = \frac{1}{8R_{cu}}\left\{\Gamma_{i-1}\left[\ell n\left(\frac{8R_{cu}}{a_{i-1}}\tan\frac{\phi_{i-1}}{4}\right) + \frac{1}{4}\right] + \Gamma_i\left[\ell n\left(\frac{8R_{cu}}{a_i}\tan\frac{\phi_i}{4}\right) + \frac{1}{4}\right]\right\}. \tag{4.121}$$

In the above equation, end points are used to define the approximate radius of curvature R_{cu} of a circular arc. In addition, Γ is the vortex strength, a is the core radius, and ϕ is the angle defined in Fig 4.48.

Eqs (4.120) and (4.121) should permit one to determine the interaction of vortex elements. However, when the vortex end point falls inside the finite core of another

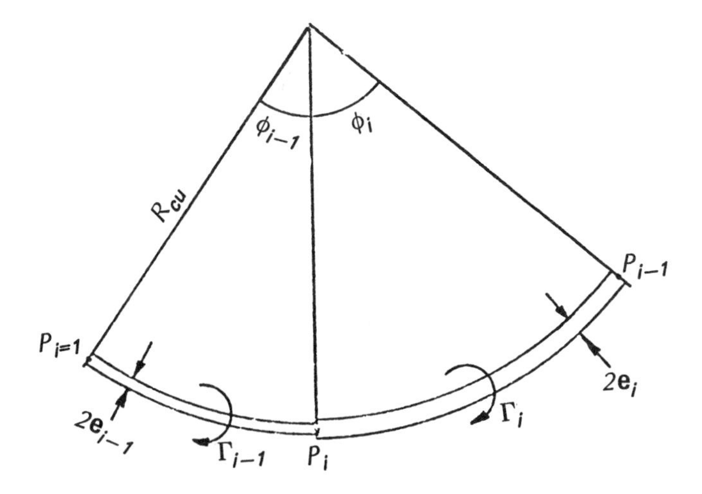

Figure 4.48 Vortex self-induced flow model

element, the theoretically predicted induced velocities may be unrealistically high. To alleviate this physically unacceptable situation, an arbitrary limit expressed as a fixed percentage of the tip speed is imposed.

Actual determination of the variation of the wake geometry is performed within a Cartesian coordinate system having its vertical axis located in the vertical plane to coincide with the rotor shaft. In Figure 4.49, the y axis is oriented along the shaft, while vector V_{fo} lies in the plane passing through the y and x axes. At this point, it should be noted that in Refs 19 and 20, the coordinate system is different from the present one.

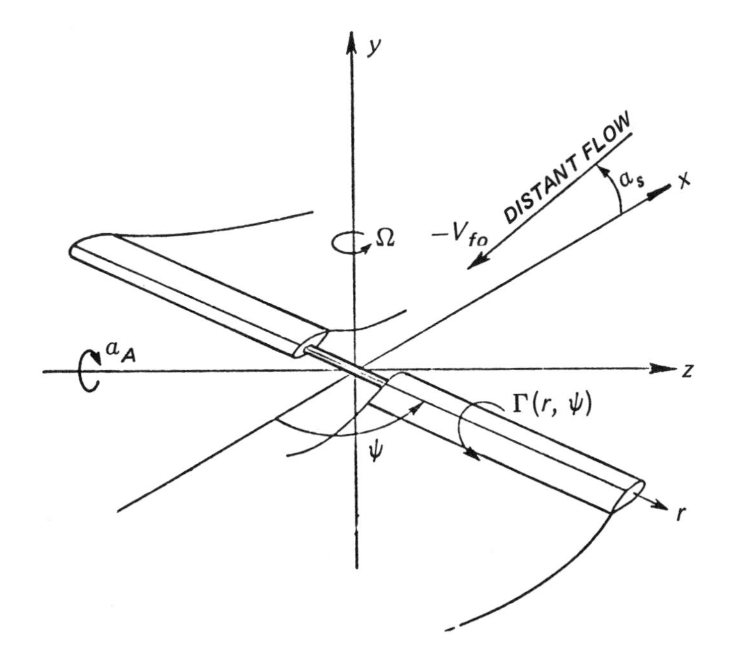

Figure 4.49 Rotor system coordinates

Blade Circulation. Circulation around the blade in the free-wake approach is computed in a manner similar to the cases previously considered. At station r, the inplane velocity component perpendicular to the blade axis is

$$U_{\perp} = \Omega r + V_{fo} \sin \psi$$

while that perpendicular to the disc plane is

$$V'_{\perp} = V_{fo}(a_s - a_\beta) + \nu \tag{4.122}$$

where a_s is the shaft axis angle with respect to the freestream, positive aft; a_β is the forward tilt of the rotor plane with respect to the shaft axis due to flapping; and ν is the induced velocity component due to the wake ($\nu = 0$ at start-up).

Combining the expressions for unit span loading according to the blade element theory with that given by the Kutta-Joukowsky law in Eq (4.16) and taking into account Eq (4.122), the following expression for blade circulation (Γ_b) at station r and azimuth ψ is obtained:

$$\Gamma_b = \tfrac{1}{2} ca[a_o(\Omega r + V_{fo} \sin \psi) + V_{fo}(a_s - a_\beta) + \nu]. \tag{4.123}$$

It is indicated in Ref 19 that "the wake-induced velocity on the blade, ν, (called w in the reference) is made up of velocities due to known circulations in the wake and to unknown circulations at the blade, and may be written in the form

$$\nu(r_i, \psi_k) = \nu_N(r_i, \psi_k) + \sum_{\ell} \sum_{j} \sigma_{\ell j}(r_i, \psi_k) \Gamma(r_\ell, \psi_j) \tag{4.124}$$

where $\nu_N(r_i, \psi_k)$ is the induced velocity due to all known wake circulations; $\Gamma(r_\ell, \psi_j)$ is the blade circulation at r_ℓ, ψ_j; and $\sigma_{\ell j}(r_i, \psi_k)$ is an influence coefficient which, when multiplied by circulation $\Gamma(r_\ell, \psi_j)$ gives the induced velocity of that element at r_i, ψ_k.

"The summations over indices ℓ and j indicate a summation over all radial sections of all blades at their respective azimuthal positions. Then a set of equations for all Γ's may be obtained, and is of the form

$$\Gamma_{ik} = (c/2)a\left[\sum_{\ell} \sum_{j} \sigma_{\ell jik} \Gamma_{\ell j} + a_0(\Omega r_i + V_{fo} \sin\psi_k) + V_{fo}(a_s - a_\beta) + \nu_N\right]. \tag{4.125}$$

"Here, Γ_{ik} is equivalent to $\Gamma(r_i, \psi_k)$, and occurs on both sides of the equation. This equation is solved with a simple iterative procedure."

Results. With blade circulation values determined for a given ψ, the vortex-induced velocities are computed at all end points of vortex elements in the wake. As the blade moves through an increment position $\Delta\psi$ during time $\Delta t = \Delta\psi/\Omega$, the end points are convected at Δt intervals at a speed resulting from the sum of the freestream flow and induced velocities. Then, the entire procedure is repeated. For details of this approach for application to predictions of blade airloads associated with the free-wake concept under steady-state flight conditions as well as in maneuvers, the reader is directed to Refs 19 and 20, respectively, Here, Fig 4.50 is reproduced from Ref 19 as an example of the path traveled by the tip vortex of a two-bladed rotor at a low μ value.

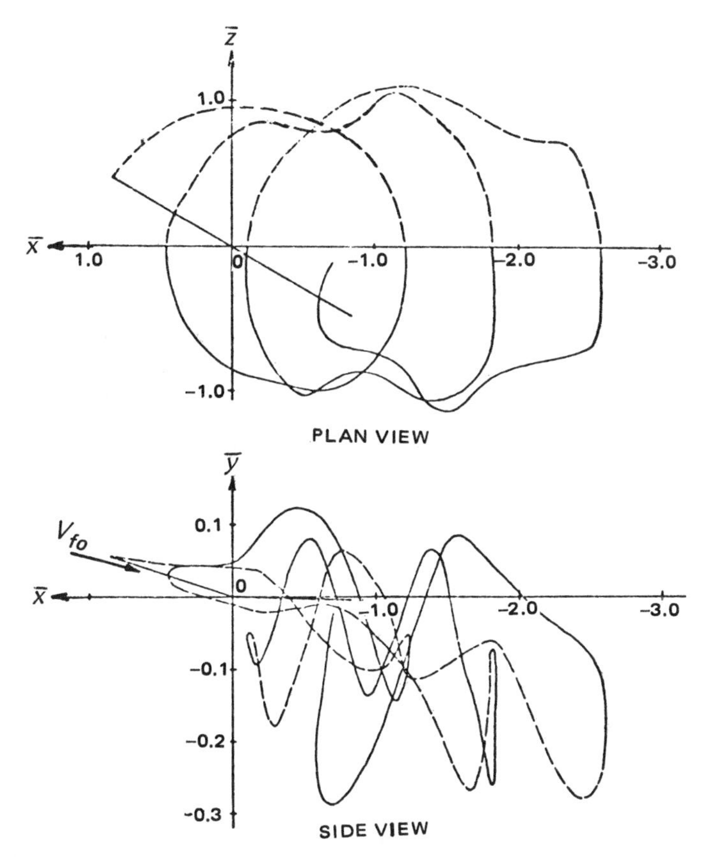

Figure 4.50 Tip vortex location of a two-bladed rotor: $\mu = 0.097$, $a_s = -3.1°$

With respect to the accuracy of predicting actual downwash by the free-wake method, it appears that under steady-state conditions, the agreement with test data is good (Fig 4.51).

The same applies to airloads (Fig 4.52). However, judging from the figures published in Ref 20, success in predicting airloads in maneuvers can be qualified as only fair.

10. COMPRESSIBILITY EFFECTS

10.1 Basic Relationships

Up to this point, compressibility effects, if considered, were applied to airfoil characteristics only (e.g., lift-curve slope, drag divergence, and maximum lift coefficient). The significance of a finite value of the speed of sound resulting in the noninstantaneous transmittal of aerodynamic signals from vortex filaments to the surrounding space has been ignored in the classic Biot-Savart law. However, the role of compressibility was considered by many investigators of aircraft propellers. It was also studied in the case of helicopter rotors by Baskin et al (Ch X of Ref 1). In the section on Basic

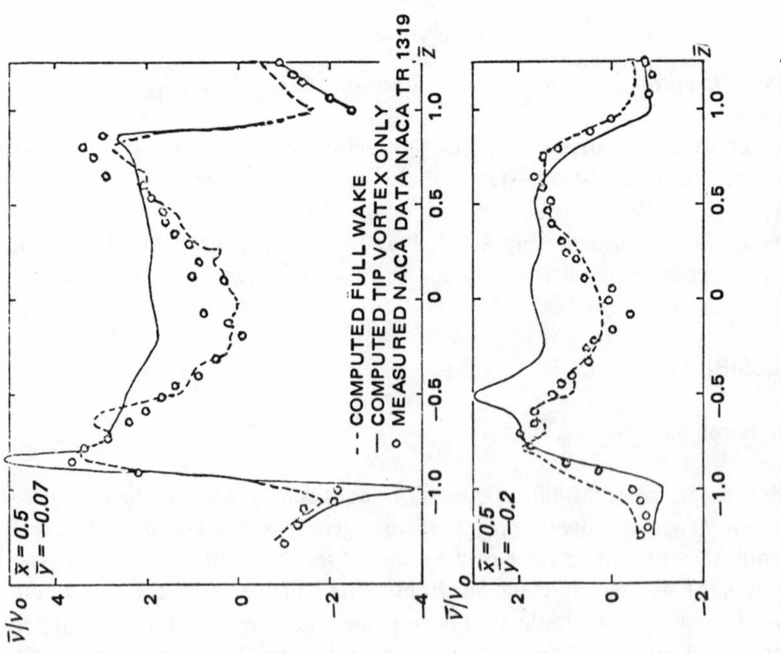

Figure 4.51 Comparison of computed and measured induced downwash below the rotor plane; $\mu = 0.14$, $C_T = 0.00371$, $x = 0.5$[19]

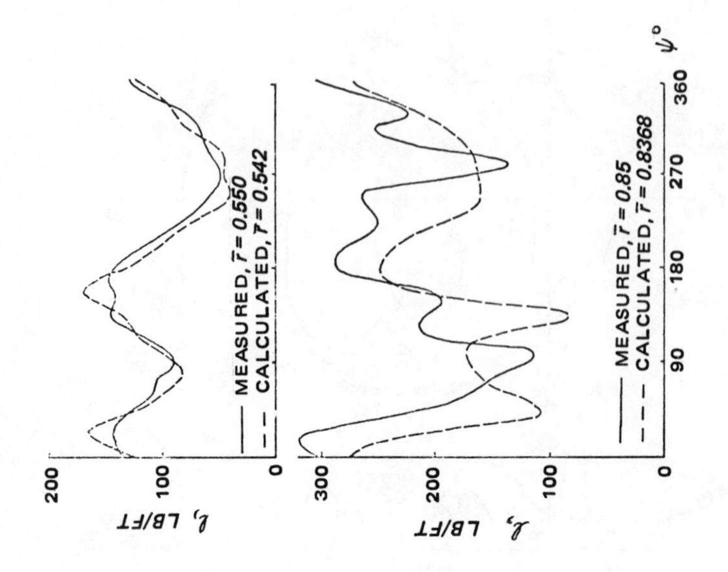

Figure 4.52 Blade airloads for H-34 helicopter; lb/ft vs azimuth: $\mu = 0.2$[19]

Assumptions and Notations, they indicate that an expansion analogous to the classic Biot-Savart vectorial relationship (4.4a) can be developed for a compressible medium.

It is assumed (1) that a field of induced velocities $\vec{v}(u,v,w)$ at a point $P(x,y,z)$ is generated by a vortex filament S moving uniformly at a speed $\vec{V}(V,0,0)$ within a large mass of compressible fluid, and (2) that $|v| \ll |V|$. Then, the vectorial relationship will be:

$$\vec{v} = \frac{\nu\Gamma}{4\pi}\int_S \frac{\vec{d} \times \vec{dS}}{d_\nu^{\,3}} \qquad (4.126)$$

where ξ, η, and ζ are coordinates of the vortex element dS,

$$d_\nu = \sqrt{\nu^2(\xi - x)^2 + (\eta - y)^2 + (\zeta - z)^2}, \quad \nu = \sqrt{1/(1 - M^2)}, \quad M = V/s$$

and s is the speed of sound in undisturbed flow.

In analogy with Eqs (4.7) to (4.9), expressions for induced velocity components are obtained:

$$u = \frac{\nu\Gamma}{4\pi}\int_{p_i}^{p_f}\left[\frac{d\eta}{dp}(z - \zeta) - \frac{d\zeta}{dp}(y - \eta)\right]\frac{dp}{d_\nu^{\,3}} \qquad (4.127)$$

$$v = \frac{\nu\Gamma}{4\pi}\int_{p_i}^{p_f}\left[\frac{d\zeta}{dp}(x - \xi) - \frac{d\xi}{dp}(z - \zeta)\right]\frac{dp}{d_\nu^{\,3}} \qquad (4.128)$$

$$w = \frac{\nu\Gamma}{4\pi}\int_{p_i}^{p_f}\left[\frac{d\xi}{dp}(y - \eta) - \frac{d\eta}{dp}(x - \xi)\right]\frac{dp}{d_\nu^{\,3}} \qquad (4.129)$$

As in Sect 3.1, it is assumed that the equation of line S is given in a parametric form with p as a parameter: $\xi = \xi(p)$; $\eta = \eta(p)$; $\zeta = \zeta(p)$. In Eqs (4.127) to (4.129), integration is performed from the initial (p_i) to the final (p_f) parameter values.

10.2 Application to a Rotor in an Oblique Flow

The rotor velocity system is similar to that considered in Sect 9, and illustrated in Fig 4.44. As previously noted, a coordinate system, xyz, is attached to the rotor which is exposed to a uniform flow of compressible fluid moving at velocity V along the X axis of another stationary Cartesian coordinate system, XYZ (Fig 4.53). Vector $\vec{V}$ makes an arbitrary angle-of-attack, a_V, with the rotor disc. The blades are modeled by segments of bound vortex filaments whose circulation varies with $\bar{r}$, $\Gamma = \Gamma(\bar{r})$, but remains constant with the azimuth: $\Gamma(\psi) = const.$ As a result of this concept, a vortex sheet composed of trailing vortices only is formed behind each blade. A parametric equation in nondimensional form for the $OXYZ$ coordinates of that vorticity surface can be written as follows:

$$\bar{\xi} = \bar{r}\cos(\psi - \vartheta)\cos a_V + \bar{V};\ \bar{\eta} = -\bar{r}\cos(\psi - \vartheta)\sin a_V;\ \bar{\zeta} = -\bar{r}\sin(\psi - \vartheta) \qquad (4.130)$$

where angle ϑ is the parameter.

Velocity $\vec{v}$ induced by the rotor can be imagined as the sum of velocity $\vec{v}_b$ due to the bound vortices, and $\vec{v}_f$ generated by the free vortices (in this case, the trailing ones

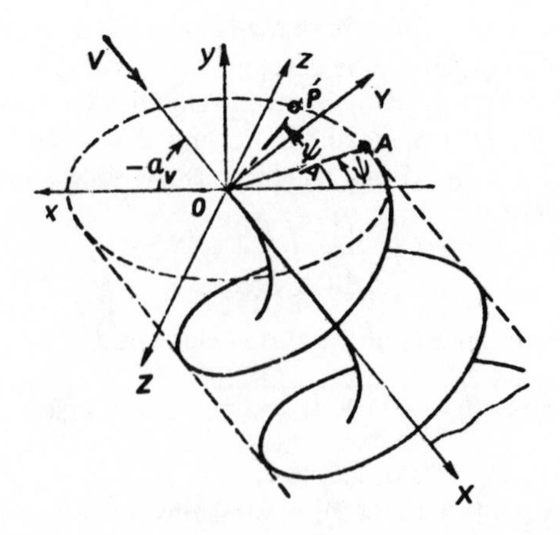

Figure 4.53 Coordinate systems

only). Using nondimensional definitions for velocities, this relationship can be expressed as follows:

$$\vec{v}(\bar{u}, \bar{v}, \bar{w}) = \vec{v}_b(\bar{u}_b, \bar{v}_b, \bar{w}_b) + \vec{v}_f(\bar{u}_f, \bar{v}_f, \bar{w}_f). \tag{4.131}$$

In order to determine the time-averaged velocity $\vec{v}_b$ induced by a *b*-bladed rotor at an arbitrary point *P* $(\bar{x}, \bar{y}, \bar{z})$ related to the *Oxyz* coordinate system, we assume that $\bar{x} = -\bar{r}_P \cos \psi_P$; $\bar{z} = \bar{r}_P \sin \psi_P$, where $\bar{r}_P$ is the radial, and ψ_P is the azimuthal location of point *P*. Next, Eq (4.130) with $\vartheta = 0$ is substituted into Eqs (4.127) through (4.129) and the results are time-averaged and then transferred to the *Oxyz* coordinate system. The results of this operation are given in Ch X of Ref 1* as

$$\bar{u}_b = \frac{-\nu b}{8\pi^2} \int_0^{2\pi} \int_{\bar{r}_o}^{1} \bar{\Gamma} \frac{\bar{y} \sin \psi}{\bar{d}_\nu^{\,3}} \, d\bar{r} \, d\psi$$

$$\bar{v}_b = \frac{-\nu b}{8\pi^2} \int_0^{2\pi} \int_{\bar{r}_o}^{1} \bar{\Gamma} \frac{\bar{r} \sin (\psi - \psi_P)}{\bar{d}_\nu^{\,3}} \, d\bar{r} \, d\psi \tag{4.132}$$

$$\bar{w}_b = \frac{-\nu b}{8\pi^2} \int_0^{2\pi} \int_{\bar{r}_o}^{1} \bar{\Gamma} \frac{\bar{y} \cos \psi}{\bar{d}_\nu^{\,3}} \, d\bar{r} \, d\psi$$

where

$$\bar{d}_\nu^{\,2} = L^2 + \bar{y}^2 + (\nu^2 + 1)\bar{d}_\xi^{\,2}; \qquad \bar{d}_\xi = \bar{y} \sin a_V - L_x \cos a_V;$$

---

* *When comparing the actual expressions in Ref 1 with those in this text, the difference in some symbol definitions should be kept in mind.*

while,

$$L = \sqrt{\bar{r}^2 + \bar{r}_p^{\,2} - 2\bar{r}\,\bar{r}_p \cos(\psi - \psi_p)};\ \text{and}\ L_x = \bar{r}\cos\psi - \bar{r}_p \cos\psi_p.$$

It can be seen from Eq (4.132) that contrary to the previously considered cases, the induced velocity $\bar{v}_b$ generated by bound vortices is not zero. However, when $\nu = 1$ (incompressible fluid), the $\bar{v}_b$ velocity field vanishes.

Velocity induced by bound vortices averaged over the entire radius $\bar{r}_p$ will be

$$\bar{v}_{b\bar{r}} = \frac{-\nu b}{16\pi^3}\int_0^{2\pi}\int_0^{2\pi}\int_0^{1} \bar{\Gamma}\,\frac{\bar{r}_p \sin(\psi - \psi_p)}{\bar{d}_\nu^{\,3}}\, d\bar{r}\, d\psi\, d\psi_p. \tag{4.133}$$

From this equation it can be seen that by changing the independent variables from ψ and ψ_p to $-\psi$ and $-\psi_p$, the $\bar{v}_{b\bar{r}}$ changes its sign. This means that the $\bar{v}_b$ field is asymmetric with respect to axis x.

The $\bar{u}_f$, $\bar{v}_f$, and $\bar{w}_f$ components due to free vortices can be determined in a similar way in the *Oxyz* coordinate system. For points on the rotor disc in particular, the following expression is obtained:

$$\bar{v}_f = \frac{b}{8\pi^2\bar{V}}\int_0^{2\pi}\int_{\bar{r}_o}^{1}\frac{\partial\bar{\Gamma}}{\partial\bar{r}}\Big[\bar{r}^2 - \bar{r}\,\bar{r}_p\cos(\psi - \psi_p) + \mu L_z$$

$$+ \frac{1}{\nu} L\bar{r}\cos\psi\cos a_v\Big]\frac{d\bar{r}\,d\psi}{L(L + \nu L_x \cos a_v)}. \tag{4.134}$$

where $\bar{V} \equiv V/V_t$ is the nondimensional velocity of distant flow which, for cases of high and even medium velocities of flight, may be considered as $\bar{V} \approx \mu$, and $L_z = \bar{r}\sin\psi - \bar{r}_p \sin\psi_p$. In axial flow, $a_v = \pm 90^\circ$ and in Eq (4.134), the second term containing ν (reflecting compressibility) disappears and that equation can be reduced to

$$\bar{v}_f = -\frac{b\bar{\Gamma}(\bar{r})}{4\pi\bar{V}}.$$

For hovering, when flow through the rotor becomes identical to the induced velocity at the disc ($\bar{V} = \bar{v} = \bar{v}_f$), the above equation can be written in a dimensional form as follows:

$$v_f = -\tfrac{1}{2}[b\Gamma(\bar{r})/v_f(2\pi/\Omega)].$$

By inspecting this equation, one would recognize that $b\,\Gamma(\bar{r})$ is the circulation transferred from radius $\bar{r}$ to the wake by all b blades, while $v_f(2\pi/\Omega)$ is the distance traveled by the trailing vortices during one rotor revolution. Consequently, it may be stated that $[b\Gamma(r)/v_f(2\pi/\Omega)] = d\Gamma/dy$, and the expression for v_f becomes identical to Eq (4.19a).

Baskin et al pointed out that the stationary field of induced velocities does not depend on the blade rotational Mach number $M_\Omega \equiv \Omega r/s$ (where s is the speed of sound), but only on $M_V \equiv V/s$ of the uniform flow along the skewed cylindrical vortex wake.

As an illustration of the influence of compressibility, comparative calculations of v_f (from Eqs (4.132) and (4.134)) were performed at $M = 0$ and $M = 0.2$ for a rotor operating at $\Omega R = 221$ *m/second*, $\mu = 0.3$, $a_v = -18$, and $C_T = 0.007$.[1] Under these conditions, compressibility effects are relatively small (Fig 4.54).

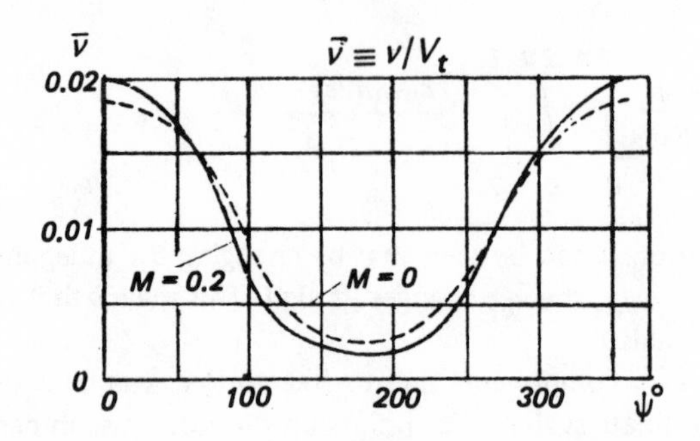

Figure 4.54 Example of compressibility effects on $\bar{v}$* *of a rotor at* $\mu = 0.3$ *and* $V_t = 221$ *m/second*

The reader interested in a more thorough discussion of the implication of the finite velocity of sound on vortices in a compressible medium, and application of those findings to airscrews, is directed to Ch XVI of Ref 1.

11. VISCOSITY EFFECTS

11.1 General Remarks

There are two important physical aspects of viscosity effects in the vortex theory. First, the existence of a finite radius core at the moment of "birth" of a vortex and second, the diffusion of that core with time. The final result of the second phenomenon may be a complete dissipation and loss of identity of the vortex structure. The process of core diffusion and complete vortex dissipation is quite complicated. However, a thorough understanding of this process and development of reliable mathematical treatment is important in predicting the decay of tip vortices of fixed-wing aircraft in terminal operations. Because of flight safety aspects, a continuous high-level effort (both analytical and experimental) may be expected in this domain[9].

As far as investigations into the diffusion of vortices of rotary-wing aircraft is concerned, the available analytical material seems to be quite limited. It also appears that outside of some initial pioneering efforts (for example, see Ref 10), there is a lack of methodical experimental studies in spite of the fact that in almost every visualization technique, the qualitative aspects of vortex dissipation are quite apparent (Figs 4.1 and 4.2).

---

* *Presumably at* $\bar{r} = 0.7$ *or* $\bar{r} = 0.75$.

With respect to the analytical treatment of the diffusion of vortices and its influence on the field of induced velocities, one may find a brief discussion of this phenomenon in Ch XV of Ref 1.

10.2 Influence of Free Vortex Diffusion on Instantaneous Induced Velocities

It should be emphasized at this point that in the vortex system of an airscrew, diffusion may affect the free (shed and trailing) vortices only, and obviously has no bearing on bound vortices.

At time zero ($t = 0$), when a free vortex separates from the straight-line vortex segment modeling the blade, circulation of the just "born" vortex is equal to Γ, while the associated vorticity field $\vec{\Upsilon}$ is enclosed inside a tubular filament. With the passage of time, a diffusion process takes place according to the following equation of dissipation.

$$(\partial\vec{\Upsilon}/\partial t) = \nu_{eq}\Delta\vec{\Upsilon} \tag{4.135}$$

where ν_{eq} is the equivalent kinematic viscosity coefficient for turbulent flow. Eq (4.135) represents an equation of heat transfer for vortex vector $\vec{\Upsilon}$.

Let it be assumed that one wants to investigate the state of diffusion of a vortex filament downstream in the wake at a location defined by angle ϑ (Fig 4.55) at the time the blade from which this filament sprang is at azimuth ψ. The time interval between the moment of "birth" of the investigated section of the vortex filament and its present position downstream at ϑ is

$$t = (\psi - \vartheta)/\Omega. \tag{4.136}$$

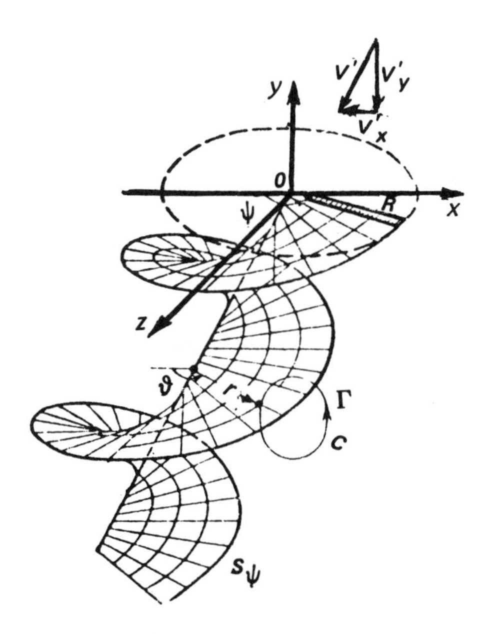

Figure 4.55 Vorticty wake of a blade

Keeping the above presented general remarks in mind, a comparison can be made between expressions determining velocities v_f induced by free vortices with, and without, consideration of the diffusion.

Without diffusion, velocity induced at a point P (x_P, y_P, z_P) whose location in space is determined by the position vector $\vec{r}_P = x_P\vec{i} + y_P\vec{j} + z_P\vec{k}$ can be obtained from the classic Biot-Savart law (Eq (4.4a)). Assuming that the n-th blade of a b-bladed rotor is at azimuth angle ψ_n, an expression for the $\vec{v}_f$ corresponding to the wake of that blade as shown in Fig 4.55 can be written as follows:

$$\vec{v}_f(\vec{r}_P, \psi_n) = \sum_{n=1}^{b} \int_0^R \int_{-\infty}^{\psi_n} \frac{\vec{d} \times D(,\Gamma,\vec{r})}{4\pi d^3}\, d\vec{r}\, d\vartheta \tag{4.137}$$

where $\vec{r} = x\vec{i} + y\vec{j} + z\vec{k}$ is the position vector locating elements of vorticity on the blade wake, $\vec{d} = \vec{r} - \vec{r}_P$, and $\psi_n = \psi + 2\pi(n-1)/b$, while

$$D(\Gamma, \vec{r}_P) = \frac{\partial\Gamma}{\partial r}\frac{\partial\vec{r}_P}{\partial\vartheta} - \frac{\partial\Gamma}{\partial\vartheta}\frac{\partial\vec{r}_P}{\partial r}$$

is the Jacobian.

When the law of dissipation as given by Eq (4.135) is taken into consideration and time is defined according to Eq (4.136), the new expression for $\vec{v}_f$ is obtained by multiplying the integrand of Eq (4.137) by $K(d)$, where

$$K(d) = (4/\pi)\int_0^{d/\delta} x^2 e^{-x^2}\, dx$$

and

$$\delta = 2\sqrt{\nu_{eq}t}.$$

Intermediate steps in the development of these relationships can be found in Ch XV of Ref 1.

At the beginning of this section, it was indicated that there was an apparent lack of intense research on the dissipation of rotor-wake vorticity. This status may be partially explained by the fact that in the past, interest was focused primarily on the flow field at the rotor disc—where induced velocities are most significantly influenced by the near wake, while vortex dissipation usually occurs further downstream. However, should this interest be expanded to include fields of flow more distant from the rotor; for instance, in the case of rotor-wake interaction with the fuselage or tail rotor, then the whole problem of vortex dissipation would acquire more practical attention.

12. CONCLUDING REMARKS

The application of vortex systems to physicomathematical modeling of both lifting and propelling airscrews offers, in principle, the most versatile tool available at present for the determination of time-average and instantaneous flow fields generated by rotary wings in unlimited space. Because of this freedom of space and time in dealing with aerodynamic phenomena, the vortex theory is attracting the attention of theoreticians as well as applied aerodynamicists and aeroelasticians, resulting in a steadily growing flow of literature. In parallel, new vortex computer programs aimed at air-load determination, aeroelastic problems, performance, and special tasks such as noise are continuously being written.

With respect to performance, it should be noted that the results have often been disappointing. In spite of large expenditures in computer time, the predicted performance figures, when compared with actual flight or wind-tunnel test results, appeared no more and sometimes less accurate than those obtained by the combined blade-element and momentum theory.

Paradoxically, these discrepancies usually result from what constitutes one of the strongest points of the vortex theory; namely, the ability to pinpoint the occurrence of very strong local induced velocities. Although the high induced-velocity regions are usually limited to the proximity of the vortex core, large loads may be generated when a strong tip vortex comes into the vicinity of the blade.

The existence of high concentrated blade loads and vibratory excitations resulting from vortex-blade interaction was investigated by many authors and confirmed both in flight (e.g., Scheiman and Ludi[22]), and in wind-tunnels (e.g., Surendraiah[23] and Ham[24]). Hence, in this case, it is clear that the ability to use the vortex theory to predict the positions of discrete vortices and indicate the existence of high induced velocities in the vicinity of their core may lead to a better understanding and more precise quantitative analysis of actual physical processes.

However, the very nature of the vortex theory which allows one to pinpoint the existence of singularities having actual physical meaning, might become a source of erroneous physical interpretation and computational errors when applied to the case of performance predictions. The classic theory, where vortex cores of infinitesimally small cross-sections are assumed, might be especially prone to errors. Here, during the execution of a computer program, physically non-existent areas of high induced velocities may appear simply as a result of the fact that, inadvertently, some of the points where induced velocity is determined happen to be too close to the vortex filaments.

For this reason, when developing computer programs based on the classic vortex theory, it is important to incorporate various safeguards against the possibility of running into mathematical singularities having no physical counterparts. Proper selection of collocation points[1] may be cited as one example of such a safeguard. Further improvement in the program may be derived from recognizing the fact that vortex cores actually have finite diameters, while vortex sheets are of finite thickness, and incorporating this into the mathematical model.

In parallel with the more meaningful modeling of individual airscrews, their assemblies, and even the whole rotorcraft through approaches derived from the vortex theory, there is a growing trend toward finding meaningful shortcuts which would allow one to reduce computer time without sacrificing basic accuracy. This was borne out by Landgrebe et al[25] in a review of the status of the vortex theory in the mid-seventies and especially, its application to aerodynamic technology of rotorcraft. For instance, in the static case, the authors came to the conclusion that "the lifting-line approach is adequate for predicting the hover performance of a wide range of conventional and advanced rotor designs." However, this approach was improved through a special program called CCHAP (Circulation Coupled Hover Analysis Program) which couples the wake geometry to load distribution. It was indicated that accuracies similar to those associated with lifting-surface methods can be achieved through CCHAP, but at a much lower computer time and cost. When predicting performance in forward flight, emphasis seems to be concentrated on the following areas: (a) use of variable inflow methods for rotor-power predictions, (b) refinements in unsteady aerodynamics and skewed flow, (c) the role of aeroelastic coupling in the design of optimized rotors, and (d) a preliminary assessment of rotor airframe interferemce[25].

Outside the framework of the vortex theory proper, there is a continuous effort to develop simpler methods, but still have the ability to determine, in both time and space, unrestricted velocity fields induced by an airscrew in various regimes of flight. The appendix to this chapter, describing the local momentum theory, is given as an example of these efforts, while in the following chapter it is shown that many basic objectives of the vortex theory can be achieved through the velocity- and acceleration-potential approach.

APPENDIX TO CH IV

LOCAL MOMENTUM THEORY

1. INTRODUCTION

The local momentum theory approach proposed by Azuma and Kawachi in Ref 26 is similar to Sect 2.2 of Ch III in that a combination of the momentum theory with the wing or blade element theories represents the main vehicle for relating wing or rotor geometry and some characteristics of their airfoil sections to the determination of induced velocity and airloads. However, in the present case, some aspects of the vortex theory are used; therefore, it seemed appropriate to include an overview of this material as an appendix to Ch IV.

The Azuma and Kawachi approach is aimed toward determination of both time-average and instantaneous fields of rotor induced velocities. The authors achieve this goal with reasonable accuracy, but without the computational complexities of the classic vortex theory. In this case, the physicomathematical model of a rotor is based on the concept of representing an actual aerodynamic load and downwash distribution of a blade by a series of n overlapping wings of decreasing size; each having an elliptical circulation along the span and therefore, producing a uniform downwash velocity.

The main task consists of finding a way of relating the geometry of a single blade (planform, twist, pitch angle, etc.) as well as that of the whole b-bladed rotor, plus its operating conditions (hovering, axial, and forward translation), to sustained loads and downwash produced by the component wings. Knowing these values, actual blade loading and downwash velocity can be obtained as a summation of the aerodynamic loads and downwash velocities produced at the point of interest by the component wings.

Since this approach is rooted in fixed-wing aerodynamics, the basic philosophy of the theory will be more appreciated by first considering its application to the non-rotating wing. However, at this point it should be stressed that practical use of the local momentum theory is more suitable for rotary-wing than for fixed-wing applications. In the latter case, a large amount of error may accumulate by not taking into account the upwash flow at both tips of the elliptic wings which model the actual wings. By contrast, the upwash flow on the root side of the elliptic wings modeling a rotor blade is not as important except perhaps, around $\psi = 90°$ in high μ forward flight.

This theory may be more helpful in solving unsteady, rather than steady, problems in rotary-wing aerodynamics. Flight maneuvers and gust response conditions may be cited as suitable areas of application because in such transient motions of the rotor, the vortex theory would require a large amount of computational effort to precisely determine the downwash distribution[20].

2. FIXED WING

The classic aerodynamic theory[11,13] indicates that an elliptically-loaded wing moving at a speed $-V$ develops a uniform downwash velocity v_o along its lifting line

(1/4-chord). Far downstream, this velocity still remains uniform, while its value becomes $2v_o$. The magnitude of the regions of upwash existing outside of the wing span can be expressed as follows (Fig A4.1)*:

$$v = v_o(1 - |\eta|)/\sqrt{\eta^2 - 1} \tag{A4.1}$$

where $\eta = y/\frac{1}{2}b$ with b being the wing span.

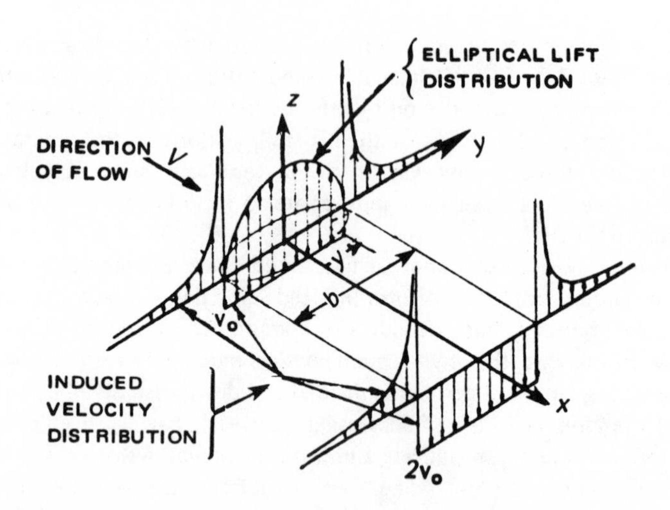

Figure A4.1 Downwash distribution of an elliptically-loaded wing

Integrated effects of upwash on the flow fields are small, and the total lift of the elliptically-loaded wing can be expressed as

$$L = 2\pi\rho(b/2)^2 Vv_o \tag{A4.2}$$

where ρ is the air density.

Load (ℓ) per unit of wing span becomes:

$$\ell = (4L/\pi b)\sqrt{1 - \eta^2}. \tag{A4.3}$$

If the spanwise wing loading is not elliptical because of planform, twist, and airfoil characteristics, it can still be approximated as a sum of the lift of a number (n) of elliptically-loaded wings. The individual contribution (L_i) of each of these wings can be expressed in a way similar to Eq (A4.2), while the total wing lift now becomes

$$L = \sum_{1}^{n} 2\pi\rho(b_i/2)^2 V\Delta v_i \tag{A4.4}$$

where the symbol Δv_i is introduced to signify the downwash velocity contribution of each wing, and b_i is the corresponding wing span.

****It should be noted that in this Appendix, a coordinate system identical to the original paper was adopted. This is different from that used in the preceding chapters.***

The contributing elliptically-loaded wings can be arranged in many ways. Examples of symmetric and one-sided arrangements are shown in Fig A4.2. If the upwash regions of the component wings are neglected, then at any nondimensional station, $\eta = y/\frac{1}{2}b$, of the actual wing, the downwash velocity will be

$$v = \sum_{1}^{n} \Delta v_i. \tag{A4.5}$$

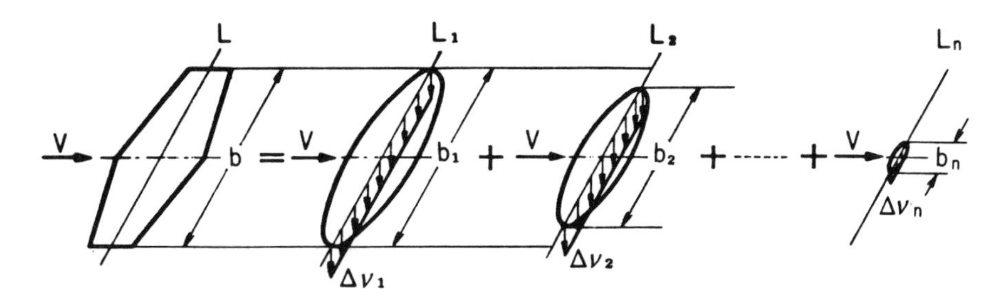

(a) SYMMETRIC ARRANGEMENT

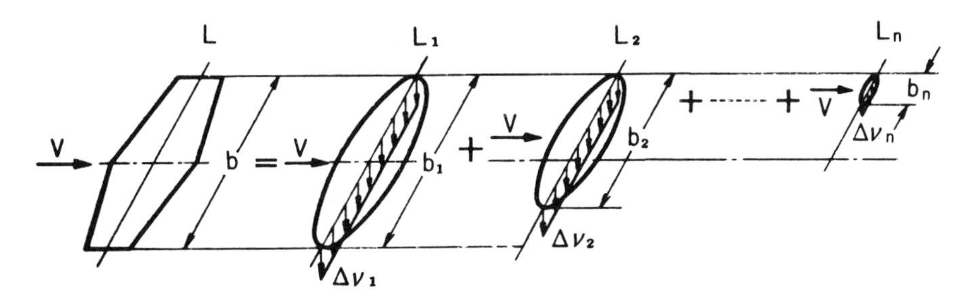

(b) ONE-SIDED ARRANGEMENT

Figure A4.2 Two types of arrangement of equivalent wings

For a symmetrical arrangement of the elliptical wings, lift (ℓ) per unit of wing span length at station ξ becomes

$$\ell = \sum_{1}^{n} \ell_i(\xi_i)$$

where $\xi_i = y_i/\frac{1}{2}b_i$ is the nondimensional span station of the i-th wing.

It was assumed that the lift distribution of the equivalent wing is elliptical; therefore, similar to Eq (A4.3), the unit span loading of the i=th wing will be

$$\ell_i = (4L_i/\pi b_i)\sqrt{1 - \xi_i^2},$$

and the expression for ℓ will now be

$$\ell = \sum_{1}^{n} (4L_i/\pi b_i)\sqrt{1 - \xi_i^2}. \tag{A4.6}$$

Since the upwash regions of the component wings have been neglected, only $|\xi_i| \leqslant 1$ are considered.

In order to express the lift supported by any single wing (say, the j) of the whole series of n equivalent wings, and to link this lift to the geometric and aerodynamic characteristics of the actual wing, the momentum and strip approaches are combined.

According to the strip approach, the lift of the portion of the actual wing between the left-hand side tips of the j and $(j + 1)$ wings (cross-hatched area in Fig A4.3) can be expressed as follows:

$$L'_j = \tfrac{1}{2}\rho V^2 \int_{-\frac{1}{2}b_j}^{-\frac{1}{2}b_{j+1}} a\left(\theta_y - \sum_i^j \Delta v_i/V\right) c_y \, dy. \tag{A4.7}$$

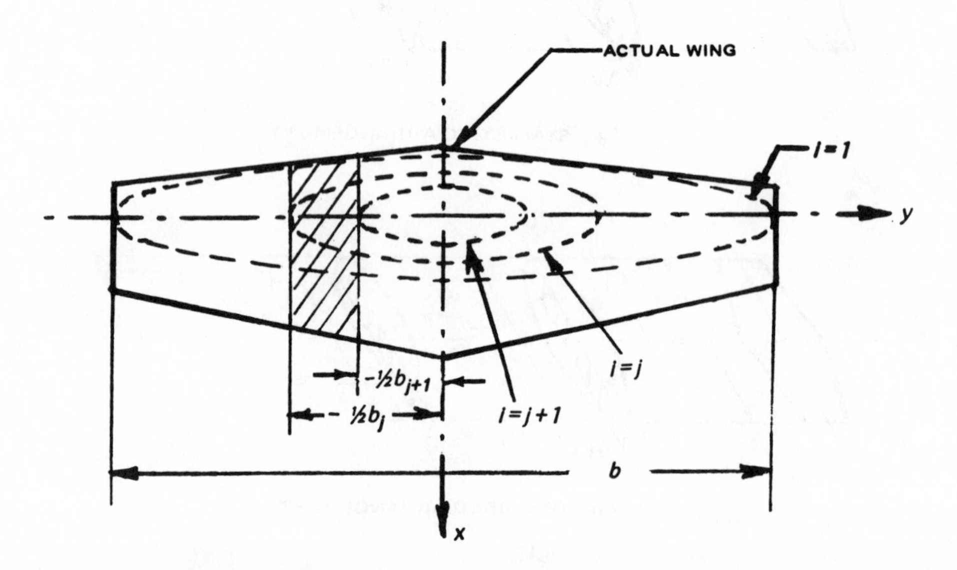

Figure A4.3 Horizontal projection of equivalent wings

where, in the most general case, the lift-curve slope a may vary (but not likely) along the wing span due to section airfoil characteristics, etc.; θ_y, the angle of incidence, may also vary along the span $\theta_y = \theta(y)$ as well as the wing chord $c_y = c(y)$. However, the value of the $\Sigma \Delta v_i/V$ sum for the i to j limits is assumed to remain constant within the interval of $y = -\tfrac{1}{2} b_j$ to $y = -\tfrac{1}{2} b_{j+1}$ (between the consecutive wing tips) and, under the small-angle assumption, represents the inflow angle which, subtracted from the local angle of incidence, gives the section angle-of-attack.

In Ref 26, the relationship expressed by Eq (A4.7) is simplified by assuming $a = const$ and taking the mean values of c and θ for the considered wing segment; namely,

$$c_j = c[-(\eta_j + \eta_{j+1})/2] \quad \text{and} \quad \theta_j = \theta[-(\eta_j + \eta_{j+1})/2]. \tag{A4.8}$$

As an alternate to Eq (A4.7), lift L'_j can be expressed as a sum of the contributions of the appropriate segments of the component elliptic wings from $n = 1$ to $n = j$.

Remembering that the resultant span loading ℓ_j of all j wings is given by Eq (A4.6), L'_j according to the momentum theory becomes

$$L_i = \int_{-\frac{1}{2}b_j}^{-\frac{1}{2}b_{j+1}} \sum_i^j \left[(4L_i/b_i)\sqrt{1 - \xi_i^2}\right] dy. \tag{A4.9}$$

Equating the right sides of Eqs (A4.7) and (A4.9), the necessary relationship for finding $v_j = \Sigma\, \Delta\, v_i$ for the 1 to i limits can be obtained. Dividing both sides of the so-developed equation by $\frac{1}{2}(b_j - b_{j+1})$, it can be rewritten in terms of the unit span loading (ℓ_j). Furthermore, replacing c_y and θ_y by their mean values according to Eq (A4.8), and switching to nondimensional span coordinates $(\eta = y/\frac{1}{2}b)$, the following is obtained:

$$\ell_i = \tfrac{1}{2}\rho V^2 \int_{-\eta_j}^{-\eta_{j+1}} c_j a(\theta_j - v_j/V)d\eta/(\eta_j - \eta_{j+1})$$

$$= \int_{-\eta_j}^{-\eta_{j+1}} \sum_1^j (4L_i/\pi b_i)\sqrt{1 - (b/b_i)^2\,\eta^2}\, d\eta/(\eta_j - \eta_{j+1}). \tag{A4.10}$$

Eq (A4.10) permits one to find v_j and ℓ_j at station j. Repeating this process for all stations from $j = 1$ to $j = n$, both the downwash and span loading along the lifting line of a fixed wing can be obtained.

3. Rotary Wing

When trying to apply the above-discussed approach to a rotary-wing, new physical facts encountered in the latter case should somehow be reflected in the development of the conceptual model. Some of the more important differences between the fixed and rotary wings are as follows: (1) in all regimes of flight, the velocity of flow varies along the blade span; consequently, for each individual blade, there is no symmetry of downwash, and (2) aerodynamic interactions between the blades should be anticipated. The induced velocity field generated by the preceding blade would influence the flow at the following blade. Furthermore, some attenuation of induced velocities may occur before the following blade enters the perturbed area.

In view of (1), Azuma and Kawachi represent the blade by a series of wings arranged to one side as shown in Fig A4.4, and assume that each of the wings has an *elliptically distributed circulation.* Due to the existence of the incoming velocity gradient $U(\bar{r}) = R\Omega\bar{r} + \sin\psi$, the spanwise load distribution along the blade will not be elliptical. Nevertheless, it is assumed that the downwash velocity along the span of each wing still remains uniform.

In order to determine the flow field at the rotor disc, a plane passing through it is divided into elements whose positions are given by coordinates ℓ and m. For the case of vertical climb and hovering, the sequence of normal velocity variation with time at some selected points of the disc are illustrated in Fig A4.5.

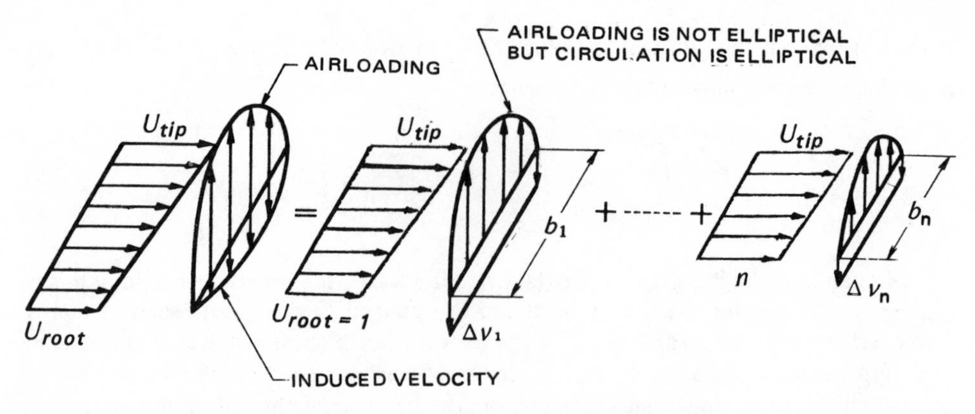

Figure A4.4 Representation of a rotary wing

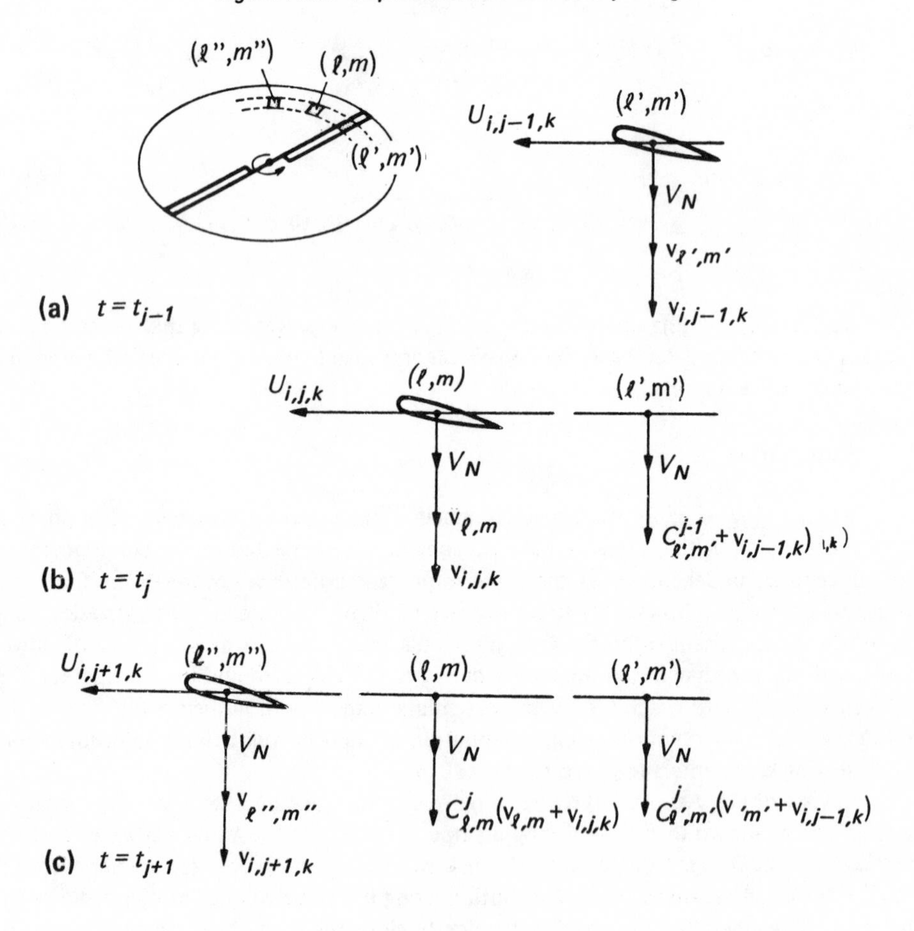

Figure A4.5 Sequence of induced velocity variation at 3 points of the disc

Let it be assumed that at some instant t_{j-1}, the blade element is located at a point (ℓ', m') where it encounters the velocity of incoming flow $U_{i,j-1,k}$. According to the convention accepted in Ref 26, "...the first subscript shows the *i-th* radial position of the blade element, the second subscript shows the time or azimuth-wise location of the blade, and the third subscript shows any quantity which is related to the *k-th* of a *b*-bladed rotor."

The total velocity component normal to the disc can be obtained as a sum of the following: (a) velocity $v_{i,j-1,k}$ induced by the blade element at the point of interest (ℓ', m') and at this particular instant $(t = t_{j-1})$; (b) velocity $v_{\zeta',\tau'}$ induced by the blade element that passed over this point one increment of time before, and (c) the normal component V_N, which may be due to axial translation of the rotor (vertical climb) and/or the flapping motion of the blade. In steady-state horizontal translation, this component would be $V_N \approx Va_d$, where a_d is the angle-of-attack of the disc plane.

At time t_j, the blade element will pass through point (ℓ, m); and at time t_{j+1}; through (ℓ'', m''). In the meantime, the velocities which had been induced at point (ℓ', m') will attenuate. This physical fact is accounted for by applying, to those velocities, a proper attenuation factor $C^j_{\ell',m'}$ corresponding to the blade position at instant $t = t_j$. At time $t = t_{j+1}$, the attenuation factor at point (ℓ', m') will be symbolized by $C^{j+1}_{\ell',m'}$. It can be seen from Fig A4.5 how this philosophy is applied to other points: (ℓ, m) and (ℓ'', m'').

Thus, as stated in Ref 26, "...$v^j_{\ell,m}$ can generally be given by the following recurrent formula:

$$v^j_{\ell,m} = C^{j-1}_{\ell,m}\left(v^{j-1}_{\ell,m} + \sum_{i=1}^{n}\sum_{k=1}^{b} v_{i,j-1,k}\,\delta_{\ell,m}\right) \tag{A4.11}$$

where the attenuation coefficient $C^{j=1}_{\ell,m}$ should be a function of the normal component of the velocity passing through the station (ℓ, m) at time t_{j-1}, and where $\delta_{\ell,m}$ should be *one* if any blade element hits station (ℓ, m) at time t_{j-1}; and otherwise, *zero*."

Similar to the previously discussed fixed-wing case, means should be found of linking blade and rotor geometry as well as its mode of operation with loads and induced velocities associated with a series of elliptical-circulation wings replacing the blade.

At this point it may be recalled that all outboard tips of the equivalent wings coincide with the blade tip. Consequently, any blade station $(\bar{r} = r/R)$ can be related to position $\bar{r}_i$ of the inboard tip of the i wing and its own nondimensionalized coordinates $\xi_i = y_i/\frac{1}{2}b$ (see Fig A4.6) in the following way:

$$\bar{r} = \tfrac{1}{2}[(1 + \bar{r}_i) + \xi(1 - \bar{r}_i)] \quad \text{or} \quad \xi = [2r - (1 + \bar{r}_i)]/(1 - \bar{r}_i). \tag{A4.12}$$

Imagining that between blade stations $r_i = \bar{r}_i R$ and at blade tip R there is a quasi-elliptically-loaded wing of semispan $R(1 - \bar{r}_i)/2$, the mass flow $\dot{m}_i$ associated with that wing would be

$$\dot{m}_i = \rho\pi[R(1 - \bar{r}_i)/2]^2 U_i \tag{A4.13}$$

where $U_i = [V \sin\psi + R\Omega(1 + \bar{r}_i)/2]$ represents the mean velocity of flow approaching the considered wing.

Similar to the previously discussed fixed-wing case, the blade geometry and rotor

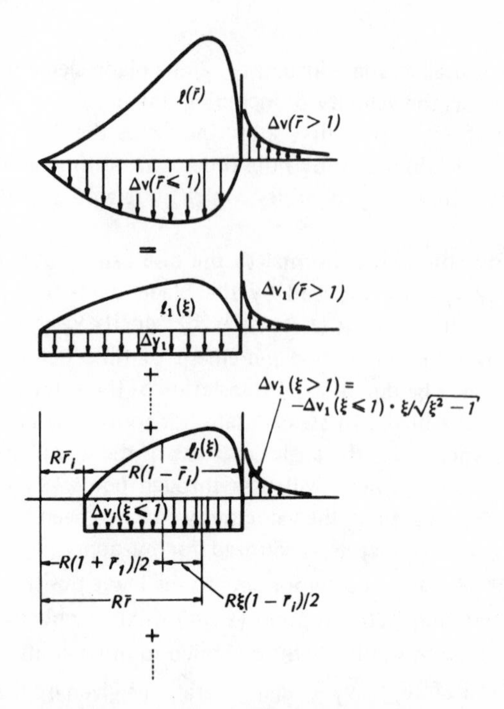

Figure A4.6 Downwash and loading of the equivalent wing

mode of operation can be linked through the blade element theory to the unit lift (ℓ_i) between blade stations $\bar{r}_i$ (position of the inboard tip of the i wing) and $\bar{r}_{i+1}$ corresponding to the inboard tip of the i+1 wing:

$$\ell_i = \tfrac{1}{2}\rho \int_{\bar{r}_i}^{\bar{r}_{i+1}} U_i^2 c_i a(\theta_i - \phi_i) d\bar{r} / (\bar{r}_{i+1} - \bar{r}_i) \qquad \text{(A4.14)}$$

where the mean values of the velocity of the incoming flow (U_i), blade chord (c_i) between stations $\bar{r}_i$ and $\bar{r}_{i+1}$, and inflow angle (ϕ_i) are given below:

$$\left.\begin{aligned} U_i &= V \sin\left(\psi_{k,o} + \sum_{\lambda=1}^{j} \Delta\psi\lambda\right) + R\Omega(\bar{r}_i + r_{i+1})/2 \\ c_i &= [c(\bar{r}_i) + c(\bar{r}_{i+1})]/2 \\ \phi_i &= (V_N + v_{\ell,m}^j + v_{i,j,k})/U_i . \end{aligned}\right\} \qquad \text{(A4.15)}$$

In the above equation, $\psi_{k,o}$ is the initial azimuth angle of the k blade, and $\Delta\psi$ the azimuthal step. In general, U_i, θ_i, and ϕ_i are $f(\psi)$ which should be reflected in full subscripts of ijk. However, for the sake of simplicity, the jk subscripts were omitted in Eq (A4.15).

The *total lift* (L_i) corresponding to the blade segment between stations $\bar{r}_i$ and $\bar{r}_{i+1}$ can be written as

$$L_i = 2\dot{m}_i \Delta v_i \tag{A4.16}$$

where $\dot{m}_i$ is given by Eq (A4.13) and Δv_i represents the downwash velocity associated with the i wing. However, additional aspects must be considered in order to establish the $\ell(\bar{r})$ relationship; i.e., to find how the total load is distributed along the blade span.

According to the Kutta-Joukowsky law, $\ell(\bar{r}) = \rho U_{\perp}(\bar{r})\, \Gamma(\bar{r})$ where $U_{\perp}(\bar{r})$ and $\Gamma(\bar{r})$ respectively, are the incoming flow velocity component perpendicular to the blade at station $\bar{r}$, and $\Gamma(\bar{r})$ is the velocity circulation at this station.

It was assumed that each of the component wings have elliptical circulation distributions. Furthermore, it was also assumed that trailing vortices are straight—perpendicular to the wing span-and extend to infinity. Consequently, the elliptical circulation can be expressed as follows:

$$\Gamma(\bar{r}) = 2R\Delta v_i(1-\bar{r}_i)\sqrt{1-\xi^2} = 2R\Delta v_i(1-\bar{r}_i)\sqrt{1-[(2\bar{r}-1-\bar{r}_i)/(1-\bar{r}_i)]^2} \tag{A4.17}$$

and the total lift of the i wing, expressed in terms of both circulation and the simple momentum relationship of Eq (A4.16) is

$$L_i = \int_{\bar{r}_i}^{r} \rho U_{\perp}(\bar{r})\, \Gamma(\bar{r})\, R\, d\bar{r} = 2\dot{m}_i \Delta v_i \tag{A4.18}$$

or, expressing $\dot{m}_i$ explicitly by Eq (A5.11), the above becomes

$$L_i = \int_{\bar{r}_i}^{1} \rho U_{\perp}(\bar{r})\Gamma(\bar{r})R\,d\bar{r} = 2\rho\pi[R(1-\bar{r}_i)/2]^2 U_i \Delta v_i \cdot \tag{A4.18a}$$

By combining Eqs (A4.17) and A4.18a), the desired expression for lift distribution along the blade span—derived from the momentum consideration—can be obtained:

$$\ell_i(\bar{r}) = [4L_i/\pi R(1-\bar{r}_i)]\,[(R\Omega\bar{r} + V\sin\psi)/U_i]\sqrt{1-[(2\bar{r}-1-\bar{r}_i)/(1-\bar{r}_i)]^2}. \tag{A4.19}$$

Equating the ℓ_i expression according to the blade element theory of Eq (A4.14) to that based on the momentum consideration of Eq (A4.19), one can obtain the equation necessary to solve the induced velocity contributions (v) appearing in Eq (A4.11).

With respect to upwash generated outside of the span of each elliptical wing, Azuma and Kawachi indicate that its effect on the inboard side can be neglected in all regimes of flight. As to the upwash at the outboard side, it may be neglected in hover and vertical climb. In the advancing rotor, however, the upwash flow left by the preceding blades is not always small on the blade-tip side and, as shown in Fig A4.7, it is necessary to take this into account when estimating the load distribution of the considered blade operating outside the tip vortices of all preceding blades. The tip-side upwash at station $\bar{r} > 1$ can be given by

$$\Delta v(\bar{r} > 1) = \sum_{1}^{n} \Delta v_i \left[1 - (2\bar{r} - 1 - r_i)/2\sqrt{(\bar{r}-1)(\bar{r}-\bar{r}_i)}\right]. \tag{A4.20}$$

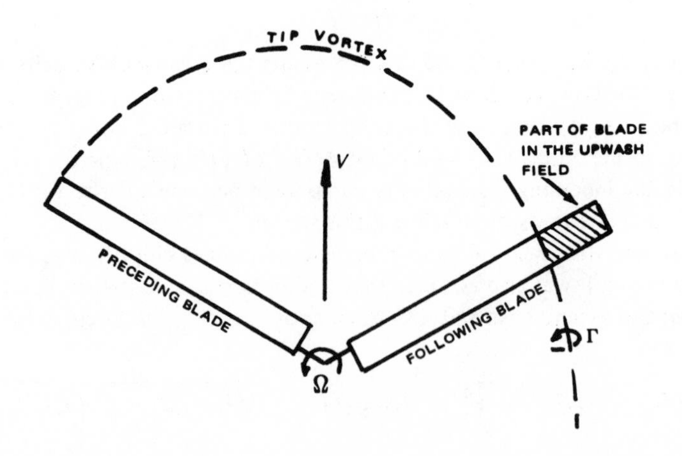

Figure A4.7 Upwash effect on the succeeding blade outside the tip vortex

The induced velocity v_j at some station (ℓ, m) must be calculated by not only taking into account the actual blade $(\bar{r} \leqslant 1)$ passing over that station, but also its hypothetical extension through the upwash region $(\bar{r} > 1)$. Similar to the downwash treatment, the same attenuation coefficients should also be applied to the upwash velocity components.

The magnitude of the attenuation coefficient, $C_{j-1} = v/v_o$ at station (ℓ,m) occupied by blade element (i, j, k), will depend on the relative position of the local station and time elapsed between passages of the blade: $\Delta t = t_j - t_{j-1}$.

Values of the attenuation coefficients computed in Ref 26 are shown in Fig A4.8a for both hover and horizontal translation. It is also indicated that in the simpler cases, the constant attenuation coefficient may be assumed $(C_{\ell,m}^{j=1} = C)$ and the representative C values for $\bar{r} = 0.75$ can be obtained for the various thrust coefficients (C_T) and number of blades shown in Fig A4.8b.

3.1 Applications

The steady-state hovering case for an articulated rotor with two untwisted blades was investigated using the above-discussed method. The spanwise partitioning of the blade was $n = 20$, and the azimuthal increment was $\Delta\psi = 2\pi/b$. The so-obtained spanwise load distributions shown in Fig A4.9 indicate a good agreement with experimental results.

Steady-state forward flight was also examined. The actual procedure as outlined by Azuma and Kawachi is as follows:

> The effect of upwash velocity observed in the outboard part of the blade must, as stated before, be included in determining the angle of attack of every succeeding blade. The trace of the upwash velocity as well as the downwash velocity left on any station in the rotor plane is stored in the computer memory and is recalled every moment, and then multiplied by the attenuation coefficient at that instant for the calculating of induced velocity.[26]

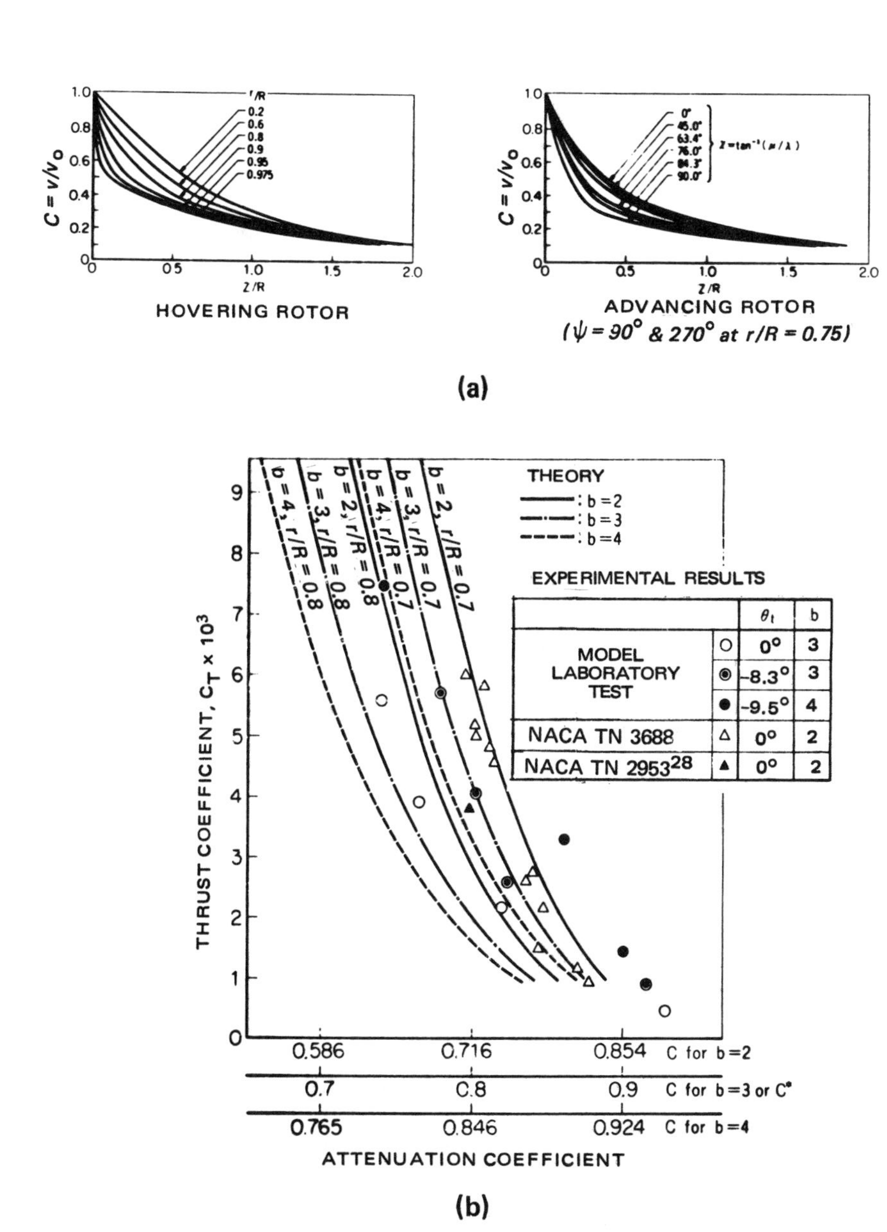

Figure A4.8 Attenuation coefficients and comparison of theoretical and experimental results of hovering rotors

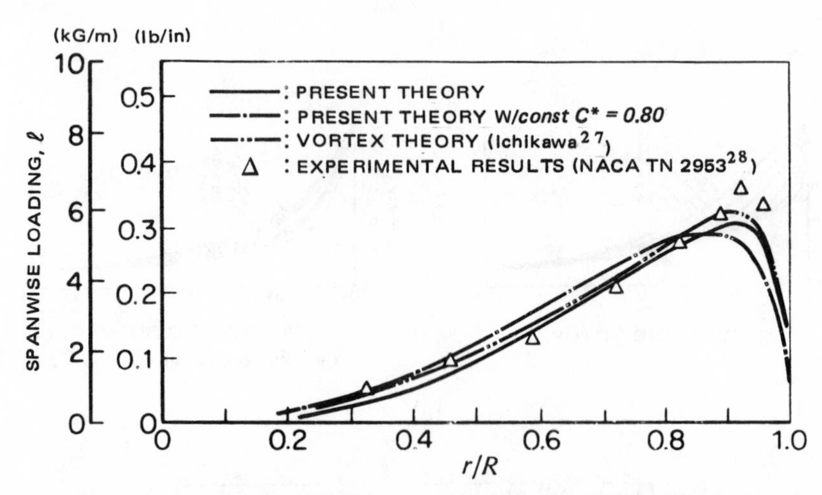

Figure A4.9 Spanwise loading of a rotor with two untwisted blades

Using this approach, spanwise airloading was computed at $\mu = 0.18$. As before, the blade was divided into $n = 20$ segments, but the azimuthal increment was reduced to $\Delta\psi = 10°$. The field was divided into a net of $R/80$ square elements, the number of which was $\ell \times m = 160 \times 320$. A lift coefficient of zero was assumed in the reversed flow region.

A comparison of predicted and experimental values of spanwise loading is shown in Fig A4.10, indicating a good agreement between analysis and tests.

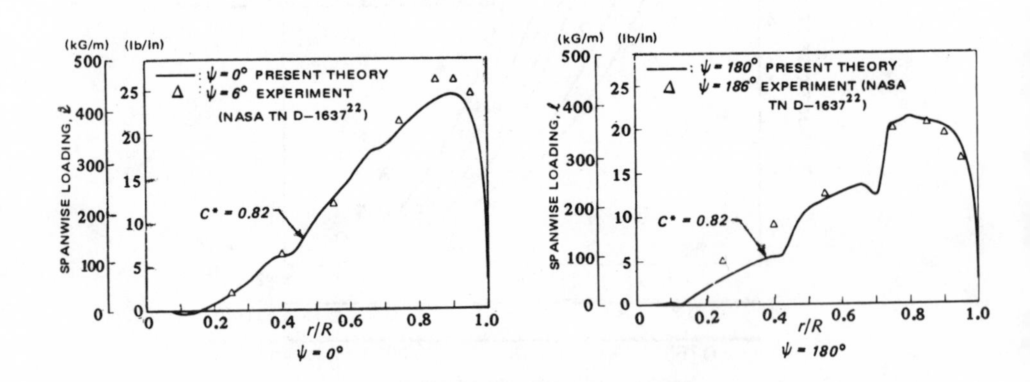

Figure 4.10 Spanwise loading in forward flight

Finally, it can be reaffirmed that this method appears especially suitable for unsteady conditions since it considerably reduces computer time. As an example of such applications, rotor response to a rapid increase in collective pitch and to sudden input of cyclic pitch is discussed in Ref 26.

References for Chapter IV

1. Baskin, V. Ye.; Vil'dgrube, L.S.; Vozhdayey, Ye. S.; and Maykapar, G.I. *Theory of Lifting Airscrews.* NASA TTF-823, 1975.

2. Goldstein, S. *On Vortex Theory of Screw Propellers.* Royal Society Proceedings (a) 123, 1929.

3. Knight, M., and Hefner, R.A. *Static Thrust Analysis of the Lifting Airscrew.* NACA TN-626, 1937.

4. Landgrebe, Anton J., and Cheney, Marvin C. J. *Rotor Wakes – Key to Performance Predictions.* AGARD-CPP 111, Sept. 1972.

5. Landgrebe, Anton J. *An Analytical and Experimental Investigation of Helicopter Rotor Hover Performance and Wake Geometry Characteristics.* USAAMRDL Tech. Report 71-24, 1971.

6. Werlé, H., and Armand, C. *Measures et Visualisation Instationnaires sur les Rotors.* ONERA T.P. No. 777, 1969.

7. Landgrebe, Anton J., and Johnson, Bruce V. *Measurement of Model Helicopter Flow Velocities with a Laser Doppler Velocimeter.* Journal of AHS, Vol. 19, No. 4, 1974.

8. Biggers, James C., and Orloff, Kenneth L. *Laser Velocimeter Measurements of the Helicopter Rotor-Induced Flow Field.* Journal of AHS, Vol. 20, No. 1, 1975.

9. McCormick, B.W.; Tangler, J.L.; and Sherrier, H.E. *Structure of Trailing Vortices.* Journal of Aircraft, Vol. 5, June 1968.

10. Gray, R.B., and Brown, G.W. *A Vortex-Wake Analysis of a Single-Bladed Hovering Rotor and a Comparison with Experimental Data.* AGARD-CPP 111, September 1972.

11. Prandtl, L., and Tietjens, O.G. *Fundamentals of Hydro- and Aeromechanics.* Translated by L. Rosenhead. Dover Publications, 1957.

12. Durand, W.F., editor-in-chief. *Aerodynamic Theory.* Durand Reprinting Committee, Div. B, Ch. III, 1943.

13. Clancy. L.J. *Aerodynamics.* Halsted Press, A Div. of John Wiley and Sons, N.Y., 1974.

14. Ormiston, Robert A. *An Actuator Disc Theory for Rotor Wake Induced Velocities.* AGARD-CPP 111, Sept. 1972.

15. Jenny, D.S.; Olson, J.R.; and Landgrebe, A.J. *A Reassessment of Rotor Hovering Performance Prediction Methods.* Journal of the AHS, Vol. 13, No. 2, April 1968.

16. Erickson, J.D., and Ordway, D.E. *A Theory for Static Propeller Performance.* CAL/USAAVLABS Symposium Proceedings, Vol. 1, June 1966.

17. Magee, J.P.; Maisel, M.D.; and Davenport, F.J. *The Design and Performance Prediction of a Propeller/Rotor for VTOL Applications.* Proceedings of the 25th Annual AHS Forum, No. 325, May 1969.

18. Brady, W.G., and Crimi, P. *Representation of Propeller Wakes by Systems of Finite Core Vortices.* CAL Report No. BB-165-1-2, February 1965.

19. Sadler, G.S. *Development and Application of a Method for Predicting Rotor Free-Wake Position and Resulting Rotor Blade Airloads.* NACA CR-1911, Vols. I and II, December 1971.

20. Sadler, G.S. *Main Rotor Free-Wake Geometry Effects on Blade Airloads and Response for Helicopters in Steady Maneuvers.* NASA CR-2110, Vols. I and II, December 1971.

21 Crimi, P. *Theoretical Predictions of the Flow in the Wake of a Helicopter Rotor.* Cornell Aero Lab Rpt. No. BB-1994-S-1, September 1965.

22. Scheiman, J, and Ludi, L.H. *Quantitative Evaluation of Effect of Helicopter Rotor-Blade Tip Vortex on Blade Airloads.* NASA TN D-1637, 1963.

23. Surendraiah, M. *An Experimental Study of Rotor Blade-Vortex Interaction.* NASA CR-1573, 1970.

24. Ham, N.D. *Some Conclusions from an Investigation of Blade-Vortex Interaction.* Journal of the AHS, Vol. 20, No. 4, October 1975.

25. Landgrebe, A.J.; Moffit, R.C.; and Clark, D.R. *Aerodynamic Technology for Advanced Rotorcraft.* Journal of the AHS, Vol. 22, Part I, No. 2, April 1977; Part II, No. 3, July 1977.

26. Azuma, Akira, and Kawachi, Keiji. *Local Momentum Theory and Its Application to the Rotary Wing.* AIAA Paper 75-865, June 1975.

27. Ichikawa, T. *Linearized Aerodynamic Theory of Rotor Blades (III).* Technical Report of National Aerospace Laboratory of Japan, NAL TR-100, 1966.

28. Meyer, J.R., Jr., and Falabella, G., Jr. *An Investigation of the Experimental Aerodynamic Loading on a Model Helicopter Rotor Blade.* NACA TN 2953, 1953.

VELOCITY AND ACCELERATION POTENTIAL THEORY

Basic formulae of the velocity potential in an incompressible fluid are recalled, and then ways of applying them to rotary-wings are indicated. Next, the same steps are taken with respect to the acceleration potential, and the links between the two potentials are shown along with ways of applying them to rotor blades moldeled by lifting lines and lifting surfaces. Considerations of the consequences of fluid compressibility completes the presentation of potential-theory fundamentals; followed by their application to such tasks as determination of the induced velocity field of an actuator disc having a prescribed area loading, and computation of blade loading in steady and unsteady flows. Finally, a brief discussion of the application of potential methods to mapping of the flow around nonrotating helicopter components concludes this chapter.

Principal notation for Chapter V

$[A]$	matrix	
C_L	area lift coefficient	
c	Blade, or wing chord	m or ft
c_ℓ	section lift coefficient	
F	load-per-unit span	N/m or lb/ft
h	elevation	m or ft
$\vec{i}, \vec{j}, \vec{k}$	unit vectors along x, y, z axes	
ℓ	distance	m or ft
m	flow doublet strength (moment)	m^4/s or ft^4/s
$\vec{n}$	vector normal to the surface	
p	pressure	N/m^2 or psf
Q	strength of flow source	m^3/s or ft^3/s
q	strength of pressure doublet	
R	radius	m or ft
r	radial distance	m or ft
$\bar{r}$	nondimensional radial distance, $\bar{r} \equiv r/R$	
S	surface	m^2 or ft^2
T	thrust	N or lb
t	time	s
U	velocity of distant flow	m/s or fps
u, v, w	velocity components along x, y, z axes	m/s or fps

Theory

V	velocity in general	m/s or fps
v	perturbation velocity	m/s or fps
W	normal velocity component	m/s or fps
X, Y, Z	Cartesian coordinates	m or ft
x, y, z	Cartesian coordinates	m or ft
Z_i, X_{ij}, Y_{ij}	coefficients	
α	angle-of-attack	rad or deg
Γ	velocity circulation	m^2/s or ft^2/s
Δ	increment	
ϵ	short distance	m or ft
θ	angle, or blade-pitch angle referred to zero-chord	rad or deg
ϑ	angle	rad or deg
Λ	angle	rad or deg
Σ	area	m^2 or ft^2
φ	velocity potential	m^2/s or ft^2/s
Ψ	acceleration potential	
ψ	azimuth angle	rad or deg
Ω	rotational velocity	rad/s
ω	solid angle	steradian
∇	del; $\nabla \equiv \vec{i}(\partial/\partial x) + \vec{j}(\partial/\partial y) + \vec{k}(\partial/\partial z)$	
∇^2	Laplacian (del^2) $\nabla^2 \equiv (\partial^2/\partial x^2) + (\partial^2/\partial y^2) + (\partial^2/\partial z^2)$	
$\{\}$	column	
[]	matrix	

Subscripts

$\parallel$	parallel
$\perp$	perpendicular
v	velocity

Superscripts

$\cdot$	time derivative
$\rightarrow$	vector

1. INTRODUCTION

From previous discussions of the various theories, the reader has probably come to the conclusion that the vortex theory offers the most precise description of the aerodynamic phenomenon of the rotor, but in practical applications, it requires the largest amount of computational effort. This is especially true if one attempts to rigorously account for all, or at least most of the interactions between vortices, such as in the case of nonrigid wakes, unsteady aerodynamics (e.g., aeroelastic phenomenona), some maneuvers, etc. Application of velocity and acceleration potentials makes it possible to determine steady and unsteady flow fields induced by the rotor in both incompressible and compressible fluids with a precision similar to that offered by the vortex theory approach but with less computational effort.

Mangler and Squire[1] were probably the first (1950s) to adapt the velocity and acceleration potential concepts to the determination of the induced velocity field of a rotor. This was done for the case of forward flight where induced velocity was considered small in comparison with that of the distant flow (reverse of the rotor translatory velocity). The accompanying pressure differential between the surfaces of the disc was assumed to be a relatively simple function of the rotor radius only (axisymmetric loading).

The Mangler and Squire approach of finding time-average induced velocities was extended by Baskin, et al[2] to include the case of a rigid wake associated with a finite number of blades. In addition to the two types of disc loading examined in Ref 1, they studied a third one; also with an axial symmetry. Finally, they investigated disc loading as a function of both rotor radius and azimuth angles.

Investigations dealing with the application of potential methods (both velocity and acceleration) to more sophisticated rotor models were initiated in the 1960s. Such aspects as unsteady aerodynamic effects and their influence on airfoil section characteristics, effects of compressibility, and finite values of the speed of sound were taken into consideration. In addition to the determination of both time-average and instantaneous induced velocities, the task of evaluating both span and chordwise loading of the blades was undertaken.

To deal with these problems, the basic methodology of potential flow was available from fixed-wing technology. However, inherent complexities resulting from rotor blade motions, as well as mutual interference between the blades makes the task of applying potential methods to rotary-wing aircraft more difficult.

When considering the adaptation of velocity and, especially acceleration potentials to unsteady rotor aerodynamics the contributions of Dat[3,4,5] and Costes[6,7] should be mentioned. van Holten[8,9] along with other investigators in this field, tend to emphasize the acceleration potential approach, while Jones and his coworkers and followers[10,11,12] built their approach around the velocity potential. Similar to the vortex theory, there is a steadily growing body of literature dealing with the application of the potential concepts to rotary-wing aircraft as exemplified by Refs 13 and 14.

Before discussing the various philosophies and methods of application presented in Refs 1-14, in more detail, some basic aspects of the potential vector field will be briefly reviewed. This will be followed by a brief resume of fundamental relationships; first of velocity and then, acceleration potentials. The presentation of actual samples, however, will be given in reverse order; starting with the acceleration potential concepts, and then the application of the velocity potential. Finally, application of the potential methods to the mapping of flow around nonrotating helicopter components will be briefly outlined.

2. VELOCITY POTENTIAL IN AN INCOMPRESSIBLE FLUID

2.1 Basic Relationships

The theory of vector fields (see any textbook on Vector Analysis; e.g., *Vector and Tensor Analysis*[15], or *Fundamentals of Aerodynamics*[16], p. 65) states that under some conditions, which will be specified later, a scalar function of the spatial (Cartesian) coordinates $\varphi(x,y,z)$ may exist, and knowledge of the $\varphi(P)$ value at an arbitrary point $P(x,y,z)$ would enable one to determine the magnitude and direction of a field vector at this point by computing its components along the x,y,z axes as: $\partial\varphi/\partial x$; $\partial\varphi/\partial y$; and $\partial\varphi/\partial z$.

Should such function φ, which will be called the velocity potential, exist in a field of flow, the components of velocity vector $\vec{v}(P)$ associated with φ can be obtained as

$$\begin{aligned} u &= \partial\varphi/\partial x \\ v &= \partial\varphi/\partial y \\ w &= \partial\varphi/\partial z. \end{aligned} \tag{5.1}$$

Velocity $v(P)$ can also be defined as a vector sum of its components as given by Eq (5.1):

$$\vec{v} = \vec{i}(\partial\varphi/\partial x) + \vec{j}(\partial\varphi/\partial y) + \vec{k}(\partial\varphi/\partial z). \tag{5.1a}$$

where $\vec{i}, \vec{j}$, and $\vec{k}$ are unit vectors along the x, y, and z axes, respectively. The relationship between $\vec{v}(P)$ and $\varphi(P)$ does not depend on the type of the coordinate system. It should be valid in the Cartesian as well as cylindrical, or any other set of coordinates. This fact can be expressed in the universal language of vector analysis as

$$\vec{v}(P) = \overrightarrow{grad}[\varphi(P)]. \tag{5.1b}$$

However, in the Cartesian system, Eq (5.1b) is often written using a vector operator called *"del"* which is symbolized by ∇:

$$\nabla \equiv \vec{i}(\partial/\partial x) + \vec{j}(\partial/\partial y) + \vec{k}(\partial/\partial z). \tag{5.1c}$$

It can easily be shown that a continuity of flow in an incompressible fluid requires that

$$(\partial u/\partial x) + (\partial v/\partial y) + (\partial w/\partial z) = 0. \tag{5.2}$$

The expression on the left side of Eq (5.2) is called the divergence, *"div"*, while the whole equation states that the divergence must be zero in order to preserve the continuity of flow in an incompressible fluid. The physical truth of this statement should be independent of the type of coordinate system. This fact is again expressed in the language of vector analysis as

$$div\,\vec{v} = 0. \tag{5.2a}$$

Since $\vec{v} = \overrightarrow{grad}\,\varphi$, the condition of continuity given by Eq (5.2a) can be written as follows:

$$div\,\overrightarrow{grad}\,\varphi = 0, \tag{5.2b}$$

which in light of Eq (5.1) leads to

$$(\partial^2\varphi/\partial x^2) + (\partial^2\varphi/\partial y^2) + (\partial^2\varphi/\partial z^2) = 0. \tag{5.3}$$

The above relationship indicates that if some function φ can fulfill the condition expressed by Eq (5.3), the vectors obtained as space gradients of φ would form a continuous field, called the potential field, while the equation itself is known as the Laplace equation, and the left side is often abbreviated through the "del^2" symbol:

$$\nabla^2 \equiv (\partial^2/\partial x^2) + (\partial^2/\partial y^2) + (\partial^2/\partial z^2).$$

In the particular case of flow fields determined by velocity vectors, the existence of a potential; i.e., a function fulfilling Eq (5.3), is synonymous with the lack of fluid rotation in such a flow.

Vector rotation (in this case, velocity $\vec{v}$) is symbolized in vector analysis as $\overrightarrow{rot}\,v$. The lack of rotation requires that $\overrightarrow{rot}\,v = 0$. In Cartesian coordinates, this can be expressed as

$$\overrightarrow{rot}\,\vec{v} = \vec{i}[(\partial v/\partial x) - (\partial u/\partial y)] + \vec{j}[(\partial u/\partial z) - (\partial w/\partial x)] + \vec{k}[(\partial w/\partial y) - (\partial v/\partial z)] = 0, \tag{5.4}$$

which, in turn, leads to the following conditions:

$$\begin{aligned}(\partial v/\partial x) - (\partial u/\partial y) &= 0\\ (\partial u/\partial z) - (\partial w/\partial x) &= 0\\ (\partial w/\partial y) - (\partial v/\partial z) &= 0.\end{aligned} \tag{5.4a}$$

Expressing u, v, and w in Eq (5.4a) according to Eq (5.1), and then summing the three expressions, will result in the Laplace equation (Eq (5.3)), thus proving that existence of the velocity potential is synonymous with the absence of rotation in the flow of a fluid.

The Laplace equation has some properties that are of importance for the application of potential methods (either velocity or acceleration) to rotary-wing aerodynamics. One of them is the fact that Eq (5.3) has an infinite number of particular solutions. This in turn provides a large latitude of function φ which can be used when selecting expressions for either velocity or acceleration potentials. In this process, however, it should be remembered that in addition to complying to Eq (5.3), these functions should also satisfy the particular boundary conditions associated with the considered problem.

Another important property of Eq (5.3) is that it is a linear equation, which means that any linear combination (addition or subtraction) of its particular solutions as represented by various potential functions will still satisfy the Laplace equation. In other words, any linear combination of potential functions also leads to a potential function, which opens the way for the use of superposition in dealing with various tasks of flow and load determination.

2.2 Application to Rotary-Wing Aircraft

Similar to the previously considered cases, it will usually be assumed that the rotor is stationary, while an infinite volume of fluid moves past it with a uniform distant velocity $(-U_\infty \equiv -U_{x_\infty})$ directed toward the negative half of the x axis of the Cartesian system of coordinates as shown in Fig 5.1.

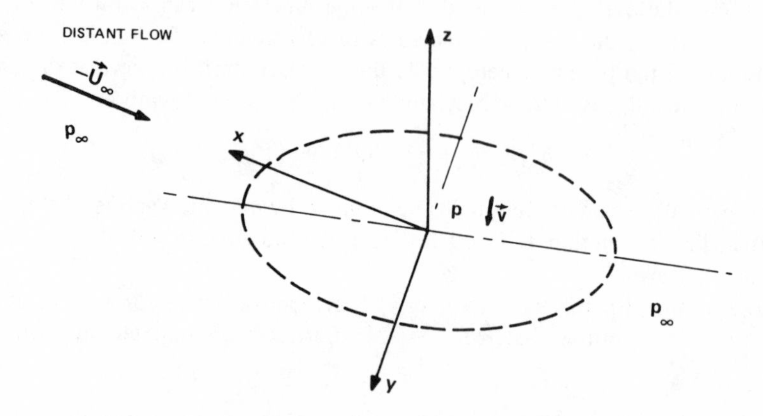

Figure 5.1 Axes of coordinates

The presence of the rotor disturbs the flow by inducing perturbation velocities $(\vec{v})$ and changing the pressure from its distant value of p_∞. Perturbation velocities

$$\vec{v} = \vec{i}u + \vec{j}v + \vec{k}w \tag{5.5}$$

will obviously modify the flow.

However, it is usually assumed that absolute values of $\vec{v}$ are small in comparison with $\vec{U}_\infty$,

$$|v| \ll |U_\infty|.$$

Consequently, it may be stated that the flow velocities in the perturbed flow will be as follows:

$$\begin{aligned} V_x &= U_\infty + u \approx U_\infty \\ V_y &= v \\ V_z &= w. \end{aligned} \tag{5.6}$$

Furthermore, the flow must satisfy some boundary conditions. The most important of which are as follows:

1. The flow is undisturbed far ahead of the rotor; hence, $\vec{v}_\infty$ vanishes.

2. Far downstream from the rotor, pressure in the fluid must return to its original value of p_∞.

3. Envelopes constituting lifting surfaces (as those of any physical nonporous body submerged in the flow) cannot be penetrated by the flow. This means that

$$\vec{V} \cdot \vec{n}_o = 0 \tag{5.7}$$

where $\vec{V}$ is the resultant velocity of the flow at the surface, and $\vec{n}_o$ is the unit vector normal to the surface.

4. The Kutta-Joukowsky condition of a smooth flow must be fulfilled at the trailing edges of the lift-producing airfoil sections.

The velocity potential approach can be used for the detemination of both time-average and instantaneous velocities induced by either a blade or a rotor. In the case of time-average problems, the potential will be expressed solely as a function of the space coordinates (spatial location of the point of interest P), $\varphi = \varphi(P)$; however, when dealing with instantaneous phenomena, the variation of the potential with time should also be considered: $\varphi = \varphi(P,t)$.

Since the existence of a potential is synonymous with the absence of rotational flow, it is obvious that potential methods cannot (at least directly) be applied to the rotor vortex system; i.e., to the rotor disc where bound vortices are present, and to the wake formed by the shed and trailing vortices. Ways of overcoming difficulties associated with this subject will be discussed later.

2.3 Expression of Velocity Potential through Doublets

In principle, there are many ways of finding functions of spatial coordinates which may express the velocity potentials of fixed-wings, rotors, or blades and still satisfy all the necessary boundary conditions. The one most often used relies on doublets of proper strength and distribution over such surfaces as the rotor disc, rotor blades, and the blade and rotor wakes.

In the case of wakes, the application of doublets would permit one to map a flow induced by such surfaces of vorticity as vortex sheets. In the vortex sheet, the flow is obviously rotational; i.e., nonpotential. However, a velocity potential usually exists outside of the sheet itself, and its value can be related to the strength and distribution of doublets located on the vortex sheet.

Flow Doublet and Its Potential. At this point it may be recalled that the concept of a doublet in the flow field of a fluid was derived as a limiting case by imagining that a source of strength Q (flow rate) and a sink of strength $-Q$ approach each other in a way such that the absolute value of Q increases without restriction as the source-sink distance ϵ decreases. At the limit when $\epsilon \to 0$, the source-sink pair becomes a doublet, while the $Q\epsilon$ product reaches a finite value: $m = Q\epsilon$. The symbol standing for m– the product of distance and flow rate–is actually the doublet moment. However, it is also called the doublet strength, and the latter, more descriptive term, is used in this text.

The velocity potential at point $P(x, y, z)$ due to a doublet located at an arbitrary point $P_o(x_o, y_o, z_o)$, and having strength m (e.g., Ref 16, p. 252) would be

$$\varphi(P) = -(\partial/\partial n_o)(m/4\pi \ell) \tag{5.8}$$

or

$$\varphi(P) = (mx/4\pi\ell^3) = (m \cos\theta/4\pi\ell^2) \tag{5.8a}$$

where $\ell = [(x - x_o)^2 + (y - y_o)^2 + (z - z_o)^2]^{1/2}$; and $\partial/\partial n_o$ signifies differentiation with respect to the doublet axis (assumed in Fig 5.2 to coincide with the x axis); while θ represents an angle between the doublet axis (x, or $\vec{n}_o$) and $\vec{\ell}$.

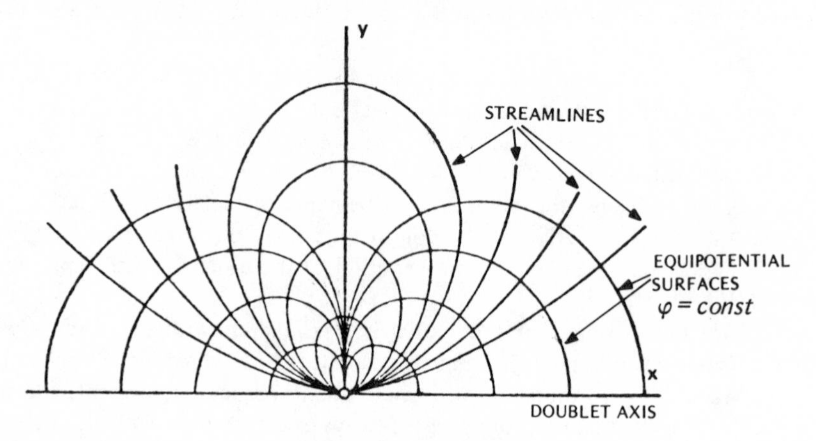

Figure 5.2 Flow doublet

Representation of a Vortex Surface through Flow Doublets. It is assumed that the point of the doublet location (P_o) lies on a surface of vorticity (say, a vortex sheet), and the doublet itself is so oriented that its axis (say, x) coincides with vector $\vec{n}_o$, normal to the surface. At point $P(x, y, z)$, this doublet would induce a velocity whose value ($v_\perp$) in a direction perpendicular to ℓ can be expressed as a proper derivative of the doublet potential given by Eq (5.8a):

$$v_\perp = (m/4\pi)(\sin\theta/\ell^3). \tag{5.9}$$

On the other hand, velocity $v_\perp$ can be expressed in terms of circulation $\Gamma(S_{v_o})$ of a vortex sheet (S_v) at the doublet location:

$$v_\perp = \int_{S_v} \frac{\Gamma(S_v)\sin\theta}{4\pi\ell^3}\, dS \tag{5.10}$$

where the subscript S_v at the integral sign signifies integration over the surface of vorticity.

Equating the right sides of Eqs (5.9) and (5.10), one finds that a relationship can be found between the strength of the doublet inducing the same velocity, and circulation value at the corresponding area of the vortex sheet:

$$m = \int_{S_v} \Gamma(S_v)dS. \tag{5.11}$$

The following expression linking velocity potential at some point P to the circulation of a vortex surface is obtained by substituting Eq (5.11) into Eq (5.8):

$$\varphi(P) = -\int_{S_v} [\Gamma(S_v)/4\pi](\partial/\partial n_o)(1/\ell)dS. \tag{5.12}$$

Velocity potential at point P may result from the influences of many areas of vorticity (S_{v_1}, S_{v_2}, ..S_{v_n}). However, due to the previously discussed linearity of the Laplace equation, this resultant potential can be simply obtained as a sum of the individual contributions of Γ_{sv_1}, Γ_{sv_2}, etc. Consequently if the shape of the whole vortex system is known and the velocity circulation values within it are related in some way to the design and operational parameters, then the following approach to the determination of the induced velocity (perturbation) field of the rotor would appear attractive. The whole vortex system could be divided into proper areas of vorticity which can be either defined by simple analytical expressions, or approximated by flat geometric figures. Knowledge of the orientation of vectors n ($\vec{n}_1, \vec{n}_2, ...\vec{n}_n$) and values ($\ell_1, \ell_2, ... \ell_n$) determining the absolute distance from those "patches" of vorticity to point P would, using Eq (5.12), permit one to calculate the individual contributions to the total velocity potential at P. As the next step, the induced velocity $\vec{v}(P)$ can be found as $\overrightarrow{grad}\ \varphi(P)$.

This approach, which in principle appears simple, unfortunately encounters considerable difficulties in calculating $\vec{v}$ at the vortex surface itself (be it vorticity represented by vortices attached to the blade, or moving freely in the wake). Looking at Eq (5.12), one would realize that with $\ell \to 0$, the expressions for the velocity potential and its gradient become meaningless.

The above result should be anticipated from the general discussion of the velocity potential which states that the potential exists only where $\overrightarrow{rot}\ v = 0$. Of course, this does not apply to the blade or rotor vortex system.

There are ways of relating velocity potential to physical quantities other than circulation. But the above-discussed difficulties in determining induced velocities at the vortex surface would still exist. For this reason, induced velocity determination based on the acceleration approach may appear advantageous[9].

3. ACCELERATION POTENTIAL IN AN INCOMPRESSIBLE FLUID

3.1 General Relationships

When discussing the velocity potential approach, only two properties of the fluid were explicitly mentioned: (1) incompressibility, which means that the speed of sound (i.e., velocity of transmitting signals) is infinite; and (2) pressure, where it was stated that in establishing boundary conditions, the pressure far downstream should return to the same value as in the distant incoming flow. Knowledge of the fluid density ρ was not required. In the acceleration potential methods, this additional characteristic of the fluid becomes important.

Acceleration of fluid—expressed by the complete derivative of velocity $\vec{v}$ with respect to time ($d\vec{v}/dt$)—can be related to the fluid pressure (p) and its density (ρ) through the Euler equation which, in vector analysis notations, is expressed as

$$d\vec{v}/dt = -(1/\rho)\,\overrightarrow{grad}\,p \tag{5.13}$$

and in Cartesian coordinates, this becomes

$$\begin{aligned}
\frac{\partial u}{\partial t} + u\frac{\partial u}{\partial x} + v\frac{\partial u}{\partial y} + w\frac{\partial u}{\partial z} &= -\frac{1}{\rho}\frac{\partial p}{\partial x} \\
\frac{\partial v}{\partial t} + u\frac{\partial v}{\partial x} + v\frac{\partial v}{\partial y} + w\frac{\partial v}{\partial z} &= -\frac{1}{\rho}\frac{\partial p}{\partial y} \\
\frac{\partial w}{\partial t} + u\frac{\partial w}{\partial x} + v\frac{\partial w}{\partial y} + w\frac{\partial w}{\partial z} &= -\frac{1}{\rho}\frac{\partial p}{\partial z}
\end{aligned} \tag{5.13a}$$

By retaining the previously accepted assumption of $|v| \ll |U_\infty|$ which led to Eq (5.6), Eq (5.13a) can be simplified to the following form:

$$\begin{aligned}
\frac{\partial u}{\partial t} + \frac{\partial u}{\partial x}U_\infty &= -\frac{1}{\rho}\frac{\partial p}{\partial x} \\
\frac{\partial v}{\partial t} + \frac{\partial v}{\partial x}U_\infty &= -\frac{1}{\rho}\frac{\partial p}{\partial y} \\
\frac{\partial w}{\partial t} + \frac{\partial w}{\partial x}U_\infty &= -\frac{1}{\rho}\frac{\partial p}{\partial z}
\end{aligned} \tag{5.14}$$

By summing up the above equations and remembering the continuity relationship expressed by Eq (5.2), it can be shown that

$$\frac{\partial^2 p}{\partial x^2} + \frac{\partial^2 p}{\partial y^2} + \frac{\partial^2 p}{\partial z^2} = 0. \tag{5.15}$$

This indicates that pressure p, which should be interpreted as the difference between p_∞ and local pressure, $\Delta p = p - p_\infty$, satisfies the Laplace equation and thus, represents a potential function.

Looking at the physical aspects of Eqs (5.13) and (5.13a), however, one would clearly see that as far as fluid acceleration is concerned, it is not the pressure differential alone, but the $(p - p_\infty)/\rho$ which influences the $d\vec{v}/dt$ values. Consequently, it appears logical to call this quantity the acceleration potential (Ψ) and define it as follows:

$$\Psi = -(p - p_\infty)/\rho. \tag{5.16}$$

By solving Eq (5.16) for p and substituting those values into Eq (5.15) it is easy to show that Ψ as defined by Eq (5.16) is also a potential function since it satisfies the Laplace equation

$$\frac{\partial^2 \Psi}{\partial x^2} + \frac{\partial^2 \Psi}{\partial y^2} + \frac{\partial^2 \Psi}{\partial z^2} = 0. \tag{5.17}$$

Similar to the velocity potential, the acceleration potential is obviously a function of the space coordinates, but it also may be a function of time; hence, in general, $\Psi(P,t)$, where P symbolizes the location of the point of interest, $P(x,y,z)$ and t is time.

3.2 Application to Lifting Surfaces

With respect to the practical application of the acceleration principle to rotary-wing aerodynamics, there is the problem, as in the case of velocity potential, of relating Ψ to some physical quantities pertaining to rotor characteristics and/or its operation. In other words, there is again a question of constructing a physicomathematical model around the basic concept of acceleration potential. In trying to find the proper mathematical expressions for $\Psi(P,t)$, care should be taken to see that the physical requirements reflected in the boundary conditions are also satisfied.

Costes[6] indicates that in dealing with load problems, it may be quite convenient to use doublets of pressure. In analogy with the case of fluid flow (Sect 2.3), it may be imagined that the doublet of pressure has a strength q which may also be a function of time, $q(t)$. Then, similar to Eq (5.8a), the acceleration potential can be expressed as

$$\Psi(P,t) = -q(t)\,\frac{\vec{\ell}\cdot\vec{n}_o}{4\pi|\ell|^3} \tag{5.18}$$

or

$$\Psi(P,t) = -q(t)\,\frac{\cos\theta}{4\pi\ell^2} \tag{5.18a}$$

where, as in the case of Eq (5.8a), ℓ signifies the distance (in this case, indicated by a vector) between the point of the doublet location $P_o(x_o, y_o, z_o)$ and the point where the potential is calculated, $P(x,y,z)$.

A physically significant dependence of the strength of the flow doublet on circulation was established in Sect 2.3 (Eq (5.11)). In the present case, another important relationship between the strength of the pressure doublets (q) and pressure differentials (Δp = aerodynamic loads) sustained by lifting surfaces can be obtained (Fig 5.3). Dat[4] accomplished this task in the following way:

With respect to a doublet located at P_o, two planes (S_1 and S_2), symmetrical with respect to P_o and perpendicular to the doublet axis n_o, are considered. These planes can be interpreted as being tangent to the two faces of a lifting surface.

Force F_1 experienced by surface Σ_1 on plane S_1 is

$$\vec{F}_1 = -\vec{n}_o \int_{(\Sigma_1)} (p - p_\infty)\,d\Sigma \tag{5.19}$$

where (Σ_1) signifies integration over the area Σ_1, $d\Sigma$ is a surface element of that area, and n_o is the unit vector. This force can also be expressed in terms of acceleration as

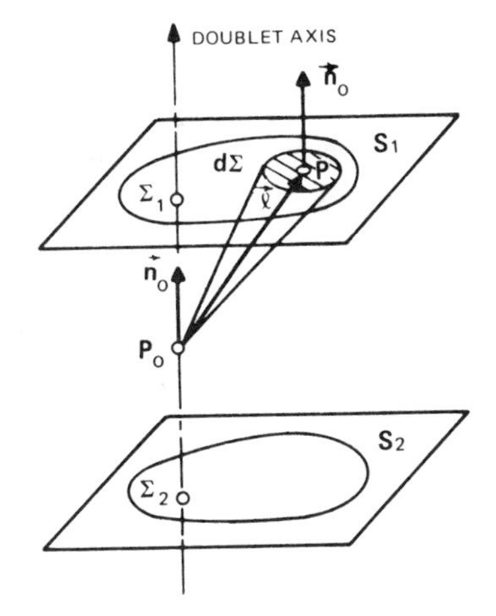

Figure 5.3 Acceleration doublet and the lifting surface

$$\vec{F}_1 = -\vec{n}_o\,\rho\,\frac{q(t)}{4\pi}\int_{(\Sigma_1)} \frac{\vec{\ell}\cdot\vec{n}_o}{|\vec{\ell}|^3}\,d\Sigma\,. \tag{5.20}$$

However,

$$\int_{d\Sigma} (\vec{\ell} \cdot \vec{n}_o / |\vec{\ell}|^3) d\Sigma$$

is the elementary solid angle $d\omega$ relating point P_o to the element $d\Sigma$. Should ω_1 represent the solid angle corresponding to the whole surface Σ_1, then

$$\vec{F}_1 = -\vec{n}_o \rho q(t) \frac{\omega_1}{4\pi}. \tag{5.21}$$

For the lower surface (S_2),

$$\vec{F}_2 = -\vec{n}_o \rho q(t) \frac{\omega_2}{4\pi}. \tag{5.22}$$

When the distance between the two surfaces becomes zero, then $\omega_1 = \omega_2 = 2\pi$ if the doublet axis pierces areas $\Sigma_1 = \Sigma_2 = \Sigma$, as in Fig 5.3. However should the doublet axis miss the Σ areas, the following relationship would hold: $\omega_1 = \omega_2 = 0$.

It can be seen, hence, that the resultant force $\vec{F} = \vec{F}_1 + \vec{F}_2$ becomes

$$\begin{aligned} \vec{F} &= -\vec{n}_o \rho q(t), \text{ if the doublet lies inside } \Sigma, \text{ and} \\ \vec{F} &= 0, \qquad \text{if the doublet lies outside } \Sigma. \end{aligned} \tag{5.23}$$

This means that the lift is concentrated only on the element of area $d\Sigma$ where the doublet is located.

On the other hand, considering that the aerodynamic load at time t on area $d\Sigma$ can be expressed in terms of the pressure differential $\Delta p(t)$ and the surface area:

$$\vec{F} = -\vec{n}_o \Delta p(t) d\Sigma, \tag{5.24}$$

and equating the right sides of Eqs (5.23) and (5.24), one finds that

$$q(t) = \Delta p(t) d\Sigma / \rho \tag{5.25}$$

which represents the sought relationship between the strength of the acceleration doublet and pressure sustained by a lifting surface.

4. RELATIONSHIPS BETWEEN VELOCITY AND ACCELERATION POTENTIAL

4.1 Integral Equations of the Velocity Potential

In actual application of the potential methods, one determines velocities (flow perturbations) induced by the blade or the rotor as a whole, by calculating the gradients of the velocity potential. However, loads sustained by the lifting surface may be given in terms of the acceleration potential. It becomes important, hence, to establish basic integral equations relating the velocity potential to the acceleration potential and eventually, to the pressure differentials experienced by the lifting surfaces.

The velocity of flow at some point P and at a given time t may be thought of as the result of acceleration acting on the fluid at that point from "all the way back" to time t. Consequently, the velocity potential φ may also be imagined as the following time integral of the acceleration potential Ψ:

$$\varphi(P,t) = \int_{-\infty}^{t} \Psi(P,t_o)\, dt_o. \tag{5.26}$$

Substituting Eq (5.18) for $\Psi(P,t_o)$, and remembering that vector $\vec{\ell}$ is the difference between the position vectors $\vec{P}$ and $\vec{P}_o(t_o)$ of points P and $P_o(t_o)$, $\varphi(P,t)$ can now be expressed in terms of the acceleration doublets:

$$\varphi(P,t) = -\frac{1}{4\pi} \int_{-\infty}^{t} \frac{q(t)[\vec{P} - \vec{P}_o(t_o)]\vec{n}_o(t_o)}{|\vec{P} - \vec{P}_o(t_o)|^3}\, dt_o. \tag{5.27}$$

Again substituting Eq (5.25) for $q(t)$ and changing $d\Sigma$ to dS in order to indicate that final values will be obtained through integration not only with respect to time, but also over a surface S, Eq (5.27) becomes

$$\varphi(P,t) = -\frac{1}{4\pi\rho} \int_{(S)} \int_{-\infty}^{t} \frac{\Delta p\,[P_o(t_o)]\,[\vec{P} - \vec{P}_o(t_o)] \cdot \vec{n}_o[\vec{P}(t_o)]\, dt_o\, dS}{|\vec{P}(t) - \vec{P}_o(t_o)|^3} \tag{5.28}$$

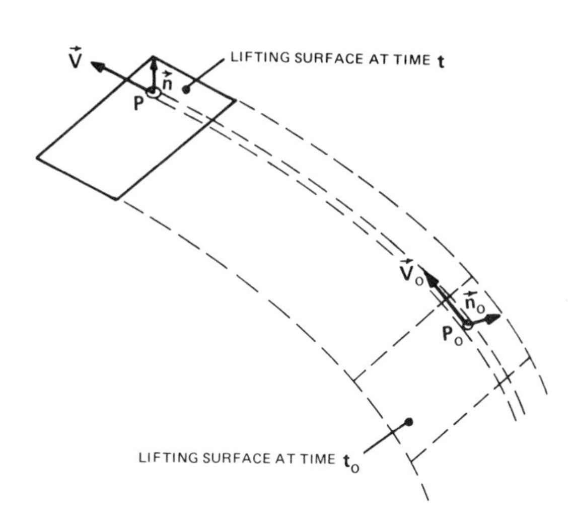

Figure 5.4 Time and space integration

In this integral, $\vec{P}_o(t_o)$, $\vec{n}_o(t_o)$, and $\Delta p(\vec{P}_o, t_o)$, respectively, denote the time history of (1) position, (2) orientation, and (3) lift, related to an element of the lifting surface[4]. With respect to the latter item, it should be noted that in Eq (5.28), Δp stands for the pressure difference between the upper and lower sides of the lifting surface: $\Delta p = p_u - p_\ell$. Consequently, negative Δp would correspond to a positive lift.

Finally, one's attention is called to the fact that the time integration indicated in Eq (5.28) using a dummy variable t_o is performed along the path of each lifting element (Fig 5.4).

4.2 Lifting-Line Case

In a particular case when the lifting surface can be represented by a lifting line located at 1/4-chord (usually an acceptable assumption for high aspect ratio surfaces such as rotor blades), Eq (5.25) giving the strength of the doublet can be written as follows[6] (Fig 5.5):

$$q(t,y) = \int_{x_2}^{x_1} \frac{\Delta p(t,y,x)}{\rho} dx \tag{5.29}$$

while the acceleration potential at any point P and time t can be expressed as:

$$\Psi(P,t) = -\frac{1}{4\pi} \int_{y_2}^{y_1} \frac{q(t,y)\cos\theta(y,t)}{\ell^2(y,t)} dy . \tag{5.30}$$

In this equation it should be noted that the integration is extended over the whole lifting line, and both ℓ and θ may vary with time and location.

As in the previously discussed three-dimensional case, the velocity potential will be obtained by substituting Eq (5.30) into Eq (5.26):

$$\varphi(P,t) = -\frac{1}{4\pi} \int_{-\infty}^{t} \int_{y_2}^{y_1} \frac{q(t_o,y)\cos\theta(y,t_o)}{\ell^2(y,t_o)} dt_o\, dy . \tag{5.31}$$

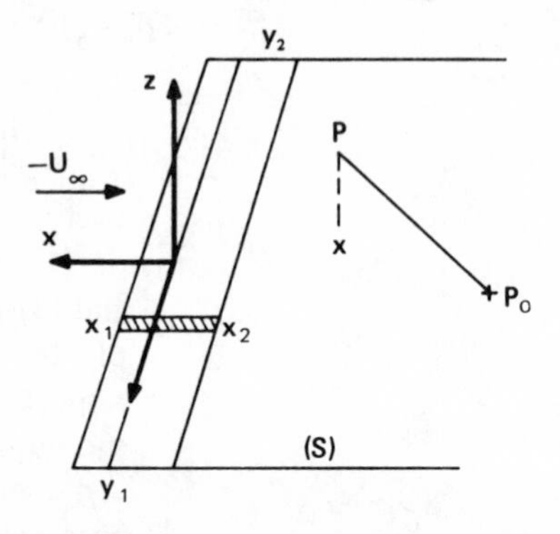

Figure 5.5 Lifting-line concept

Similar to the more general case of the lifting surface, it is necessary to know the path (S) traveled by the lifting line (Fig 5.5) in order to perform the integration indicated in Eq (5.31). For instance, for a wing moving horizontally at a speed U_∞ the coordinate x can be expressed as a function of time: $x = U_\infty(t - t_o)$, and Eq (5.31) can be rewritten in terms of variables related to spatial coordinates only[6] :

$$\varphi(P,t) = -\frac{1}{4\pi U_\infty} \int_{y_2}^{y_1} \int_{-\infty}^{0} \frac{q(P_o)\cos\theta(P_oP)}{\ell^2(P_oP)} dy\, dx. \tag{5.32}$$

Eq (5.32) represents an integration performed over a plane surface swept by the lifting line up to the instant t.

When $P \to P_o$ and thus, $|\ell| \to 0$, it can be seen that Eq (5.32) becomes meaningless. In order to investigate this case, Costes[6] considers two points, P_1 and P_2, located symmetrically with respect to S (Fig 5.6). According to Eq (5.18a), acceleration potentials $\Psi(P_1)$ and $\Psi(P_2)$ due to a doublet at P_o will be of the same magnitude at P_2 and P_1, but of opposite signs. The same will be true of their velocity potentials: $\varphi(P_2) = -\varphi(P_1)$. This means that velocity potential φ is antisymmetric and discontinuous with respect to S.

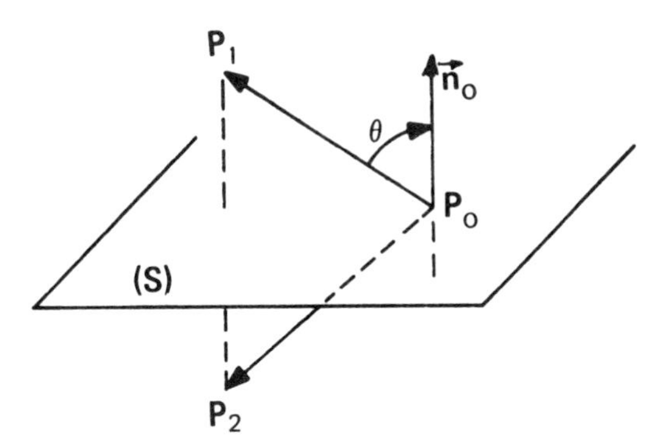

Figure 5.6 Points symmetrical with respect to S

An investigation of the resultant velocities ($\vec{v} = \overrightarrow{grad}\,\varphi$) would show that their components parallel to the plane S would be opposite to each other for the points located on opposite sides of that plane. By contrast, the normal component for the points above and below S would be the same. Consequently, plane S or, more generally, surface S to which the considered plane may be assumed as a tangent, appears as a surface of discontinuity for the velocities tangent to the surface. The discontinuity of velocity components parallel to the surface was considered in Sect 3.1 of Ch IV, where it was shown that this phenomenon occurs in the presence of a vortex sheet. The same approach can now be extended to include the present case, where surface S, along which the integration indicated by Eq (5.32) should be performed, may be interpreted as a vortex sheet of free vortices forming the wing wake. 'Reduction to practice' of this idea will become clearer when examples of the application of the potential approach to the solution of various problems of rotary-wing aerodynamics are discussed later in Sect 6.

5. ACCELERATION AND VELOCITY POTENTIALS IN COMPRESSIBLE FLUIDS

In acoustics, the velocity potential of the three-dimensional sound propagation satisfies the following wave equation[17]:

$$\frac{\partial^2 \varphi}{\partial x^2} + \frac{\partial^2 \varphi}{\partial y^2} + \frac{\partial^2 \varphi}{\partial z^2} - \frac{1}{s^2}\frac{\partial^2 \varphi}{\partial t^2} = 0. \tag{5.33}$$

where s is the velocity of sound.

Just as in the case of incompressible flow where both the velocity (Eq 5.3) and the acceleration (Eq 5.17) potentials satisfied the Laplace equation, the acceleration potential must satisfy a wave equation of the type given by Eq (5.33):

$$\frac{\partial^2\Psi}{\partial x^2} + \frac{\partial^2\Psi}{\partial y^2} + \frac{\partial^2\Psi}{\partial z^2} - \frac{1}{s^2}\frac{\partial^2\Psi}{\partial t^2} = 0. \tag{5.34}$$

It appears that the acceleration potential approach is especially suited for a proper interpretation of the physical aspects of the flow phenomena in a compressible medium. This is due to the link existing between acceleration potential and pressure variations which, traveling at the speed of sound, represent a physical means of transmitting aerodynamic signals in a compressible medium.

In the incompressible case, it was shown that the acceleration can be directly related to the pressure differential through Eq (5.16). This relationship must now be slightly modified to indicate that fluid density ρ in the area of the occurrence of pressure variation may no longer be the same as that in distant flow (ρ_∞). Consequently, Eq (5.16) for the compressible flow medium should be written as

$$\Psi = -(p - p_\infty)/\rho_\infty. \tag{5.35}$$

Using p instead of the longer terms, ($p - p_\infty$) or Δp, to denote pressure fluctuations, and substituting the so-defined p from Eq (5.35) into Eq (5.34), the following equation is obtained for pressure potential in a compressible fluid:

$$\frac{\partial^2 p}{\partial x^2} + \frac{\partial^2 p}{\partial y^2} + \frac{\partial^2 p}{\partial z^2} - \frac{1}{s^2}\frac{\partial^2 p}{\partial t^2} = 0. \tag{5.36}$$

It should be emphasized at this point that here, as in the case of incompressible flow, the linearity of the basic differential equations permits one to obtain a complete solution by superimposing particular solutions.

Time Lag in Signal Transmittance. It is obvious from Eq (5.36) that since pressure signals directly related to the acceleration potential are transmitted through the medium at a finite velocity s (speed of sound), a time delay develops between the instant τ when the signal is emitted from point P_o and the time (t) it is received at point P:

$$t - \tau = |\vec{P} - \vec{P}_o|/s \tag{5.37}$$

where $\vec{P}_o$ and $\vec{P}$ are radius vectors (having a common origin) which locate points P_o and P respectively, in the space. Eq (5.37) can also be written as follows:

$$t - \tau = |\ell|/s. \tag{5.37a}$$

The physical fact of the $t - \tau$ time delay in transmitting pressure signals leads to the concept of retarded potential.

Velocity Potential. As in the incompressible case, a doublet of acceleration may be considered as representing an element of the lifting surface. However, in order to find a mechanism for computing induced velocities resulting from the lifting surface loading, relationships establishing the dependence of velocity potential on the strength of the acceleration doublets as given in Eq (5.25) must be developed first.

This was done by Dat[3], who derived the velocity potential at a point P and time t generated by the moving doublet. His approach in this case was similar to that taken for incompressible fluid. Acceleration potential for both positive and negative (sink) forces was written in terms of the strength of the acceleration doublet, except that in this case, the retarded potential aspects were introduced by replacing $q(t)$ with its Dirac distribution:

$$q(t) = \int_{-\infty}^{\infty} q(\tau)\delta(t-\tau)d\tau .$$

In Sect 2.3 of this chapter, it was indicated that the strength of a doublet represents a product of the source strength and distance ϵ between the source and the sink. It may be stated hence, that the strength of an acceleration source would be $q(t)/\epsilon$ and the potential of the source would be:

$$\psi(P,t) = -\int_{-\infty}^{\infty} \frac{q(\tau)\delta\left(t-\tau-\frac{|\vec{\ell}(\tau)|}{s}\right)}{4\pi|\vec{\ell}(\tau)|\epsilon} d\tau$$

where $\vec{\ell}(\tau) = \overrightarrow{P_o(\tau)P(t)}$.

An expression for the potential of an acceleration doublet was derived—as in the incompressible case—by adding the potentials of a source and a sink, and assuming that the distance along the vector n_o representing the doublet axis decreases to zero: $\epsilon \to 0$.

The so-obtained potential for an acceleration doublet is integrated with respect to time, thus yielding the velocity potential expressed in terms of the strength of the acceleration doublet.

Assuming that the doublet moves at a velocity smaller than s (subsonic), the velocity potential due to this doublet may be written as follows:

$$\varphi(P,t) = \frac{q(\tau_1)\vec{\ell}(\tau_1)\cdot\vec{n}_o(\tau_1)}{4\pi s|\vec{\ell}(\tau_1)|^2\left[1-\frac{\vec{\ell}(\tau_1)\cdot\vec{V}_o(\tau_1)}{s|\vec{\ell}(\tau_1)|}\right]} + \int_{-\infty}^{\tau_1} \frac{q(\tau)\vec{\ell}(\tau)\cdot\vec{n}_o(\tau)}{4\pi|\vec{\ell}(\tau)|\epsilon} d\tau \quad (5.38)$$

where

$\tau_1 = t - \frac{|\vec{\ell}(\tau_1)|}{s}$, and referring to Fig 5.7,

$\vec{P}_o(\tau)$ defines the time history of the doublet position

$\vec{V}_o(\tau)$ defines the time history of the doublet velocity

$\vec{n}_o(\tau)$ defines the time history of the doublet orientation

$q(\tau)$ defines the time history of the doublet strength.

The physical reason for making τ_1 instead of t as the upper limit of integration is the fact that signals (acoustic waves) emitted during the time interval $t-\tau_1$ have not had time to reach point P.

Similar to the previously considered case of incompressible fluid, it can be seen that as point P approaches P_o; i.e., as $|\ell| \to 0$, singularities will appear in Eq (5.38) which, as before, means that a line or surface on which doublets are distributed represents a surface of discontinuity for tangential velocity components.

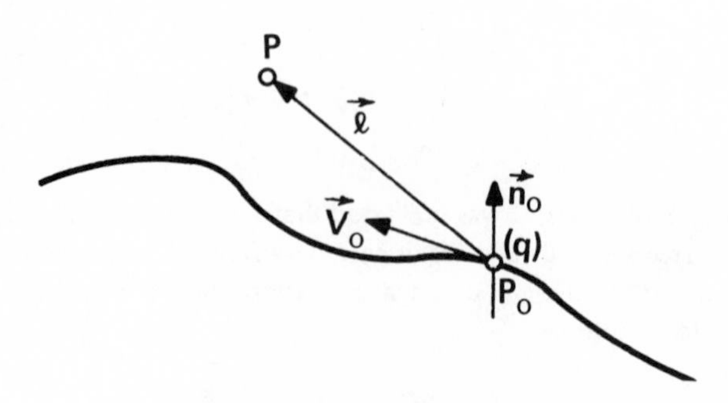

Figure 5.7 Path of a doublet

Application to Lifting Surfaces. In order to get the velocity potential due to the whole lifting surface, Eq (5.38) must be integrated over a surface S on which doublets are distributed. Again, as in Section 4.2, vortex sheets forming the wake of a wing, blade, or the whole rotor can be interpreted as the surfaces over which integration should be performed. It should also be noted that the condition of discontinuity of tangential velocities requires that vectors $\vec{n}_o$ (doublet axes) be perpendicular to the surface. One should also recall that vortex filaments move with the fluid. Consequently, there should be no flow through the vortex sheet; which can be expressed mathematically as $\vec{V}_o \cdot \vec{n}_o = 0$ where $\vec{V}_o$ is the resultant velocity of flow at point P_o; i.e., at surface S.

A helicopter rotor or a propeller where the blades are modeled by lifting lines can be used as an example of determining the velocity potential. At time t under the $\vec{V}_o \cdot \vec{n}_o = 0$ condition, the sought velocity at point M (over blade No. 1) is now expressed by the following:

$$\varphi(P,t) = \int_{R_o}^{R_1} \frac{q(r,\tau)\vec{\ell} \cdot n_o}{4\pi s|\ell|^2\left[1 - \dfrac{\vec{V}_o \cdot \vec{\ell}}{s|\ell|}\right]} dr + \int_{R_o}^{R_1} \int_{-\infty}^{\tau_1(r)} \frac{q(r,\tau)\vec{\ell}\cdot\vec{n}_o}{4\pi|\ell|^3} d\tau\, dr \tag{5.39}$$

This equation was obtained from Eq (5.38) in a way similar to Eq (5.32).

It is obvious that for the helicopter rotor shown in Fig 5.8, it is necessary to perform the integration on the wakes of all three blades. In this way, the mutual interaction of the blade is taken into consideration. In Eq (5.39), τ_1 depends on r, hence the integration is performed within that part of the vortex wake which, in the neighborhood of the blade, is limited by the $\tau_1(r)$ line after which it extends into infinity.

Theoretical considerations of the potential concept presented thus far in this chapter should have provided the reader with the basic information needed to better grasp the application of this principle to various problems of rotary-wing aerodynamics.

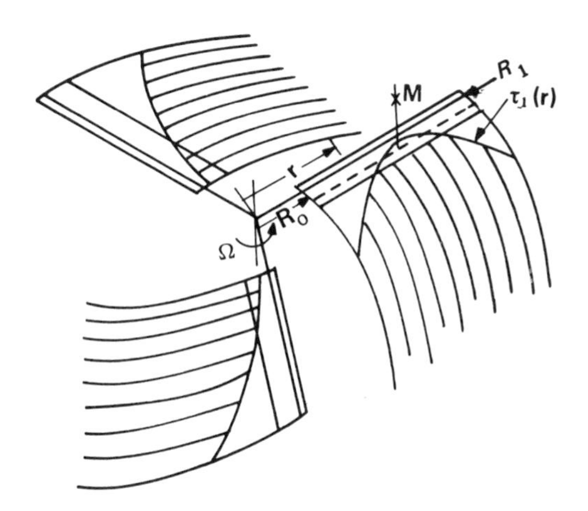

Figure 5.8 Zones of integration

6. EXAMPLES OF POTENTIAL METHODS

6.1 Actuator Disc with Axisymmetrical Loading

Assumptions. Mangler and Squire[1] developed a rather simple way of relating acceleration and then velocity potentials to the character of load (pressure) distribution over an actuator disc which is assumed to model a rotor in forward flight at velocity U_∞ (Fig 5.9). The time-averaged induced velocity field $(\vec{v} = \vec{i}u + \vec{j}v + \vec{k}w)$ is obtained from the velocity potential under the assumption that $|\vec{v}| \ll |\vec{U}_\infty|$. It is postulated that the rotor is lightly loaded in order to assure compliance with this condition.

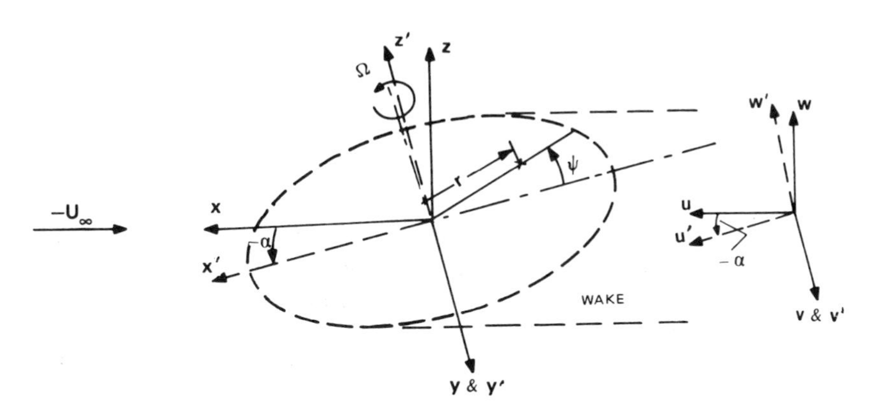

Figure 5.9 Wind (x,y,z) and disc-plane (x', y', z')

This is satisfied by the requirement that the rotor disc lift coefficient (C_L), based on U_∞, is

$$C_L \equiv T/\tfrac{1}{2}\rho U_\infty^2 \pi R^2 \ll 1.0. \tag{5.40}$$

As the fluid moves in the $-x$ direction at velocity $-U_\infty$, it experiences an acceleration which, assuming that $\partial u/\partial t$ is small in comparison with $(\partial u/\partial x)U_\infty$, etc., can be expressed from Eq (5.14) as follows:

$$-U_\infty(\partial\vec{v}/\partial x) = -(\overrightarrow{grad}\, p)/\rho, \tag{5.41}$$

where the function expressing the spatial variation of p satisfies the Laplace equation (Eq 5.25).

Velocity $\vec{v}$ induced by the rotor at points having their abscissa equal to x can now be obtained by integrating Eq (5.41) from ∞, where both the pressure differential and induced velocity vanish to x. When expressed in the nondimensional form of $(\vec{v}/U_\infty)$, it becomes

$$\vec{v}/U_\infty = (1/\rho U_\infty^2)\int_\infty^x \overrightarrow{grad}\, p\, dx. \tag{5.42}$$

Since Eq (5.42) represents potential flow, it is, in principle, valid everywhere except inside the rotor wake which contains vorticity and thus, $rot\,\vec{v} \neq 0$. Neglecting deviations due to induced velocities, the wake may be assumed to be represented by the half cylinder extending downstream from the rotor disc with its generatrices being parallel to $-\vec{U}_\infty$. The flow through the disc is continuous; consequently, the induced velocity component perpendicular to the disc is the same both above and below its surface. But $\vec{v}$ components parallel to the disc must also be the same both above and below the disc surface. Otherwise, this type of velocity discontinuity (shear) would imply the existence of inplane forces, which are not assumed in the actuator disc concept.

Induced Velocities. Under the above assumptions, the induced velocity component u, or its nondimensional value (u/U_∞) along the x axis can be obtained by integrating Eq (5.42). Outside of the wake, this would give

$$u/U_\infty = (p - p_\infty)/\rho U_\infty^2 \tag{5.43}$$

and inside the wake,

$$u/U_\infty = [(p - p_\infty) + \Delta p]/\rho U_\infty^2 \tag{5.44}$$

where p_∞ is the pressure in undisturbed flow and $\Delta p \equiv (p_u - p_o)$ is the pressure rise across the disc directly ahead of the point under consideration.

Eqs (5.43) and (5.44) permit one to determine only one component of the induced velocity. In order to determine the whole field of induced velocities, Mangler and Squire solved the Laplace equation for the pressure potential (Eq (5.15)). They did this In terms of Legendre functions of the elliptic coordinates associated with the disc, which are discontinuous between two faces of the disc, but continuous everywhere else. The induced velocity components are then determined from Eq (5.42) for some simple axisymmetrical types of nondimensionalized load distribution over the disc, expressed as $\Delta p/\rho U_\infty^2 C_L = f(\bar{r})$ where $\bar{r} \equiv r/R$ (Fig 5.10).

The most important induced velocity components; namely, those appearing at the disc surface in a perpendicular direction (w'), as well as those perpendicular to the distant flow far downstream $(w_{-\infty})$ are presented in graphical form in Figs 5.11 to 5.14. Figs 5.11

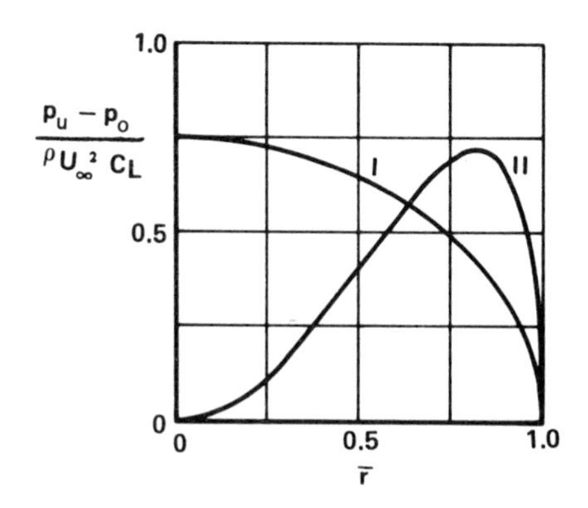

DISTRIBUTION I: $\dfrac{p_u - p_o}{\rho U_\infty^2 C_L} = \dfrac{3}{4}\left(1 - \bar{r}^2\right)^{1/2}$

DISTRIBUTION II: $\dfrac{p_u - p_o}{U_\infty^2 C_L} = \dfrac{15}{8}\,\bar{r}^2\left(1 - \bar{r}^2\right)^{1/2}$

Figure 5.10 Pressure distribution on the rotor disc

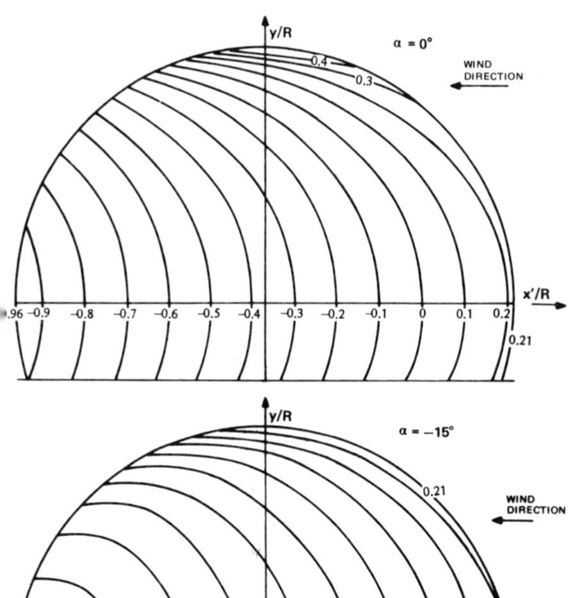

Figure 5.11 Contours of nondimensionalized induced velocity $w'/U_\infty C_L$ at the disc for Type I pressure distribution $\alpha = 0°$ and $\alpha = -15°$

and 5.13 should enable one to calculate downwash distribution at the disc for both types of the assumed disc-area-loading, while Figs 5.12 and 5.14 would provide downwash distribution far downstream from the rotor. Additional figures may be found in Ref 1.

Harmonic Representation. The nondimensional downwash distribution at the disc may be expanded into a Fourier series in $\bar{r}$ and ψ. As indicated in Ref 1, it will contain the cosine terms only, because of the symmetry of loading with respect to the fore-and-aft diameter:

$$(w'/U_\infty C_L) = \tfrac{1}{2}a_0(\bar{r},\alpha) + \sum_{n=1}^{\infty} a_n(\bar{r},\alpha)\cos n\psi. \tag{5.45}$$

The expression for the zero-harmonic coefficient for Type I loading will be

$$a_{I_0} = \Delta p/\rho U_\infty^2 C_L = \tfrac{3}{4}(1 - \bar{r}^2)^{1/2} \tag{5.46}$$

and for type II,

$$a_{II_0} = \Delta p/\rho U_\infty^2 C_L = (15/8)\bar{r}^2(1 - \bar{r}^2)^{1/2}. \tag{5.47}$$

Values of the higher harmonic coefficients can be found in Ref 1.

Mean Induced Velocity at the Disc. It is interesting to figure out the value of induced velocity w' averaged over the disc ($\overline{w}'_{av}$):

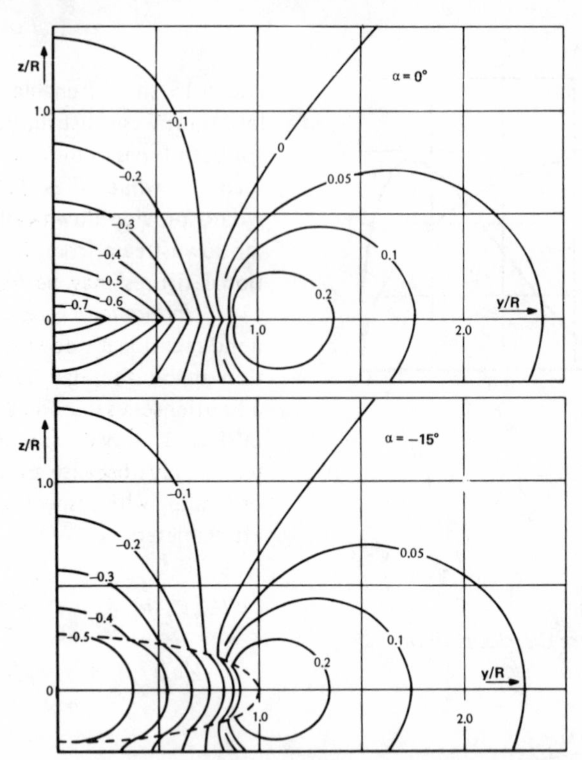

Figure 5.12 Contours of nondimensionalized induced velocity $w_{-\infty}/U_\infty C_L$ for Type I pressure distribution far downstream $\alpha = 0°$ and $\alpha = -15°$

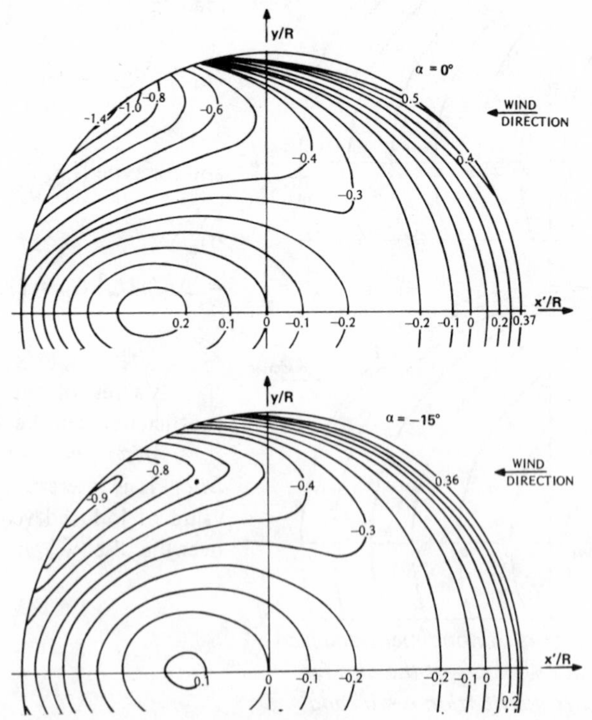

Figure 5.13 Contours of nondimensionalized induced velocity $w'/U_\infty C_L$ for Type II pressure distribution at the disc $\alpha = 0°$ and $\alpha = -15°$

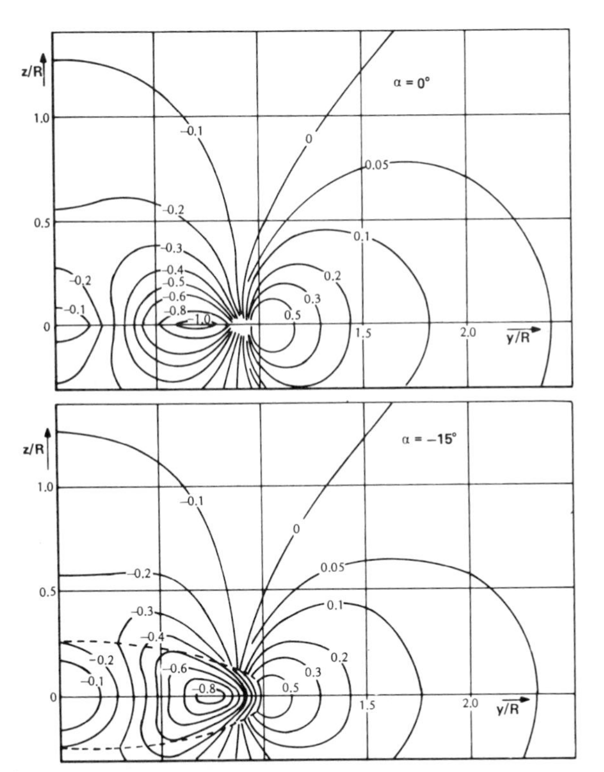

Figure 5.14 Contours of nondimensionalized induced velocity $w_{-\infty}/U_{\infty}C_L$ for Type II pressure distribution $\alpha = 0°$ and $\alpha = -15°$

$$\bar{w}'_{av} \equiv w'_{av}/U_{\infty}C_L = (1/\pi R^2)\int_{-\pi}^{\pi}\int_{0}^{R}(w'/U_{\infty}C_L)r\,dr\,d\psi. \tag{5.48}$$

It is easily determined that the harmonics higher than zero will vanish in the above integration; thus leading to

$$\bar{w}'_{av} = \int_{0}^{1} a_{1_0}\bar{r}\,d\bar{r} = \int_{0}^{1} a_{11_0}\bar{r}\,d\bar{r}. \tag{5.49}$$

Substituting the zero-harmonic values from Eqs (5.46) and (5.47) into Eq (5.49), and performing the indicated integration, it is found that both (Types I and II) loadings lead to

$$\bar{w}'_{av} = 1/4, \tag{5.50}$$

which can be rewritten in a dimensional form for w':

$$w' = T/2\pi R^2 \rho U_{\infty}. \tag{5.51}$$

This will be immediately recognized as the Glauert formula for induced velocity of an ideal rotor in horizontal flight when $v_{id} = w'$ is small in comparison with U_{∞} (Eq 2.31). In Ch II, this equation was given simply—without additional proof—as a logical extension of axial translation, and as being analogous to the elliptically-loaded fixed-wing. The derivation of Eqs (5.50) and (5.51) may be considered as more rigorous proof of this basic relationship; at least for the case of lightly-loaded rotors ($w' \ll U_\infty$) modeled by an actuator disc.

6.2 General Remarks re Blade-Loading Examples

The works of Dat and his associates[3-7] formed the foundation for what may be called the ONERA School of Application of the Acceleration Potential Principle to Various Aerodynamic and Aeroelastic Phenomena of both Fixed and Rotary-Wings in Incompressible and Compressible Flow. As an example of the application of the ONERA approach, the case of predicting instantaneous blade loads on helicopter blades in an incompressible unstalled flow will be considered. Here, the blade will be represented by a lifting line located at the first one-quarter chord. As a second example, the case of determination of helicopter blade loads, taking into consideration the unsteady-aerodynamic phenomena and stall, will be outlined.

6.3 Instantaneous Blade Loads in Unstalled, Incompressible Flow

Statement of the Problem. The helicopter rotor is assumed to be in a steady-state translation; hence, load sustained by the blade and blade motions become functions of the azimuth angle only. A linearized approach is applied to this case, which allows one to express the load through some basic functions and in turn, to determine perturbation velocities generated by each of these functions. The next step is to find a combination of functions which would satisfy the requirement of nonpenetration of the lifting surface by the flow. This condition is verified at some selected points on the blade surface at various azimuth angles.

It will later be shown that perturbation velocities corresponding to each of the basic functions depends only on the translational and rotational velocities of the rotor, as well as the planform of the blades. They are not, however, a function of other characteristics of the blade, such as cyclic and collective pitch.

Costes points out that the required combination of the basic functions which would satisfy various boundary conditions can be obtained with little additional calculation.

Presentation of Blade Loads. An airscrew blade is represented by a lifting line located at the first one-quarter chord. The load F per unit span of the blade is a function of both time t and location on the blade as given by the radial distance r: $F = F(t, r)$.

The load also varies along the blade chord; hence, $F(t, r)$ must be determined as follows:

$$F(t,r) = -\int_0^c \Delta p \, dc \tag{5.52}$$

where Δp is the pressure differential between the upper and lower blade surfaces along chord c at time t, and blade station r. It may be recalled (Sect 4.1 of this chapter) that negative Δp corresponds to positive lift as reflected in the minus sign in Eq (5.52).

As indicated by Eq (5.25), a lifting-blade segment can be represented by a doublet whose strength depends on the average pressure differential sustained by the surface of this segment. However, one can imagine that this surface is reduced to a lifing line; thus, the surface distribution becomes a line distribution of doublets per unit-span. This intensity is the sum of the doublet surface intensity along the chord.

$$q(t, r) = \int_{o}^{c} \frac{\Delta p}{\rho_o} dc$$

but according to Eq (5.52), the integral of $\Delta p\, dc$ is equal to $-F(t, r)$. Consequently,

$$q(t, r) = -F(t, r)/\rho_o . \tag{5.53}$$

Under steady-state conditions at $r = const$, $F(t, r)$ assumes the same value after every rotor revolution. In other words, it becomes a periodic function of time t which, in turn, can be related to the local azimuth angle ψ_ℓ (Fig 5.15) by the following relationship:

$$\psi_\ell = \Omega t + \vartheta(r) \tag{5.54}$$

where $\vartheta(r)$ represents a time lag permitting one to take into account a possible stagger in the time counting, beginning with zero at various points along the lifting line.

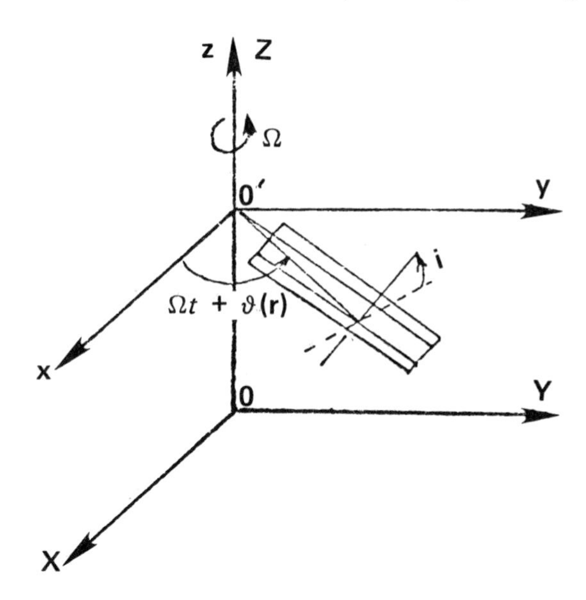

Figure 5.15 Definition of the local azimuth angle ψ

The n points located at r_i $(i = 1, 2, ...n)$ are selected along the blade span, and the force distribution will be determined at $(2m + 1)$ azimuthal positions: ψ_j $(j = 1, 2, ...m)$.

Variation of the load F at a given blade station r_i is periodic with time. Consequently, the variation $F(t)_{r_i = const}$ can be expanded into the following Fourier Series;

$$F(t)_{r_i} = Z_i + \sum_{j=1}^{m} X_{ij} \cos j(\Omega t + \vartheta_{r_i}) + \sum_{j=1}^{m} Y_{ij} \sin j(\Omega t + \vartheta_{r_i}) \tag{5.55}$$

where X_{ij}, Y_{ij}, and Z_i, are coefficients of expansion, and must be determined.

For the lifting line passing through the axis of rotation, ϑ_{r_i} will obviously remain constant for all values of n.

Eq (5.55) represents the load-per-unit span at discrete points ($i = 1, 2, \dots n$) of the blade span. In order to obtain—with a limited number of points—the best possible determination of the F values along the blade span, values of r_i were selected as the Gauss points for the weighing function $\sqrt{1 - \eta^2}$, where $\eta = (2r_i - R_1 - R_o)/(R_1 - R_o)$, as shown in Fig 5.16(a). The blade span loading at an instant t can be obtained by interpolation of the $F(t = const)_{r_i}$ values between the i points.

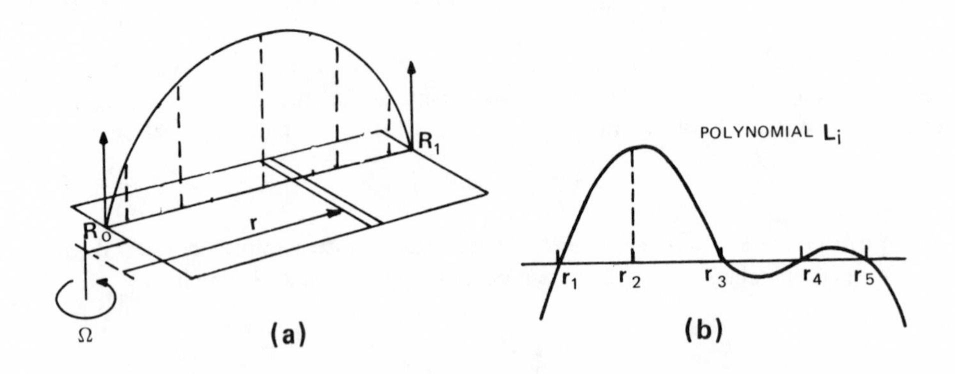

Figure 5.16 (a) Load distribution along the blade, and (b) interpolating Lagrange polynomial for $i = 2$

This is done with the help of the Lagrange polynomial $L_i(r)$ which has a zero value at points r_k for $k \neq 1$, and is equal to one for r_i. Now the variation of span loading with both time and blade radius becomes

$$F(t, r) = \sum_{i=1}^{n} L_i(r)\, F(t)_{r_i}. \tag{5.56}$$

In order to explicitly show the variation of F with time, Eq (5.55) for $F(t)_{r_i}$ is substituted into Eq (5.56):

$$F(t,r) = \sum_{i=1}^{n} Z_i\, L_i(r) + \sum_{i=1}^{n} \sum_{j=1}^{m} X_{ij} \cos j\big(\Omega t + \vartheta(r)\big) L_i(r)$$

$$+ \sum_{i=1}^{n} \sum_{j=1}^{m} Y_{ij} \sin j\big(\Omega t + \vartheta(r)\big) L_i(r). \tag{5.57}$$

Furthermore, ϑ_{r_i} is replaced by $\vartheta(r)$ in Eq (5.57) to better account for the geometry of the lifting line. For $r = r_i$, the values of $F(t,r)$ as given by Eq (5.57) become identical to the $F(t)_{r_i}$ values obtained from Eq (5.55).

In order to cope with the abrupt variations of loading at the inboard and tip extremities of the blade, a large number of stations r_i would be required. This situation, however, can be remedied by introducing into the expansion of $F(t, r)$, some functions reflecting the expected behavior of the blade span loading at both extremities of the blade.

The theory of finite-span wings indicates that the drop-off of loading at the tips should be proportional to the square root of the distance from the tips. Assuming, the same as Costes, that a similar drop-off would occur at the blade tip as well as at its roots, Eq (5.57) is multiplied by $\sqrt{1-\eta^2}$, where $\eta = (2r - R_1 - R_o)/(R_1 - R_o)$.

In this way, fewer points r_i—smaller n—would be required for a realistic determination of the blade span loading. This would result in fewer equations for computing the unknown coefficients Z_i, X_{ij}, and Y_{ij}.

Keeping in mind Eq (5.25), and using Eq (5.57) as refined by the tip-shape function $\sqrt{1-\eta^2}$, the strength of the doublets located along the lifting line can be expressed as follows:

$$q(r,t) = \left(\frac{\sqrt{1-\eta^2}}{\rho_\infty}\right) \sum_{i=1}^{n} Z_i L_i(r) + \sum_{i=1}^{n} \sum_{j=1}^{m} X_{ij} \cos j\left(\Omega t + \vartheta(r)\right) L_i(r)$$
$$+ \sum_{i=1}^{n} \sum_{j=1}^{m} Y_{ij} \sin j\left(\Omega t + \vartheta(r)\right) L_i(r). \qquad (5.58)$$

Collocation Method. Eq (5.57), in principle, would permit one to determine the span blade loading at any given time t—equivalent to a blade azimuthal angle ψ—but it contains the unknown coefficients Z_i, X_{ij}, and Y_{ij}. In order to compute them, advantage will be taken of the fact that having $q(r,t)$ as given by Eq (5.58), expressions for acceleration and velocity potential can be written for both incompressible and compressible cases. Knowing the velocity potential, the associated induced (perturbation) velocities $\vec{v}$ can be determined at any point in the space surrounding the rotor. The method of collocation as originally developed for fixed wings allows a selection of the number of points best suited for the task of finding coefficients Z_i, X_{ij}, and Y_{ij}.

In multibladed rotors, the interaction between the blades is taken into consideration by performing the summations indicated by Eq (5.31) for incompressible, or Eq (5.39) for compressible, fluids over the wakes of each blade, while maintaining the proper value for $\vartheta(r)$:

For instance, for a three-bladed rotor:

Blade No. 1: $\vartheta_1(r)$

Blade No. 2: $\vartheta_2(r) = (2\pi/3) + \vartheta_1(r)$

Blade No. 3: $\vartheta_3(r) = (4\pi/3) + \vartheta_1(r)$.

As to the location of the collocation points, similar to fixed-wing practice, the induced velocity is determined along the three-quarter chord line of the blade chord, while the lifting line is assumed to extend along the one-quarter chord. Points of collocation are located at distances of r_i from the axis of rotation.

In order to retain the linearity of the problem, it is assumed that the wakes trail independently from each of the blades, and that the blade motions in the direction normal to the wake are small. The second condition implies that the velocity components normal to the blade are small in comparison with tangential ones.

In principle, the induced velocity at any point can be calculated if the velocity potential around the rotor is known. Of special interest would be the induced velocities v_n which are located along the lifting line, and are perpendicular to the blade. However, it can be seen that for both the incompressible medium as noted in Eq (5.31) and the compressible medium in Eq (5.39), the presence of the $|\ell|^2 = 0$ term in the denominators would lead to uncertainty. The following finite difference approach[18] was suggested in order to overcome this difficulty.

Velocity v_n in a direction perpendicular to the blade is computed at a collocation point as:

$$v_n(h, \Delta h) = [\varphi(P_1) - \varphi(P_2)]/\Delta h \tag{5.59}$$

where, as shown in Fig 5.17, h is the normal elevation of point P_1 over the blade wake, and Δh is the distance between P_2 and P_1. It was indicated in Ref 6 that the values $h \approx 0.035c$ and $\Delta h \approx 0.005c$ where c is the blade chord, can be used for practical calculations.

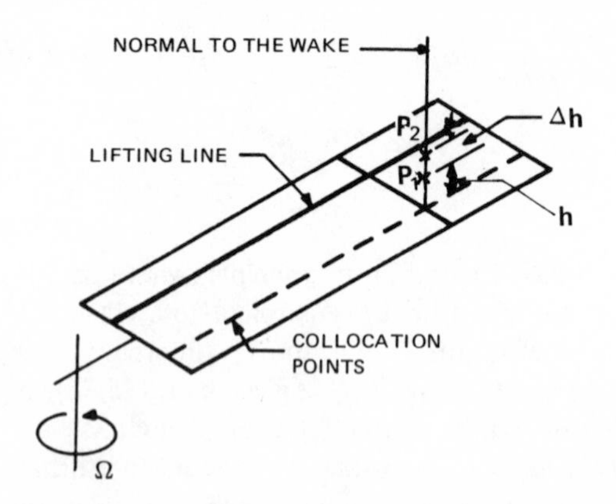

Figure 5.17 Location of velocity potential points

Computational Procedure. Velocity potential φ at a point P at time t can be determined for the incompressible case from Eq (5.31), and from Eq (5.39) for the compressible case by performing an integration with respect to r and τ. Before doing this, however, Eq (5.58) must be substituted for $q(r,\tau)$ appearing in Eqs (5.31) and (5.39), while the vectors $\vec{\ell} = \overrightarrow{P_oP}$, $\vec{V}_o$ are determined for the trajectory of each blade depending on r and τ and elevation h, or $h + \Delta h$, above the blade wake.

By substituting Eq (5.58) into Eqs (5.31) and (5.39), a relationship can be established between the velocity potential and the unknown coefficients Z_i, X_{ij}, and Y_{ij}. The normal components of induced velocity v_n at the collocation point (P_k) and instant t_1 can now be obtained by following the procedure leading to Eq (5.59), which now enables one to express $v_n(P_k, t_1)$ under the following algebraic form;

$$v_n(P_k, t_1) = \sum_{i=1}^{n} Z_i w_{io}^{(0)}(P_k, t_1)$$

$$+ \sum_{i=1}^{n} \sum_{j=1}^{m} X_{ij} w_{ij}^{(1)}(P_k, t_1) + \sum_{i=1}^{n} \sum_{j=1}^{m} Y_{ij} w_{ij}^{(2)}(P_k, t_1). \tag{5.60}$$

Costes[6] explains that coefficients w_{ij} are integrals in r and τ: $w_{io}^{(0)}(P_k, t_1)$ is the velocity induced at point P_k, and at an instant t_1 when the strength of the doublet is given by

$$q(r) = \left[\sqrt{1 - \eta^2(r)} / \rho_\infty\right] L_i(r);$$

$w_{ij}^{(1)}(P_k, t_1)$ is the induced velocity when the strength of the doublets is given by

$$q(r,\tau) = \left[\sqrt{1 - \eta^2(r)} / \rho_\infty\right] L_i(r) \cos j[\Omega t + \vartheta(r)];$$

and $w_{ij}^{(2)}(P_k, t_1)$ is the induced velocity when the strength of the doublets is given by

$$q(r,\tau) = \left[\sqrt{1 - \eta^2(r)} / \rho_\infty\right] L_i(r) \sin j[\Omega t + \vartheta(r)].$$

Numerical calculations consist of finding $w_{ij}^{(0)}$, $w_{ij}^{(1)}$, $w_{ij}^{(2)}$ through integration over the assembly of wakes represented by more or less complicated, but known, functions.

Eq (5.60) expresses $v_n(P_k, t_1)$ as a function of Z_i, X_{ij}, Y_{ij} through a matrix relationship. By considering the same number of points of collocation and instances of time t_1 as the number of unknowns, w_{ij} becomes a regular matrix, and after inverting it, a matrix which expresses the unknown Z_i, X_{ij}, Y_{ij} as a function of the normal velocities of $v_n(P_k, t_1)$ is obtained. The latter may also be directly expressed as a function of parameters which determine movement and deformation of the blades, while the condition of the tangency of flow to the lifting surface is maintained.

Condition of Tangency of Flow to the Blade. The requirement of nonpenetration of the blade by the flow can be expressed for any point P on the blade surface and time t on the condition that the normal component $v_n(P,t)$ of the velocity induced by the wake must be equal to the normal projection of the velocity of that point with respect to the air at rest. This means that the velocity experienced by point P must be expressed in reference to a fixed system of coordinates $OXYZ$. For the case of a rotor blade articulated in flapping, but otherwise rigid, the procedure would be as follows:

Another system of coordinates $O'xyz$ is considered, assuming that O' coincides with the rotor hub center and axes x,y,z are parallel to X, Y, Z. The position of O' with respect to O is determined by the magnitude of velocity U_∞ and its angle Λ in the OXZ plane (Fig 5.18). In this figure, the location of the flapping axis O_1 is given by r_o and angle $\Omega t + \vartheta_o$.

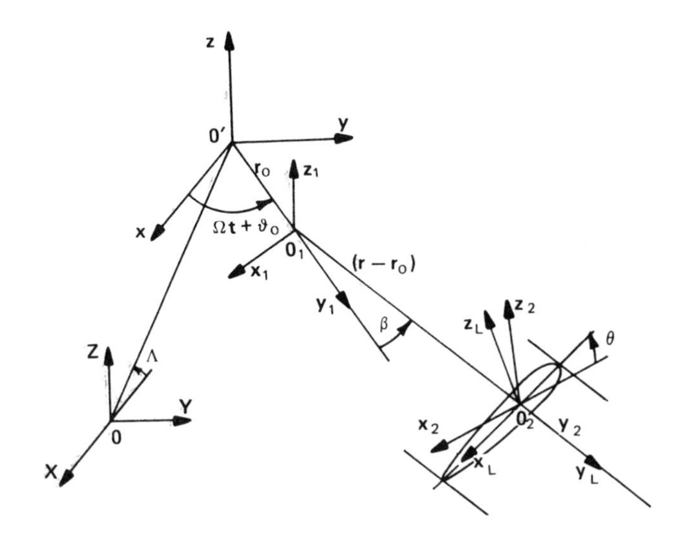

Figure 5.18 Blade section coordinate system

Flapping (β) is determined with respect to the $O_1x_1y_1z_1$ system where it is assumed that axes $O_1y_1z_1$ are in the plane $O'xy$. The $O_2x_2y_2z_2$ system is related to the blade station being considered and finally, the $O_2x_Ly_Lz_L$ system is obtained from the $O_2x_2y_2z_2$ by assuming that the O_2x_L axis coincides with the zero chord airfoil section which makes an angle θ (resulting from collective and cyclic inputs) with the O_2x_2 axis.

The blade is assumed to be infinitely thin and thus may be represented by a surface—or for untwisted blades, by a plane—formed by zero-lift chords. Any point P on the so-modeled blade would have the coordinates x_L, $0,0$ in the $O_2x_Ly_Lz_L$ coordinate system. Knowing r, $\beta(t)$, $\Omega t + \vartheta_o$ and U_∞, it is possible to determine the velocity V_p of point P in the $OXYZ$ system.

$$\vec{V}_P = (d/dt)\overrightarrow{OP} = (d/dt)\overrightarrow{OO'} + (d/dt)\overrightarrow{O_1O_2} + (d/dt)\overrightarrow{O_2P}. \tag{5.61}$$

A normal to the blade at point P would be parallel to the O_2z_L axis, and the condition of nonpenetration of the blade by the flow can be expressed as

$$v_n(P,t) = \vec{n}_P \cdot \vec{V}_P. \tag{5.62}$$

Presenting the vectorial relationships expressed by Eq (5.61) in Cartesian coordinates, the condition expressed by Eq (5.62) can be written as follows[6]:

$$\begin{aligned} v_n(P,t) = &- U_\infty \cos\Lambda[\sin\theta \sin(\Omega t + \vartheta_o) - \sin\beta(t)\cos\theta\cos(\Omega t + \vartheta_o)] \\ &+ U_\infty \sin\Lambda \cos\beta(t)\cos\theta \\ &- \Omega\sin\theta[r_o + (r - r_o)\cos\beta(t)] - \sin\beta(t)\Omega x_L + (r - r_o)\dot{\beta}(t)\cos\theta. \end{aligned} \tag{5.63}$$

Wake Geometry. In order to calculate coefficients $w_{ij}^{(k)}$, the geometry of the wakes streaming behind the blades must be known. In the linearized approach, as in the vortex theory, the blade wakes are assumed to be rigid—no contraction, hence no change in distance between vortex sheets of individual blades. This approach is based on the supposition that the wake shape in the immediate proximity of the rotor has the most significant influence on the structure of the induced velocity field at the disc. In order to still further improve the similarity between the real wake and its physicomathematical model represented by the rigid wake, the influence of the ideal induced velocity at the disc is taken into consideration. It is assumed that the wake of each blade is formed by the trailing vorticity moving in a direction perpendicular to the rotor disc at a velocity of $U_\infty \sin\Lambda + v_{id}$, while being transferred at a speed of $U_\infty \cos\Lambda$ in a direction parallel to the disc. It should be noted that v_{id} does not enter into the calculation of $v_n(P_k, t)$ according to Eq (5.63). The reader's attention should also be called to the importance of small blade movements and deformation in the determination of v_n values in Eq (5.63) versus the calculation of the $w_{ij}^{(k)}$ coefficients.

In the first case, flapping, blade twist, and both collective and cyclic pitch influence the values of the normal velocity components at the points of collocation. In the second case where $w_{ij}^{(k)}$ coefficients are determined from the wake geometry, the wake shape is influenced by blade flapping motion, pitching, and torsional deformation. Under steady-state conditions due to the periodicity of those occurrences, however, the relative distance between the individual blade wakes would remain the same since all of them are affected in the same manner. Consequently, vector $\vec{\ell} = \overrightarrow{P_oP}$ appearing in Eq (5.39) can be determined solely by the translatory velocity of the rotor, its rotational speed, and its inclination with respect to the distant flow. As indicated in Ref 6, this further simplifies calculations of $w_{ij}^{(k)}$ coefficients by assuming that $\beta(t) \equiv 0$ for rotors with a low coning angle.

Test Comparisons. Using potential methods, airloads were calculated for the rotor shown in Figure 5.19 for which wind-tunnel tests were available[6].

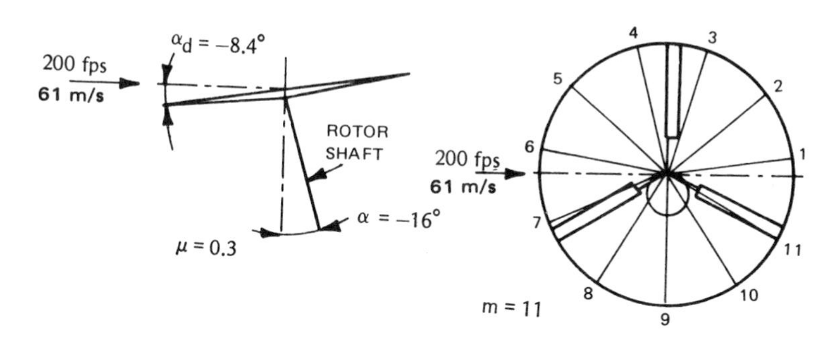

Figure 5.19 Definition of a rotor for tests and calculations

In the analytical predictions, five blade stations ($n = 5$) and eleven azimuthal positions ($m = 11$) were considered in Eq (3.58) for determination of the doublet strength. The selected values of n and m results in 55 unknown coefficients Z_i, X_{ij}, Y_{ij}, whose determination would require 55 equations. As each position of the blade permits one to write 5 equations, it is necessary to calculate induced velocities for 11 positions of the blade, which are presumed to be as shown in Fig 5.19.

Strictly speaking, in the inboard regions of the blade, and especially in the reversed velocity areas, induced velocities are no longer small in comparison with the incoming flow experienced by the blades. This, of course, violates the assumption of small perturbations which is essential for the linearized approach. However, since the air loads are also quite low because of the low resultant velocities, the condition of nonpenetration of the blade by the flow was replaced by the condition of zero lift within the reversed velocity area[6]. In other applications of the nonpenetration rule, the blade was considered as rigid, while its flapping motion was directly taken from test measurements which gave $\beta = 1.22° - 7.55° \cos \psi$.

The computational problem consists of solving a system of 55 linear equations with 55 unknowns. The air loads at selected blade stations obtained in this way were plotted versus azimuth angle and then compared with tests (Fig 5.20). In general, there is good agreement. The areas of largest discrepancies are around $\psi = 180°$. This is explained in Ref 6 as follows:

1. *The relative position of the lifting line and the collocation line are related to the pressure distribution along the airfoil; it is possible that in the neighborhood of 0° and 180°, the radial velocity component varies and thus introduces a systematic error.*

2. *The velocity generated by the hub and blade attachments is ignored. However, this may be important for $\psi = 0°$.*

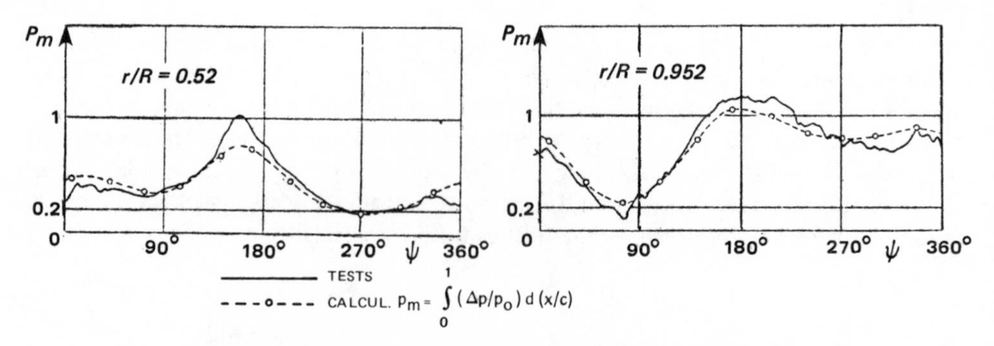

Figure 5.20 Examples of measured and calculated airloads

The potential method can also be applied to detail performance calculations. However, this particular application is not likely to be used on a routine basis.

6.4 Aerodynamic Loads on Helicopter Blades in Unsteady Flow

Statement of the Problem. In contrast to the preceding case of determining aerodynamic loads under linear assumptions of unstalled airfoils, the approach here will be the prediction of aerodynamic forces on helicopter blades during forward flight when nonlinear aerodynamic effects must be considered. This problem was examined by Dat[5] and Costes[20]. The basic principles of this method will be explained using Dat's approach, which is based on the linear theory of lifting surfaces and, for a long time, was used in dealing with aeroelastic problems of fixed wings. Since the method uses the potential theory, some difficulty may be expected in the case of stall. However, this difficulty is diminished by proposing a physicomathematical model where the vorticity zone associated with stall is limited to a relatively thin layer of rotational flow, while outside of this zone, the flow may be considered as being potential.

Linear Theory of a Lifting Surface in an Arbitrary Motion. A method, based on the principle of the integral equation used in the flutter analysis of fixed-wings, was developed by ONERA in dealing with helicopter blades modeled by lifting surfaces. However, in order to make this approach applicable to helicopters, a more general equation dealing with the arbitrary motions of a lifting surface were developed.

The lifting surface is modeled by a surface sustaining a discontinuity of pressure. It is imagined that the surface moves within a great volume of fluid thus generating small perturbations which can be expressed as gradients of velocity potential φ. Each element $d\Sigma$ of the surface is assumed to support a doublet of pressure $\Delta p d\Sigma$ with its axis along the normal $\vec{n}_o$. Similar to the development of Eq (5.39), an integral equation expressing velocity potential in terms of the distribution of pressure differential Δp and time history (from $-\infty$ to the present τ) can be obtained[3] (Figure 5.21):

$$\varphi(P,t) = \iint_{(A)} \frac{\Delta p[\vec{P}_o(\tau)]\,[\vec{P}-\vec{P}_o(\tau)]\cdot\vec{n}_o(\tau)d\Sigma_o}{4\pi\rho_\infty s|\vec{P}-\vec{P}_o(\tau)|^2\left[1-\dfrac{[\vec{P}-\vec{P}_o(\tau)]\,\vec{V}_o(\tau)}{s|\vec{P}-\vec{P}_o(\tau)|}\right]}$$

$$+ \iint_{(A)} \int_{-\infty}^{\tau_o} \frac{\Delta p[\vec{P}_o(\tau_o)]\,[\vec{P}-\vec{P}_o(\tau_o)]\,\vec{n}_o(\tau_o)}{4\pi\rho_\infty|\vec{P}-\vec{P}_o(\tau_o)|^3}\, d\tau_o\, d\Sigma_o \quad (5.64)$$

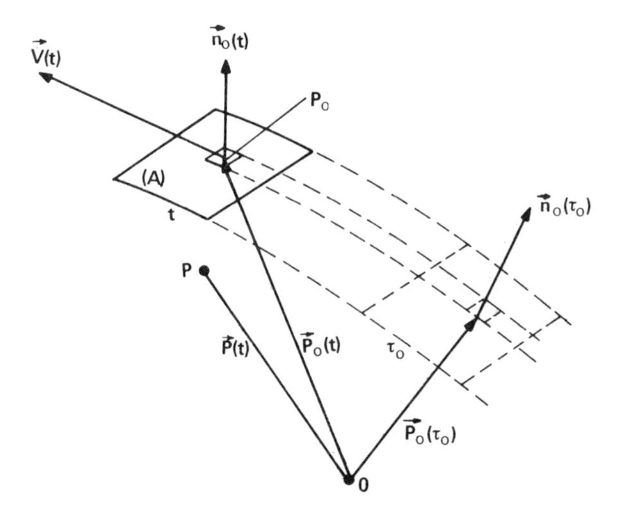

Figure 5.21 Notations for Eq (5.64)

where as in Eq (5.37), τ_o is determined from the equation reflecting retardation of the acoustic signal $t - \tau_o = [\vec{P} - \vec{P}_o(t)]/s$, speed $V_o = d\vec{P}_o/d\tau_o$, and s is the speed of sound.

Eq (5.64) can be applied to a single surface, or to several surfaces, as long as the condition of small perturbation is fulfilled. As in the case in Sect 6.3, the requirements of nonpenetration of flow and the Kutta-Joukowski condition should also be fulfilled.

As long as there is no stall, the vortex sheet forming the wake of each blade is relatively thin. Hence, it may be treated as a surface of discontinuity where rotation exists; outside of which the flow is potential. When a blade is stalled, this turbulent zone where *rot* $\vec{v} \neq 0$ becomes thicker. However, it may be assumed that at some point P which is sufficiently removed from the turbulent zone, the flow is potential and the assumption of small perturbations is valid (Fig 5.22). Therefore, at that point P where the velocity potential exists, perturbations generated by a stalled lifting surface will be the same as those induced by an unstalled lifting surface having the same lift as the one in stall.

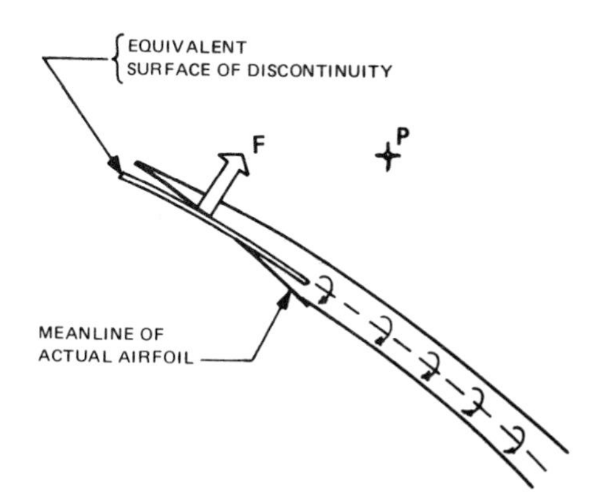

Figure 5.22 Modeling of a stalled airfoil

The derivative of the potential in a direction perpendicular to the lifting surface, which enters the boundary condition of nonpenetration of the lifting surface by the resultant flow, cannot be determined at the surface itself. However, for points P which are located sufficiently far from the layer of turbulence, one may use Eq (5.64) which relates velocity potential to the lifting surface load. Thus, this approach may also be used in case of stalled blades.

Loads on Helicopter Blades in Steady Flight. Blades (1,2,3,...) are modeled by infinitesimally thin lifting surfaces. The points on these surfaces are determined by non-dimensional spanwise $\eta = (2r - R - R_o)/(R - R_o)$ and chordwise $(\xi = x/\tfrac{1}{2}c)$ coordinates (Figure 5.23). It should be noted that at the inboard station of the blade $\eta = -1$ and at the tip, $\eta = 1$; while at the leading edge, $\xi = -1$; and at the trailing edge, $\xi = 1$.

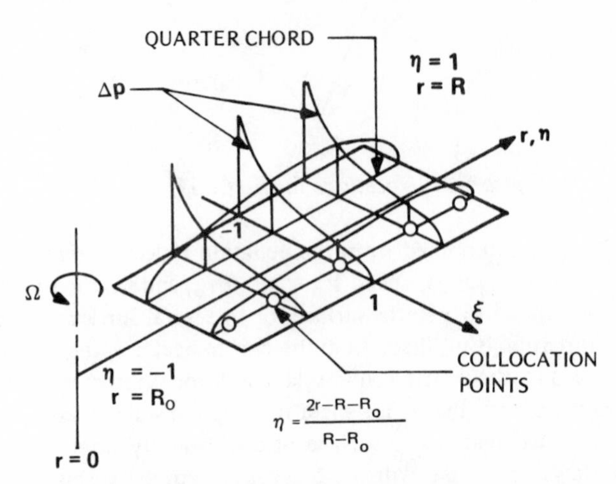

Figure 5.23 Pressure distribution on a blade

The task consists of determining differential pressure distribution Δp as a function of blade coordinates and time: $\Delta p(\xi, \eta, t)$, assuming that the blade motion resulting from helicopter velocity of flight, rotor rotation, and oscillatory and vibratory movements of the blade is known.

In steady-state flight, Δp should be a periodic function of time $T = 2\pi/\Omega$. Consequently, the pressure differential on any blade (say, blade No. 1 on a three-bladed rotor) can be expressed in terms of time and blade coordinates as follows:

$$\Delta p_1(\xi,\eta,t) = \sqrt{(1-\xi)/(1+\xi)}\sqrt{1-\eta^2}\left[\sum_{i=1}^{M}\sum_{j=1}^{N}\sum_{p=0}^{R} \cos p\Omega t P_i(\xi)Q_j(\eta)X'_{ijp} + \sum_{i=1}^{M}\sum_{j=1}^{N}\sum_{p=1}^{R} \sin p\Omega t P_i(\xi)Q_j(\eta)X''_{ijp}\right]. \quad (5.65)$$

As in the case considered in Sect 5.3, the multiplier $\sqrt{1-\eta^2}$ appearing in Eq (5.65) assures a drop-off of the blade aerodynamic load at both spanwise extremities. The role of the other multiplier $\sqrt{(1-\xi)/(1+\xi)}$ is to guarantee that the chordwise load is zero at the trailing edge which, in turn, assures compliance with the Kutta-Joukowsky law.

Functions $P_i(\xi)$ and $Q_i(\eta)$ are independent polynomials, and X'_{ijp} and X''_{ijp} make an assembly of the $M \times N \times (2R+1)$ unknowns.

For the other two blades (Nos 2 and 3), the corresponding pressure differentials Δp_2 and Δp_3 can be expressed as:

$$\Delta p_2(\xi, \eta, t) = \Delta p_1[\xi, \eta, t + (2\pi/3\Omega)] \tag{5.66}$$

and

$$\Delta p_3(\xi, \eta, t) = \Delta p_1[\xi, \eta, t - (2\pi/3\Omega)] \,. \tag{5.67}$$

By replacing Δp in Eq (5.65) with Eqs (5.66) and (5.67) and taking into consideration the trajectories of points P_o of the three blades, it becomes possible to express potential φ at a given point and a given time as a function of the unknown X'_{ijp} and X''_{ijp}. Using the same procedures discussed in Sect 6.3, the perturbation velocity component perpendicular to the blade is determined from the finite differences in the potential values (Fig 5.24). At the time t for which the normal velocity component is determined, the position of point P on the disc is known.

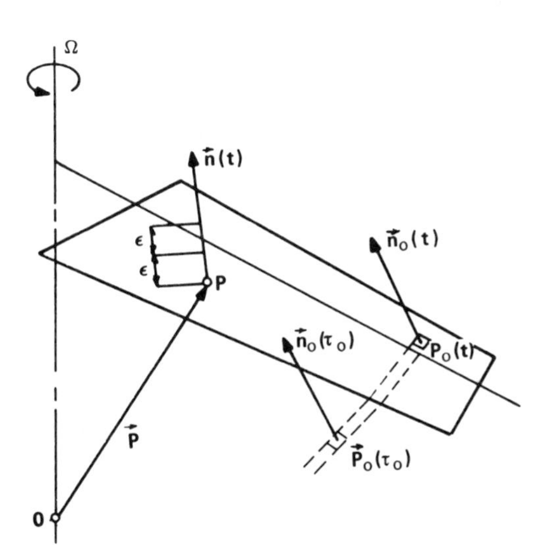

Figure 5.24 Notations for normal velocity calculations as finite differences

By taking $M \times N$ points on the same blade and considering $2R + 1$ as different instants during the same period, we indicate the $M \times N \times (2R + 1)$ points on the disc corresponding to the points of the normal velocity components, or points of collocation.

The process described above would make it possible to write a system of algebraic equations relating the velocity component normal to the blade W at those points as a function of the unknown X'_{ijp} and X''_{ijp}. In matrix notations, the equations may be expressed as

$$\{W\} = [A]\{X\} \tag{5.68}$$

where

$\{W\}$ is the column of normal velocities

$\{X\}$ is the column of the unknown X'_{ijp} and X''_{ijp}, and

$[A]$ is a matrix resulting from the numerical calculation of Eq (5.65) and the establishment of derivatives along a normal to the blade through a finite difference process.

Column $\{W\}$ can be determined if the motion of the blade with respect to air is known.

Pressure points, where discontinuities of pressure Δp are related to aerodynamic forces, correspond to the points of normal velocity. Positions of points of pressure differential and those of normal velocities are selected in such a way that maximum accuracy in the sequence of integration will be assured. In order to reduce the time required to calculate matrix A, it is necessary to consider a relatively small number of points. For each blade segment, this number is usually limited to one point for velocity and one for pressure. In such a case, the representation of a lifting surface is reduced to a single lifting line and normal velocity is calculated at the three-quarter chordline, while the linear spanwise blade load (F) is assumed to be concentrated at the 1/4-chordline. Since this blade load is linearly related to the unknown coefficients X'_{ijp} and X''_{ijp}, it is possible to write a matrix relationship of the form

$$\{F\} = [S]\{X\} \tag{5.69}$$

where $\{F\}$ is the column of airloads on the blade segments at the considered instant (each segment contains one point of pressure and one point of normal velocity).

$[S]$ is a matrix whose coefficients are determined by the basic functions of the expression given by Eq (5.65). From Eqs (5.68) and (5.69), one obtains

$$W_{k\ell} = \sum_{mn} G_{k\ell,mn} F_{mn} \tag{5.70}$$

where G is the matrix product, $G = AS^{-1}$; $W_{k\ell}$ is the normal velocity component at 3/4-chord of the blade at radial station η_k at the moment of the blade passage through the azimuthal coordinate ψ_ℓ. F_{mn} is the load at the 1/4-chord in section η_m when the blade passes through an azimuth ψ_n and when the blade motion with respect to the air is known.

In the absence of stall, normal velocities can be found from the condition of nonpenetration of the blade by the flow. Then it is enough to invert matrix G in order to obtain airloads F_{mn}.

However, even if the blade is stalled, it is still possible to find a nonstalled flow which, for the points outside of the stalled zone and its wake, is the same as the real flow. This equivalent flow can be obtained by replacing the real blades with fictitious ones having the same lift, but which were developed at below-stall angles-of-attack. Eq (5.70) remains valid for such hypothetical blades, but since the blade motion is not known á priori, the normal velocity components are also unknown. In order to find them, it is assumed that each blade segment behaves as an airfoil oscillating in a two-dimensional flow. The behavior of these airfoils at high angles-of-attack will be deduced from experimental wind-tunnel data and the equivalent nonstalled flow. These inputs will be used only for the determination of induced velocity at each segment to account for unsteady effects.

Representation of Airfoils in Two-Dimensional Flow. Actual lift coefficients are lower at high angles-of-attack than those given by the linear theory. In those cases where the difference between the linear theory and experiments are large, however, it is still possible to use the previously outlined method by replacing the actual airfoil by a hypothetical one at an equivalent angle-of-attack a^* as shown in Fig 5.25. This approach can be justified by the physical fact that for points distant from the wake, the two-dimensional flow around the unstalled airfoil at a^* is equivalent to the real stalled flow.

Figure 5.25 also shows how the effective angle-of-attack can be determined from the known $c_\ell(a)$ curve obtained under static conditions. However, as the airfoil oscillates, points representing the lift coefficient do not follow the static curve, but form an hysteresis loop* (Fig 5.26).

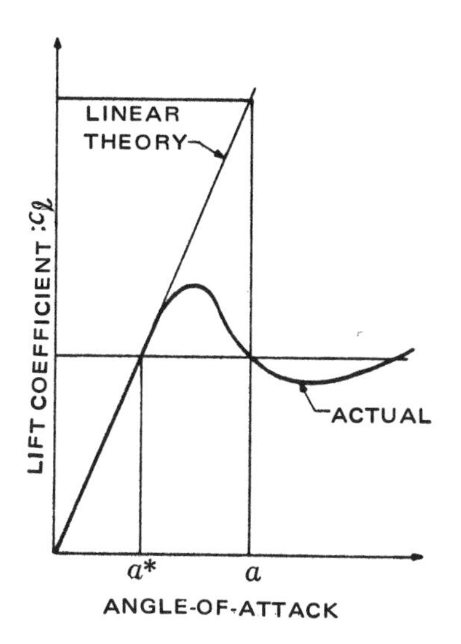

Figure 5.25 Determination of the effective angle-of-attack

Figure 5.26 Lift hysteresis loop of an oscillating airfoil

Cross and Harris[19] proposed a method for determining the loop if variation of the angle-of-attack versus lift is known. They define the reference angle-of-attack (a_{ref}) as

$$a_{ref} = a - \gamma\sqrt{|c\dot{a}/2V|}\, sin\,(\dot{a}) \tag{5.71}$$

where γ is a function determined empirically from wind-tunnel measurements of oscillating airfoils.

The instantaneous lift coefficient c_{ℓ_i} is given as

$$c_{\ell_i} = (a/a_{ref})c_\ell\,(a_{ref}) \tag{5.72}$$

where $c_\ell\,(a_{ref})$ is the lift coefficient corresponding to a_{ref} under stationary conditions.

This method is depicted in Fig 5.26. In this case, it is possible to assume that real flow is equivalent to a nonstalled flow corresponding to an effective angle-of-attack a^*.

*This subject is also discussed in Ch VI.

In this way, it is possible to develop a physicomathematical model which will explicitly determine the equivalent angle-of-attack as a function of the actual angle-of-attack a and its time derivative $\dot{a}$. Denoting the operator which defines this model as g, we have

$$a^* = g(a,\dot{a}). \tag{5.73}$$

Operator g depends on R_e, M, and reduced frequency of oscillation $\sqrt{c\dot{a}/2V}$. It can be determined only on the basis of extensive wind-tunnel tests.

Rotors Encountering High Angles-of-Attack. The potential theory allows one to determine blade loads and local angles-of-attack, even in the presence of stall. However, to do this, it is necessary to consider the unknown motion of a hypothetical blade. Eq (5.70) can be solved for the unstalled case because of the fact that $W_{k\ell}$ could be obtained from the known blade motion and the condition of nonpenetration of the lifting surface by the flow. Now, the same condition should be fulfilled for the equivalent blade; thus, in analogy to Eq (5.70), the following can be written

$$W^*_{k\ell} = \sum_{mn} G_{k\ell\, mn} F_{mn} \tag{5.74}$$

where $W^*_{k\ell}$ is the normal component of the velocity of flow at point $k\ell$ for the fictitious blade.

It was stated before that each segment of the blade behaves as an airfoil in a two-dimensional flow. However, the angle-of-attack of the section is not known a' priori, as the direction of flow appraching each segment is also influenced by the velocity v, induced by the three-dimensional effects of the remaining blades, as well as other segments of the considered blade (Fig 5.27). Yet, the lift is proportional to the equivalent angle-of-attack, where the coefficient of proportionality is determined by the two-dimensional linear theory. Hence, it is possible to write the following equation:

$$F_{mn} = D_{mn}\, a^*_{mn} \tag{5.75}$$

where D_{mn} is a coefficient which depends on air density, local Mach number, and velocity of the airfoil movement $|\vec{V}_{mn}|$ (projection of the resultant velocity on the plane as defined by the blade chord and the normal $\vec{n}$).

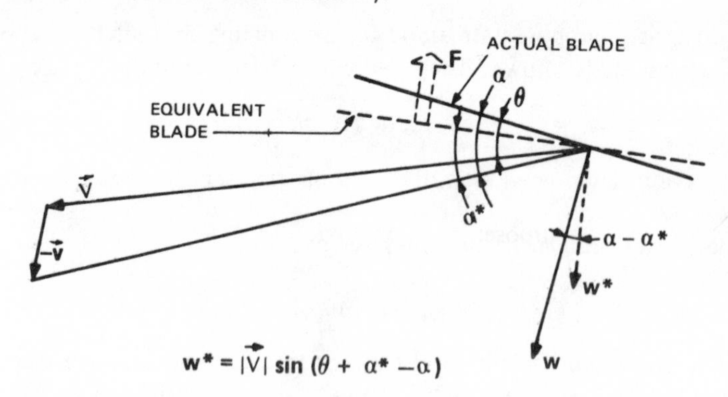

Figure 5.27 Determination of velocity component normal to the equivalent blade

On the other hand, the equivalent angle-of-attack a^* can be expressed in terms of the real angle-of-attack and its time derivative. Thus,

$$a^*_{mn} = g(a_{mn}, a_{mn}). \tag{5.76}$$

Finally, it should be realized that the difference between the angles-of-attack at the considered section of the real and hypothetical blades is equal to $a_{mn} - a^*_{mn}$ so that if velocity $\vec{V}_{mn}$ makes an angle θ_{mn} with the chord, the velocity component normal to the hypothetical blade is

$$W^*_{mn} = |\vec{V}_{mn}| \sin(\theta_{mn} + a^*_{mn} - a_{mn}). \tag{5.77}$$

When the blade movement is given, $\vec{V}_{mn}$ and θ_{mn} are known.

If we eliminate W^* and F from Eqs (5.74), (5.75), and (5.77), the following system remains:

$$|V_{k\ell}| \sin(\theta_{k\ell} + a^*_{k\ell} - a_{k\ell}) = \sum_{mn} G_{k\ell mn} D_{mn} a^*_{mn} \tag{5.78}$$

$$a^*_{mn} = g(a_{mn}, a^*_{mn}). \tag{5.78a}$$

Eqs (5.78) and (5.78a) form a system of nonlinear equations from which the real $(a_{k\ell})$ and hypothetical $(a^*_{k\ell})$ angles-of-attack can be obtained.

The aerodynamic load F can now be computed from Eq (5.75). The solution of Eq (5.78) is obtained by iteration.

6.5 Actual Calculations of Blade Loads and Angles-of-Attack

Costes[20] shows how the ideas discussed in general terms on the preceding pages can be reduced to practice and applied to such problems as determination of blade loads and angles-of-attack under both stalled and unstalled conditions. This is done under the following assumptions:

(1) The blade movement is a known periodic function of time.
(2) The blade is represented by a lifting line located along the 25-percent chord.
(3) The nonpenetration condition is applied to the 75-percent chord station, and
(4) The lift per unit of blade span is a linear combination of the time (azimuth) and blade radius functions.

The latter condition can be expressed by the following formula representing Eq (5.65) adapted to the present purpose:

$$F(r,t) = \sqrt{1-\eta^2}\left\{\sum_{i=1}^{n} Z_i L_i(r) + \sum_{i=1}^{n}\sum_{j=1}^{m} X_{ij} L_i(r)\cos j\psi + \sum_{i=1}^{n}\sum_{j=1}^{m} Y_{ij} L_i(r)\sin j\psi\right\} \tag{5.79}$$

where $L_i(r)$ is a Lagrange polynomial, r is the radial coordinate of a blade section, and Z_i, X_{ij}, and Y_{ij} are unknown coefficients, to be determined by application of conditions 1 and 3. Other symbols are as previously defined.

In Ref 20, five radial stations ($n = 5$) and six azimuthal positions ($m = 6$) were assumed; thus resulting in a total of 65 unknown coefficients (Z, X, Y). Now by time integration, the total induced velocity generated by each lift function $L_i(r)$, $L_i(r) \cos j\psi$, and $L_i(r) \sin j\psi$ is computed. Integration is performed for each of the 65 functions on 65 points P_o, evenly distributed on the rotor disc.

Because of the linearity assumption, the induced velocity may be expressed as a sum of the velocities induced by each of the L functions with the appropriate values of coefficients Z_i, X_{ij}, and Y_{ij}. This allows one to write a set of algebraic equations which express the relationship between the velocity at point P_o and the Z coefficients. This can be written in matrix form as follows:

$$[A] \begin{bmatrix} Z_{ij} \\ X_{ij} \\ Y_{ij} \end{bmatrix} = \{W\} \tag{5.80}$$

where $[A]$ is the matrix resulting from the integration—each coefficient of A giving the induced velocity at a particular point by a particular lift function; the Z, X, Y column represents the unknown coefficients; and $\{W\}$ is the column of the total induced velocity for each point P_o.

When any of the blade control points (P_o) passes through the rotor disc point defined by the combination of i and j values, the total velocity at point P_o with respect to the air can be computed from the known motion of the blade. In addition, the geometry of the blade section passing through point P_o is also determined; say, by the shape of the airfoil mean line at the blade station i. Finally, the total pitch of the blade section at i when the blade is at an azimuthal position j should also be known. Now, at point $P_o(i)$, it becomes possible to calculate the component of total air velocity perpendicular to the representative blade lifting surface formed by airfoil meanlines, or represented by an assembly of the airfoil zero-lift chords.

In order to satisfy the requirement of nonpenetration of the lifting surface by the flow, absolute values of the total induced velocity W appearing in Eq (5.80) must be equal to that resulting from the blade motion. The system of Eq (5.80) can now be solved; giving values of coefficients Z, X, and Y which, in turn, would permit one to calculate (from Eq (5.79)) the blade spanwise load distriabution at various azimuthal positions.

Determination of the Blade Angle-of-Attack. The corresponding angle-of-attack at disc points (r, ψ) can readily be obtained from the known blade spanwise values. For the non-stall condition, it may be assumed that the slope of the lift curve, corrected for compressibility effects* is

$$a = 2\pi/\sqrt{1 - M^2}$$

where $M \equiv U_{\perp}/s$ is the Mach number of a blade section at disc location $P(r, \psi)$ when the air velocity component perpendicular to the blade axis is $U_{\perp}$. Now, the angle-of-attack of a blade section at (r, ψ) becomes

$$a_p = F\sqrt{1 - M^2}/\pi\rho U_{\perp}^2 c. \tag{5.81}$$

The approach described above is usually satisfactory for the determination of blade section angles-of-attack of practical helicopters because of the low frequency of blade

---

****These corrections will be discussed in Ch VI.***

oscillations in pitch. The determined a_p values can now be used for computation of blade drag forces and in mapping the stall regions.

However, a more refined approach would require recognition of the fact that at high angles-of-attack, the blade sections would experience an unsteady-aerodynamic phenomena governed by the reduced frequency of oscillations of the blade sections, as well as by its Mach and Reynolds numbers.

The spanwise unit blade load for unsteady conditions (F_{un}) can be found by following the general approach described in Sect 6.4, and the effective angle-of-attack can be determined using a formula similar to Eq (5.81):

$$a_{eff} = F_{un}(r,t)\sqrt{1 - M^2}/\pi \rho U_{\perp}^2 c. \tag{5.82}$$

Usually,

$$|F_{un}(r,t)| \leqslant |F(r,t);$$

therefore,

$$|a_{eff}| \leqslant |a_p|.$$

To simplify the problem, the surface formed by zero-lift airfoil chords instead of that consisting of airfoil mean lines is assumed to represent the lifting surface. With this simplification, the procedure of actual calculations for the unsteady nonlinear case can be outlined as follows[20]: Similar to the linear case, forces acting on the rotor disc can be determined using procedures based on the assumption of nonpenetration of the lifting surface by the resultant flow; except that the real airfoil meanlines are now replaced by fictitious ones. It is still assumed that wake-induced velocities are determined by the same linear relationships as in the linear case with the exception that the formerly linear forces are now replaced by nonlinear ones. In principle, Eq (5.80) remains valid except that the right-hand term is now an unknown; however, it can be considered as velocity perpendicular to the equivalent zero lift chord of the stalled airfoil. The following step-by-step approach is used by Costes to compute the nonlinear forces: (1) The linear approach is applied to calculate a_p values. (2) The equivalent angle of attack ($a_{p_{eff}}$) is obtained from Eq (5.76). (3) Finally, the $a_{p_{eff}}$ values are used in Eq (5.82) to determine the new right-hand values in Eq (5.80), as well as the corresponding forces, F_{un}. The problem is equivalent to solving a nonlinear system of 65 equations having 65 unknown parameters.

Resolution of the Nonlinear System. For convenience, the aerodynamic incidences a_p are chosen as unknown parameters to be determined at the 65 P_o points of the rotor disc. Once the incidences are determined, the effective angle-of-attack ($a_{p_{eff}} \equiv a^*$) is computed from Eq (5.73), and the actual nonlinear forces F on the 65 P_o points are obtained from Eq (5.82). The interpolation of $F(r,\psi)$ over the entire rotor disc is made —as for the linear computation—according to Eq (5.79). Because of the large number of unknown a_p parameters (currently 65), Newton's generalized method, which converges fast enough if the nonlinearities are not too severe, is used. The starting point is provided by the solution of the associated linear system. The convergence is not likely to be difficult for strongly nonlinear cases; that is to say, for heavily stalled rotors. Fortunately, the stalled region remains restricted to the side of the retreating blade which, in fact, reduces the number of nonlinear equations.

A comparison of the blade section angle-of-attack computed by linear and nonlinear methods are shown in Fig 5.28(a), while Fig 5.28(b) gives a comparison of the

CONDITIONS: $\mu = 0.3$; $C_T/\sigma = 0.1138$; $V_t = 200$ m/s; $\alpha_{shaft} = -15.94°$;
$\theta_{0.75} = 13.94°$; No Feathering

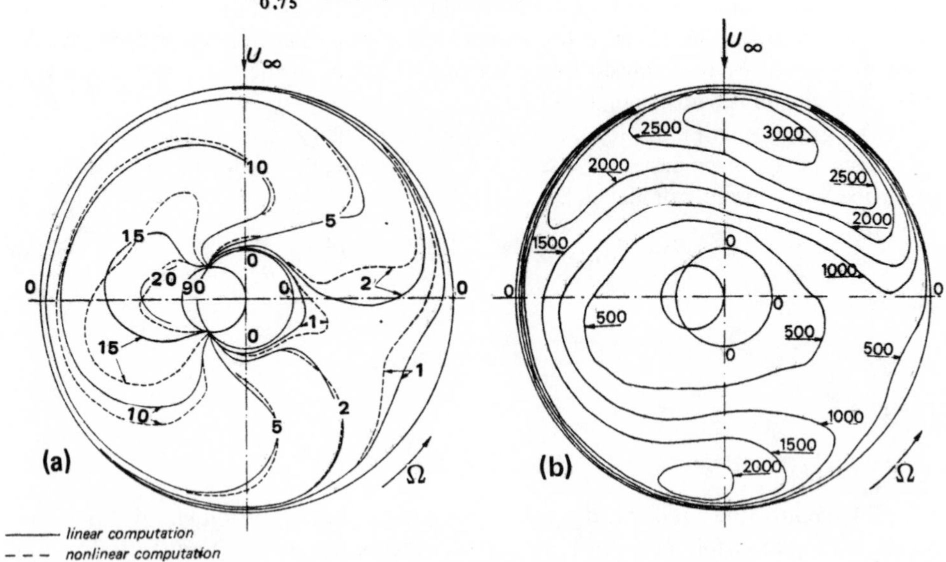

Figure 5.28 A comparison of (a) angles-of-attack, and (b) blade span loading, computed by linear and nonlinear methods

corresponding unit span lift. In both diagrams, it can be seen that outside of the stall region (around $\psi = 270°$) the difference between the linear and nonlinear values of α_P and F are small.

A further comparison was made between analytical load predictions and experimental results obtained from pressure transducers. These tests were run In the Modane Wind Tunnel on an articulated rotor with very rigid blades. The conditions, representing heavy stall under which the tests were run, are specified in Fig 5.28. The comparison of $r/R = 0.520$ and $r/R = 0.855$ shown in Fig 5.29 indicates that the use of the nonlinear approach improves the accuracy of the blade load predictions.

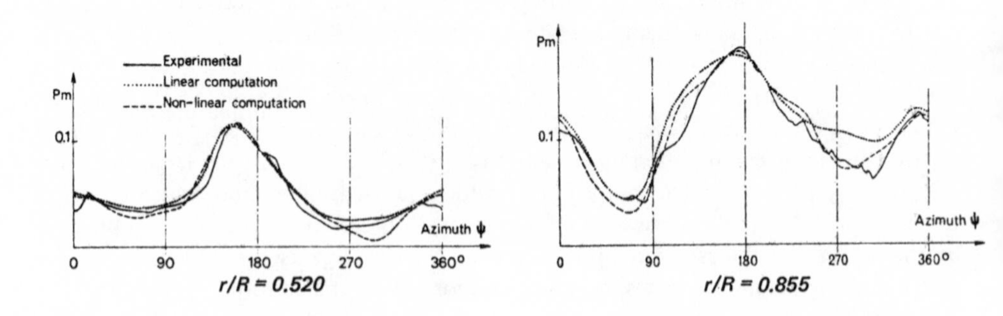

Figure 5.29 A comparison of predicted and measured air loads

7. DETERMINATION OF FLOW AROUND THREE–DIMENSIONAL, NONROTATING BODIES

7.1 Introductory Remarks

Body-Generated Flow Fields. Similar to the design process of fixed-wing aircraft, one needs to develop the shapes of nonrotating components of rotary-wing aircraft in such a way that they would offer the lowest possible drag and create a minimum of interference with other nonlifting and lifting components. However, in order to properly attack this problem, it is necessary to know the flow patterns at the body surface; strictly speaking, just outside the boundary layer, as well as in the surrounding space.

If the geometry of the body is already fixed, its aerodynamic characteristics can be obtained from the flow field at its surface by first calculating the pressure distributions, and then integrating the pressures in order to obtain forces and moments. Furthermore, the character of pressure variation on the body surface may indicate areas either susceptible to, or actually affected by, the flow separation. Knowledge of the flow fields more distant than just on the surface is essential in evaluating aerodynamic interactions.

Once the flow is determined, the designer–guided by intuition and experience–may modify the geometry of the body or bodies, and possibly their arrangements, in order to improve aerodynamic characteristics. This process is called *analysis.*

However, the problem can be formulated differently; namely, as a requirement that an optimally shaped body will be defined within geometric constraints. This is known as the *design* approach.

Until the Sixties, flow analysis could be accomplished almost exclusively through such flow visualization techniques as tufts, smoke, and gas bubbles; while streamline velocities were measured using hot-wire and pitot-static anemometers. Tests were either performed in flight (having the advantage of true Reynolds and Mach numbers, but still expensive and time consuming), or in wind or water tunnels. In addition to flow visualization, the wind-tunnel approach has an obvious advantage of providing direct force and moment measurements, plus pressure distributions. It also permits one to make quicker shape modifications than in flight tests. One disadvantage of this approach lies in the uncertainties which may result from the fact that the Reynolds and Mach numbers do not usually correspond to actual operating conditions.

As for obtaining optimum body shapes within given geometric constraints, tunnel and flight test techniques could only help indirectly by providing experimental data as a guide for the designer's intuition.

Potential Methods. Flow potential methods use various singularities (sources, doublets, and vortices) properly distributed on the body surface, inside the body itself, and in the wake; thus making it possible to map flow fields around the bodies or a combination of bodies. In principle, this approach has been known since the beginning of this century. However, because of computational difficulties, application was restricted to geometrically-simple shapes such as bodies of revolution in the non-lifting, and thin airfoils, in the lifting case.

With the advent of high-speed computers, it became possible to adapt potential methods to the development of analytical programs; thus enabling the practicing engineer to determine flow fields around arbitrary-shaped, three-dimensional bodies and their combinations. It also became possible to attack the problem of designing bodies of optimal aerodynamic shapes.

The pioneering efforts of Hess and Smith[21] in the early Sixties can be cited as one of the first attempts to develop engineering applications of these analytical techniques. By the mid and late Sixties, complete computer programs dealing with these problems were already in existence. In the U.S., the works of Hess and his group were being paralleled by Ruppert and his coworkers. All of these efforts resulted in two programs which, continuously modified and improved even as of this writing, are probably the most widely used in this country. The program based on Hess's approach is also called the Douglas-McDonald program[22], while that of Rubbet and his coworkers is referred to as the Boeing program[23,24]. Both programs permit one to deal not only with such tasks as fuselage shapes, fuselage-wings, and fuselage-empennage combinations, but also to study flow both outside and inside of such components as engine nacelles and ducts. Typical problems of fixed-wing aircraft which can be solved using these potential flow modeling methods are indicated in Fig 5.30[25].

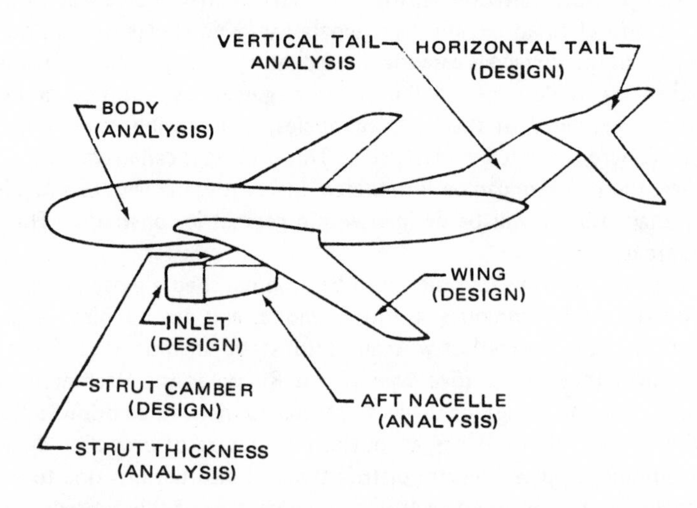

Figure 5.30 Typical problems of practical interest for airplane design

With respect to rotary-wing aircraft; in addition to studies of the flow around the fuselage, engine nacelles, and lifting surfaces, the influence on the flow patterns of such items as the transmission cooling-ducts, inlets and exits, and other similar systems can be investigated. Even the interaction of nonrotating aircraft components and the flow field of the rotor can be determined.

It can be seen, hence, that there is a wide field of practical applications of the flow potential methods. There are even some enthusiasts who believe that computer-based flow analysis may eliminate the need (unlikely) or at least reduce the scope (more probable) of wind or water-tunnel testing. Consequently, the number of actual programs and the volume of literature dealing with analytical predictions of flow fields is continuously growing. It is because of this rapid development that only essential aspects of the application of potential methods to flow determination about three-dimensional bodies representative of helicopter nonrotating components can be discussed here. However, the so-acquired basic knowledge should enable the reader to follow more advanced original works on that subject.

7.2 Development of Conceptual Model

Statement of the Problem. For the sake of simplicity, it will be postulated that the flow velocity in the neighborhood of, as well as far from, the examined body is sufficiently low that the flow itself may be assumed to be incompressible. Under this assumption, the main task in the development of the conceptual model of a single body or a combination of bodies consists of selecting the proper singularities and then distributing them both on and inside the body itself and perhaps, in the wake, in such a way that the resulting flow would comply with all the essential boundary conditions. It may be stated hence, that this task represents a boundary value problem.

Once the proper selected boundary conditions are met, one may have a high degree of confidence that the predicted flow patterns at the body surface, and in the surrounding space, will closely resemble those which would actually exist under the same conditions.

With respect to the material needed for the construction of a model, one would find the necessary "building blocks" under the form of such singularities as sources (both positive and negative), doublets, and vortices (filaments and sheets). Which of these should be used depends on the physical nature of the investigative problem.

Nonlifting Three-Dimensional Body without Flow Separation. In this case, the flow is entirely potential, as vorticity is nonexistent. Streamlines follow the body contour and upon leaving the body, assume the same direction as that of the distant incoming flow. No surface of velocity discontinuity is present anywhere.

Under these conditions the body can be represented exclusively by sources of proper strength and sign, appropriately distributed on the body surface.

Nonlifting Three-Dimensional Body with Flow Separation. Should flow separation be present, then recognition of this physical fact would require a modified approach.

It will be assumed that the line along which the flow separates from the body is known (Fig 5.31). The part of the body with unseparated flow can be modeled, as in the preceding case, through sources. By contrast, the separated wake where rotation is present, would be modeled through the vorticity contained in its volume. This should be done

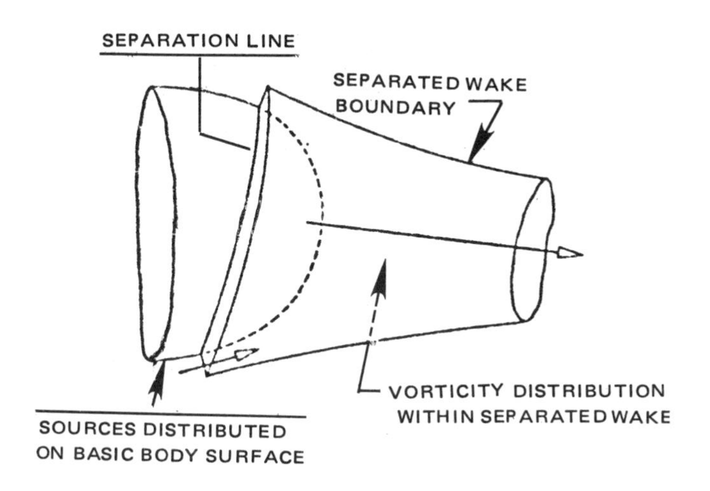

Figure 5.31 Conceptual model for body with separated flow

in such a way that the flow resulting from these vortices as well as sources on the non-separated body would produce streamlines forming a surface boundary around the separated (rotational) wake. The flow outside of the wake will be irrotational; thus, potential methods can be applied for mapping the flow.

As to the determination of the separation line, this is usually done by first assuming that the flow around the body is unseparated. On this basis, velocities and pressure variations are computed along the streamlines. Regions of the body surface where the adverse pressure gradient is high are designated as areas having a high probability of flow separation. The actual procedure for determining the line of separation is described by Stricker and Polz[27].

Combination of Nonlifting and Lifting Bodies. A combination of nonlifting (e.g., fuselage) and lifting (e.g., wings and empennage) bodies represents still another modeling problem. Even assuming for simplicity that there is no flow separation on either of these components, sources can not be used as the sole building blocks of the conceptual model. This is due to the fact that the existence of lift requires the presence of circulation which. in turn, can be modeled by either vortices or doublets. Fig 5.32[23] illustrates a model of the fuselage-wing combination.

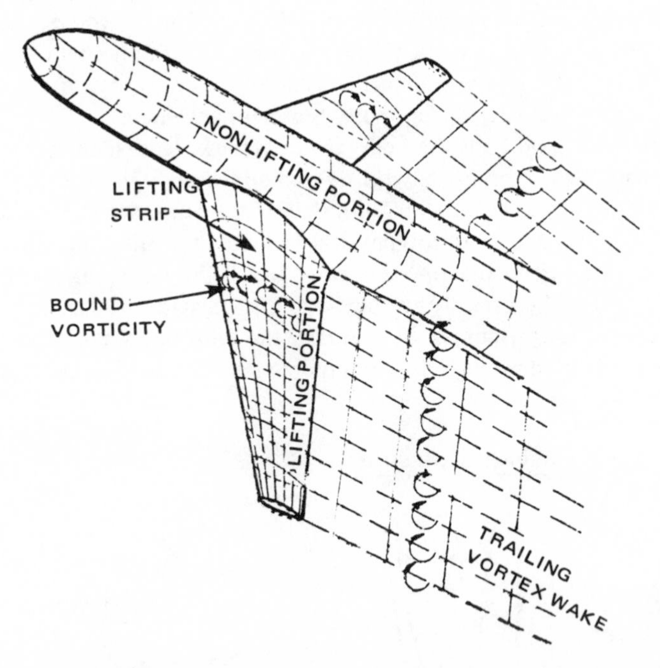

Figure 5.32 Example of a conceptual model representing a combination of lifting and nonlifting bodies

It can be seen from this figure that vortices properly distributed on the wing and in the wake were used as building blocks in determining the lifting component of the fuselage-wing combination.

Selection of the Type of Flow for Further Studies. The above discussion represents only a very rough sketch of the techniques used in the development of conceptual models. In order to give the reader more insight into the actual technique of flow determination, the case of a nonlifting body with no separation will be considered in some detail. The selection of this particular illustrative example is justified not only by its relative simplicity, but also by the fact that many flow aspects related to nonrotating components of classical helicopters can by analyzed using this approach.

7.3 Determination of Nonseparated Flow about Nonlifting Bodies

Geometric Approximation of the Body Surface. The shapes of three-dimensional nonrotating bodies encountered in aircraft design may be quite complicated. Consequently, finding the required distribution of sources on such shapes would represent an insolvable problem, even with high-capacity computers, unless the body shape is approximated in such a way that the computational procedure is discretized.

This is usually done through a special subroutine in the program by approximating the actual body by an assembly of four-sided flat panels on which sources of a uniform strength (m) per unit area are distributed. The number of quadrilateral panels would vary, depending on the accuracy desired and complexity of the body. For instance, for flow studies around isolated helicopter fuselages, anywhere from 250 to 350 source panels may be used to serve as a compromise between the desired accuracy and the required computer time[28]. The sizes of the panels are usually not equal and in those regions where a particular phenomenon is more thoroughly investigated, the number of panels per unit of body surface increases (e.g., inlet to the nacelle in Fig 5.33).

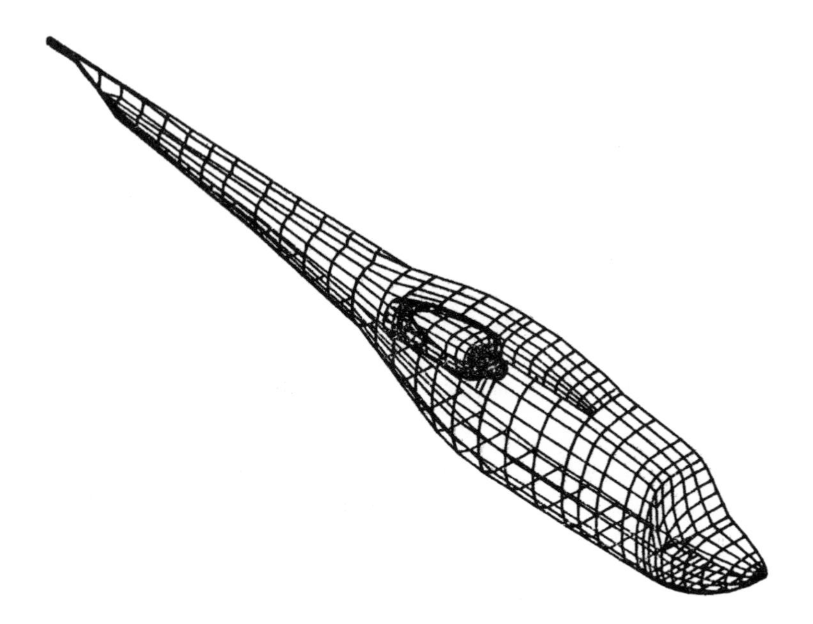

Figure 5.33 Panel representation of a helicopter fuselage-nacelle combination

It may be anticipated that it would be advantageous to approximate the body geometry through an assembly of three-dimensional, rather than flat, panels. In addition, if the strength of the source, instead of remaining constant, could vary over the panel according to some rule (say linearly), then fewer panel elements would be required to model an actual body. In fact, procedures along these lines are being developed. However, at this writing, programs incorporating these advanced ideas (such as those by Rubbert and Saaris) are still in a preliminary stage.

Velocity of Flow. The velocity of flow (V_P) at some point $P(x,y,z)$ located either on the body surface or in the surrounding space can be expressed as follows:

$$\vec{V}_P = \vec{U}_\infty + \vec{v}_P \tag{5.83}$$

where $\vec{U}_\infty$ represents the velocity of flow far from the body, and $\vec{v}_P$ is the velocity induced by singularities (in this case, sources) distributed over the body surface.

Since the flow is potential, the total velocity at P; i.e., $\vec{V}(x,y,z)$ can be expressed as a space derivative of the velocity potential Φ:

$$\vec{V}_P = \nabla\Phi. \tag{5.84}$$

Rubbert et al[24], indicate that it is convenient to represent Φ as the sum of the distant flow potential, $\vec{U}_\infty \cdot (\vec{i}x + \vec{j}y + \vec{k}z)$, and the disturbance potential φ:

$$\Phi = \vec{U}_\infty \cdot (\vec{i}x + \vec{j}y + \vec{k}z) + \varphi \tag{5.85}$$

Far away from the body, the velocity of flow becomes equal to $\vec{U}_\infty$, which means that

$$\varphi \to 0, \quad \text{as } x,y,z \to \infty. \tag{5.86}$$

Boundary Conditions. As far as the behavior of the unseparated flow at the body surface is concerned, two possibilities are expressed by the so-called Neuman conditions:

(1) The surface is impermeable (e.g., the walls of the fuselage), which means that no fluid can either enter or leave the body. This situation, similar to those considered in the preceding sections, can be expressed as follows:

$$\vec{V}_P \cdot \vec{n}_P = 0 \tag{5.87}$$

where, as before, $\vec{n}_P$ is a unit vector perpendicular to the body surface and directed outward.

(2) Some areas on the body surface can either allow the fluid to enter (e.g., various inlets and intakes), or allow it to leave (e.g., exhausts and outlets). Such body areas are considered to be permeable, and this condition can be expressed as follows:

$$\vec{V}_P \cdot \vec{n}_P \neq 0, \tag{5.88}$$

while the above product value can be specified a priori.

In view of Eqs (5.84) and (5.85), the condition of nonpenetration as expressed by Eq (5.87) can be rewritten as follows:

$$(\partial\varphi/\partial n)_P = -\vec{n} \cdot \vec{U}_\infty \tag{5.89}$$

or, in the case of a permeable surface,

$$(\partial\varphi/\partial n)_P = -\vec{n} \cdot (\vec{U}_\infty - \vec{v}) \tag{5.90}$$

where the value of the $(\partial\varphi/\partial n)_P$ derivative can be specified.

As to the location of the points (P) where boundary conditions must be met, Rubbert and Saaris[26] indicate that for the case of constant strength source panels, these control points are usually selected at the centroid of the panel (a and b in Fig 5.34). However, they can also be located outside of the panel where in addition to the coordinates of the control points, direction cosines of the unit vector and magnitude of the velocity component along n are also specified (Fig 5.34c).

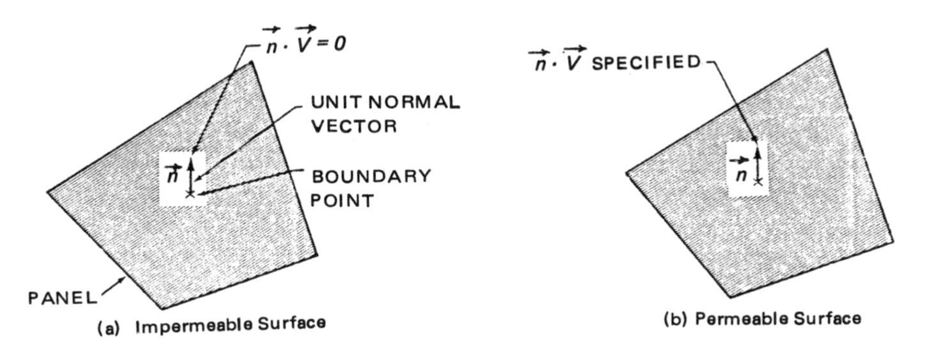

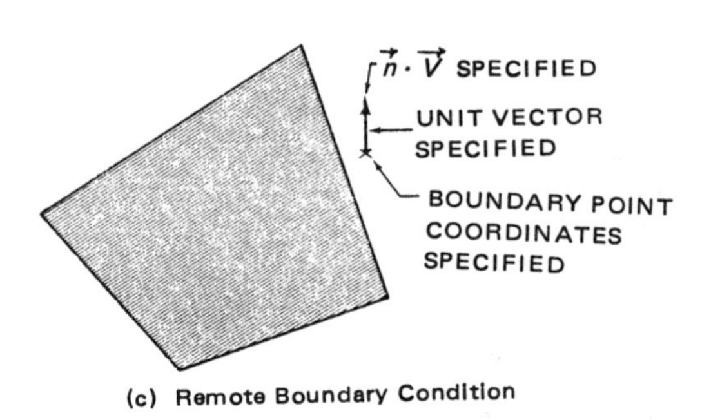

Figure 5.34 Examples of boundary condition specification

7.4 Governing Equations

Once the boundary conditions are specified, the solution to the boundary-value problem is obtained by expressing φ in terms of distribution of sources over the whole surface (S) of the considered body or bodies.

It is indicated in Ref 24 that since the disturbance potential satisfies the Laplace equation, φ can be expressed as a solution to that equation under the following form:

$$\varphi_P = -(1/4\pi) \iint_S (1/\ell)\,(\partial\varphi/\partial n)\, ds \tag{5.91}$$

where S under the double integral sign indicates that integration is performed over the entire body area, and ℓ is the distance from point P to surface S.

But

$$\partial\varphi/\partial n = m(S)$$

where $m(S)$ is the source strength per unit area.

Substituting $m(S)$ for $\partial\varphi/\partial n$ in Eq (5.91), the following relationship between the disturbance velocity potential at P and the source strength at various elementary panels representing the considered body is obtained:

$$\varphi_P = -(1/4\pi) \iint_S [m(S)/\ell]\, ds. \qquad (5.92)$$

Assuming that φ_P is known, $\vec{v}_P$ values can be computed as $\vec{v}_P = \nabla\varphi$ and then the total flow vector at that point can be calculated from Eq (5.83).

However, before Eq (5.92) can be used, the source strength-per-unit area must be computed, taking advantage of the boundary conditions generally specified by Eqs (5.88) and (5.89). To do this, an integral equation–to be solved numerically–is constructed by allowing the field point P in Eq (5.92) approach boundary S, and differentiating the equation with respect to the surface normal at P[24]:

$$\left.\frac{\partial\varphi}{\partial n}\right|_S = -\vec{n}\cdot(\vec{U}_\infty - \vec{v}) = \frac{1}{4\pi}\frac{\partial}{\partial n}\iint_S \frac{m(S)}{\ell} \qquad (5.93)$$

The left side of Eq (5.93) is known at every point of the body surface from the specified boundary conditions. However, the actual numerical solution requires a special computer program.

Examples. A computer program, developed at Boeing along the above lines, was used by C.N. Keys (guided by Saaris) to solve the flow problems around the fuselage-nacelle combination schematized in Fig 5.33. One of the results of their efforts is shown in Fig 5.35(a) where the flow velocity vectors are marked around the nacelle-body combination and the hub fairing, while the resulting streamlines are traced in Fig 5.35(b).

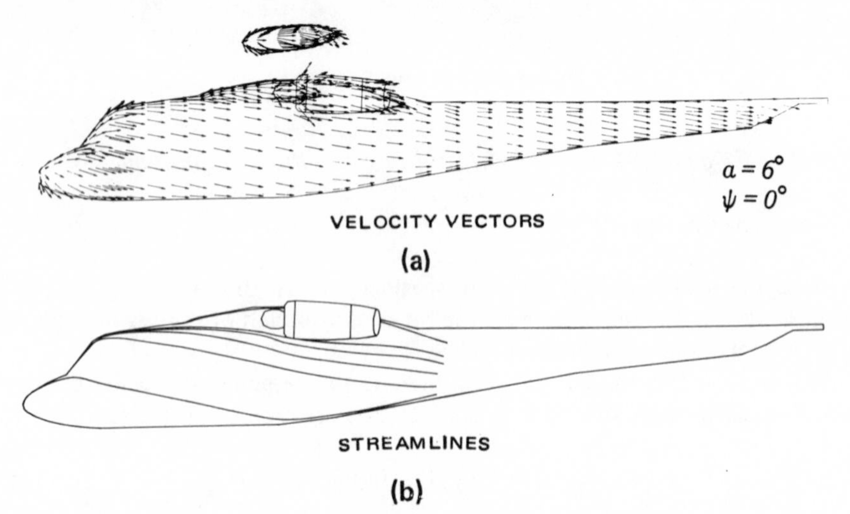

Figure 5.35 Example of flow studies for the fuselage-nacelle-hub fairing combination

In studies of aerodynamic interaction, potential methods may be applied to determine the flow in the space surrounding the body. One example of this application is shown in Fig 5.36, wherein Keys depicts the upwash angles at the rotor disc caused by the presence of the fuselage.

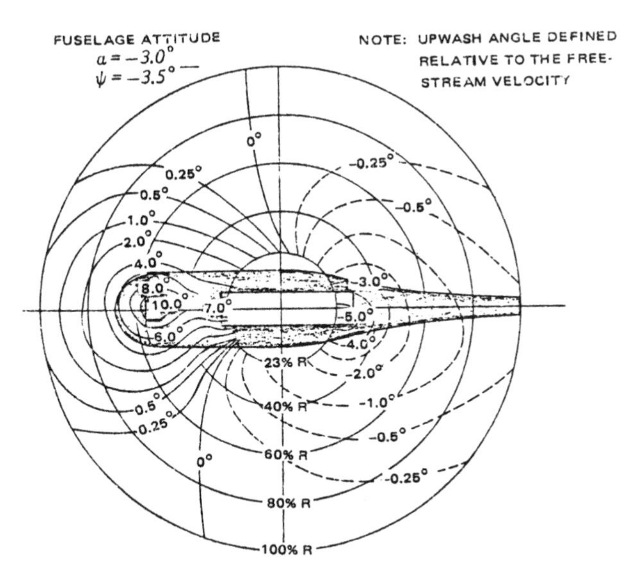

Figure 5.36 Upwash angles at the rotor disc due to the presence of the fuselage

8. CONCLUDING REMARKS

Potential methods in helicopter analysis are primarily used for the following tasks: (1) Determination of blade airloads (both time average and instantaneous), (2) Prediction of flow fields resulting from the presence of a loaded airscrew, and (3) Mapping the flow fields around nonrotating three-dimensional single bodies and their combinations. The first two of the the above tasks were previously discussed in the chapter on Vortex Theory which in a broad sense, belongs in the domain of potential methods, for outside of the vortex filament all flow is irrotational (potential). However, the difference between the considered approaches in Chs IV and V consists of the fact that in Ch IV, singularities based on the vortex concept were used exclusively as building blocks for conceptual models of airscrews, and vortex-generated velocities were directly calculated using the Biot-Savart law. In Ch V, the number of building blocks has been increased by adding sources and doublets, while determination of velocity and acceleration potentials became an intermediate step in analytical procedures.

The use of the doublet approach allowed one to develop methods of blade load and local angle-of-attack predictions which in some cases may result in more simplified computational procedures than those offered by the vortex theory. The use of sources alone in the case of nonlifting bodies and nonseparated flow; and in combination with doublets and vortices for the case of lifting bodies, or separated flow, made it possible to map the flow at body surfaces and in the surrounding space. This opened the way toward a better understanding of the generative process of aerodynamic forces (especially drag) and moments on various nonrotating helicopter components as well as their assemblies.

It can be seen hence, that potential methods offer the helicopter designer an important tool to assist him in the task of developing more efficient rotorcraft.

References for Chapter V

1. Mangler, K.W. and Squire, H.B. *The Induced Velocity Field of a Rotor.* R&M No. 2642, May 1950.

2. Baskin, V.E., Vil'dgrube, L.S., Vozhdayev, E.S., and Maykapar, G.I. *Theory of the Lifting Airscrew.* NASA-TT F-823, February 1976.

3. Dat, Rolland. *La Théorie de la Surface Portante Appliquée à l'Aile Fixe et à l'Hélice.* ONERA T.P. No. 1298, 1973; also Recherche Aerospatiale No. 1973-74.

4. Dat, R. *Aeroelasticity of Rotary-Wing Aircraft.* AGARD Lecture Series No. 63 on Helicopter Aerodynamics and Dynamics, 1973; also ONERA T.P. No. 1198, 1973.

5. Dat, R. *Aérodynamique Instationnaire des Pales d'Helicoptère.* Paper presented at AGARD/FMP Symposium, Gottingen, May 30, 1975.

6. Costes, J.J. *Calcul des Forces Aérodynamiques Instationnaires sur les Pales d'un Rotor d'Helicoptère.* AGARD Report 595. English translation, NASA TTF 15039, 1973.

7. Costes, J.J. *Introduction du Décollement Instationnaire dans la Théorie du Potential d'Acceleration. Application à l'Helicoptère.* Recherche Aérospatiale No. 1975.

8. van Holten, Th. *The Computation of Aerodynamic Loads on Helicopter Blades in Forward Flight Using the Method of Acceleration Potential.* Dr. of Engineering Thesis, Delft Tech. University, March 1975.

9. van Holten, Th. *Computation of Aerodynamic Loads on Helicopter Rotor Blades in Forward Flight Using the Method of Acceleration Potential.* IGAS Paper 74-54, Haiffa, August 1974.

10. Jones, W.P. and Moore, Jimmie A. *Simplified Aerodynamic Theory of Oscillating Thin Surfaces in Subsonic Flow.* AIAA Journal, Vol. II, No. 9, September 1973.

11. Rao, B.M. and Jones, W.P. *Application to Rotary-Wings of a Simplified Aerodynamic Lifting Surface Theory for Unsteady Compressible Flow.* AHS/NASA Specialists' Meeting on Rotorcraft Dynamics, February 1974.

12. Schatzel, Paul R. *A Simplified Numerical Lifting Surface Theory Applied to Rotary Wings in Steady Incompressible Flow.* AIAA Paper 75-218, January 1975.

13. Caradonna, Francis X. and Isom, Morris P. *Subsonic and Transonic Flow over Helicopter Rotor Blades.* AIAA Journal, Vol. 10, No. 12, December 1972.

14. Summa, Michael J. *Potential Flow about Three-Dimensional Lifting Configurations with Application to Wings and Rotors.* AIAA Paper 75-126, January 1975.

15. Coburn, Nathaniel. *Vector and Tensor Analysis.* Dover Publications, New York, 1970.

16. Durand, William Frederick. *Aerodynamic Theory.* Durand Reprinting Committee, CAL Tech, January 1943.

17. Wood, Alexander. *Acoustics.* Dover Publications, New York, N.Y. 1966.

18. Dat R. *Représentation d'une Ligne Portante Animée d'un Mouvement Vibratoire par une Ligne de Doublets d'Accélération.* Recherche Aérospatiale No. 133, November-December 1969.

20. Costes, J.J. *Rotor Response Prediction with Non-Linear Aerodynamic Loads on the Resulting Blade.* Second European Rotorcraft and Powered Lift Aircraft Forum, Buckeburg, West Germany, Paper No. 22, September 1976.

21. Hess, J.L. and Smith, A.M.O. *Calculation of Nonlifting Potential Flow About Arbitrary Three-Dimensional Bodies.* ES 40622, Douglas Aircraft Co., 1962.

22. Hess, J.L. and Smith, A.M.O. *Calculation of Potential Flow About Arbitrary Bodies.* Progress in Aeronautical Sciences, Vol. 8, Perganon Press, 1967.

23. Hess, J.L. *Calculation of Potential Flow About Arbitrary Three-Dimensional Lifting Bodies.* McDonnell Douglas Rpt No. MDC J5679-01, NADC Contract No. N00019-71-C-0524, October 1972.

24. Rubbert, P.E.; Saaris, G.R.; Scholey, M.B.; Standen, N.M.; and Wallace, R.E. *A General Method for Determining the Aerodynamic Characteristics of Fan-in-Wing Configurations.* Vol. 1 – *Theory and Application,* USAAVLABS Tech. Rpt. 67-61A, 1967.

25. Hink, G.R.; Gilbert, R.F.; and Sundstrom. *A General Method for Determining the Aerodynamic Characteristics of Fan-in-Wing Configurations.* Vol II – *Computer Program Description.* USAAVLABS Tech. Rpt. 67-61B, 1967.

26. Rubbert, P.E. and Saaris, G.R. *Review and Evaluation of a Three-Dimensional Lifting Potential Flow Analysis Method for Arbitrary Configurations.* AIAA Paper No. 72-188, 1972.

27. Stricker, R. and Polz, G. *Calculation of the Viscous Flow around Helicopter Bodies.* Third European Rotorcraft and Powered Lift Forum, Aix-en-Provence, France, September 1977.

28 Gillespie, J., Jr. and Windsor, R.I. *An Experimental and Analytical Investigation of the Potential Flow Field, Boundary Layer, and Drag of Various Helicopter Fuselage Configurations.* USAAMRDL TN 13, January 1974.

AIRFOILS FOR ROTARY–WING AIRCRAFT

Basic definitions and working formulae used in applied airfoil aerodynamics are reviewed first, followed by brief presentations of thin and finite-thickness airfoil theories. Considerations of viscosity, compressibility, and unsteady aerodynamic effects conclude a review of the fundamentals. An analysis of the particular operational environment of airfoils applied to rotary-wing aircraft, a discussion of the contribution of airfoil characteristics to rotor performance, and a summary of requirements for airfoils best suited for helicopter applications represent sections forming a link to the design aspects.

Principal notation for Chapter VI

A	coefficient in Glauert's expansion	
AR	wing aspect ratio: $AR \equiv b^2/S$	
a	lift-curve slope: $a \equiv dc_\ell/d\alpha$	rad^{-1}, or deg^{-1}
a	cylinder radius	m, or ft
a	constant	
b	wing span, or airfoil half-chord	m, or ft
C	speed of molecular motion	m/s, or fps
C_D	wing drag coefficient: $C_D \equiv D/\frac{1}{2}\rho V^2 S$	
C_L	wing lift coefficient: $C_L \equiv L/\frac{1}{2}\rho V^2 S$	
C_M	wing moment coefficient: $C_M \equiv M/\frac{1}{2}\rho V^2 S\bar{c}$	
C_p	pressure coefficient: $C_p \equiv (p_u - p_\ell)/\frac{1}{2}\rho V^2$	
c	airfoil chord	m, or ft
c_c	airfoil chord-force coefficient	
c_d	airfoil drag coefficient: $c_d \equiv d/\frac{1}{2}Vc\rho V^2$	
c_ℓ	airfoil lift coefficient: $c_\ell \equiv \ell/\frac{1}{2}c\rho V^2$	
c_m	airfoil moment coefficient: $c_m \equiv m/\frac{1}{2}\rho V^2 c^2$	
c_n	airfoil normal-force coefficient	
d	drag per unit length	N/m, or lb/ft
i	$i \equiv \sqrt{-1}$	
ℓ	lift per unit length	N/m, or lb/ft
ℓ	length	m, or ft
M	Mach number: $M \equiv V/s$	
M	moment	N-m, or ft-lb
m	moment per unit length	N/m, or lb/ft

m	airfoil camber; $y_{m\ell}/c$	
R_e	Reynolds number: $R_e \equiv Vc/\nu$	
S	surface	m^2, or ft^2
s	speed of sound	m/s, or fps
u,v,w	velocity components along x, y, z axes	m/s, or fps
V	velocity of flow in general	m/s, or fps
v	induced velocity	m/s, or fps
W	complex potential: $W \equiv \phi + i\psi$	
x,y,z	Cartesian coordinates	m, or ft
z	complex variable: $z \equiv x + iy$	
α	angle of attack	rad, or deg
Γ	circulation	m^2/s, or ft^2/s
γ	circulation per unit length	m/s, or fps
ζ	complex variable in the ζ plane: $\zeta \equiv \eta + i\xi$	
η,ξ	coordinates in the ζ plane	m, or ft
θ	angle	rad, or deg
μ	rotor advance ratio: $\mu \equiv V/V_t$	
μ	viscosity coefficient	kg/m-s, or slugs/ft-s
ν	kinematic coefficient of viscosity: $\nu \equiv \mu/\rho$	
ρ	air density	kg/m^3, or $slugs/ft^3$
φ	velocity potential	m^2/s, or ft^2/s
ψ	stream function	
ψ	blade azimuth angle	rad, or deg
Ω	angular velocity	rad/s, or deg/s

Subscripts

DD	drag divergence
a.c	aerodynamic center
com	compressible
c.p	center of pressure
hs	high speed
i	ideal
inc	incompressible
ℓs	low speed
ℓo	local

Theory

m	meanline
n	indication of numerical sequence
t	tip
$\|\|$	parallel
$\perp$	perpendicular

Superscripts

$\cdot$	time derivative
—	average (mean)
$\rightarrow$	vector

1. INTRODUCTION

The significance of rotor-blade airfoil characteristics was indicated in the preceding three chapters, especially in Ch III where the influence on helicopter performance of such airfoil characteristics as c_d vs c_ℓ, $c_{\ell_{max}}$, compressibility effects, and sensitivity to blade surface roughness was discussed. By this time, the reader is probably aware that improvements in a set of airfoil characteristics, while beneficial in one regime of flight, may prove to be detrimental in another.

This divergency in the requirements of airfoils suitable for helicopter rotor blades chiefly stem from the particular environment in which they operate. Consequently, the designer when confronted with the problem of either selecting the most suitable blade airfoil section from a catalog or attempting to develop new ones, must seek a compromise between often conflicting requirements.

A knowledge of airfoil theories should contribute to a better judgement of the possibilities, as well as the limitations, regarding rotorcraft performance gains which can be achieved through the application of the most suitable blade airfoil sections. In addition, these theories should contribute to a better understanding of physical phenomena and interpretation of experimental results. They may also provide a tool for the development of sections which should exhibit a priori defined aerodynamic characteristics.

There are many texts (both classic and modern) which are either devoted to subsonic airfoils[1], or at least contain large sections on this subject[2,3,4,5,6]. There is also a recent publication, called *DATCOM*[7], which deals exclusively with airfoils suitable for helicopter applications. In addition to some theoretical discussions, it contains a catalog compendium of airfoil sections most widely used in helicopter design along with those potentially well-suited for rotary-wing applications.

The material presented in this chapter was selected in a way that should enable the reader on one hand, to grasp the theoretical aspects of airfoil analysis and development; while on the other, to help apply this knowledge to the aerodynamic design of rotors. The above goals are accomplished by (1) a review of the most important aspects of two-dimensional airfoil section theory, (2) an analysis of the particular aerodynamic environment in which the blade sections operate, and (3) a determination of the airfoil characteristics most desirable for various regimes of flight and operational applications.

2. REVIEW OF BASIC RELATIONSHIPS

2.1 Airfoil Characteristics

Lift, Drag, and Moment Coefficients. Airfoil data are usually presented under the form of nondimensional force and moment coefficients. For wings of finite aspect ratios, these characteristics are expressed by the following relationships: $C_L(a)$, $C_D(a)$, $C_D(C_L)$, $C_m(a)$, and $C_m(C_L)$; and for the two-dimensional case ($AR = \infty$), they become $c_\ell(a)$, $c_d(a)$, $c_d(c_\ell)$, $c_m(a)$, and $c_m(c_\ell)$.

Airfoil section characteristics as given by the two-dimensional data are almost exclusively used for rotary-wing applications. However, it may happen that only results of three-dimensional tests of a wing of a finite aspect ratio are available. In this case, an approximation of the section-lift characteristics can be computed from the following relationships based on the classic Prandtl lifting-line theory of finite-span wings.

The two-dimensional lift-curve slope ($a \equiv dc_\ell/da$) is related to that of a finite aspect ratio wing ($a' \equiv dC_L/da$) as follows (Fig 6.1a):

$$1/a = 1/a' - 1/\pi AR_e \tag{6.1}$$

where AR_e is the wing equivalent aspect ratio—usually given with the test data. The two-dimensional profile drag coefficient (c_d) at a lift coefficient (c_ℓ)—assumed equal to that of the 3-D wing ($c_\ell = C_L$)—can be obtained from the total drag coefficient (C_D) of a finite span wing as (Fig 6.1b):

$$c_d = C_D - C_L^2/\pi AR_e. \tag{6.2}$$

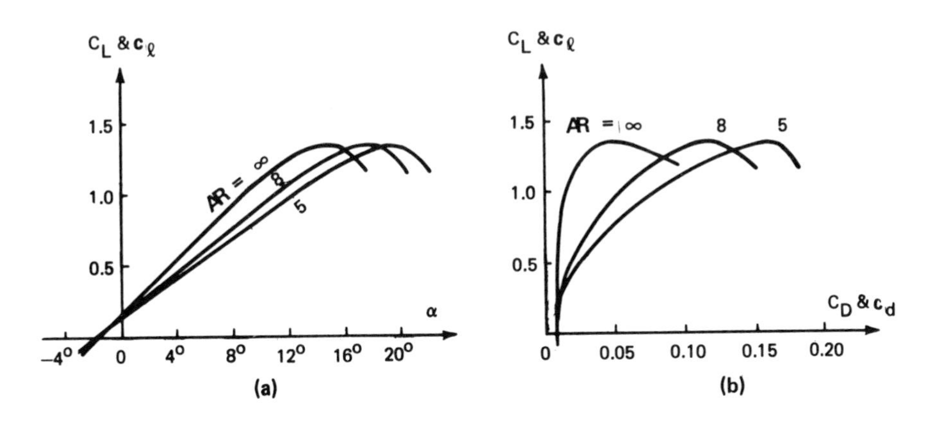

Figure 6.1 Examples of (a) lift slope and (b) drag coefficient variations for three aspect ratios

Normal (c_n) and Chordwise (c_c) Force Coefficients. Knowing c_ℓ and c_d, the normal and chordwise force coefficients (Fig 6.2) can be calculated from the following relationships:

$$c_n = c_\ell \cos a + c_d \sin a.$$
$$c_c = c_d \cos a - c_\ell \sin a. \tag{6.2}$$

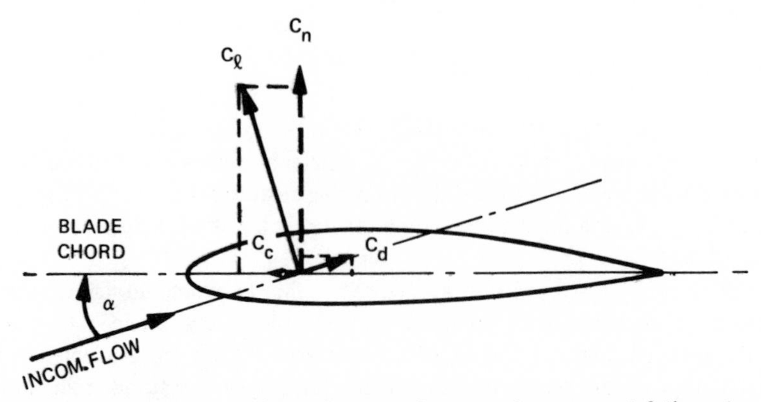

Figure 6.2 Components of aerodynamic forces acting on an airfoil section

It is usually assumed that $c_n \approx c_\ell$. By contrast, c_c may be quite different from c_d as far as magnitude and sign are concerned. For instance, at high angles of attack (prior to stall), the c_c component may be directed forward; thus providing a driving force in an autorotating rotor.

Pitching Moment Coefficient (c_m) and Aerodynamic Center (a.c). As long as there is no lifting-line sweep of a finite aspect ratio wing, it may be assumed that the c_m (based on the average chord of the wing) obtained from three-dimensional wing tests will be the same as that for the two-dimensional case.

In principle, the pitching moment coefficient can be referred to any axis perpendicular to the plane of the airfoil section (say, passing through the leading or trailing edges). However, following past NACA practice, this reference axis is usually assumed to pass through the so-called aerodynamic center *(a.c)*. In contrast to all other possible reference points, the moment coefficient referred to an axis passing through the *a.c* of an airfoil section ($c_{m_{a.c}}$), when plotted vs c_ℓ, remains constant up to the inception of stall (Fig 6.3).

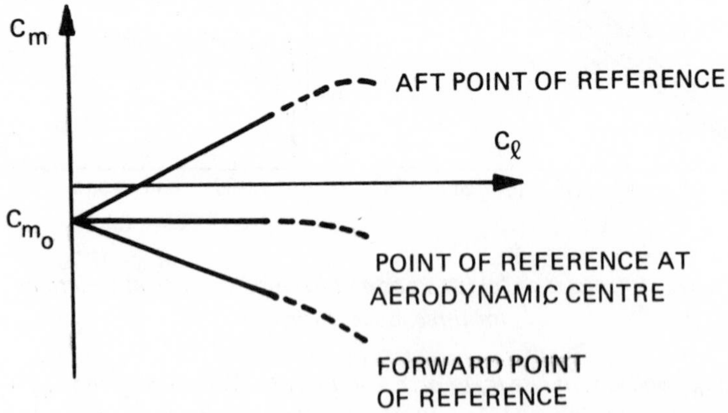

Figure 6.3 Moment coefficients about various reference points

When dealing with experimental results, the position of the aerodynamic center (usually close to the 1/4-airfoil chord point) is often determined by the "cut and try" process. However, for theoretically developed airfoil sections, both thin[1] and those of finite thickness[4], the $a.c$ location can be obtained analytically.

Center-of-Pressure Position. The drag, or more strictly, the chordwise component of the aerodynamic force contributes little to the moment about the $a.c$. Consequently, it may be assumed that the pitching moment about the $a.c$ is entirely due to the normal component, approximately equal to the lift. Under this assumption, the location of the center-of-pressure ($c.p$) can be determined as follows:

$$x_{c.p}/c = c_{m_{a.c}}/c_n \tag{6.3}$$

or, since $c_n \approx c_\ell$:

$$x_{c.p}/c = c_{m_{a.c}}/c_\ell \tag{6.3a}$$

where $x_{c.p}$ is the chordwise distance between the point of application of the normal force and the $a.c$; $x_{c.p}$ being positive when measured ahead of the $a.c$ and negative, aft.

It can be seen from Eq (6.3a) that for $c_\ell > 0$ airfoils having $c_{m_{a.c}} < 0$, the center-of-pressure is located aft of the $a.c$ and would exhibit a rearward movement for decreasing α and hence c_ℓ values. By contrast, for airfoils having $c_{m_{a.c}} > 0$, the center-of-pressure would stay ahead of the $a.c$ and would move forward when α—and hence, c_ℓ also—decreases. For $c_{m_{a.c}} = 0$, the center-of-pressure remains fixed at the aerodynamic center; at least as long as there is no stall, or compressibility effects.

The sign of $c_{m_{a.c}}$ is important from the point-of-view of blade aeroelastic phenomena. Since the $c.p$ movement associated with the angle-of-attack variation has a stabilizing effect by shifting in the direction that generates a moment opposing torsional deformations of the blade, $c_{m_{a.c}} > 0$ proves advantageous. By contrast, $c_{m_{a.c}} < 0$ is somewhat more difficult to handle. For this reason, early successful helicopters employed symmetrical airfoil sections with $c_{m_{a.c}} = 0$ (no $c.p$ movement when the angle-of-attack varies); while attempts to develop new airfoil sections specifically suited for helicopter blades were directed toward shapes exhibiting $c_{m_{a.c}} > 0$.

An increasing understanding of the blade aeroelastic phoenomena in the Sixties enabled rotor designers to use cambered blade airfoil sections in spite of the fact that they have $c_{m_{a.c}} < 0$. The attractiveness of these cambered airfoils lies in the higher $c_{\ell_{max}}$ values than those achievable in symmetrical airfoil sections ($c_{m_{a.c}} = 0$), or the S-shaped ones with their trailing edges deflected up ($c_{m_{a.c}} > 0$).

Meanline, Shape, and Thickness Distribution. It becomes clear—even from the above cursory discussion—that it is important to establish a dependable relationship between the airfoil shape and its aerodynamic characteristics. In particular, it would be desirable to know how these characteristics are influenced by modifications in the shape of various elements constituting an airfoil. It will be shown later that the two classical theories; namely, of thin and of finite-thickness airfoils, provide some analytic guidance regarding the dependence of aerodynamic characteristics on airfoil shapes. However, there are also other approaches to this subject; for instance, that of NACA, where theoretical guidance was combined with experimental inputs, thus leading to a systematic development of families of airfoils, known as the four- to the six-digit series.

In this approach, an airfoil is visualized as being constructed of two basic components: the meanline, constituting its skeleton, and the shape of the thickness distribution, forming an outer envelope "skewed" on the meanline (Fig 6.4).

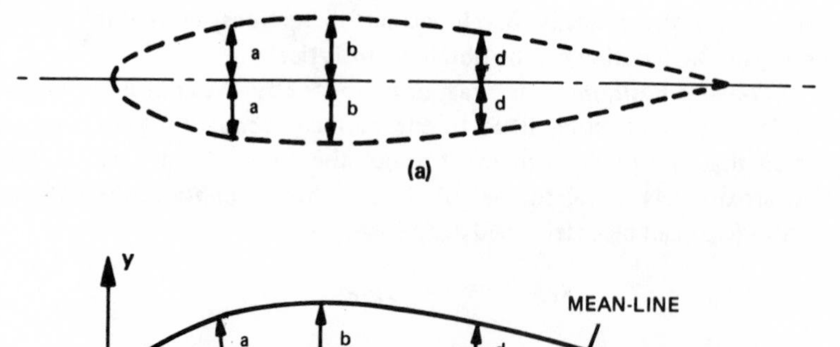

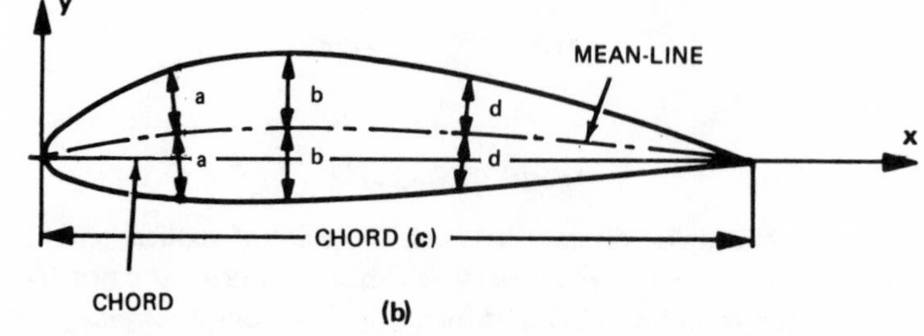

Figure 6.4 (a) Thickness distribution, and (b) complete airfoil section

The spheres of influence of these two basic airfoil components can not be completely separated. However, it may be assumed that the shape of the meanline mostly governs pitching moment characteristics and $c_{\ell_{max}}$ levels, while the thickness distribution influences profile drag and character of stall (the latter in combination with the meanline shape). The best insight into the development philosophy and experimental data on the NACA family of airfoils in general can be obtained from Ref 1, while current data as applied to helicopters are given in Ref 7.

3. THEORY OF THIN AIRFOIL SECTIONS IN IDEAL FLUID

3.1 Basic Approach

Abbott and von Doenhoff[1] indicate that the theory of thin airfoil sections may be quite helpful in understanding many of the important airfoil characteristics governed by the shape of the airfoil meanline; for instance:

(1) chordwise load distribution
(2) value of the zero-lift angle
(3) magnitude of the pitching moment
(4) value of the lift-curve slope, and
(5) approximate location of the *a.c.*

This theory will be reviewed, following the classical Glauert[2] approach in which the airfoil meanline is modeled by an infinitely long vortex sheet having a width equal to the chord length, c; while its generatrix is perpendicular to the incoming flow (Fig 6.5).

The intensity of vorticity per unit length of the meanline is γ and may vary with x: $\gamma(x)$. As explained in Sect 3.1 of Ch IV, a vortex sheet represents a surface of discontinuity of velocities tangential to the surface just above and below that surface, while γ (Eq 4.3) is numerically equal to the velocity difference. Total circulation Γ about the entire vorticity surface would be

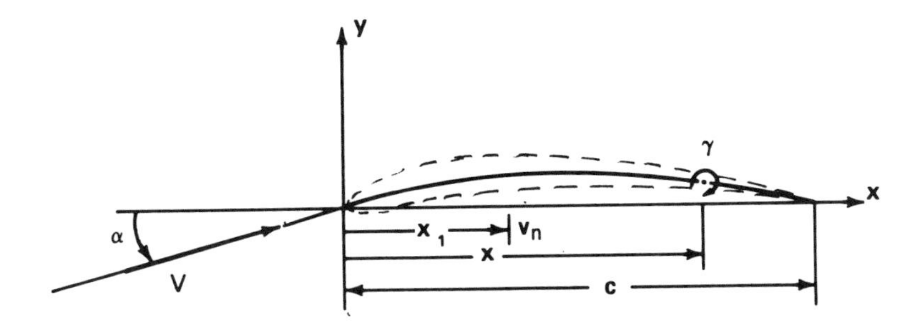

Figure 6.5 Conceptual model of a thin airfoil section

$$\Gamma = \int_0^c \gamma dx. \tag{6.4}$$

The actual integration indicated by Eq (6.4) should be accomplished along the meanline. However, since the camber is small in comparison with the chord length, it is conceivable to assume that $c \approx \ell_{m\ell}$ where $\ell_{m\ell}$ is the length of the meanline.

Also assuming that a dx-long element of vorticity located at an abscissa x is small enough to be treated as a single vortex filament, the velocity component (dv_n) perpendicular to the chord can be expressed according to equation (4.1) as follows:

$$dv_n = \frac{\gamma dx}{2\pi(x - x_1)}, \tag{6.5}$$

while the total component induced by the whole sheet becomes

$$v_n = \int_0^c \frac{\gamma dx}{2\pi(x - x_1)} dx. \tag{6.6}$$

The principle of nonpenetration of the lifting surface by the flow, often discussed in the previous chapter, in this case can be expressed under the small-angle assumption as follows:

$$Va + v_n - V(dy/dx) = 0 \tag{6.7}$$

where dy/dx expresses the slope of the meanline.

In principle, Eqs (6.6) and (6.7) offer the possibility of determining aerodynamic characteristics of a given thin airfoil; and also give the opportunity to study how some characteristics of a thick airfoil are influenced by the shape of the meanline. It should be noted, however, that similar to other cases discussed in Chs IV and V, mathematical procedures directed toward a practical determination of airfoil characteristics involve difficulties associated with undetermined values of Eq (6.6) when $x \to x_1$. Various methods of dealing with this problem can be found in texts devoted to subsonic aerodynamics. However, most of the methods are usually based on Glauert's original approach of expressing x with the help of a new independent variable; namely, angle θ:

$$x = (c/2)(1 - \cos\theta).$$

It is obvious that x varies within the interval $0 \leqslant x \leqslant c$; while θ is included within $0 \leqslant \theta \leqslant \pi$. The functional dependence of γ on x; $\gamma(x)$, can now be expressed through the following Fourier series:

$$\gamma(x) = \gamma(\theta) = 2V(A_0 \operatorname{ctn}(\theta/2) + \sum_{n=1}^{\infty} A_n \sin n\theta) \quad (6.8)$$

where coefficients $A_0, A_1, \ldots, A_n$ should be so-selected that the condition of nonpenetration of the lifting surface expressed jointly by Eqs (6.6) and (6.7) is fulfilled.

Eq (6.6), when rewritten in terms of the new variable θ becomes:

$$v_n(\theta_1) = (1/2\pi) \int_0^{\pi} \frac{\gamma(\theta) \sin\theta \, d\theta}{\cos\theta_1 - \cos\theta}. \quad (6.9)$$

Multiplying the right side of Eq (6.8) by $\sin\theta$; substituting this product for $\gamma(\theta)$ $\sin(\theta)$ in Eq (6.9), and finding the simple integrals appearing in Eq (6.9), the following is obtained:

$$v_n(\theta_1) = V\left[-A_0 + \sum_{n=1}^{\infty} A_n \cos n\theta_1\right]. \quad (6.10)$$

The condition of nonpenetration given by Eq (6.7) can now be written as

$$\alpha - A_0 + \sum_{n=1}^{\infty} A_n \cos n\theta = (dy/dx). \quad (6.11)$$

The coefficients in Eq (6.11) can be obtained by multiplying, in turn, both sides of the equation by $\cos\theta$, $\cos 2\theta, \ldots$ and performing the indicated integrations from $\theta = 0$ to $\theta = \pi$, with $(c/2)(1 - \cos\theta)$ substituted for x. This results in the following:

$$A_0 = \alpha - (1/\pi) \int_0^{\pi} (dy/dx) d\theta$$

$$A_n = (2/\pi) \int_0^{\pi} (dy/dx) \cos n\theta \, d\theta. \quad (6.12)$$

The lift-per-unit span is given in Eq (4.14) as

$$\ell = \rho V \Gamma.$$

Substituting the right side of Eq (6.4) for Γ in the above equation; expressing $\gamma(x)$ according to Eq (6.8); and then integrating within 0 to π limits, the following is obtained:

$$\ell = \rho V^2 c \pi [A_0 + (A_1/2)] \quad (6.13)$$

and

$$c_\ell = 2\pi(A_0 + \tfrac{1}{2}A_1). \quad (6.13a)$$

Treating an element of vorticity $\gamma(x)dx$ as a single vortex filament; the elementary lift $d\ell$ generated by this filament becomes $d\ell = \rho V \gamma(x) dx$.

Now, the moment about the leading edge can be expressed as:

$$M = -\rho V \int_0^c \gamma(x) x \, dx.$$

Following a procedure similar to that used for the determination of ℓ, it can be found that

$$M = -(\pi/4)\rho V^2 c^2 [A_0 + A_1 - (A_2/2)], \tag{6.14}$$

and the moment coefficient about the leading edge becomes:

$$c_m = -(\pi/2)(A_0 + A_1 - \tfrac{1}{2}A_2). \tag{6.14a}$$

Eqs (6.13a) and (6.14a) indicate that both lift and moment coefficients are influenced by the three first-harmonic coefficients (A_0, A_1, A_2) only. Higher harmonics are of no significance.

Substituting the expressions for A as given by Eq (6.12) into Eqs (6.13a) and (6.14a), the following is obtained:

$$c_\ell = 2\pi(a + \epsilon_0) \tag{6.15}$$

$$c_m = 2[\mu_0 - (\pi/4)\epsilon_0] - \tfrac{1}{4}c_\ell. \tag{6.16}$$

where

$$\epsilon_0 = -(1/\pi) \int_0^\pi (dy/dx)(1 - \cos\theta) d\theta$$

and

$$\mu_0 = -\tfrac{1}{4} \int_0^\pi (dy/dx)(1 - \cos\theta) d\theta.$$

It can be seen from Eq (6.15) that when $a = -\epsilon_0$, $c_\ell = 0$. This means that $-\epsilon_0$ represents the zero-lift angle-of-attack. Eq (6.15) also indicates that the lift-curve slope of thin airfoils is $dc_\ell/da = 2\pi$.

Eq (6.16) can be rewritten in a form more suitable for practical application. Calling c_{m_0} the moment coefficient corresponding to $c_\ell = 0$, and noting from Eq (6.16) that $c_{m_0} = 2[\mu_0 - (\pi/4)\epsilon_0]$, Eq (6.16) becomes

$$c_m = c_{m_0} - \tfrac{1}{4}c_\ell. \tag{6.16a}$$

3.2 Meanline Shape for Uniform Loads

The development of a procedure for determining the mean-line shape assuring a uniform chordwise load distribution is considered here as an illustrative example of (a) a way of giving the airfoil some desired characteristics, and (b) a method of dealing with singularities.

In order to have a uniform chordwise load distribution, $\gamma(x)$ must be constant; consequently, Eq (6.6) can be rewritten as follows:

$$v_n(x_1) = (\gamma/2\pi) \int_0^c [dx/(x - x_1)]. \tag{6.17}$$

Again treating each vorticity element $\gamma\, dx$ as a vortex filament; the lift-per-unit span (ℓ) can be written per Eq (4.14) as

$$\ell = \int_0^c \rho V \gamma dx = \rho V \gamma c \tag{6.18}$$

or, in a coefficient form, the ideal lift coefficient ($c_{\ell i}$) corresponding to the desired uniform load distribution will be

$$c_{\ell i} = 2\ell/\rho V^2 c = 2\gamma/V \tag{6.19}$$

and

$$\gamma = \tfrac{1}{2} V c_{\ell i}. \tag{6.20}$$

By substituting Eq (6.20) into Eq (6.17) and making x and x_1 nondimensional, as x/c and x_1/c_1, one obtains

$$\frac{v_n}{V}\left(\frac{x_1}{c}\right) = \frac{c_{\ell i}}{4\pi} \int_0^1 \frac{d(x/c)}{(x/c) - (x_1/c)}. \tag{6.21}$$

To avoid undetermined values as $x \to x_1$, the integration indicated in Eq (6.21) is performed in the following way:[1]

$$\frac{v_n}{V}\left(\frac{x_1}{c}\right) = \frac{c_{\ell i}}{4\pi} \lim_{\epsilon \to 0} \left[\int_0^{(x_1/c)-\epsilon} \frac{d(x/c)}{(x/c) - (x_1/c)} + \int_{(x_1/c)+\epsilon}^1 \frac{d(x/c)}{(x/c) - (x_1/c)}\right]$$

$$= \frac{c_{\ell i}}{4\pi} \lim_{\epsilon \to 0} \left\{\left[\ln\left(\frac{x}{c} - \frac{x_1}{c}\right)\right]_0^{x_1 - \epsilon} + \left[\ln\left(\frac{x}{c} - \frac{x_1}{c}\right)\right]_{x_1+\epsilon}^1\right\} \tag{6.22}$$

$$= \frac{c_{\ell i}}{4\pi}\left[\ln\left(1 - \frac{x_1}{c}\right) - \ln\left(\frac{x_1}{c}\right)\right].$$

It is indicated in Ref 1 that uniform load distribution is obviously symmetrical with respect to the chord mid-point; consequently, the meanline will also be symmetrical with respect to that point, while the angle-of-attack at which this uniform load is realized would be zero. Now, substituting Eq (6.22) for the v_n/V ratio, Eq (6.7) becomes

$$dy/dx = \frac{c_{\ell i}}{4\pi}\left[\ln\left(1 - \frac{x_1}{c}\right) - \ln\left(\frac{x_1}{c}\right)\right] \tag{6.23}$$

and integration results in the following equation for the meanline:

$$y/c = -\frac{c_{\ell i}}{4\pi}\left[\left(1-\frac{x}{c}\right)\ell n\left(1-\frac{x}{c}\right)+\frac{x}{c}\,\ell n\,\frac{x}{c}\right]. \tag{6.24}$$

In Eq (6.24), constants of integration were selected to make $y(o) = 0$, and $y(c) = 0$.

Similar procedures can be applied to obtain shapes of the meanline corresponding to other desired forms of chordwise load distributions at a specified design (also called ideal) lift coefficient, $c_{\ell i}$. One such shape acquired wide recognition through its application to laminar flow airfoils (described in Sect 5.2) developed by NACA in the late Thirties. This shape has a uniform chordwise load from $x/c = 0$ to $x/c = a$, and then decreases linearly to zero (Fig 6.6).

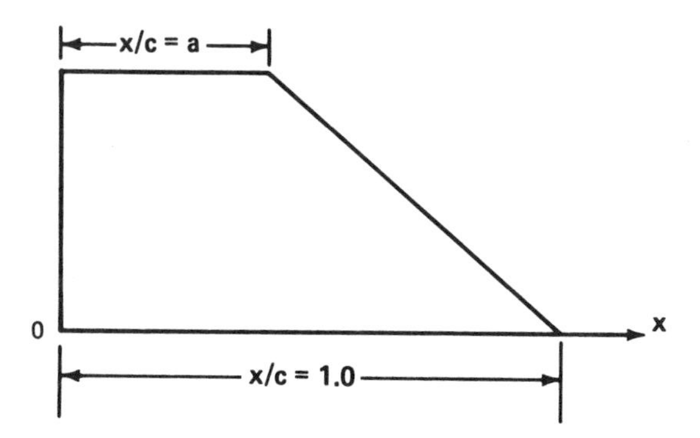

Figure 6.6 Partially-uniform chordwise load distribution

The expression for ordinates of the meanline assuring the type of load shown in Fig 6.6 at $c_\ell = c_{\ell i}$ is given as follows[1]:

$$\frac{y}{c} = \frac{c_{\ell i}}{2\pi(a+1)}\left\{\frac{1}{1-a}\left[\frac{1}{2}\left(a-\frac{x}{c}\right)^2 \ell n\left|a-\frac{x}{c}\right| - \frac{1}{2}\left(1-\frac{x}{c}\right)^2 \ell n\left(1-\frac{x}{c}\right) + \frac{1}{4}\left(1-\frac{x}{c}\right)^2 - \frac{1}{4}\left(a-\frac{x}{c}\right)^2\right] - \frac{x}{c}\,\ell n\,\frac{x}{c} + g - h\frac{x}{c}\right\}. \tag{6.25}$$

where

$$g = \frac{-1}{1-a}\left[a^2\left(\frac{1}{2}\,\ell n a - \frac{1}{4}\right) + \frac{1}{4}\right]$$

$$h = \frac{1}{1-a}\left[\frac{1}{2}\left(1-a\right)^2 \ell n\left(1-a\right) - \frac{1}{4}\left(1-a\right)^2\right] + g.$$

The ideal angle-of-attack for these meanlines is

$$a_i = \frac{c_{\ell i}\, h}{2\pi(a+1)} \tag{6.26}$$

At this point it should be recalled that the thin airfoil theory outlined in this section incorporates the linear assumptions reflected in Eqs (6.1) through (6.3), where it was assumed that the angle-of-attack of the incoming flow was small, and that the straight-line chord could be substituted for the curved mean line.

The advantage of linearization is obviously the ability to obtain new solutions through superposition of already existing ones. For instance, if a new chordwise loading can be broken down into simpler forms for which the corresponding meanline shapes have already been established, then the ordinates of the mean line needed for this new load distribution will be obtained by simply adding meanline ordinates associated with the simpler loading forms. However, the limitations of the linearized approach should also be realized. An inspection of Eqs (6.24) and (6.25) indicates that y/c ordinates are proportional to $c_{\ell i}$, which would also be true for other shapes of the chordwise loading. This means that as $c_{\ell i}$ values increase, the required mean lines would have more camber and thus, the assumption of $\ell_{m\ell} \approx c$ becomes questionable. Furthermore, Eq (6.26) in-indicates that the angle-of-attack corresponding to the increasing $c_{\ell i}$ would also increase, thus placing the small-angle assumptions in doubt. It may be expected hence, that for higher $c_{\ell i}$ values, the predictions of aerodynamic characteristics based on the linear thin airfoil theory may be less accurate.

4. THEORY OF FINITE–THICKNESS AIRFOILS IN AN IDEAL FLUID

4.1 Flow around a Cylinder with Circulation

The prime purpose of airfoils is the generation of lift. The simplest conceptual model explaining lift generation in a two-dimensional flow of ideal fluid is that of an infinite cylinder around which exists a circulation of strength Γ, while an infinite mass of fluid flows with a uniform velocity V in a direction perpendicular to the cylinder axis. As can be recalled from Eq (4.14), lift generated per unit length of such a cylinder will be $\ell = \rho V \Gamma$, and the flow pattern can be conveniently described through the use of the so-called complex potential (w):

$$w = \varphi + i\psi \tag{6.27}$$

where φ is the velocity potential, ψ is the stream function, and $i \equiv \sqrt{-1}$. Eq (6.27) as a whole can be represented as a function of the complex variable $z \equiv x + iy$;

$$w = f(z).$$

Once the above relationship is given, the velocity of flow at any point z can be obtained as

$$dw/dz = u + iv \tag{6.28}$$

where u is the velocity component in the x, and v in the y, direction.

In almost all textbooks one can find that the complex potential of uniform flow around a cylinder with circulation is as follows:

$$w(z) = V_\infty[z + (a^2/z)] + i(\Gamma/2\pi)\ell n\ z \tag{6.29}$$

where a is the cylinder radius.

By performing the differentiation indicated in Eq (6.28), the well-known flow patterns shown in Fig 6.7 can be obtained from Eq (6.29).

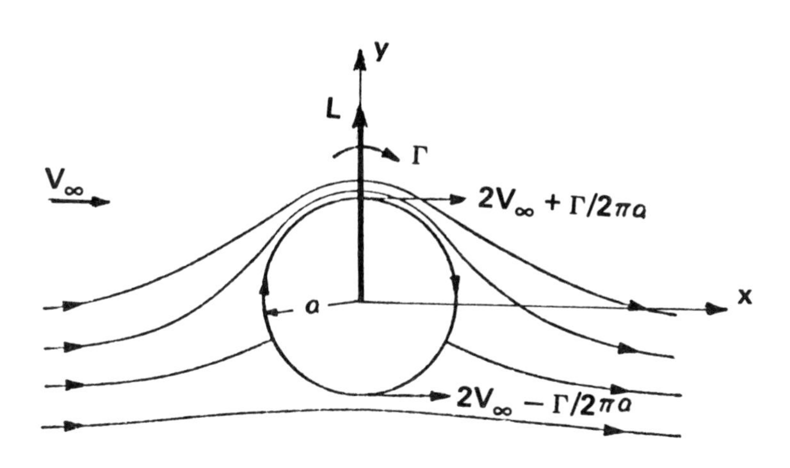

Figure 6.7 Flow pattern around a cylinder with circulation

In particular, at the top of the cylinder where $z = 0 + ia$, the resultant velocity $V_u = 2V_\infty + \Gamma/2\pi a$; and at the bottom where $z = 0 - ia$, $V_b = 2V_\infty - \Gamma/2\pi a$.

It should also be noted that so-called stagnation points are located at those elements of the cylinder surface where the incoming flow has only a radial velocity component.

Looking at Fig 6.7, one sees that streamlines are more closely concentrated at the top of the cylinder than at the bottom; thus indicating that the velocity of local flow in the upper region is larger, and smaller in the lower region, than that of the distant flow (V_∞). This means that according to the Bernoulli equation, the pressure at the top of the cylinder surface will be lower, and at the bottom, higher than in the distant fluid. Consequently, an upward-directed lift force is generated. However, due to the symmetry of flow with respect to the y axis, no drag is present in this scheme.

Lift Generation with Circular Airfoils. Attempts were made to reduce to practice the aerodynamic-force-generating scheme shown in Fig 6.7. This was done in the case of ships by replacing sails with rotating cylinders (Flettner). However, the work of Cheeseman[8] should be of direct interest to students of rotary-wing aircraft, since he attempted to make lifting rotor blades out of circular tubes where circulation was produced by blowing through properly located slots (Fig 6.8). This scheme appeared especially attractive for convertible aircraft, as the lift-generating ability of the blades would instantly be "killed" by simply switching off the blowers; thus eliminating the most difficult problem of stopping the blade in flight.

The concept of the cylinder with circulation has not found direct application to practical designs of lifting rotor blades because of the compressibility and viscous effects encountered in such real fluids as air. However, enormous indirect benefits to aviation in general, including rotary-wing aircraft, resulted from the theoretical exploration of the type of flow represented by the circular cylinder with circulation exposed to a uniform

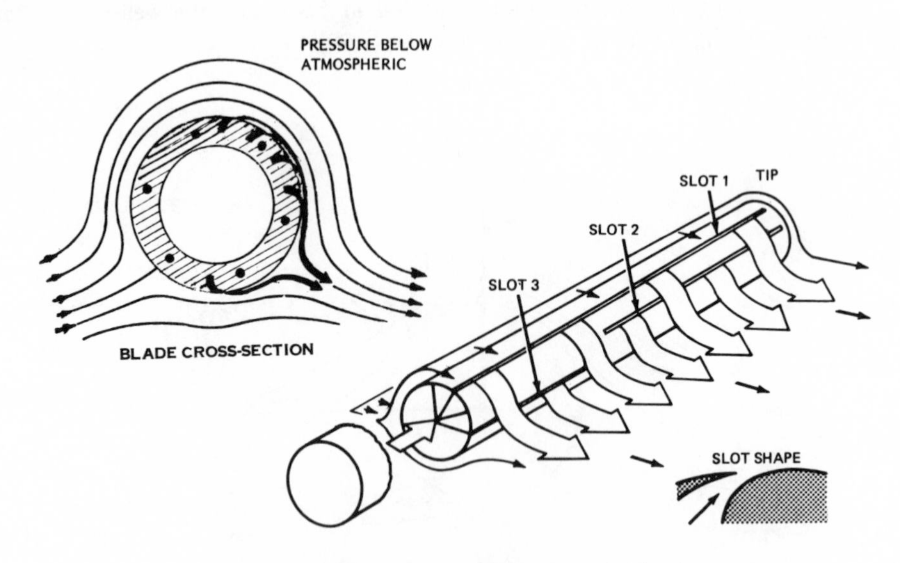

Figure 6.8 Cylindrical rotor blade with blowing

flow; as this concept formed the basis for the development of the finite-thickness airfoil theory. The initial step in this development was taken in the beginning of this century by Joukowsky who, by using the so-called conformal transformations, pointed out a way of transforming a circle into streamlined shapes which we recognize as airfoils, while the flow about a cylinder with circulation is also transformed into a corresponding flow around the airfoil.

4.2 Conformal Transformation

The conformal transformation consists basically of "methods whereby a geometrical field 1, characterized by an assemblage of points and lines may be transformed into another field 2, point by point, and line by line, in such a manner that the infinitely small element of area in field 1 shall transform, in field 2, into an element of similar geometrical form and proportions, while at the same time, the aggregate in field 2 may be quite different from that in field 1" (Durand, Ref 3, Vol 1, p 89).

Use of the Complex Variable. The conformal transformation can be performed either graphically or analytically. The analytical transformation is usually done with the help of functions of the complex variable.

The use of the complex variable can be considered as a method of representing a vector in a plane or determining the position of a point in the plane. For instance, in plane z, the position of point P (Fig 6.9a) can be determined by the value of z:

$$z = x + iy.$$

Similarly, in plane ζ, the position of point P' (Figure 6.9b) can also be determined by the value of ζ:

$$\zeta = \xi + i\eta.$$

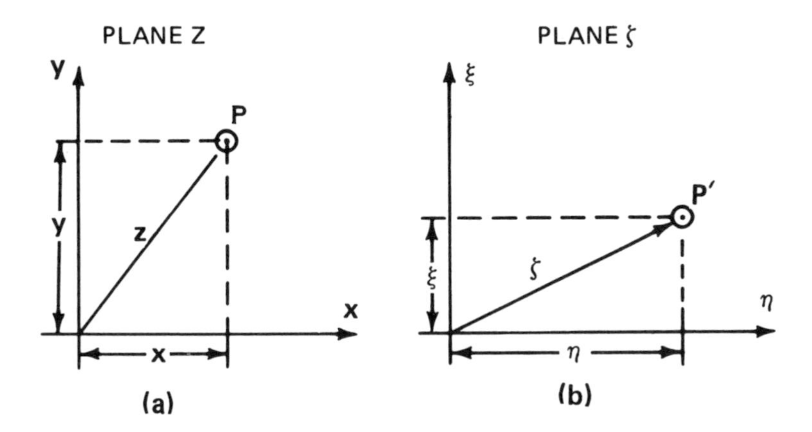

Figure 6.9 The two planes of conformal transformation

If the relationship between ξ and z is expressed as a real, continuous algebraic function,

$$\zeta = f(z), \tag{6.30}$$

then for an assembly of points in the z plane, a corresponding assembly of points in the ζ' plane can be found. Furthermore, under the above conditions, the requirement for conformal transformation is automatically fulfilled[3]. THis indicates that there may be a great variety of $\zeta(z)$ functions which can be used in conformal transformation.

Transformation of Circles into Airfoils. The importance of conformal transformation of a circle into an airfoil-like shape results from the fact that the velocity of flow tangent to the contour of the cylinder at some point P in plane z will also be tangent to the airfoil shape at the corresponding point P' in plane ζ (Fig 6.9). By the same token, stagnation points on the circle in plane z will also represent the stagnation points on the airfoil in plane ζ.

Since the location of the stagnation point on a circle representing a cross-section of a cylinder exposed to a uniform flow with velocity V_∞ depends on the strength of circulation, its value should be so-selected that the location of the aft stagnation point coincides with the trailing edge of the airfoil obtained in transformation. This automatically leads to the fulfillment of the Kutta-Joukowsky condition of a smooth flow at the trailing edge.

The actual transformation is accomplished by having two circles in plane z; one with radius a ($a = 1$ is assumed in Fig 6.10) having its center at the origin of coordinates, and another intersecting the smaller circle at two points. The transformation is obtained through the following Joukowsky transform formula:

$$\zeta = z + (a^2/z). \tag{6.31}$$

Now the small circle becomes a segment of a straight line $4a$ long (in this case, simply having four a-unit lengths). This segment is located along the ξ axis with its center at the origin of coordinates in the ζ plane.

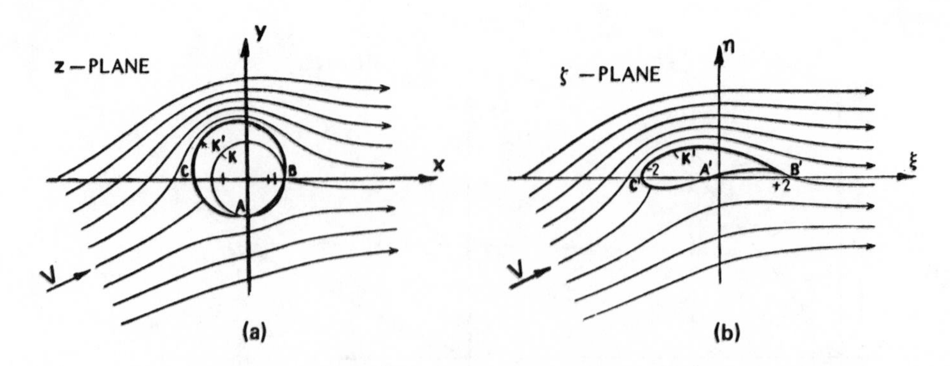

Figure 6.10 Joukowsky transformation of a circle into a profile

In order to transform the larger circle in the z plane and the flow around it into an airfoil and its corresponding field of flow, the following should be noted:

The complex potential for the flow around the larger circle, when the velocity at infinity is inclined at an angle a_o toward the x axis, can now be expressed (Ref 1, p 52) as

$$w = V\left[(z+\epsilon)e^{-ia_o} + \frac{(a+\epsilon)^2 e^{ia_o}}{z+\epsilon}\right] + \frac{i\Gamma}{2\pi}\,\ell n\,\frac{(z+\epsilon)e^{-ia_o}}{a+\epsilon} \tag{6.32}$$

where ϵ is the displacement of the larger circle from the center of coordinates.

Because of the nature of conformal transformation, the distant-flow velocity vector V_∞ will also form an angle a_o with the ξ axis in the ζ plane. The potential of the flow around the airfoil can be obtained by substituting $z(\zeta)$ from Eq (6.31) into Eq (6.32). However, it should be pointed out that this would lead to complicated expressions. A simpler way, indicated by Abbott and von Doenhoff, consists of selecting the values of z representing various points on the larger circle and finding the correspondng ζ points through the use of Eq (6.31). The velocities at these so-transformed points can be found by remembering that in analogy to Eq (6.28), velocities in the ζ plane can be expressed as

$$dw/d\zeta = (dw/dz)(dz/d\zeta).$$

As indicated by the above, Eq (6.32) is differentiated with respect to z, while $dz/d\zeta$ is obtained from Eq (6.31); thus,

$$dw/d\zeta = \left[V\left(e^{-ia_o} - \frac{(a+z)^2 e^{ia_o}}{(z+\epsilon)^2}\right) + \frac{i\Gamma}{2\pi(z+\epsilon)}\right]\left(\frac{z^2}{z^2-a^2}\right). \tag{6.33}$$

It can be seen from Eq (6.33) that for point $z = a$; i.e., point B in the z plane which transforms into the trailing edge (point B' in the ζ plane), the velocity of flow would be infinite—thus violating the Kutta-Joukowsky conditions—unless the expression in the square brackets of Eq (6.33) is zero. Writing that equality and solving the so-obtained equation for Γ, one would find that the circulation satisfying the Kutta-Jowsky condition is:

$$\Gamma = 4\pi(a+\epsilon)\,V \sin a_o. \tag{6.34}$$

It can be shown that as long as $\epsilon \ll a$, the chord of the airfoil will be $c = 4a$ (Fig 6.10). Remembering that lift per unit span is $\ell = \rho V_\infty \Gamma$ and substituting Eq (6.34) for Γ, the following expression for the section-lift coefficient is obtained:

$$c_\ell = 2\pi[1 + (\epsilon/4a)] \sin a_o. \tag{6.35}$$

Under the small-angle assumption, $\sin a_o \approx a_o$, and it can be seen from Eq (6.34) that the lift-curve slope is somewhat higher than $2\pi/rad$, as obtained in the thin airfoil theory. It should be noted however, that ϵ is related to the airfoil thickness as follows[1]: $(t/c) \approx (3\sqrt{3}/4)(\epsilon/a)$. This means that for a 12-percent thick airfoil, the theoretical lift-curve slope would only be 2.5 percent higher than for the zero-thickness section.

Since the above-discussed transformation gives not only the shape of the airfoil section, but also the velocity of flow about its contours, pressure distribution at the airfoil contour can be computed using Bernoulli's equation. Having obtained these pressures, lift and moment coefficients can be calculated. Numerous comparisons of measured and predicted pressure distributions of airfoils transformed from circles usually indicated a good agreement between theory and tests, at least for unstalled conditions.

Flow Around Arbitrarily-Shaped Airfoils. A problem opposite to that discussed on the preceding pages is of special interest to aerodynamicists since it represents the prediction of the flow pattern and hence, the pressure distribution around any arbitrarily-shaped airfoil section. Given such knowledge, it would be possible to study, without recourse to tests, the lift and moment characteristics of an airfoil section which appears promising on paper. The task would be easy if a function(s) transforming the circle into this given airfoil were known. Consequently, the attack on solving this inverse problem may proceed along lines tracing the works of von Karman and Treffetz (Ref 3, Vol II, p 80), Theodorsen[9, 10], and others.

In all of these approaches, a given airfoil is first transformed into a curve of nearly circular shape. Then another attempt is made to distort this shape into a circle or at least, as closely as possible approximating a circle. Once these functions of transformation have been established, the flow around the circle can be transformed into that around the airfoil; thus obtaining the required pressure distribution.

Even more interesting would be the task of finding the shape of an airfoil section which would produce a given (desired) pressure distribution. This problem was briefly discussed for thin airfoils in the preceding section. However, for airfoils of finite thickness, the first satisfactory solutions were found in the prewar years and much progress was made during World War II[11]. Here, again, the desired solutions were obtained through the method of conformal transformation, which proved once more its aerodynamic usefulness.

5. EFFECTS OF VISCOSITY

5.1 Shearing Forces in Gases

Transfer of Momentum. In the preceding sections of this chapter, air was treated as an ideal, non-viscous, incompressible fluid. But in reality, air is viscous, and bodies moving through it are subjected to tangential (shearing) forces caused by viscosity. The existence of viscous (shearing) forces in gases can be explained by the transfer of momentum by gas particles traveling between layers of gas moving at relatively different speeds.

When the flow is turbulent—visually detectable through smoke or other means—small masses of air are moving in a direction perpendicular to the main flow. Hence, it is obvious that if there is a difference in speed between the adjacent air layers, then these small traveling masses provide a means for exchange of momentum, resulting in shearing forces.

However, when the flow is laminar, there are no visually detectable masses of air traveling between the adjacent layers which, in the case of differences in speed between them, can be imagined as sliding on each other like solid bodies. But even in this case, shearing forces are still present. Their existence can be explained by the kinetic theory of gases which shows that the momentum is transferred by free molecules of gas which are exchanged between the gas layers moving at different relative speeds (Ref 12, pp 156-184).

It is clear that the necessary condition for the existence of shearing forces in gases is the difference in speeds of gas layers or, in other words, the presence of a velocity gradient across the flow.

It may be assumed that the magnitude of the shearing forces per unit area (τ) is proportional to the velocity gradient for both laminar and turbulent flow:

$$\tau = \mu(\partial u/\partial y). \tag{6.36}$$

The coefficient of proportionality μ is known as the coefficient of viscosity. For air at standard sea-level conditions, $p_o = 101{,}320\ N/m^2$, $T_o = 288.2\,K$ ($p_o = 14.7\ psi$, $T_o = 518.7^\circ R$), and μ becomes

SI $\qquad \mu_o = 1.79 \times 10^{-5}\ kg/ms$

Eng $\qquad \mu_o = 3.37 \times 10^{-7}\ slugs/ft\ sec.$

Dividing μ by the air density ρ, we obtain the so-called kinematic coefficient of viscosity ν whose value under sea-level conditions, where $\rho_o = 1.225\ kg/m^3$, is

SI $\qquad \nu_o = \mu_o/\rho_o = 1.46 \times 10^{-5}\ m^2/s$

Eng $\qquad \nu_o = \mu_o/\rho_o = 1/6380\ ft^2/sec.$

Reynolds Number. It was shown that such aerodynamic forces as drag may be present due to the fluid viscosity. However, it may be anticipated that not only drag, but other aerodynamic forces as well may be influenced in some way by viscosity. Once such an assumption is made, it can be proven (by means of dimensional analysis) that the magnitude of the aerodynamic forces and hence, of the nondimensional coefficients C_L, c_ℓ, C_D, c_d, etc., will in turn depend on the value of a special ratio called the Reynolds number (e.g., Ref 3, Vol I, p. 30).

$$R_e = \rho \ell V/\mu, \tag{6.37}$$

where V is the velocity of air, ℓ is a characteristic linear dimension of the body, ρ is the air density, and μ is the coefficient of viscosity. For such bodies as wings or blades, the chord length c represents the characteristic dimension as far as flow is concerned. In addition, remembering that $\mu/\rho = \nu$, Eq (6.37) can be rewritten as follows:

$$R_e = Vc/\nu, \tag{6.37a}$$

For sea-level conditions, using the SI units of m and m/s, Reynolds number becomes

$$R_{e_o} = 68{,}493\ Vc. \tag{6.37b}$$

or, in English units, ft and ft/sec,

$$R_{e_o} = 6380\ Vc.$$

The dependence of aerodynamic forces on Reynolds number is also known as *scale effect*.

5.2 Boundary Layer

Definition. When air is flowing past a body, the air layer immediately adjacent to the body has zero speed (adhesion), while at some distance from the body surface, the velocity reaches its full magnitude U (Fig 6.11). This means that within this layer of air where the speed grows from zero to U, there is a velocity gradient $\partial u/\partial y > 0$ and hence, according to Eq (6.36), shearing forces exist; but their presence is limited to this layer only, since outside of it $\partial u/\partial y = 0$, and the flow may be considered as inviscous. This layer of air, where the speed grows from zero to its full value and where all shearing forces acting on a body are generated, is generally called the boundary layer or sometimes, Prandtl's layer. It is usually very thin in comparison with the characteristic dimensions of the body. Consequently, in most practical cases, the existence of the boundary layer may be ignored when determining the potential flow field around a body.

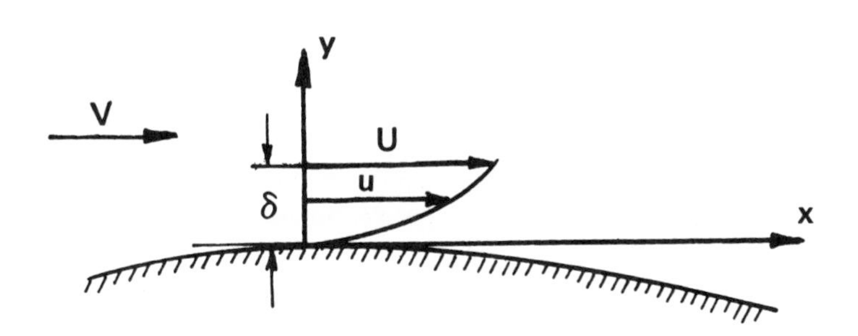

Figure 6.11 Boundary layer scheme

Only a few topics referring to boundary layer will be discussed here. A deeper insight into basic aspects of boundary-layer pheonomena can be obtained from such books as those of Abbott and v.Doenhoff[1] and Schlichting[13], while references to both theory and application (until the early Sixties) can be found in a text edited by Lachman[14], and the Proceedings of the NYAS Congress on Subsonic Aeronautics[15].

Boundary Layer of a Flat Plate. As an illustration of boudary-layer phenomena, the flow past a flat plate will be considered (Fig 6.12).

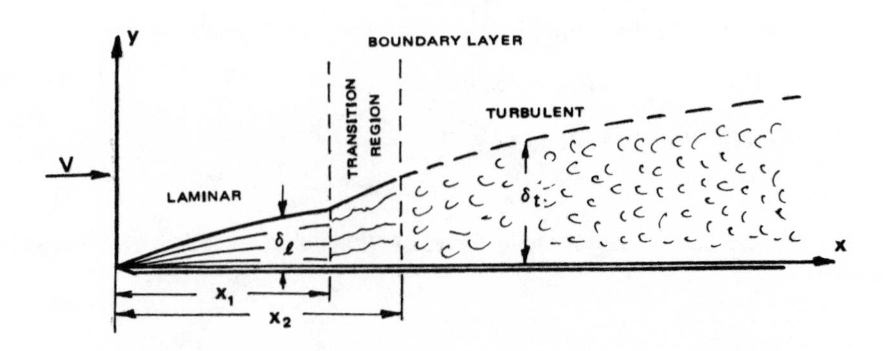

Figure 6.12 Development of boundary layer along a flat plate

The flow in the boundary layer at the leading edge is laminar and the boundary layer is relatively thin. But moving downstream, the boundary layer thickness, δ_ℓ, grows, as can be seen from Blasius' expression for this state of flow:

$$\delta_\ell = 5.48x/\sqrt{R_x}, \tag{6.38}$$

where x is the distance measured from the leading edge, and R_x is the local Reynolds number, $R_x = Vx/\nu$.

At some distance from the leading edge; say, between points x_1 and x_2, laminar flow begins to break down and beyond point x_2, the flow in the boundary layer is fully turbulent.

The thickness of the boundary layer in the turbulent state, δ_t, can be expressed by von Karman's formula,

$$\delta_t = 0.377x/\sqrt[5]{R_x}, \tag{6.39}$$

where all symbols are the same as those in Eq (6.38).

The mechanics of transition from laminar to turbulent flow and the causes producing it were not entirely clear until Dryden's studies[16]. Prior to this, there were two different schools trying to explain the physics of transition. Representatives of the so-called Gottingen School believed that infinitesimal disturbances may cause a transition by developing into unsteady wave motions at some Reynolds numbers (laminar boundary-layer oscillations). Adherents of the British school tried to explain the transition as the result of turbulence already present in the free airstream.

Experimental results indicate that depending on the circumstances, the causes singled out by both schools may start the transition. In addition, such factors as surface roughness or an adverse pressure gradient also facilitate the transition. On the other hand, transition may be delayed by such actions as removing a part of the boundary-layer air, favorable pressure gradient, etc. For a better understanding of the origin of turbulence and the nature of transition, the reader is directed to Schlichting's considerations (Ref 13, pp 308-367) and Tollmien's & Grahne's analysis (Ref 14, pp 602-636).

Aside from transition, separation is another very important phenomenon of boundary-layer mechanics. According to Tetervin[17], "the separation point is the point on the surface of the body at which the surface friction is zero. Upstream of the point, the direction of flow in the boundary layer next to the surface is downstream; and downstream of the point, the direction of flow in the boundary layer next to the surface is upstream" (Fig 6.13).

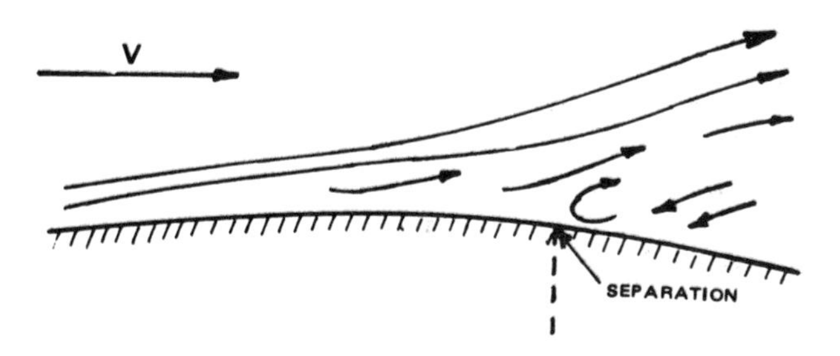

Figure 6.13 Boundary layer separation

In the presence of an adverse pressure gradient (pressure increasing downstream), both laminar and turbulent boundary layers will eventually separate. However, the turbulent boundary layer would show a greater degree of stability than the laminar one. This is because there is a greater mixing of air between the layers in turbulent flow and hence, more particles of the outer layers having a higher kinematic energy will penetrate the inner layers, thus increasing their momentum; i.e., the ability of air particles to move against the increasing pressure without appreciably slowing down.

Skin Friction Drag. Knowing the velocity gradient within the boundary layer; i.e., the so-called velocity profile, we can calculate the friction drag of a plate and other bodies such as wings, etc.

Since the friction drag depends on the contact area between body and fluid, it seems logical to use the wetted area rather than the projected or cross-sectional area when calculating this type of drag. Hence, the friction drag coefficient C_f is defined as

$$C_f = D_s/qA, \tag{6.40}$$

where D_s is the skin friction drag, q is the dynamic pressure of the flow ($\frac{1}{2}\rho V^2$) far from the body, and A is the wetted area. In the case of airfoils, the wetted area will obviously be slightly larger than two times the projected area.

The friction drag coefficient of a laminar flow will be different from that of the turbulent one because the velocity profile for the turbulent boundary layer differs markedly from the velocity profile of the laminar boundary layer (Fig 6.14). "At large boundary-layer Reynolds number the turbulent velocity profile shows an extremely rapid rise in velocity in a very short distance. This large slope at the wall causes the turbulent skin friction coefficient to be higher than the laminar skin friction coefficient."[17]

Blasius gives the following expression for the skin drag coefficient of the laminar flow:

$$C_f = 1.33/\sqrt{R_e} \tag{6.41}$$

where R_e is the Reynolds number as given in Eq (6.37b).

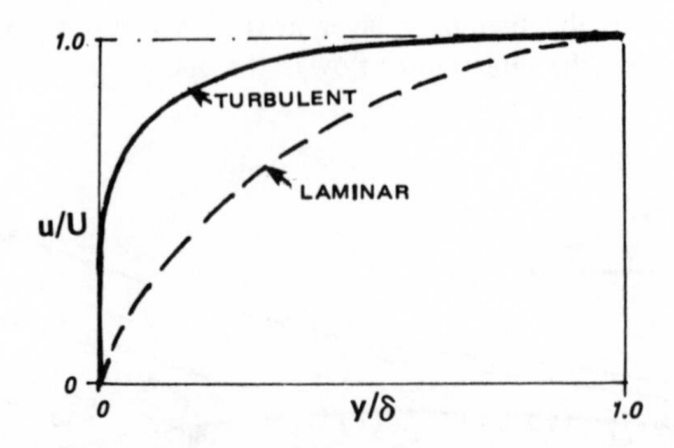

Figure 6.14 Profiles of laminar and turbulent boundary layers

In the case of the turbulent boundary layer, the best known expressions for C_f may be credited to Falkner:

$$C_{f_t} = 0.0306/\sqrt[7]{R_e}, \tag{6.42}$$

and von Karman:

$$C_{f_t} = 0.072/\sqrt[5]{R_e}. \tag{6.43}$$

For instance, for $R_e = 6 \times 10^6$, Eq (6.42) gives $C_{f_t} = 0.0033$; and for Eq (6.43) gives $C_{f_t} = 0.0032$.

When friction coefficients for laminar and turbulent flow in the boundary layer are plotted against Reynolds number (Fig 6.15), it becomes obvious that for the same R_{e_x} values, the drag coefficient for turbulent flow is much higher than the drag coefficient for laminar flow.

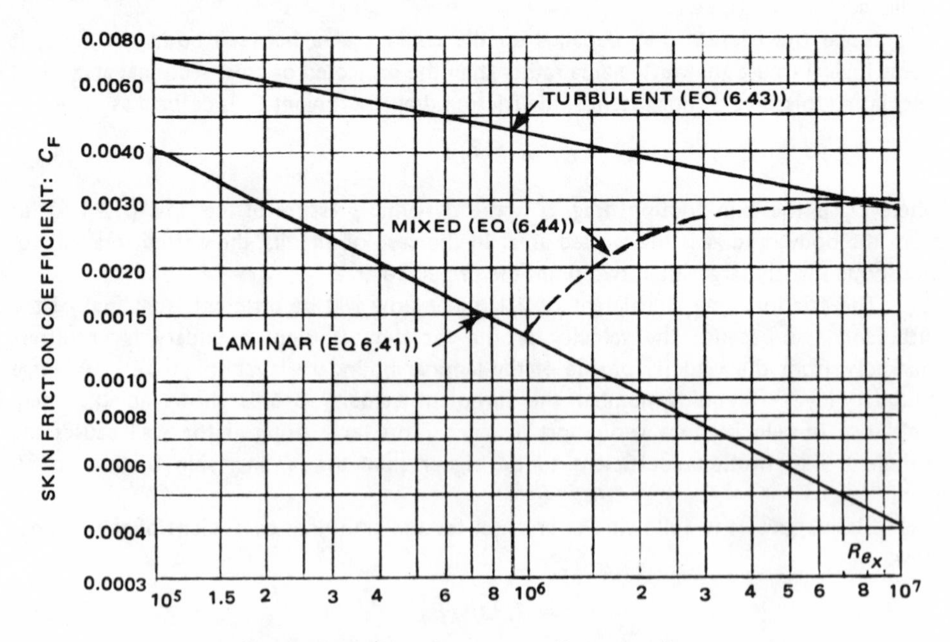

Figure 6.15 Skin friction coefficient vs R_{e_x}

A transition region where the flow is mixed—partially laminar and partially turbulent—often exists in the flow around the wings and other bodies. Assuming that transition occurs at $R_e = 10^6$, Clancy[5] proposes the following formula for C_f in the mixed flow:

$$C_f = (0.0306/\sqrt[7]{R_e}) - 2924/R_e. \tag{6.44}$$

Values of C_f computed from Eqs (6.41), (6.43), and (6.44) are shown in Fig 6.15.

Large differences between the skin friction coefficients for laminar and turbulent flows became apparent shortly after the presentation of the boundary-layer theory by Prandtl in 1903. Consequently, various attempts have been made to obtain laminar flow over the largest possible areas of wings and other bodies exposed to the air flow.

Removal of the Boundary-Layer Air. One of the long-recognized possible means of delaying transition—thus reducing the drag—is removal of the boundary layer by means of suction through slots, porous surfaces, or perforated skin. In spite of promising possibilities[15]; as yet, the application of this method of drag reduction to practical fixed-wing aircraft has met with little success. The task of preventing transition in rotary-wing aircraft may be even more difficult, since the tendency of the laminar flow to break down increases with the turbulence level of the oncoming fluid. This is especially true in forward flight where the rotor blade usually encounters increased air turbulence caused by the vorticity left by all rotor blades, as well as hubs and other components.

Favorable Pressure Gradient. The creation of a favorable pressure gradient as demonstrated by the use of low-drag, or laminar, airfoil sections probably represents the most practical and widespread method of obtaining laminar flow over large sections of an airfoil. In order to better understand this approach, the mechanics of transition on conventional sections will be discussed first.

At positive lift coefficients, conventional airfoils exhibit a pressure distribution with a high suction area at the leading edge, followed by a marked drop in suction; ie, increase in pressure in the downstream direction (Fig 6.16). This means that the air particles in the boundary layer on the top of the airfoil have to move against the increasing pressure, or in other words, against an adverse pressure gradient, which facilitates transition.

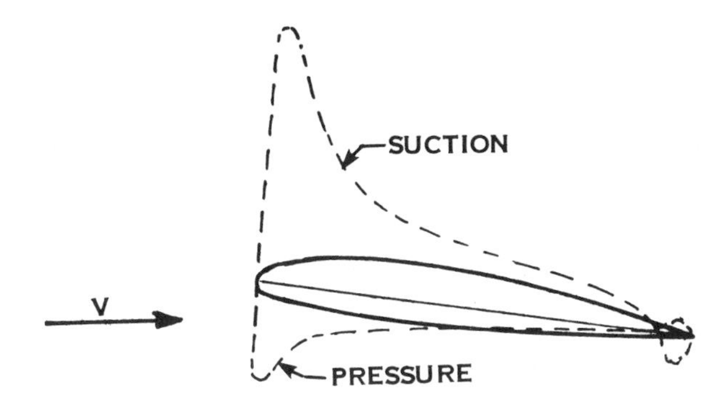

Figure 6.16 Pressure distribution over a conventional airfoil at a high angle-of-attack

In order to obtain a low-drag airfoil section; ie, an airfoil with laminar flow extending as far downstream as possible, it is necessary to create a decreasing pressure downstream or, at least to hold it constant for a large part of the chord. This has been achieved through the design of airfoil sections having a desirable pressure distribution (Sect 3.2 of this chapter.

It should be stressed that it is usually impossible to obtain a favorable pressure distribution over an airfoil section for the entire range of practical values of c_ℓ. Therefore, the low drag is limited to a certain range of c_ℓ and outside this region, laminar-flow airfoil sections may exhibit even higher profile drag coefficients than conventional sections (Fig 6.17(a)). It should also be mentioned that low-drag airfoils are extremely sensitive to such factors as surface roughness, deviation of the real shape from the theoretical one (manufacturing tolerances), and waves and notches. In this respect, conventional airfoil sections show more desirable characteristics (Fig 6.17(b)). However, it should be noted that it is possible to obtain a forced transition of the whole boundary layer of both low-drag and conventional airfoils by making the leading edge rough; for instance, by covering it with a material such as carborundum. This procedure is used for obtaining airfoil characteristics of so-called standard roughness levels. More information about various aspects of roughness is contained in Vol II, Ch 3, Sect 1, *Drag Estimate.* The reader is also referred to Ref 14, pp 637-747, where Tani discusses the effects of two-dimensional and isolated roughness on laminar flow; v. Doenhoff and Braslow consider the problem from the point-of-view of distributed surface roughness; and finally, Coleman analyzes the influence of roughness due to insects. Specific data on the effect of roughness on various types of airfoil sections can be found in Ref 1, Ch 7.

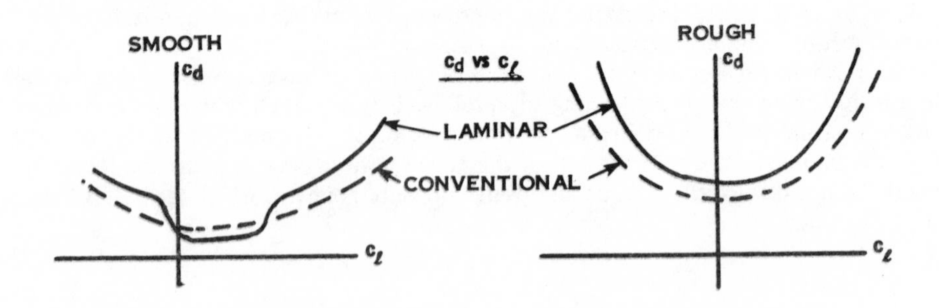

Figure 6.17 Character of the drag polar for conventional and laminar airfoils

Stall. Stall—a condition that manifests itself as a more or less sudden decrease in lift associated with increasing angle-of-attack and a large increase in profile drag (Fig 6.18)—is a phenomenon caused by separation of the boundary layer when the adverse pressure gradient becomes too high.

Three basic types of stall are recognized: (1) Trailing-edge stall that occurs when the flow begins to separate at the trailing edge, and with increasing angle-of-attack, the separation gradually progresses toward the leading edge (Fig 6.18(a)). This is considered as a gentle type of stall and more desirable than other types, since the lift decreases gradually from its maximum value. (2) Leading-edge stall which begins as a short bubble is formed in the neighborhood of leading edge. When it bursts, a rapid change of flow over the upper surface of the airfoil occurs, resulting in both a sudden drop in lift and an increase in the profile drag. (3) Thin-airfoil stall, which begins with a long "stable" bubble which elongates gradually and eventually bursts (Fig 6.18(c)).

The type of stall is strongly influenced by the geometry of the front part of the airfoil section within 10-15 percent from the leading edge; the most important factors being the shape of the mean-line curve between 0 and 15 percent c, and the leading-edge radius [18,19].

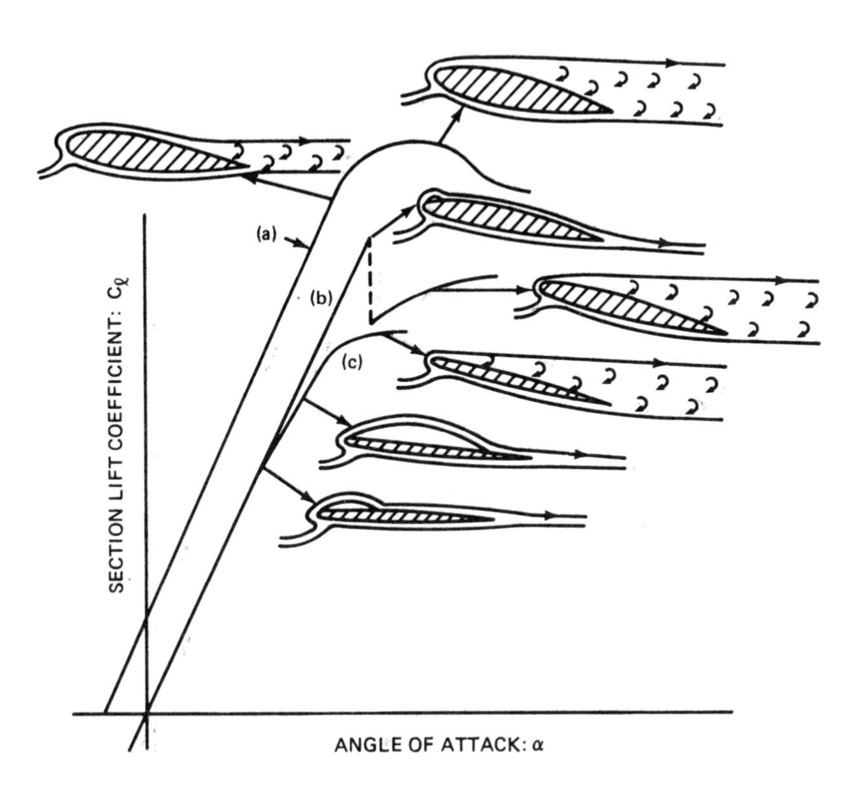

Figure 6.18 Three types of stall: (a) trailing-edge, (b) leading-edge, and (c) thin-airfoil

The airfoils should be shaped so that the tangent to the mean line at the trailing edge would make the smallest possible angle with the streamline of the incoming flow. The leading-edge radius should be large enough to facilitate the flow from the stagnation point located on the lower surface of the airfoil (Fig 6.18(a)) to the upper surface without the high centrifugal accelerations which would enhance the flow separation in this area. It becomes obvious that, in general, thick airfoil sections which usually have large leading-edge radii would be less likely to exhibit the leading edge stall than thin sections; especially, symmetric ones having no favorable camber in the leading-edge area.

The loss of lift of stalled airfoils can be explained through the reasoning based on the Kutta-Joukowski law (Eq (4.11)):

When the boundary layer separates, the character of flow around the airfoil, as sketched in Fig 6.19(a), changes to that shown in Fig 6.19(b). It is obvious from Fig 6.19(b) that the circulation decreases around an airfoil in the stalled position. Between points A–B in Fig 6.19(a), there is always some velocity component tangent to the path along which the circulation is computed.

When stall occurs, the length of the A–B line in Fig 6.19(b) passes through a turbulent wake which contributes little or nothing to the circulation. Thus, the circulation around the stalled airfoil is reduced and consequently, lift is decreased.

The boundary-layer separation on airfoils (stall) and other aerodynamic shapes can be prevented, or at least delayed, through both suction and blowing.

Removal of the boundary-layer air through distributed suction may be cited as a means of preventing flow separation of airfoils; thus leading to higher values of the maximum lift coefficients. The technical feasibility of this approach was demonstrated long ago by Raspet and his followers[20]. However, such practical aspects as icing and dust-clogging of the suction passages represent some of the obstacles to operational applicability of this concept.

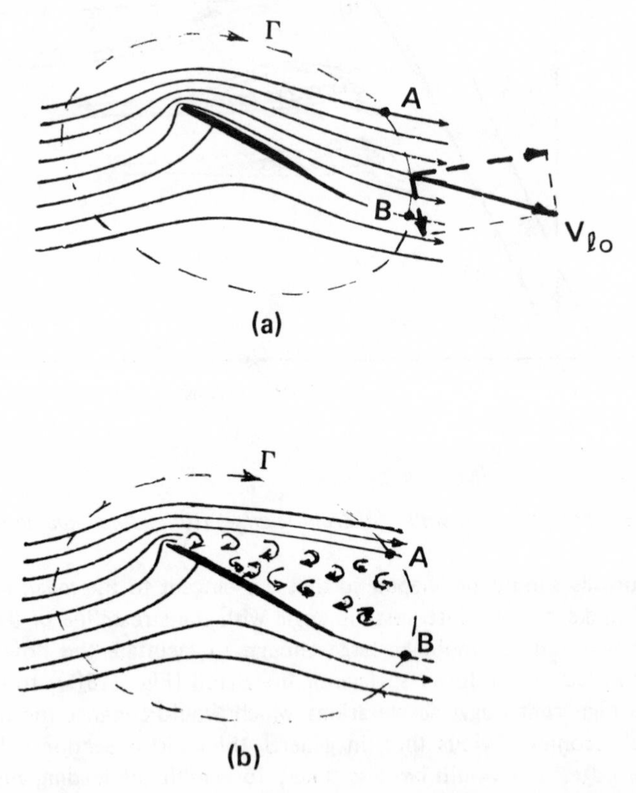

Figure 6.19 Circulation around an airfoil just before and after leading-edge stall

Because of the above-mentioned operational disadvantages, the application of suction to rotor blades attracts little interest at the present time. It should be realized, however, that lift augmentation may become an important means for improving maximum speeds of helicopters by delaying stall of the retreating blade, and allow operation at higher average lift coefficients in hover, thus leading to a reduction in the rotor blade area which, in turn, would contribute not only to a reduction in the power required in hovering, but also to an improvement of the rotor L/D_e values in forward flight.

For this reason, other means of delaying boundary-layer separation by energizing it through blowing should not be neglected since they appear—at least in principle—better suited for practical applications. The feasibility of blowing as a means of stall delay and $c_{\ell_{max}}$ augmentation has been demonstrated in full-scale wind-tunnel rotor tests[21].

Readers more interested in the whole field of blade boundary-layer control are directed to the work of Wuest (Ref 14, pp 196-208), and Schlichting's analysis (Ref 13, pp 229-241) as far as suction aspects are concerned. In Ref 14 (pp 209-231), Carriere and Eichelbrenner present a theory of flow reattachment by a tangential jet discharged against a strong adverse pressure gradient.

Scale Effects. Wind-tunnel tests of airfoils are seldom performed at exactly the same, or even similar, Reynolds number values as would occur in actual flight. Consequently, a methodology is required to predict the R_e scale effects on such airfoil characteristics as profile drag coefficient, $c_d(R_e)$; maximum lift coefficient values, $c_{\ell_{max}}(R_e)$; lift-slope levels, $a(R_e)$; and pitching moment coefficient about a geometrically-fixed point; say $c_{m_{c/4}}(R_e)$, since the position of the aerodynamic center may depend on R_e. Detailed procedures for dealing with these effects can be found in Ref 1, Ch 7, and in Ref 7, Sect 1.2.60. It should be noted that up to the incipience of stall, it may be assumed—for practical purposes—that $a(R_e) = const$; while $c_{m_{c/4}}(R_e)$ varies little with R_e. Consequently, only the two most pronounced scale effects for conventional airfoils will be touched upon to illustrate the trend; ie, $c_{\ell_{max}}(R_e)$ and $c_d(R_e)$. Some basic relationships relative to this subject were established by Jacobs and Sherman[22]. Most of the conventional airfoil characteristics were given by NACA at the standard Reynolds number, $R_{e_{std}} = 8 \times 10^6$. The maximum section-lift coefficient, $c_{\ell_{max}}$, at R_e other than 8×10^6 can be found from that obtained at the standard Reynolds number by adding or subtracting a correction term, $\Delta c_{\ell_{max}}$:

$$c_{\ell_{max}} = (c_{\ell_{max}})_{std} + \Delta c_{\ell_{max}}. \tag{6.45}$$

The value of $\Delta c_{\ell_{max}}$ for different types of airfoil sections with various R_e values can be found in graphs such as those found in Ref 22. The trend for $c_{\ell_{max}}(R_e)$ for a symmetrical (NACA 0012) and cambered (NACA 23012 and 13021) airfoils is illustrated in Fig 6.20, which shows that symmetrical airfoils are much more sensitive to R_e values than cambered ones as far as $c_{\ell_{max}}$ is concerned.

When the maximum section-lift coefficient, $c_{\ell_{max}}$ of an airfoil at a given R_e is known, the section-drag coefficient c_d at different c_ℓ values can be estimated as follows:

$$c_d = c_{d_{min}} + \Delta c_d \tag{6.46}$$

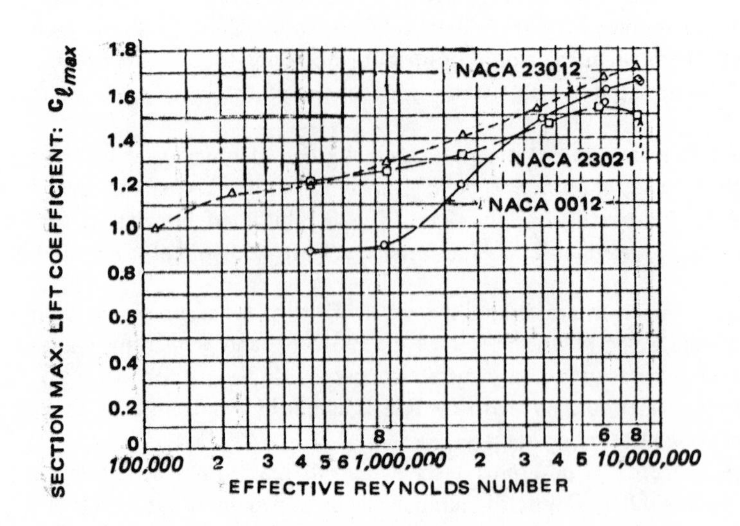

Figure 6.20 Example of $c_{\ell_{max}}(R_e)$ for symmetrical and cambered airfoils

where $c_{d_{min}}$ represents the profile-drag component whose value depends directly on R_e, while Δc_d is the increase in profile drag which depends on the relation of the actual c_ℓ to the $c_{\ell\,opt}$ and $c_{\ell max}$, where $c_{\ell\,opt}$ is the c_ℓ corresponding to $c_{d_{min}}$.

For many conventional airfoil sections, the variation of $c_{d_{min}}$ with R_e can be expressed as

$$c_{d_{min}} = (c_{d_{min}})_{std}(8 \times 10^6/R_e)^{0.11} \tag{6.47}$$

where $(c_{d_{min}})_{std}$ is the minimum profile drag coefficient obtained at standard $R_e = 8 \times 10^6$, and R_e represents the value for which $c_{d_{min}}$ is calculated. However, it must be pointed out that Eq (6.47), although usually correct for R_e greater than 8×10^6 and conventional airfoils, is not always true for $R_e < 8 \times 10^6$ even for conventional airfoils, and is seldom correct for laminar sections. In these two cases, in finding $c_{d_{min}}$ at a given R_e, it is always better to refer to experimental data obtained directly at the desired R_e. Fig 2.5 in Vol II shows the trend of $c_{d_{min}}(R_e)$ for several airfoils suitable for rotary-wing applications.

The correction factor, Δc_d, shown in Fig 6.21, can be expressed as a function of the following ratio:

$$(c_\ell - c_{\ell_{opt}})/(c_{\ell_{max}} - c_{\ell\,opt}),$$

where c_ℓ is the section-lift coefficient for which Δc_d is computed, $c_{\ell_{max}}$ is the maximum lift coefficient at a given R_e, and $c_{\ell_{opt}}$ is the c_ℓ at which c_d reaches its minimum value.

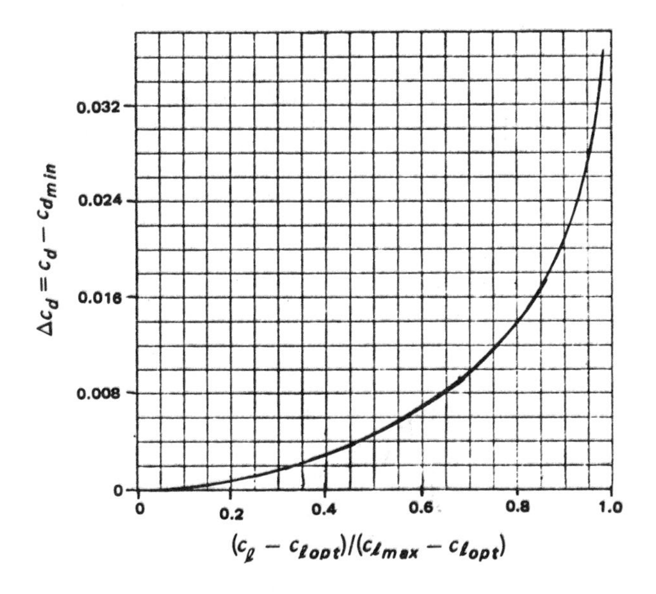

Figure 6.21 Determination of the Δc_d correction term in Eq (6.45)

Three-Dimensional Boundary Layer. The three-dimensional boundary layer is characterized by the presence of a cross-flow, which can be defined as a flow in the boundary layer in a direction normal to that of the outer streamlines.

Yawed and swept wings, even when exposed to a steady flow, develop a considerable cross-flow (especially on the suction side), and thus the boundary layer can no longer be treated as two-dimensional. It may be expected hence, that for the helicopter rotor in forward flight, or for any other airscrew experiencing an inplane velocity component, flow conditions and thus, the boundary-layer behavior on the blade itself will be three-dimensional and, probably, more complicated than for yawed or swept fixed wings. This assumption is based on the fact that in addition to the time-variable blade yaw angle with respect to the incoming flow, the flow itself is also time-dependent. Finally, the boundary layer is subjected to the centrifugal acceleration from the rotation about the rotor axis.

Despite the possibility that the three-dimensional phenomena occurring in the blade boundary layer may have a strong influence on rotor performance in forward flight, theoretical and experimental knowledge on this subject is quite limited. This is in contrast with the fixed-wing case where a considerable amount of literature exists on this subject; for instance, Schlichting's considerations of the phenomena associated with a yawed cylinder in steady motion (Ref 13, pp 169-179); studies of Linfield et al, of approximate methods for calculating three-dimensional boundary-layer flow on wings (Ref 14, pp 842-912); Brown's investigations of stability criteria for three-dimensional boundary layers (Ref 14, pp 914-923); and Eichelbrenner's paper on three-dimensional boundary layers (laminar and turbulent) where an attempt was made to develop a unified, three-dimensional, boundary-layer theory[15].

However, initial steps toward a better understanding of the behavior of the boundary layer of rotary-wings have already been taken. In that respect, it should be mentioned that three-dimensional effects in the blade boundary layer were studied, both experimentally and analytically, in the late Sixties or early Seventies by Tanner and Yaggy[23], Velkoff and Blaser[24], Dwyer and McCroskey[25], and others.

McCroskey's measurements of boundary-layer transition, separation, and streamline direction on rotating blades are especially interesting[26]. "For the suction surface of typical helicopter rotors, the transition from laminar to turbulent flow is dominated by the chordwise adverse pressure gradient and a conventional laminar separation bubble, rather than by Reynolds number, rotational, or cross-flow effects. The cross-flow in the unseparated boundary layer is small with respect to the undisturbed potential flow streamlines, whether the flow is laminar or turbulent. Therefore, the boundary-layer flow on a wide class of rotating blades bears a marked resemblance to its counterpart on conventional wings.

"There are, of course some significant differences in rotor and fixed-wing boundary layers. Though unimportant in a practical sense, centrifugal forces move the fluid significantly outward in separated regions such as the laminar separation bubble or in trailing-edge separated flow. On the more practical side, high-aspect ratio helicopter rotors in forward flight and low aspect ratio propeller blades are known to develop much greater thrust without stalling than would be expected on the basis of two-dimensional boundary-layer behavior. This investigation completely ignored any unsteady effects, and little was learned about the actual mechanisms of stall on the rotating blades.

"A better understanding was gained into the laminar flow on contemporary helicopter rotors as well as some initial insight into the nature of turbulent flow. Purely rotational effects appear to be relatively unimportant, and the major three-dimensional and unsteady effects on the stall characteristics of actual rotors probably occur in the turbulent and separated regions of the boundary layer.

"Since the laminar flow on the upper surface separates so near the leading edge for large α, the correct flow model apparently is a small region of essentially two-dimensional quasi-steady laminar flow, followed by a three-dimensional unsteady turbulent flow. This turbulent boundary layer has its starting or initial conditions in the x-direction determined by a classical short separation bubble, and its subsequent behavior is predominantly influenced by the chordwise pressure distribution. Another important feature is that the boundary layer cross-flow is largely determined by the spanwise flow at the outer edge of the boundary layer, which is approximately $V_\infty \cos \psi$ in high-speed forward flight. This new knowledge, it is hoped, will serve as a useful guide for further theoretical and experimental investigations."

6. COMPRESSIBILITY EFFECTS

6.1 Physical Aspects of Compressibility

Speed of Sound. Compressibility effects stem from the physical fact that in such real fluids as air, the velocity of the propagation of pressure signals (sound) is finite. The mechanism of pressure propagation relies on the collision of molecules. Thus, it may be expected that speed of sound (s) would be proportional to the mean value of the speed of molecular motion ($\bar{C}$). In this respect, the kinetic theory of gases[12] indicates that

$$s = 0.742\bar{C}. \tag{6.48}$$

where $\bar{C}$ is the mean value of the speed of the molecules.

On the other hand, since the mean velocity of the molecules of a given gas depends only on its absolute temperature, it is evident that the speed of sound in air is a function of the absolute temperature only. In fact, the speed of sound in air can be expressed (in m/s) as

$$s = 20.05\sqrt{T} \tag{6.49}$$

where T is the absolute temperature in K; or in mph, as

$$s = 33.35\sqrt{T} \tag{6.49a}$$

where T is in °Rankine.

Airfoils in Compressible Medium. Many papers (e.g., Ref 27) deal with airfoils suitable for high subsonic speeds. However, here, only major aspects of compressibility effects are discussed. For freestream velocities smaller than $M = 0.4$ where M is the Mach number of the distant flow, compressiblity effects can usually be ignored, but as the freestream Mach number becomes higher, the aerodynamic characteristics of bodies, including airfoils, undergo changes which can be better clarified by the introduction of such terms as critical Mach number, subcritical and supercritical speed, and supersonic speed.

When the flow over some part of an airfoil (or any other body) equals the speed of sound (local Mach number $M_{\ell o}$ becomes unity), it is said that the Mach number of the distant flow has reached its critical value, M_{cr}. Speeds for Mach numbers lower than M_{cr} are called subcritical and those higher, supercritical. When the speed of sound is exceeded at some point or points on the body; i.e., when the local Mach number becomes $M_{\ell o} \geqslant 1.0$, a zone of supersonic flow appears, which usually terminates as a clearly defined shock wave. Downstream of the shock wave, the flow again becomes subsonic. The above-described type of flow is schematically indicated in Fig 6.22(a), while the detailed flow field around an airfoil at $M \approx 0.78$ is shown in Fig 6.22(b)[28].

As M increases, the region of supersonic flow covers larger and larger portions of both the upper and lower surfaces of the airfoil. Eventually, even though M is still less than 1.0, a situation may develop where practically the whole flow over the airfoil is supersonic (Fig 6.23(a)).

When the freestream M becomes $M > 1.0$, the flow over subsonic airfoils having rounded leading edges will be characterized by the existence of a detached shock wave slightly ahead of the L.E. A small region of subsonic flow will exist behind the shock wave, while the flow over practically the whole airfoil surface will be supersonic (Fig 6.23(b)).

It should be noted that compressibility effects begin to appear even at subcritical speeds. The physical mechanism of compressibility effects can be better understood if one visualizes that the speed at which pressure signals from the airfoil are being sent upstream become lower and lower as the Mach number of the incoming flow increases,

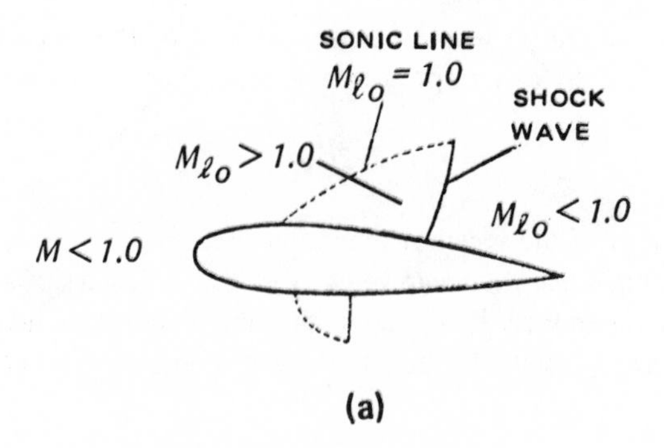

(a)

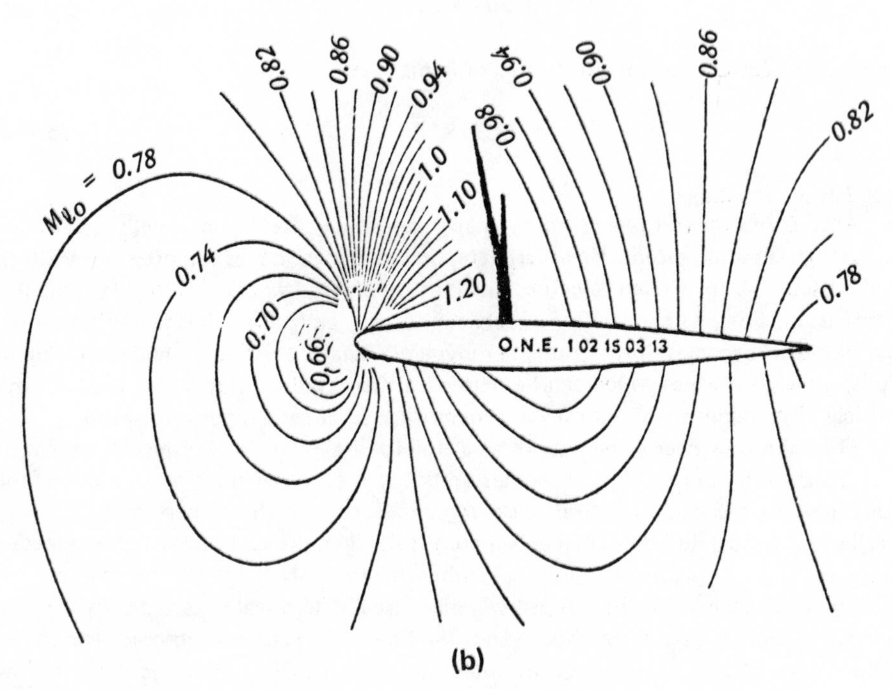

(b)

Figure 6.22 Field of flow and position of the shockwave on an airfoil

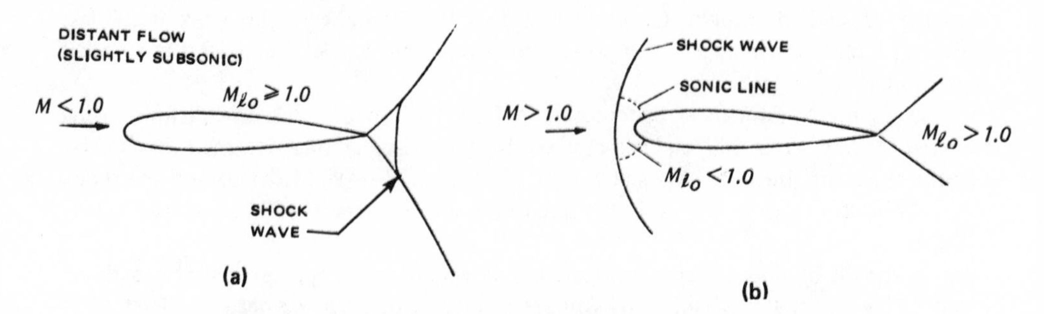

Figure 6.23 Shock waves of high subsonic and supersonic speeds on nonsupersonic airfoils

until finally, at $M \geqslant 1.0$, no signal can be sent upstream of the airfoil. Due to this retardation of the signals directed upstream at $M \to 1.0$, but still being <1.0, the streamlines of the approaching flow receive signals at a shorter distance ahead of the airfoil. Consequently, having a shorter path in which to adjust themselves to the flow pattern around the airfoil, they approach the wing at a steeper angle (Fig 6.24). This has the equivalent effect of an increased angle-of-attack[5].

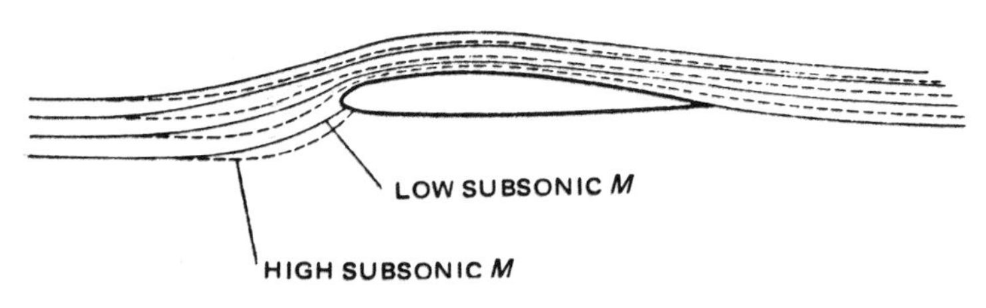

Figure 6.24 Streamline at both low and higher Mach numbers[5]

The influence of these effects on some airfoil characteristics in two-dimensional flow will be discussed briefly in the following paragraphs. The reader who is more interested in compressibility phenomena is directed to such texts as Ref 1, Ch 9, Ref 5, Ch 11; and Ref 7, Chs 10-12.

Critical Mach Number. M_{cr} can be found from relations established by von Karman and Tsien when the pressure distribution around a body in an incompressible (low-speed) flow is known:

$$\frac{p_{cr} - p_o}{q} = \frac{2}{\gamma M_{cr}^2}\left[\left(\frac{2}{\gamma + 1} + \frac{\gamma - 1}{\gamma + 1} M_{cr}^2\right)^{\frac{\gamma}{\gamma - 1}} - 1\right], \qquad (6.50)$$

where p_o is the static pressure in the freestream, p_{cr} is the pressure at the point of maximum suction (highest speed according to Bernoulli's equation), q is the dynamic pressure of the freestream, and γ is the specific heat ratio: $C_p/C_V = 1.4$.

Eq (6.50) can be presented in a graphical form as $M_{cr} = f[(p_{cr} - p_o)/q]$, where $(p_{cr} - p_o)/q$ is the maximum suction coefficient at the surface of an airfoil. For airfoil sections, however, it is more convenient for practical purposes to represent the critical Mach number as a function of the low-speed section-lift coefficient $(c_\ell)_{\ell s}$ (Fig 6.25).

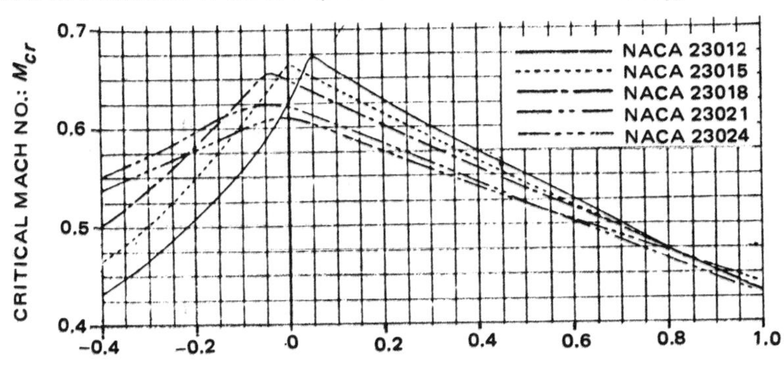

Figure 6.25 Typical presentation of M_{cr}

It is now easy to read the critical Mach number corresponding to any value of the section-lift coefficient experienced by the airfoil at low speeds.

Drag Coefficients. Drag coefficients of airfoils are influenced by compressibility, and there are several theoretical and semi-empirical ways of estimating this influence[7] One, which may be called the classical way of accounting for compressibility effects on the airfoil section drag at subcritical speeds in two-dimensional flow, consists of the application of the Glauert-Prandtl correction factor, $1/\sqrt{1 - M^2}$. This is done following von Karman's approach which showed that the profile drag coefficient in compressible flow of an airfoil having a thickness ratio t and camber m, and being at an angle-of-attack α, is equivalent to the drag coefficient in incompressible flow (low Mach number) of this airfoil having α, t, and m multiplied by the Glauert factor. The practical importance of this statement lies in the fact that if the low-speed profile drag coefficients are known at different angles-of-attack for a family of airfoils characterized by various thickness ratios and cambers, then the drag coefficient at high Mach number can be found by multiplying t, m, and α of the actual airfoil by $1/(1 - M^2)^{1/2}$ and finding the low-speed profile drag of the sections modified in this way. Therefore, the low Mach number wind-tunnel measurements could be substituted to some extent for the high-speed tests. However, it must be emphasized that when Mach numbers exceed or even approach their critical drag values, von Karman's correction will result in an underestimation of the actual drag rise (Fig 6.26).

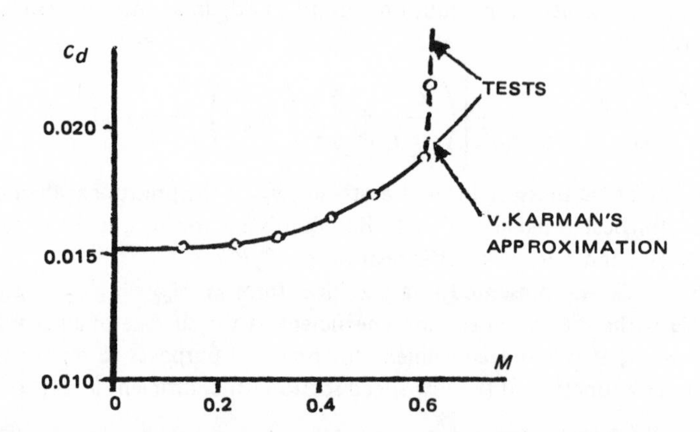

Figure 6.26 Example of the validity of von Karman corrections for airfoil section drag

When $M \ll M_{cr}$, the increase of drag over that found for incompressible flow (obtained at low speed) is obviously negligible. However, when the ratio M/M_{cr} approaches unity, the drag starts to increase rapidly until finally, when the M/M_{cr} ratio exceeds a value of 1.05 to 1.15, the drag increase is quite rapid. The freestream Mach number corresponding to the point marking the beginning of a rapid drag increase is called the drag divergence Mach number (M_{DD}) which may be defined as an M value where the slope of the profile drag curve vs freestream Mach number becomes $dc_d/dM = 0.1$.

The drag increase associated with high Mach number can also be represented as a ratio of the compressible drag coefficient ($c_{d_{com}}$) to its incompressible value ($c_{d_{inc}}$) occurring at either the fixed c_ℓ, or α value.

A typical variation of that ratio vs M/M_{cr} is shown in Fig 6.27.

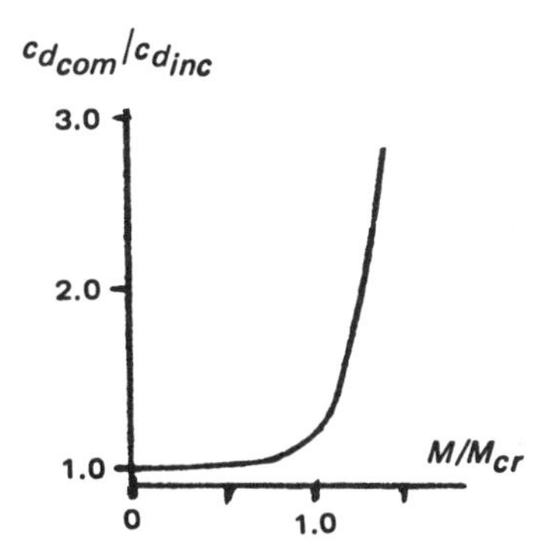

Figure 6.27 Typical variation of $c_{d_{com}}/c_{d_{inc}} = f(M/M_{cr})$

Unfortunately, even with the known M_{cr}, it is impossible to develop a universal curve similar to Fig 6.27 which would permit one to predict drag increases resulting from high M values. According to Nitzberg and Crandall[29], this is because "...for some types of airfoil sections the drag rises rapidly as soon as the freestream Mach number exceeds the critical, whereas for other types no appreciable drag rise occurs until the Mach number of the freestream is considerably above the critical." They further suggest that a good measure of the freestream Mach number at which abrupt supercritical drag rise begins can be found at the Mach number at which sonic local velocity occurs at the crest of the airfoil.

Lift Slope. As indicated in Fig 6.24 and the accompanying discussion, compressibility effects in subsonic flow cause an apparent increase in the angle-of-attack of the flow approaching an airfoil. It may be expected hence that the lift-slope coefficient at high subsonic speeds (a_{com}) should be higher than at low speeds. The simplest correction is given by the Glauert-Prandtl formula,

$$a_{com} = a_{inc}/\sqrt{1 - M^2} \tag{6.51}$$

where a_{inc} is the lift slope for incompressible flow (low speeds). There are also more refined formulas such as Kaplan's correction[7], but for relatively low Mach numbers (up to $M = 0.6$ to 0.7), Eq (6.51) usually shows a good agreement with experimental values. Since there is no reliable source from which to obtain values for higher Mach numbers, this information can be gained only through the process of experimentation.

Fig 6.28 illustrates the $a(M)$ dependence by showing the variation of c_ℓ vs M at a constant angle-of-incidence[5].

Interaction between Shocks and Boundary Layer. The appearance of a shock wave may, and often does, induce separation of the boundary layer downstream from the shock which, in turn, would increase the drag. On the other hand, the character of the boundary layer influences the structure of the shocks themselves.

Schlichting[13] pp 297-307, indicates that the understanding of mutual interaction between shock waves and the boundary layer is complicated by the fact that behavior

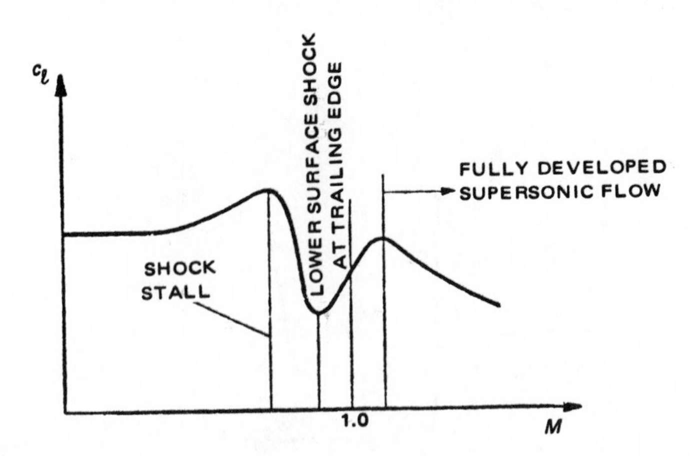

Figure 6.28 Character of variation of c_ℓ vs M at α = const[5]

of the boundary layer depends on Reynolds number, while that of the shock waves depends on the Mach number. This creates difficulties for analytical treatment as well as for experimental separation of these two influences.

It may be stated however that the appearance of the shock will be different when the boundary layer is laminar from that when it is turbulent. In the first case, there are usually several individual shock waves: a weak oblique shock located far upstream is followed by a stronger normal shock or shocks. In the presence of a turbulent boundary layer, there is only one normal shock wave.

The behavior of the boundary layer in the presence of the shock wave can also be quite irregular. For instance, the upstream laminar boundary layer may maintain its character downstream from the shock. However, this boundary layer could separate upstream from the shock due to the adverse pressure gradient, and then reattach itself to the surface beyond the shock, either in the laminar (original), or in the turbulent state. However, it may not reattach at all. Furthermore, the upstream turbulent boundary layer may either remain attached beyond the shock or it could become separated.

In view of all these possibilities, it is important to develop proper techniques for the prevention of shock-induced separation, both by design and by boundary-layer control. The interested reader is directed to Pearcey's discussion of this subject on pp 1166-1339 of Ref 14.

The so-called "peaky airfoils" may be exemplified as an attempt to create a favorable interaction between shocks and boundary layer by developing shapes with well-rounded leading edges where at zero angle-of-attack, the "peak" of supervelocity appears in the vicinity of the leading edge. Such airfoils seem to delay transonic difficulties at high speeds. At lower subsonic speeds however, they appear inferior (drag-wise) to conventional airfoil sections (say, NACA 00 Series), but even more so to the laminar ones[28].

Maximum Section-Lift Coefficient. Normally, $c_{\ell_{max}}$ decreases with increasing M (Fig 6.29). As indicated in the preceding section, this can be attributed to the formation of shock waves on the upper surface of the airfoil, and the resulting flow separation. But with the Mach number still increasing, the $c_{\ell_{max}}$ may go up again, due chiefly to the presence of supersonic unseparated flow over some portion of the airfoil. Thus, it is evident that, depending on the character of the boundary layer and speed distribution at the surface, different airfoils may behave differently regarding the $c_{\ell_{max}}(M)$ relationship. This behavior is well illustrated by Fig 6.29[30] where $c_{\ell_{max}}$ values for several airfoil sections suitable for rotary-wing application are plotted vs M. Coordinates and aerodynamic characteristics of the NRL 7223-62 airfoil can be found in Ref 30, while those of other airfoils, in Ref 7.

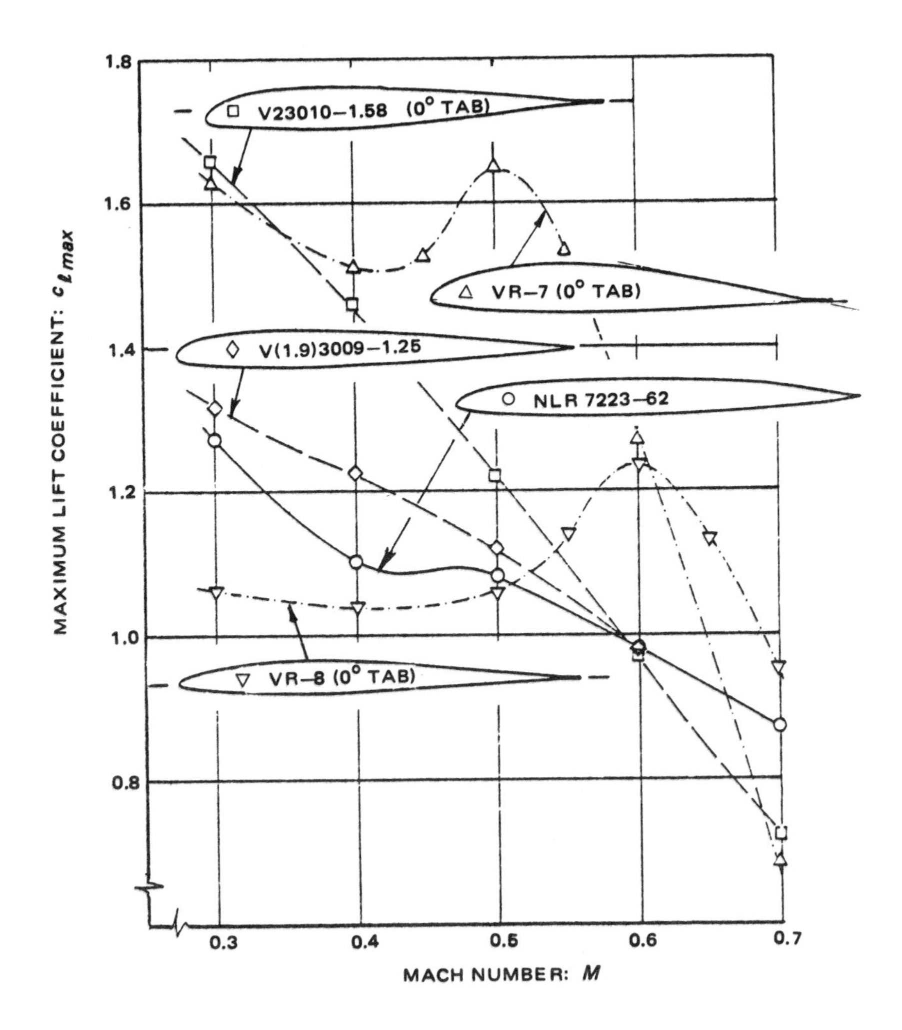

Figure 6.29 Comparison of $c_{\ell_{max}}(M)$ for several airfoil sections

Center-of-Pressure Travel. The position of the airfoil center-of-pressure also varies as M increases. The available theoretical and experimental material is not sufficient to establish universal expressions for precise quantitative computations of these variations. It can only be stated that the aerodynamic centers of very thin airfoil sections (5 to 6 percent) have a tendency to move aft with increasing Mach number. The opposite is true for thicker sections ($t/c \geqslant 12$ percent). Here, a forward c.p. shift of increasing subsonic speed may be observed, followed by a rearward motion. In any case, at $M \gg 1.0$, the center-of-pressure approaches the mid-chord point.

Some idea about physical phenomena influencing the c.p. travel at high subsonic Mach numbers can be obtained from Figs 6.30 and 6.31. In the first of these figures, the character of the chordwise pressure distribution ($C_p(c)$) at the high subcritical flow is compared with that at $M > M_{cr}$.

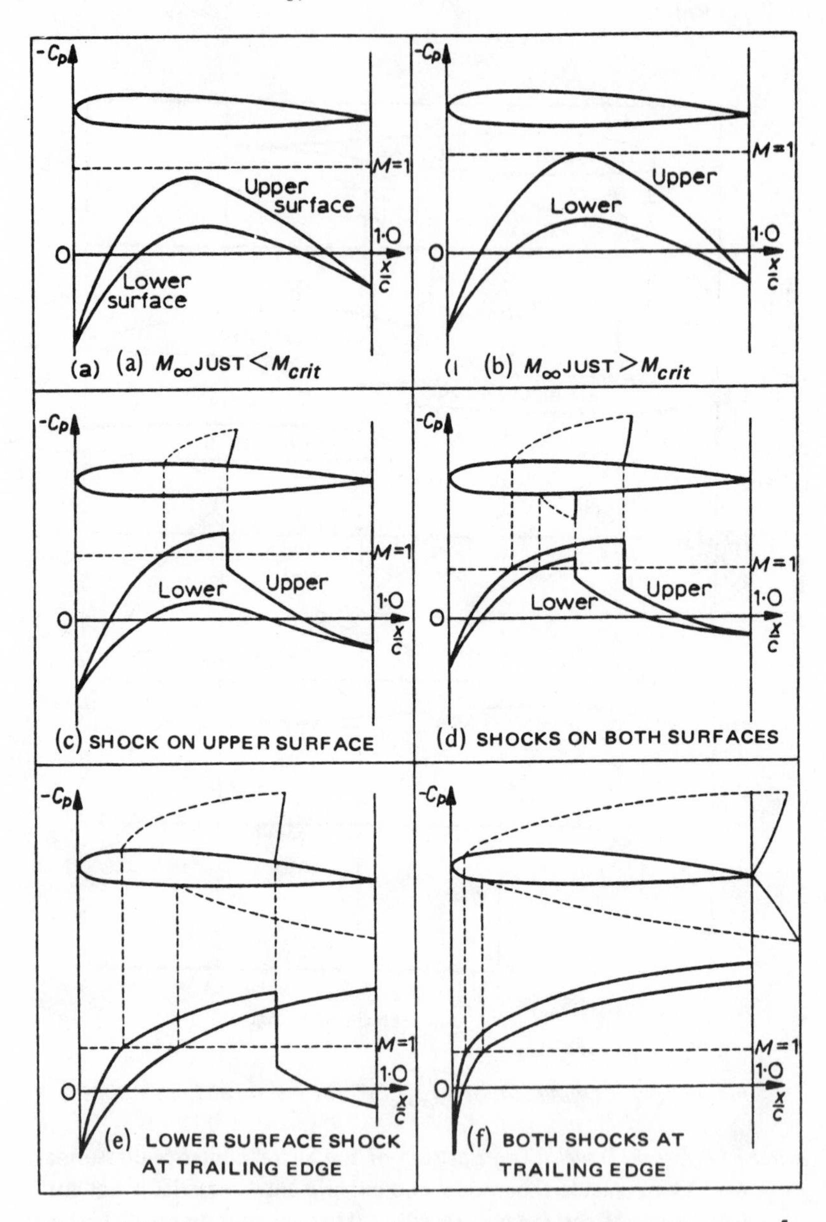

Figure 6.30 Character of chordwise pressure distribution at $M \approx M_{crit}$[5]

Figure 6.31, representing a cross-plot of the wind-tunnel measurements given in Ref 30, seems to confirm the trend of events depicted in Fig 6.31. Just before the M_{cr} is reached, the c.p. moves forward, and then as the freestream Mach number becomes higher than M_{cr}, the center of pressure begins to move toward the mid-chord position.

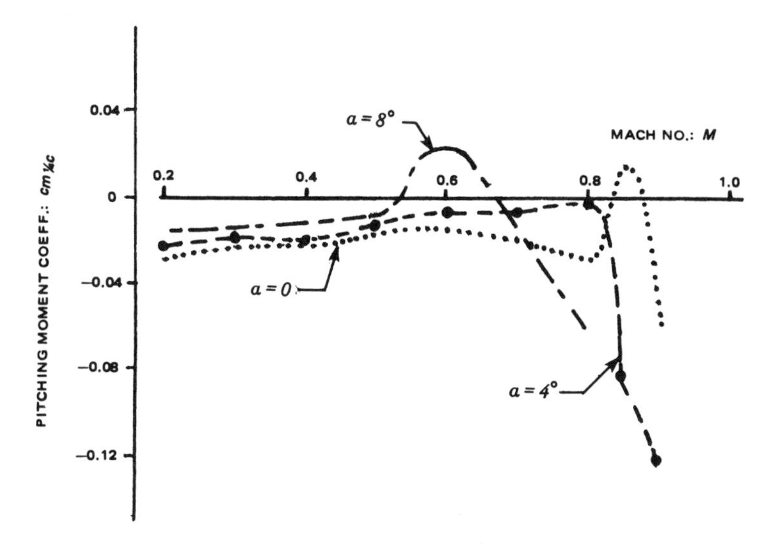

Figure 6.31 Example of c_m variation vs M (NLR 7223-62 Airfoil)[30]

7. UNSTEADY AERODYNAMIC EFFECTS

7.1 General Considerations

Unsteady Aerodynamic Effects Encountered by Airfoil Sections in Fixed Wings and Rotors. Unsteady aerodynamic effects alter airfoil characteristics of both fixed and rotary-wing aircraft, and thus are of importance to both types. However, in the case of airplanes, the unsteady aerodynamic phenomena are primarily linked to aeroelastic problems. As a matter of fact, catastrophic manifestations of wing flutter of airplanes in the early Thirties spurred analytical studies of unsteady aerodynamic phenomena by such researchers as Theodorsen[31], whose efforts were paralleled by the experimental investigations of Silverstein and Joyner[32]. In the Forties and Fifties, considerable literature was published on this subject, including such textbooks as that of Bisplinghoff, et al[33]. Through the Sixties and Seventies, a number of publications related to unsteady aerodynamic effects on airfoils of fixed wings continued to grow (including various aspects of transonic and supersonic speeds, and hypersonic flow) but, as can be seen from a more recent textbook by Fung[34], they are still chiefly related to aeroelastic problems.

By contrast, as far as helicopter rotors are concerned, interest in unsteady aerodynamics of airfoils stem not only from its importance to aeroelastic problems, but performance as well.

This latter aspect became evident in the early Sixties when the attention of researchers (e.g., Harris and Pruyn[35]) was focused on the fact that the boundaries of rotor stall in forward flight were more favorable than those predicted on the basis of static two-dimensional airfoil data. It appeared that differences in airfoil characteristics resulting from unsteady aerodynamic phenomena may have been responsible for the discrepancies. The importance of unsteady aerodynamic phenomena to the understanding and analysis of blade aeroelastic problems provided an additional incentive for experimental and analytical studies of unsteady flow. Experimental studies by Liiva, et al[36] represent one of the first efforts directed exclusively toward unsteady aerodynamic problems of blade airfoil sections.

It is obvious that unsteady fluid dynamic phenomena are not limited to airfoil sections of either fixed or rotating wings, but as McCroskey[37] points out, cover a wide range of fluid flows of practical importance, which include flows around slender and blunt bodies as well as their arrays.

Unsteady flow dynamics is a fast growing field as witnessed by the fact that most of the 290 literature positions mentioned in Ref 37 were published in the Seventies. In this text, however, attention will be focused on an area limited to unsteady aerodynamic effects of airfoils in two-dimensional subsonic flow.

Blade Section Angle-of-Attack Variation. Rotor blade sections, especially in forward flight, experience time-dependent variations of their angles-of-attack which, in turn, may lead to the manifestation of various unsteady-aerodynamic phenomena.

In Fig 6.32, the velocities encountered by a blade airfoil section in forward flight are shown with respect to the stationary air. From this figure, it can be seen that

$$a = f(\theta, U_\perp, U_\parallel)$$

where θ is the section pitch angle, while $U_\perp$ and $U_\parallel$, respectively, are blade section velocity components perpendicular and parallel to the tip-path plane.

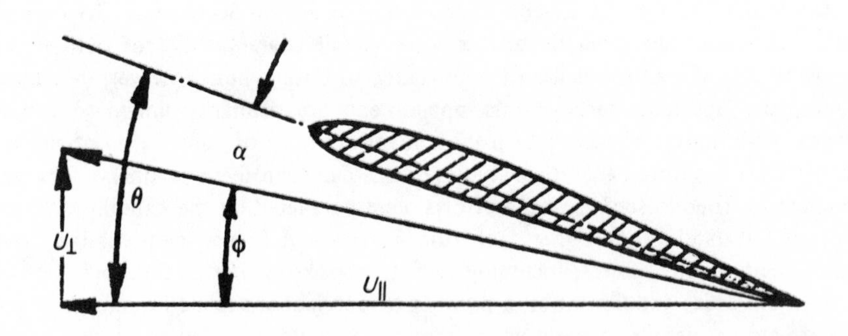

Figure 6.32 Blade section velocities with respect to air

As Liiva, et al[36] point out, variations of the θ, $U_\perp$, and $U_\parallel$ parameters are periodic, since

θ varies:

a. at 1/rev because of cyclic pitch and/or flapping motion (accepting the classical equivalence between flapping and feathering), and

b. at higher frequencies (4/rev or higher) because of blade torsional motion.

$U_\perp$ does not experience any 1/rev variation since axes are referred to the plane of no-flapping, but varies:

a. at frequencies of 2/rev and higher because of flap bending at appreciable amplitude, and

b. irregularly, because of vortex effects.

$U_\parallel$ varies:

a. sinusoidally at 1/rev because of aircraft forward speed,

b. at higher frequencies (but very low amplitude) because of chord bending, and

c. irregularly, because of vortex effects.

Wind-Tunnel Simulation. When experimental programs for investigation of unsteady aerodynamic effects on rotor blades were formulated in the Sixties, the question was asked as to what extent the above-discussed variation of the three parameters could be simulated in a wind tunnel. It is obvious that it would be impossible to simulate the $\alpha(U_\parallel)$ function directly in existing wind tunnels, since it would require sinusoidal variation of the magnitude of flow velocity. This means that for practical purposes, the $\alpha(U_\parallel)$ dependence must be approximated by such other means as airfoil oscillation ($\alpha(\theta)$) about its pitching axis (usually ¼-chord) and through plunging motions ($\alpha(U_\perp)$)[36].

The $\alpha(\theta)$ and $\alpha(U_\perp)$ relationship can be directly simulated in a wind tunnel using arrangements similar to those shown in Figs 33 and 34.

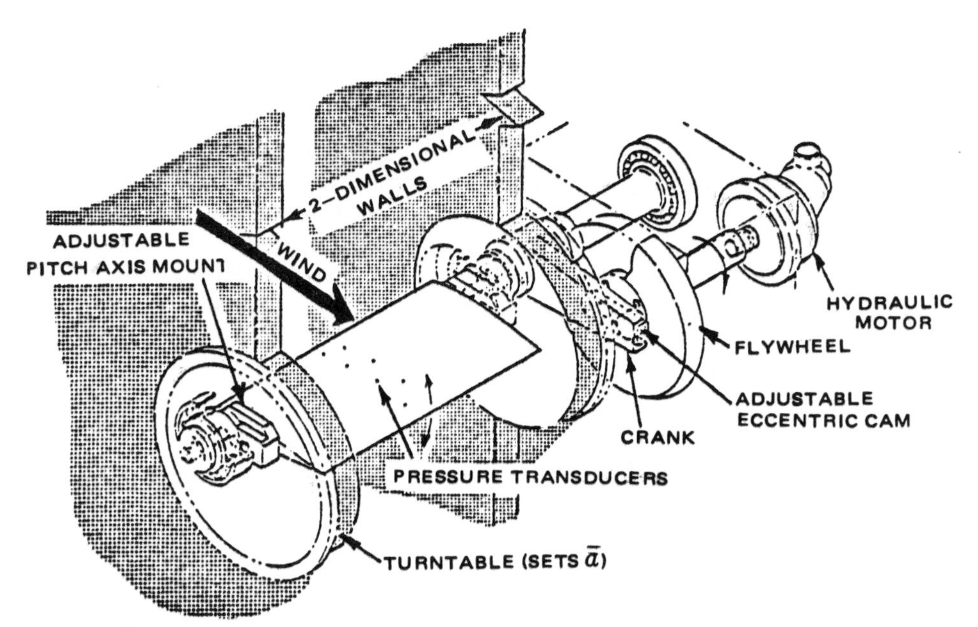

Figure 6.33 An example of a wind-tunnel arrangement for simulation of airfoil pitching oscillations[30]

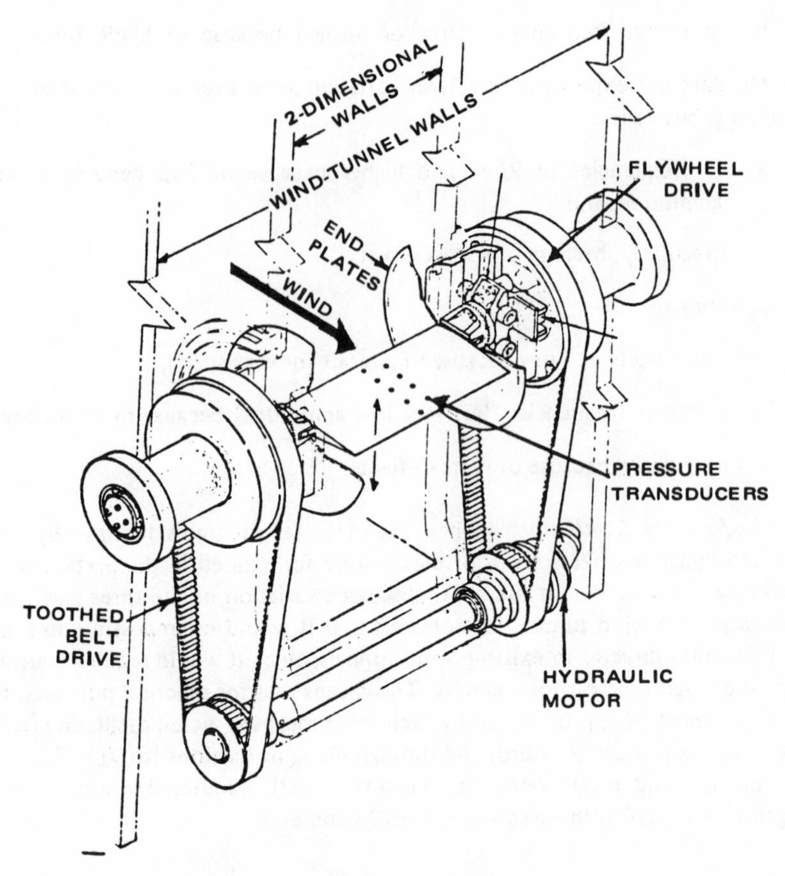

Figure 6.34 An example of wind-tunnel arrangement for simulation of airfoil plunging oscillations[36]

In the case of pitch, the airfoil executes sinusoidal oscillations of half amplitude, $\pm \Delta \alpha$, around some mean angle-of-attack, $\bar{\alpha}$. In plunging tests, the airfoil is preset at an angle-of-attack α and then is subjected to sinusoidal vertical oscillations of a given fraction of the chord ($\pm y/c$). In both cases, pressure transducers register the pressure distribution on the surface from which normal forces and pitching moments are calculated, while drag is usually obtained from the wake momentum loss measurements.

7.2 Physics of Unsteady Flow

Small Oscillations Below Stall. It may be anticipated that at low angles-of-attack (below stall) and at small airfoil oscillations for both pitch and plunging, the behavior of the airfoil can be interpreted as being within the framework of potential theories; thus, the linear approach would apply.

It can be seen from Figs 6.35(a) and 6.35(b)—taken from flow visualization studies by Werlé and Armand[38]—that the boundary layer remains attached which, according to Jones[39], signifies that viscous (nonlinear) effects may be ignored.

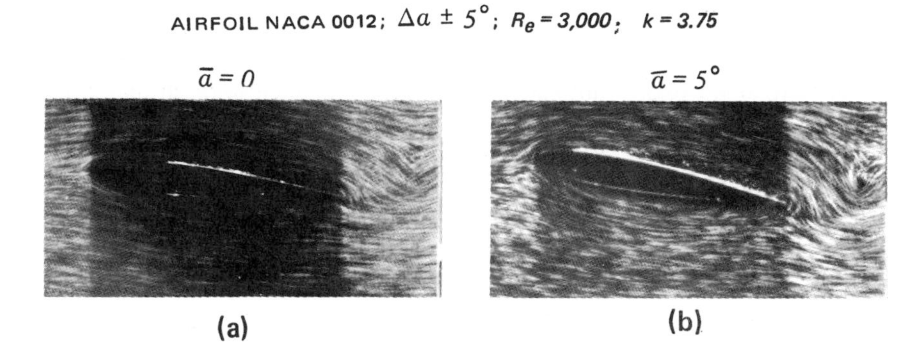

Figure 6.35 Flow visualization around airfoils oscillating in pitch about 0.25c at low angles-of-attack[38]

Under these circumstances, the most physically significant phenomenon would consist of the formation of a sinusoidal vorticity wake (Fig 6.36). This wake would obviously be generated behind the oscillating-in-pitch and plunging airfoils due to the variation in their angle-of-attack being synonymous with the fluctuations of lift (ℓ) sustained by a unit span of the airfoil. Referring back to Eq (4.14), this means that circulation (Γ) around the airfoil would also vary. Every change ($\Delta\Gamma$) in Γ will be accompanied by a vorticity shedding equal to $-\Delta\Gamma$. Consequently, a vortex wake is generated by an airfoil undergoing periodic variations in lift. This wake, in turn, would influence the forces and moments acting on the airfoil. In the particular case of a uniform flow in which an airfoil is undergoing a sino-type variation of its angle-of-attack, a sinusoidal wake would form. This wake may be imagined as being composed of discrete vortex filaments of varying strength and direction of rotation (Fig 6.36)

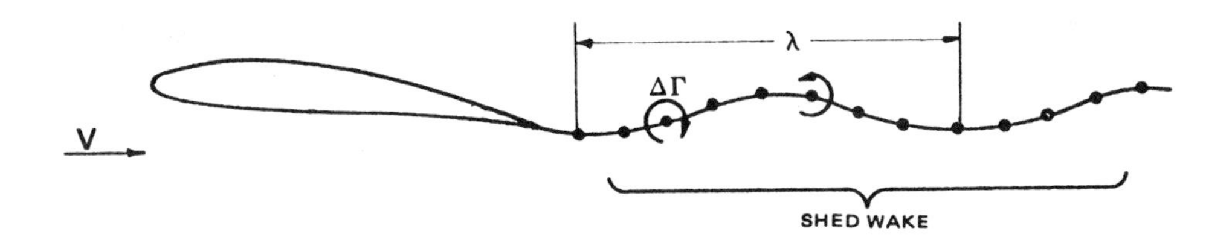

Figure 6.36 Scheme of the vortex wake of an oscillating airfoil

It may be expected that the ratio of the airfoil chord (c) to the wave length (λ) of the wake would represent a physically significant parameter. Remembering that λ is the distance traveled by fluid at the distant-flow velocity V during the time corresponding to one complete oscillation at cycle frequency f (in Hz), the sought ratio becomes:

$$c/\lambda = fc/V. \tag{6.52}$$

Noting that cycle frequency is related to circular frequency (ω in rad/s), i.e., $f = \omega/2\pi$; Eq (6.52) can be rewritten as follows-

$$c/\lambda = \omega c/2\pi V. \tag{6.53}$$

McCroskey[37] indicates that in the broad class of problems which can be treated by linear theory, unsteady effects become important when $\omega c/V$ is of the order of 1 or greater. In light of Eq (6.53), this statement can be interpreted to mean that unsteady effects begin to be significant when $\lambda/c \leqslant 2\pi$.

It is apparent that the fc/V ratio has a definite physical meaning, and all expressions for the so-called reduced frequency (k) contain that ratio. For instance, Liiva[35] and Dadone[30] express reduced frequency in Hz as

$$k = \pi fc/V. \tag{6.54}$$

McCroskey refers it to circular frequency (in rad/s) and defines it as

$$k = \omega c/2V. \tag{6.54a}$$

There are other authors (e.g., Warlé and Armand[38]) who define reduced frequency as

$$k = 2\pi fc/V. \tag{6.54b}$$

The practical importance of reduced frequency lies in the fact that as long as the nonlinear effects caused by viscosity and compressibility are not significant, or are taken care of in tests by maintaining Reynolds and Mach numbers at the full-scale level, the wind-tunnel simulation of unsteady aerodynamic effects would be truly representative of the full-scale phenomena as long as the test k values are the same as those of simulated airfoils.

Experimental results show that oscillations around low angles-of-attack (below stall) produce some hysteresis as far as both forces (chiefly the normal component, enumerated through the c_n value) and moments (characterized by the c_m coefficient) acting on the airfoil are concerned (Fig 6.37a, where $a_o = 7.33$). However, it can be seen from this figure that in this case, hysteresis loops are narrow and c_n or c_m values forming them do not deviate significantly from the statically-obtained values.

Although the moment loop in this case is flat, the reader's attention is called to the general significance of the $c_m(a)$ relationship. This is important because, depending on the character of that relationship, either positive or nagative damping in pitch will be generated. It may be generally stated that if within some interval of the angles-of-attack (say, between a_1 and a_2, where $a_1 < a_2$), the following relationship occurs:

$$\int_{a_1}^{a_2} [c_m(a_{incr}) - c_m(a_{decr})]\, da < 0, \tag{6.55}$$

then increments of aerodynamic moments are generated in such a way that they oppose the angle-of-attack variation. This obviously is synonymous with the existence of positive damping (see Fig 6.37(a), where $a_o = 7.33°$). Should the inequality given by Eq (6.55) be positive, then negative damping will be present (see parts of the moment coefficient loop corresponding to $a_o = 14.92°$).

Oscillation Close to Stall. Oscillations close to stall are of special significance to performance prediction, as they influence the values of various parameters (e.g., advance

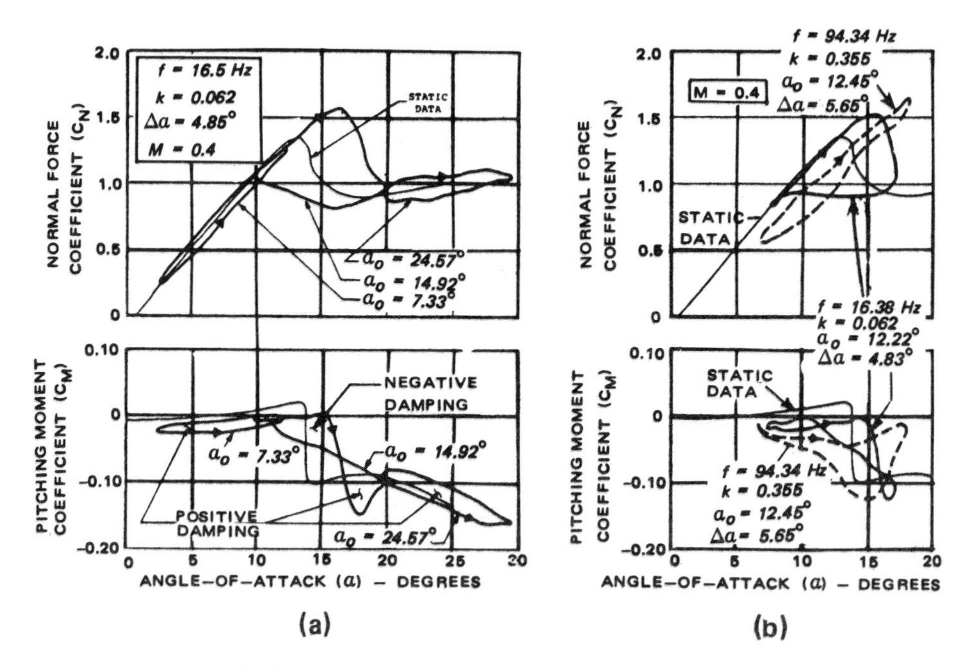

Figure 6.37 Typical force and moment data of oscillating airfoils[35]

ratio, thrust coefficient, and thrust inclination) which determine stall boundaries for helicopter operations.

As far as the physics of this type of oscillation is concerned, it is anticipated that in addition to the previously discussed influence of reduced frequency, viscous effects (with all associated nonlinearities) will predominate. Also, the influence of compressibility which may trigger radical changes in the behavior of the boundary layer (e.g., separation) could be stronger here than in the case of oscillation around low angles-of-attack (Fig 6.38)

The strong role of viscosity and compressibility (when present) make the development of a comprehensive picture of physical events taking place in the case of dynamic stall more difficult. McCroskey[37] points out that although during the last decade a better insight into dynamic stall phenomena has been obtained, a complete understanding of that event is still lacking.

The fact that dynamic stall more radically alters static-type flow around airfoils than in the case of oscillations around low angles-of-attack is quite visible from experimental measurements. It can be seen from Fig 6.37(a) where $a = 14.92^\circ$, and (b) where $a = 12.45^\circ$, that considerable departures from the static test airfoil characteristics are now encountered, as large hysteresis loops are generated for both the $c_n(a)$ and $c_m(a)$ relationships.

In order to obtain at least some insight into the physics of dynamic stall, let us start with an examination of the flow visualization pictures obtained by Werlé and Armand[38] in a water tunnel at low Reynolds numbers (Fig 6.39).

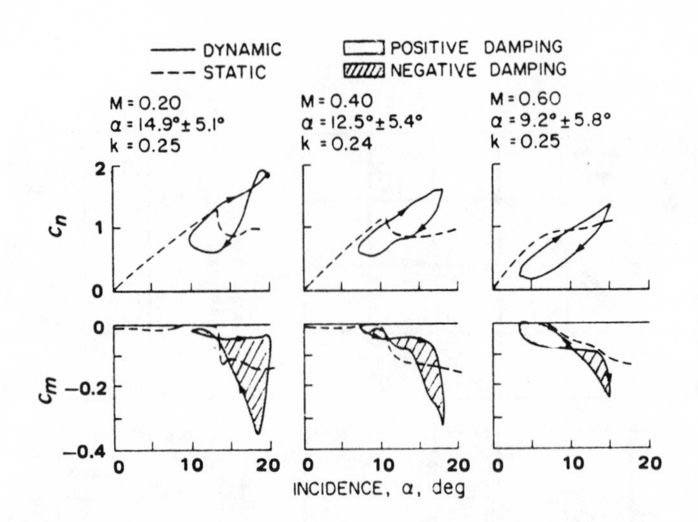

Figure 6.38 The influence of Mach numbers on dynamic stall for sinusoidal pitch oscillations

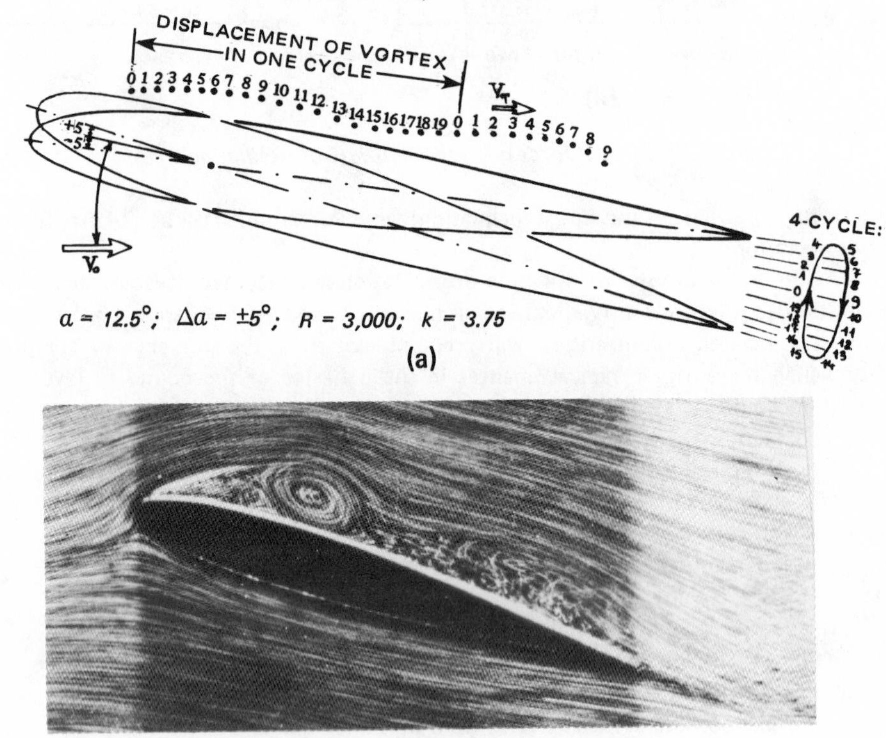

$a = 12.5°$; $\Delta a = \pm 5°$; $R = 6{,}000$; $k = 0.9$

(b)

Figure 6.39 (a) NACA 0012 airfoil oscillating about a ¼-chord axis, and (b) flow visualization around it at maximum angle-of-attack

It can also be seen from Fig 6.39(a) that when the airfoil is close to the static stall angle, a vortex-type disturbance forms close to the leading edge (position 0 on the upper surface of the airfoil), and moves downstream along the surface following the path noted on the sketch. In Fig 6.39(b), the position of the vortex is shown as the airfoil is approximately at its highest angle-of-attack value. Further downstream, a somewhat weaker vortex can be noted.

The flow along the lower airfoil surface appears to be attached, and is probably not much different from that encountered under static conditions at high angles-of-attack. Consquently, events occurring at the upper surface are chiefly responsible for the differences in airfoil characteristics associated with static and dynamic stall.

Physical aspects of the events occurring along the upper surface can be even better appreciated by looking at Fig 6.40[37], where an analytically-predicted flow is compared with that obtained by flow visualization.

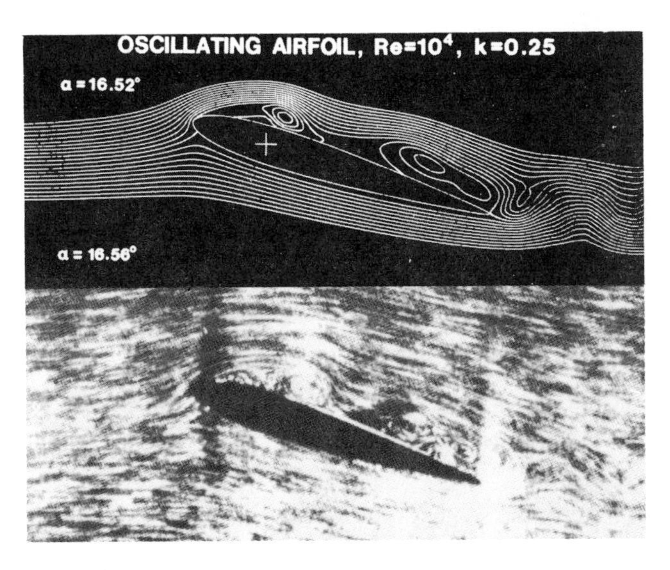

Figure 6.40 Navier-Stokes calculations by U.B. Mehta and flow visualization by H. Werlé[37]

It can be seen from this figure that the vortex-like formation seems to exercise flow stabilizing effects. Unlike the case of static stall (Fig 6.19), the streamlines outside of the vortices are not broken, but remain continuous. Using the reasoning outlined in Sect 5.2, it may be anticipated that the overall circulation around the airfoil and therefore, the lift as well, at high angles-of-attack above the static-stall angle will, in general, remain high. However, at still higher angles-of-attack, the flow will eventually break down and cause stall. These changes in the flow pattern are obviously accompanied by a corresponding variation in pressure distribution along the upper surface of the airfoil, as shown in Fig 6.41.

The above-discussed events are summarized in Fig 6.42[37] which originally appeared in NASA TN D-8382 by Carr, McAllister and McCroskey.

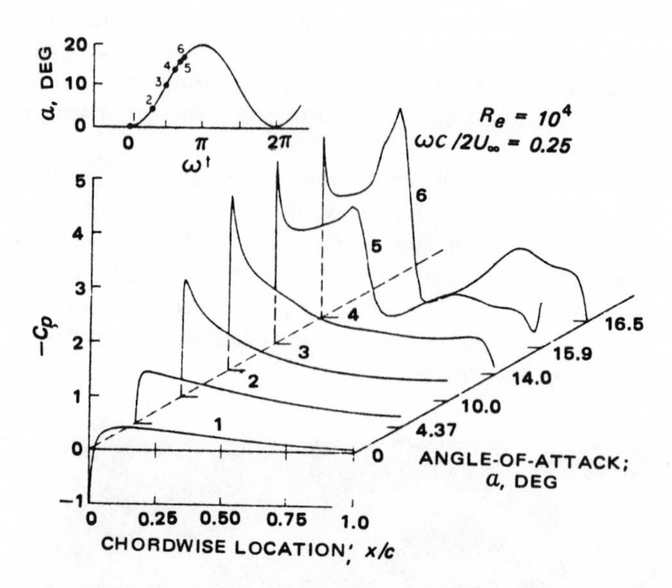

Figure 6.41 Calculations of the upper-surface pressure distribution[37]

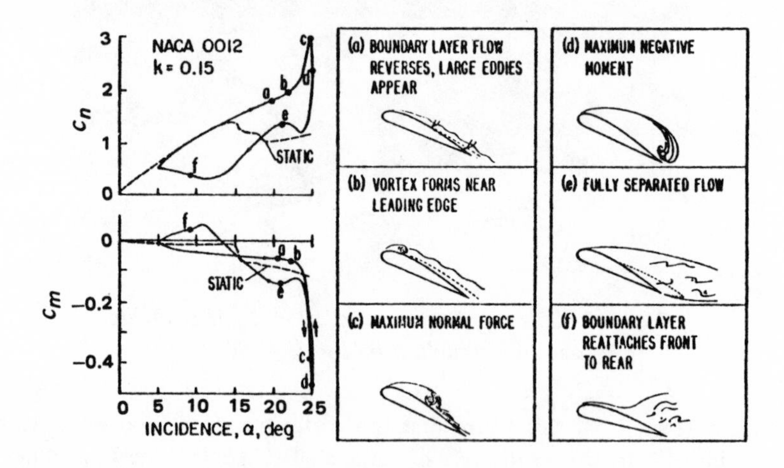

Figure 6.42 The important dynamic stall events on an oscillating airfoil at $R_e = 2.5 \times 10^6$

It is apparent that vortices forming near the leading edge and then traveling along the upper airfoil surface constitutes one of the main tools for producing unsteady aerodynamic effects. It may be expected hence, that reduced frequency, which influences the distance between the vortices with respect to the chord length (Fig 6.43[38]) should be considered as one of the most important parameters.

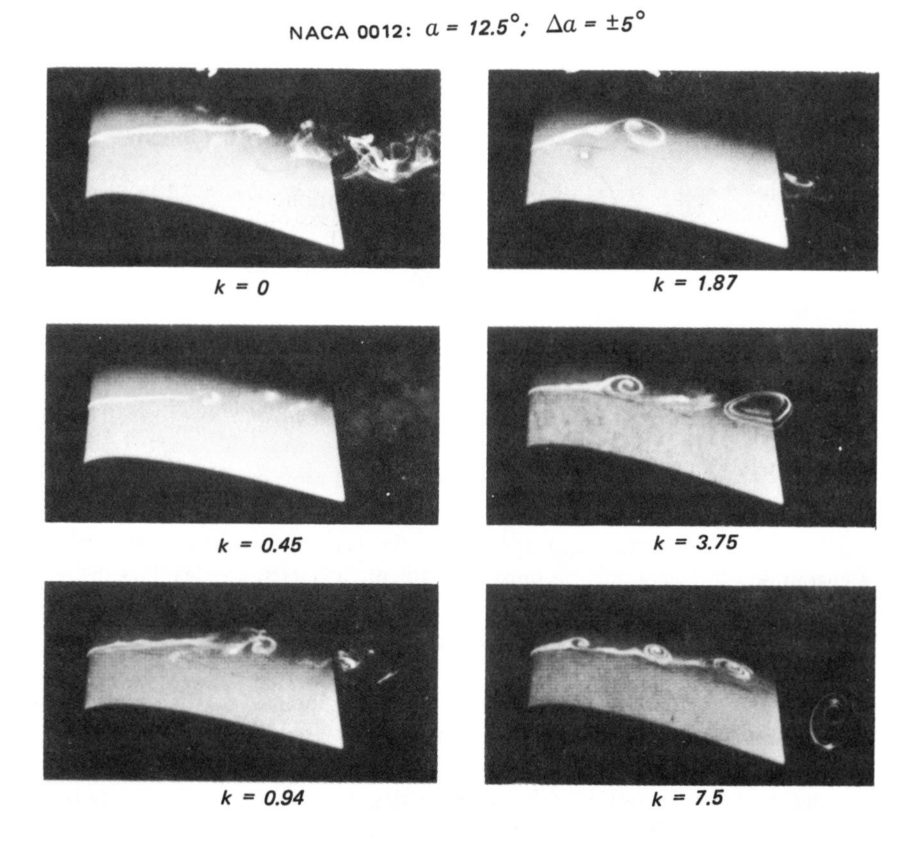

Figure 6.43 Influence of frequency on the character of flow around an airfoil oscillating in pitch[38]

The actual mechanism producing these vortices and the factors which influence them, as well as the behavior of the boundary layer, including its eventual separation and then delayed reattachment, pose questions which still remain unanswered.

This problem was extensively studied in the late Sixties and early Seventies by, among others, Johnson and Ham[41], and McCroskey et al[42]. Based on the analytical and experimental results depicted in Fig 6.44, the events occurring at the leading edge are described in Ref 41 as follows: "...the boundary layer is laminar initially, separates when it encounters the adverse pressure gradient in the vicinity of the airfoil leading edge, becomes turbulent, and reattaches, forming a separation bubble."

As the angle-of-attack increases, the separation point moves forward and eventually reaches a stagnation point. It should be noted that this forward motion of the separation point toward the stagnation point occurs under static conditions as well as when the airfoil is oscillating in pitch at various frequencies and at various locations of the pitching axis.

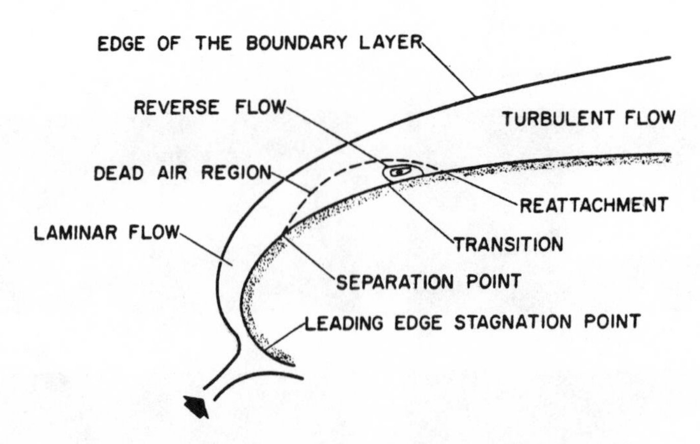

Figure 6.44 Bubble structure on a thin airfoil[41]

However, as the value of $\dot{\alpha}$ increases, the forward movement of the turbulent reattachment point is retarded, suggesting that the bubble elongates. Finally, the bubble contracts as the reattachment point approaches the leading edge and, as the shortened bubble encounters the large adverse pressure gradient, it bursts, and leading-edge vortex shedding commences. From now on, the flow and hence, forces and moments, acting on the airfoil are influenced by the vortex movement.

Eventually, as the vortex approaches the trailing edge, the flow in that area separates (possibly due to the large adverse pressure gradient induced by the vortex), a counter vortex is shed from the trailing edge, and the flow progresses to the fully stalled state.

Among the factors affecting stall characteristics, the role of compressibility is somewhat easier to grasp. Going back to Fig 6.38, it may be quoted from McCroskey[37] that

> ...progressively smaller mean angles were selected because of the decrease in the static-stall angle with increasing Mach number. The similarity of the static-stall characteristics at $M_\infty = 0.2$ and 0.4 suggests that transonic shockwave formation does not play a role in either case, but the static $M_\infty = 0.6$ data show clear evidence of shock-induced separation and stall. The dynamic data at $M_\infty = 0.6$ suggest that the formation of shock waves somehow inhibits the development of the vortex shedding process, although some vestiges of the phenomenon remain. New experiments similar to the types that have been done at low speeds are needed to resolve this question further.

Pitching vs Plunging. Finally, at the conclusion of this discussion of physical and experimental aspects of oscillating airfoils, a question should be asked regarding the extent to which pitching can be considered as equivalent to plunging. Experimental results of two-dimensional airfoil tests in both pitching and plunging were reported by Liiva et al[36], but no comparative analysis between the two types of oscillation were made. However, using the basic data of Ref 36, this aspect was investigated by Fukishima and Dadone[40] who came to the following conclusions:

1. Leading and trailing edge pressures (suction) during stall are generally lower in pitch than in plunging.

2. In general, the rate at which stall propagates chordwise is initially greater in vertical translation.

3. Only weak secondary stall phenomena have been observed; vertical translation appears to be somewhat more conducive to secondary stall events.

4. For pitch and translation oscillations, similar changes occur in the chordwise progression of stall with variations in frequency and mean angle-of-attack.

5. Within the limitations of the available data, no significant differences have been observed in the mechanism of stall recovery with respect to effect on the normal force.

It appears hence, that for practical engineering application, unsteady aerodynamic characteristics—especially those related to dynamic stall—obtained from airfoil tests in pitching motions can be assumed as representative of both this and plunging types of oscillations.

7.3 Analytical Approach to Oscillations of Thin, Flat Airfoils Below Stall

Application of Potential Methods. It was indicated in the preceding section that at small oscillations below stall and before occurrence of compressibility effects, the fluid may be assumed as inviscid and incompressible; thus, the flow around the airfoils is considered as potential. This opens wide possibilities for analytical treatment of unsteady aerodynamic problems since, through application of the superposition principle, the velocity potential of the flow around an airfoil can be presented as the sum of potentials associated with various aspects of both the characteristics and the motion of the airfoil.

Once the total velocity potential (φ) of flow around the airfoil is established, the difference in pressure (p) at any point P on the airfoil surface from that at infinity can be computed using the Bernoulli theorem:

$$p_P = -\rho[(V_P{}^2/2) + (\partial\varphi_P/\partial t)] \tag{6.56}$$

where V_P is the velocity of flow at point P, and φ_P is the total velocity potential at the point which, in turn, can be expressed as the sum of the potentials representing various aspects of the oscillating airfoil:

$$\varphi_P = \varphi_1 + \varphi_2 + \ldots . \tag{6.57}$$

It is obvious that normal forces and moments can be easily computed once the pressure distribution along the airfoil surface is known.

Theodorsen's Approach. Theodorsen studied aerodynamic forces and moments acting on a wing-aileron system oscillating as a whole about a pitching axis and executing vertical motions while the aileron is free to move about its hinge[31]. In this approach, the aerodynamic surfaces were modeled by flat plates. If one assumes that aileron deflection becomes the angle-of-attack of the whole airfoil, the so-simplified wing-aileron system could be used as a mathematical model for pitching and plunging oscillations of airfoil sections (Fig 6.45).

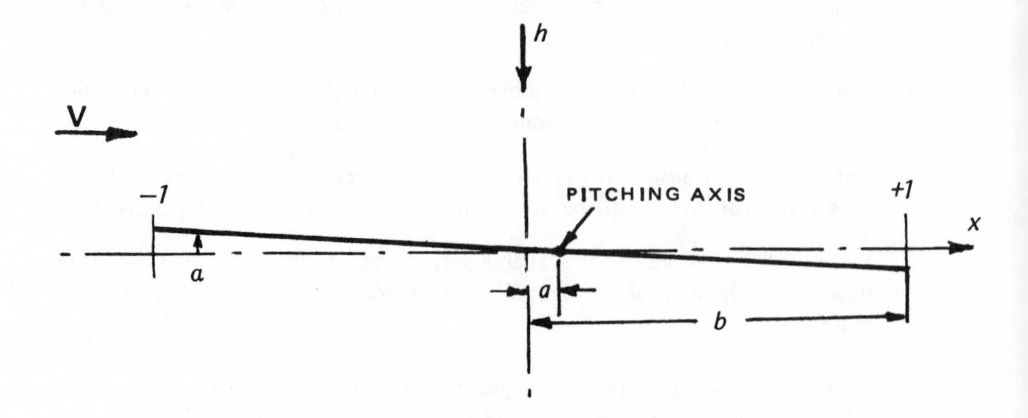

Figure 6.45 Basic notations for Theodorsen's mathematical model

To further simplify the problem, it was postulated that the oscillations were infinitesimal. Consequently, the vortex wake of the airfoil would not be wavy as shown in Fig 6.36, but would be flat.

In order to obtain the total velocity potential needed for pressure computations, Theodorsen assumes that it is composed of two classes of potentials: (a) one, associated with noncirculating flow which can be expressed through sources and sinks, and (2) the other, linked to circulatory flow related to the flat vorticity surface extending from $x = 1$ to $x = \infty$.

Velocity Potential of the Noncirculatory Flow. The velocity potentials of the noncirculatory flow are developed by transforming a segment of a straight line (representing the airfoil section) $2b$ long (where b is the half-chord) into a circle of radius b (Fig 6.46).

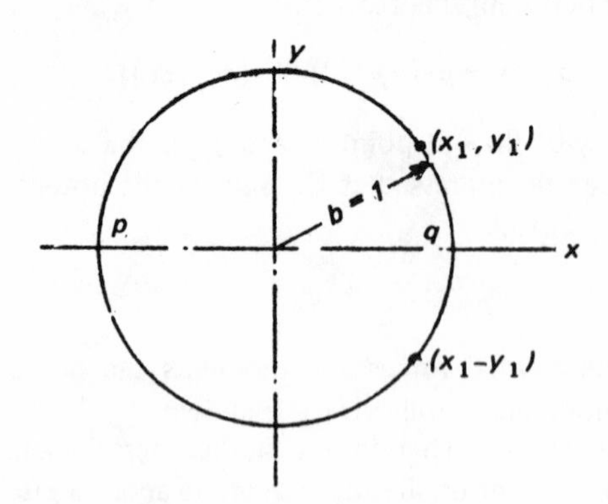

Figure 6.46 Conformal transformation of an airfoil in noncirculatory flow

The potential of a source ϵ at (x_1, y_1) on the circle is

$$\varphi = (\epsilon/4\pi)\ log\,[(x - x_1)^2 + (y - y_1)^2]. \tag{6.58}$$

By locating the double source 2ϵ at (x_1, y_1), and a corresponding sink -2ϵ at $(x_1, -y_1)$, the following is obtained for the potential of the flow around a circle:

$$\varphi = \frac{\epsilon}{4\pi}\, log \frac{(x - x_1)^2 + (y - y_1)^2}{(x - x_1)^2 + (y + y_1)^2}. \tag{6.59}$$

But, as Theodorsen indicates, the function φ on the circle directly gives the surface potential of a straight line, pq. Now, $y = \sqrt{1 - x^2}$ and thus, φ is a function of x only.

Substituting $-Vab$ for ϵ and integrating Eq (6.59) with respect to x along the whole airfoil chord; i.e., from -1 to 1, the following expression for the potential due to angle-of-attack (φ_a) is obtained:

$$\varphi_a = Vab\sqrt{1 - x^2}. \tag{6.60}$$

The potential $\varphi_{\dot{h}}$, associated with the plunging motion at vertical velocity $\dot{h}$, is directly obtained from Eq (6.60) by noting that $a = \dot{h}/V$ for small angles. Hence,

$$\varphi_{\dot{h}} = \dot{h}b\sqrt{1 - x^2}. \tag{6.61}$$

Finally, the potential $\varphi_{\dot{a}}$ resulting from the airfoil pitching velocity a can also be derived from φ_a by realizing that the rotation about some pitching axis located at a (Fig 6.45) is equivalent to the rotation about the leading edge ($x = -1$) at an angular velocity $\dot{a}$, plus a vertical motion with a linear velocity $-\dot{a}(1 + a)b$. This leads to

$$\varphi_{\dot{a}} = \dot{a}b^2(½x - a)\sqrt{1 - x^2}. \tag{6.62}$$

The total potential related to the noncirculatory flow (φ_{nc}) now becomes

$$\varphi_{nc} = \varphi_a + \varphi_{\dot{a}} + \varphi_{\dot{h}}. \tag{6.63}$$

Eq (6.56) can now be used for the determination of pressure differentials (Δp) between the upper and lower surfaces of the airfoil. Realizing that the total velocity at a point P can be expressed as $V_p = V + \partial\varphi/\partial x$, the following is obtained:

$$\Delta p = -2\rho[V(\partial\varphi/\partial x) + (\partial\varphi/\partial t)]. \tag{6.64}$$

Velocity Potential of Circulatory Flow. Using the conformal transformation as in the preceding case, it is assumed that the vortex sheet of unit strength γ extends from $x = b$ (where $b = 1$) to infinity (Fig 6.47). The velocity potential of the flow around the circle associated with a discrete vortex element $-\Delta\Gamma = -\gamma\Delta x$ located at $X_0, 0$ is:

$$\varphi_\Gamma = \frac{\Delta\Gamma}{2\pi}\left[tan^{-1}\frac{Y}{X - X_0} - tan^{-1}\frac{Y}{X - (1/X_0)}\right]. \tag{6.65}$$

where X, Y are coordinates of the circle and X_0 gives the location of $-\Delta\Gamma$ on the x axis.

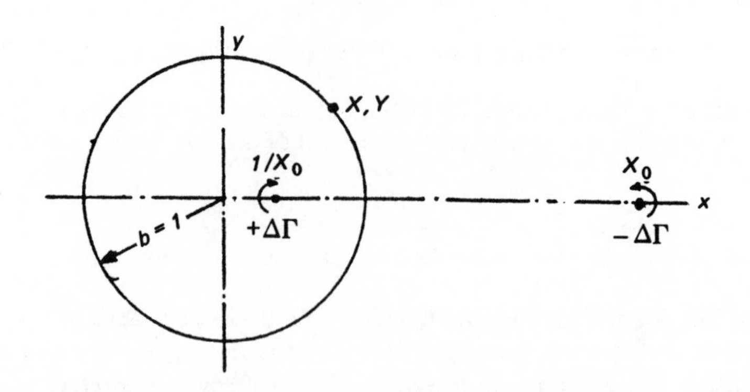

Figure 6.47 Conformal transformation of an airfoil in circulatory flow

Defining $X_0 + (1/X_0) = 2x_0$ and noting that $X = x$ and $Y = \sqrt{1-x^2}$, the velocity potential at point x due to the elementary vortex $-\Delta\Gamma$ at x_0 becomes

$$\varphi_{xx_0} = -\frac{\Delta\Gamma}{2\pi}\tan^{-1}\frac{\sqrt{1-x^2}\,\sqrt{x_0^{\,2}-1}}{1-xx_0}. \tag{6.66}$$

Realizing that the vortex element $-\Delta\Gamma$ moves away from the airfoil along the positive x axis at a speed V, the time derivative appearing in Eq (6.64) can now be expressed as $\partial\varphi/\partial t = V\partial\varphi/\partial x_0$. Hence, Δp_Γ, due to the circulatory flow, becomes

$$\Delta p_\Gamma = -2\rho V[(\partial\varphi/\partial x) + (\partial\varphi/\partial x_0)]. \tag{6.67}$$

However,

$$\frac{2\pi}{\Delta\Gamma}\frac{\partial\varphi}{\partial x} = \frac{\sqrt{x_0^{\,2}-1}}{\sqrt{1-x^2}}\,\frac{1}{x_0-x}$$

and

$$\frac{2\pi}{\Delta\Gamma}\frac{\partial\varphi}{\partial x_0} = \frac{\sqrt{1-x^2}}{\sqrt{x_0^{\,2}-1}}\,\frac{1}{x_0-x}.$$

Substituting $\partial\varphi/\partial x$ and $\partial\varphi/\partial x_0$ from the above equations into Eq (6.67), Δp_Γ is expressed as follows:

$$\Delta p_\Gamma = -\rho V\frac{\Delta\Gamma}{\pi}\,\frac{x_0+x}{\sqrt{1-x^2}\,\sqrt{x_0^{\,2}-1}}. \tag{6.67a}$$

The increment of force (F) on the whole airfoil, resulting from the elementary vortex $\Delta\Gamma$ located at x_0 will be

$$\Delta F_\Gamma = \int_{-1}^{1}\Delta p_\Gamma\,dx = -\rho Vb\,\frac{x_0}{\sqrt{x_0^{\,2}-1}}\,\Delta\Gamma. \tag{6.68}$$

On the other hand, $\Delta\Gamma = \gamma dx_0$ and thus, the total force on the airfoil resulting from the vortex wake extending from $x = 1$ to ∞ will be

$$F_\Gamma = -\rho Vb \int_1^\infty \frac{x_0}{\sqrt{x_0{}^2 - 1}} \gamma dx_0 \qquad (6.69)$$

Using a similar approach, the moment on the whole airfoil (caused by the vortex wake) about an axis located at a becomes:

$$M_\Gamma = -\rho Vb^2 \int_1^\infty \left[\frac{1}{2}\sqrt{\frac{x_0 + 1}{x_0 - 1}} - (a + ½) \frac{x_0}{\sqrt{x_0^2 - 1}}\right] \gamma dx_0 \,. \qquad (6.70)$$

It should be noted that since the elements of vorticity move with the fluid with a velocity V, their values would remain constant when referred to a system moving with the fluid. By contrast, for a fixed system related to the airfoil, it may be stated that

$$\gamma = f(Vt - x_0)\,,$$

where t is the elapsed time since the beginning of the motion.

The strength of circulation is determined by applying the Kutta condition of finite velocity at the trailing edge or, in other words, at $x = 1$:

$$V_{x=1} = \partial\varphi/\partial x = (\partial/\partial x)(\varphi_\Gamma + \varphi_a + \varphi_{\dot a} + \varphi_{\dot h}) = \textit{finite.}$$

Substituting the values of noncirculatory and circulatory potentials from Eqs (6.60), (6.61), (6.62), and (6.66) into the above formula, the following relationship is obtained:

$$(1/2\pi) \int_1^\infty \left[\left(\sqrt{x_0 + 1}\right)/\left(\sqrt{x_0 - 1}\right)\right] \gamma dx_0 = Va + \dot h + b(½ - a)\dot a = Q\,. \qquad (6.71)$$

Using Eqs (6.69), (6.70), and (6.71), the following expressions are obtained for the force (F) and moment (M_a) resulting from the circulatory potential:

$$F_\Gamma = -2\pi\rho VbQC \qquad (6.72)$$

and

$$M_{a\Gamma} = -2\pi\rho Vb^2 Q[½ - (a + ½)]C. \qquad (6.73)$$

where C is a special quantity defined by Theodorsen as a ratio of the following integrals:

$$C = \int_1^\infty \left[\left(x_0/\sqrt{x_0 - 1}\right) \gamma dx_0\right] \Big/ \int_1^\infty \sqrt{\left[\left(x_0 + 1\right)/\left(x_0 - 1\right)\right.} \left.\gamma dx_0\right] \qquad (6.74)$$

which is later expressed as a function of the positive constant k determining the wave length. This is done by assuming that

$$\gamma = \gamma_0 e^{i\left[\left(k\frac{s}{b} - x_0 + \varphi\right)\right]}$$

where $s = Vt$ is the distance from the first vortex element to the airfoil ($s \to \infty$). Now,

$$C(k) = \int_1^\infty (x_0/\sqrt{x_0^2 - 1})\, e^{-ikx_0}\, dx_0 \Big/ \int_1^\infty \left[(x_0 + 1)/\sqrt{x_0^2 - 1}\right] e^{-ikx_0}\, dx_0 . \tag{6.75}$$

Upon solving these integrals, $C(k)$ is expressed in terms of the functions E and G*
as follows:

$$C(k) = E + iG. \tag{6.75a}$$

These values are graphically shown in Figure 6.48.

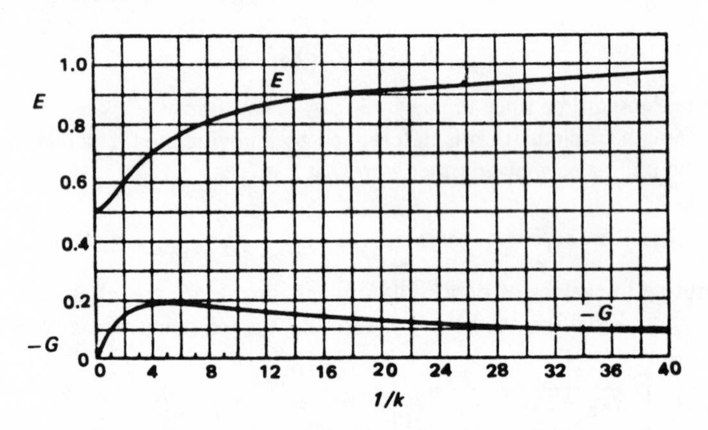

Figure 6.48 Functions E and C vs 1/k[31]

The total force and moment resulting from the noncirculatory and circulatory flows are now given as functions of various parameters characterizing an oscillating airfoil motion:

$$F = -\rho b^2 (V\pi\dot{a} + \pi\ddot{h} - \pi b a\ddot{a}) - 2\pi\rho b C[Va + \dot{h} + b(\tfrac{1}{2} - a)\dot{a}] - 2\pi \tag{6.76}$$

and

$$M_a = -\rho b^2 [\pi(\tfrac{1}{2} - a)Vb\dot{a} + \pi b^2(1/8 + a^2)\ddot{a} - a\pi b\ddot{h}]$$

$$+ 2\rho V b^2 \pi(\tfrac{1}{2} + a)C[Va + \dot{h} + b(\tfrac{1}{2} - a)\dot{a}]. \tag{6.77}$$

---

*Expressed as F and G in the original notations of Theodorsen.

7.4 Analytical Approach to Oscillations of Airfoils with Thickness and Camber

Oscillations Below Stall. The analytical approach to oscillations of flat plates as discussed in the preceding section can be extended to airfoils having finite thickness and camber. This problem will be briefly discussed by following the McCroskey approach[43].

Similar to the preceding case, it is assumed that the fluid is inviscid and incompressible, while displacements of the airfoil—whether pitching or plunging—are small. Under this assumption, the problem may be considered as linear; thus, the velocity potential of the flow around airfoils (φ) can be expressed as a sum of potentials due to thickness (φ_t), camber (φ_c), and angle-of-attack resulting from the rotation about a pitching axis located at a (φ_a):

$$\varphi = \varphi_t + \varphi_c + \varphi_a. \tag{6.78}$$

The first two components of the total potential are independent of time; therefore, they are the same as for a stationary airfoil. The third component, due to unsteady motion, can be interpreted as being the same as for the oscillating flat plate as discussed in Sect 7.3

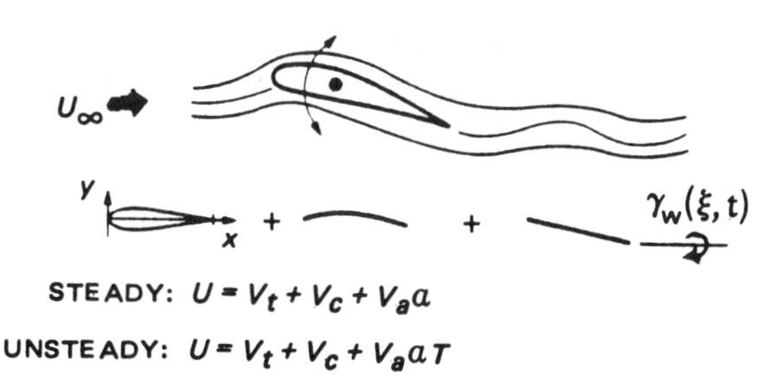

Figure 6.49 Interpretation of the flow around an airfoil as a sum of contributions from thickness, camber, and oscillating flat plate[43]

In light of Eq (6.78), the nondimensionalized chordwise velocity on the airfoil surface ($\bar{U} \equiv U/V_\infty$) existing at the airfoil chordwise coordinate x at time t, can be expressed as follows:

$$\bar{U}(x,t) = \bar{V}_t + \bar{V}_c + \bar{V}_a\, aT \tag{6.79}$$

where aT is the local instantaneous angle-of-attack, and $\bar{V}_t$, $\bar{V}_c$, and $\bar{V}_a$ are velocity components nondimensionalized with respect to V_∞.

T is assumed to be a function of both x and t, and is defined in Ref 43 as follows:

> The function $T(x,t)$, which contains *all* the unsteady effects, is the ratio of the unsteady to quasi-steady solutions for a flat plate at angle of attack a. In other words, we have merely substituted aT in place of a in the steady theory to obtain an unsteady solution of comparable quality.

From the unsteady Bernoulli equation, the chordwise pressure distribution is

$$C_p = 1 - U^2 - (2c/V_\infty)(\partial\varphi_a/\partial t) \tag{6.80}$$

where U is given by Eq (6.79) and $\partial\varphi_a/\partial t$ can be developed from the appropriate thin airfoil solution for an unsteady flat plate.

In order to obtain a better insight into the dependence of C_p on various parameters characterizing oscillatory motion of an airfoil, McCroskey linearized Eq (6.80). This was done by first squaring Eq (6.79):

$\overline{U}^2 = 2(\overline{V}_t + \overline{V}_c + \overline{V}_a \alpha T) - 1$ + *second-order terms.* When the second-order terms are neglected, and the so-obtained linearized expressions for $\overline{U}^2$ are substituted into Eq (6.80), a linearized expression for C_p (called $C_{p_{lin}}$) is obtained:

$$C_{p_{lin}} = 2(1 - \overline{V}_t - \overline{V}_c - \overline{V}_a \alpha T) - (2c/V_\infty)(\partial\varphi_a/\partial t).$$

This equation can be rewritten as follows:

$$\tfrac{1}{2}C_{p_{lin}} = 1 - \overline{V}_t - \overline{V}_c - \overline{V}_a \alpha R. \tag{6.81}$$

where the quantity $R = R(x,t)$ is analogous to $T(x,t)$ appearing in Eq (6.79), except that this time R represents all the unsteady effects for pressure distribution and similar to the determination of T, R can be defined as the ratio for unsteady to quasi-steady solutions for a flat plate at an angle-of-attack α.

In general, it may be stated that for sinusoidal oscillations, Theodorsen's function $C(k)$ is used to determine R and T, while for such other manifestations of unsteady aerodynamic phenomena as a step change in α or sharp-edge gusts, advantage is taken of other functions (i.e., those of Wagner & Kussner).

The above-indicated expressions for velocities and pressures at the surface of oscillating airfoils permit one to get at least an insight into some of the phenomena associated with dynamic stall. For instance, the unsteady phase lag and attenuation of the inviscid pressure gradients near the leading edge explain the dynamic delay in laminar boundary separation on oscillating airfoils. In this case, the inviscid theory correctly indicates the trends in the onset of large-scale boundary-layer separation, but not the quantitative delay above the angle-of-attack for static stall.

In order to achieve the goal of precise analysis of dynamic stall, the viscid-inviscid interactions, which became clearly apparent during the discussion of the physics of dynamic stall, must be incorporated into the mathematical model. Unfortunately, formulation and solution of the problem is extremely difficult. Consequently, attempts have been made to develop semi-empirical engineering methods for predicting airfoil dynamic stall characteristics.

An Example of Engineering Analysis of Dynamic Stall. The method of Erickson and Reding[44] for predicting dynamic stall characteristics using static data can be given as an example of an engineering analysis of the problem for airfoils oscillating in pitch at relative low reduced frequencies when $\omega c/V_\infty < 0.5$. in this approach it is assumed that there is equivalence between boundary-layer improvement due to pitch-rate-induced effects and increasing Reynolds number. Consequently, the maximum lift coefficient achievable during dynamic stall can not exceed $c_{\ell_{max}}$ corresponding to $R_e = \infty$.

"The technique is semiempirical and uses experimental data to determine certain critical proportionality constants for the effects of pitch amplitude and frequency on the dynamic-stall characteristics."

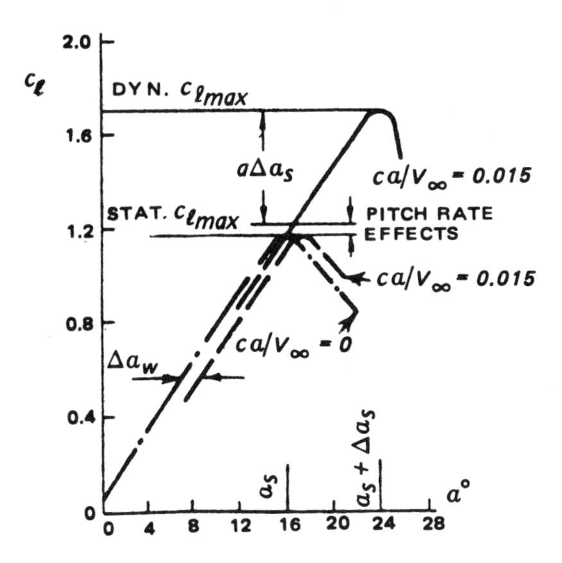

Figure 6.50 Composition of static $c_{\ell max}$ overshoot[44]

To illustrate the proposed approach, the reader's attention is called to the determination of the dynamic $c_{\ell max}$ overshoot over its static value (Fig 6.50).

First of all, it should be noted that the dynamic $c_\ell(a)$ curve is displaced to the right through an angle Δa_w, where the subscript w indicates that this displacement is associated with the wake effect. The displacement is the result of the existence of a time lag occurring in the circulation buildup of the oscillating airfoil. Using the v. Karman-Sears approach, this circulation lag is expressed as follows:

$$\Delta a_w = \xi_w c\dot{a}/V_\infty \tag{6.82}$$

where the value of the coefficient ξ_w is assumed to be *1.0*.

It is indicated in Ref 44 that the experimentally-determined Δa_w is somewhat lower than that given by Eq (6.82).

It can be seen from Fig 6.50 that the most important effect is that of Δa_s, representing an increase in the stall angle resulting from dynamic effects. This Δa_s increment is expressed as:

$$\Delta a_s = K_a c\dot{a}/V_\infty \tag{6.83}$$

where the K_a coefficient for airfoils oscillating about the quarter-chord axis is assumed to be $K_a = 3.0$.

Once the extension of $c_\ell(a)$ under unsteady conditions to the dynamic $c_{\ell max}$ value is established, the hysteresis loop for the $c_n(a)$ curve should be established. Details of this procedure can be found in Ref 45.

8. OPERATIONAL ENVIRONMENT OF ROTOR-BLADE AIRFOILS

The operational environment of the blade airfoil in the helicopter regime of flight is, in general, quite different from that of airplane wings. Furthermore, in nonaxial translation (forward flight), it also differs from that encountered by propellers. Most peculiarities existing in helicopter-type forward flight stem from the asymmetry of the incoming flow encountered by the blade in the rotor-disc plane. It should be emphasized, however, that although the helicopter-type forward flight generates a particularly difficult environment for blade airfoils, they may also encounter detrimental working conditions in axial translation (vertical climb and descent), and even in hovering.

To give the reader some quantitative insight into the operating conditions of blade airfoils during one complete rotor revolution in forward flight, the following cases are considered:

Variation of Inplane Velocity Componets. Expressing the inplane velocity component perpendicular to the blade axis $(U_{\perp})$, in nondimensional form as $\overline{U}_{\perp} \equiv U_{\perp}/V_t$, Eq (3.62) can be rewritten as follows:

$$\overline{U}_{\perp} = \overline{r} + \mu \sin \psi. \tag{6.84}$$

It can be seen from this equation that for $|-\mu \sin \psi| > \overline{r}$, $\overline{U}_{\perp}$ becomes negative or, in other words, airfoils encounter flow from the trailing edge. The extent of the regions of this reversed flow within the disc area can be judged from Fig 6.51.

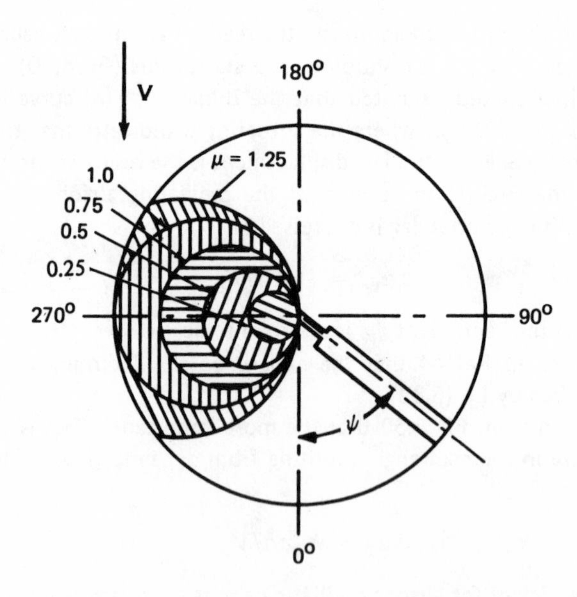

Figure 6.51 Regions of reversed flow encountered at various μ values

Figure 6.52 supplements Fig 6.51 by showing the magnitude of the $\overline{U}_{\perp}(\psi)$ variation at some selected blade stations and various μ values. It can be seen from this figure that at the so-called representative blade station $\overline{r} = 0.75$ and $\mu = 0.4$, the normal velocity component during one rotor revolution would fluctuate from $\bar{U}_{\perp} = 1.15$ at $\psi = 90°$, to $\bar{U}_{\perp} = 0.35$ at $\psi = 270°$.

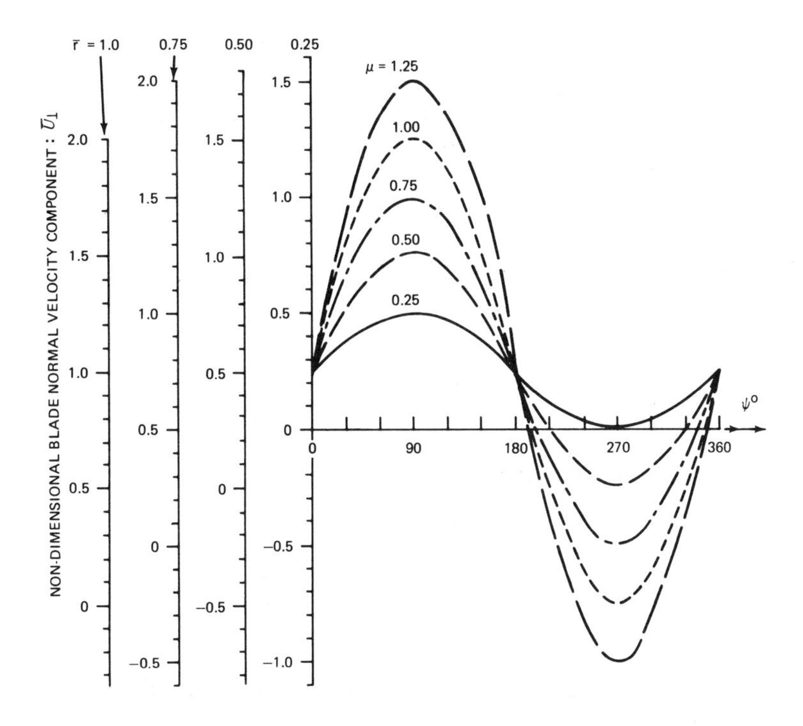

Figure 6.52 Variation of $\bar{U}_{\perp}$ vs ψ for various μ values at four blade stations

Presented in a nondimensional form, the inplane velocity component ($U_{\parallel}$) parallel to the blade axis ($\bar{U}_{\parallel} \equiv U_{\parallel}/V_t$) becomes

$$\bar{U}_{\parallel} = \mu \cos \psi. \tag{6.85}$$

Because of the $\bar{U}_{\perp}$ and $\bar{U}_{\parallel}$ variations, the sweep angle (Λ) existing at various blade stations would, in forward flight, also undergo periodic variations as the blade moves through the azimuth angle. Combining Eqs (6.84) and (6.85), the sweep angle becomes

$$\Lambda = \tan^{-1} \cos \psi/[(\bar{r}/\mu) + \sin \psi]. \tag{6.86}$$

Fig 6.53 gives an example of the sweep-angle variation for two values of the $\bar{r}/\mu$ parameter $\bar{r}/\mu = 0.5$, representing the blade station locations of the center of the reversed velocity circle, and $\bar{r}/\mu = 2.0$, corresponding to the blade station twice removed from the outer border of the reversed velocity circle. For instance, for $\mu = 0.4$, $\bar{r}/\mu = 0.5$ would represent $\bar{r} = 0.2$; and for $\bar{r}/\mu = 2.0$, $\bar{r} = 0.8$.

It can also be seen from this figure that at $\bar{r} = 0.2$, the blade airfoil sections during one revolution are subjected to a complete variation of the sweep angle including flow along the blade axis in both directions ($\Lambda = -90°$ and $\Lambda = 270°$) and flow from the trailing edge ($-90° > \Lambda > -270°$). By contrast, the sweep angle at the $\bar{r} = 0.8$ station would oscillate within much lower limits of $-30° < \Lambda < 30°$.

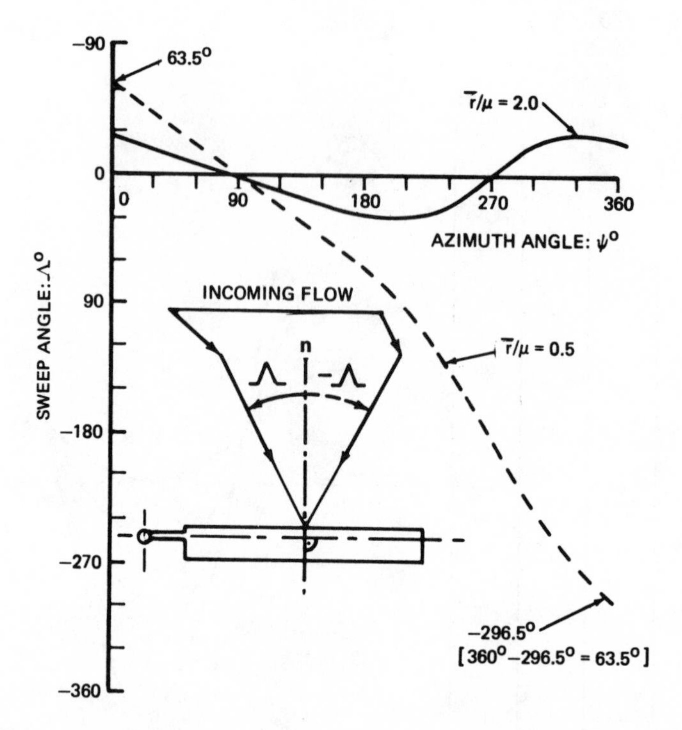

Figure 6.53 Examples of variation of sweep angles inside ($\bar{r}/\mu = 0.5$) and outside ($\bar{r}/\mu = 2.0$) of the reversed velocity circle

Periodic Angle-of-Attack Variation. Periodic variation of the blade section angle-of-attack represents another important characteristic of the operational environment of rotor blade airfoil sections.

The frequency of oscillations (f, in Hz) associated with the first-harmonic flapping and pitch control inputs will obviously be equal to the rotor rps. However, higher oscillatory frequency may be encountered due to higher blade flapwise bending modes, elastic torsional oscillations, or higher harmonic control inputs. This latter contribution is, as yet, seldom found in practice. Once the f values are known, reduced frequencies can be determined from the equations given in Sect 7.2.

With respect to amplitudes associated with blade oscillations, the highest are usually associated with the frequencies resulting from the first-harmonic inputs. For instance, the angles-of-attack of the outboard blade sections may vary from $a \leqslant 0$ to a's higher than those corresponding to static stall.

In addition to frequencies of of oscillations and the associated amplitudes, there is another important parameter; namely the $c\dot{a}/V_{\infty}$ ratio. It can be seen from Eq (6.83) that this quantity appears in the expression for the increment of the dynamic stall angle over its static value.

For this reason, an expression containing $c\dot{a}/V_{\infty}$ which, for the case of blades becomes $c\dot{x}/U_{\perp}$, is defined as the stall delay function (also called gamma function):

$$\gamma = c\dot{a}/2U_{\perp}. \tag{6.87}$$

Assuming, for simplicity, that the angle-of-attack of a blade station varies in a sinusoidal manner between its lowest, (a_{min}) at $\psi = 90°$, and its maximum, (a_{max}) at $\psi = 270°$, values so that $(\Delta a)_{max} \equiv a_{max} - a_{min}$; the incremental variation of $\Delta a(\psi)$ can be expressed as follows:

$$\Delta a = ½(\Delta a)_{max}(1 - sin\ \psi); \tag{6.88}$$

and consequently, $\dot{a}$ becomes

$$\dot{a} = -½\Omega(\Delta a)_{max}\ cos\ \Omega \tag{6.89}$$

where $\Omega \equiv d\psi/dt$.

The other quantities appearing in Eq (6.87) can be expressed as follows: $c = \sigma\pi R/b$, where σ is the rotor solidity, R is the radius, b is the number of blades; and $U_\perp = \Omega R(\bar{r} + \mu\ sin\ \psi)$. Thus,

$$\gamma = -(\pi/4)[(\Delta a)_{max}(\sigma/b)\ cos\ \psi]\ /\ (\bar{r} + \mu\ sin\ \psi). \tag{6.90}$$

From Eq (6.90) it becomes apparent that values of the parameter γ are independent of Ω and consequently, are not influenced by the rotor diameter and its tip speed, but by the rotor geometry only (number of blades and solidity). This finding is of practical importance, since it permits one to transfer empirical results—within proper limits of Reynolds and Mach numbers—of the stall limit extension obtained from geometrically similar models to full-scale rotors operating at the same μ values.

Acceleration of Boundary Layer. Another phenomenon—absent in fixed-wing aircraft, but encountered by airscrew blades—is the centrifugal acceleration field acting on the boundary layer. At present, the significance of this acceleration and its infuence on airfoil characteristics is not fully understood. Nevertheless, it should be realized that (1) the boundary layer of all airscrews are subjected to high accelerations, and (2) for the same tip-speed value, this magnitude increases inversely proportionally to the airscrew radius R. This can be readily seen from the expression for acceleration at a blade station $\bar{r}$ (a_r in m/s^2 or fps^2):

$$a_r = V_t\bar{r}(V_t/R). \tag{6.91}$$

Acceleration at the tip (expressed in g's) is shown in Fig 6.54 as an example of the function of R for the three selected levels of V_t. For instance, for a blade having a radius of 20 ft and a tip speed of 700 fps, the tip acceleration would be $a_t = 760\ g's$. However, for a 1/5-scale model, $a_t = 3800\ g's$.

Surface Roughness and Flow Turbulence. Finally, it should be emphasized that because of their operational environment, the blades of rotary-wing aircraft are frequently exposed to sand and rain erosion. Consequently, a higher degree of surface roughness (especially in the leading-edge area) can be expected than on fixed-wing aircraft.

It should also be noted that due to the presence of vorticity in the blade wake —especially in forward flight—the airfoils encounter much more turbulent incoming flow than airfoils of fixed-wing aircraft. Both surface roughness and incoming flow turbulence should be taken into consideration when selecting or developing new airfoil sections for rotor blades.

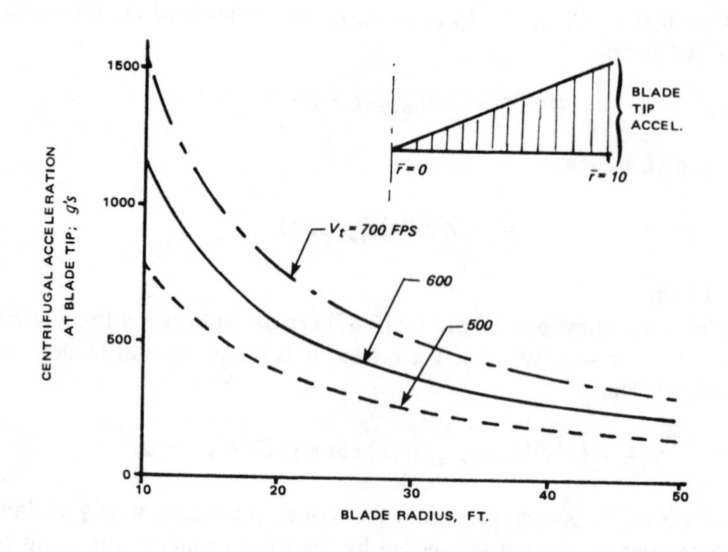

Figure 6.54 Centrifugal acceleration acting on the boundary layer at the blade tip

9. CONTRIBUTION OF AIRFOIL CHARACTERISTICS TO PERFORMANCE

In order to give the reader some idea regarding the extent to which improvements in airfoils may contribute to the efficiency of helicopter performance in hover and forward flight, the influence of the $c_d(c_\ell)$ on figure-of-merit (FM) and L/D_e are briefly discussed.

9.1 Hover

As far as hover is concerned, the influence of airfoil characteristics on performance would manifest itself through the dependence of the rotor figure-of-merit on the $c_d(c_\ell)$ function.

To facilitate this investigation, the expression for $FM \equiv \eta_{eh}$ (Eq 3.57) is rewritten in the following terms:

$$FM = 1/[k_{ind} + 2.60/\sqrt{\sigma}(\bar{c}_\ell^{\,3/2}/\bar{c}_d)] \tag{6.92}$$

where k_{ind} is the ratio of actual induced rotor power to the ideal induced rotor power $k_{ind} = RP_{ind}/RP_{id}$; $\bar{c}_\ell$ is the average rotor lift coefficient, $\bar{c}_\ell = 6w/\sigma\rho V_t^2$; $\bar{c}_d$ is the average profile drag coefficient, $\bar{c}_d \equiv 8RP_{pr}/\sigma\pi R^2\rho V_t^3$; and σ is the rotor solidity.

The rotor solidity is strongly governed by structural weight considerations and other design factors; however, its level is also influenced by such airfoil characteristics as absolute value of the airfoil $c_{\ell_{max}}$ and the character of the $c_d(c_\ell)$ relationship in the high c_ℓ region. It is obvious that high values of $c_{\ell_{max}}$ accompanied by gentle airfoil stall characteristics would allow the designer to reduce rotor solidity. Nevertheless, for the present consideration, it will be assumed that $\sigma = const$ in Eq (6.92).

It becomes clear that in order to maximize the figure-of-merit, k_{ind} should be made as low as possible, while $\bar{c}_\ell^{\,3/2}/\bar{c}_d$ should be as high as possible. The k_{ind} is minimized by first attacking the induced power level through such means as blade chord and twist distribution; and then, the geometry of the blade arrangement within the rotor. Such airfoil characteristics as prevention of stall, and the steepness of the lift-curve slope also contribute somewhat to the k_{ind} level.

With respect to the second term in the square brackets of Eq (6.92), the task would consist of the development of invariable airfoils showing the highest possible $c_\ell^{3/2}/c_d$ levels. It is, in principle, also possible to improve this ratio for the case of hover through the proper variation of the airfoil shape (e.g., deflection of the trailing- or leading-edge segments), and through such means as boundary-layer or circulation control.

Assuming that $\sigma = 0.10$ is typical of contemporary designs, Fig 6.55 was prepared in order to give some idea of the combinations of k_{ind}'s and $\bar{c}_\ell^{\,3/2}/\bar{c}_d$'s required for $0.7 \leqslant FM \leqslant 0.85$.

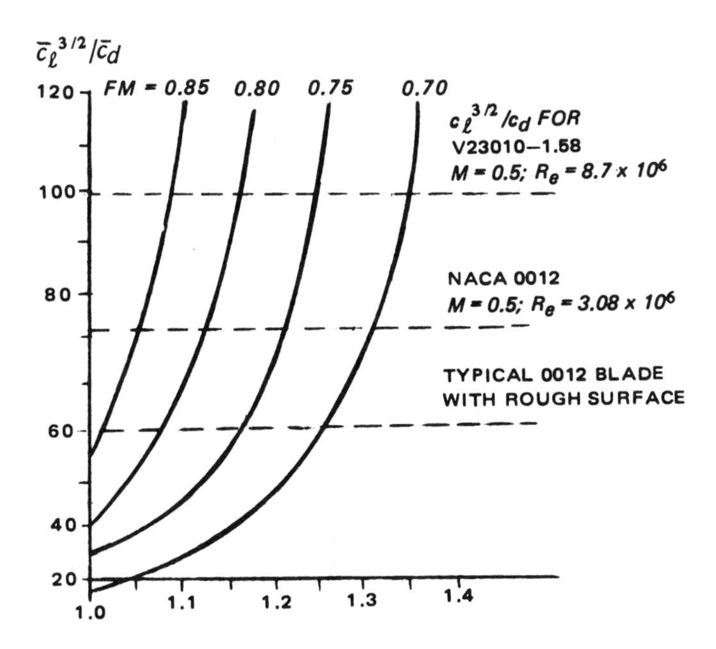

Figure 6.55 Required $\bar{c}_\ell^{\,3/2}/\bar{c}_d$ vs k_{ind} for given FM values

In addition, typical values of $(c_\ell^{3/2}/c_d)_{max}$, which are obtainable with the symmetrical 0012 and cambered V23010-1.58 airfoils, are also noted on this figure. The two higher values are based on wind-tunnel tests of smooth models[7] (Fig 6.56), while the lowest one may be considered as representative of symmetrical airfoils of blades having surfaces roughened by erosion as encountered in actual operations.

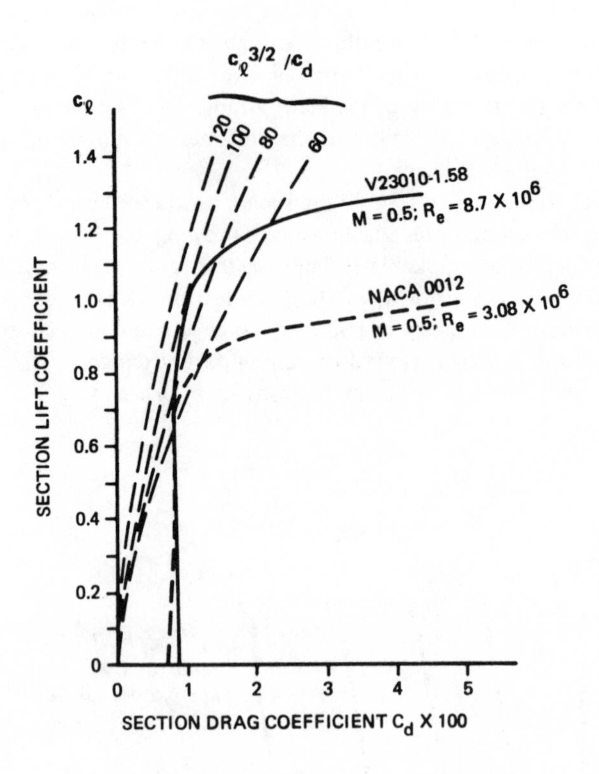

Figure 6.56 Examples of cambered (V23010-1.58) and symmetrical (NACA 0012) airfoil sections

Eq (6.92) and Fig 6.55 relate the figure-of-merit to the $\bar{c}_\ell^{\,3/2}/\bar{c}_d$ ratios based on the average values of lift and profile drag coefficients. This means that in order to maximize these ratios, the blade planform and/or blade twist distribution should be such that particular blade sections operate at c_ℓ's as close as possible to those corresponding to $(c_\ell^{\,3/2}/c_d)_{max}$. However, it would probably be impossible to reach this maximum at all blade stations simultaneously. Consequently, the maximum $\bar{c}_\ell^{\,3/2}/\bar{c}_d$ value for the blade as a whole should be expected to be lower than the $(c_\ell^{\,3/2}/c_d)_{max}$ found from wind-tunnel tests of smooth models of a particular airfoil section.

It is also apparent from Fig 6.55 that once $\bar{c}_\ell^{\,3/2}/\bar{c}_d$ becomes higher than 100 for $\sigma \approx 0.1$, the *FM* curves become less sensitive to further improvements in this ratio. It appears hence, that from the *FM* point-of-view, airfoil development has already reached a level where further improvements in their $c_\ell^{\,3/2}/c_d$ levels would contribute little to an improvement in the figure-of-merit for rotors having $\sigma \approx 0.1$. Thus, a reduction of the k_{ind} values becomes the way for increasing the figure-of-merit.

It should also be noted from Fig 6.56 that maximum values of $c_\ell^{\,3/2}/c_d$ occur at c_ℓ's close to the incipience of stall. The need of providing a sufficient margin for maneuvers may force designers to select the average operational rotor-lift coefficient in hover—generally, at lower values than those corresponding to $(\bar{c}_\ell^{\,3/2}/\bar{c}_d)_{max}$.

9.2 Forward Flight

It is intended to give the reader only a general idea about the influence of blade airfoil section characteristics on forward flight performance. To achieve this goal, expressions for the lift-to-the-equivalent-drag ratio of the rotor $(L/D_e)_R$ alone—with no hub contribution—are obtained by assuming, in Eq (3.106), that the rotor lift (L) is substituted for the weight (W), the parasite drag term is equal to zero, and the download coefficient is equal to 1.0. The so-obtained expression for HP_R is then substituted into Eq (3.109); thus, $(L/D_e)_R$ becomes

$$(L/D_e)_R^{-1} = \frac{wk_{ind}}{2\rho V_{ho}^2} + \frac{1}{8}\frac{\sigma\pi R^2 \rho V_t^3(1+4.7\mu^2)\bar{c}_d}{LV_{ho}}. \tag{6.93}$$

Remembering that $6w/\sigma\rho V_t^2 = 6L/\sigma\pi R^2 \rho V_t^2 = \bar{c}_\ell$, where the so-defined $\bar{c}_\ell$ is the average lift coefficient, Eq (6.93) becomes

$$(L/D_e)_R^{-1} = (1/12\mu^2)\sigma k_{ind}\bar{c}_\ell + (3/4\mu)(1+4.7\mu^2)(\bar{c}_d/\bar{c}_\ell) \tag{6.94}$$

where k_{ind}, in the present case, represents the induced power coefficient in forward flight; and μ is the advance ratio. The remaining symbols are as defined in the preceding section.

Differentiating Eq (6.94) with respect to $\bar{c}_\ell$, and solving the $\partial(L/D_e)_R/\partial\bar{c}_\ell = 0$ equation, an expression for the $\bar{c}_\ell$ value ($\bar{c}_{\ell\,opt}$) minimizing $(L/D_e)_R^{-1}$; i.e., maximizing (L/D_e), is obtained:

$$\bar{c}_{\ell\,opt} = 3\left[\sqrt{\mu(1+4.7\mu^2)/\sigma k_{ind}}\right]\sqrt{\bar{c}_d}. \tag{6.95}$$

Taking the optimum value of $\bar{c}_\ell$ from Eq (6.95) and substituting this value into Eq (6.94), we obtain the following expression for the inverse of the maximum $(L/D_e)_R$:

$$\left[(L/D_e)_R\right]_{max}^{-1} = \left[\sqrt{\sigma k_{ind}}/2\mu\right]\left[\sqrt{(1/\mu)(1+4.7\mu^2)}\right]\sqrt{c_d}. \tag{6.96}$$

It is again assumed, as in Sect 9.1, that the rotor solidity is fixed ($\sigma = const$) as a result of design considerations, and that k_{ind} is known.

Eqs (6.95) and (6.96) clearly indicate that for a given μ, the average blade profile drag coefficient ($\bar{c}_d$) is the parameter governing both the level of optimum operational average rotor lift coefficient and maximum value of the rotor lift-to-the-equivalent-drag ratio. The influence of $\bar{c}_d$ can be seen from Fig 6.57 where $\bar{c}_{\ell\,opt}$ and $[(L/D_e)_R]_{max}$ are shown as functions of $\bar{c}_d$ for several μ values. Auxiliary $\bar{c}_\ell/\bar{c}_d = const$ lines should enable the reader to appreciate the order of magnitude of the airfoil lift-to-drag ratios associated with the $\bar{c}_{\ell\,opt}$ values.

Figure 6.58 was prepared in order to indicate how deviations in the $\bar{c}_\ell$ value from its optimal level of $\bar{c}_{\ell\,opt}$ would affect the decrease of $(L/D_e)_R$ from $[(L/D_e)_R]_{max}$. It can be seen that the character of $(L/De)_R = f(\bar{c}_\ell)$ is such that deviation from $\bar{c}_{\ell\,opt}$ by as much as $\Delta\bar{c}_\ell \pm 0.2$ leads to a relative small decrease in $(L/D_e)_R$ from $[(L/D_e)_R]_{max}$ as

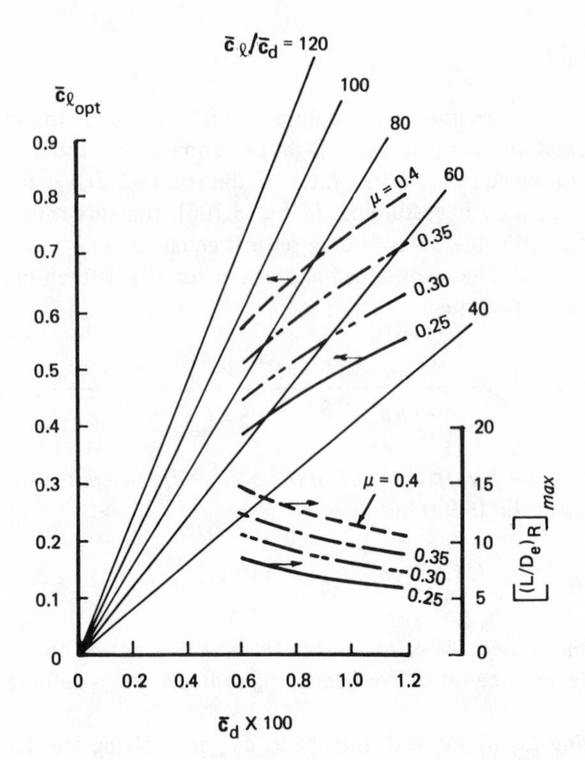

Figure 6.57 $\bar{c}_{\ell_{opt}}$ *vs* $\bar{c}_d$ *and corresponding* $[(L/D_e)_R]_{max}$ *for* $k_{ind} = 1.25$ *and* $\sigma = 0.1$

long as $\bar{c}_d$ remains constant. This means that from the point of view of the blade airfoil characteristics, it is important to have airfoils which retain low c_d values both above and below $c_\ell \approx \bar{c}_{\ell_{opt}}$.

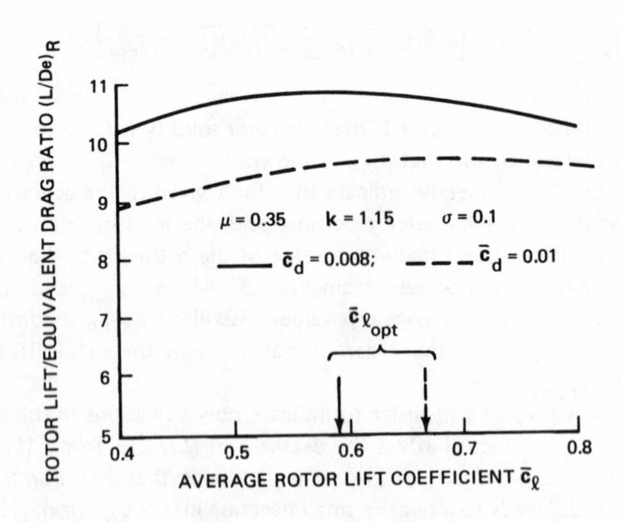

Figure 6.58 $(L/D_e)_R$ *values as affected by the deviation of* $\bar{c}_\ell$*'s from their optimum level*

It can be seen from Eq (6.96) that the product of rotor solidity (σ) and the k_{ind} factor can be treated as a single parameter—the lower the value, the higher the maximum lift-to-the-equivalent-drag ratio at a given μ.

In order to show the combination of σk_{ind} and $\bar{c}_d$ required to achieve various $[(L/D_e)_R]_{max}$ values, Fig 6.59 was prepared giving this relationship at $\mu = 0.35$.

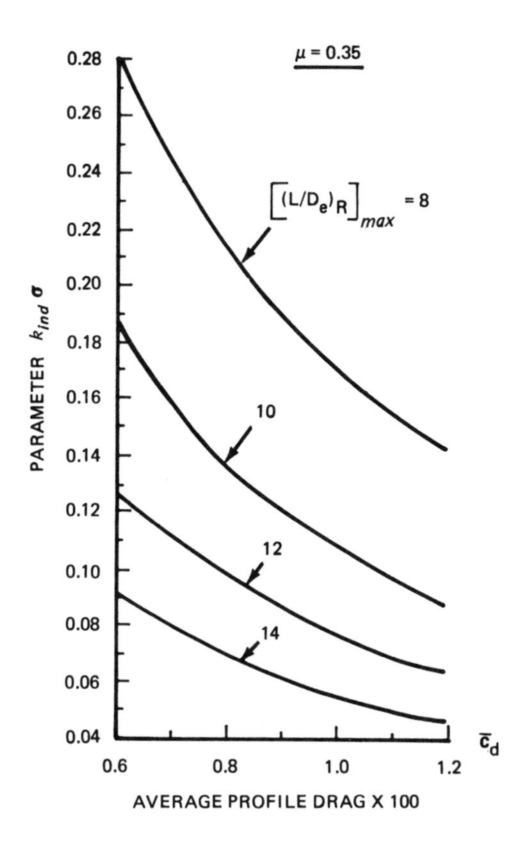

Figure 6.59 Combinations of $\bar{c}_d$ and the σk_{ind} parameter required to achieve given $[(L/D_e)_R]_{max}$ values

10. CONCLUDING REMARKS

For decades, the NACA 00 airfoil family and especially, the 0012, appeared to be the only blade section suitable for helicopter application.

In the late Fifties, it became apparent that incorporation of a camber (although still through modifications of the 0012 airfoil) resulted in performance improvements in all regimes of flight. This, in turn, provided an incentive for a more vigorous search for airfoil shapes that would still further contribute to performance gains and/or improvements in flying qualities. However, this task is not easy since conflicting airfoil characteristics may be considered as optimum, depending on the regime of flight, or even the azimuthal position of the blade.

Hover. It was indicated that a high $c_\ell^{3/2}/c_d$, combined with low rotor solidity ratios, is required in order to maximize the figure-of-merit. But, in turn, the solidity ratio is inversely proportional to the average lift coefficient at which the rotor operates. This obviously means that airfoil sections best suited for hovering conditions should possess the following characteristics: (a) ability to operate continuously at high section-lift coefficients, (b) still provide a lift coefficient margin for maneuvers and gusts at high operational c_ℓ's, and (c) have the highest $c_\ell^{3/2}/c_d$ ratios possible. Finally, operations at these high lift coefficients should be as free as possible from compressibility effects, thus allowing the designer more freedom in selecting higher rotor tip speeds, should this become desirable for overall design optimization.

To satisfy all these demands, highly cambered airfoils, about 12-percent thick, appear as the most suitable. However, even in this case, a conflict in these requirements may now develop since high camber is usually accompanied by elevated negative c_m values which may be unacceptable because of aeroelastic instabilities and high control loads. An upward-reflected trailing-edge section may be required to alleviate these problems. However, this usually leads to a reduction in the ability to achieve high section-lift coefficients.

Forward Flight. During high-speed forward flight, an obvious conflict exists in the blade characteristic requirements on the advancing and retreating sides. Because of the combination of very low section-lift coefficients and high Mach numbers encountered by the outer sector of the advancing blade, one would like to see airfoils exhibiting high drag and moment divergence Mach numbers around $c_\ell = 0$. This, of course, would favor either symmetrical or slightly cambered thin airfoil sections.

On the retreating side, the operational environment is just the opposite; here, relatively low Mach numbers are encountered, but airfoil section-lift coefficients should be high and the $c_\ell(\alpha)$ relationship should be characterized by gentle stall. Consequently, moderately thick (or thick) cambered airfoil sections appear as the most suitable for this situation.

For the azimuth angles around the upwind and downwind positions of the blade, the operational conditions of airfoils are again different from those on the advancing and retreating sides. Here, there is the possibility that the blade sections may encounter a combination of relatively high Mach numbers with high section-lift coefficients. Under these circumstances, it would be desirable to have blade airfoil sections that would tolerate the moderately high $M - c_\ell$ combinations with no detrimental compressibility effects. Moderately thin cambered airfoils would fulfill these requirements.

At present, only invariable blade airfoil sections are used in practical helicopter designs. Consequently, the only possible solution consists of selecting airfoil shapes in such a way that a compromise can be reached between the often conflicting demands for aerodynamic characteristics. However, one can imagine that in the future, practical means of varying the airfoil geometry for different regimes of flight or even as a function of the azimuth angle, may be developed. It is also possible that some goals of the variable geometry airfoil sections could be achieved through boundary layer or circulation controls. Should any of these stipulations become a reality, new gains in performance of conventional helicopters would be possible.

The fundamentals of airfoil theory presented in this chapter should provide the reader with a better insight into the possibilities and problems of finding or developing blade sections representing a good compromise within the invariable geometry constraints or to establish guidelines for the development of variable geometry blade sections or flow-control schemes. The latter would enable one to develop blade sections whose aerodynamic characteristics could be adapted to various operating conditions.

This chapter on airfoil theory concludes the presentation of basic theories of rotary-wing aerodynamics.

References for Chapter VI

1. Abbott, I.H. and v. Doenhoff, A.E. *Theory of Wing Sections.* Dover Publications, New York, 1958.
2. Glauert, H. *The Elements of Aerofoil and Airscrew Theory.* Cambridge University Press, London, 1926.
3. Durand. *Aerodynamic Theory.* Durand Reprinting Committee, Cal Tech., 1943.
4. Rauscher, M. *Introduction to Aeronautical Dynamics.* John Wiley & Sons, Inc., New York, 1953.
5. Clancy, L.J. *Aerodynamics.* John Wiley & Sons, Inc., New York, 1975.
6. Hoerner, S.F. and Borst, H.V. *Fluid-Dynamic Lift.* Henry Borst Assoc., Ardmore, Pa. 19003, 1975.
7. Dadone, L.V. *U.S. Army Helicopter Design DATCOM,* Vol 1 – *Airfoils.* USAAMRDL CR 76-2, Moffett Field, Ca. 94025, 1976.
8. Cheeseman, I.C. and Seed, A.R. *The Application of Circulation Control by Blowing to Helicopter Rotors.* Journal of the Royal Aeronautical Society, Vol 71, July 1967.
9. Theodorsen, Theodore. *Theory of Wing Sections of Arbitrary Shape.* NACA TR No. 411, 1931.
10. Theodorsen, Theodore and Garrick, I.E. *General Potential Theory of Arbitrary Wing Sections.* NACA TR No. 452, 1933.
11. Peebles, G.H. *A Method for Calculating Airfoil Sections from Specification on the Pressure Distribution.* Journal of Aero. Sciences, Vol 14, No 8, Aug. 1947.
12. Jeans, James. *An Introduction to the Kinetic Theory of Gases.* Cambridge University Press, 1940.
13. Schlichting, H. *Boundary-Layer Theory.* McGraw Hill Co., New York, Sixth Edition, 1968.
14. Lachman, V. (Editor). *Boundary Layer and Flow Control, Its Principle and Application.* Vols I and II, Pergamon Press, New York, London, Paris, 1961.
15. *Proceedings of the International Congress of Subsonic Aeronautics.* The New York Academy of Sciences, New York, Apr. 1967.

16. Dryden, Hugh L. *Recent Advances in the Mechanics of Boundary Layer Flow. Advances in Applied Mechanics.* Edited by R.v. Mises and T. v. Karman, Vol 1, Academic Press, Inc., New York, 1948.

17. Tetervin, Neal. *A Review of Boundary Layer Literature.* NACA TN No 1348, 1947.

18. Gault, D.E. *A Correlation of Low-Speed, Airfoil-Section Stalling Characteristics with Reynolds Number and Airfoil Geometry.* NACA TN 3963, March 1957.

19. van den Berg, B. *Reynolds Number and Mach Number Effects on the Maximum Lift and the Stalling Characteristics of Wings at Low Speeds.* NLR TR 69025U, March 1969.

20. Raspet, August. *Control of the Boundary Layer on Sailplanes.* Mississippi State College, Mi., 1952.

21. McCloud, J.L., III, Hall, L.P., and Brady, J.A. *Full-Scale Wind-Tunnel Tests of Blowing Boundary-Layer Control Applied to Helicopter Rotors.* NASA TN D-335, Sept. 1960.

22. Jacobs, E.N., and Sherman, A. *Airfoil Section Characteristics as Affected by Variations of the Reynolds Number.* NACA TR No 586, 1937.

23. Tanner, W.H. and Yaggy, P.F. *Experimental Boundary Layer Study on Hovering Rotors.* J. Amer. Helicopter Soc., Vol 11, No 3, 1966, pp. 22-37.

24. Velkoff, H.R., Blaser, D.A., and Jones, K.M. *Boundary Layer Discontinuity on a Helicopter Rotor Blade in Hovering.* AIAA Paper 69-197, 1969.

25. Dwyer, H.A. and McCroskey, W.J. *Crossflow and Unsteady Boundary-Layer Effects on Rotating Blades.* AIAA Paper 70-50, 1970.

26. McCroskey, W.J. *Measurements of Boundary Layer Transition, Separation and Streamline Direction on Rotating Blades.* NASA TN D-6321, Apr. 1971.

27. Sloof, J.W., Wortman, F.X., and Duhnon, J.H. *The Development of Transonic Airfoils for Helicopters.* Preprint 901 of the 31st AHS National Forum, May 1975.

28. Vincent de Paul, M. and Dyment, A. *Récherches sur profils' d'alles en écoulement subsonique élevé.* ONERA TP No 815, 1970.

29. Nitzberg, G.E. and Crandall, S. *A Study of Flow Changes Associated with Airfoil Section Drag Rise at Supercritical Speeds.* NACA TN No 1813, Febr. 1949.

30. Dadone, Leo. *Two-Dimensional Wind-Tunnel Tests of an Oscillating Rotor Airfoil.* Vol 1, NASA CR 2914, Dec. 1977.

31. Theodorsen, T. *General Theory of Aerodynamic Instability and the Mechanism of Flutter.* NACA TR No 496, 1935.

32. Silverstein, A. and Joyner, U.T. *Experimental Verification of the Theory of Oscillating Airfoils.* NACA TR 673, 1939.

33. Bisplinghoff, R., Ashley, H., and Halfan, R. *Aeroelasticity.* Addison-Wesley, 1955.

34. Fung, Y.C. *The Theory of Aeroelasticity.* Dover Publications, New York, 1969.

35. Harris, F.D. and Pruyn, R.R. *Blade Stall—Half Fact, Half Fiction.* Proceedings of the 23rd Annual National AHS Forum, No. 101, May 1967.

36. Liiva, J., Davenport, F., Gray, L., and Walton, I. *Two-Dimensional Tests of Airfoils Oscillating near Stall.* U.S. Army Aviation Materiel Lab., Fort Eustis, Va., Tech. Report 68-13, April 1968; also J. of Aircraft, Vol 6, No 1, Jan. 1969, pp. 46-51.

37. McCroskey, W.J. *Some Current Research in Unsteady Fluid Dynamics.* J. of Fluid Engineering, Vol 99, March 1977.

38. Werle' H. and Armand, C. *Mesures et visualisations instationnaires sur les rotors.* ONERA TP No 777, 1969.

39. Jones, W.P. *Research on Unsteady Flow.* J. of the Aerospace Sciences, Vol 29, No 3, March 1962.

40. Fukishima, T. and Dadone, L.V. *Comparison of Dynamic Stall Phenomena for Pitching and Vertical Translation Motions.* NASA CR 2793, Feb. 1977.

41. Johnson, W. and Ham, N.D. *On the Mechanism of Dynamic Stall.* J. of AHS, Vol 17, No 4, Oct. 1972.

42. McCroskey, W.J., Carr, L.W., and McAlister, K.W. *Dynamic Stall Experiments on Oscillating Airfoils.* AIAA Journal, Vol 14, No 1, Jan. 1976, pp. 57-63.

43. McCroskey, W.J. *The Inviscid Flowfield of an Unsteady Airfoil.* AIAA Paper No 72-681, June 1972.

44. Ericsson, L.E. and Reding, J.P. *Dynamic Stall Analysis in Light of Recent Numerical & Experimental Results.* J. of Aircraft, Vol 13, No 4, Apr. 1976.

45. Ericsson, L.E. and Reding, J.P. *Analytic Prediction of Dynamic Stall Characteristics.* AIAA Paper 72-682, Boston, Mass., June 1972.

INDEX

VOLUME II:

PERFORMANCE PREDICTION OF HELICOPTERS

PREFACE

It is generally recognized that the educational value of a textbook is enhanced when numerical examples are included in the text. The readers and students not only become acquainted with computational procedures, but they also acquire an awareness regarding the magnitude of various values encountered in practice.

This need for illustrating theories by showing their practical application through numerical examples and special problems can be satisfied through two approaches.

(1) *The classroom approach*, used in many technical textbooks, is based on the incorporation of mutually unrelated, or only loosely related, short problems—quite often with answers—usually presented at the end of chapters or even shorter sections. This philosophy may be especially appealing to professors and instructors as being better suited for purely academic applications.

(2) *The total project approach* represents another way of providing the necessary illustrative material. Here, the submitted example is patterned on the actual industrial practice of dealing with a complete task which, in this case, is the prediction of helicopter performance. Various phases of performance calculations are related to suitable theoretical counterparts, thus providing examples for their reduction to practice. In addition to the purely illustrative aspect, a unified picture of the application of aerodynamic theory to performance predictions along with the computational methods used in industry can be presented.

Since the completed text is destined not only for classroom use, but also is intended to be of some help to the practicing engineer, the second approach was selected. Consequently, this volume was written to complement the rotary-wing aerodynamic theories discussed in Volume I, and contains complete and detailed performance calculations for conventional single-rotor, winged, and tandem-rotor helicopters.

Volume II is divided into five chapters and two appendices. Chapters I, II, and III describe detailed performance techniques for a single-rotor helicopter in hover, vertical ascent, and forward flight. Winged and tandem-rotor helicopter performance calculations are presented in Chapters IV and V as extensions and modifications of single-rotor methodology. The Appendices deal with the following special problems: (a) determination of guaranteed performance values based on both theory and test data , and (b) techniques of "growing" an aircraft to offset unprojected increases in weight empty.

Many of the sample calculations presented in Volume II employ computers to integrate the blade element expressions derived in Volume I. Computer data based on the vortex theory is compared with the approximate results obtained from the simplified momentum theory and blade element solution. In many cases correction factors or adjustments to the expressions are determined from these comparisons, and are often used to develop practical short-cut, but sufficiently accurate, prediction methods.

The calculations reflect up-to-date practices used in industry. Although the methods are chiefly based on those used by Boeing Vertol Company, they may be considered typical of the techniques used by a majority of helicopter manufacturers. This premise was borne out factually and enhanced through extensive reviews.

The presented text was first critically examined by Mr. A. Morse and Dr. F. H. Schmitz of Ames Directorate of AMRDL. Then, to further assure that the material was in compliance with generally accepted computational methods, the manuscript was submited to representatives of research institutions, industry, and universities as suggested by Dr. I. Statler of AMRDL. Many valuable technical and editorial inputs resulted from the reviews, and most or them were incorporated by the editors into a revised version.

The editors regret that manuscript deadlines prevented conversion of this volume to the SI (International Metric System) units; however, all formulae presented in Volume I are given in both SI and English units.

The author and editors wish to express their sincere appreciation to all those who devoted their time and effort in reviewing the text, and especially to Dr. I. Statler, Mr. A. Morse, and Dr. F. H. Schmitz of AMRDL; Dr. Andrew Z. Lemnios and staff of Kaman Aerospace Corporation; Professor Barnes W. McCormick, Dept. of Aerospace Engineering, The Pennsylvania State University; and personnel of Langley Directorate, AMRDL; Bell Helicopter Textron; Hughes Helicopters; and Sikorsky Aircraft.

W. Z. Stepniewski

TABLE OF CONTENTS

INTRODUCTORY CONSIDERATIONS

This chapter contains a description of the hypothetical single-rotor helicopter configuration, performance summary, engine performance characteristics, and the standard-day atmosphere relationships used to define ambient pressure, temperature, and density ratio.

Principal notation for Chapter I

c	chord	ft
c_d	airfoil section drag coefficient	
c_ℓ	airfoil section lift coefficient	
c_m	airfoil section moment coefficient	
DW	design gross weight	lb
h	altitude	ft
FUL	fixed useful load	lb
f_e	equivalent drag flate plate area	ft^2
INT	intermediate	
M	Mach number	
N	rotational speed	rpm, or rps
p	pressure	lb/ft^2, or in. of Hg
R	gas constant	ft/°C
R	rotor radius	ft
R_e	Reynolds number	
r	radial distance from rotor axis	ft
SL	sea level	
STD	standard	
sfc	specific fuel consumption	lb/hp;hr
V	velocity	fps or kn
W	weight, or gross weight	lb
WE	weight empty	lb
$T = 273.16 + t°C$	absolute temperature	K
t	temperature	°C, or °F
x	abscissa	in, or ft
y	ordinate	in, or ft
$\delta = p/p_o$	pressure ratio	
$\theta = T/288.16$	temperature ratio	
ρ	air density	$slugs/ft^3$
$\sigma_\rho = \rho/\rho_o$	density ratio	

Performance

Subscripts

F	fuel
o	initial, or SL/STD
p	pressure
t	tip
tr	tail rotor
ρ	density

Superscripts

	derivative with respect to time	per s, or hr

1. DESCRIPTION OF THE HYPOTHETICAL HELICOPTER CONFIGURATION

A typical 15,000-lb gross weight aircraft with a 50-ft diameter main rotor was chosen to illustrate the techniques used to predict single-rotor helicopter performance. To make the aircraft as realistic as possible, the configuration design is similar to one of the studies of the Utility Tactical Transport Aircraft System (UTTAS) helicopters. A detailed description of this configuration is given in Table I-1, and a 3-view drawing of the aircraft is shown in Fig 1.1. A brief discussion of the most important features of this design is presented below.

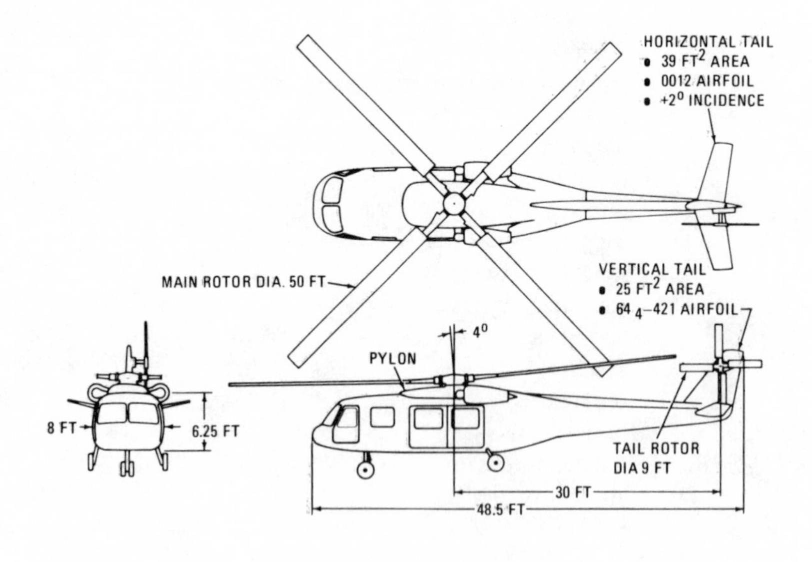

Figure 1.1 Three-view drawing of the hypothetical single-rotor helicopter

WEIGHTS	
Maximum Gross Weight	18,000 lb
Design Gross Weight	15,000 lb
Disc Loading @ Design Gross Weight	7.64 lb/ft^2
Weight Empty	9,450 lb
Weight Empty/Design Gross Weight	0.630
Fixed Useful Load (2 Crew @ 200 lb ea, and 30 lb trapped liquid)	430 lb
Fuel Capacity (354 gal, JP-4)	2,300 lb
MAIN ROTOR	
Diameter	50 ft
Chord	24 in
Solidity	0.102
Tip Speed	700 fps
Number of Blades	4
Airfoil	V23010-1.58
Twist	-10°
Cutout (r/R)	20%
RPM	267.4
TAIL ROTOR	
Diameter	9 ft
Chord	9 in
Solidity	0.212
Tip Speed	700 fps
Number of Blades	4
Airfoil	V23010-1.58
Twist	-8°
Cutout r/R	20%
Type	Pusher
AIRFRAME	
Parasite Drag	19.1 ft^2
Landing Gear	fixed
ENGINES (HYPOTHETICAL)	
Number	2
Rating SL/STD (INT/Max Cont)	1600/1300
Lapse Rate (Sea Level)	6.0 hp/°F
Installation Losses	1%
TRANSMISSION RATINGS	
Dual Engine (SL/84°F–INT Power)	2900 SHP
Single Engine (SL/STD–INT Power)	1600 SHP

TABLE I-1 CONFIGURATION DEFINITION

1.1 Weights

As shown in Table I-1, the helicopter has a design gross weight *(DW)* of 15,000 lb, and a maximum gross weight (W_{max}) of 18,000 lb. The weight empty *(WE)* is 9,450 lb, or 63 percent of the *DW*. Subtracting the full fuel weight (W_F = 2,300 lb), fixed useful load *(FUL = 430 lb)*, and weight empty from the design gross weight results in a payload capability of *PL = 2,820 lb.* This is equivalent to taking off with a full load of fuel and approximately 14 passengers.

1.2 Main Rotor and Tail Rotor Geometry

The main rotor is a four-bladed hingeless rotor design having a 50-ft diameter and 2-ft chord. The dimensions of the main rotor are close to figures obtained during an optimization process aimed at a minimization of weight and costs which was performed during preliminary design studies of an actual UTTAS-type helicopter. The selection of the hingeless configuration was also the result of comparative design studies of various rotor types – chiefly articulated and hingeless. In this process, advantages and disadvantages were weighed, and both concepts were evaluated, taking into account such criteria as performance (parasite drag), controllability, permissible limits for c.g. travel, vibration, maintenance time, and dimensions affecting air transportability. However, in this text dealing with performance, the lower drag of the hingeless configuration, at least in principle, served as sufficient justification for preferring it over the completely articulated one.

The hingeless feature of this rotor means that there are no lead-lag or flapping hinges; however, flapping and lead-lag motions still occur through bending of the entire blade. The resulting blade motion is similar to the flapping characteristics of an articulated rotor with a relatively large hinge offset[1]. For this reason, rotor performance and stability evaluations are often conducted using an articulated rotor analysis while assuming a virtual or equivalent hinge offset as shown in Fig 1.2. A more detailed discussion is contained in Sect 4.3, Ch I, Vol I. This simulation gives the correct trim attitude for fuselage drag and download calculations.

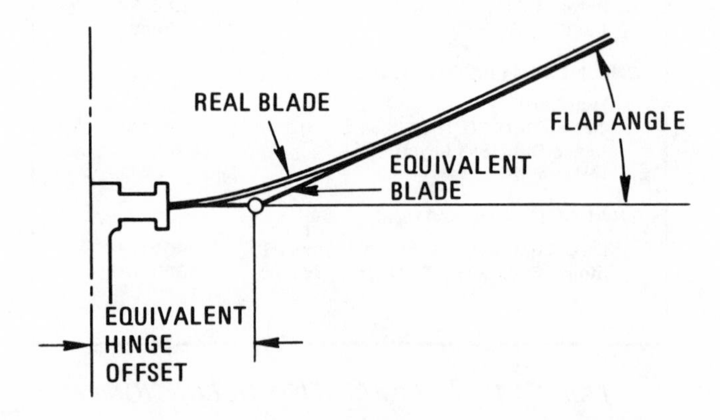

Figure 1.2 Equivalent hinge offset representation of a rigidly-attached blade

By analyzing the forces at the virtual flapping hinge, it can be shown that a hub moment is created when the tip-path plane deviates from the rotor-disc plane. This moment is one advantage of the hingeless rotor because it provides a more rapid response to control movements than fully-articulated rotors having a small (2 to 3 percent) or no flapping hinge offset which leads to improved maneuverability. In actual design practice, however, this and other previously mentioned advantages of the hingeless configuration must be weighed against dynamic couplings which are usually more complicated, resulting in vibratory problems more difficult to solve than those encountered in the low-offset, articulated rotors.

Other characteristics of the hypothetical helicopter main rotor given in Table 1-1 show that the rotor operates at a tip speed of $V_t = 700\ fps$, and has a cambered V23010-1.58 airfoil section similar to the NACA 23010 series except that the leading-edge radius is increased to 1.58 percent of the chord. The cambered airfoil was selected because of the following advantages over symmetrical sections: (1) higher $c_\ell^{3/2}/c_d$ values; resulting in an improved figure-of-merit in hover (Sect 9, Ch VI, Vol 1); and (2) higher $c_{\ell_{max}}$ coefficients, leading to the retreating-blade-stall retardation in forward flight.

The coordinates of the V23010-1.58 airfoil are presented in Table 1-2. It should be noted that a trailing edge tap extends over 4 percent of the chord. In order to reduce the pitching moment coefficient resulting from the camber, the tab is deflected up 3° relative to the Vertol reference line as described in Ref 2 (1.7° relative to the NACA chordline). The basic characteristics of this airfoil as they appear in the Airfoil DATCOM[2] with tab deflected 3° up are shown in Fig 1.3. Further details of this family of airfoils and methods for interpolation of the data are included in this reference.

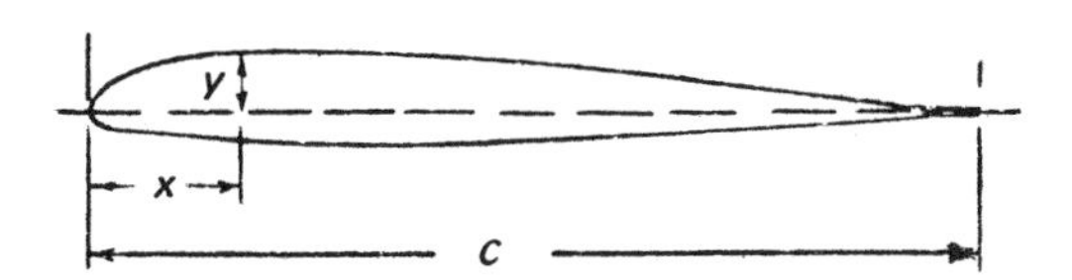

x/c	y/c_u	y/c_ℓ	x/c	y/c_u	y/c_ℓ
0.0	-0.0225	-0.0225	0.39	0.048	-0.0505
0.005	-0.0078	-0.0329	0.43	0.0465	-0.0487
0.01	-0.0024	-0.0362	0.47	0.0446	-0.0468
0.015	0.0019	-0.0378	0.51	0.0424	-0.044
0.025	0.0096	-0.0394	0.55	0.0397	-0.0412
0.035	0.0155	-0.0404	0.59	0.0369	-0.038
0.047	0.0214	-0.0412	0.63	0.0336	-0.0346
0.06	0.0265	-0.042	0.67	0.0301	-0.0308
0.08	0.0327	-0.0434	0.71	0.0263	-0.0269
0.11	0.0396	-0.0449	0.75	0.0223	-0.0226
0.15	0.0455	-0.0471	0.79	0.0181	-0.0182
0.19	0.0489	-0.0494	0.83	0.0137	-0.0136
0.23	0.0499	-0.0513	0.87	0.0093	-0.0093
0.27	0.0499	-0.0522	0.91	0.0056	-0.0057
0.31	0.0497	-0.05215	0.96	0.00235	-0.00235
0.35	0.049	-0.0517	1.0	0.00445	-0.00025

NOTES: Coordinates defined in the Vertol reference system, where the reference line approximately bisects the aft 50% of an airfoil.
Thickness, t/c = 0.102.
Leading edge radius, r/c = 0.0158.
Center of leading edge circle at x/c = 0.0158; y/c = −0.0225.
Trailing edge tab from x/c = 0.96 to x/c = 1.0.

TABLE 1-2 V23010-1.58 AIRFOIL COORDINATES WITH –3° T.E. TAB

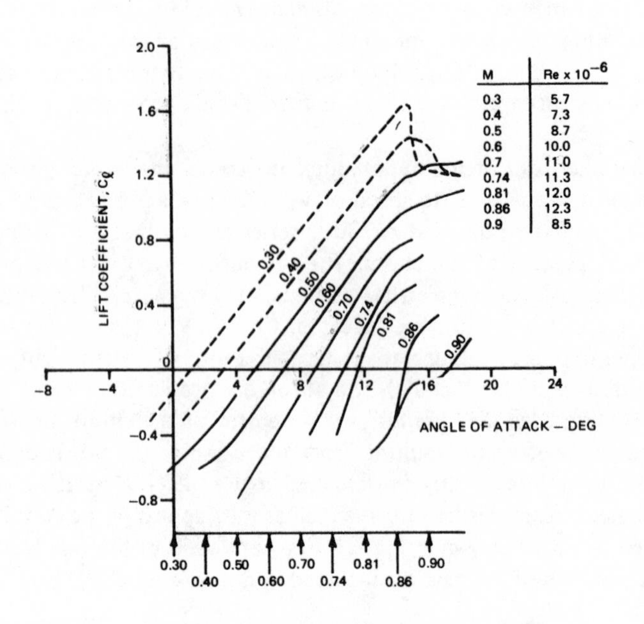

Figure 1.3a

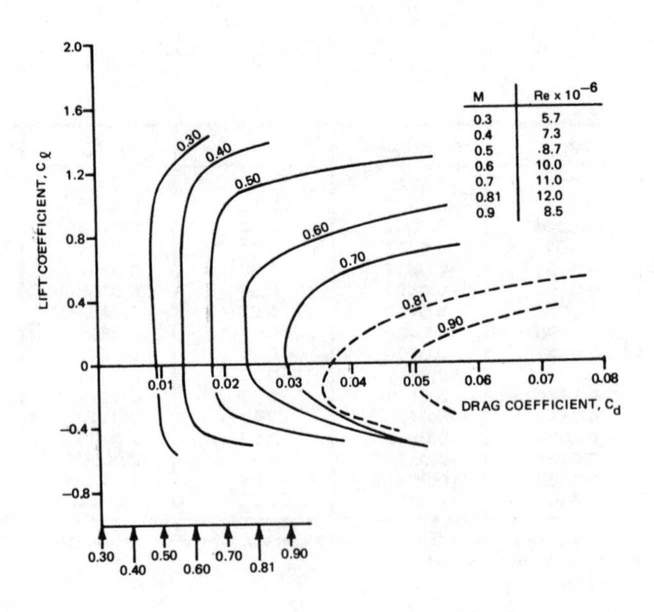

Figure 1.3b

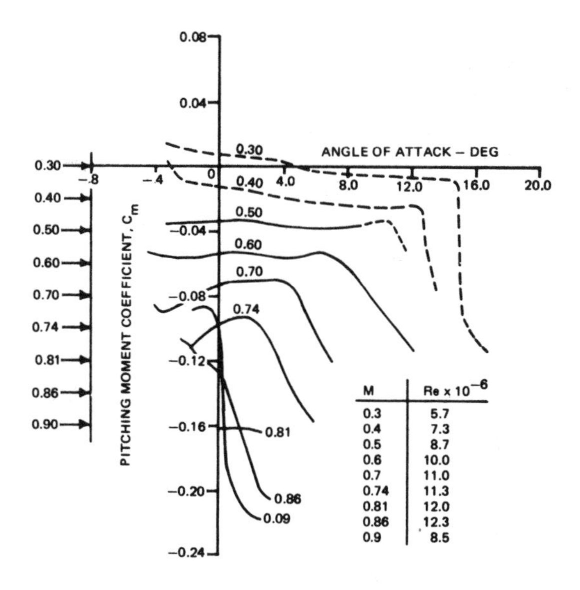

Figure 1.3c

Figure 1.3 Basic aerodynamic characteristics of V23010-1.58 airfoil with T.E. tab deflected 3° up

The tail rotor of the hypothetical aircraft is a 9-ft diameter, four-blade design having a 9-in chord and a V23010-1.58 airfoil section. The tail rotor is sized to provide adequate directional control in hover and forward flight and operates at a nominal tip speed of $V_{t_{tr}} = 700$ *fps.*

1.3 Transmission Limits

Helicopter transmissions are primarily limited by stress considerations corresponding to a given torque level. For an aircraft designed to operate at one rotor speed, as assumed for the hypothetical example, the torque limit becomes synonymous with a specific power-available limitation. As indicated in Table I-1, a dual engine transmission limit of 2900 hp was selected, which provides sufficient hover and vertical climb capability at low altitudes and high gross weights corresponding to INT power at SL/84°F. The selection of a single-engine gearbox rating corresponding to intermediate power at SL/STD atmosphere ($SHP_{int} = 1600$) was dictated by the consideration that with one engine out, the other can still be used to its maximum capability without exceeding the gearbox limits.

1.4 Airframe Configuration

The heliocopter fuselage selected for the sample calculations is similar to a version of the UTTAS design which was tested in a wind tunnel. The availability of tunnel data is the primary reason for using this particular airframe configuration, since it affords an opportunity to verify the accuracy of drag prediction techniques and provides some insight into the order-of-magnitude of the drag of various components. Parasite drag estimates, which will be discussed in detail later in Chapter III, show that the airframe shape is aerodynamically clean, as expressed by the ratio of the gross weight *(W)* to the equivalent flat plate area (f_e), when compared with current production helicopters (Fig 1.4).

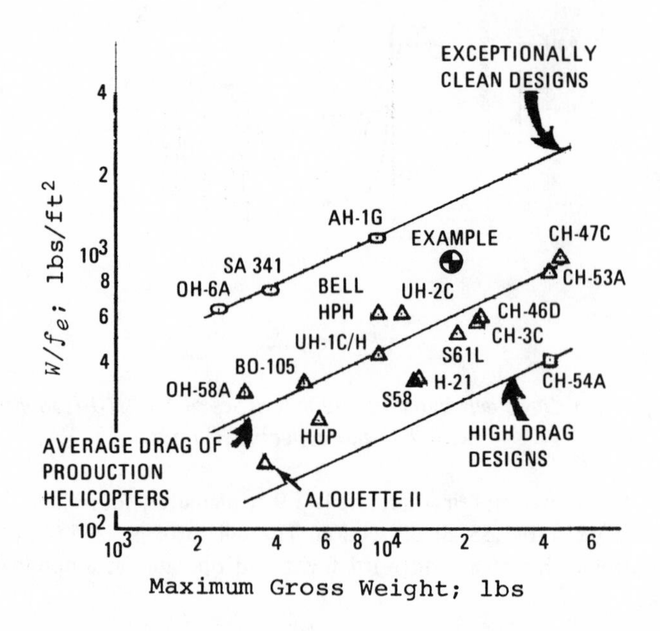

Figure 1.4 Helicopter drag trends

2. PERFORMANCE SUMMARY

An introductory discussion of a new helicopter design would not be complete without some reference to its performance capability. Therefore, a summary of the hypothetical helicopter performance is presented in Table I-3 for *DW = 15,000 lb.* The information shown in this table is typical of the type of performance calculations described in the text and includes hover, vertical climb and forward flight data at both standard day and 4000 ft/95°F conditions. The latter represents a primary design condition for new U.S. Army helicopters.

SPEED CAPABILITY Maximum continuous power, 4000 ft/95°F	161 kn
RANGE CAPABILITY Full fuel, 2-min. warmup, 10% fuel reserve, 4000 ft/95°F	330 n.mi.
ENDURANCE CAPABILITY Full fuel, 2-min. warmup, 10% fuel reserve, loiter at minimum power speed, 4000 ft/95°F	2.9 hr
HOVER CEILING Out-of-ground effect, 95°F, intermediate power	5700 ft
HOVER CEILING In-ground effect (5-ft wheel height), 95°F, intermediate power	9800 ft
VERTICAL CLIMB CAPABILITY 4000 ft/95°F, intermediate power	800 fpm
FORWARD FLIGHT CLIMB CAPABILITY Maximum continuous power, 4000 ft/95°F, dual engine	1900 fpm
SINGLE ENGINE SERVICE CEILING Standard day, intermediate power	13,700 ft

TABLE I-3 PERFORMANCE SUMMARY AT 15,000–LB DESIGN GROSS WEIGHT

3. ENGINE PERFORMANCE

3.1 General Characteristics of Shaft Turbines

Shaft turbines are continuous gas-flow engines (Fig 1.5). Consequently the power available at the shaft depends on (1) the amount of heat energy introduced in the form of fuel per pound of ambient air, and (2) rate of air flow ($\dot{W}_A$; lb/s) through the engine.

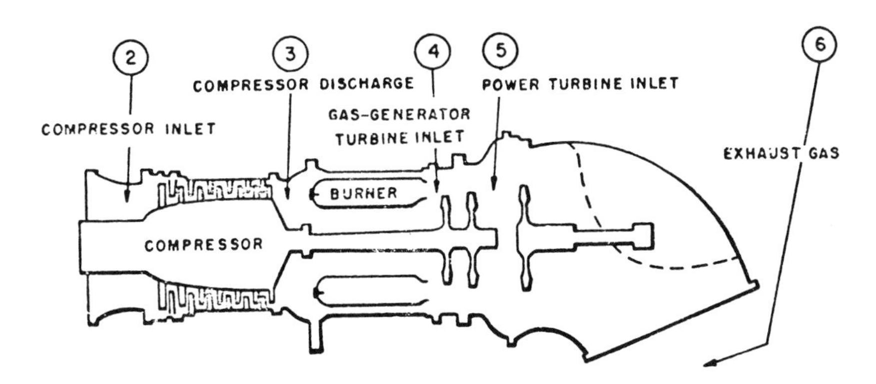

Figure 1.5 Typical gas flow diagram showing stations used in performance analysis

The amount of heat energy per pound of air (ΔE) will be

$$\Delta E = c_p(T_4 - T_3) \tag{1.1}$$

where T_3 and T_4 are absolute temperatures at the burner entrance and gas-generator inlet respectively, while c_p is the specific heat at constant pressure.

It becomes clear from Eq (1.1) that T_4 should be as high as possible in order to maximize the ΔE value. Understandably, however, the T_4 is limited by the state of technology as to the ability of turbine blades to withstand high operational temperatures. To assure that T_4 does not exceed values endangering the structural integrity of the turbine, the gas temperature must be monitored. Because of the ease of installing thermocouples, this temperature monitoring is usually done at the power turbine inlet, and the permissible T_5 values—instead of the T_4 which are obviously higher—are specified in operational manuals.

It should be realized from Eq (1.1) that since T_4 is limited, and also since T_3 increases with ambient temperature for a given compression ratio, an increase in the ambient temperature would reduce the amount of energy per pound of ingested air.

The rate of air flow ($\dot{W}_A$) is dependent on both ambient pressure and temperature. The effect of air pressure is straightforward. As the ambient pressure drops, the air density also drops. Consequently,

$$\dot{W}_A \sim p,$$

however, the influence of the ambient temperature is somewhat more complicated. Air density is inversely proportional to the absolute ambient temperature *(T);* but the speed of molecular motion is proportional to $\sqrt{T}$ (see Vol I, Ch.VI). This effect contributes to the increase in the speed of flow through the duct represented by the powerplant as a whole. The combined effect of these two influences is such that the rate of air flow through the engine would vary inversely proportional to $\sqrt{T}$. Now the air flow under a given ambient T and p condition becomes:

$$\dot{W}_A \sim p/\sqrt{T}. \tag{1.2}$$

It can be seen from Eqs (1.1) and (1.2) that the amount of heat energy ($\Delta E\ \dot{W}_A$) introduced each second into the engine cycle is directly related to the ambient temperature and pressure. It may be expected hence, that the shaft power available, which is proportional to the $\Delta E\ \dot{W}_A$ product, will also be influenced by the ambient condition.

3.2 Power Ratings and Effects of Ambient Conditions on Engine Characteristics

From the preceding discussion, it can be seen that power ratings of a turbine-type engine are related to the gas temperature level (as expressed by the T_5 values). For instance, $T_5 \approx 850°C$ allows the engine to operate for a limited stretch of time of no more than 30 minutes. Consequently, the corresponding *intermediate* rating (SHP_{int}) is used for takeoff, or emergency situations. However, when the gas temperature is sufficiently lowered (say, $T_5 \approx 750°C$), the engine can be operated for an unlimited time, and the corresponding rating becomes *maximum continuous* ($SHP_{m.c.}$).

In Armed Forces Specifications, the two above-mentioned power settings are also often called *military* (SHP_{mil}) and *normal rated* ($SHP_{n.r.}$) ratings.

It is now clear that turbine engine characteristics are such that the relationship of $SHP/\delta\sqrt{\theta}$ vs T_5/θ essentially results in a single curve. Then, having this curve and T_5 for a given power rating, the available power for any ambient condition can be determined. In addition to power, other engine characteristics are also corrected in order to relate them to SL/STD conditions.:

Power:	$SHP_o = SHP/\delta$*	(a)	
Fuel Flow:	$\dot{W}_{F_o} = \dot{W}_F/\delta\sqrt{\theta}$	(b)	
Air Flow:	$\dot{W}_{A_o} = \dot{W}_A\sqrt{\theta}/\delta$	(c)	(1.3)
Gas Generator, or Power Turbine Speed:	$N_o = N/\sqrt{\theta}$	(d)	

* For given rating and ambient temperature.

3.3 Powerplant for Hypothetical Helicopter

The powerplant used in the sample calculation consists of two hypothetical turbo-shaft engines with the following power ratings at SL/STD: intermediate, $SHP_{int} = 1600$ *hp;* and maximum continuous, $SHP_{m.c.} = 1300$ *hp.*

The variation of power available versus altitude for standard and 95° conditions is shown in Fig 1.6. Performance was established using the generalized power available shown in Fig 1.7. Altitude effects are taken into account according to Eq (1.3a).

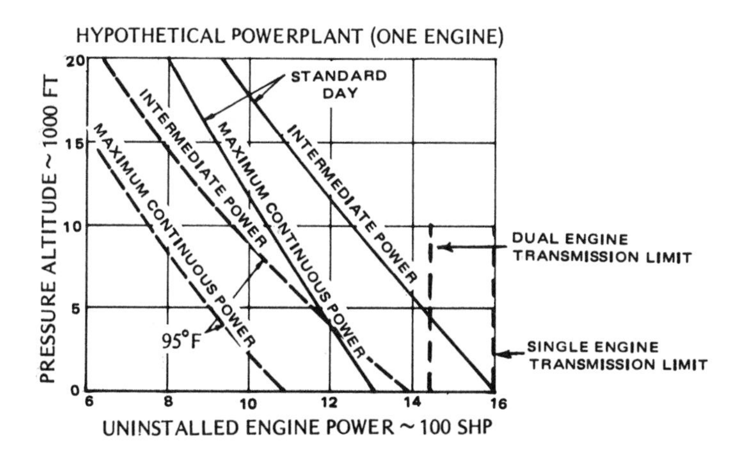

Figure 1.6 Uninstalled power available

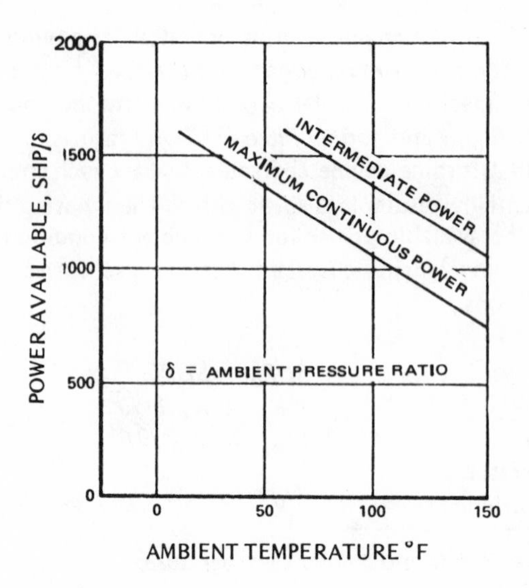

Figure 1.7 Generalized plot of power available

The slope of the intermediate and maximum continuous power lines at sea level ($\delta = 1.0$) is *-6 hp/°F*, which is typical of lapse rates for current engines of similar size.

The hypothetical engine fuel flow characteristics are shown in Fig 1.8 in terms of the parameters $\dot{W}_F/\delta\sqrt{\theta}$ (Eq (1.3b)) and $SHP/\delta\sqrt{\theta}$.

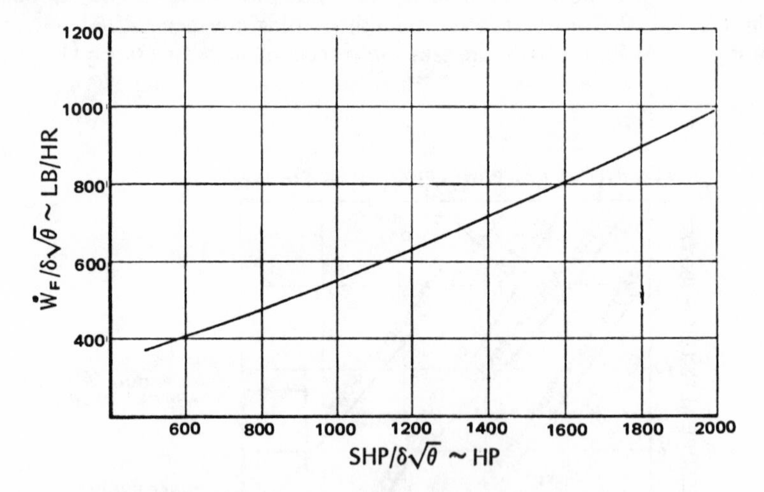

Figure 1.8 Fuel flow for hypothetical engine

This relationship was developed on the basis of trend curves for similar size engines, and results in $sfc = (\dot{W}_F/SHP)$ versus percent of military power as illustrated in Fig 1.9. As noted, at 60 percent of intermediate power, which corresponds to a cruise power setting, *sfc = 0.56 lb/hr SHP.*

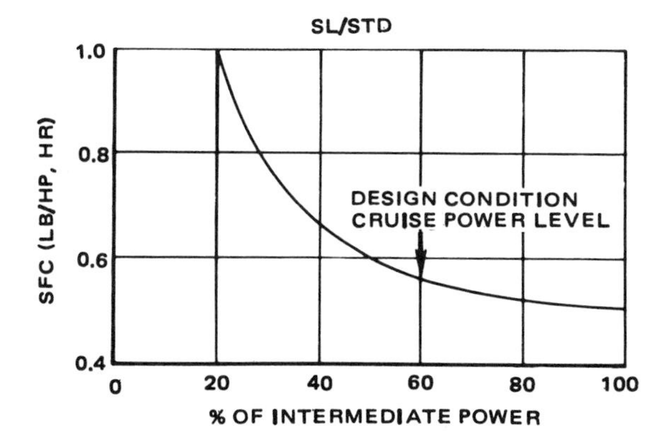

Figure 1.9 Hypothetical engine specific fuel consumption

3.4 Engine Power Constraints

Maximum power available is enclosed within an envelope formed by various constraints. For example, engine *SHP* vs ambient temperature ($p = p_o$) is shown in Fig 1.10. For $t > t_o$ and gas temperature (T_5) fixed, the power lapse rate with ambient temperature forms one of the constraining boundaries. As ambient temperature becomes lower than t_o, one might expect that engine power would increase, maintaining the same *SHP* vs t slope as for the $t > t_o$ region. However, the full benefit of this potential power increase usually cannot be realized because of the gas generator speed limit (N_1). At still lower ambient temperatures, a new stronger constraint in the form of the fuel-flow limit appears.

Finally, for an engine installed on a helicopter, two additional constraints associated with strength of the transmission for both dual and single-engine operations may appear.

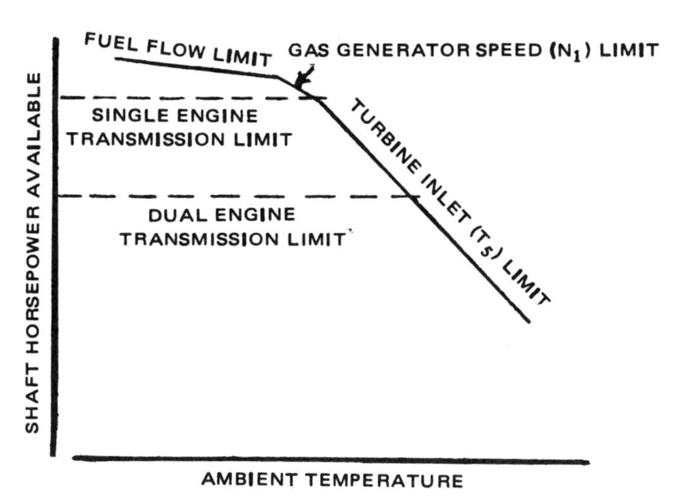

Figure 1.10 Typical constraints for engine power available

3.5 Engine Installation Losses

Installation of the engines on the airframe generally results in a decrease in performance when compared to the engine manufacturer's performance specifications. Losses associated with engine installation can be divided into (1) inlet losses, (2) exhaust losses, and (3) losses due to bleed air extraction. A brief discussion of each of these items is presented in the following paragraphs.

Inlet Losses. Inlet losses result from either a rise in temperature or a pressure drop at the inlet. In hover, the predominant effect is the temperature rise due to the recirculation of hot exhaust gases which occurs primarily in ground effect. A rise in inlet air temperature may also occur for installations with the gearbox located in front of the inlet. Pressure losses generally result from flow disturbances or separation at, or ahead of, the inlet. These effects are especially noticeable in forward flight, where flow separation may result in sizeable pressure losses; however, these losses are often offset by a decrease in flow velocity and and an increase in air pressure as it enters the inlet (ram recovery). When particle separators or screens are installed, additional sizeable losses may occur both in hover and forward flight.

Exhaust Losses. Exhaust losses are caused by backpressure normally resulting from a redirection or rerouting of the exhaust flow, from the installation of equipment such as an infrared suppressor, or from nozzeling to reduce parasite drag.

Extracting Bleed Air. Additional losses are incurred if bleed air is extracted from the compressor for anti-ice protection of the engine inlets when operating under cold ambient temperatures or for cabin or cockpit air-conditioning systems under hot ambient conditions.

For designs having podded engines, as assumed for the hypothetical aircraft, the engine installation losses are minimized because the engines are essentially detached from the airframe. Based on flight test experience, the power losses for this type of installation are generally less than one percent. Therefore, the one-percent loss assumed for all sample calculations is conservative. In addition, it is conservatively assumed that there is no increase in the power available due to ram recovery effects in forward flight.

Loss of power due to the inlet pressure drop (as a result of engine installation) also leads to a corresponding decrease in fuel flow (Eq (1.3b)). Typically, a loss in pressure will result in a reduction in fuel flow of 0.5 percent or less for each one-percent decrease in power available, thus resulting in a net increase in sfc. By contrast, a temperature rise will produce approximately equal power and fuel flow reductions with no net sfc change. Assuming, for the hypothetical helicopter, that a reduction in fuel flow amounts to 0.5 to 1 percent for 1 percent of power decrement, the installed fuel flow versus power relationship would remain within 0.5 percent of the uninstalled curve in Fig 1.8. Since these tolerances are small, this figure will be used as a basis for the installed fuel flow. However, in calculating performance, the sfc is increased 5 percent[3,4] over the values resulting from Fig 1.8. This increase accounts for (1) engine and airframe deterioration, and (2) nonoptimum piloting techniques.

4. STANDARD DAY ATMOSPHERE

The performance capability of an aircraft depends on the density of the surrounding air which, in turn, is a function of the local ambient conditions (temperature and pressure). This makes it difficult to compare the performance of various aircraft, or even the same aircraft from one day to another, unless the data is reduced to some standard conditions. An international standard atmosphere has been established for this purpose with air density varying with altitude as shown in Fig 1.11.

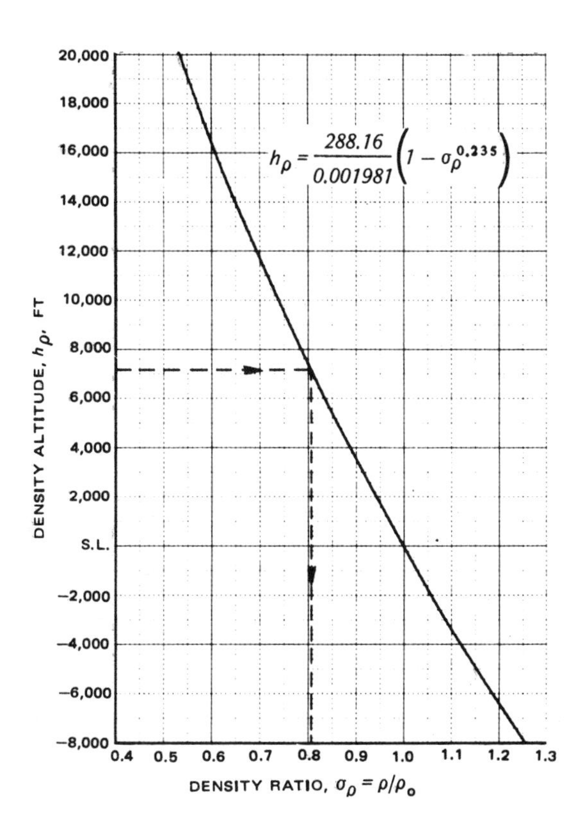

Figure 1.11 Density altitude vs density ratio (standard atmosphere)

In this case, altitude is not a "tape-measured" elevation over sea level, but is a hypothetical height referred to as *density altitude* based on the following criteria for standard atmosphere[5]:

(1) The air is assumed to be a perfect dry gas having a constant of $R = 96.04$ *ft/°C.*

(2) The pressure at sea level is $p_o = 29.92$ *Hg.*

(3) The temperature at sea level is $t = 15°C$ *(59° F).*

(4) The temperature varies linearly with altitude according to the expression $t = 15° - 0.001981h$; where altitude h is expressed in feet, and temperature in °C.

The linear temperature gradient or lapse rate assumed for standard day conditions approximates the average actual year-round temperature variation with altitude which occurs in North America at about 40° latitude[6,7,8].

Pressure altitude, defined as "the altitude at which a given pressure p is found in the standard atmosphere"[8], is again a "non-tape-measured" hypothetical height more frequently used than density altitude in performance calculations.

Using the equation of state for an ideal gas, and accounting for gravitational effects, the following expression for pressure altitude (h_p) in feet can be developed[7].

$$h_p = (288.16/0.001981)\,[1 - (p/p_o)^{0.1903}]. \tag{1.4}$$

Pressure altitude, therefore, is solely a function of the ambient pressure ratio. The practical advantage of this concept lies in the fact that pressure altitude is directly measured by aircraft altimeters which are essentially barometers calibrated according to Eq (1.4). By contrast, density altitude must be computed from altimeter and temperature readings. Consequently, performance data is generally quoted for a given pressure altitude and ambient temperature rather than density altitude. If an altitude is not qualified, then it is generally assumed to be pressure altitude.

Knowing the pressure altitude in feet and temperature in °C, the density ratio $\sigma_p = \rho/\rho_o$; pressure ratio $\delta = p/p_o$, and temperature ratio $\theta = T/T_o$ can be computed using the following relationships:

$$\delta = [1 - (0.001981\,h_p/288.16)]^{5.256} \tag{1.5}$$

$$\theta = (t + 273.16)/288.16 \tag{1.6}$$

$$\sigma_p = \delta/\theta = 288.16/(t + 273.16)[1 - (0.001981\,h_p/288.16)]^{5.256} \tag{1.7}$$

where t is in °C.

The ratios defined by Eqs (1.5) through (1.7) are used throughout this textbook to reduce performance predictions to SL/STD atmosphere conditions. In actual calculations, however, it is generally more convenient to use tabulations of δ and σ_p values as defined in Refs 5 and 7, or to obtain these ratios from charts (Figs 1.11 through 1.13). For instance, values of the pressure ratio δ, which are often needed in engine performance, can be more easily obtained from a graph such as the one shown in Fig 1.12 rather than computing them from Eq (1.5).

Computations of the absolute temperature ratio (Eq (1.6)) are so simple that no graphical help is required.

Graphs are also quite useful in determining values of density ratios (σ_p) which are required in calculations of such aerodynamic quantities as lift, drag, and induced velocity. For example. to compute the density ratio at 4000-ft pressure altitude/95°F, a density altitude of 7100 feet is read from Fig 1.13. The density ratio is then found from Fig 1.11 ($\sigma_p = 0.81$). The exact value from Eq (1.7) is $\sigma_p = 0.8076$, where $h_p = 7123$ *ft.*

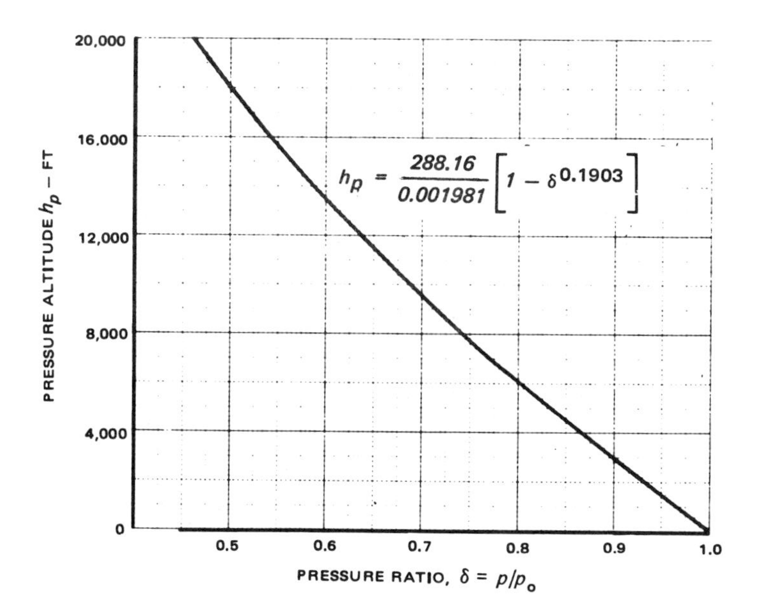

Figure 1.12 Ambient pressure ratio δ

The density ratio, σ_ρ, is also used in the preparation of flight manuals. Here, it is needed to convert true airspeed (ground speed in zero wind) to that indicated in the cockpit because airspeed measurement is, in practice, a measure of dynamic pressure ($\frac{1}{2}\rho V^2$). For more detail on this subject, see Appendix A. Supplement 3.

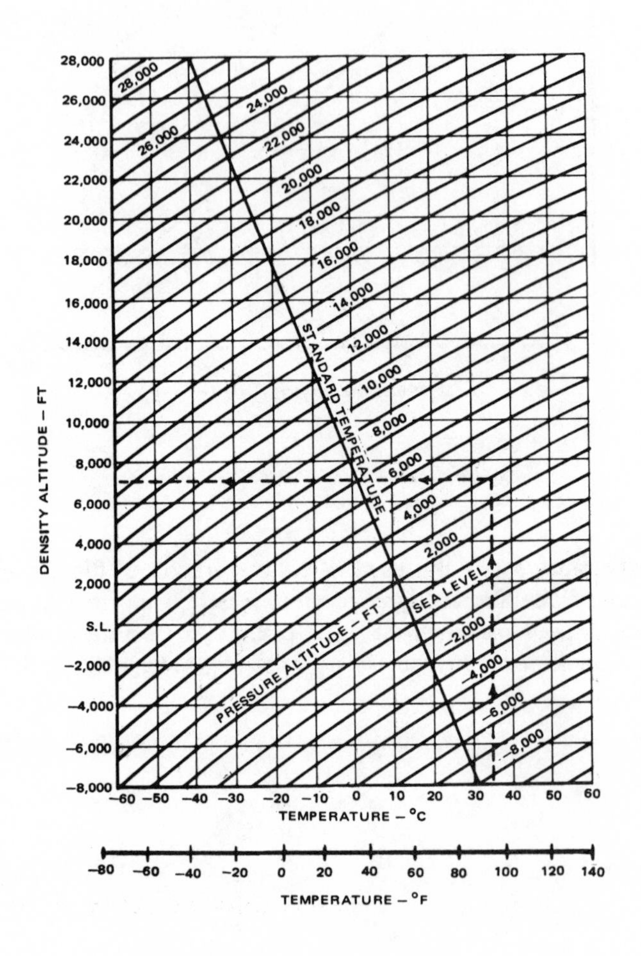

Figure 1.13 Density altitude chart

References for Chapter I

1. Reichert, G. *Basic Dynamics of Rotor Control and Stability of Rotary-Wing Aircraft Aerodynamics and Dynamics of Advanced Rotary-Wing Configurations.* AGARD Lecture Series No. 63, March 1973.

2. Dadone, L. U. *U.S. Army Helicopter Design DATCOM – Volume 1, Airfoils.* Contract NAS2-8637, May 1976.

3. MIL-M-7700A. *Military Specification – Manuals, Flight.* 14 Feb. 1958.

4. MIL-C-5011A. *Military Specification – Charts: Standard Aircraft Characteristics and Performance, Piloted Aircraft,* 5 Nov. 1951.

5. International Civil Aviation Organization. *Manual of the ICAO Standard Atmosphere Calculations by the NACA,* NACA TN 3182, May 1954.

6. Gregg, Willis Ray. *Standard Atmosphere Supplements.* NACA Report 147, 1922.

7. *U.S. Standard Atmosphere Supplements.* NASA CR-88870, 1966.

8. von Mises, Richard. *Theory of Flight.* Published by Dover, 1959 Edition.

SINGLE ROTOR HELICOPTER HOVER AND VERTICAL CLIMB PERFORMANCE

Hover out-of ground effect (OGE) capability, in-ground effect (IGE), and vertical climb prediction techniques are discussed in this chapter. Detailed calculations for the single-rotor hypothetical helicopter are presented to illustrate these techniques.

Principal Notation for Chapter II

A	area	ft^2
a $(a_o = 1116.4\ fps)$	speed of sound	fps
b	number of blades	
C_D	body drag coefficient	
$C_P = P/\rho\pi R^2 V_t^3$	rotor power coefficient	
$C_Q = Q/\rho\pi R^3 V_t^2$	rotor torque coefficient	
$C_T = T/\rho\pi R^2 V_t^2$	rotor thrust coefficient	
c	chord	ft
c_d	profile drag coefficient	
c_ℓ	profile lift coefficient	
$c_{\ell a}$	lift curve slope	rad^{-1} or deg^{-1}
D	drag	lb
d	rotor diameter	ft
H	height	ft
H_z	rotor height above fuselage	ft
IGE	in-ground effect	
INT	intermediate (power rating)	
k_d	downwash development factor	
k_g	download correction factor	
$k_{ind} = RP_{ind}/RP_{id}$	induced power correction factor	
k_p	climb efficiency factor	
k_v	integrated downwash velocity factor	
ℓ	length, or distance	ft
$M = V/a$	Mach number	
N	rotational speed	rpm
OGE	out-of-ground effect	
P	power	ft-lb/s, or hp
P_T	thrust power	ft-lb/s, or hp
Q	torque	ft-lb
R	rotor radius	ft
RHP	rotor horsepower	hp
RP	rotor power	ft-lb/s, or hp

R_e	Reynolds number	
r	radial distance	ft
S_v	vertical fin area	ft^2
SHP	engine shaft horsepower	hp
SL/STD	sea-level standard	
SP	engine shaft power	ft-lb/s, or hp
s	tail rotor fin separation	ft
T	rotor thrust	lb
$T = 273.16 + t°C$	absolute temperature	K
t	temperature	°C, or °F
V	velocity in general	fps
v	induced velocity	fps
W	weight	lb
w	width	ft
α	airfoil angle-of-attack relative to chord in NASA Reference System	rad, or deg
$\delta = p/p_o$	pressure ratio	
ζ	distance from rotor disc leading edge	ft
$\theta = T/T_o$	absolute temperature ratio	
η	efficiency	
$\sigma = bcR/\pi R^2$	rotor solidity	
$\sigma_\rho = \rho/\rho_o$	density ratio	

Subscripts

a	accessory
av	available
c	climb, or compressible
d	divergent, or downwash
e	equivalent
f	fuselage
h	hover
i	incompressible
id	ideal
ind	induced
mr	main rotor
n	indicator of numerical order
o	initial, or SL/STD
pr	profile
r	rotor
ref	referred
t	tip
tm	transmission
tr	tail rotor
v	vertical

Superscripts

average

1. HOVER OUT-OF-GROUND EFFECT PERFORMANCE

1.1 General Procedure

The hover OGE performance calculation procedure, in principle, consists of comparing helicopter shaft power required for a given ambient condition with the engine installed power available.

As diagrammatically shown in Fig 2.1, several intermediate calculations must be performed in order to determine these two quantities.

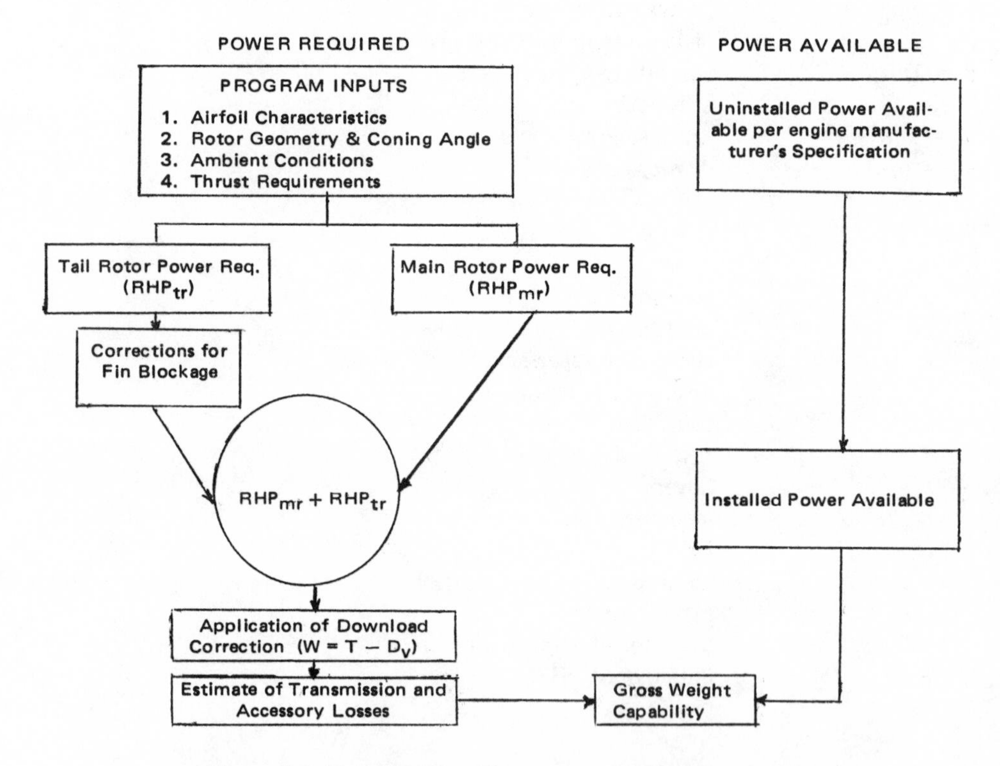

Figure 2.1 Hover OGE performance calculation procedure

Power Required. Airfoil characteristics, rotor and blade geometry, ambient conditions, and assumed thrust of the main rotor (close to the anticipated gross weight) represent initial inputs. Next, both the main and tail rotor power required are computed either manually or through the use of computer programs. These calculations may be based on any acceptable theory relating rotor thrust to rotor power required (i.e., combined blade element and momentum theory, vortex, local momentum, or potential theories). However, the computer programs most frequently used in industrial practice are based on the vortex theory.

After corrections to power required by the tail rotor due to the aerodynamic interference of the vertical fin are made, the total power required by the main and tail rotors is determined.

The gross weight (W) corresponding to the originally assumed thrust (T) can be resolved through computation of the vertical drag (D_v). Application of transmission and accessory losses allows one to establish a relationship between gross weight and shaft power required: $SHP_{req}(W)$.

Power Available. Determination of the shaft power available begins with uninstalled SHP as set by the engine manufacturers for assumed ambient conditions. Accounting for installation losses leads to the establishment of the shaft horsepower actually available (SHP_{av}).

A comparison of $SHP_{req}(W)$ with SHP_{av} permits one to determine the gross weight of the aircraft in hover OGE under given ambient conditions.

1.2 Power vs Thrust Calculations

An actual computer program of one of the helicopter manufacturing companies was used in performance calculations of the hypothetical helicopter.

An isolated rotor nonuniform downwash analysis described as the *Explicit Vortex Influence Technique*[1] was used to predict both the main and tail rotor power required. This represents the prescribed wake approach, which is discussed in more detail in Ch 4 of Vol I. Only the more salient features of this technique are recalled here. The technique is basically an extension of the fixed-wing, lifting-line theory where each blade is represented by a lifting line and trailing vortex wake. This wake is composed of an infinite number of weak vortex filaments which the theory mathematically approximates by a finite number of vortices streaming from various radial locations. The positioning of the vortices below the rotor is indicated by the semi-empirical prescribed rate of wake contraction since the vortex filaments must travel at the velocity of the surrounding fluid. The contraction rate, specified as a function of the thrust coefficient $C_T = T/\rho\pi R^2 V_t^2$, is determined by analytical studies of finite-core vortex ring flows and by correlation of calculated and measured propeller static performance. As the wake is defined empirically rather than allowing it to form freely, this type of model is defined as a prescribed wake. This method is generally preferred over the free-wake method throughout the industry in order to obtain reasonable computer run time (Vol I, Ch IV.6).

The strength of the vortices is determined by the section lift (c_ℓ) distribution using the Kutta-Joukowski theorem. The angle-of-attack and hence, the c_ℓ distribution is determined by the downwash velocity induced by the vortices defined by the Biot-Savart law. An iterative technique is used to obtain a mutually consistent c_ℓ and downwash distribution. Once an agreement is achieved, the c_ℓ and section drag (c_d) distributions are integrated taking into consideration the local downwash angle, thus thrust and power required are obtained. If the computed thrust and the desired thrust do not agree, the collective pitch angle setting is changed and the entire process is repeated.

The iterative calaculations described above require the use of a high-speed computer. A simplified block diagram of the computer program is presented in Fig 2.2. As shown in this figure, the inputs required are:

- airfoil section c_ℓ and c_d characteristics
- rotor geometry
- ambient condition
- required thrust.

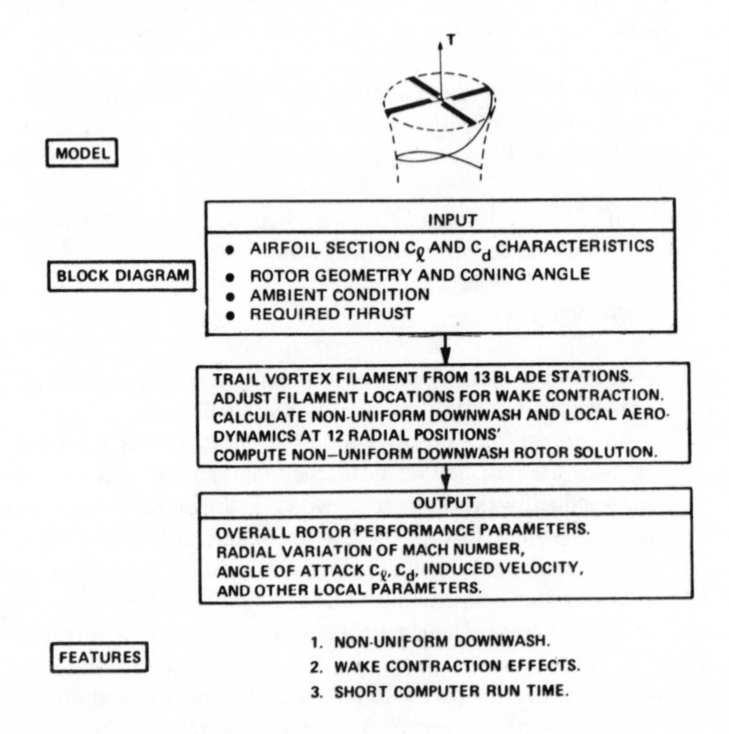

Figure 2.2 Hover performance analysis computer program (explicit vortex influence technique)

A brief description of the specific computer program input parameters for the hypothetical aircraft are presented below.

Airfoil Section Aerodynamic Characteristics. As noted in Ch I, the hypothetical aircraft design uses V23010-1.58 airfoil sections for both the main and tail rotors. The lift characteristics of this airfoil are illustrated in Fig 2.3, where lift coefficient versus angle-of-attack is shown at Mach numbers from 0.3 to 0.9. This data is based on wind-tunnel testing conducted in the Boeing two-dimensional wind tunnel. The high Mach number data has lower $c_{\ell_{max}}$ values and higher lift-curve slopes ($c_{\ell a}$). The actual variation of the lift-curve slope with Mach number (for the anticipated M-value range) agrees

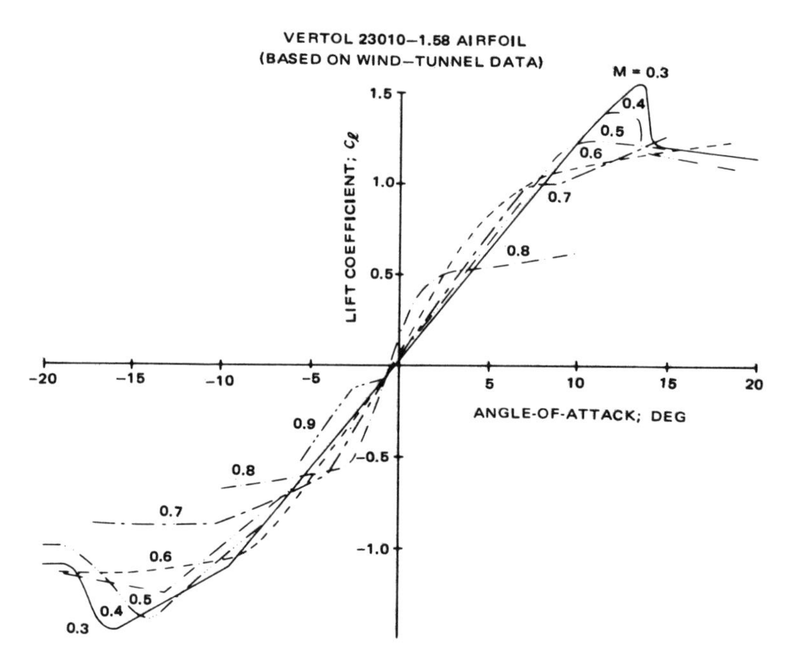

Figure 2.3 Variation of section-lift characteristics with Mach number

well with the Prandtl-Glauert expression (see Vol I, Ch VI)

$$c_{\ell a_c} = c_{\ell a_i}/\sqrt{1 - M^2}$$

where $c_{\ell a_c}$ is the slope corrected for compressibility effects ($M > 0$), and $c_{\ell a_i}$ corresponds to the incompressible case ($M = 0$).

The V23010-1.58 section drag characteristics are presented in Fig 2.4 as a function of Mach number for various angle-of-attack settings. The airfoil c_d varies from 0.008 to 0.018 at Mach numbers lower than the drag divergent Mach number (M_d). M_d is defined as the Mach number where the the slope $\Delta c_d/\Delta M = 0.1$. It represents the Mach number at which a weak oblique shock forms on the crestline (tangential point of freestream velocity with the upper surface of the airfoil) resulting in separation of the boundary layer. As noted in Fig 2.4, the drag divergence Mach number decreases with increasing angle-of-attack. This variation is due to an increase in local velocity on the upper surface, and movement of the crestline chordwise location forward into the high velocity region near the airfoil leading edge.

The data presented in Figs 2.3 and 2.4 was obtained at full-scale Reynolds number (R_e) corresponding to a CH-47C helicopter rotor blade, with a chord of 25.25 inches, operating at a tip speed of approximately 750 fps. A Reynolds number drag correction must be applied to use this data for the hypothetical aircraft. The variation in section c_d with Reynolds number is shown in Fig 2.5. This data is based on two-dimensional

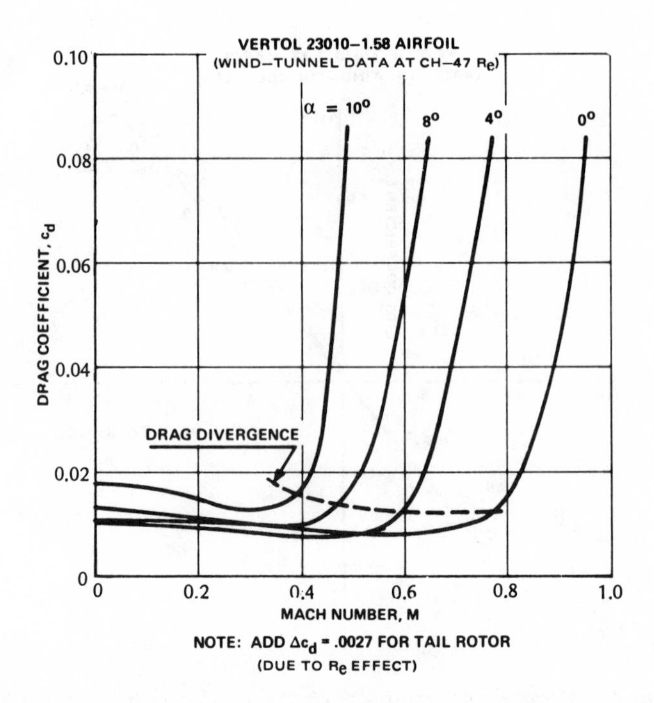

Figure 2.4 Variation of section-drag characteristics with Mach number

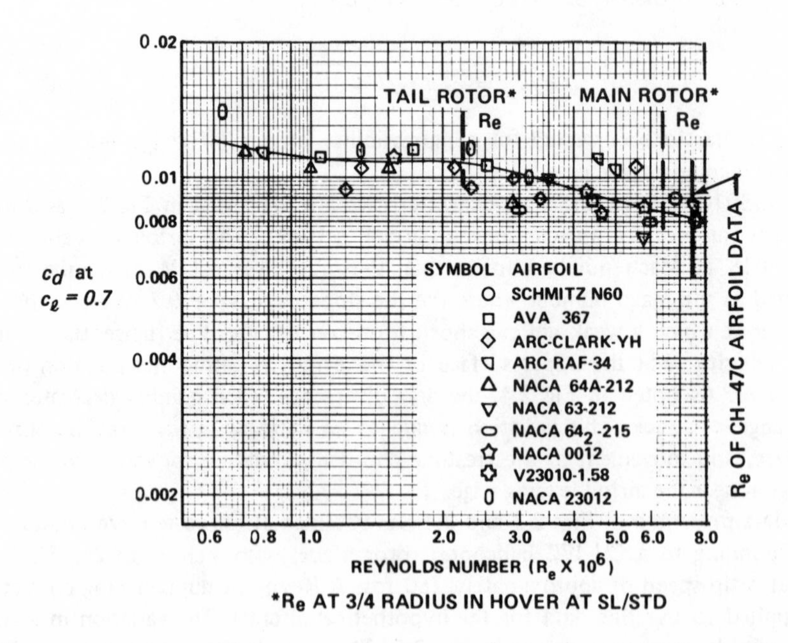

Figure 2.5 Reynolds number effect on airfoil section drag

testing and the trend was confirmed by model rotor data using the 3/4-radius location to define the average blade Reynolds number. Since the R_e values for the CH-47C rotor blades and those of the main rotor of the hypothetical helicopter are quite close, the CH-47C airfoil c_d data is sufficiently accurate for estimating the main rotor c_d values. It is not suitable, however, for use in defining tail rotor drag coefficients. The c_d increment due to the difference between the tail rotor Reynolds number and the airfoil data Reynolds number is $\Delta c_d = 0.0027$. Therefore, this Δc_d correction, representing a 34-percent increase in drag, is applied to all tail rotor performance predictions in this text.

Rotor Geometry and Coning Angle. The geometry of both the main and tail rotors is defined in Ch I. The elastic coning angles, which exclude any built-in coning, may be approximately computed from the relationships given in Vol I, Ch I, using the virtual flapping-hinge concept. This can also be done utilizing the simplified calculations discussed in Ref 2. Most industrial organizations have in-house trim analysis computer programs from which values of the coning angle can be found.

Ambient Conditions. Hover performance is a function of the air density and ambient temperature. The air density is either computed from inputs of pressure altitude and ambient temperature as defined by the equations found in Ch I, or determined from appropriate graphs. The number of computed conditions can be kept to a minimum by nondimensionalizing the power required and rotor thrust for air density as described later in this chapter. The only stipulation is that the blade tip Mach number must be correct. For an aircraft which is designed to operate at one rotor speed, such as the hypothetical helicopter, the tip Mach number variation can be satisfied by considering a range of ambient temperatures.

Rotor Thrust. Thrust values ranging from the minimum flying weight (weight empty, fixed useful load, and fuel reserve) to the maximum gross weight were inputted into the computerized program used in the case of the hypothetical helicopter. Additional calculations extending to zero thrust were also obtained to provide a comparison with the momentum theory predictions at low thrust levels.

1.3 Example of the Main Rotor Power Required

The hypothetical helicopter nondimensionalized main rotor power required is shown in Fig 2.6. Thrust coefficient $C_T = T/\rho\pi R^2 V_t^2$ is shown as a function of power coefficient $C_P = (RHP \times 550)/\rho\pi R^2 V_t^3$ for various tip Mach numbers, M_t. It should be noted that the fan shape formed by the C_P/C_T curves for various Mach number values is due to compressibility effects which become negligible at $M_t \leqslant 0.606$ and $C_T/\sigma < 0.08$. At 700 fps tip speed, $M_t < 0.606$ corresponds to ambient temperatures above 95°F.

Presenting the rotor performance in a nondimensional form is quite convenient during aircraft concept definition and preliminary design phases when a number of configurations must be evaluated and compared. However, the values of C_P and C_T are small; typically, $C_P = 0.0005$ and $C_T = 0.005$, which many find difficult to interpret and cumbersome to use when computing detailed performance for a given aircraft. For this reason, once a design is finalized, the nondimensional method of presenting power required is often replaced by a dimensional method known as the *referred power/referred thrust* (weight) *method.* Referred is based on the fact that at a given set of C_P

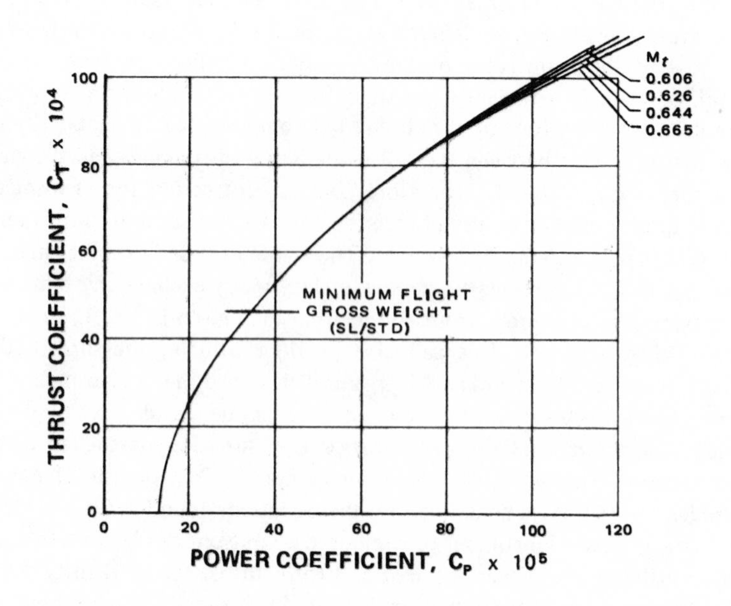

Figure 2.6 Main rotor hover out-of-ground effect power required

and C_T values, power is proportional to $\rho \pi R^2 V_t^3$ and thrust is proportional to $\rho \pi R^2 V_t^2$; therefore, the rotor horsepower (RHP) and thrust (T) at any altitude and tip speed can be referred to the equivalent SL/STD density altitude conditions as shown below.

$$RHP_{ref} = (RHP/\sigma_\rho)(V_{t_{ref}}/V_t)^3 = C_P \rho_o \pi R^2 V_{t_{ref}}^{\;3}/550$$
$$T_{ref} = (T/\sigma_\rho)(V_{t_{ref}}/V_t)^2 = C_T \rho_o \pi R^2 V_{t_{ref}}^{\;2} \tag{2.1}$$

where $\sigma_\rho = \rho/\rho_o$
RHP_{ref} = power required at SL/STD
T_{ref} = thrust at SL/STD
$V_{t_{ref}}$ = reference operating tip speed.

Referred thrust for the sample problem aircraft is plotted as a function of referred power in Fig 2.7, assuming a reference tip speed of $V_{t_{ref}} = 700$ *fps*. Compressibility effects must be accounted for by referring along lines of constant Mach number. Tip Mach number can be related to tip speed in a more convenient way by noting that

$$M_t = V_t/a_o\sqrt{\theta}$$

where

a_o = speed of sound at 59°F (15°C); fps.

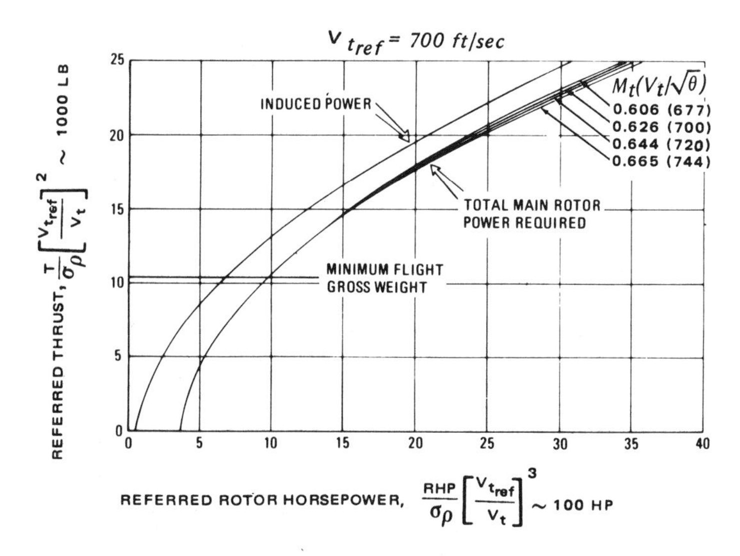

Figure 2.7 Main rotor power required (hover OGE)

The lines of constant Mach number in Fig 2.7, therefore, are also lines of constant $V_t/\sqrt{\theta}$. It should be noted that for a known rotor radius (in this case, $R = 25\ ft$), V_t can be replaced by an equivalent expression containing rpm, since $V_t = \Omega R = \pi\, rpm\, R/30 = 700\ fps$.

The induced power component of the main rotor power required is also shown in this figure. In the case of a complete computer program based on the vortex theory, the induced power can be computed by making an input of the section $c_d = 0$. It can be seen that under SL/STD atmosphere conditions, the induced power amounts to approximately 80 percent of the total RHP at $T = 15{,}000\ lb$. The power difference between induced and total power represents the profile power, including the compressibility penalty.

A detailed breakdown of the profile power and induced power under SL/STD atmosphere is presented in Fig 2.8. The ideal induced power and simplified blade element theory profile power required (assuming $\bar{c}_d = 0.008$) are also shown for comparison with vortex theory results. The vortex theory induced power is considerably higher than the ideal power given by the momentum theory. At $T = 15{,}000\ lb$, for example, the vortex theory induced power is 15 percent greater than its ideal value. This difference is primarily due to the fact that nonuniform downwash effects and tip losses are not taken into account when determining ideal power. The induced power correction factor k_{ind_h}, defined as the ratio of the actual induced power to the ideal induced power (Vol I, Ch II) is also shown in Fig 2.8. It can be seen that k_{ind_h} increases from $k_{ind_h} = 1.08$ at $T = 10{,}000\ lb$, to $k_{ind_h} = 1.22$ at $T = 20{,}000\ lb$.

The general trend of k_{ind_h} increasing with T or, more strictly speaking, with C_T, is probably correct. However, it should be emphasized that the trend shown in Fig 2.8 is the result of a program based on the prescribed wake approach where wake contraction

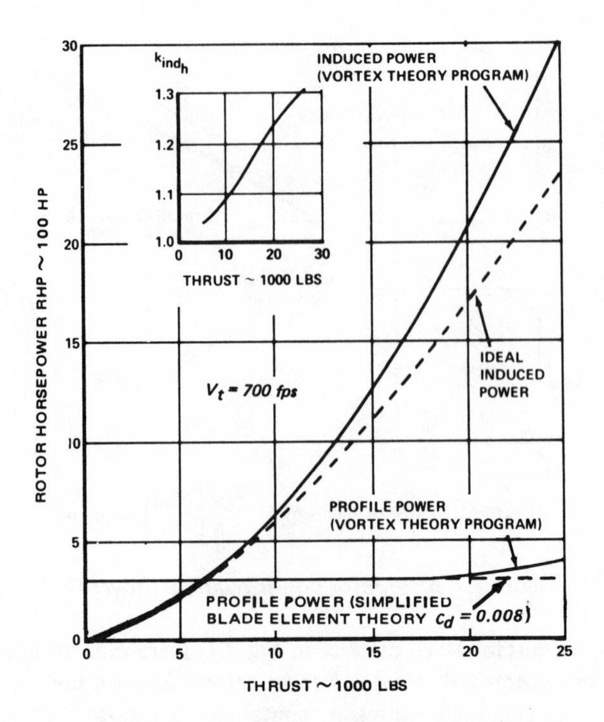

Figure 2.8 Main rotor induced and profile power at SL, STD

was empirically determined as a function of C_T. Consequently, an extrapolation to C_T values higher than actually tested may lead to error. Furthermore, it should also be remembered that blade section lift coefficients also become higher with increasing C_T and thus, c_d values also increase–sometimes quite rapidly, due to adverse $c_\ell - M$ combinations. Hence, caution must be exercised in the "bookkeeping" so that the induced and profile power effects are properly registered.

With these words of warning, let it be assumed that the relationship shown in Fig 2.8 is correct. Therefore, the thrust scale of the k_{ind_h} chart can be nondimensionalized in terms of C_T to predict the performance of other rotor configurations. The only stipulations are that the designs have four blades and –10° linear twist. Due to such airfoil characteristics as camber, lift-curve slope, and trailing-edge tab setting, some variation in k_{ind_h} may occur when changes are made in the airfoil section.

The variation of the induced power factor versus C_T, linear twist, and number of blades is illustrated in Figs 2.9 and 2.10[3]. In Fig 2.9, it can be seen that the benefits of twist begin to decrease at twist values $>|-10°|$. For typical rotor designs having C_T/σ levels (*0.06* to *0.08*) and twist values between *–5* and *–15°*, the induced power varies approximately 1 percent per degree of twist.

As shown in Fig 2.10, the induced power correction factor decreases with increasing number of blades. This trend may be expected because as the number of blades increase, the rotor approaches an actuator disc consisting of an infinite number of blades

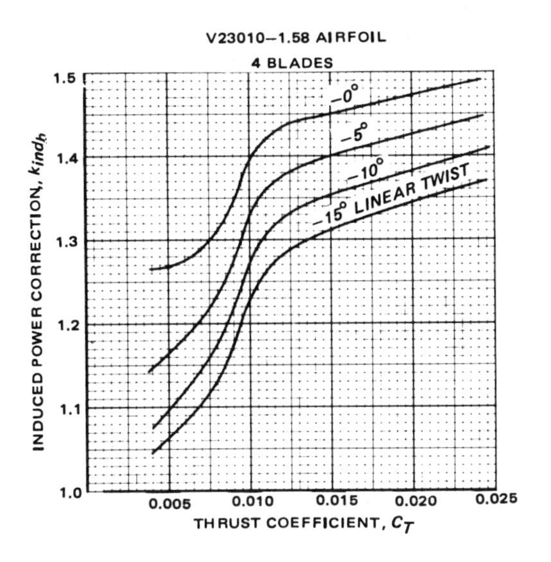

Figure 2.9 Variation of induced power factor with thrust coefficient and blade twist

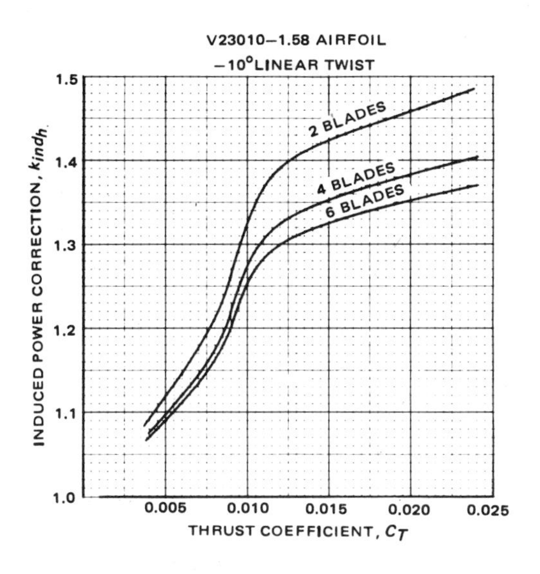

Figure 2.10 Variation of induced power factor with thrust coefficient and blade number

of infinite aspect ratio. However, the actual rate of improvement becomes lower as the number of blades increases. This is due to the fact that as the distance between the blades decreases, the trailing vortices from the preceding blade move closer to the following blade; causing adverse changes in its lift distribution.

In contrast to the above-discussed comparison of induced power predicted by the simple momentum versus vortex theories, predictions of profile power based on the simple element theory using a constant $\bar{c}_d$ value show good correlation with those obtained from the vortex theory (Fig 2.8). Both methods agree for thrust levels up to $T \leqslant 15{,}000$ *lb*. For $T > 15{,}000$ *lb*, the vortex theory predictions of profile power show gradually increasing values caused by compressibility effects, as well as an increase in c_d values due to the increasing level of local blade lift coefficient (c_ℓ).

The profile power (RHP_{pr}) shown by the dashed line in Fig 2.8 was determined using the following expression derived in Vol I, Ch III.1.3.

$$RHP_{pr} = (1/4400)\sigma\pi R^2 \rho \bar{c}_d V_t^3. \tag{2.2}$$

The term $\bar{c}_d$ in this equation is the average blade airfoil section drag coefficient. It is determined using airfoil section data at the average section lift coefficient $\bar{c}_\ell$ and representative Mach number $\bar{M}$ of the rotor where $\bar{c}_\ell$ is defined from such blade element considerations as: $\bar{c}_\ell = 6C_T/\sigma$, and $\bar{M} = 0.75V_t/a$.

For the hypothetical helicopter, the airfoil section data given in Figs 2.3 and 2.4 was used. For profile power predictions based on blade element theory, it is generally more convenient to present this data in drag polar form as shown in Fig 2.11. As noted in this figure, $\bar{c}_\ell$ goes up to *0.64* for a thrust level of 25,000 lb at SL/STD atmosphere conditions and the corresponding Mach number (at 75-percent radius) is $M = 0.47$. At this Mach number, the average drag coefficient would be *0.008,* and would not vary significantly with lift coefficient for c_ℓ values up to *0.7*.

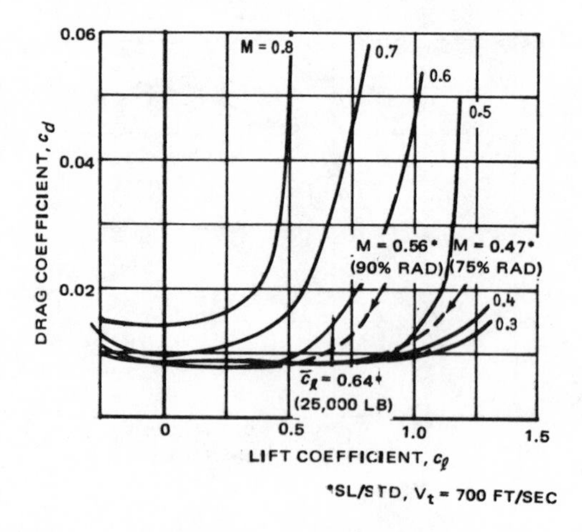

Figure 2.11 Vertol 23010-1.58 airfoil section drag polars

However, due to compressibility effects which cause a rise in drag at the blade tip at high thrust levels, the assumption that the representative Mach number occurs at the 3/4-blade radius location becomes invalid. The two profile power predictions shown in Fig 2.8 can be made to agree reasonably well if the average drag coefficient and associated Mach number are assumed to occur at 90 percent instead of 75 percent of the blade radius. As shown in Fig 2.11, at $T = 25{,}000\ lb$ and SL/STD conditions, c_d increases from $c_d = 0.008$ to $c_d = 0.0102$ as the representative Mach number increases from $M = 0.47$ at the 75-percent radius to $M = 0.56$ at the 90-percent radius.

A more rigorous method of accounting for compressibility is to apply profile power corrections obtained from the computer program based on the vortex theory. An example of this correction for the hypothetical helicopter is shown in Fig 2.12 as $\Delta C_P/\sigma$ versus tip Mach number for various C_T/σ values. The same correction can be used as an approximation for other airfoil sections provided they have a thickness ratio of from 10 to 12 percent.

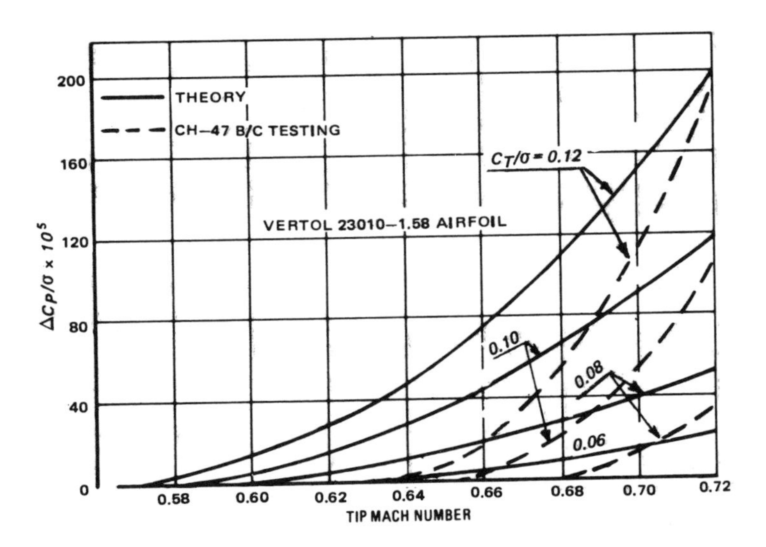

Figure 2.12 Profile power compressibility correction for hover OGE

The theoretical compressibility shown in this figure was found to be conservative when compared to flight test measurements. Test data generally indicates that compressibility effects in hover for a V23010-1.58 airfoil are not significant for tip Mach numbers of $M_t \leqslant 0.64$. In contrast, the theoretical compressibility power divergence occurs at $M_t = 0.60$ to 0.62. This discrepancy between test and theory is attributed to relieving blade-tip effects[4,5,6]. In the theoretical hover analysis, two-dimensional airfoil data were used to predict power required; however, it did not account for the reduced local velocities that occur at the blade tip due to three-dimensional flow effects.

The induced and profile power discussion presented above indicates that calculations based on simple momentum and blade element theories can be modified to account for nonuniform downwash and compressibility effects so that reliance on large and expensive high-speed computers is not necessary. These shortcut techniques are particularly valuable for preliminary design studies where many different configurations must be evaluated within a limited budget. Additional information on shortcut methods can be found in the HESCOMP User's Manual (Helicopter Sizing and Performance Computer Program)[3].

1.4 Tail Rotor Power Required

The solidity of tail rotors is generally larger than that of the main rotors in order to obtain acceptable average lift coefficients or C_T/σ values both in equilibrium and in maneuvers.

The tail rotor must produce sufficient thrust in equilibrium to compensate the main rotor torque in hover. This trim condition requirement can be expressed as

$$T_{tr_{net}}\ell_{tr} = Q \tag{2.3}$$

and

$$Q = 5252\,RHP_{mr}/N \tag{2.4}$$

where

Q = main rotor torque; ft-lb
$T_{tr_{net}}$ = net tail rotor thrust
ℓ_{tr} = moment arm (line from tail rotor shaft perpendicular to main rotor shaft); ft
N = main rotor rpm.

Substituting Eq (2.4) into Eq (2.3) gives

$$T_{tr_{net}} = 5252\,RHP_{mr}/N\ell_{tr}. \tag{2.5}$$

For the hypothetical helicopter,

$$T_{tr_{net}} = 5252\,RHP_{mr}/(267.4 \times 30) = 0.655\,RHP_{mr}.$$

The tail rotor power required corresponding to this net tail rotor thrust level was determined for the hypothetical helicopter by applying the same vortex theory analysis technique used to predict main rotor hover performance. However, in order to determine the net tail rotor thrust, an additional correction must be applied to account for fin blockage effects which, due the vertical fin side force, decreases the thrust of the isolated rotor. A discussion of isolated tail rotor power required calculations and fin blockage effects is presented below.

Isolated Tail Rotor Power Required. The isolated tail rotor performance, as in the case of the main rotor, can be determined using any appropriate theory. In this particular case, the existing hover analysis computer program based on the explicit vortex influence technique was used. The computer input requirements are similar to those defined for the main rotor (Fig 2.1). It should be emphasized at this point that airfoil drag characteristics were significantly different from those of the main rotor; therefore an increment of

$\Delta c_d = 0.0027$ was added to the main rotor section drag data to account for R_e effects (Fig 2.5).

The power required for the hypothetical aircraft tail rotor for SL/STD atmosphere is presented in Fig 2.13. Total power, induced power, and profile power obtained from the vortex theory program are compared to the ideal induced power and simplified blade element theory profile power predictions. As shown, the total rotor thrust required to hover at $W = 15{,}000\ lb$ is $T_{tr} = 1{,}170\ lb$. The corresponding power predicted by the vortex theory is $RHP_{tr} = 210\ hp$, consisting of 87 percent induced and 13 percent profile power. The vortex theory induced power is 30 to 50 percent higher than that predicted by the simple momentum theory. This percentage difference is almost twice as large as the values noted in the main rotor discussion because the tail rotor operates at much higher disc loadings or thrust coefficients where nonuniform downwash effects become

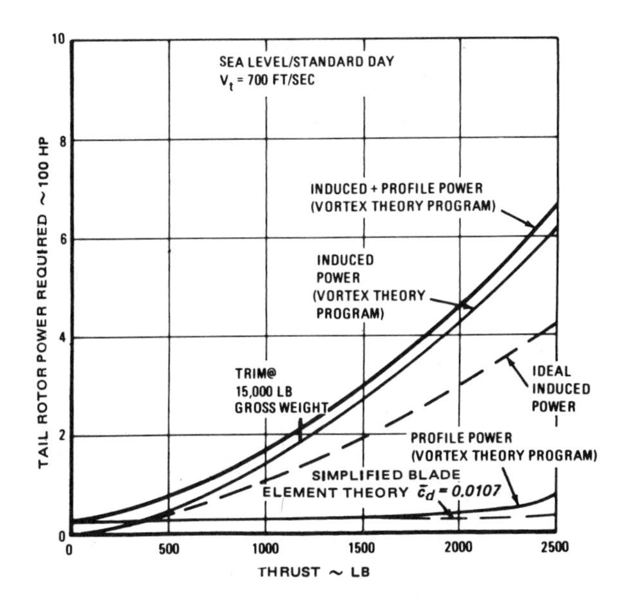

Figure 2.13 Tail rotor induced and profile power

more significant. Wind tunnel test data obtained with 2 and 4-bladed tail rotors generally confirm the large differences between vortex theory and ideal induced powers. The tail rotor induced power factor k_{ind_h} varies from 1.3 to 1.5; depending on the value of C_T. For tail rotor applications where $C_T > 0.01$, a mean value of $k_{ind_h} = 1.4$ can be used to estimate the tail rotor performance as shown in Figs 2.9 and 2.10.

Comparisons of profile power required are also shown in Fig 2.13. The tail rotor power required, as predicted by the blade element theory, agrees with the vortex theory computer program estimates at thrust levels up to approximately $T_{tr} = 1500\ lb$. Above this value, the vortex theory shows higher power required due to compressibility effects. As described in the main rotor analysis, better agreement can be achieved at higher thrust levels if a 90-percent radius representative section Mach number instead of the 3/4-radius value is utilized or, if the correction shown in Fig 2.12 is applied. In general, however, the tail rotor operates in trimmed hover conditions at C_T/σ values sufficiently low that the compressibility power increment represents a relatively small percentage of the total power required and therefore, can be neglected for most preliminary design studies.

Vertical Fin Blockage Effects. Since this is assumed to be a pusher-type tail rotor, the inflow, in hover, is blocked by the vertical tail and a fin force is generated which acts to reduce the net thrust available for antitorque purposes. The tractor-type tail rotor downwash impinges on the vertical tail; again, creating a fin force reducing the net rotor thrust. In either case, isolated tail-rotor performance must be adjusted for the blockage effect by increasing the thrust required as shown in Fig 2.14. This data was obtained by measuring the fin force and thrust of various model fin and tail rotor configurations[7]. The configurations tested included both pusher and tractor-type tail rotors located at varying distances from the vertical fin (s). The ratio of the tail rotor thrust to net thrust is a function of the fin/rotor separation (s/R_{tr}) and the fin area to rotor-disc area ratio ($S_v/\pi R_{tr}^2$). Utilizing the thrust ratio from Fig 2.14 and Eq (2.5), it can be shown that the isolated rotor thrust required to trim the aircraft is as follows:

$$T_{tr} = \left(\frac{T_{tr}}{T_{tr_{net}}}\right)\left(\frac{5252\, RHP_{mr}}{N \ell_{tr}}\right). \tag{2.6}$$

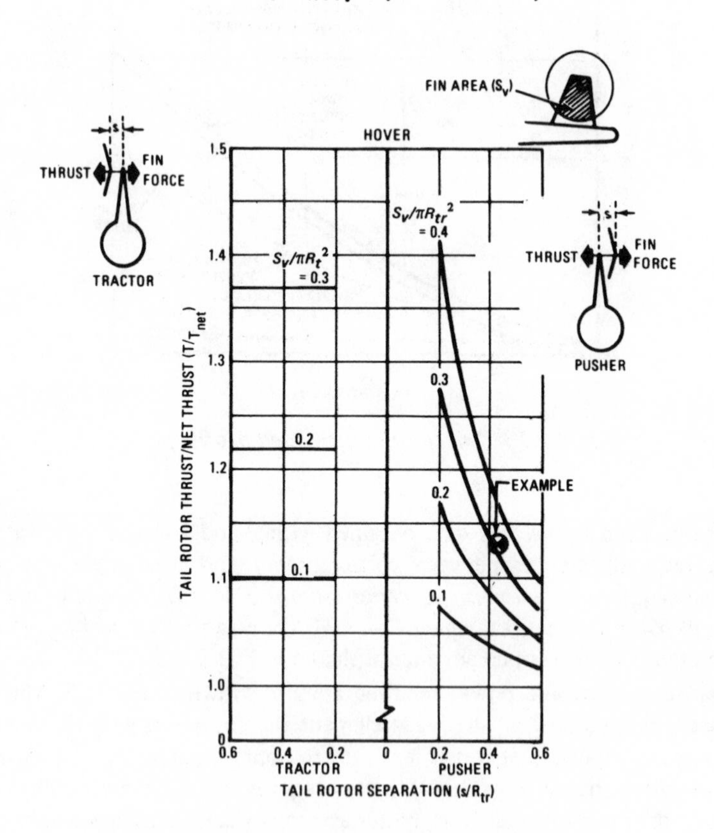

Figure 2.14 Vertical tail blockage correction

The hypothetical aircraft has a pusher-type tail rotor with a fin separation ratio $s/R_{tr} = 0.426$ and a fin area to rotor-disc area ratio $S_v/\pi R_{tr}^2 = 0.32$. As noted in Fig 2.14, the thrust ratio for this configuration is $T/T_{net} = 1.13$. The effect of vertical tail blockage on tail rotor power required is shown in Fig 2.15, where net thrust is presented as a function of power required with and without the vertical tail installed. It can be seen that the blockage correction increases the tail rotor power required at the $T = 15{,}000\text{-}lb$ trim point by 18 percent, which is equivalent to approximately a 2-percent increase in total aircraft power required. For preliminary design studies, this penalty may be neglected since it is relatively small; however, for detailed performance predictions, it should be included in the calculations.

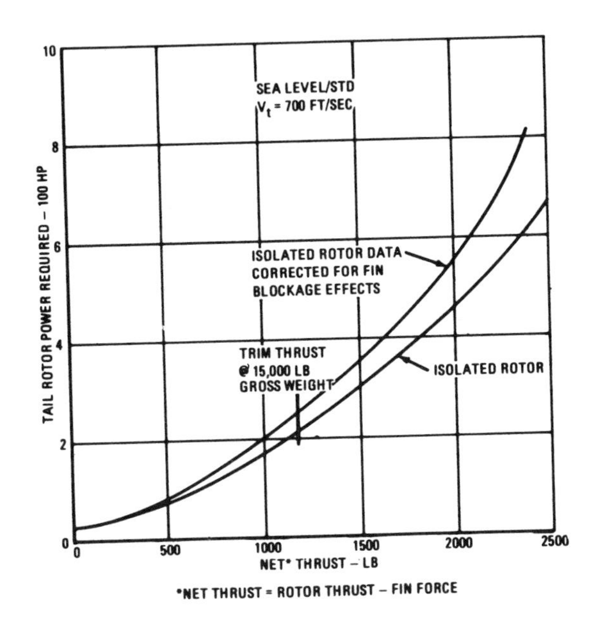

Figure 2.15 Effect of vertical tail blockage on tail rotor power required

1.5 Fuselage Download

Having determined main rotor and tail rotor power required as a function of main rotor thrust, the next step is to adjust the main rotor thrust values for download effects to obtain gross weight. The total thrust required by the main rotor in hover is equal to the gross weight plus the vertical drag (D_v) or download on the fuselage caused by the rotor downwash velocity ($T = W + D_v$). Vertical drag is calculated by combining the estimated fuselage vertical drag coefficients with downwash velocity distributions based, preferably, on wind-tunnel testing; or analytical predictions, if test results are lacking. The calculation procedure involves dividing the fuselage into segments and computing the drag increments of each segment. For example, the incremental download ΔD_v of segment n to $n+1$ is

$$\Delta D_{v_n} = \int_{\zeta_n}^{\zeta_{(n+1)}} C_{D_v} \tfrac{1}{2}\rho v^2 \, w_n \, d\zeta \qquad (2.7)$$

where w_n is the average segment width, C_{D_v} is the local drag coefficient, and v is the downwash velocity acting on the area $w_n\,d\zeta$. The parameter ζ is the distance from the forward edge of the rotor disc to segment n, and is measured along the fuselage centerline shown in Fig 2.16.

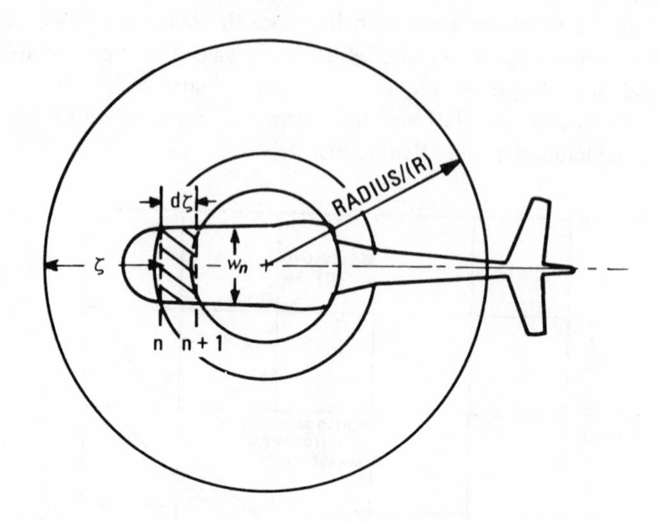

Figure 2.16 Incremental fuselage area

Assuming that w_n and $C_{D_{v_n}}$ are constants for a given segment, Eq (2.7) can be rewritten as

$$\Delta D_{v_n} = C_{D_{v_n}}\, \tfrac{1}{2}\rho w_n \int_{\zeta_n}^{\zeta_{n+1}} v^2\, d\zeta. \tag{2.8}$$

From momentum theory,

$$T = 2\rho\pi R^2\, v_{id}{}^2. \tag{2.9}$$

Dividing Eq (2.8) by (2.9) results in the nondimensional download expression

$$\Delta D_{v_n}/T = C_{D_{v_n}}\, W_n/4\pi R \int_{(\zeta/R)_n}^{(\zeta/R)_{n+1}} (v/v_{id})^2\, d(\zeta/R) \tag{2.10}$$

where $\Delta D_v/T$ and ζ/R are defined as percentages.

The term $\int(v/v_{id})^2\, d(\zeta/R)$ in Eq (2.10) is equal to the area under a plot of $(v/v_{id})^2$ versus (ζ/R) between stations n and $n+1$. To simplify the integration procedure, a variable

$$k_{v_n} = \int_0^{(\zeta/R)_n} (v/v_{id})^2 d(\zeta/R)$$

is introduced. Substituting this expression in Eq (2.10) gives:

$$\Delta D_{v_n}/T = (C_{D_{v_n}} w_n/4\pi R)\ (k_{v_{n+1}} - k_{v_n}). \tag{2.11}$$

To illustrate the download prediction technique outlined above, detailed sample calculations for the hypothetical helicopter are presented in Table II-1. The fuselage is

STEP	①	②	③	④	⑤	⑥	⑦	⑧	⑨
ITEM	ζ (ft)	ζ/R (%)	$k_{v_{n+1}}$	k_{v_n}	Δk_v	C_{D_v}	w (ft)	$C_{D_v}w/4\pi R$	$\Delta D_v/T$ (%)
CALCULATION PROCEDURE		①/25 100	See Fig 2.19		③ − ④	Fig 2.17	Segment Width	⑥ × ⑦/4π × 25	⑧ × ⑤
SEGMENT (n)									
1-2 COCKPIT	09.3	037	138	087	51	0.5	6.00	0.00955	0.49
2-3 CABIN	12.7	051	205	138	67	0.4	8.00	0.01020	0.68
3-4 NACELLE	25.9	104	210	205	05	1.2	8.65	0.03300	0.17
4-5 AFTERBODY									
	31.1	124	225	210	15	0.5	6.50	0.01035	0.16
	33.6	134	252	225	27	0.5	5.34	0.00850	0.23
	36.1	144	298	252	46	0.5	4.18	0.00665	0.31
5-6 TAILBOOM									
	39.5	158	385	298	87	0.5	2.66	0.00423	0.37
	44.6	178	410	385	25	0.5	2.00	0.00318	0.08
	50.0	200		410					
									D_v/T = 2.49% D_v/W = 2.55%

TABLE II-1 DOWNLOAD CALCULATIONS

divided into five segments. The cross-section shape of each segment and the corresponding C_{D_v} are defined in Fig 2.17. The drag coefficients are based on wind-tunnel pressure and force measurements obtained on numerous fuselage shapes.

Download drag coefficients for typical fuselage section shapes based on wind-tunnel tests are illustrated in Fig 2.18. Additional drag data applicable to download predictions can be found in Ref 8; however, this data applies to two-dimensional shapes and must be adjusted for three-dimensional effects if it is to be used for fuselage sections near the cockpit. An estimate of three-dimensional effects can be obtained from Ref 9.

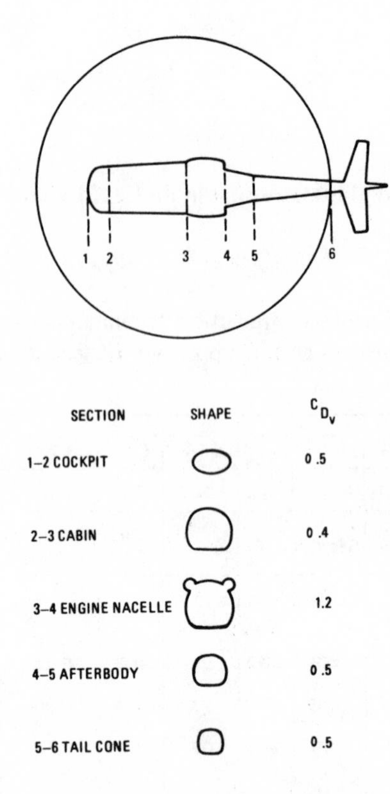

Figure 2.17 Hypothetical helicopter vertical drag coefficients

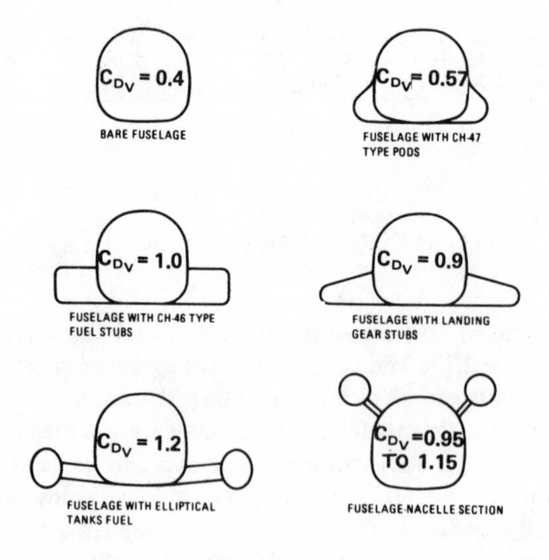

Figure 2.18 Typical helicopter fuselage section vertical drag coefficients

The download increment for each of the five fuselage segments is then calculated by combining the drag coefficient with the downwash velocity. The downwash profile applicable to single-rotor helicopters is illustrated in Fig 2.19. The nondimensional velocity ratios, v/v_{id}, $(v/v_{id})^2$ and k_v values in this figure are plotted as a function of distance from the forward tip of the rotor in percent radius ζ/R. This data is based on Universal Helicopter Model (UHM)* measurements, with the front rotor removed.

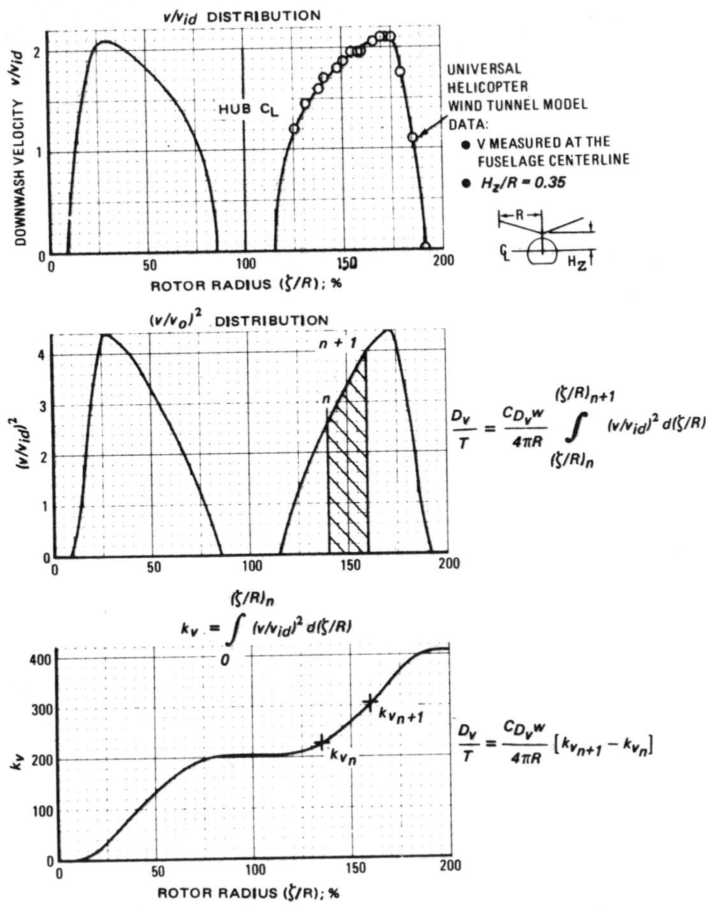

Figure 2.19 Hover downwash velocity distribution

A photograph of the UHM tandem configuration installed in the tunnel is shown in Fig 2.20. The model tested was a 5.35-ft diameter rotor having -9° linear twist, a solidity of 0.0619 and a rotor height above the fuselage of $H_z/R = 0.35$. Of these parameters, the rotor height has the most significant effect on download. For rotor configurations with H_z/R values lower than that of the model—such as the hypothetical helicopter which has a $H_z/R = 0.2$—the model data represents a conservative estimate of downwash velocity since the downwash velocity decreases with decreasing rotor-to-fuselage clearance.

*The name of this model reflects its versatility, permitting one to test tandem rotors in various geometric configurations, or to obtain measurements on one rotor only.

Figure 2.20 Universal Helicopter wind-tunnel model

The last step in the calculations is the summation of the $\Delta D_v/T$ increments. As noted at the bottom of Table II-1, the download (D_v) of the hypothetical helicopter is 2.49 percent of thrust–or 2.55 percent of gross weight. The relationship between D_v/T and D_v/W is

$$D_v/W = (D_v/T)/(1 - D_v/T).$$

The sample problem main rotor thrust versus rotor power required data presented in previous sections can now be converted to gross weight versus power required, using the D_v/W value derived above:

$$W = T/(1 + D_v/W). \tag{2.12}$$

The hypothetical helicopter combined main and tail rotor power required as a function of gross weight is shown in Fig 2.21. The tail rotor power required–including compressibility effects–is also shown in this figure. As noted, at 15,000 lb gross weight

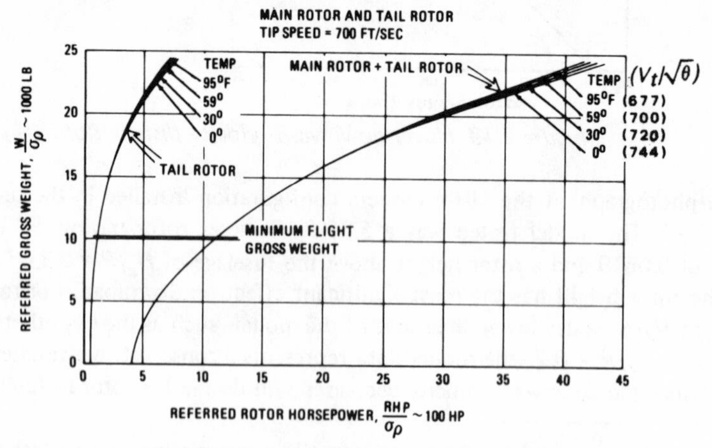

Figure 2.21 Total hover (OGE) power required

under SL/STD conditions, the tail rotor power required amounts to 12 percent of the total required RHP. The percentage varies from 10 percent at $W = 10{,}000\ lb$ to 14 percent at $W = 18{,}000\ lb$. Compressibility effects do not significantly influence these percentages for referred weights below 17,000 lb.

The download prediction methodology described above will give slightly conservative answers because of thrust recovery effects. The main rotor operates in partial ground effect, provided by the upper surface of the airframe. This results in an increased thrust capability at a fixed power available level. The amount of gain is a function of blade twist, blade cutout, and rotor/fuselage clearance[10]. It should be noted that a similar phenomenon is also encountered in the tail operation; hence, the predicted vertical fin blockage corrections may be conservative.

1.6 Transmission and Accessory Losses

The rotor horsepower required (*RHP*) shown in Fig 2.21 was computed at the rotor shaft, and does not take into account losses which occur in the transmission of power from the engine shafts, nor the additional power required to operate accessories. The total shaft horsepower (*SHP*) required of the engines is

$$SHP = RHP/\eta_t + P_a \tag{2.13}$$

where

η_t = transmission efficiency,

P_a = accessory power.

A detailed discussion of transmission and accessory losses is presented below.

Transmission Losses. Neglecting the fact that even at zero power there are some transmission losses due to gear movement in the air/oil environment (windage), it is assumed for simplicity that they represent a fixed percentage of the input SHP. The losses are then determined by totaling the estimated losses for each gear mesh in the drive system. The following loss-per-mesh values were used for the sample calculations.

(1) 0.5 percent for low and intermediate-speed bevel and planetary gears;

(2) 1.0 percent for high-speed bevel gears.

Estimated losses for the hypothetical helicopter at the 2900 SHP transmission limit power setting are presented in Table II-2. Transmission losses amount to two percent of the power available. Tail rotor transmission losses are so small that they are not considered in these calculations.

Accessory Losses. Accessory losses include power extraction for items such as engine and transmission cooling blowers, electrical power generation, and hydraulic power supplies. For the hypothetical helicopter, accessory losses are assumed to be 30 hp, or approximately one percent of the transmission limit. Therefore, the total transmission plus accessory losses incurred at the transmission limit is about three percent. This value is typical of losses measured during flight test evaluations of current production aircraft. The final expression for converting RHP to SHP for the hypothetical aircraft is

$$SHP = (RHP/0.98) + 30.$$

ASSUMED NUMBER AND TYPE OF TRANSMISSION	POWER AT GEAR MESH	% POWER LOSS/MESH	POWER LOSS (ΔHP)
ENGINE GEARBOXES:			
2 High-Speed Bevels	1450	1.0	29.0
MAIN ROTOR:			
2 Intermediate Speed Bevels	1450	0.5	14.5
1 Planetary	2900	0.5	14.5
	TOTAL		58.0
	PERCENTAGE OF 2900 SHP		2.0

TABLE II-2 TRANSMISSION LOSS AT THE TRANSMISSION LIMIT POWER SETTING

2. HOVER IN–GROUND–EFFECT (IGE) PERFORMANCE

When helicopters hover close to the ground, the power/thrust relationships undergo changes, depending on the relative height (H_r/d) of the rotor disc above the ground. This is due to the fact that induced velocity decreases close to the ground, causing a drop in induced power. By the same token, at a fixed power setting, the main rotor thrust increases, and the fuselage download decreases. Because of the mathematical complexity of the problem, semi-empirical methods are usually used to evaluate the ground effect. This approach is also applied to the prediction of IGE performance capability of the hypothetical helicopter.

2.1 Rotor Thrust Variation, IGE

Correction factors to out-of-ground effect values were developed from flight test and model rotor data. The ratio of thrust IGE at constant power settings for various single-rotor aircraft (based on flight-test measurements[11]) as a function of H_r/d is presented in Fig 2.22. It can be seen that the inception of ground effect occurs at $H_r/d < 1.3$. For the hypothetical helicopter, the thrust ratio at a wheel height of 5 feet from the ground–corresponding to $H_r/d = 0.3$–is $T_{IGE}/T_{OGE} = 1.14$ (Fig 2.23).

Other possible methods of correcting OGE performance for IGE effects can be based on the SHP_{IGE}/SHP_{OGE} at a fixed thrust level. However, the approach based on thrust-ratio data at constant SHP provides more useful generalized trends because it eliminates the influence of variation in profile power, tail rotor power, and transmission accessory power which, at constant SHP, remain the same regardless of the thrust value. To separate induced power from other power components, generalized power required data must be analyzed by comparing $C_P^{2/3}$ versus C_T.

2.2 IGE Download

When using the thrust ratio in determining in-ground-effect hover capability, it is necessary to apply another empirical correction to account for the reduction in vertical

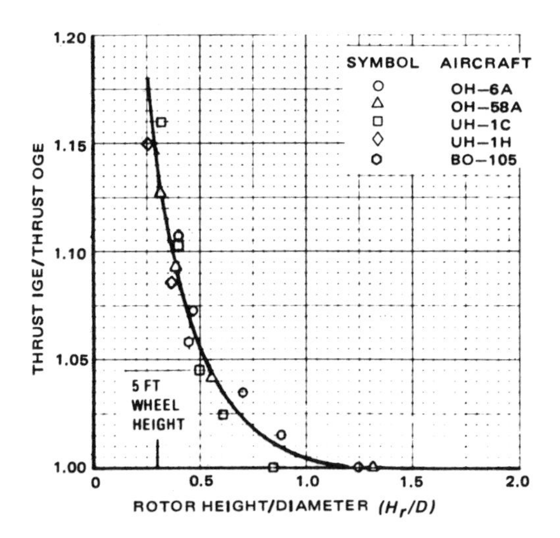

Figure 2.22 Thrust variation in-ground-effect at constant power

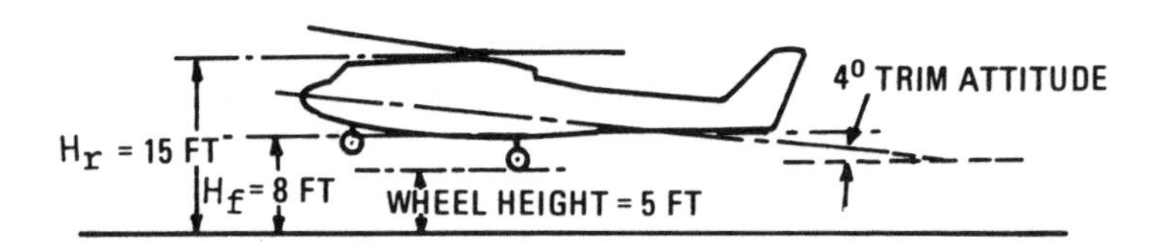

Figure 2.23 Hypothetical helicopter geometric parameters (IGE)

drag (D_v) which occurs in this regime of flight. This decrease in IGE download results from favorable interference effects between the lower surface of the airframe and the ground, as the pressure on the lower half of the fuselage increases due to higher static pressure in the surrounding downwash field.

The download correction factor (k_g), derived from the model rotor test data shown in Fig 2.24, is defined as

$$k_g = D_{v_{IGE}}/D_{v_{OGE}}.$$

In this figure, k_g is presented as a function of the ratio of average distance between the fuselage lower surface and the ground (H_f), and rotor diameter (d). For the hypothetical helicopter having a 5-ft wheel height, $k_g = 0.09$, and the download becomes zero at a wheel height of 3.5 ft, or $H_f/d = 0.13$.

Since k_g is used to compute the gross weight ratio W_{IGE}/W_{OGE} from the thrust ratio data presented in Fig 2.22 by noting that $T = W + D_v$; and since $D_{v_{IGE}} = k_g D_{v_{OGE}}$, then

$$\frac{T_{IGE}}{T_{OGE}} = \frac{W_{IGE}}{W_{OGE}}\left[\frac{1 + k_g(D_v/W)_{OGE}}{1 + (D_v/W)_{OGE}}\right]. \qquad (2.14)$$

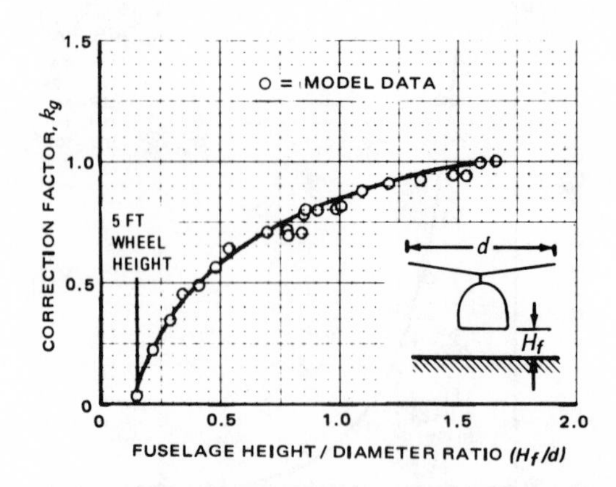

Figure 2.24 Effect of ground proximity on hover download

It should be noted that although the thrust curve shown in Fig 2.22 is based on the gross weight data presented in Ref 11, corrections for download effects (utilizing Eq (2.14)) were applied, thus obtaining a relationship between the IGE and OGE rotor thrust values.

As shown in Fig 2.23, the hypothetical helicopter rotor and fuselage height parameters at a typical ground-effect wheel height of 5 ft are $H_r/d = 0.3$ and $H_f/d = 0.16$. The average height of the cabin and tail boom above the ground at the nominal hover attitude of 4° (nose up) was used to define H_f; however, the distance between the bottom of the cabin and the ground at 0° pitch attitude can also be used with no significant reduction in accuracy. The thrust augmentation in-ground-effect and the download correction factors obtained from Figs 2.22 and 2.24 are $T_{IGE}/T_{OGE} = 1.14$ and $k_g = 0.09$. Substituting these values in Eq (2.14) and noting that $(D_v/W)_{OGE} = 0.0255$ for the hypothetical helicopter, we obtain $W_{IGE}/W_{OGE} = 1.17$. Therefore, the IGE gross weight capability of the hypothetical helicopter at a 5-ft wheel height is 17 percent higher than when out-of-ground effect, assuming constant power available. This analysis also assumes zero wind. In general, wind tends to reduce IGE thrust augmentation by deflecting the rotor downwash.

3. HOVER CEILING OGE AND IGE

Hover ceiling capability is calculated by matching the power available (see Ch I) with the power required for a range of operational altitudes. OGE and IGE hover ceilings for the hypothetical aircraft are presented in Fig 2.25 under standard atmosphere and 95° F ambient conditions. As shown, the helicopter has a 16,000-lb OGE gross weight capability at 4000 ft/95° F. The IGE capability is restricted—due to the maximum weight limit of the aircraft—to $W = 18{,}000$ *lb*. Detailed sample calculations for these two points are presented below to illustrate the calculation procedure. The main rotor tip speed is $V_t = 700$ *fps*, and the ambient constants at 4000 ft/95° F are $\sigma_\rho = 0.8076$ and $\sqrt{\theta} = 1.034$ (as defined in Ch I).

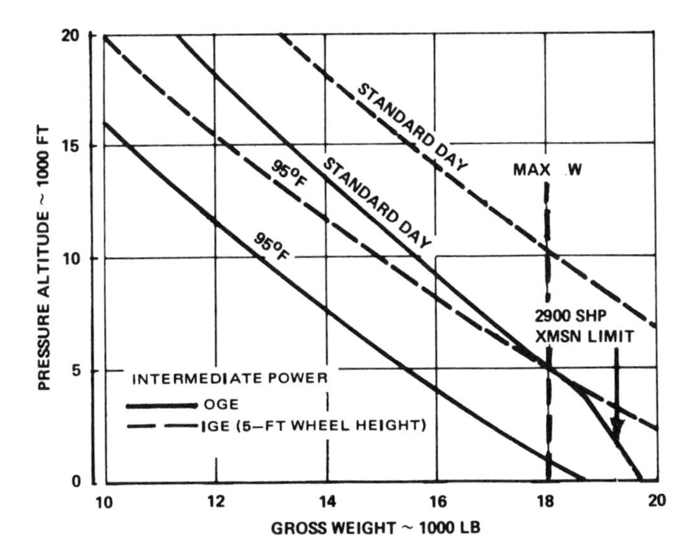

Figure 2.25 Hover ceiling in-and-out-of-ground effect

Starting with the above inputs, further calculations are performed using the following steps:

(1) Determine engine power available from Fig 1.6:

$SHP = 1196\ hp$ (one engine, intermediate (INT) power).

(2) Correct for installation losses (-1 percent):

$SHP = 0.99(1196 \times 2) = 2368\ hp$ (two engines).

(3) Convert SHP to RHP:

$RHP = 0.98(SHP - 30)$

$RHP = 0.98(2368 - 30) = 2291\ hp.$

(4) Calculate referred power at $V_t = 700\ fps$:

$RHP_{ref} = (RHP/\sigma_p)(700/Vt)^2$

$RHP_{ref} = (2291/0.8076) = 2835\ hp.$

(5) Compute $V_t/\sqrt{\theta}$:

$V_t/\sqrt{\theta} = 700/1.034 = 677\ fps.$

(6) From Fig 2.21, at $V_t/\sqrt{\theta} = 677\ fps$; the referred gross weight,

$W_{ref} = 19{,}820\ lb.$

(7) Calculate HOGE gross weight:

$W_{OGE} = W_{ref} \times \sigma_p(V_t/700)^2$

$W_{OGE} = 19820 \times 0.8076 = 16{,}010\ lb.$

(8) Calculate IGE weight from the OGE weight:

$W_{IGE} = 1.17\ W_{OGE}$

$W_{IGE} = 18{,}730\ lb.$

Since this weight exceeds the maximum operational weight limitation, the HIGE capability is restricted to a gross weight of 18,000 lb.

4. VERTICAL CLIMB CAPABILITY

The hypothetical helicopter vertical climb capability as a function of gross weight for maximum continuous and intermediate power settings is presented in Fig 2.26 at SL/STD and 4000 ft/95°F ambient conditions. For a design gross weight of 15,000 lb, V_c = *900 fpm* at intermediate power and 4000 ft/95°F conditions. A description of the method used to compute this performance data, including detailed basic sample calculations, is presented in the following paragraphs. In addition, a simplified method of computing vertical climb performance using potential energy considerations is provided.

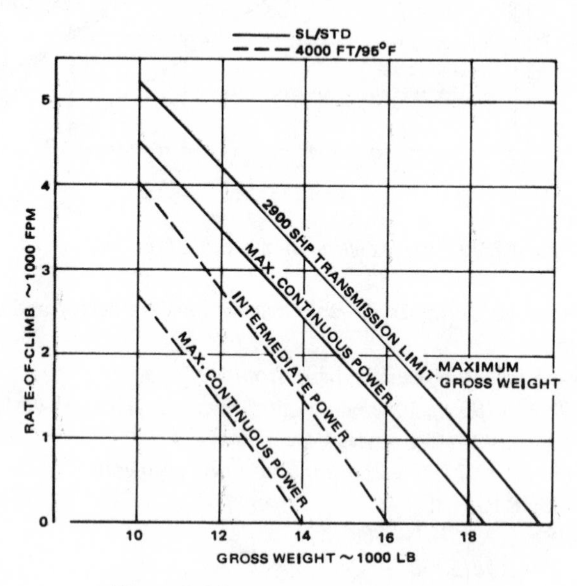

Figure 2.26 Vertical climb capability

4.1 Detailed Analysis

The relationships used to calculate vertical climb performance are based on the momentum theory expressions developed in Vol I, Ch II.4. To account for nonuniform downwash effects, the ideal induced velocity (v_{id}) and ideal thrust horsepower [$P_{T_{id}} = T(V_c + v_{id})/550$] defined by the momentum theory can be replaced by an equivalent induced velocity $v_e = k_{ind}\, v_{id}$ and the *actual* thrust horsepower (P_T) determined from the vortex theory. This procedure is shown below:

$$T = \rho \pi R^2 (V_c + v_e) 2 v_e \tag{2.15}$$

and

$$P_T = T(V_c + v_e)/550 \tag{2.16}$$

where

V_c = climb velocity; fps

T = main rotor thrust; lbs.

Rearranging Eq (2.16) gives:

$$V_c = (550P_T/T) - v_e. \tag{2.17}$$

Vertical rate of climb, therefore, is a function of (1) main rotor thrust power, (2) thrust, and (3) equivalent induced velocity. Knowing the main rotor power available ($RHP_{av_{mr}}$), the corresponding available rotor thrust power ($P_{T_{av}}$) can be determined as shown below:

$$P_{TAV} = RHP_{av_{mr}} - P_{pr} \tag{2.18}$$

where P_{pr} is the main rotor profile power, and

$$RHP_{av_{mr}} = (SHP_{av} - P_a)\eta_t - P_{tr};$$

SHP_{av} being engine shaft power available; P_{tr}, the tail rotor power required; and P_a is the accessory power.

In this relationship, the tail rotor power requirements in climb are assumed to be equal to the value needed to trim the aircraft in hover at $RHP_{av_{mr}}$, and the main rotor profile power in climb is assumed to be equal to the hover profile power for the same thrust level ($P_{prc} = P_{prh}$ if $T_h = T_c$). The tail rotor power and main rotor profile power variation between hover and climb conditions is usually small enough at low-to-moderate climb rates to justify these assumptions.

The remaining unknown parameter in Eq (2.17) is the equivalent induced velocity. This term is determined by noting that if the thrust in hover is equal to the thrust in climb, $T_h = T_c$; and that $V_c = 0$ for hover, then from Eq (2.15),

$$\rho\pi R^2 v_{e_h} 2v_{e_h} = \rho\pi R^2 (V_c + v_{e_c}) 2v_{e_c}. \tag{2.19}$$

Defining ($V_c + v_e c$) as U, Eq (2.19) becomes

$$v_{e_c} = v_{e_h}^{\,2}/U. \tag{2.20}$$

Knowing the actual induced main rotor hp in hovering (P_{ind_h}), and having thrust power available in climb (Eq (2.17)), it can be seen from Eq (2.16) that v_{e_h} and U in Eq (2.20) are

$$v_{e_h} = 550\,P_{ind_h}/T \tag{2.21}$$

and

$$U = 550\,P_{T_{av}}/T. \tag{2.22}$$

Substituting Eqs (2.21) and (2.22) in Eq (2.20) and then substituting the resulting equation in Eq (2.17) gives the vertical rate of climb in fps:

$$V_c = (550/T)[P_{T_{av}} - (P_{ind_h}^{\,2}/P_{T_{av}})]. \tag{2.23}$$

The method described above is based on rotor thrust which should be corrected for download effects. Download or vertical drag in climb is estimated by adjusting the hover download for the inflow velocity variations that occur in axial translation because of the change in induced velocity and the vertical climb velocity component.

The download adjustment procedure consists of first dividing the airframe into segments as described in the hover download discussion in Ch II.1.5. The download of each segment is

$$D_v = \tfrac{1}{2}\rho V'^2 A C_{D_v} \tag{2.24}$$

where

V' = total vertical velocity at the fuselage; fps

A = planform area of the segment; ft^2.

The velocity V' can be expressed as a function of vertical climb velocity, and the equivalent induced velocity at the rotor as follows:

$$V' = V_c + k_d v_e. \tag{2.25}$$

The parameter k_d in this equation is the downwash development factor defined as the ratio of the induced velocity at the fuselage to the induced velocity at the rotor disc. For fully-developed flow, $k_d = 2.0$ as defined by the momentum theory. However, most airframes are located sufficiently close to the rotor so that the downwash is less than its fully-developed value.

To determine the degree of downwash development at the fuselage, model rotor measurements (Fig 2.19) were compared with theoretical induced velocity predictions at the rotor disc. The latter was based on the vortex theory. These comparisons indicate that for fuselages located within a distance of $H_z/R \approx 0.3$, the average $k_d \approx 1.6$. For configurations with $H_z/R > 0.3$, the $k_d = 1.6$ value will result in optimistic download estimates, and for aircraft, such as the hypothetical helicopter and most of today's aircraft having $H_z < 0.3$, the 1.6 value will give a conservative estimate of download.

Knowing k_d, C_{D_v}, and the area A of each segment, the vertical climb download (D_{v_c}) can be expressed as a function of the hover download (D_{v_h}):

$$D_{v_c} = \left[\frac{\Sigma C_{D_v} A (V_c')^2}{\Sigma C_{D_v} A (V_h')^2}\right] D_{v_h}. \tag{2.26}$$

Substituting Eq (2.25) in Eq (2.26), D_{v_c} becomes

$$D_{v_c} = k_1 \left[\frac{k_2 U^2 + v_h^2 + k_3 v_e^2}{V_h^2}\right] D_{v_h}. \tag{2.27}$$

The terms k_1, k_2, and k_3 in this expression are constants.

In order to simplify the above described calculation procedure, a first approximation to download in climb can be obtained by assuming an average fuselage vertical drag coefficient ($\overline{C}_{D_v}$), and neglecting contraction effects. Then,

$$D_{v_c} = ½\rho A_f U^2 \bar{C}_{D_v} \tag{2.28}$$

where A_f is the total fuselage planform area. Substituting Eq (2.22) in Eq (2.28) gives

$$D_{v_c} = ½\rho A_f (550 P_{T_{av}}/T)^2 \bar{C}_{D_v}. \tag{2.29}$$

The corresponding hover equation is

$$D_{v_c} = ½ A_f (550 P_{ind_h}/T)^2 \bar{C}_{D_v}. \tag{2.30}$$

Dividing Eq (2.30) by (2.29) gives

$$D_{v_c}/D_{v_h} = (P_{T_{av}}/P_{ind_h})^2. \tag{2.31}$$

Step-by-step sample calculations are presented in Tables II-3 and II-4 for the hypothetical single-rotor helicopter operating at $SHP = 2900\ hp$ (transmission limit) and SL/STD atmosphere ambient conditions. Although, in principle, intermediate power rating could have been used for climb, this cannot be done since the aircraft is transmission limited at this ambient condition.

The initial calculations are shown in Table II-3 where the rate-of-climb is computed for five thrust levels. Eqs (2.15) through (2.23) serve as the basis for steps (1) through (11) in this table. The thrust values were then corrected for climb download effects. These computations are based on Eqs (2.24) through (2.27). The constants k_1, k_2, and k_3 used in the calculations were obtained from Table II-4. As shown in this table, the fuselage is divided into eight segments. The downwash development factor for airframe segments in the rotor downwash is $k_d = 1.6$, and a value of $k_d = 0$ is used for areas not located in the rotor wake, such as the fuselage sections under the cutout region of the rotor disc and the horizontal tail.

A first approximation to the download in climb can be determined by substituting the tabulated values of induced power shown in steps (4) and (6) into Eq (2.31). This procedure replaces the lengthy calculations in Tables II-3 and II-4. The results of the so-abbreviated calculations, together with the more detailed method, are shown in Fig 2.27, where performance is shown for two cases: (1) where download is constant as in the case of hover ($D_{v_h} = 0.025W$), and (2) where D_v varies with the rate of climb: $D_v = f(V_c)$. In the latter case, download estimates based on both the detailed calculation method and the first approximations are shown. It can be seen that the climb download correction is negligible at rates of climb less than approximately 1000 fpm. At higher climb rates, either the detailed calculation method or the first approximation of climb effects should be used. This figure indicates that the first approximation method will give a slightly conservative estimate of climb capability when compared with the detailed technique; primarily due to neglecting wake contraction effects. Wake contraction, being more pronounced in hover, contributes more to the increase of download in hover than in vertical climb.

STEP NO.	PROCEDURE	THRUST (T): LBS				
		10260	13330	16410	19490	20200
(1)	W_h (lb): $T/[1 + (D_v/W)]$	10000	13000	16000	19000	19700
(2)	RHP_{mr}: Required in Hover	968	1318	1750	2275	2420
(3)	$RHP_{av_{mr}}$	2420	2420	2420	2420	2420
(4)	P_{ind_h} (RHP)	670	1020	1460	1960	2090
(5)	P_{pr} (RHP): (2) – (4)	298	298	290	315	330
(6)	$P_{T_{av}}$ (RHP: (3) – (5)	2122	2122	2130	2105	2090
(7)	v_h (fps): ((4) x 550)/T	35.92	2.09	48.93	55.31	56.90
(8)	v_h^2 (fps²)	1290	17701	2394	3059	3238
(9)	U (fps): ((6) x 550)/T	113.75	87.55	71.39	59.40	56.90
(10)	v_e (fps): (8) / (9)	11.34	20.22	35.53	51.50	56.90
(11)	V_c (fpm): 60((9) – (10))	6144	4038	2272	747	0
(12)	D_{v_h} (lb): T – (1)	260	330	410	490	500
(13)	U^2 (fps²): (9)²	12940	7665	5097	3528	3238
(14)	v_e^2 (fpm²): (10)²	128.6	408.8	1124	2652	3238
(15)	$k_2 U^2 = -1.91$ (13) (fps²)	–24715	–14640	–9735	–6738	–6185
(16)	$k_3 v_e^2 = -1.35$ (14) (fps²)	–173.6	–551.9	–1518	–3580	–4371
(17)	(15) + (16) + (8) (fps²)	–23599	–13422	–8859	–7259	–7318
(18)	(17) / (8) (fps²)	–18.29	–7.58	–3.70	–2.37	–2.26
(19)	k_1 (18) $= -0.442$ (18)	8.08	3.35	1.64	1.04	1.00
(20)	D_{v_c} = (19) x (12) (lv)	2101	1106	672	510	500
(21)	Climb Gross Weight (lb): T – (20)	8159	12224	15738	18980	19700

TABLE II-3 DETAILED VERTICAL CLIMB SAMPLE CALCULATION

FUSELAGE SEGMENT	A (FT^2)	C_{D_v}	$C_{D_v}A$	k_d	V' $U+(k_d-1)v_e$	$(V')^2$	$C_{D_v}A(V')^2$
1. NOSE	20.4	0.5	10.2	1.6	$U+0.6v_e$	$U^2+1.2Uv_e+0.36v_e^2$	$10.2U^2+12.24Uv_e+3.67v_e^2$
2. CABIN	61.1	0.4	24.4	1.6	$U+0.6v_e$	$U^2+1.2Uv_e+0.36v_e^2$	$24.4U^2+29.28Uv_e+8.78v_e^2$
3. CUTOUT	71.1	0.8	56.8	0	$U-v_e$	$U^2-2Uv_e+v_e^2$	$56.8U^2-113.6Uv_e+56.8v_e^2$
4. NACELLE	18.0	1.2	21.6	1.6	$U+0.6v_e$	$U^2-1.2Uv_e+0.36v_e^2$	$21.6U^2+25.92Uv_e+7.78v_e^2$
5. AFTERBODY	43.8	0.5	21.9	1.6	$U+0.6v_e$	$U^2-1.2Uv_e+0.36v_e^2$	$21.9U^2+26.28Uv_e+7.88v_e^2$
TAILBOOM:							
6. UNDER ROTOR	24.4	0.5	12.2	1.6	$U+0.6v_e$	$U^2+1.2Uv_e+0.36v_e^2$	$12.2U^2+14.64Uv_e+4.39v_e^2$
7. AFT OF ROTOR	3.3	0.5	1.7	0	$U-v_e$	$U^2-2Uv_e+v_e^2$	$1.7U^2-3.4v_e+1.7v_e^2$
8. HORIZONTAL TAIL	39.0	1.2	46.8	0	$U-v_e$	$U^2-2Uv_e+v_e^2$	$46.8U^2-93.6Uv_e+46.8v_e^2$

$$[\Sigma C_{D_v}A(V')^2]_c = 195.6U^2 - 102.24Uv_e + 137.81v_e^2$$

$$[\Sigma C_{D_v}A(V')^2]_h = 231.17v_h^2$$

DERIVATION OF CONSTANTS:

$$D_{v_c} = \left[\frac{[\Sigma C_{D_v}A(V')^2]_c}{[\Sigma C_{D_v}A(V')^2]_h}\right]D_{v_h}$$

$$= \left[\frac{0.846U^2 - 0.442v_h^2 + 0.596v_e^2}{v_h^2}\right]D_{v_h} \quad \text{where } v_h^2 = Uv_e$$

$$D_{v_c} = 0.0442\left[\frac{-1.91U^2 + v_h^2 - 1.35v_e^2}{v_h^2}\right]D_{v_h} \quad \text{where } k_1 = -0.0442;\ k_2 = -1.91;\ k_3 = -1.35$$

TABLE II-4 DETERMINATION OF VERTICAL CLIMB DOWNLOAD CONSTANTS

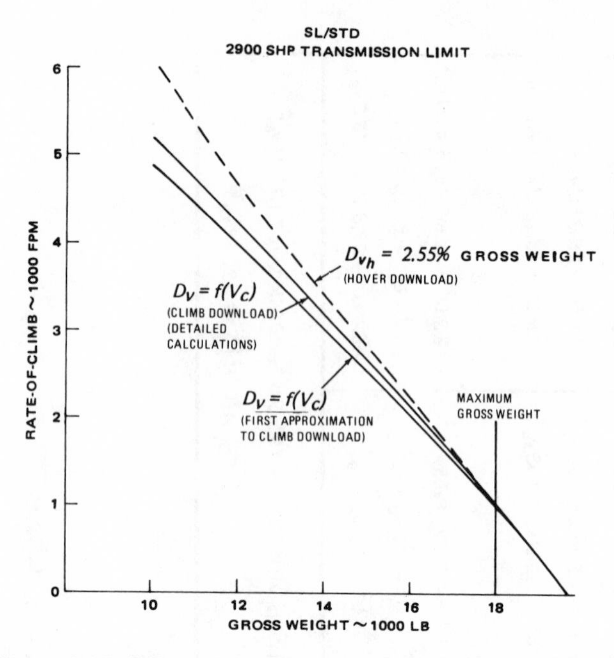

Figure 2.27 Effect of download on vertical climb performance

4.2 Simplified Vertical Climb Predictions

The calculation procedure described above is generally too detailed and time-consuming to be used for preliminary design performance estimates. A simplified method of estimating vertical climb performance can be developed by assuming that the excess shaft horsepower over that required in hover ($\Delta SHP = SHP_{av} - SHP_h$) times a correction factor is used entirely for moving the gross weight (W) against the pull of gravity:

$$550\,\Delta SHP\,k_p = V_c W \tag{2.32}$$

where

V_c = rate of climb; fps
k_p = climb efficiency factor
$\Delta SHP = SHP_{av} - SHP_h$
SHP_h = hover OGE shaft horsepower required.

The rate-of-climb in fpm can now be obtained from Eq (2.32):

$$V_c = \frac{33000\,\Delta SHP\,k_p}{W}. \tag{2.33}$$

For a given gross weight and available engine power, climb capability can be estimated if the climb efficiency factor, k_p, is known. On one hand, this factor should reflect the power losses—transmission efficiency, nonuniform downwash, tip losses, and tail and accessory power requirements—contributing to the difference between the ideal power available at the rotor and SHP. On the other hand, one should take into account the gains resulting from a reduction in induced power due to the increase in inflow velocity resulting from the rate of climb. It should be remembered, however, that an increase in the rate of climb usually leads to higher download values.

The k_p factor for the hypothetical helicopter shown in Fig 2.28 was computed on the basis of the detailed calculations discussed in the previous section. From this figure, it can be noted that kp varies from 1.5 at a 500 fpm rate of climb, to 1.0 at 4170 fpm.

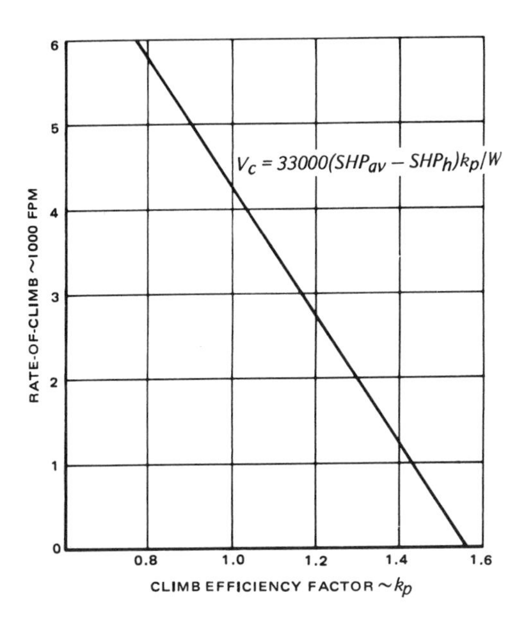

Figure 2.28 Vertical climb efficiency factor

It can be seen from Eq (2.11), Vol I, that k_p values are generally expected to be greater than 1.0 because of the reduction in induced velocity occurring as the result of an increased rotor inflow in vertical climb. However, at high rates of climb, this improvement is offset by increased download and the fact that the TV_c term in Eq (2.17) representing the power associated with working against gravity constitutes the major portion of the total power required in climb.

For preliminary design purposes, the k_p vs V_c relationship shown in Fig 2.28 can be used to predict the vertical climb capability of other helicopters similar to the hypothetical aircraft. However, for more rigorous evaluations, the procedure described in Sect 4.1 should be followed.

References for Ch II

1. Magee, J. P.; Maisel, M. D.; and Davenport, F. J. *The Design and Performance Prediction of a Propeller/Rotor for VTOL Applications.* Proceedings of the 25th Annual AHS Forum, No. 325. May 1969.

2. Gessow, Alfred; and Myers, Garry C. Jr. *Aerodynamics of the Helicopter.* Frederick Ungar Publishing Co., New York. Originally published in 1951 and republished in 1967.

3. Davis, S. Jon; and Wisniewski, J. S. *User's Manual for HESCOMP, the Helicopter Sizing and Performance Computer Program.* Report D210-10699-2 (Contract NAS2-6107) by Boeing Vertol Co. for NASA-Ames Research Center, Moffett Field, Ca. Nov. 1974.

4. Prouty, R. *Tip Relief for Drag Divergence.* Journal of the American Helicopter Society, Vol. 16, No. 4.

5. LeNard, John M.; and Boehler, Gabriel. *Inclusion of Tip Relief in the Prediction of Compressibility Effects on Helicopter Rotor Performance.* USAAMRDL Tech. Report 73-7, Dec. 1973.

6. LeNard, John M. *A Theoretical Analysis of the Tip Relief Effect on Helicopter Rotor Performance.* USAAMRDL Tech. Report 72-7, Aug. 1972.

7. Wiesner, Wayne; and Kohler, Gary. *Design Guidelines for Tail Rotors.* USAAMRDL Tech. Report D210-10687-1, Sept. 1973.

8. Delany, N. K.; and Sorensen, N. E. *Low-Speed Drag of Cylinder of Various Shapes.* NACA TN-3038, Nov. 1953.

9. Hoerner, S. F. *Fluid-Dynamic Drag.* Published by the author, 1965.

10. Fradenburgh, E.A. *Aerodynamic Factors Influencing Overall Hover Performance.* Paper No. 7 presented at AGARD Specialists Meeting on Aerodynamics of Rotary Wings, Marseille, France. Sept. 1972.

11. Lewis, Richard B., II. *Army Helicopter Performance Trends.* Journal of the American Helicopter Society, Vol 17, No. 2. April 1972.

FORWARD FLIGHT PERFORMANCE

The procedures for estimating airframe drag in forward flight are presented first. This is followed by an explanation of the method of determining power required in horizontal flight; illustrated by examples based on the hypothetical helicopter. Predictions of forward flight climb and descent, as well as the level-flight maneuver envelope, are also discussed. A complete presentation of forward-flight performance capabilities of the hypothetical helicopter concludes this chapter.

Principal notation for Chapter III

A	area	ft^2
AR	aspect ratio	
b	number of blades	
C	correction term	
C_D	body drag coefficient	
$C_P = 550\ HP/\pi R^2 \rho V_t^3$	rotor power coefficient	
$C_T = T/\pi R^2 \rho V_t^2$	rotor thrust coefficient	
$C'_T = L/\pi R^2 \rho V_t^2$	rotor lift coefficient	
$C_Q = Q/\pi R^3 \rho V_t^2$	rotor torque coefficient	
C_f	skin friction drag coefficient	
c	blade chord	ft
c_d	section drag coefficient	
$c_{\ell a}$	lift-curve slope	per radian
D	drag	lb
D_n	nacelle diameter	ft
f_e	equivalent flat-plate area	ft^2
g	acceleration of gravity	32.2 fps^2
h	altitude	ft
I	interference factor	
I	moment of inertia	slugs-ft^2
IGE	in-ground-effect	
k	grain size	in, or ft
k_c	cooling system design factor	
k_p	vertical flight correction factor	
k_{pt}	pressure drag factor	
k_s	supervelocity correction factor	
$k_{3\text{-}D}$	three-dimensional correction factor	
k_v	download factor	
L	rotor lift (rotor thrust component $\perp$ to distant flow)	ft
ℓ	length	ft
$M = V/a$	Mach number	
M	moment	ft.lb

Performance

m	mass	slugs
N	rotationsl speed	rpm
NUD	nonuniform downwash	
OEI	one-engine-inoperative	
P	power	ft.lb/s
PL	payload	lb
Q	torque	ft.lb
QSP	empirical stall parameter	
q	dynamic pressure	lb/ft^2
R	rotor radius	ft
$R_e = V\ell/\nu$	Reynolds number	
r	radial distance	ft
SR	specific range	n.mi/lb
T	rotor thrust	lb
T	absolute temperature	K
$TOGW$	takeoff gross weight	lb
t	time	s, or hr
U	velocity	fps
V	velocity of distant flow, or speed of flight	fps, or kn
W	weight (gross weight in particular)	lb
w	width	ft
X	rotor propulsive force	lb
$x = r/R$	nondimensional radial distance	
y	lateral distance	ft
z	hub elevation over pylon	ft
α	angle-of-attack	deg, or rad
β	sideslip angle	deg, or rad
$\gamma = \rho c_{\ell a} c R^4/I_b$	blade Lock number	
Δ	increment	
$\delta = p/p_o$	ambient pressure ratio	
η	efficiency	
η_t	transmission efficiency	
$\theta = T/T_o$	ambient temperature ratio	
ϑ	angle	
$\mu = V_{\|\|}/V_t$	rotor advance ratio	
ν	kinematic coefficient of viscosity ($\nu_o = 1/6380$)	ft^2/s
ρ	air density	slugs/ft^3
$\sigma = bcR/\pi R^2$	rotor solidity ratio	
$\sigma_\rho = \rho/\rho_o$	ambient density ratio	
ψ	azimuth angle ($\psi = 0$ for downwind position)	deg, or rad

Subscripts

a	air
b	blade
c	contraction

c	compressible
c	climb
d	divergent
d	descent
E	empty
e	equivalent
ex	exit, or exhaust
F	fuel
f	fuselage
f	forward
h	hover
i	initial
id	ideal
ind	induced
$inst$	installed
mr	main rotor
nu	nonuniform
o	initial
o	SL/STD
p	pylon
p	propulsive
s	stall
t	tail (horizontal)
tr	tail rotor
u	uniform
w	wetted
$\bullet$	cross-section
$\parallel$	parallel
$\perp$	perpendicular

Superscripts

$\bar{}$	average
$\bar{}$	nondimensional
$\dot{}$	derivative with respect to time

I. INTRODUCTORY REMARKS

The forward flight performance of a helicopter is primarily composed of (1) speed capability, (2) range and endurance levels, (3) rate of climb, (4) service ceiling, and (5) autorotational characteristics. To compute these items, power required as a function of airspeed, weight, and altitude must be determined. The following step-by-step procedures should be executed when estimating power required or when computing the forward flight performance of new helicopter designs:

1. Estimate the airframe drag, lift, side-force, and pitching and rolling moments.

2. Determine the complete level-flight power-required curve through the following intermediate steps:

   a. Calculate power required for aircraft trimmed conditions using a uniform downwash approach. (Note: An existing trim analysis computer program was used for the hypothetical helicopter.)
   b. Correct power required for nonuniform downwash effects.
   c. Apply parasite power correction.
   d. Define low-speed power required.

3. Determine climb and descent power required through the use of climb efficiency and descent calculation factors.

4. Apply the structural airspeed limitations associated with rotor stall.

5. Calculate performance capability by matching engine performance with speed.

Each step of the techniques used for performance predictions is explained in this chapter with detailed sample calculations for the hypothetical single-rotor aircraft. It should be noted that all airspeeds are considered as true airspeed which, under no-wind conditions, is equal to the ground speed.

2. DRAG ESTIMATES

The total parasite drag of the hypothetical helicopter can be determined by adding the incremental equivalent flat-plate area (Δf_e) of each of the components given in Table III-1.

ITEM	ESTIMATE Δf_e; ft^2		WIND TUNNEL TEST RESULTS Δf_e; ft^2	
BASIC FUSELAGE & PYLON		2.35		2.5
LANDING GEAR		4.56		4.9
MAIN	*(2.82)*		*(2.9)*	
NOSE	*(1.74)*		*(2.0)*	
MAIN ROTOR HUB		5.22		4.3
ENGINE NACELLES		1.09		1.7
VERTICAL & HORIZONTAL TAIL		0.83		0.7
TAIL ROTOR HUB ASSEMBLY		1.19		1.4
TRIM DRAG		0.80		0.7
SUBTOTAL		16.04		16.2
ESTIMATED ITEMS:				
ROUGHNESS & LEAKAGE	*(1.0)*		*(1.0)*	
PROTUBERANCES	*(1.6)*		*(1.6)*	
COOLING LOSSES	*(0.3)*	2.90	*(0.3)*	2.90
GRAND TOTAL		18.94		19.1

TABLE III-1 HYPOTHETICAL HELICOPTER PARASITE DRAG ESTIMATES

Both wind-tunnel test results and predicted drag values are shown in this table. Test data is available because the hypothetical helicopter fuselage is similar to an early version of a UTTAS prototype aircraft evaluated in the tunnel. The similarity of the two airframes is evident in the drawings and photograph presented in Fig 3.1.

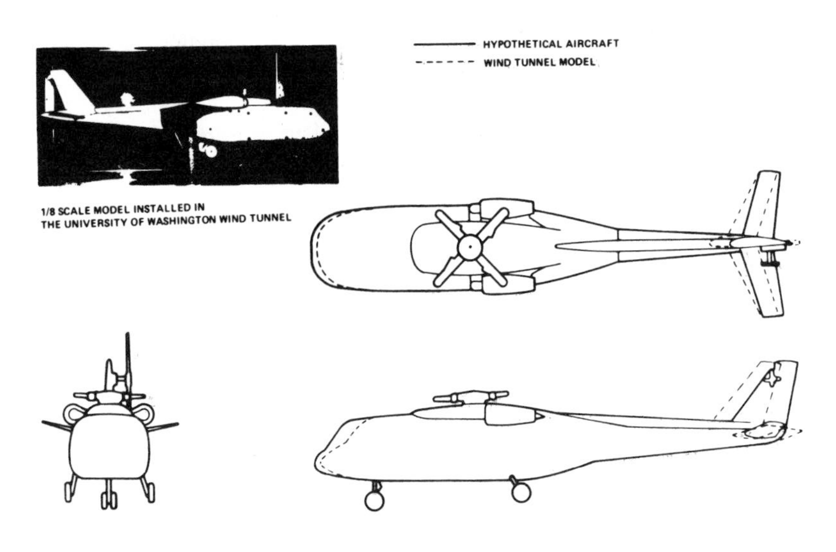

Figure 3.1 Fuselage configuration comparison

One may notice that the estimated and wind-tunnel measured subtotal drag values shown in Table III-1 are very close. A further examination of the table indicates that this is partially due to a random averaging of differences existing in predicted and measured drag values of individual items. However, even these individual differences are not too high, which may be due to the availability of wide wind-tunnel based general information on drag of various components. Without this background material, much larger discrepancies between predicted and measured drag levels of components and the total airframe may be expected.

The drag of the components shown in Table III-1 reflects values representative of streamlined items and nonstreamlined bodies. It also includes trim drag due to fuselage angle-of-attack effects and miscellaneous items resulting from roughness due to rivets, skin waviness, protuberances (antennas, lights, etc.), leakage, and cooling air momentum losses.

The total wind-tunnel value, $f_e = 19.1\ ft^2$, will be used for all forward flight performance of the hypothetical helicopter in this volume. A discussion of the differences in the accuracy of drag predictions for individual items such as the hub and engine nacelles, including prediction techniques, scale model effects, and an evaluation of the differences is presented below.

2.1 Drag of Streamlined Components

Although the streamlined components which include the basic fuselage (nose, cabin, and tail boom) pylon or crown area, tails, and engine nacelles are larger in size, they account for only 26 percent of the total drag of the hypothetical helicopter (Table III-1). This is due to the fact that the drag of these items consists primarily of skin friction resulting from shearing stresses developed in the boundary layer of the fluid.

The method of predicting the drag of the streamlined components consists of estimating the skin friction drag coefficient C_f, corresponding to fully turbulent flow over a flat plate area at the same Reynolds number as that of the part itself, and then applying correction factors to account for three-dimensional and mutual interference effects between the components[1,2]. A fuselage angle-of-attack of zero degrees is assumed for these calculations. The Δf_e is then computed using the wetted, or exposed, surface area A_w as a reference.

$$\Delta f_e = C_f A_w (1 + k_{3\text{-}D}) I \tag{3.1}$$

where

$k_{3\text{-}D}$ = three-dimensional correction

I = interference factor.

The skin friction coefficient in Eq (3.1) is based on the assumption of a fully turbulent boundary layer and thus, varies with Reynolds number and surface roughness[1] as shown in Fig 3.2. The use of C_f values based on this state of flow is valid because the

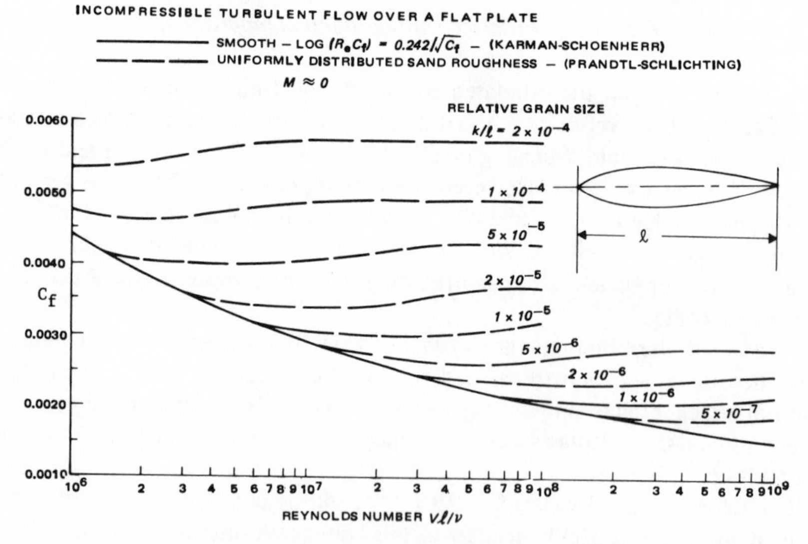

Figure 3.2 Average skin friction coefficient

surface is generally sufficiently rough due to rivets, seams, skin waves, etc., to cause the boundary layer to transition near the component leading edge. The data shown in this figure can also be used to correct wind-tunnel results for Reynolds number effects, provided transition strips were used to fix the model boundary layer transition near the leading edge.

There are two categories of roughness in the drag accounting system; either surface or discrete, depending on size. Surface roughness refers to the grain size of the paint or surface finish. As noted in Fig 3.2, grain size is specified in terms of the equivalent average grain size (k) to body length (ℓ) ratio. An equivalent value is required to utilize the experimental data. A typical value of k for mass production spray-painted finishes is 1.2×10^{-3} (Ref 2). Surface roughness does not include larger surface discontinuities such as rivet seams, or waviness, which is defined as discrete roughness.

Three-Dimensional Effects. For three-dimensional bodies, additional corrections ($k_{3\text{-}D}$) must be applied to the flat-plate skin-friction drag estimates to account for the following:

Supervelocity effect — due to local speed of flow exceeding the freestream value.

Pressure drag — resulting from the loss of momentum in the boundary layer.

Additional drag increase — resulting from the fact that the body surface is not a flat plate, but usually resembles a cylinder. The three-dimensional boundary layer of a cylinder is thinner than that of the flat-plate at the same R_e values.

The parameter $k_{3\text{-}D}$ can be determined from Ref 2 for optimum streamlined bodies of revolution and for wing and tail surfaces. Minor adjustments to these expressions are required to predict the drag of helicopter fuselage shapes because they are generally not bodies of revolution. The equation for $k_{3\text{-}D}$, including these adjustments, is presented below.

$$k_{3\text{-}D} = \underbrace{0.001(\ell/d)}_{\substack{\text{skin friction}\\ \text{drag due to}\\ \text{warping effect}}} + \underbrace{1.5(d/\ell)^{3/2}}_{\substack{\text{supervelocity}\\ \text{effect}}} + \underbrace{8.4(d/\ell)^{3}}_{\substack{\text{pressure}\\ \text{drag}}} + C \tag{3.2}$$

where

ℓ/d = effective length-to-diameter ratio (with d/ℓ as its reciprocal).

C = correction factor for noncircular cross-section shapes ($C = 0.05$ is a typical value for helicopters).

Eq (3.2) applies to the basic fuselage and engine nacelles. For tail surfaces, the expression for $k_{3\text{-}D}$ is

$$k_{3\text{-}D\,(tails)} = k_s(t/c) + k_{p_t}(t/c)^4 \tag{3.3}$$

where

t/c = average thickness in percent of chord

k_s = supervelocity factor

k_{p_t} = pressure drag factor.

Additional details concerning $k_{3\text{-}D\,(tails)}$ can be found in Ch IV, Sect 4.1 (Fig 4.14).

The last term in Eq (3.1) is the interference factor I. This factor accounts for the mutual interference drag which occurs when one body is placed in the vicinity of, or attached to, another. The increase in velocity and/or separation at the juncture point causes the combined drag of the two or more bodies to be greater than the sum of their individual values. The magnitude of interference effects are difficult to assess because they vary greatly with the basic shape of the two bodies and location of one relative to the other. The 0.6 ft^2 difference between the engine nacelle predictions and test results shown in Table III-1 illustrates this point.

The hypothetical helicopter nacelle interference drag may be reduced by locating the nacelles further outboard of the airframe. As shown by the CH-47 data in Fig 3.3, interference drag is minimized when the nacelles are located at least one diameter away from the basic fuselage[3]. The effect of different fillets on such an installation are almost impossible to evaluate accurately without wind-tunnel test data.

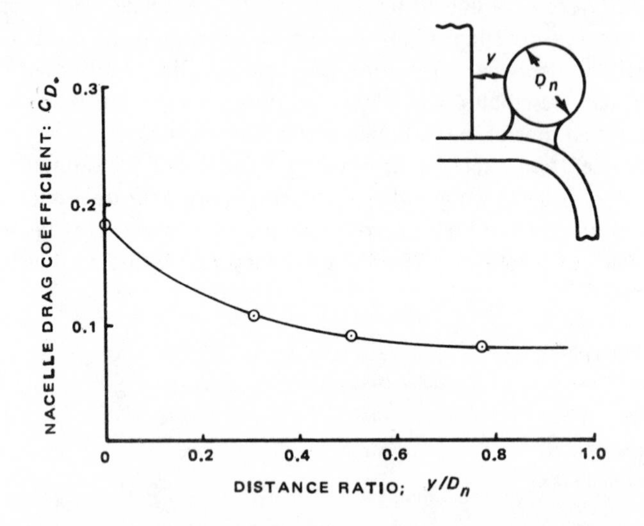

Figure 3.3 Effect of nacelle location on interference drag

The interference factor is usually determined from wind-tunnel tests or from information presented in Refs 1 and 2. In the absence of more reliable experimental data, a minimum value of 1.2 times the isolated component drag can be used to account for a typical level of interference between helicopter components and the basic fuselage. To facilitate "bookkeeping", interference drag should be included in the component drag values rather than charged to the fuselage.

The discussion presented above applies to streamlined fuselage shapes having no significant afterbody or aft section separation. For fuselage configurations employing the abrupt afterbody contraction required for rear-loading designs, additional drag terms must be added to Eq (3.1). In this case, based on Boeing Vertol test data[2,3], drag increments resulting from separation due to the adverse pressure gradient must be accounted for by such corrections as those shown in Fig 3.4. For afterbodies with symmetrical shapes in the side-view, and little or no lateral contraction, the correction expressed in terms of equivalent flatplate area is as follows:

$$\Delta f_{e_{cont}} = 0.008[6(d_e/\ell_c)^{5/2} - 1]A_{\bullet}. \tag{3.4}$$

However, for afterbodies having their side-view mean-line turned up (cambered) as shown in Fig 3.5 (but still having little, or no lateral contraction), another pressure drag correction must be applied because of still stronger three-dimensional flow effects.

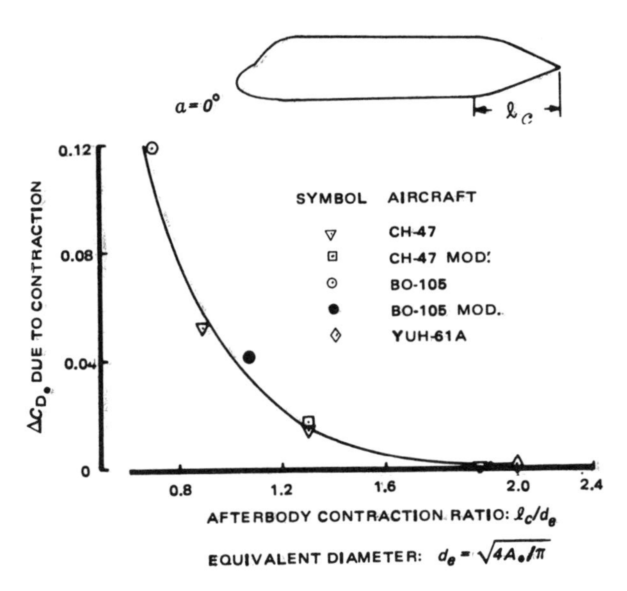

Figure 3.4 Effect of afterbody contraction ratio on drag

This drag increment ($\Delta f_{e_{camb}}$) can be approximated for afterbody shapes such as the CH-47 aircraft as follows:

$$\Delta f_{e_{camb}} = 0.0959(x/d_e)A_e \tag{3.5}$$

where

x/d_e = afterbody camber ratio .

A less abrupt contraction (high ℓ_c/d_e values) will, to some extent, counterbalance the effects of camber. However, more data is required to totally define this interaction. For example, the graph shown in Fig 3.5 indicates that a lateral contraction ratio of 1.3 or higher is required for the CH-47-type afterbody to minimize the drag due to negative camber.

2.2 Drag of Nonstreamlined Components

The major nonstreamlined components are the main rotor hub, tail rotor hub, and landing gear. As illustrated in Table III-1, the hubs and landing gear account for over 50 percent of the total aircraft drag; 30 percent of the total drag is due to the hubs. The drag of these components consists primarily of pressure drag resulting from large separated areas at the base of the component. The technique for estimating this drag

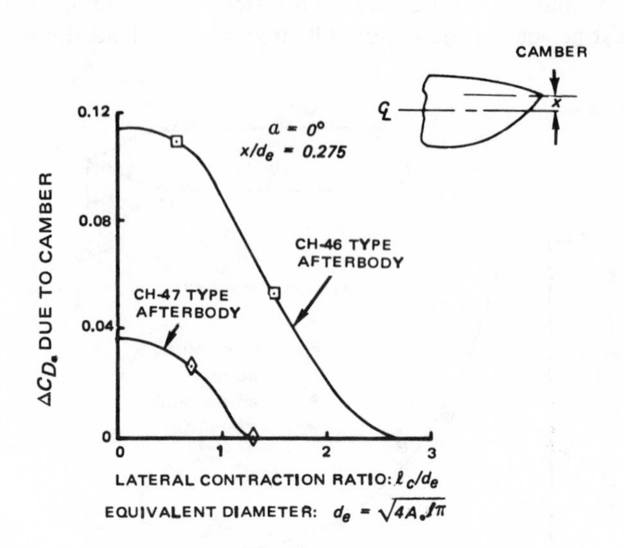

Figure 3.5 Effect of camber on drag

consists of obtaining representative drag coefficients and interference factors from Hoerner[1] or from past wind-tunnel test results and computing the equivalent flat plate area $\Delta f_e = C_{D_\bullet} A_\bullet I$ where $C_{D_\bullet}$ is based on frontal area $A_\bullet$ and I is the interference factor.

Landing Gear Drag. If wind-tunnel results are not available, the drag coefficient of the landing gear can usually be computed using the data from Ref 1. If test results are available, care should be taken to account for Reynolds number effects when applying the model results to the full-scale aircraft. Wind-tunnel model landing gear struts are usually tested at subcritical Reynolds numbers, while those of the full-scale aircraft operate in cruise at supercritical Reynolds number. Due to this correction, the drag of the full-scale landing gear would be about 10 to 15 percent lower than that indicated by wind-tunnel results for typical models (scale about 1:8).

Skid gears, rather than wheeled ones, are often utilized on single-rotor aircraft. A comparison of skid and wheeled gear drag trends is presented in Fig 3.6. For a helicopter of 20,000-lb gross weight, the skid drag is approximately 40 percent lower than that of the wheeled gear. Also, even lower values of drag can be achieved by streamlining the skids and support structure[3].

The wind-tunnel-determined landing gear drag for the hypothetical helicopter (4.9 ft^2) appears slightly lower than indicated by the wheeled-gear trend curve. This is due to such design aspects as location of the front-wheel torque scissors behind the main strut, and other similar design details.

Hub Drag. A slightly different approach must be taken to predict hub drag because of rotational and interference effects due to the proximity of the hub to the airframe. A summary of the calculation procedure described in detail in Ref 4 is presented below.

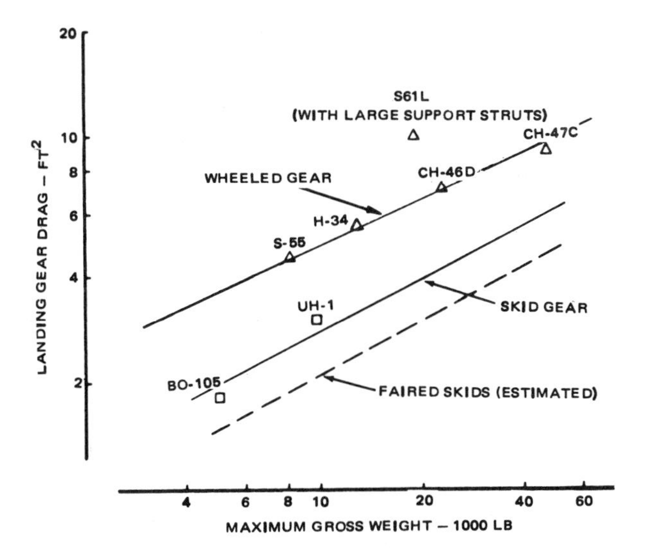

Figure 3.6 Landing gear drag trends

The initial step for predicting hub drag is to divide the hub into basic components consisting of shanks, blade attachment fittings, pitch housing and center-section as shown in Fig 3.7. The drag of the blade shank, attachment fittings, and pitch housing can be computed using the two-dimensional data published in Refs 1 and 5, and making

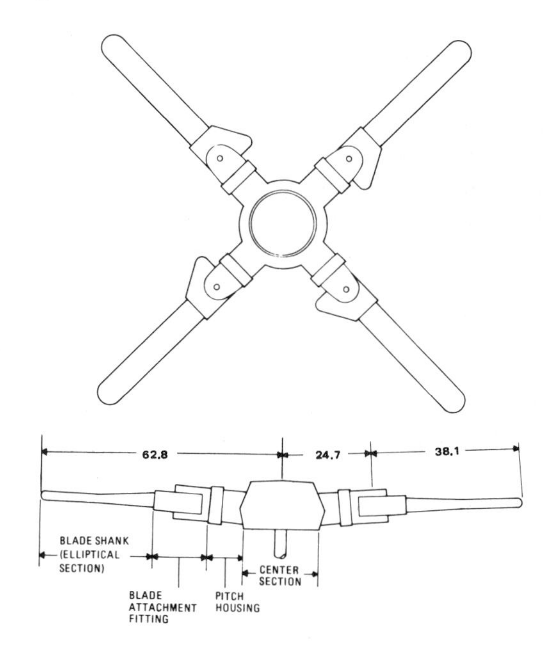

Figure 3.7 Hypothetical helicopter hub and blade shank

corrections for rotational effects. These corrections can be developed by noting that as each blade shank moves around the rotor azimuth, it encounters variations in dynamic pressure and projected frontal area. In the fore and aft position, this area may be assumed to be zero. At the advancing ($\psi = 90^\circ$) and retreating ($\psi = 270^\circ$) azimuthal positions, the frontal area is maximum, and the average dynamic pressure $\bar{q} = (q_{90^\circ} + q_{270^\circ})/2$. The Δf_e of the shank averaged over one revolution, is

$$\Delta f_e = \Delta D/q_o = c_{d_\bullet} A_\bullet (\bar{q}/q_\bullet)/2 \tag{3.6}$$

where

$c_{d_\bullet}$ = shank section drag coefficient

$A_\bullet$ = frontal area of section at $\psi = 90^\circ$ and 270°

q_o = freestream dynamic pressure.

$\bar{q}$ is obtained by integrating the local velocities along the shank extending from radial station m to station n, selected in such a way that $c_{d_\bullet}$ may be considered constant within their limits.

$$\bar{q} = q_o \left\{ 1 + \left[\frac{(1/\mu_n)^3 - (1/\mu_m)^3}{3(1/\mu_n - 1/\mu_m)} \right] \right\} \tag{3.7}$$

Here, μ_m and μ_n are the advance ratios at radial stations m and n, respectively. Therefore, the total equivalent flat-plate area of the shanks for a hub with b blades is

$$\Delta f_e = D/q_o = (b/2) c_{d_\bullet} A_\bullet / (\bar{q}/q_o). \tag{3.8}$$

Using this approach, the drag of the shanks was charged to the airframe parasite drag thus leading to a clearer understanding of the influence of blade airfoil section characteristics and blade geometry on rotor performance. There are, however, aerodynamicists who prefer a different "bookkeeping" method where both drag and torque of blade shanks are accounted for in the rotor performance calculations.

The hub center-section drag contribution can be computed using drag coefficients based on wind-tunnel test results[4]. As shown in this report, the hub center-section drag coefficients based on the rotating projected frontal area at 0° angle-of-attack vary from $C_{D_\bullet} = 0.55$ to 0.65. The tested configurations included two-bladed teetering rotors and three-bladed articulated hubs.

The $C_{D_\bullet}$ values noted above are for isolated hubs and do not include interference effects on the fuselage. However, for practical designs, interference drag can be significant, as shown in Fig 3.8. Here, interference factors based on wind-tunnel tests are presented as a function of the hub-gap to pylon-width ratio for various angles-of-attack. This data can be used to predict the interference effect of the outboard and center sections of the hub. For components located outboard of a circle defined by the width of the pylon, interference drag may be neglected. For example, for the hub center-section of the hypothetical helicopter, the gap is zero and the interference factor at zero degrees fuselage angle-of-attack is 1.97.

The technique outlined above was used to define the hypothetical helicopter hub drag shown in Table III-1. As noted in this table, the predicted hub-drag values are

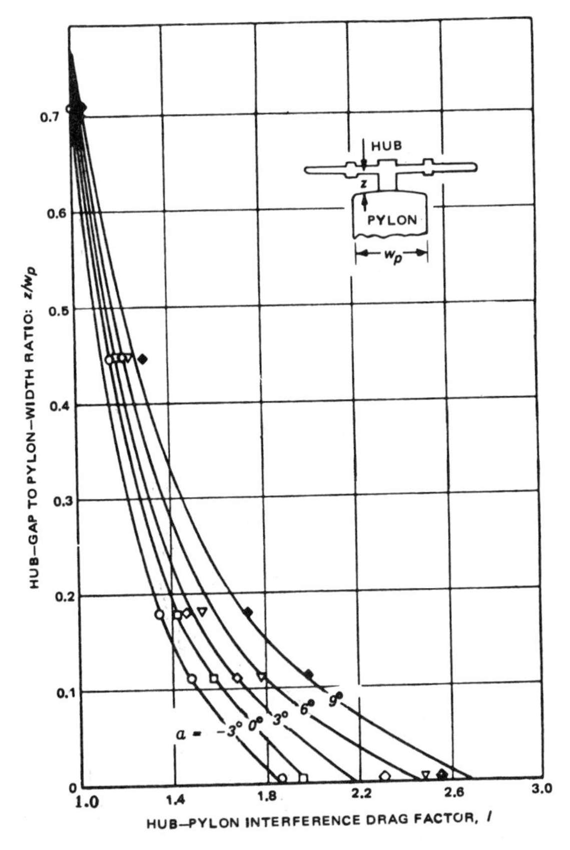

Figure 3.8 Effect of hub/fuselage gap on interference drag

0.9 ft² higher than the experimental results. This discrepancy is attributed to the use of component drag coefficients associated with articulated rotors having discrete hinges, and to variations in the interference drag. It can be seen from Fig 3.9 that articulated designs with discrete hinges are aerodynamically dirtier than the corresponding hingeless rotor or elastomeric hubs. This is due, in part, to lead-lag and flapping hinges, and lag dampers. An additional hub drag reduction can be achieved by utilizing flex-strap and elastomeric designs which eliminate the need for large pitch-bearing housings.

For preliminary estimates of hub drag when details of the hub components have not been defined, the trends shown in Fig 3.9 can be used. This data is based primarily on scale model tests, with no Reynolds number corrections. Comparisons of model and full-scale results show no significant effect of R_e on unfaired hub drag, although the local model shank and pitch-arm operate at Reynolds numbers below critical values. Additional testing is required to fully understand the effect of R_e on both faired and unfaired hub configurations.

2.3 Trim Drag

The component drag estimates discussed above are generally calculated at zero angles of attack unless information for other angles is available. A correction must be

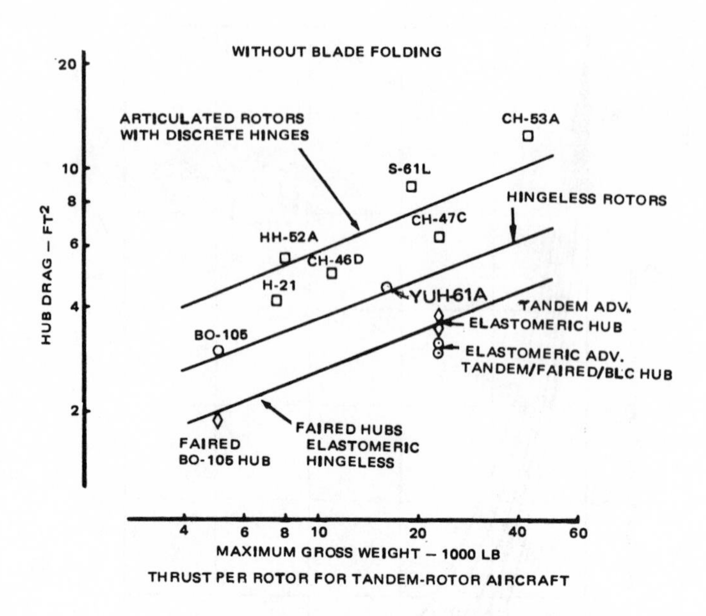

Figure 3.9 Hub drag trends

applied to the aircraft as a whole for the variation of data between zero degrees and cruise angle-of-attack which, for helicopter fuselages, is typically $\alpha_f = -4°$ to $-6°$ (nose down). For example, for the hypothetical helicopter, $\alpha_f = -5.2°$, which consists of a pitch or trim attitude of $-3.2°$ and $-2°$ of downwash angle at the following level flight trim conditions: *V = 140 kn; W = 15,000 lb; SL/STD day; 8-inch fwd c.g. position;* and *sideslip angle,* $\beta = 0°$.

The variation of fuselage drag with angle-of-attack is due primarily to the induced drag of the basic fuselage and stabilizer. There is no systematic data for determining the effect of trim attitude on drag for helicopter fuselage shapes. The only recourse is to rely on previous wind-tunnel tests for somewhat similar configurations. The trim drag of tail surfaces can be evaluated by the same method as was used for fixed wings[6,8]. The variation of drag with angle-of-attack of other components such as the hub, gear, etc., is generally negligible. The contribution of fuselage download—as is usually the case; or lift, which is seldom in horizontal flight—should also be taken into account.

The data presented in Ref 7 was used to determine the predicted hypothetical helicopter trim drag increment of 0.8 ft² between $\alpha_f = 0°$ and $-5°$. This drag value does not include any horizontal tail trim drag because the horizontal stabilizer angle-of-attack in cruise is $\alpha_t = -3°$; and at $\alpha_f = 0°$, $\alpha_t = 2°$. The difference in induced drag between these angles for the horizontal empennage is negligible.

To obtain the aircraft pitch attitude, the level flight fuselage angle-of-attack used in trim calculations is determined by solving the equations of motion for a force and moment balance about the c.g., and then adding the rotor downwash effects. The solutions to the complete set of trim equations, including a detailed analysis of rotor forces, was obtained for the hypothetical helicopter through the use of a computer as described later in this chapter. However, this can also be done through hand calculations by selecting a fixed speed in horizontal flight, and then writing an equilibrium equation for moments about the pitching axis; and another set of equilibrium equations for forces acting along the horizontal and vertical axes. Next, assuming a few (at least three) values of the fuselage angle-of-attack find, by interpolation, a a_f value for which moment and force equilibrium is reached.

When computing a_f, the effect of c.g., gross weight, and airspeed should be taken into account (Fig 3.10). The effect of c.g. on the hypothetical helicopter angle-of-attack

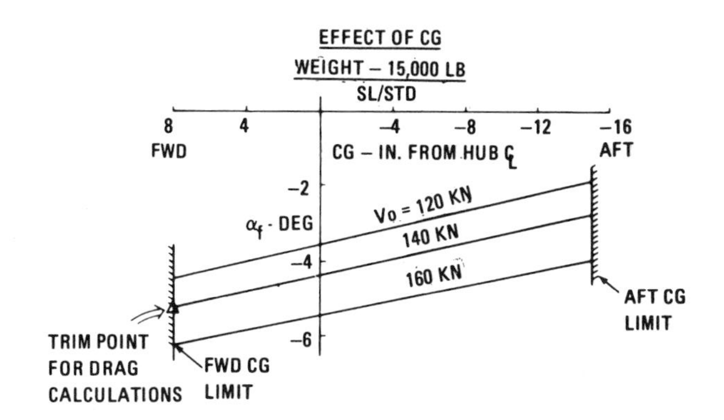

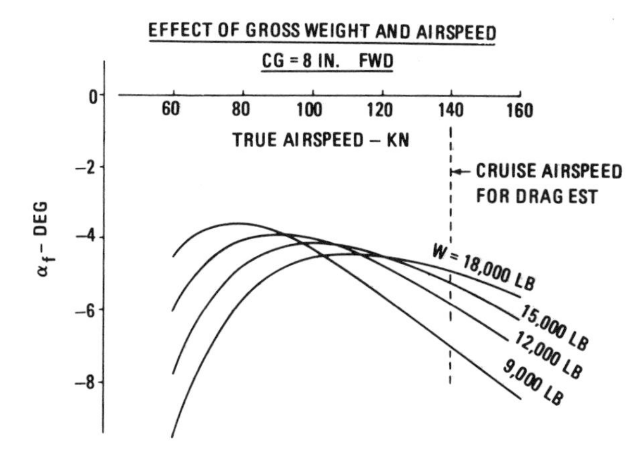

Figure 3.10 Effect of c.g., gross weight, and airspeed on fuselage angle-of-attack

is presented in the upper half of this figure. The most forward c.g. (8 inches forward) results in the largest negative angle-of-attack. At these c.g. positions, the aircraft must be trimmed more nose-down in order to achieve moment equilibrium.

Between the 15-inch aft and the 8-inch forward c.g. positions, there is a 2.5° difference in angle-of-attack. Performance data is typically presented at the most forward c.g. because this is the most adverse condition for single-rotor helicopters. All hypothetical helicopter performance presented in this text is based on the 8-inch forward c.g. position.

The effect of airspeed and gross weight on α_f is illustrated in the lower half of Fig 3.10. It can be seen that the airframe nose-down angle-of-attack increases with airspeed, due to the higher propulsive force requirements resulting from increased fuselage drag. This variation of cruise angle with airspeed is most prominant at the lower gross weights, since the thrust vector tilt required to achieve a given propulsive force is larger at these conditions. At speeds below approximately 80 kn, it should be noted that α_f becomes more negative as the speed decreases. This trend is due to higher rotor downwash and occurs regardless of the gross weight of the aircraft.

Trim analysis computations for the hypothetical helicopter at 140 kn indicate that approximately 4° of sideslip is required to trim at 0° roll attitude. Higher angles are required at lower airspeeds. The drag penalty associated with the $\beta = 4°$ trim change attitude is 0.6 ft² based on wind-tunnel test results. However, to simplify the sample calculations, this penalty is not included in the performance predictions.

2.4 Miscellaneous Items

Items included in the following discussion refer to the discrete roughness and leakage, protuberances, and cooling air momentum indicated at the bottom of Table III-1.

Discrete Roughness and Leakage. As noted previously, discrete roughness drag includes surface irregularities such as rivet heads, seams, waviness in the skin, etc. Leakage drag results from air that enters or exits the fuselage around cowlings, access doors, windows, etc. Data compiled in Ref 2 indicates that roughness and leakage effects increase the basic skin friction drag of current aircraft by 20 percent; 10 percent of which is due to roughness and 10 percent to leakage. A 20-percent increase in skin friction drag for the hypothetical helicopter is equivalent to a 5-percent increment of the total drag value.

Protuberances. Protuberances are represented by larger external items such as antennas, vents, drains, and anticollision lights. If detailed drawings are available to locate and define all of the protuberances, drag estimates can be obtained using data presented in Ref 1; however, for preliminary design work where such details are not available, the protuberances are generally accounted for by increasing the aircraft drag by 5 to 10 percent as suggested in Refs 2 and 6. A fixed amount of protuberances installed on a relatively clean fuselage such as that of the hypothetical design would lead to a larger relative drag increase than in the case of a "dirty" design. Therefore, a value of a 10-percent increment of the total aircraft drag (excluding the miscellaneous items) was assumed.

Cooling Momentum Drag. This drag results from the loss of air momentum as it enters and exits the cooling system of the hydraulic power supply, engines, and transmissions. Losses also occur due to air entering the heating and ventilation systems.

If details of the aircraft cooling system are available so that the mass flow through the system and the area of the inlet and exit ducts are defined, the cooling drag can be computed using the simple momentum relationship.

$$D = \dot{m}_a(V_o - V_{ex} \cos \vartheta_{ex}) \tag{3.9}$$

where

$\dot{m}_a = \dot{W}_a/g$; mass flow ($W_a \equiv$ weight of air-flow in lb/s, and $g \equiv$ acceleration of gravity)

V_o = forward speed; fps

V_{ex} = exit velocity

ϑ_{ex} = exit angle relative to the freestream velocity.

With properly designed inlets and exits, much of the momentum loss that occurs at the inlet could be recovered at the exit. The only problem is that helicopter cooling systems have to be designed for hover, where the highest power and resulting operating temperatures occur. To satisfy these requirements, inlets and exits are usually designed as large openings located on top of the fuselage where the exit airflow cannot be readily directed aft. Cooling systems depend on the total pressure in the downwash, or free-stream air flow, combined with pressure differential generated by blowers to move the air through the system. Typical exhaust velocities for blower installations are on the order of 60 to 80 fps to provide adequate cooling and to minimize power loss.

The procedure for estimating momentum drag may be simplified by taking the approach suggested in Ref 2, which indicates that the cooling momentum drag is proportional to the installed power available as shown below:

$$\Delta f_e = 2.5 \times 10^{-5}(SHP_{inst})k_c \quad (3.10)$$

where

k_c = cooling system design factor

SHP_{inst} = installed shaft horsepower available.

Depending on the design of their cooling systems, typical production helicopters have k_c values ranging from 4 to 6. A value of 4 is used for the hypothetical helicopter design. Therefore, the cooling air momentum losses at the 2900 SHP transmission limit results in a $\Delta f_e = 0.3\ ft^2$ drag penalty.

2.5 Net Engine Thrust

Turboshaft engines produce a net thrust, or drag, depending on the engine mass flow, exhaust direction and velocity, and flight speed. For typical installations, the air-flow velocity originally decreases at the inlet; but is increased at the exhaust to provide thrust. The net resultant thrust, T_{net} is

$$T_{net} = \dot{m}(V_{ex} \cos \vartheta_{ex} - V_o) \quad (3.11$$

where

$\dot{m}$ = engine mass flow; slug/s

V_{ex} = exhaust velocity; fps

ϑ_{ex} = exhaust velocity cant angle; deg.

For typical turboshaft installations, the engines provide a net thrust at speeds up to 140 to 150 kn; however, a momentum drag appears at higher speeds (Fig 3.11). The magnitude of this thrust or drag in cruise is normally small enough to be omitted for most performance calculations, but care must be taken when making exhaust modifications—

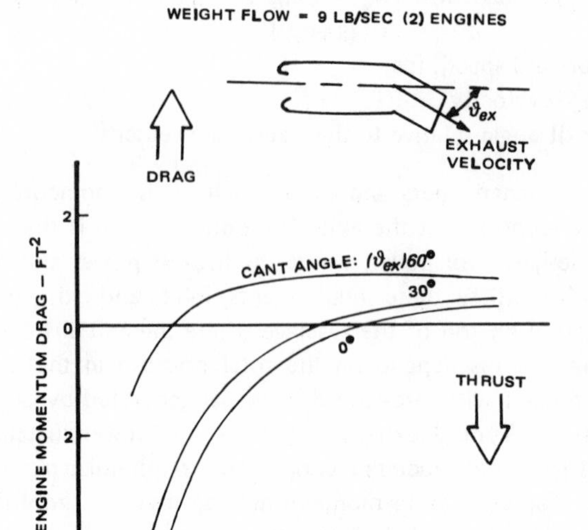

Figure 3.11 Example of engine momentum drag/thrust

for instance, installations of ejector shrouds and IRS suppressors, as they tend to reduce the exit velocity.

2.6 Detailed Sample Drag Calculations for the Hypothetical Helicopter

Details of the hypothetical helicopter drag calculations, including reference areas, drag coefficients, and interference factors are shown in Table III-2. The results were summarized and compared with wind-tunnel measurements in Table III-1. To further illustrate the procedure for estimating the drag of various components, the following step-by-step calculations are shown for streamlined (basic fuselage) and nonstreamlined (nose gear) items.

Streamlined Component – Calculation of the hypothetical helicopter basic fuselage drag (nose, cabin, and tail boom drag).

1. Determine skin friction drag coefficient, C_f, from Fig 3.2, where C_f is defined as a function of Reynolds number R_e, ($R_e = V\ell/\nu$), and equivalent roughness k/ℓ, where k is the grain size, ℓ is the characteristic body length, and ν is the kinematic coefficient of viscosity.

At 150 kn, SL/STD,

$$R_e = \frac{140 \times 1.689 \times 47}{1.567 \times 10^{-4}} = 7.1 \times 10^7$$

Assume an equivalent grain size, $k = 1.2 \times 10^{-3}$, for mass production spray-painted surfaces[2]:

ITEM	WETTED AREA (ft^2)	FRONT AREA (ft^2)	C_D; C_f	INTERFERENCE FACTOR (I)	Δf_e (ft^2)	COMMENTS
BASIC FUSELAGE					1.92	*See Note 3*
Skin Friction (flat Plate)	685		0.00235		1.61	
3-Dimensional Effects					0.31	
PYLON					0.43	*See Note 3*
Skin Friction (Flat Plate)	63		0.0027	2.40	0.41	
3-Dimensional Effects					0.02	
NOSE GEAR					1.74	*Strut airflow is supercritical*
Oleo Strut		0.9	0.5	1.25	0.56	
Scissors		0.65	0.6		0.39	
Wheels (2)		1.75	0.3	1.50	0.79	
MAIN GEAR					2.82	*Trailing arm airflow is supercritical*
Trailing Arm Strut (2)		1.48	0.5	1.25	0.92	
Wheels (2)		2.54	0.5	1.50	1.90	
MAIN ROTOR HUB					5.22	*C_D values and frontal areas defined at 0° shaft angle. Blade shanks have elliptical cross-section*
Center Section		1.65	0.55	1.97	1.79	
Pitch Housing		0.52(1)	0.4	1.70/1.02*	0.72	
Blade Attachment Fittings		0.75(1)	1.0	1.57/1.06*	2.50	
Blade Shanks		0.58(1)	0.15	1.00/1.20*	0.21	
ENGINE NACELLES (2)					1.09	*See Note 3*
Skin Friction (Flat Plate)	54		0.0034	2.0	0.37	
3-Dimensional Effects					0.40	
Base Drag (Fairing)					0.32	
VERTICAL TAIL					0.39	*See Note 3*
Skin Friction (Flat Plate)	53		0.0035	1.05	0.19	
3-Dimensional Effects					0.10	
Induced					0.10	
HORIZONTAL TAIL					0.44	*See Note 3*
Skin Friction (Flat Plate)	72		0.00385	1.05	0.29	
3-Dimensional Effects					0.08	
Induced					0.07	
TAIL ROTOR HUB ASSEMBLY					1.19	*Interference includes mutual interference between parts and effect on vertical tail*
Shaft		0.28	0.5	1.50	0.21	
Hub and Slider		0.61	1.0	1.50	0.92	
Base Fairing		0.21	0.2	1.50	0.06	
TRIM DRAG (Δ Drag between 0° and -5° α_f)					0.80	*Function of planform area and pressure drag*
				SUBTOTAL	(16.04)	

ITEM	Δf_e	COMMENTS
ROUGHNESS AND LEAKAGE	1.0	*Calculated as 20% of skin friction drag (10% roughness and 10% leakage; or 6% of subtotal)*
PROTUBERANCES	1.6	*Estimated to be 10% of subtotal*
COOLING LOSSES (Transmission and Engine)	0.3	*Proportional to installed power available*
SUBTOTAL	(2.90)	
GRAND TOTAL	18.94	

NOTES:
1. $R_e/ft = 1.51 \times 10^6$ *(140 kn, SL/STD)*
2. All items evaluated at $\alpha_f = 0°$
3. Surface roughness equivalent grain size $k = 1.2 \times 10^{-3}$ inches

* Rotational Effects $\bar{q}/q_o$; where $\Delta D/q = (b/2) C_{d_o} A \cdot I (\bar{q}/q_o)$

TABLE III-2 PARASITE DRAG ESTIMATE

$$k/\ell = \frac{1.2 \times 10^{-3}}{47 \times 12} = 2.13 \times 10^{-6}$$

Hence, $C_f = 0.00235$ (Fig 3.2).

2. Calculate fuselage wetted area (A_w):

A quick method of estimating A_w based on a three-view drawing is presented below. This method assumes that the cross-section shape does not vary along the body length.

$$A_w = 2 \times \kappa(A_{top} + A_{side}) - A_{wp} \tag{3.12}$$

where

κ = circumference/2 (height + width) of cross-section
A_{top} = planform view area
A_{side} = side-view area
A_{wp} = pylon juncture wetted area.

Assuming the cross-section shown in Fig 3.1 is representative of the entire fuselage,

$$\kappa = 24/2(8 + 6.3) = 0.84,$$

and

$A_{top} = 230\ ft^2$
$A_{side} = 214\ ft^2$
$A_{wp} = 61\ ft^2$.

Now

$$A_w = 2 \times 0.84(230 + 214) - 61 = 685\ ft^2.$$

The fuselage should be divided into smaller segments if the shape of the cross-section varies considerably along the length.

3. Determine $k_{3\text{-}D}$ from Eq (3.2) as a function of ℓ/d and C, where

$\ell = 47\ ft$
$d_e = \sqrt{A_\bullet 4/\pi} = \sqrt{(46 \times 4)/\pi} = 7.64\ ft,$
$\ell/d_e = 47/7.64 = 6.15$
$C = 0.05.$

Substituting these values in Eq (3.2) gives:

$$k_{3\text{-}D} = 0.142 + 0.05 = 0.192.$$

4. Compute basic fuselage Δf_e per Eq (3.1), assuming no significant afterbody separation:

$$\Delta f_e = C_f A_w (1 + k_{3\text{-}D})$$

$$\Delta f_e = 0.00235 \times 685(1 + 0.192) = 1.92 \; ft^2.$$

Nonstreamlined Component – Calculation of the hypothetical helicopter nose-gear drag.

1. Using Fig 3.12, compute oleo strut and axle drag:

At 140 kn, SL/STD,

$$R_e = (140 \times 1.689 \times 3)/(12 \times 1.567 \times 10^{-4}) = 3.77 \times 10^5.$$

Depending on the relative roughness, $C_{D_\bullet} = 0.3$ to 0.5 (Hoerner[1], pg 3-10). The higher value of 0.5 is used to provide a degree of conservatism and to account for discontinuities along the length of the strut. Assuming an interference factor of 25 percent (Ref 1, pp 8-9 and 8-19),

$$\Delta f_e = C_{D_\bullet} A_\bullet \times 1.25$$

$$\Delta f_e = 0.5 \times 0.9 \times 1.25 = 0.56 \; ft^2$$

where

$$A_\bullet = [(16)(1) + (4.5)(7.5) + (20)(3) + (4)(5)] / 144 = 0.9 \; ft^2.$$

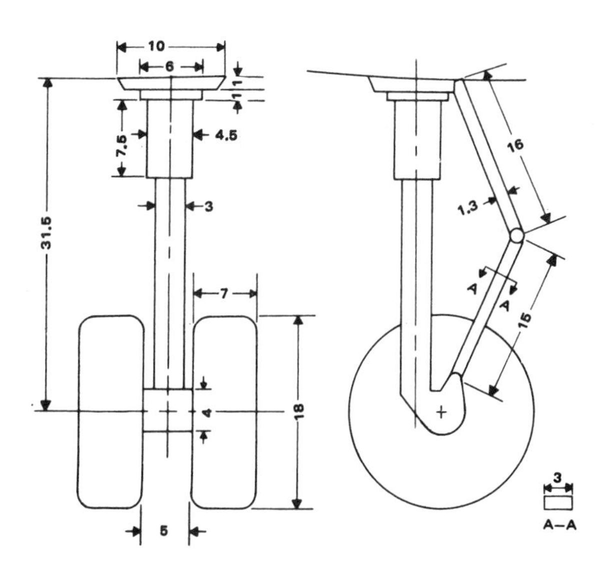

Figure 3.12 Nose-gear configuration (extended on flight position)

2. Calculate torque scissors drag:

Assuming $C_{D_\bullet} = 1.2$ (flat plate), and local $q = 0.5q_o$ (freestream) due to strut wake,

$$\Delta f_e = 1.2 \times 0.65 \times 0.5 = 0.39\ ft^2$$

where

$$A_\bullet = (3)(31)/144 = 0.65\ ft^2.$$

3. Estimate wheel drag:

According to Ref 1 (pg 13-14), $C_D = 0.3$, when based on an area (A) defined by tire width x maximum diameter:

$$A = (18 \times 7)/144 = 0.875\ ft^2.$$

Interference between the wheels and the struts is assumed to be 50 percent of wheel drag. This is based on previous test-versus-theory comparisons conducted on the CH-46 aircraft which has a similar nose-gear arrangement. Therefore, the drag of the two wheels is:

$$\Delta f_e = 2(0.3 \times 0.875 \times 1.5) = 0.79\ ft^2.$$

The total estimated gear drag is:

Oleo Strut	$\Delta f_e = 0.56$
Torque Scissors	$\Delta f_e = 0.39$
(2) Wheels	$\Delta f_e = 0.79$
TOTAL	$\Delta f_e = 1.74\ ft^2.$

3. METHODOLOGY FOR DETERMINING LEVEL FLIGHT POWER REQUIRED

The most widely used practical means of determining power required in horizontal flight consist of performing detailed computations for (1) flying speeds higher than about 60 kn, and (2) hovering. The $SHP = f(V_o)$ relationship for the $0 \leqslant V_o < 60$ interval is usually established by "guided" interpolation. Such a guide (determining the basic shape of the power-required curve within the low-speed region) can be provided by the simple momentum theory, taking into account the complete flow velocity through the disc $(\vec{V}' = \vec{V}_o + \vec{v})$ as indicated in Ch II of Vol I. It can also be obtained from vortex theory methods adapted to the low-speed region (Ch IV of Vol I).

With respect to flying speeds higher than ≈ 60 kn, the combined momentum and blade element theory (Ch III of Vol I) should provide a sufficiently accurate basis for routine engineering practice. This is especially true if performance predictions obtained in this manner are checked against flight test results, or more accurate (at least in principle) theoretical methods such as those based on vortex theories (Ch IV of Vol I). Should discrepancies—between either flight tests, or more sophisticated analytical procedures—be pinpointed with respect to the area of their occurrence (say, induced or profile power predictions), then appropriate correction factors could be established and applied.

In the particular case of the hypothetical helicopter, the level flight power required for speeds above ≈60 kn was determined using an available trim analysis computer program. The program employs a combination of blade element and momentum theory to compute the rotor and fuselage trim forces. The accuracy of the induced power prediction is improved by applying a nonuniform downwash correction to values obtained on the basis of uniform downwash. This correction is derived from comparisons of power levels theoretically predicted for forward flight; using uniform and nonuniform downwash analyses. In addition, an empirical correction based on wind-tunnel test results is applied to account for discrepancies between theoretical predictions and wind-tunnel measurements of parasite power.

Low-speed power required below 60 kn is determined by utilizing either the trim analysis program, or the basic momentum theory induced power relationships to define the shape of the low-speed power-required curve. When the uniform downwash theory is used, adjustments are required to have the hover point agree with the values given by the vortex theory, and the 60-kn point with nonuniform downwash predictions.

Details of the trim program and low-speed power required prediction techniques are presented on the following pages. Power required data and calculations for the hypothetical helicopter are provided to illustrate the methodology. At the end of this section, the trim program predictions are compared with the simplified power required expressions derived in Vol I.

3.1 Trim Analysis Computer Program

Aircraft trim for forward-flight conditions is determined by solving the six steady-state equations of motion developed from a force and moment balance about the center-of-gravity. The computer program for the hypothetical helicopter calculations was formulated in such a way that the flight conditions indicated in the input box in Fig 3.13 contained the following items: gross weight, speed of flight (horizontal), and sideslip angle. The results consisted of: fuselage pitch attitude (in this program, symbolized by θ), fuselage roll angle (ϕ), and longitudinal cyclic pitch angle.

In Sect 2.3 of this chapter, it was indicated that it is desirable to have the ϕ angle close to zero. This means that should the trim analysis indicate that ϕ does not equal zero (say $\phi > |0.1°|$), new sideslip angles must be assumed; until the desired $\phi \approx 0$ is obtained. In this respect, it would be more desirable to develop a computer trim analysis program where $\phi = 0$ would be input as one of the flight conditions, and the output would be the fuselage yaw angle required to achieve $\phi = 0$.

As to the actual process of solving equations of equilibrium, iterative techniques are required because of the complexity of the rotor analysis needed to compute the rotor forces and moments. The rotor analysis is a subroutine in itself and uses a numerical approach for solving the rotor flapping and force equations. Blade stall, reverse flow, and compressibility effects are taken into account by the use of two-dimensional airfoil section data (see Ch II). However, in order to simplify this analysis, the following assumptions were made.

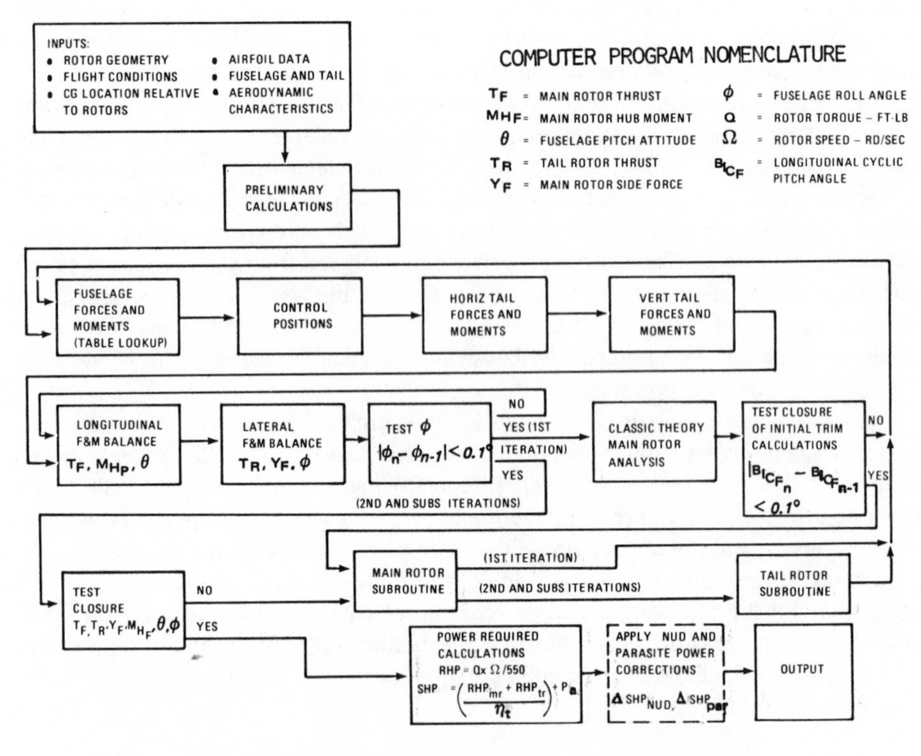

Figure 3.13 Block diagram for single-rotor trim analysis

1. Induced velocity distribution is assumed to be uniform.
2. Blade lag and all elastic degrees of freedom are neglected.
3. Unsteady aerodynamic and spanwise flow effects are ignored.
4. Three-dimensional compressibility effects at the blade tip are not considered.

Once the trim has been established, corrections to the power required for non-uniform downwash (NUD) and parasite power are added to the basic trim power required predictions. A discussion of these corrections is presented later in this chapter.

A complete block diagram of the single-rotor helicopter trim analysis is presented in Fig 3.13. The initial step in the iterative procedure is to compute the fuselage and tail forces and moments. A matrix of fuselage forces and moments is input into the program as a function of angle-of-attack ($\alpha_f = \pm 90°$), and sideslip angle ($\beta_f = \pm 90°$) based on wind-tunnel testing. An example of longitudinal fuselage and horizontal tail forces and moments used for the hypothetical helicopter is presented in Fig 3.14.

If wind-tunnel data is not available, the fuselage characteristics must be estimated. This involves predicting basic tail-off airframe characteristics plus the aerodynamic contriubtions of the tail surfaces. It can be seen from Fig 3.14 that the tails have a significant effect on the fuselage moments and forces. Their effects can be estimated from basic wing theory as, for instance, that presented in Ref 6. However, precise methods of estimating tail-off fuselage characteristics for helicopter-type shapes are not available. Consequently, previous wind-tunnel test results for similar configurations must be used as described in the discussion of trim drag (Sect 2.3).

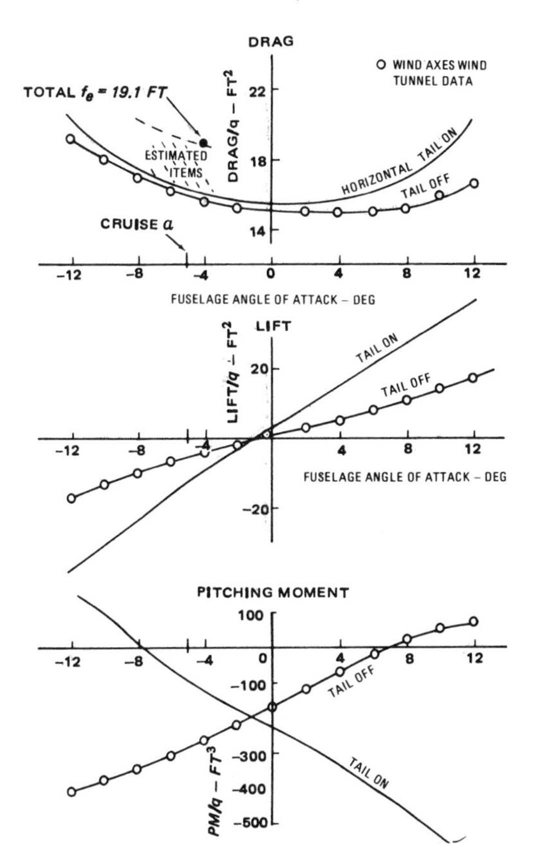

Figure 3.14 Fuselage aerodynamic characteristics

When computing the fuselage characteristics, attention should be focused on the drag, lift, pitching moment and side-force components since these forces and moments have the most significant effect on trim attitude and power required. As shown in Fig 3.14, the fuselage lift in level flight is negative due to the nose-down cruise attitude. The download increases the main rotor thrust needed to trim the aircraft, which results in an increase in power required. Similarly, the fuselage side-force characteristics determine the trim sideslip angle, which also contributes to an increase in parasite drag.

The main rotor forces and moments used in the trim iteration procedure outlined in Fig 3.13 are based on the classical blade element and momentum theory relationships for the first iteration (see Chs II and III of Vol I). For the second or subsequent iterations, a main and tail rotor subroutine is used to compute both the main and tail rotor trim forces (Fig 3.15). The subroutine uses collective pitch angle from the previous iteration to begin the computations. The next step is to define the blade flapping motion by summing the blade moments about the real or virtual flapping hinge. This involves solving a differential equation by numerical methods and incorporation of two-dimensional airfoil data, including blade stall and compressibility effects.

Having defined the blade motion, the next step in the rotor subroutine is to compute the rotor forces and moments by summing the elemental forces and moments of a

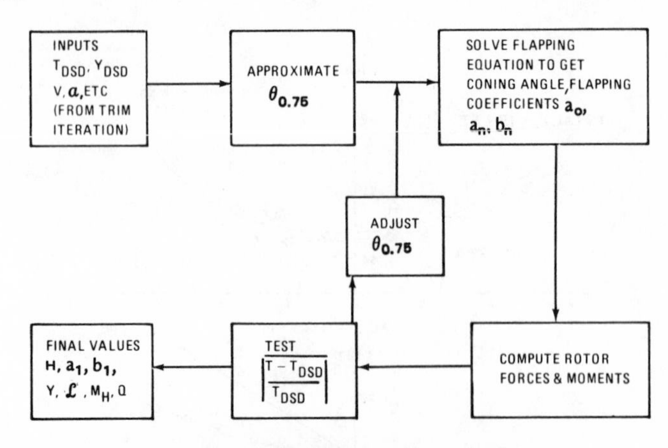

COMPUTER PROGRAM NOMENCLATURE

$\theta_{0.75}$ = COLLECTIVE PITCH ANGLE MEASURED AT 3/4 RADIUS
H = ROTOR H FORCE
a_n = nth HARMONIC OF LONGITUDINAL FLAPPING
a_o = CONING ANGLE
b_n = nth HARMONIC OF LATERAL FLAPPING
$\mathcal{L}$ = ROTOR ROLLING MOMENT
Q = ROTOR TORQUE
T_{DSD} = THRUST DESIRED
Y_{DSD} = SIDEFORCE DESIRED
V = AIRSPEED
a = ROTOR DISC ANGLE OF ATTACK

Figure 3.15 Rotor subroutine flow chart

specific number of discrete blade elements located at various radial positions, and then averaging these values at equally-spaced azimuthal locations around the disc. The thrust resulting from analysis is then compared with the desired value in a test for closure. If they don't agree, the collective pitch angle is adjusted accordingly and the rotor subroutine procedure is repeated until the thrust values are within 1 percent of the desired value. The remaining rotor forces and moments are then fed back into the main trim iteration routine shown in Fig 3.13 and the iteration is repeated until the six control variables defined in this figure (T_F, T_R, Y_F, MH_F, θ, ϕ) converge within the prescribed tolerances.

The last step in the trim analysis computations shown in Fig 3.13 is to compute the shaft horsepower required, including transmission and accessory losses. The nonuniform downwash and parasite power corrections are also applied at this stage of the analysis. A detailed discussion of each of these corrections is presented in the following sections.

3.2 Nonuniform Downwash (NUD) Correction

A nonuniform downwash correction was developed by comparing power predictions for an isolated rotor obtained by a computer program based on the application of the vortex theory with another, based on the simple momentum theory.

In the vortex theory approach, it is assumed that the wake is rigid (Ch IV, Vol I); that is, there is no allowance for the wake contraction which occurs relatively far downstream of the rotor and thus, has little effect on performance. The rapid movement of the wake away from the rotor in forward flight also permits the use of other simplifying calculation techniques, such as assuming that the trailing vortices roll up into a concentrated tip vortex system after 45° of blade rotation. These assumptions reduce the computer run time substantially with no significant reduction in the accuracy of performance predictions.

Within this general approach, local induced velocities are determined from a trailed vortex system. Values of velocities induced by the vortex filaments are used to compute the blade loads which, in turn, serve as an input in recalculating the corresponding induced velocities. The iteration is continued until the airloads and induced velocities are mutually consistent.

A summary block diagram and list of features of the nonuniform downwash analysis used in the hypothetical helicopter calculations is presented in Fig 3.16. As noted, the program includes optional yawed flow, elasticity, and unsteady aerodynamic features.

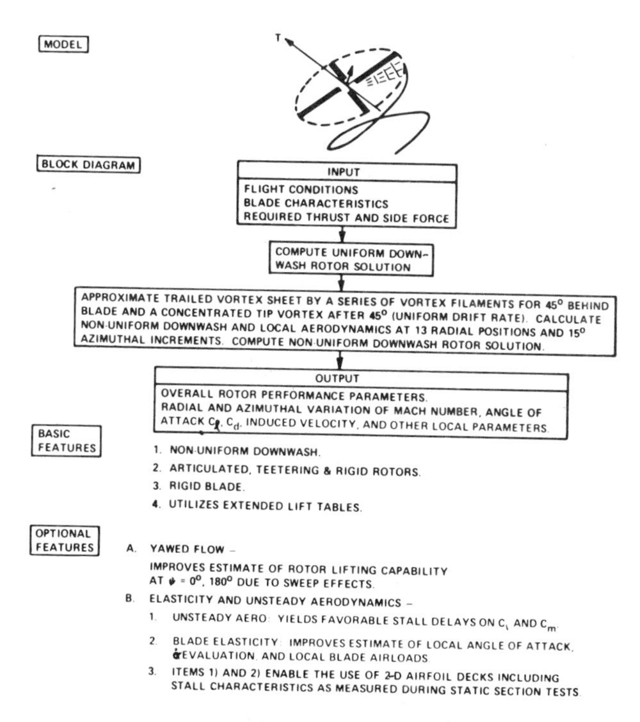

Figure 3.16 Rotor airloads and performance analysis with nonuniform induced inflow

Regardless of the use of a considerable number of simplifying assumptions, the nonuniform downwash computer run time is still considerably higher and costlier than the uniform downwash program. For this reason, a simplified nonuniform downwash (NUD) correction was developed:

A flight velocity coordinate system was used to define rotor lift (L) and rotor equivalent drag (D_{e_r}) forces. The rotor lift is represented by the rotor-thrust component perpendicular to the flight velocity factor. This is presented nondimensionally under the form of the following coefficient:

$$C'_T = L/\rho A V_t^2, \text{ or } C'_T/\sigma = L/\rho A V_t^2 \sigma$$

The equivalent drag is interpreted as

$$D_{e_r} = 550\, RHP/V_o - X$$

where $550\ RHP/V_o$ represents the equivalent drag of the entire helicopter (based on RHP and not on SHP), and X is the rotor propulsive force.

In trimmed flight, X is equal to the fuselage drag, hence

$$D_{e_r} = (P_{ind} + P_{pr})/V_o.$$

Here, D_{e_r} combines the induced and profile power (P_{ind} and P_{pr}) into a synthetic drag and thus, converts rotor power required into a fundamental fixed-wing type of parameter. D_{e_r} can also be nondimensionalized by dividing it by $q_o d^2 \sigma$, where q_o is the velocity of flight dynamic pressure, and d is the rotor diameter.

At this point, it should be emphasized that although the D_{e_r} value actually depends on both P_{ind} and P_{pr}, it is assumed that differences between D_{e_r} values obtained by the uniform downwash approach and those obtained either through tests or more refined (nonuniform downwash) calculations are chiefly due to the induced power discrepancies. Consequently, all of the corrective action for D_{e_r} values is contained under the common name of NUD corrections.

These so-defined NUD corrections can be obtained as follows: (1) plot the isolated rotor performance, incorporating nonuniform downwash effects, in terms of C'_T/σ vs $D_{e_r}/q_o d^2 \sigma$ for selected values of μ and given rotor geometry (Fig 3.17), and (2) compare the so-obtained graphs with $C'_T/\sigma = f(D_{e_r}/q_o d^2 \sigma)$ derived under the uniform downwash assumption.

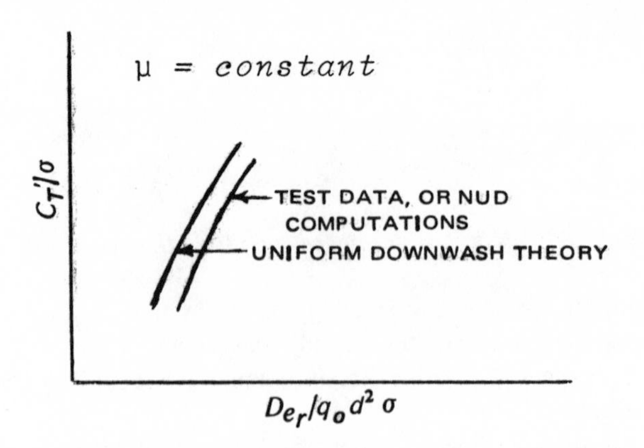

Figure 3.17 Effect of NUD on power required

It should also be pointed out that the De_r concept is also useful for evaluating rotor efficiency in terms of lift-to-drag ratio, L/De_r; while here in the NUD calculations, it is used as a means of isolating the induced and profile power from parasite components.

On the basis of a graph such as the one depicted in Fig 3.17, nondimensional equivalent rotor drag increments due to NUD effects $(\Delta D_e/q_o d^2 \sigma)$ can be plotted as a function of rotor-lift coefficient C'_T/σ (Fig 3.18). The incremental data appeared to be reasonably linear for $C'_T/\sigma \leqslant 0.08$; thus a linear fairing passing through the origin was established for each value of μ.

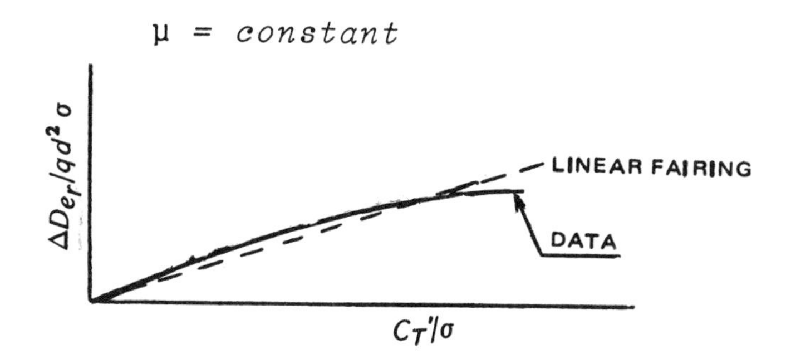

Figure 3.18 Incremental NUD power required

The slopes, $\partial(\Delta D_e/q_o d^2 \sigma)/(\partial(C'_T/\sigma)$, of these linear fairings were then plotted as a function of the variable $1/bAR_b$, where b = number of blades, and AR_b = blade aspect ratio, $(R/\bar{c})$. This relationship was again found to be linear, and the lines pass through the origin as shown in Fig 3.19.

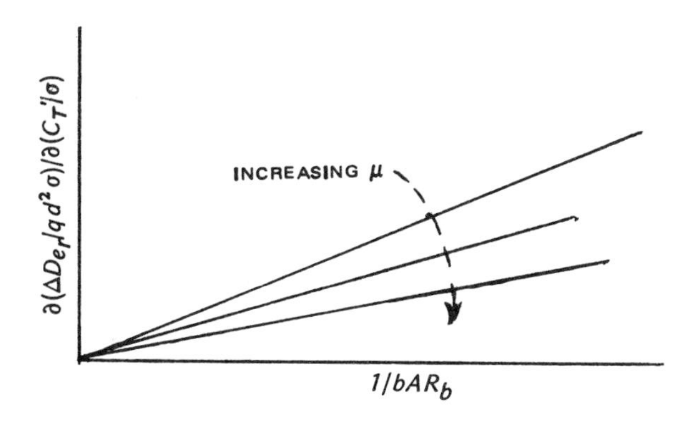

Figure 3.19 Effect of blade aspect ratio and number on NUD power increment

Based on the nondimensional data shown in Fig 3.19, an NUD incremental power penalty (ΔRHP_{nu}) can be estimated for any rotor geometry:

$$\Delta RHP_{nu} = k_n L \bar{c} V_t / 432\, bd \tag{3.13}$$

where

$\bar{c}$ = average blade chord; ft
d = rotor diameter; ft
b = number of blades
L = rotor lift,

and

$$k_n = \{[\partial(\Delta D_e/qd^2\sigma)/\partial(C'_T/\sigma)]bAR_b\}\mu^3.$$

The variable k_n represents the slope of the lines in Fig 3.19 times μ^3. The variation of k_n with advance ratio is illustrated in Fig 3.20, and is considered valid for computing performance at $0.15 \leqslant \mu \leqslant 0.4$. At μ values outside of the above boundaries, this method gives optimistic results. The k_n factor can be used to compute both main rotor and tail rotor NUD effects; however, the tail rotor correction is generally small enough to be neglected.

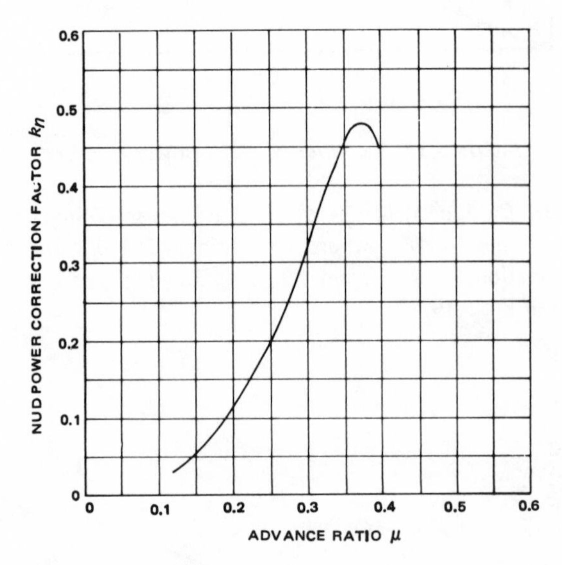

Figure 3.20 Nonuniform downwash incremental power correlation factor

It is interesting to note that k_n and the corresponding ΔRHP correction increases rapidly with advance ratio up to $\mu = 0.37$. This large variation in induced power is attributed to the fact that in forward flight, the rotor downwash varies with both the blade azimuth and radius. However, with increasing advance ratios, the azimuthal variation becomes more significant.

At this point, it should be emphasized that the above-described procedure for NUD corrections represents a method used by one company, and is not necessarily used throughout the industry. It is obviously possible to devise other approaches to this problem, but the experience of the company seems to support the practical validity of this approach.

To illustrate the NUD procedure, sample calculations for the hypothetical helicopter at $W = 15{,}000\ lb$ and $V = 150\ kn$ are presented below:

1. The required rotor geometry parameters are:

   $\bar{c} = 2\ ft;\ d = 50\ ft;\ b = 4;\ V_t = 700\ fps;$

   $\mu = V_o/V_t = 0.36$; and $L \approx 15{,}000\ lb.$

2. From Fig 3.20, $k_n = 0.475$ at $\mu = 0.36$,

3. Substituting these values in Eq (3.13 gives:

$$\Delta RHP_{nu} = \frac{k_n L \bar{c} V_t}{432\, bd} = \frac{0.475 \times 15{,}000 \times 2 \times 700}{432 \times 4 \times 50}$$

$$\Delta RHP_{nu} = 115.5\ hp.$$

3.3 Parasite Power Correction

A comparison of model rotor parasite power measurements with theoretical uniform downwash performance predictions indicates that the theory underpredicts the power required to generate a given propulsive force. The ratio of theoretically predicted parasite power ($XV_o/550$) to that actually measured–called *propulsive efficiency* (η_p) is a function of advanced ratio (Fig 3.21). It should be noted that at advance ratios of

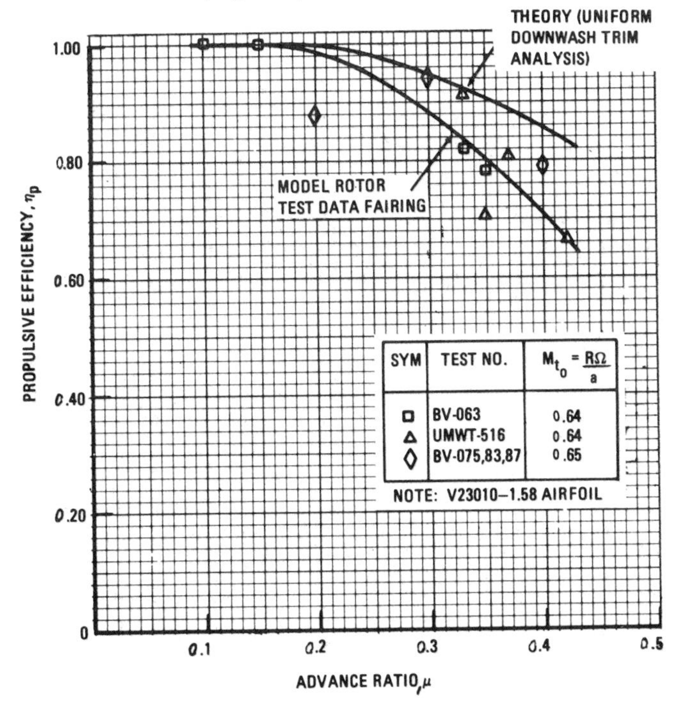

Figure 3.21 Parasite power correction

$\mu > 0.15$, the actual parasite power from model tests is higher than its theoretical value ($\eta_{p\,test} < \eta_{p\,theory}$). This disagreement in parasite power levels is not completely understood; however, it could be due to blade contributions to the total helicopter parasite drag (see Ch III, Vol I) resulting from local separation on the retreating blade tip, as well as in the reverse flow region. It can also mask nonuniform downwash or aeroelastic effects. If it is stall related, then differences between the Reynolds numbers used in theoretical predictions and those corresponding to the actual model test data shown in Fig 3.21 may be a factor. More experimental and theoretical investigations of this phenomena are required. Nevertheless, even without a complete understanding of the phenomena, a practical way of dealing with it must be developed. One approach in that respect is outlined below. However, no special claims are made as to its universal merits.

Using the data presented in Fig 3.21, the trim analysis parasite power is corrected to the test level by adding the following increment to the total power required:

$$\Delta RHP_{par} = \frac{XV_o}{550}\left(\frac{1}{\eta_{p\,test}} - \frac{1}{\eta_{p\,theory}}\right). \tag{3.14}$$

For example, neglecting the rotor profile drag component, the hypothetical helicopter propulsive force required to balance the airframe drag at 150 kn under SL/STD conditions is

$$X = drag = f_e \times q = 19.1\ ft^2 \times 76.4;$$

$$X = 1460\ lb$$

$$\mu = V_o/V_t = 0.362.$$

From Fig 3.21, $\eta_{p\,test} = 0.78$ and $\eta_{p\,theory} = 0.89$. The ΔRHP_{par} correction, therefore, is

$$\Delta RHP_{par} = \frac{1460 \times 253.5}{550}\left(\frac{1}{0.78} - \frac{1}{0.89}\right);$$

$$\Delta RHP_{par} = 106\ hp.$$

3.4 Determination of Low-Speed Power Required

Determination of the power required in the 0 to 60-kn speed range is necessary primarily to analyze takeoff and landing capability, and to determine the effect of wind on hover performance. Furthermore, this regime of flight is important, due to the increasing interest of the military in low-speed nap-of-the-earth (NOE) operations. In this speed range (approaching hover), wake contractions become significant. This may require development of a technique permitting a transition from the hover analysis employing an empirical wake contraction technique, to a forward-flight nonuniform downwash analysis based on a rigid wake. In addition, flow visualization studies[27] have shown that two concentrated wing-tip vortices form at the edge of the wake, resulting in an extremely complex wake structure.

Until practical vortex theory analyses applicable to predictions of low-speed performance are developed, the power required in this speed regime will continue to be defined by the shape of the $RHP = f(V)$ curve based on simple momentum theory (uniform downwash). It was shown that power required trends in this speed range, predicted by momentum theory, usually agree with wind-tunnel measurements. However, incremental power corrections are needed in order to match the hover and 60-kn power required points determined by the nonuniform downwash approach. This correction is based on the assumption that the power adjustment at low speeds is proportional to that in hover, and follows the trend given by the momentum theory. The procedure for this calculation can be explained with the aid of the hypothetical helicopter power required shown in Fig 3.22. In this figure, uniform downwash and nonuniform downwash power required are shown at SL/STD conditions for a gross weight of 15,000 lb. Designating the values of the hover and 60-kn power required by a, b, c, and d as noted in this figure, the nonuniform downwash correction (ΔRHP_{nu}) for airspeeds of 0 to 60 kn is:

$$\Delta RHP_{nu} = \left(\frac{(a-b)-(c-d)}{(b-d)}\right)\left(RHP_u - d\right) + \left(c-d\right). \qquad (3.15)$$

In this equation, RHP_u is the uniform downwash power required at the airspeed at which ΔRHP is being computed. The total power required, therefore, is

$$RHP = RHP_u + \Delta RHP_{nu}. \qquad (3.16)$$

This technique can be further refined by removing the parametric power prior to using Eq (3.15).

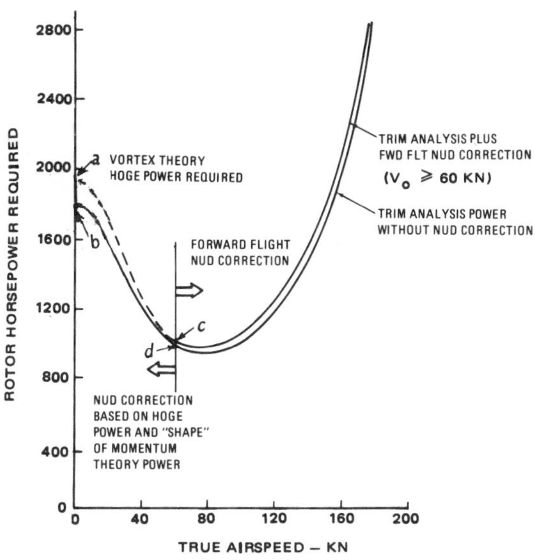

Figure 3.22 Determination of low-speed power required

To illustrate the above procedure, detailed calculations for the 25-kn point in Fig 3.22 are presented below.

1. From this figure:

$$
\begin{aligned}
a &= 1813\ hp \\
b &= 1661 \\
c &= 896 \\
d &= 884 \\
RHP_u &= 1362
\end{aligned}
$$

2. Substituting these values in Eqs (3.15) and (3.16),

$$
\begin{aligned}
RHP &= 1362 + 98 \\
RHP &= 1460.
\end{aligned}
$$

4. EXAMPLES OF LEVEL FLIGHT POWER REQUIRED PREDICTIONS FOR THE HYPOTHETICAL SINGLE–ROTOR AIRCRAFT

A discussion of the hypothetical helicopter forward flight power required based on the calculation procedure described above is presented in this section of the text. The discussion includes an evaluation of the referred and generalized data presentation techniques used to account for the effect of air density and compressibility at various ambient conditions. In addition, a comparison of trim analysis computer program predictions versus simplified calculations is provided.

4.1 Power Required Based on the Trim Analysis

Level flight power required is presented in Figs 3.23 and 3.24 for SL/STD and 4000 ft, 95°F ambient conditions. Speed power polars based on the trim analysis and correction factors discussed in the previous section are presented for gross weights from 9,000 to 18,000 lb. Increasing gross weights result in increased power and higher minimum-power airspeed. Due to the fact that the induced power changes become more rapid at low speeds, significant power required increments are most apparent in the speed range from $V = 0$ to that corresponding to minimum power. In contrast, at higher speeds, the induced power continues to decrease, resulting in a convergence of the speed power polars.

Engine power available and transmission power limits are also shown in Figs 3.23 and 3.24. For SL/STD conditions, it should be noted that the speed capability at 15,000 lb gross weight is 172 kn at max. cont. power and 178 kn at the transmission limit. A complete discussion of performance capabilities is presented in Sect 7 of this chapter.

The difference between the SL/STD and 4000 ft/95°F power required are due to effects of air density and compressibility.

Air Density. At 4000 ft/95°F, the air density is 0.808 of the SL/STD value. This tends to increase the hover and low-speed power required because of increased induced

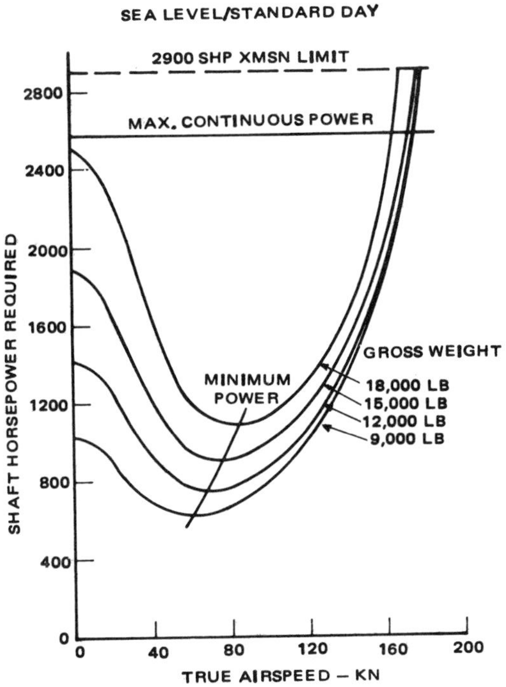

SEA LEVEL/STANDARD DAY
2900 SHP XMSN LIMIT
MAX. CONTINUOUS POWER
SHAFT HORSEPOWER REQUIRED
0
400
800
1200
1600
2000
2400
2800
MINIMUM POWER
GROSS WEIGHT
18,000 LB
15,000 LB
12,000 LB
9,000 LB
0
40
80
120
160
200
TRUE AIRSPEED – KN

Figure 3.23 Level flight power required at SL/STD

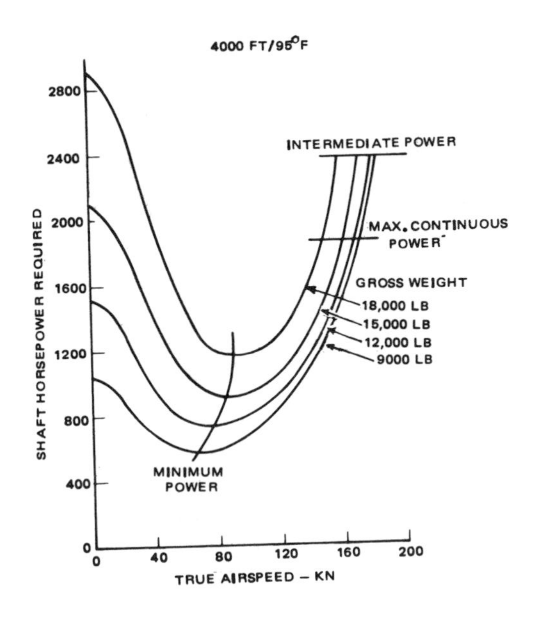

Figure 3.24 Level flight power required at 4000 ft/95° F

power. At high speeds, the lower air density reduces the total power required by decreasing the parasite drag, unless the rotor lift coefficient becomes sufficiently high to cause a divergence in power required due to rotor stall. The effect of density variations can be accounted for by dividing the rotor horsepower required (*RHP*) and weight (*W*) by the density ratio (σ_ρ), as illustrated in Fig 3.25. The parameters RHP/σ_ρ and W/σ_ρ are called referred power and referred weight, and they are developed from the basic C_p, C_T nondimensional relationships.

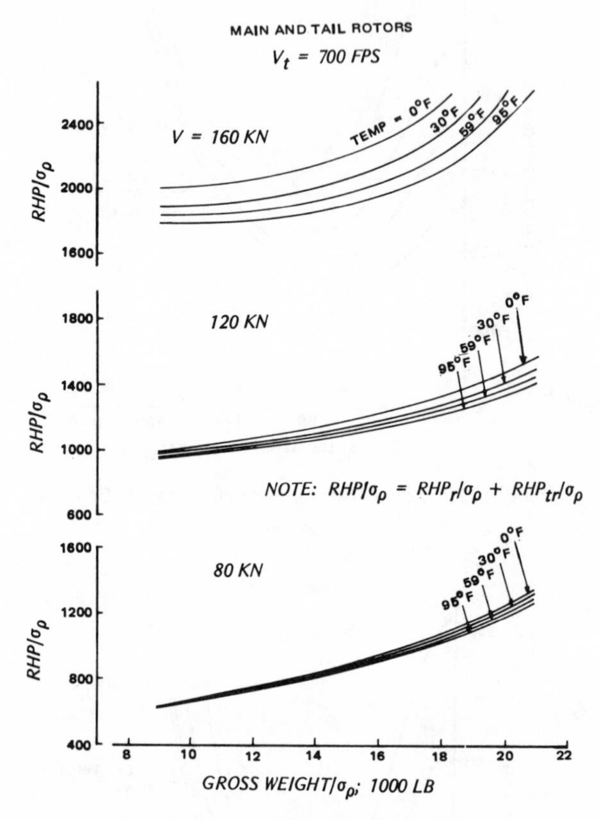

Figure 3.25 Forward-flight compressibility power

Compressibility Effects. The graphs shown in Fig 3.25 also incorporate compressibility effects based on the two-dimensional airfoil characteristics shown in Figs 2.3 and 2.4. The variation in referred power with temperature is due primarily to the effect of compressibility on the main rotor. This effect on the tail rotor can be neglected since, at the speeds indicated in Fig 3.25, the tail rotor contributes only 3 to 5 percent of the total power required. Compressibility effects vary with increasing airspeed, gross weight, and temperature, since these parameters are a function of average rotor c_ℓ and advancing blade tip Mach number $M_{(t)90}$. The variation of $M_{(t)90}$ with forward speed and ambient temperature is apparent in the following chart:

TEMPERATURE	$M_{(t)90}$ @ 80 Kn	$M_{(t)90}$ @ 160 Kn
0°F	0.795	0.923
95°F	0.723	0.841

The effect of varying rotor rpm can be accounted for by taking advantage of the referred data in Fig 3.25. This is accomplished by using $(V_{t_o}/V_t)^2$ and $(V_{t_o}/V_t)^3$ values in addition to the W/σ_ρ and RHP/σ_ρ parameters as discussed in Ch II. Knowing the advance ratio μ, the flight speed should then be converted to referred airspeed $V_{rfd} = V(V_{t_o}/V_t)$ where V_{t_o} is the design tip speed. Compressibility effects due to RPM variations can be accounted for by changing the constant temperature lines in Fig 3.25 to lines of constant average Mach number $\bar{M}_t = \bar{V}_t/a$, where a is the speed of sound, and $\bar{M}_t$ is a function of $M_{(t)90}$:

$$\bar{M}_t = M_{(t)90}/(1 + \mu). \tag{3.17}$$

The referred weight and power parameters are more convenient to work with than the nondimensional coefficients C_T and C_P which have low values on the order of 0.005 and 0.0005, respectively.

A third method of determining power required, called the *generalized method*, is often used to present flight test data. This method converts C_P, C_T, μ, and M to the dimensional form of W/δ, $V/\sqrt{\theta}$, $N/\sqrt{\theta}$, and $SHP/\delta\sqrt{\theta}$ as shown below:

$$C_T = \left(\frac{30^2}{\rho_o A \pi^2 R^2}\right)\left(\frac{W/\delta}{(N/\sqrt{\theta})^2}\right)$$

$$C_P = \left(\frac{550(30)^3}{\rho_o A \pi^3 R^3}\right)\left(\frac{SHP/\delta\sqrt{\theta}}{(N/\sqrt{\theta})^3}\right)$$

$$\mu = \left(\frac{30}{\pi R}\right)\left(\frac{V/\sqrt{\theta}}{N/\sqrt{\theta}}\right)$$

$$M_{(t)90} = \left(\frac{\pi R}{30 a_o}\right)\left(\frac{N}{\sqrt{\theta}}\right)\left[1 + \left(\frac{30}{\pi R}\right)\left(\frac{V/\sqrt{\theta}}{N/\sqrt{\theta}}\right)\right]$$

where

a_o = speed of sound at SL/STD conditions
ρ_o = air density at SL/STD conditions
δ = ambient pressure ratio
θ = ambient temperature ratio
A = rotor area

N = rotor rpm
R = rotor radius.

Maintaining a constant W/δ and $N/\sqrt{\theta}$ is equivalent to a constant C_T; and a constant $V/\sqrt{\theta}$ and $N/\sqrt{\theta}$ is equivalent to a constant $M_{(t)90}$.

Even though the three methods are interchangeable, the generalized method is becoming increasingly popular, particularly for presenting flight test data where compressibility effects are significant. During performance test programs, it is easier to fly constant W/δ and $N/\sqrt{\theta}$ than it is to maintain a constant C_T and average Mach number, because the generalized parameters are readily determined from cockpit measurements of pressure altitude, rotor speed, and ambient temperature. Intermediate density altitude calculations are not necessary. When presenting generalized data, speed power polars for various W/δ levels are required for a range of $N/\sqrt{\theta}$ values.

The results of the trim analysis program (Fig 3.25) were based on two-dimensional airfoil characteristics. However, studies using the potential flow theory[9,10,11] indicate that drag divergence occurs at higher Mach numbers than given by the two-dimensional data. This three-dimensional *tip relief effect* is due to the spanwise movement of streamlines located within approximately one chord-length of the tip. Consequently, the flow encounters an apparent decrease in airfoil thickness, and local velocity at the thickest portion of the airfoil is reduced. This effect is similar to that encountered in swept fixed-wings.

4.2 Simplified SHP_{req} Estimates vs Trim Analysis Program

The theoretical relationships developed in Vol I, Chs II and III, and the trim analysis computer program used for the sample problem calculations employ the same fundamental momentum theory and blade element concepts for computing power required. The primary differences between the two techniques are: (1) the procedure used to integrate the blade element forces over the rotor disc, and (2) the trim analysis which inherently accounts for the effects of helicopter attitude on performance. In the classical method (without the use of computers), the rotor torque is determined with the help of an average rotor drag coefficient $\overline{c}_d$. The trim analysis uses numerical techniques programmed on the computer to average the force contributions of discrete blade sections located at various radial and azimuthal positions. Local effects such as compressibility and stall are accounted for in this manner.

In Fig 3.26, the power required computed by a simplified method is compared with the results obtained from trim analysis. Here, SHP_{req} at SL/STD was computed for 15,000-lb gross weight. To simplify this comparison, the NUD and parasite power corrections were not applied. The following equations from Vol I were reproduced for application to the hypothetical aircraft (all velocities are in fps):

$$SHP = (RHP_{mr} + RHP_{tr})/\eta_t + \Delta SHP$$

where

ΔSHP = accessory losses, and η_t = transmission efficiency.

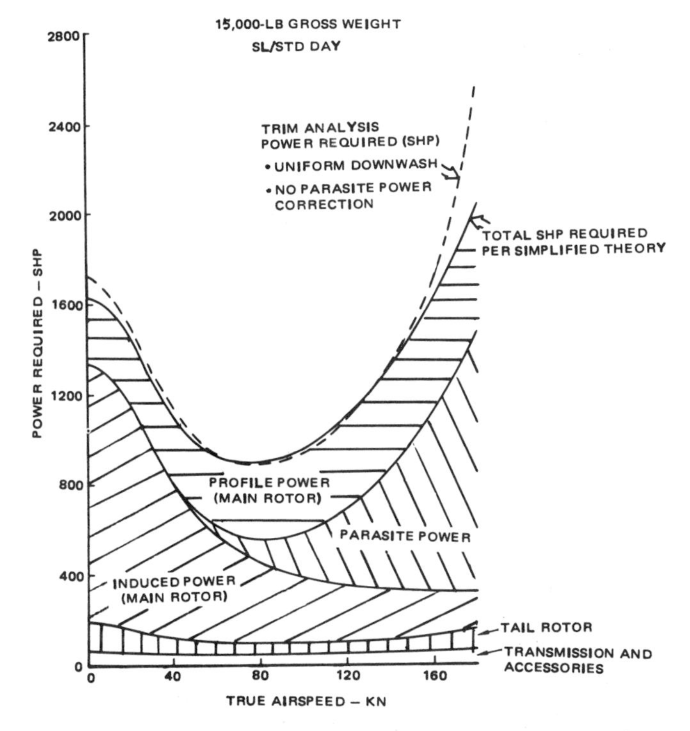

Figure 3.26 Simplified approach to power required

$$RHP_{mr} = \underbrace{\frac{T v_{if}}{550}}_{\text{Induced Power}} + \underbrace{\frac{\sigma \bar{c}_d (1 + 4.7\mu^2)\rho \pi R^2 V_t^3}{4400}}_{\text{Profile Power}} + \underbrace{\frac{f_e \rho V^3}{1100}}_{\text{Parasite Power}}$$

and

$$RHP_{tr} = \frac{T_{tr} v_{if_{tr}}}{550} + \frac{\sigma \bar{c}_d (1 + 4.7\mu^2)\rho \pi R_{tr}^2 V_{tr}^3}{4400}$$

The above induced velocities (v_{if} and $v_{if_{tr}}$, respectively) are computed as follows:

$$v_{if} = v_o \left[\sqrt{-(V/v_o)^2/2 + \sqrt{(V/v_o)^4/4 + 1}} \right]$$

where

$$v_o = \sqrt{T/2\rho \pi R^2 \bar{r}_e^2}$$

and

v_o = hover induced velocity per momentum theory

$\bar{r}_e$ = r_e/R, effective nondimensional rotor radius which extends from blade root cutout to $0.97R$, where tip losses become significant (see Vol I, Ch III for tip loss discussion).

These relationships can be further simplified by neglecting the fuselage download effects and tail rotor unloading due to the cambered vertical tail assumed in the sample problem. Therefore,

$$T_{mr} \approx W$$

and

$$T_{tr} \approx 5252\ RHP_{mr}/\ell_{tr}N_{mr}$$

where

ℓ_{tr} ≈ tail rotor moment arm
N_{mr} ≈ main rotor rpm.

In addition, it is assumed that

$\eta_t = 0.98$

$\bar{r}_e = 0.95 = \sqrt{(0.97)^2 - (0.2)^2}$, where cutout = $0.2R$ and effective rotor radius = $0.97R$

$\ell_{tr} = 30\ ft$

main rotor $\bar{c}_d = 0.008$, tail rotor $\bar{c}_d = 0.0107$ — For derivation, see Hover Section (Ch II, Sect 1.2 and 1.3)

$f_e = 19.1\ ft^2$ (fuselage cruise angle-of-attack = $-5°$)

$\Delta SHP = 30\ hp$ (accessory drive power)

σ, R, V_t, *etc.*, see Table I-1.

Based on the above assumptions, the total power required calculated by the simplified method agrees well with the trim analysis results up to 140 kn—the only exception being small differences in the low-speed range (Fig 3.26). These differences are primarily due to the omission of the effect of the fuselage download on the main rotor induced power. The two approaches would result in an agreement if the 2.55 percent of gross weight hover download is included in the induced power required estimates for airspeeds up to approximately 60 kn. At speeds above 60 kn, cruise download can be approximately estimated from the fuselage aerodynamic characteristics presented previously in Fig 3.14, assuming a constant cruise angle-of-attack of $\alpha_f = -5°$; thus neglecting variations of the actual α_f values due to changing main rotor downwash.

As shown in Fig 3.10, the angle-of-attack at 15,000 lb is within $\approx 1°$ of this angle for airspeeds between 60 and 140 kn.

Because of compressibility and stall effects above 140 kn, the simplified calculations under-predict the power required. However, theoretical stall and compressibility corrections can be added to the basic profile power expressions based on the equations developed by McCormick[13] and modified to agree with the vortex theory analysis[12]. For example, for the V23010-1.58 airfoil, an increment in the average rotor section drag coefficient due to compressibility ($\Delta\bar{c}_{d_c}$) is

$$\Delta\bar{c}_{d_c} = 0.2(M_{(t)90} - M_d)^3 + 0.0085(M_{(t)90} - M_d) \tag{3.18}$$

where M_d is the drag divergence Mach number defined empirically as

$$M_d = 0.82 - 2.4(C_T/\sigma). \tag{3.19}$$

In Eq (3.18), $M_{(t)90}$ is the advancing blade tip Mach number; therefore, the total profile power (RHP_{pr}) becomes

$$RHP_{pr} = \frac{\sigma(\bar{c}_d + \Delta\bar{c}_{d_c})(1 + 4.7\mu^2)}{4400} \rho\pi R^2 V_t^3 \tag{3.20}$$

For a rotor with the V23010-1.58 airfoil section, a similar average rotor blade section drag increment can be developed for predicting the increase in power due to stall, ($\Delta\bar{c}_{d_s}$).

$$\Delta\bar{c}_{d_s} = 18.3(1 - \mu)^2 F^3 \tag{3.21}$$

where

$$F = \left[\frac{C_T/\sigma}{(1 - \mu)^2}\right]\left[1 + \frac{f_{eq}}{W}\right] - 0.1375. \tag{3.22}$$

Equations for the NACA 0012, as well as advanced, airfoils can be found in Ref 12.

The agreement between calculations based on the simplified approach and the more sophisticated computerized procedure indicates that the first method can be used to reduce computer costs and increase turnaround time if compressibility, stall, nonuniform downwash and propulsive efficiency effects are accounted for. For preliminary design work, the simplified relationships are particularly useful because they provide a cost-effective means of conducting parametric studies on many different helicopter configurations.

The simplified calculation techniques also give details concerning the power-required breakdown which provides some insight into means of optimizing a design for a specific mission. This breakdown was obtained for the hypothetical helicopter by applying the download, compressibility, NUD and parasite power corrections to the appropriate values of the power-required components predicted by the simplified approach and shown in Fig 3.26. Data for hover and airspeeds of 80 and 150 kn are presented in Table III-3.

TRUE AIRSPEED	HOVER	80 KN	150 KN
	% SHP		
Induced Power (Main Rotor)	72	40	19
Profile Power (Main Rotor)	16	38	30
Parasite Power	0	12	43
Tail Rotor Power	8	5	4
Transmission & Accessories	4	5	4

TABLE III-3 POWER REQUIRED BREAKDOWN (PERCENT)

It can be seen that in hover, the induced power accounts for 72 percent, and profile power for 16 percent, of the total SHP required. Therefore, to design a helicopter having a low energy consumption in hover and/or maximum vertical takeoff performance capability, the induced power should be kept to a minimum. Selection of a low disc loading and minimization of induced power losses through optimization of the blade twist and planform represent some of the design approaches which may be applied in this case.

At 80 kn, which is about the minimum power condition, the induced and profile powers are each approximately equal to 40 percent of the total. One-engine-inoperative (OEI) speed at which SHP_{req} is minimum, and that corresponding to maximum endurance usually coincide*. Therefore, to provide maximum OEI performance capability and maximum loiter time for endurance missions, more attention should be given to profile power; thus, tip speed and blade area should be optimized to minimize this quantity. However, since the induced power is also sizeable—decreasing the disc loading and reducing the nonuniformity of the rotor downwash (see Ch III, Vol I) will also significantly contribute to an improvement in performance.

At $V_o = 80$ *kn*, parasite power accounts for only 12 percent of the total SHP. Thus, even a sizeable drag reduction would provide only a limited benefit in this regime of flight. By contrast, at a cruise speed of 150 kn, the parasite drag is responsible for 43 percent of the total SHP required, while the profile and induced power represent 30 and 19 percent of the total, respectively. Therefore, means of reducing parasite drag should be explored in order to provide maximum speed and range capability.

5. POWER REQUIRED IN CLIMB AND DESCENT

Power required in climb or descent at airspeeds equal to, or exceeding, the minimum power speed can be computed using the trim analysis program; however, this can result in considerable run time. It is generally less costly and easier to take advantage of the approximate climb prediction method based on the excess of shaft horsepower available over that required for forward flight at a particular flight speed. In analogy to Eq (2.32) derived in the preceding chapter, the following expression can be written.

*Small deviations may occur due to the variation of specific fuel consumption with SHP, with the result that SHP_{min} does not coincide with minimum fuel flow in forward flight.

$$\Delta SHP = W(V_c)/33{,}000k_{p_c} \tag{3.23}$$

where

$\Delta SHP = SHP_{climb} - SHP_{level\ flight}$ = incremental climb power

V_c = rate of climb in fpm

k_{p_c} = climb efficiency factor.

The climb efficiency factor, which can be derived from flight test data, wind-tunnel tests, or more precise analytical calculations (e.g., trim analysis program), accounts for such factors as fuselage lift and drag, induced power, and tail rotor power being different in climb than in horizontal flight. It also covers transmission efficiency—reducing the excess power actually available for climb.

5.1 Climb Power Required

Assuming that profile power in climb remains the same as in horizontal flight, the most important influence of the variables that affect the additional power required to climb can be discussed in light of the momentum theory considerations presented in Ch II of Vol I. Using this approach, and assuming that the rotor plane is positioned almost horizontally,

$$\Delta SHP = \frac{W(V_c)}{33000}\left[\frac{k_{vf}}{\eta_t}\left(1 + \frac{v_c - v}{V_c}\right)\right] \tag{3.24}$$

where

k_{vf} = forward flight download factor

η_t = transmission efficiency

v_c = induced velocity in climb

v = induced velocity in forward flight.

Comparing Eqs (3.23) and (3.24), it can be seen that the variables in the brackets of Eq (3.24) are equal to the reciprocal of the climb efficiency factor k_{p_c}. It should also be noted that the efficiency factor is made up of the download factor k_{vf} which accounts for the increase in negative lift and rise in trim drag of the fuselage in climb and η_t. In addition, the term, $1 + (v_c - v)/V_c$, accounts for the variation in downwash velocity in climb. This is less than 1.0 for positive rates of climb, since $v_c < v$ in this flight regime. Therefore, it is apparent that the climb efficiency factor would be greater than 1.0 if there were no increase in download or transmission power losses.

In contrast to paired main rotor systems, the increase in tail rotor power required in climb for single-rotor configurations causes an additional reduction in the climb efficiency factor. Typically, $0.85 \leqslant k_{p_c} \leqslant 0.95$ for tandems; while for single-rotor aircraft, $0.80 \leqslant k_{p_c} \leqslant 0.90$.

Average k_{pc} values for a specific range of weights, airspeed, and rates of climb are often used to reduce computer and calculation time. For example, computed values of k_{pc} are presented in Fig 3.27 as a function of gross weight for airspeeds of 80 to 120 kn. Here, it is shown that k_{pc} increases with decreasing speed. As discussed in Ch II, values on the order of 1.5 are achieved in hover. At 80 kn, $0.8 < k_{pc} \leqslant 0.9$, depending on the rate of climb. Since most forward flight climb performance is calculated at the minimum power speed of $V \approx 80\ kn$ (where rate of climb is highest), an average value of $k_{pc} = 0.85$ was selected for subsequent climb performance calculations.

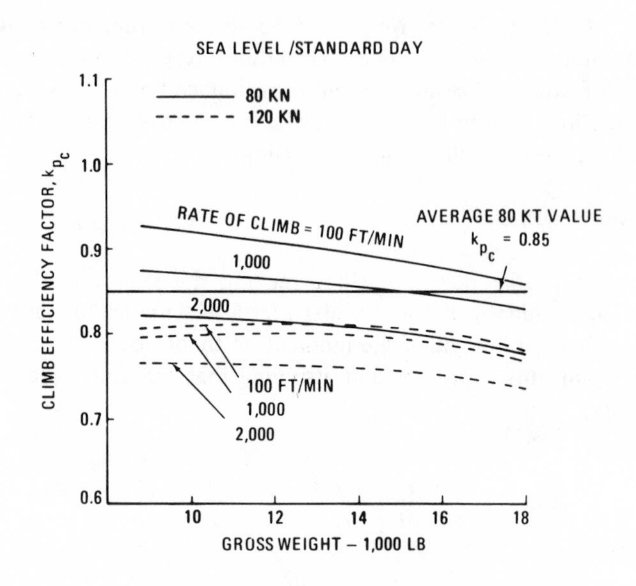

Figure 3.27 Climb efficiency factor

Power required at 100 fpm rate of climb is presented as a function of gross weight in Fig 3.28 to illustrate the application of the climb efficiency factor for minimum power speed (60 to 80 kn) in level flight. Here, it is shown that the incremental climb power at 15,000 lb is approximately 50 hp. The data is presented in the referred system for various ambient temperatures in order to illustrate compressibility effects at minimum power airspeeds, which can have an effect on service-ceiling calculations. For example, dual-engine operations at altitudes close to the service ceiling are often performed at low air densities and temperatures. This results in high rotor coefficients (C_T/σ) combined with increased advancing-blade Mach numbers.

5.2 Descent Power Required

Descent performance is treated as a negative rate-of-climb calculation. In descent, the induced velocity term in Eq (3.24) is less than 1.0, and the download and tail rotor power required are lower than in level flight. The result is that at autorotational rates of descent, $k_p \approx 1.0$ or even higher. For this reason, instead of being called an *efficiency* factor, it will be referred to as a descent *correction* factor (k_{pd}) when applied to the potential energy expression in subsequent discussions.

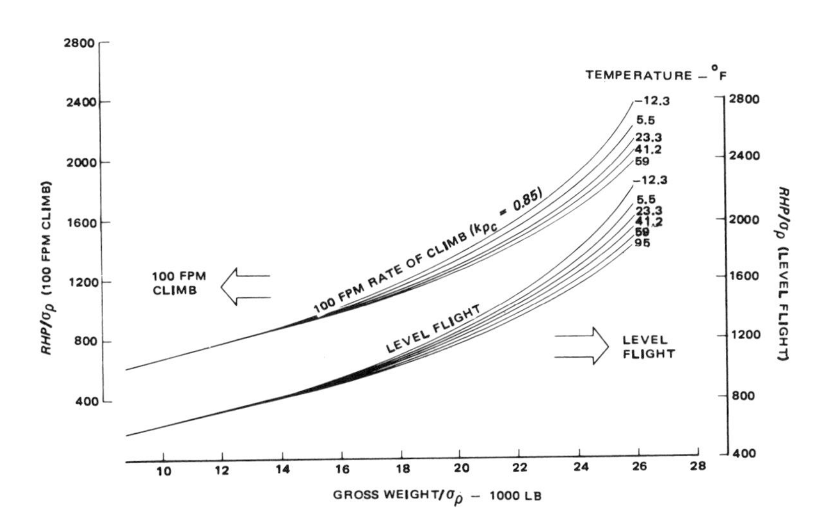

Figure 3.28 Hypothetical helicopter minimum power required

Based on the trim analysis program, sample calculations of the descent factor as a function of gross weight for the hypothetical helicopter at airspeeds of 80 to 120 kn are presented in Fig 3.29. Here, $k_{pd} = f(W)$ is shown for partial-power descent at rates of 500 and 2000 fpm. Boundaries defining autorotational rates of descent are also indicated. The autorotational rate of descent used to compute k_{pd} values varies with weight, and is determined for a given weight by plotting power required as a function of the descent rate and noting its value at zero power required. At 15,000 lb gross weight, the factor varies from $k_{pd} = 0.9$ at 500 fpm, to $k_{pd} = 1.0$ at 2000 fpm. Below a descent rate of 500 fpm, the same values as those for the average climb efficiency factor; i.e., $k_{pc} = 0.85$, can be used with reasonable accuracy to compute descent performance. At autorotational rates of descent, $0.9 \leqslant k_{pd} \leqslant 1.0$, and does not change significantly with airspeed. It appears that $k_{pd} = 1.0$ represents a good autorotational value and has been used successfully to predict the autorotational performance of tandems as well as single-rotor aircraft, as verified by flight-test measurements.

Empirical data indicates that k_{pd} increases rapidly as forward speed approaches zero. Values of well over 2.0 were recorded in vertical descent for isolated rotors. These high values are primarily due to the adverse effects of operating in the vortex ring state[14].

6. LEVEL FLIGHT AND MANEUVER AIRSPEED ENVELOPE

The level flight speed and maneuver capability of a helicopter is usually limited by one or more of the following items:

(1) power available
(2) transmission (torque) limits
(3) excess vibration/structural limits not related to stall
(4) compressibility effects on advancing blade
(5) stall inception.

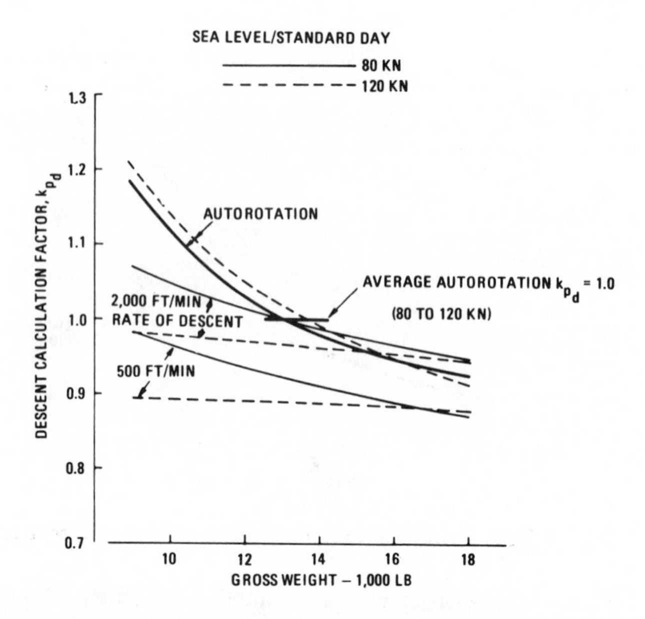

Figure 3.29 Descent calculation procedure

The power available and transmission limits were discussed briefly in Ch I, and further details will be provided in Sect 7 (Performance Capability). The third item, excess vibration/structural limits, is not a consideration for new designs because the dynamic system and airframe are (theoretically) designed to avoid this type of restriction.

Compressibility effects on the advancing blade pitching moment is another potential structural limitation. As pointed out by Dadone[15]

> *Only relatively recently the growth of pitching moments with Mach number has become a significant parameter. This has occurred with the introduction of cambered airfoils and structurally softer blades. The phenomenon has been referred to as "pitching moment break" or, borrowing the term from fixed-wing terminology, "Mach tuck." Essentially, this growth in pitching moment coincides with the onset of transonic flow conditions and it is associated with both a rearward shift in the aerodynamic center and an increase in pitching moments about the aerodynamic center.*

The development of pitching moment break along with other compressibility effects can be delayed by reducing the relative thickness and camber of the tip airfoil section, while deflecting the trailing edge tab upward would cause a general reduction of the negative pitching moment (see Ch VI, Vol I).

The remaining airspeed limitation criterion (rotor limits due to stall inception) is the primary constraint of the airspeed-altitude envelope for new helicopters. It is associated with the high alternating control system loads, and deterioration in stability and control resulting from retreating blade stall. For single-rotor helicopters such as the hypothetical aircraft, the effect of stall on component stress levels, rather than flying quality deterioration which may appear at the occurrance of stall, is generally the limiting criterion. Tandems are more prone to flying quality stall limitations since they chiefly depend on differential thrust between the forward and aft rotors for longitudinal control.

The methodology used to estimate rotor flying speed limitations resulting from control loads is discussed in the following sections. Included are sample calculations of the level-flight envelope and maneuver capability of the hypothetical helicopter based on model rotor test data.

6.1 Rotor Stall Limits Methodology

Level-flight rotor limits are encountered because of the development of stall on the retreating blade ($240 \leqslant \psi \leqslant 300°$). As the blade enters and leaves this region, the section pitching moment decreases abruptly due to moment stall of the airfoil section, and thus induces the aeroelastic phenomena known as stall flutter. Moment stall causes spikes in the blade torsional waveform as illustrated in Fig 3.30. The rate of growth of the peak alternating control loads is also shown in this figure. Therefore, operating the rotor at thrust or airspeeds beyond stall inception results in loads which build up quickly to the fatigue or endurance limit of the rotor control system because of the high alternating torsional loads feed directly into the system. This limitation has been the primary factor in defining helicopter structural envelopes. The rapid growth of the control loads after stall inception also restricts the amount by which the original envelope can be enlarged by strengthening the control system. For this reason, the structural flight envelopes of growth aircraft are often inside their power limits[16].

The endurance limit is defined as the maximum alternating load that can be sustained by a component for an indefinite number of cycles without fatigue failure. For instance, the CH-47C limit corresponds to approximately three times the unstalled alternating control load (Fig 3.30). However, with the increased requirements for highly maneuverable aircraft, a finite component life (typically, 4000 to 5000 hours) is accepted, thus allowing operation for a given percentage of time beyond the endurance limit.

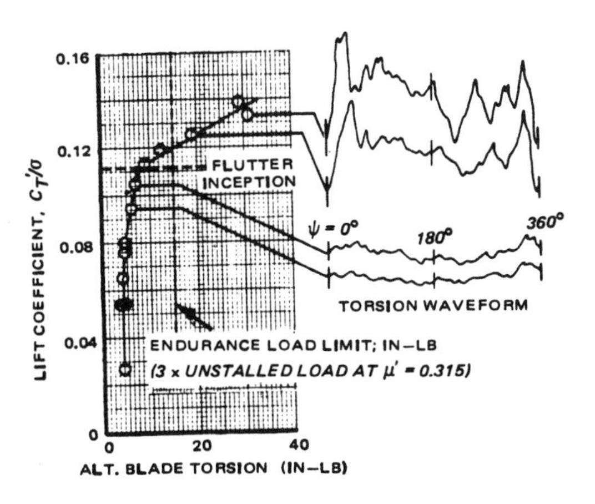

Figure 3.30 Stall flutter boundary determination

Various methods are used throughout the industry to define the rotor stall limits. These techniques include (1) nondimensionalizing flight test or wind-tunnel test data, and (2) pure theoretical rotor system loads and performance prediction programs. Currently, most companies rely on flight-test or wind-tunnel empirical techniques, since purely analytical stall load predictions have not been sufficiently developed to warrant their use as the primary means of determining rotor limits.

Use of Flight Test Data to Define Rotor Limits. Flight-test measurements of control system loads are used to detect stall inception and to define the rate of load growth in stall. The thrust and airspeed envelope determined by either the stall inception or endurance limit criteria can be nondimensionalized in terms of rotor-lift coefficient C_T'/σ and μ where $C_T'/\sigma = W/\rho\pi R^2\ V_t^2\,\sigma$. These two parameters can be used to predict the stall limits of other aircraft, provided the new helicopter design has a similar airfoil section, nondimensional propulsive force $X = f_e/d^2x$, tip speed, and blade number. An example of the nondimensional flight envelopes of some current production aircraft based on test data is presented in Fig 3.31. The data is based on level-flight airspeed limits published in military manuals for SL/STD conditions.

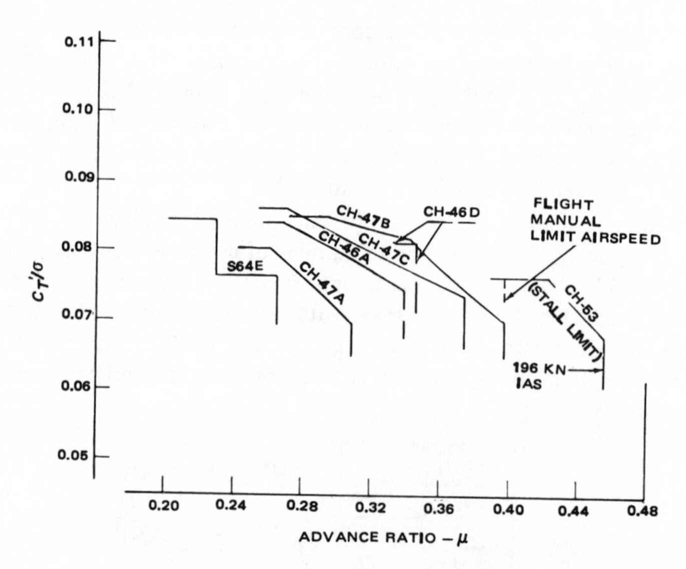

Figure 3.31 Nondimensional flight envelope limitations

The C_T'/σ versus μ method of nondimensionalizing test data does not account for differences in blade torsional characteristics. To account for these effects, a second technique, generalizing test-measured stall inception boundaries[17], was developed. Basically, this method establishes the stall limits in terms of a stall flutter parameter (SFP) and the retreating blade angle-of-attack, $a_{(t)270}$. The SFP is an empirical quantity relating blade torsional properties to $a_{(t)270}$ at given μ and V_t as well as air density corresponding to stall inception of the test aircraft. Here, $a_{(t)270}$ is determined from the trim analysis computer program. SFP is also used to relate the stall limits of one aircraft to the rotor systems of other aircraft having the same airfoil. The advantage of this approach is that it automatically accounts for variations in $\overline{X}$; however, based on recent studies, the validity of the torsional parameter is questionable as this approach is very sensitive to the aerodynamic model used to represent the blade.

A third method of generalizing test data is to correlate the inception of stall observed in flight test with predicted increases in inplane torque levels which occur at particular azimuthal positions[18]. The inplane torque per blade rises abruptly as the blade enters the stall region, due to increased drag on the outboard section of the blade. By comparing test and theoretical torque predictions, a nondimensional empirical stall parameter (*QSP*) can be defined as shown below:

$$QSP = bC_{Q_d}/\sigma \tag{3.25}$$

where

b = number of blades

σ = rotor solidity

$$C_{Q_d} = (\sigma/2b)\int_0^{1.0} c_d \overline{U}_\perp^2\, \overline{r}\, d\overline{r} = \text{profile drag torque coefficient per blade}$$

$\overline{U}_\perp = U_\perp/R\Omega$ = component, perpendicular to the blade span, of the resultant inplane velocity at station $\overline{r}$, where $\overline{r} = r/R$.

This parameter can then be combined with theoretical rotor analyses to predict the stall boundaries of other rotor designs. A value of $0.0035 \leqslant QSP \leqslant 0.004$ has been shown to agree with test measurements of stall inception. One advantage of this technique is that it accounts for the stall and compressibility effects that occur in the third quadrant ($180^\circ \leqslant \psi \leqslant 270^\circ$) and is not limited to one azimuth angle ($\psi = 270^\circ$) as in the SFP approach.

Development of Rotor Limits from Wind-Tunnel Test Data. Model data is often used to evaluate incremental variations in rotor limits due to blade geometry changes, airfoil section modifications, etc. In this respect, trends in performance improvements detected in the wind tunnel will also probably be found in full-scale aircraft in spite of the difference in Reynolds number values. However, a more careful interpretation of the influence of the Reynolds number levels should be applied when model tests are used to define the absolute stall flutter limits of new rotor designs.

The primary advantage of model testing is that the model can be "flown" well into stall under controlled conditions to provide sufficient data to accurately define both stall inception and endurance limits. The amount of full-scale flight-test data obtained beyond stall is generally limited by safety, vibration, weight, and power available restrictions.

The methods of nondimensionalizing wind-tunnel test data are identical to the flight-test techniques outlined in the previous section. Blade torsional loads or pitch-link loads are recorded until stall is observed. The inception points are then nondimensionalized; i.e., generalized, in terms of (1) C_T'/σ and μ, (2) stall flutter parameter, or (3) the inplane torque parameter. It should be noted that when using wind-tunnel data to define absolute stall boundaries, no corrections to the model data for Reynolds number effects are currently applied; thus providing a degree of conservatism. As shown in Fig 3.32, comparisons of stall flutter boundaries based on full-scale flight tests of the 60-ft diameter CH-47C and 44-ft diameter H-21 rotor with those obtained from a 6-ft diameter model tested at the same tip speed show no significant Reynolds number effect for advance ratios of 0.2 to 0.4[16]. These results are attributed to the unsteady turbulent aerodynamic environment in forward flight which increases the effective Reynolds number of the model rotor. However, additional analyses and testing are required to fully understand Reynolds number effects.

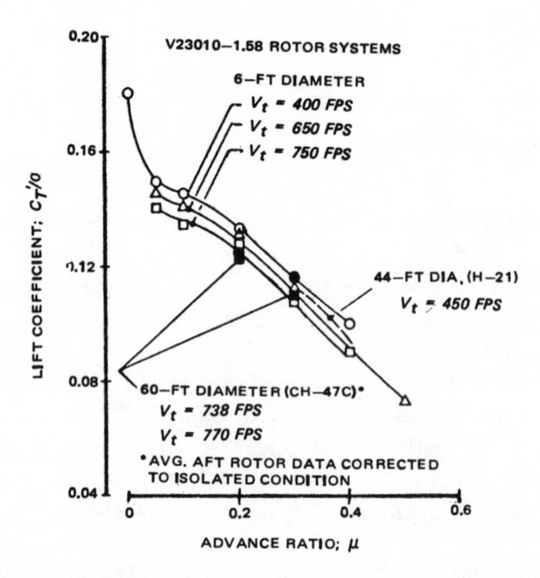

Figure 3.32 Stall flutter boundary — effect of scale

To illustrate the use of wind-tunnel data to predict rotor limits, the 6-ft diameter model data presented in Fig 3.33 will be used to determine the hypothetical helicopter level-flight structural envelope and maneuver capability. Stall inception and hypothetical endurance limit boundaries are shown as a function of μ. The endurance limit is defined as the C_T'/σ value where the load equals three times the unstalled alternating loads (based on CH-47 test data). The stall boundary of the model rotor is corrected for propulsive force and tip speed differences between the model and full-scale rotor configurations. The level-flight structural envelopes developed in Sect 6.2 are based on the endurance limit boundary reduced 10 percent (thrust decrease at constant μ) in order to provide some margin for turbulence encountered in normal operation.

An additional boundary defined for a 2-g banked turn maneuver is also shown in Fig 3.33. The difference between the 1-g and 2-g boundary is the effect of pitch rate alleviation. The pitch rate generated in a banked turn permits a small amount of lift offset to occur, thus unloading the retreating blades slightly and extending the boundary. More details concerning the calculation of pitch rate effects will be presented later in Sect 6.3.

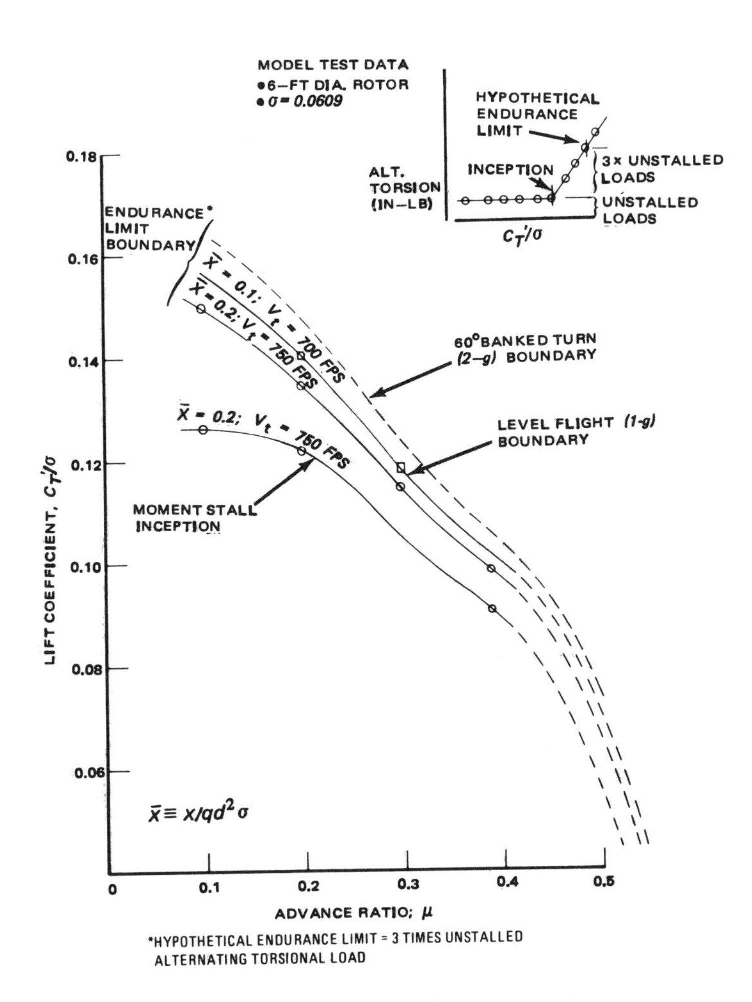

Figure 3.33 Hypothetical helicopter moment stall limits

Theoretical Loads Prediction Computer Programs. It should be emphasized that theoretical methods of determining rotor stall boundaries from blade-load analysis have not, as yet, reached the state-of-the-art where they can be utilized as the sole means of defining structural flight-envelope limits. Consequently, the empirical methods described above are the primary means used to establish these boundaries for new aircraft. The difficulty in developing theoretical loads analysis lies in the establishment of a truly representative mathematical model, reflecting the complex unsteady aerodynamic environment and its interaction with blade elastic properties[19]. Although much work remains to be done, some progress has been made[20,21].

The latest loads prediction computer programs include shed and trailing vortex wake representations as well as elastic blade effects. Typically, a skewed helical trailing vortex system is used to account for the spanwise variation in lift. Shed vorticity (Fig 3.34) due to the azimuthal variation of lift is defined mathematically by applying Theodorsen's relationships of unsteady aerodynamics[22] to the helicopter rotor.

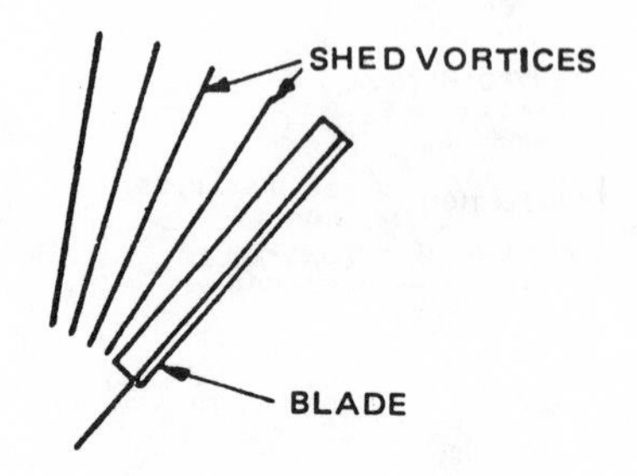

Two additional factors having a significant effect on stall inception and the resulting unsteady aerodynamic load growth are dynamic stall and spanwise flow. Consequently, they have been incorporated in the loads prediction programs.

Figure 3.34 Shed wake pattern

Dynamic effects alter the static airfoil-section stall characteristics. Due to blade flapping or pitching motion, the local blade section angle-of-attack and pitch rate vary periodically with the azimuth angle (Fig 3.35).

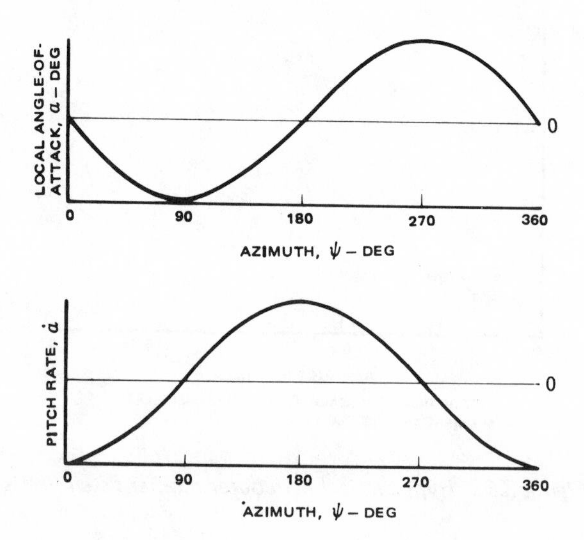

Figure 3.35 Variation of local angle-of-attack and pitch rate with the azimuth

Positive pitch rates are developed over the aft section of the rotor disc, and negative pitch rates over the forward section. Two-dimensional oscillating airfoil wind-tunnel tests[22] have shown that as the blade section angle-of-attack approaches stall, a hysteresis effect occurs which delays the stall at positive pitch rates ($\dot{\alpha}$) and promotes stall at negative values as shown in Fig 3.36.

The increase in two-dimensional airfoil stall angle-of-attack for both moment stall and lift stall for the V23010-1.58 airfoil section at $M = 0.4$ is shown in Fig 3.37 as a function of the nondimensional pitch rate factor $\sqrt{c\dot{\alpha}/2V}$. A physical interpretation of $\sqrt{c\dot{\alpha}/2V}$ can be obtained by noting that c/V is approximately the time it takes for a particle of

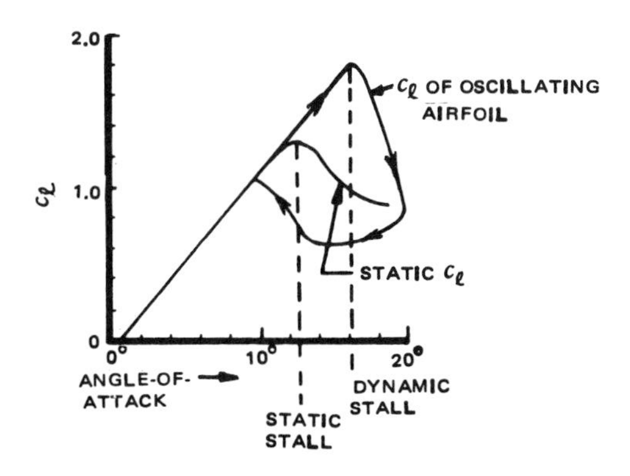

Figure 3.36 Effect of pitch oscillation on stall

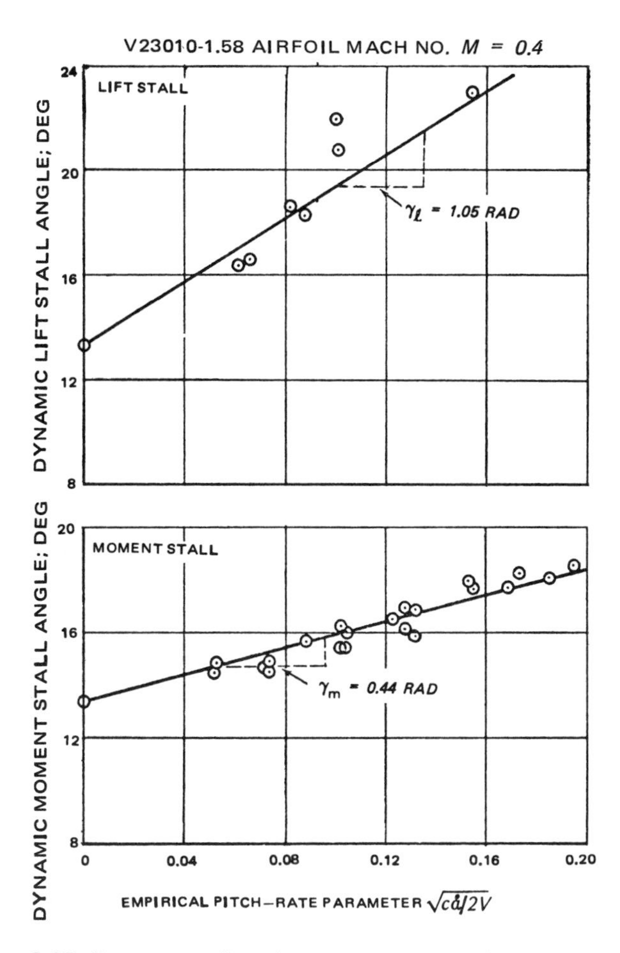

Figure 3.37 Dynamic stall angle vs nondimensional pitch rate factor

of air to travel from the leading edge to the trailing edge of the airfoil and is therefore, a measure of the time it takes for stall to fully develop. Hence, the term $\dot{a}(c/V)$ is the Δa that can occur before stall effects become significant. The pitch rate has more effect on lift stall than on the moment stall which casues high control loads. Furthermore, the beneficial effects of $\dot{a}$ on retarding the stall are reduced at higher Mach numbers as illustrated in Fig 3.38, where the stall delay function $\gamma = \Delta\dot{a}_{stall}/\Delta\sqrt{c\dot{a}/2V}$ is a function of

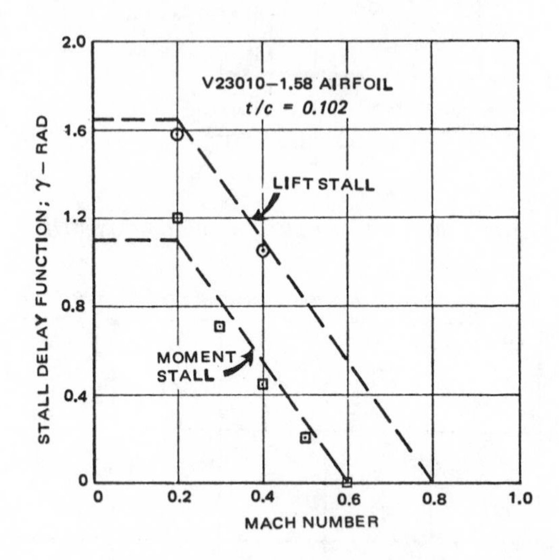

Figure 3.38 Effect of compressibility on stall delay function

Mach number. The parameter γ is simply the slope of the graph presented in Fig 3.37 which is applied to two-dimensional static airfoil data in order to incorporate oscillating airfoil effects in existing rotor analyses, as described in Ref 22.

The other factor identified as having a significant effect on rotor stall is spanwise flow. The radial component of freestream velocity has usually been ignored in the past. However, the total velocity at each blade element is actually larger than the normal component, and it places the blades at an equivalent yaw angle. There are two basic effects of yawed or radial flow. First, the increase in actual velocity augments the section drag as explained in Ch III, Vol I. The second effect is to make the section-lift coefficients, $c_{\ell_{max}}$, referenced to inplane velocities normal to the blade span, appear higher than their actual three-dimensional values.

The effects of dynamic stall and yawed flow on thrust and power required predictions—discussed in detail in Ref 22—are illustrated in Fig 3.39[23]. The baseline theory, using static airfoil data, shows much higher thrust and power required penalties due to stall than indicated by the test data. Incorporating dynamic stall and yawed flow corrections in the analysis significantly improved the high thrust-level correlation. However, theoretical and experimental gaps still remain; for instance, an understanding of the effect of spanwise flow on the hysteresis loop of the unsteady airfoil data.

Improvements in the prediction of torsional load growth in stall have also been obtained. Fig 3.40 shows the correlation of the U.S. Government Program (C-81) with

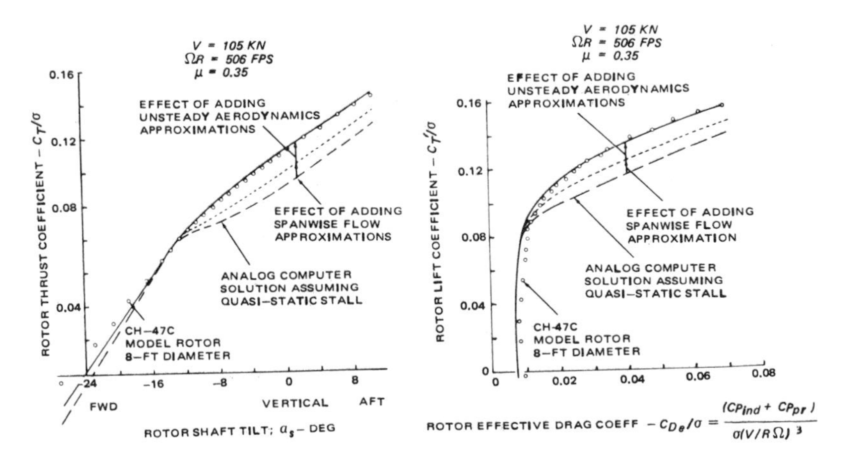

Figure 3.39 Predictions and tests of rotor thrust and lift coefficients

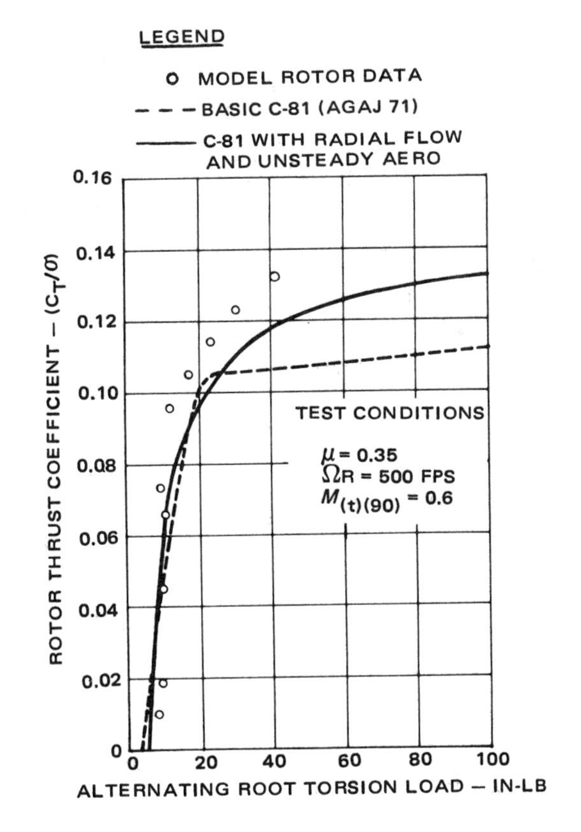

Figure 3.40 Correlation of theoretical torsional load prediction with model rotor data

model data, and a comparison of the Vertol loads program (C-60) with flight-test data is presented in Figs 3.41 and 3.42. Additional discussions of these analytical techniques can be found in Refs 19, 20, and 21.

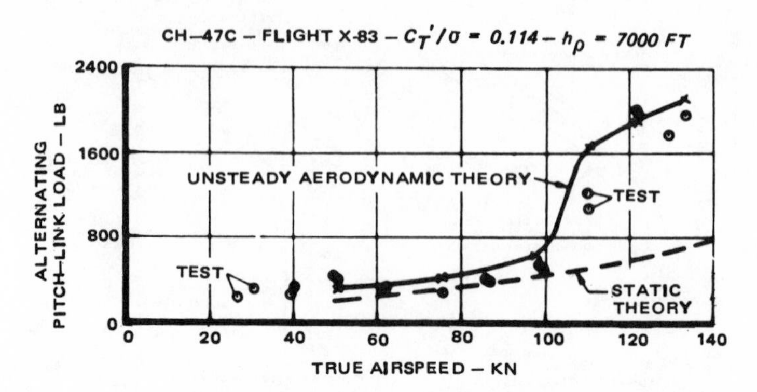

Figure 3.41 Comparison of test and computed pitch-link loads for an airspeed sweep

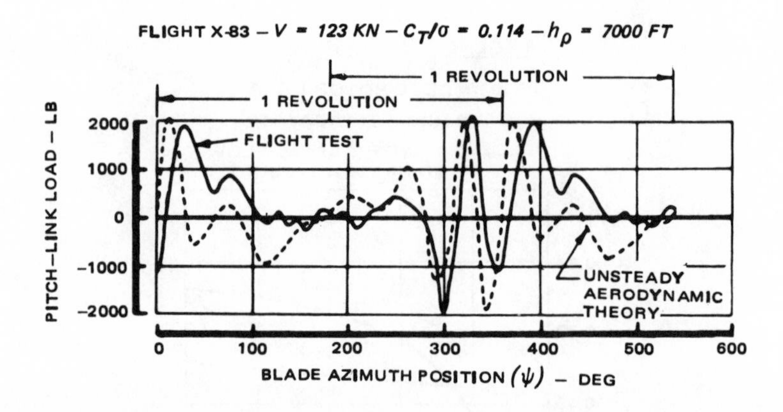

Figure 3.42 Pitch-link load waveform correlation using unsteady aerodynamic theory

6.2 Level-Flight Airspeed/Altitude Structural Flight Envelope

The structural envelopes of helicopters are generally presented as a function of density altitude as shown for the hypothetical helicopter in Fig 3.43. This envelope is based on the wind-tunnel model endurance limit boundaries given in Fig 3.33. The data is given with and without a 10-percent thrust margin to account for turbulence effects. It can be seen that the airspeed envelopes decrease rapidly with altitude due to the increase in rotor $\bar{c}_\ell$ for a given gross weight. Maximum continuous and intermediate power speed capability points at 15,000-lb gross weight are indicated for SL/STD and 4000 ft, 95°F ambient conditions. The aircraft does not have sufficient power in level flight to exceed the structural envelope at these conditions. This relationship between power

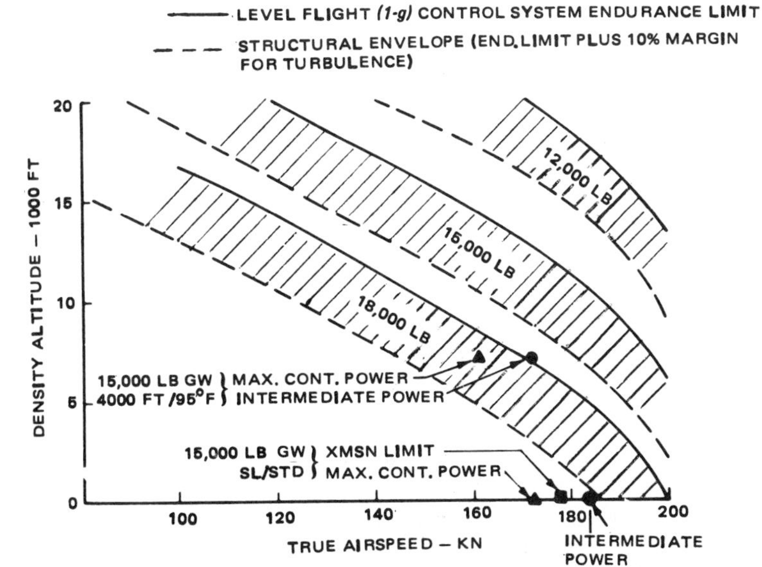

Figure 3.43 Hypothetical helicopter structural envelope

limits and structural limits is a desirable design feature as it reduces the risk of inadvertently operating beyond the structural envelope in level flight.

At high altitudes, the aircraft has sufficient continuous power to exceed the structural airspeed limits as shown in Fig 3.44 for standard day conditions and $W = 15{,}000$ *lb*. It is apparent that the structural airspeed decreases more rapidly with altitude than the corresponding power limit envelope.

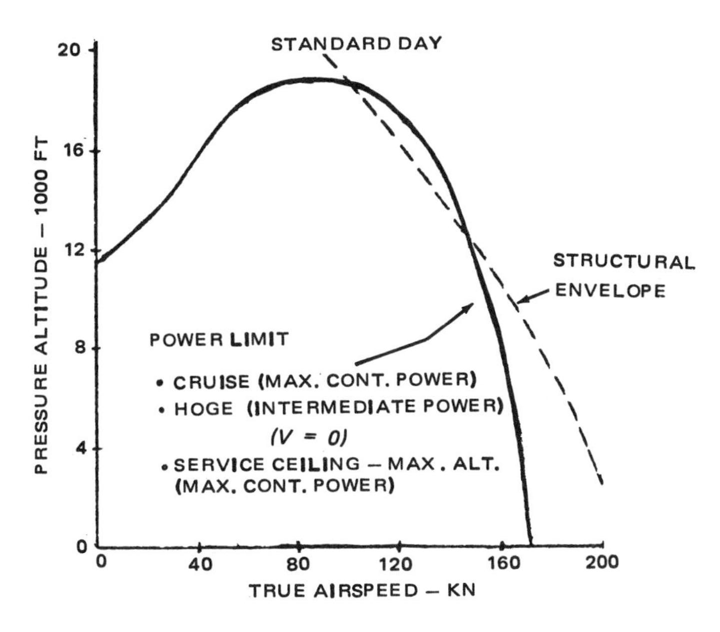

Figure 3.44 Flight envelope at 15,000-lb gross weight

6.3 Maneuver Capability

The current generation of military helicopters is designed to meet specific maneuver requirements such as contour flying where the aircraft operating close to the ground (50–100 ft altitude) must follow the contour of the terrain. In the past, much less emphasis was placed on maneuver capability. Consequently, rotors were sized for hover and cruise efficiency.

A maneuver is defined as *accelerated flight when acceleration* $a = (n - 1)32.2$, where n is the thrust-to-weight ratio (T/W) expressing the load factor, and $32.2\ ft/sec^2$ is the acceleration of gravity. If the acceleration occurs normal to the freestream velocity, then the flight path will be curved, resulting in a pull-up or banked-turn type of maneuver. The acceleration in this case is centripetal, and is equal to V^2/R where R is the radius of the turn, or pull-up.

The hypothetical helicopter banked-turn maneuver capability at 15,000-lb gross weight and 4000 ft, 95°F is illustrated in Fig 3.45. It is assumed that structural loads during maneuvers cannot exceed the endurance limit (Fig 3.33). However, most of the currently designed helicopters have sufficiently severe maneuver requirements that the life of the components must be defined on the basis of a given percentage of time for operation beyond the endurance limit. The hypothetical helicopter has a 1.35-g rotor limit capability at 150 kn. However, the engine rating does not provide sufficient power to maintain the required thrust during maneuvers. Consequently, the extra energy must be provided by one of the following means:

(1) descent (potential energy)
(2) decreased rotor speed (rotational kinetic energy)
(3) deceleration (translational kinetic energy).

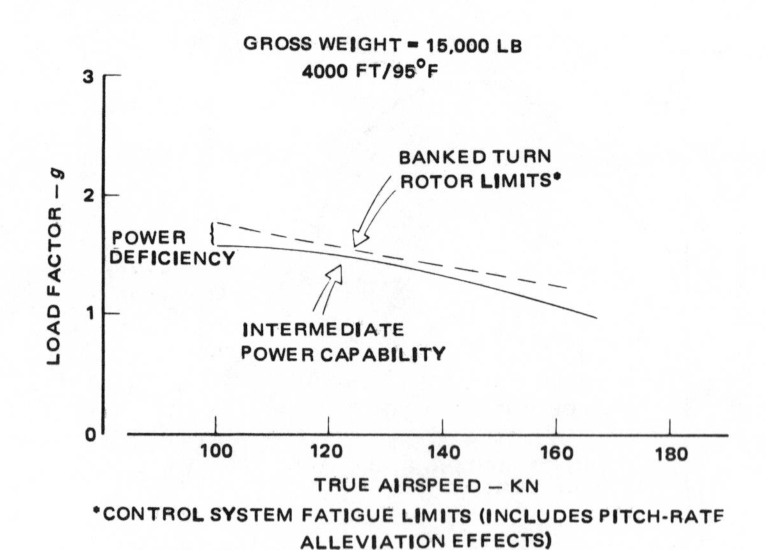

Figure 3.45 Maneuver capability

During contour-flying operations conducted at low altitudes, obviously neither descent nor rotor speed decay are acceptable. Consequently, deceleration must be used in maneuvering beyond the power limits. This will require reducing the forward speed by using cyclic control to tilt the tip-path plane aft. The deficiency in power available (ΔRHP) is compensated by a reduction in airspeed from the initial speed (V_i) to the final speed (V_f) as shown in Eq (3.26).

$$\Delta RHP = (1/1100)(W/g)(V_f^2 - V_i^2)/\Delta t \tag{3.26}$$

where

Δt = time increment of the maneuver; sec

V = velocity; fps.

Utilizing this type of analysis, time histories of aircraft deceleration, rotor decay, altitude, etc., can be developed. A detailed discussion of the energy tradeoffs during maneuvers is presented in Ref 24.

The rotor limits presented in Fig 3.45 are based on the maneuver C_T'/σ boundaries given in Fig 3.33. The difference in level flight and banked-turn maneuver rotor limits is due to pitch rate alleviation (*PRA*) effects. As stated in Ref 25, positive helicopter pitch rates developed during symmetrical pull-ups or banked turns result in a favorable gyroscopic moment acting on the rotor system. This moment affects the blade flapping motion in such a way as to unload the retreating blade and delay the onset of stall. The blade flapping equations can be solved with and without PRA to obtain the corresponding incremental decrease in retreating blade tip angle-of-attack $\Delta a_{(t)270}$.

$$\Delta a_{(t)270} = \frac{-16 q_r R}{\gamma(V_t - V)} \tag{3.27}$$

where

γ = blade Lock number $(\rho c_{\ell a} c R^4)/I_b$

q_r = pitch rate (body axes reference system); rad/s

V_t = tip speed fps.

In this equation, the aircraft pitch rate during a banked turn (q_{r_b}) or cyclic pull-up (q_{r_p}) can be approximated by the following expressions[6,25]:

$$q_{r_b} = (32.2/V)\,[n - (1/n)] \tag{3.28}$$

$$q_{r_p} = (32.2/V)(n - 1) \tag{3.29}$$

where

n = load factor (g's)

V = airspeed; fps

Eqs (3.27)–(3.29) indicate that the stall benefits resulting from pitch rate decrease with increasing forward speed. In addition, the pitch rate and $\Delta a_{(t)270}$ developed for a given load factor are larger for a banked turn than for cyclic pull-up. For example, the pitch rate generated during a 2g banked turn is 1.5 times the corresponding pull-up pitch rate.

The $\Delta a_{(t)270}$ due to pitch rate effects must be combined with trim analysis $a_{(t)270}$ predictions to obtain the incremental load factor increase (Δn) due to PRA, assuming that stall will occur at a given $a_{(t)270}$ value during the maneuver. The graphical procedure for obtaining Δn is shown in Fig 3.46.

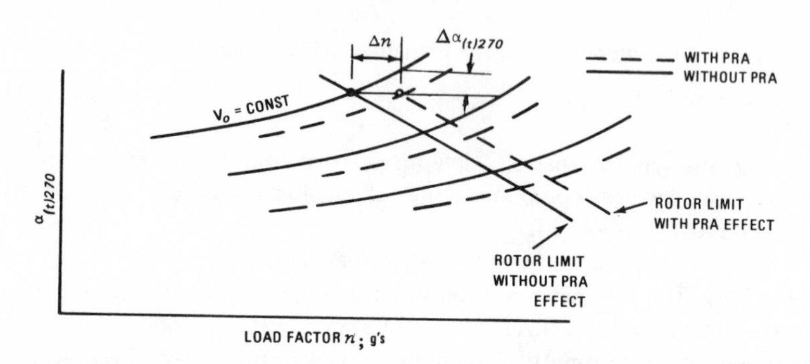

Figure 3.46 Determination of pitch rate alleviation (PRA) effects from trim analysis

The basic gyroscopic moment effects can also be included in the trim program flapping equations in order to account for power required as well as stall limit variations due to pitch rate effects.

7. CALCULATION OF FORWARD–FLIGHT PERFORMANCE CAPABILITY

The methodology for computing forward-flight performance capability as described in this section, applies to the following items:

(1) mission performance analysis
(2) payload/range
(3) payload/endurance
(4) ferry range
(5) speed capability
(6) climb capability
(7) service ceiling
(8) autorotation capability.

7.1 Mission Performance

Mission performance calculations involve computing the payload (cargo, passengers, equipment, etc.) that can be carried during a given mission. The payload (PL) is obtained by subtracting the weight of fuel (W_F) required, plus the weight empty (W_E), and fixed useful load (FUL) determined by the weight of crew plus trapped oil and fuel, from the takeoff gross weight ($TOGW$):

$$PL = TOGW - (W_E + FUL + W_F). \tag{3.30}$$

The maximum practical TOGW should be used to determine the maximum payload capability of an aircraft. The criteria used to determine this weight is a function of the type of takeoff site assumed for the mission. For example, if the takeoff area is surrounded by trees, a vertical climb capability is required; however, if the area is an open space, then the takeoff can be conducted from an IGE wheel height with a gradual transition to forward flight. Operating from helipads located on tops of buildings, oil rig platforms, forest clearings, or other confined areas may require alternate takeoff criteria. For most military missions, TOGW is based either on hover OGE capability at intermediate power, or a specified vertical rate-of-climb level at either intermediate or 95 percent of that power rating. These criteria provide a sufficient performance margin for defining a realistic takeoff weight.

The weight of fuel in Eq (3.30) depends on the type of mission being evaluated. Using the range mission as an example, the various segments of the profile shown in Fig 3.47 are as follows:

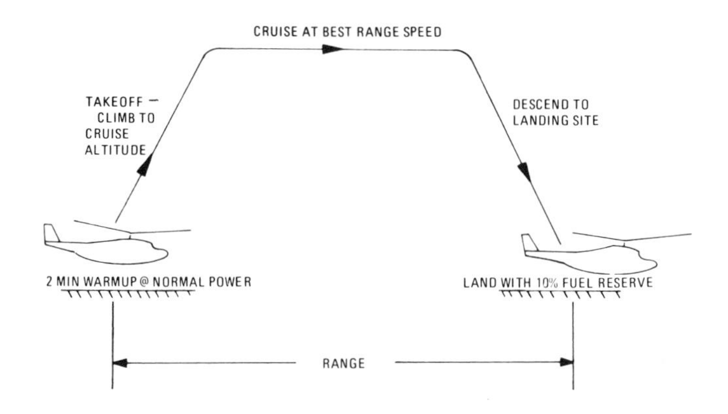

Figure 3.47 Mission profile

(1) *Warmup* — Includes fuel consumed to start and check out the aircraft. A fuel allowance of two to five minutes at maximum continuous power is typical.

(2) *Takeoff* — The fuel required for takeoff and transition to forward flight is generally small enough to be neglected when calculating missions where the cargo or payload is carried internally. During missions where the cargo is carried externally, the hover fuel required for the hookup of the load can be significant.

(3) *Climb to Cruise Altitude* — For comparative performance calculations and for missions flown 1000–2000 ft above the takeoff site, climb fuel and distance are usually neglected. Cruise at higher altitudes will require consideration of the climb effect on fuel.

(4) *Cruise at Constant Altitude* — The cruise portion of the mission is generally conducted at airspeeds which provide the maximum range for a given quantity of fuel,

unless the aircraft is limited by power available or structural considerations. This speed is referred to as the *best range speed.*

(5) *Descent to Landing Site* – Because of the lower power settings in descent, the fuel used during this stage of the mission is considered negligible. In addition, no allowance is taken for distance traveled in descent, unless the cruise altitude is significant.

(6) *Landing with Fuel Reserve* – For internal cargo missions where no hover time is required for detaching loads prior to landing, the fuel required to land is negligible. However, it is assumed that the aircraft lands with a specific quantity of fuel reserve, which is typically 10 percent of the initial fuel quantity, or an allowance of sufficient fuel to either cruise at best range speed, or loiter at minimum power speed, for 20 to 30 minutes. The loiter time reserve is generally used for short missions where a 10-percent reserve would be insufficient.

The objective of the range mission is to maximize the distance traveled one way per given quantity of fuel. These missions are generally computed at constant altitude and optimum airspeed conditions; however, aircraft altitude as well as speed can be optimized to further increase the distance traveled. The altitude and airspeed optimization is usually reserved for *ferry-range missions,* discussed later in this section.

Maximum endurance represents another basic mission aimed at maximizing endurance or time on station rather than distance. This mission is typical of search and surveillance operations and is flown at minimum-power speeds (70 to 90 kn), where fuel consumption per unit of time is lowest.

There are many other derivatives of the range and endurance missions such as *radius missions* (cruise out, land, and unload part or all of the payload, and return), *antisubmarine (ASW), mine countermeasures (MCM),* and *vertical replenishments.* However, the basic calculation methods are essentially the same as the range and endurance procedures which are applied to each segment of the mission, while careful account of the fuel is maintained. For missions with more than one cruise segment, some iteration on the distribution of fuel may be required.

In this section, only the typical features of range (constant altitude), endurance, and ferry-range mission performance prediction techniques are presented, using the hypothetical helicopter to illustrate the methods. For a more detailed analysis, the reader is referred to a design handbook[26].

7.2 Payload/Range Capability

Weight empty (W_E) and fixed useful load (FUL) are functions of basic aircraft configurations and number of crew. For the hypothetical helicopter,

$$W_E = 9450\ lb$$

$$FUL = 430\ lb$$ (2 crew @ 200 lb ea + 30 lb of trapped oil and fuel).

Assuming 4000 ft/95°F ambient conditions, $TOGW_{max} = 16{,}000\ lb$, based on hover OGE capability at intermediate power (Fig 2.25). Substituting these values into Eq (3.30) gives

$$PL = 6120 - W_F.$$

The payload, therefore, decreases as the fuel weight increases until the internal fuel capacity is reached. Further increases in fuel capacity require the addition of auxiliary tanks, which increases the W_E as discussed in the Ferry Range section (7.4).

The tradeoff of payload for fuel and its subsequent influence on the payload-range relationship can be determined by first computing the *specific range* (n.mi/lb of fuel). The specific range is simply the cruise speed V in kn divided by the average fuel consumption $\dot{W}_F$ in units of lb/hr. Then, the incremental range (dR) becomes

$$dR = (V/\dot{W}_F)dW \tag{3.31}$$

where

V = airspeed (kn)

dW = incremental change in weight due to fuel burnoff (lb).

The *total range* in n.mi for a given quantity of cruise fuel (lb) is obtained from Eq (3.31) as

$$R = \int_{W_2}^{W_1} (V/\dot{W}_F)dW \tag{3.32}$$

where

W_1 = initial gross weight

$W_2 = W_1 -$ *fuel burnoff.*

The integral in Eq (3.32) is normally evaluated by graphical methods. The range is equal to the area under a plot of specific range (SR) vs gross weight between the initial W_1 and the final W_2 gross weight after fuel burnoff (Fig 3.48). If the specific range data is reasonably linear between W_1 and W_2, this area can be computed by using the average SR. The $SR = f(W)$ plots are developed from the power required and the engine fuel flow

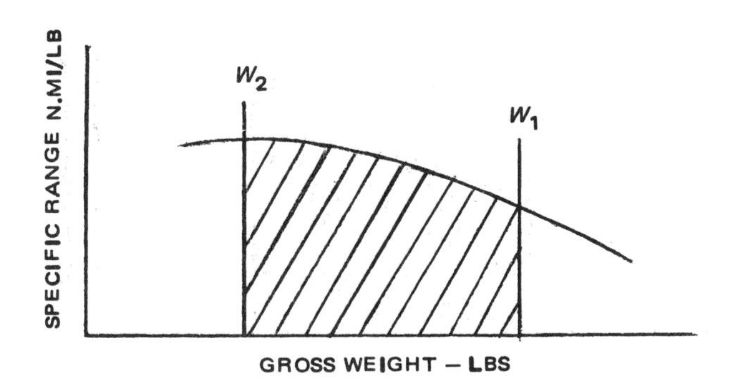

Figure 3.48 Specific range vs gross weight

characteristics at the desired cruise ambient conditions and cruise speeds. Range missions are usually computed at the speed for maximum range. The procedure for computing the best range speed and associated specific range is illustrated in Fig 3.49 (reproduced from Ref 26), and outlined below.

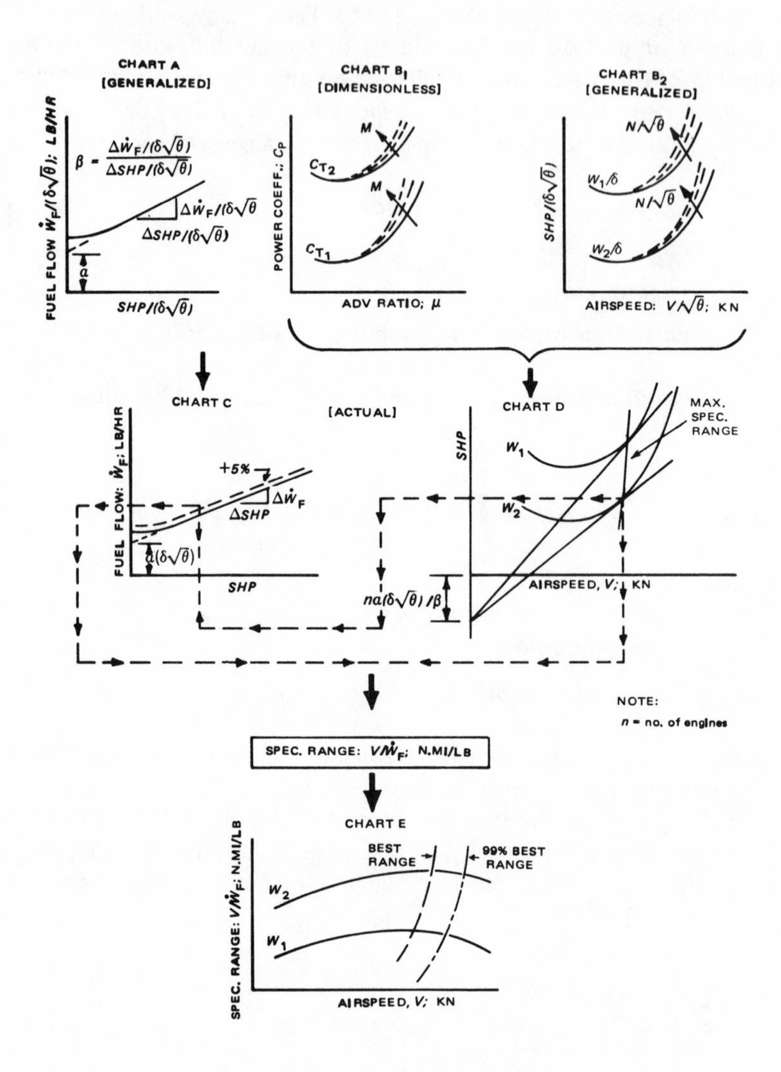

Figure 3.49 Determination of specific range

(1) *Determination of Required SHP* – The required SHP for level flight can be based on analysis or test data and should include compressibility effects. Speed-power polars are needed for at least three to four gross weights ranging from the aircraft minimum flying weight to takeoff weight. Power required may be defined in generalized, referred, or nondimensional form.

(2) *Calculation of Fuel Flow vs Airspeed* – For two-engine aircraft, the SHP required is assumed to be equally divided between the engines. The total fuel flow is calculated using the engine manufacturers' fuel flow vs power relationships at the correct ambient conditions which are usually accounted for by generalized fuel flow ($\dot{W}_F/\delta\sqrt{\theta}$) and SHP ($SHP/\delta\sqrt{\theta}$) data (Ch I, Sect 3). For many applications, the engine manufacturers' fuel flow is increased 5 percent to account for differences between production engines, variations due to different pilot operating techniques, and engine performance degradation in service. The 5-percent fuel flow adjustment may also be employed when power required estimates are based entirely on theory.

(3) *Computation of SR = f(V)* – The specific range, $SR = V/\dot{W}_F$ (n.mi/lb), is computed for each weight and plotted as a function of airspeed. The resulting curves reach a maximum, or best range value, at one airspeed. However, the mission cruise speed is generally defined as that corresponding to $0.99(SR)_{max}$. Since the specific range curve is relatively flat near the best range speed, the one-percent loss in range will result in a six or seven-percent increase in cruise speed and a corresponding reduction in mission time. For this reason, almost all best range performance is actually defined at the 99-percent best range conditions.

(4) *Check of Power Available and Structural Airspeed Limits* – The maximum continuous power, transmission, stall, and structural envelope limitations can be indicated on the $SR = f(V)$ graph to determine whether the best range speeds are within the aircraft flying envelope. Many times, an aircraft can fly at best range speeds at low gross weights, but at high weights, is limited to lower speeds.

(5) *Determination of the Final Specific Range vs Gross Weight* – This is done by cross-plotting the specific range data as a function of gross weight.

Step-by-step calculations of specific range for the hypothetical helicopter at 15,000 lb gross weight and 4000 ft/95°F cruise condition is shown in Table III-4. The graphs presented in Figs 3.50 and 3.51 were developed by repeating this procedure for three other weights.

① AIRSPEED V; kn	80	100	120	140	160	170
② SHP_{req} (Fig 3.24)	915	930	1060	1320	1810	2260
③ SHP per engine =②/2	458	465	530	660	905	1130
④ $SHP/\delta\sqrt{\theta}$ = ③ $/\delta\sqrt{\theta}$*	513	521	593	739	1019	1265
⑤ $\dot{W}_F/\delta\sqrt{\theta}$; lb/hr (Fig 1.7)	381	382	408	456	560	668
⑥ $\dot{W}_F$ per engine =⑤ x $\delta\sqrt{\theta}$; lb/hr	340	341	365	407	500	596
⑦ $\dot{W}_F$ (2 engines + 5% sfc increase) ⑥ x 2.1; lb/hr	714	716	766	854	1049	1252
⑧ Specific Range SR = ①/⑦; n.mi/lb	0.112	0.140	0.158	0.164	0.152	0.136
NOTE: * $\delta\sqrt{\theta}$ = 0.8932						

TABLE III-4 CALCULATIONS OF SPECIFIC RANGE AT 15,000 LB; 4000 FT/95°F

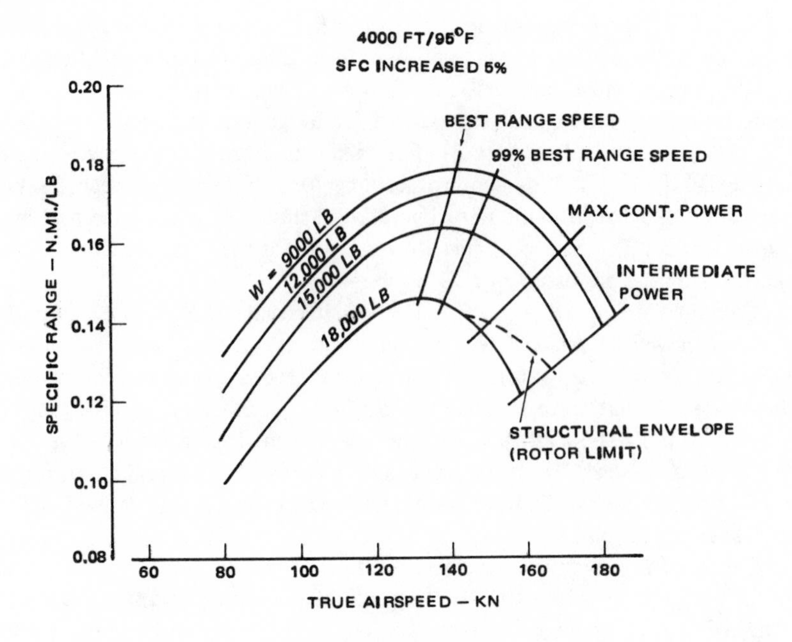

Figure 3.50 Specific range

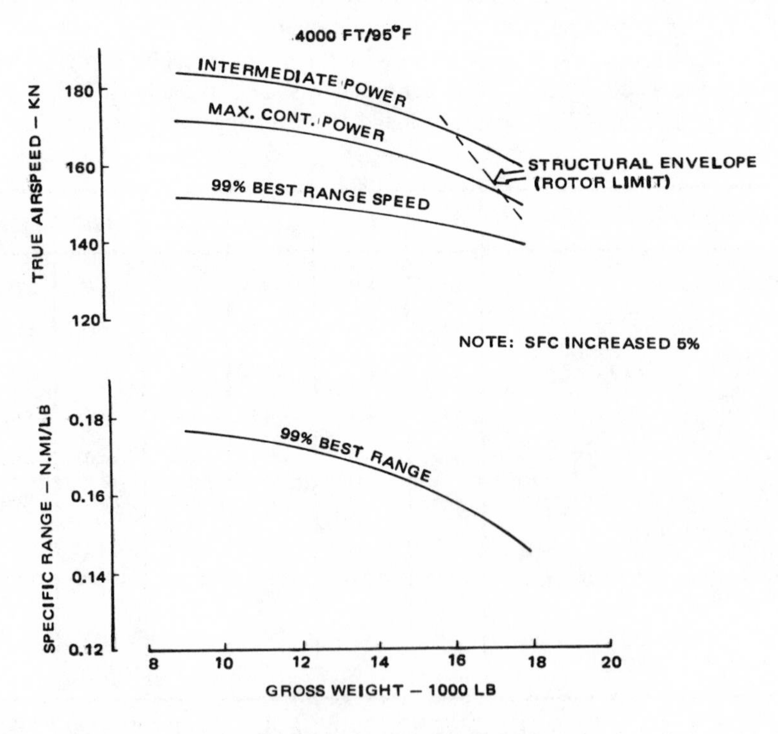

Figure 3.51 Cruise performance

The airspeed for 99-percent best range speed is approximately 7 percent higher than the 100-percent best range values, and is considerably below normal power and structural envelope limits.

Utilizing the specific range curve in Fig 3.51, the mission fuel in Eq (3.30) can be converted to range; leading to a plot of payload vs range (Fig 3.52). Here, the maximum payload at zero range is approximately 6100 lb, and decreases as the fuel required increases with range until the permanent tank fuel capacity is reached. Then, the TOGW must be reduced, or auxiliary tanks must be added to increase the range. The payload and range capability can be determined from this plot for any combination of gross weight and fuel quantity up to the maximum takeoff weight and internal fuel capacity.

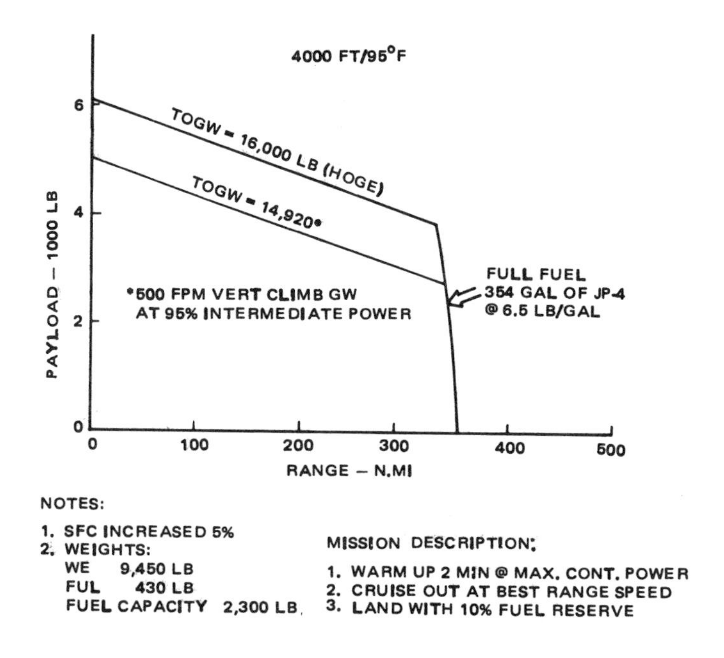

Figure 3.52 Hypothetical helicopter payload/range capability

A line representing a 500-fpm vertical rate of climb takeoff criteria at 95-percent intermediate power corresponds to $TOGW \approx 15{,}000$ *lb*. This takeoff performance was used to determine the hypothetical helicopter design gross weight since it is representative of current U.S. Army takeoff criteria.

The payload/range curve is determined by first computing the zero and full-fuel points at the maximum takeoff gross weight of 16,000 lb, and connecting the points with a straight line. Intermediate points should be computed using an iterative technique, since neither the payload nor the range is fixed for these calculations. These points, however, do not deviate sufficiently from a straight-line interpolation to justify the additional calculations. The last step is to compute two additional points along the full-fuel line; one intermediate and another, at the zero payload point.

The calculation of payload and range for these points requires a step-by-step accounting system to keep track of the gross weight and fuel weight throughout the mission. Generally, a detailed tabulation sheet similar to the one shown in Table III-5 is used. The sample calculations shown on this sheet use the specific range data presented in Fig 3.51 for the mission described in Fig 3.52.

PROCEDURE	ZERO RANGE	FULL FUEL CONSIDERATIONS		
	PL_{max} ($TOGW_{max}$)	PL_{max} ($TOGW_{max}$)	ZERO PL	$PL = 0.5PL_{max}$
PAYLOAD CALCULATIONS; lb				
1 Weight Empty	9,450	9,450	9,450	9,450
2 Fixed Useful Load	430	430	430	430
3 Total Fuel[1] = *13*	40	2,300	2,300	2,300
4 PAYLOAD = 5 – (1 + 2 + 3)	6,080	3,820	0	1,910
5 Takeoff Gross Weight	16,000	16,000	12,180	14,090
6 Warmup, 2 min @ max cont[2]	36	36	36	36
7 Gross Weight = *5 – 6*	15,964	15,964	12,144	14,054
8 1/2 Cruise Fuel = *16/2*	0	1,017	1,017	1,017
9 Avg Gross Wt = *7 – 8*	15,964	14,947	11,127	13,037
10 Remaining Cruise Fuel = *16/2*	0	1,017	1,017	1,017
11 Landing Weight = *9 – 10*	15,964	13,930	10,110	12,020
12 WE + FUL = 11 – (4 + 15)	9,880	9,880	9,880	9,880
FUEL ANALYSIS; lb				
13 Total Fuel	40	2,300	2,300	2,300
14 Warmup Fuel	36	36	36	36
15 Reserve Fuel = 0.1 x *13*	4	230	230	230
16 Cruise Fuel = *13 – (14 + 15)*	0	2,034	2,034	2,034
RANGE CALCULATION:				
17 SR @ Avg *W* (Fig 3.50) (n.mi/lb)	–	0.1627	0.1736	0.1690
18 RANGE = 17 x 16 (n.mi)	0	331	353	344

NOTES: 1. Specific Fuel Consumption increased 5 percent

2. Mission Description:

a. Warmup: 2 min @ maximum continuous power
b. Cruise out at best ferry range speed
c. Land with 10 percent fuel reserve

[1] *Full Fuel Weight = 354 gal x 6.5 lb/gal = 2300 lb*

[2] *Warmup fuel calculation at 400 ft/95°F: W_F = 1070 lb/hr (2 engines @ max cont Power)*
Fuel for 2 min = 36 lb

TABLE III-5 PAYLOAD–RANGE CALCULATIONS @ 4000 FT/95°F

The trend of the specific range data with weight is assumed to be sufficiently linear to utilize the mission mid-point weight and associated specific range to compute the total range capability – an exception to this approach would be when constraints such as rotor stall reduces specific range at high gross weights early in the mission; in which case, a segmented mission analysis should be used. The calculations continue until the landing weight is obtained. The fuel reserve and payload is then subtracted from the landing weight as shown in Step 12 of Table III-5. This weight should be equal to $W_E + FUL$.

7.3 Payload/Endurance Capability

Endurance missions require maximum time on station for the purpose of surveillance, loiter, search and destroy, etc. The payload for these missions usually consists of electrical equipment or external armament (torpedoes, missiles, etc.).

Maximum endurance is obtained at the minimum engine fuel flow ($\dot{W}_{F_{min}}$). By calculating $dSHP/dV = 0$ and solving for V, it can be shown that minimum fuel flow vs SHP curve is linear in the range of SHP considered. In most cases, the engine fuel flow is essentially linear over the small range of power required defining the speed power polar "bucket." The exact expression for maximum endurance (t_{max}) is

$$t_{max} = \int_{W_2}^{W_1} (1/\dot{W}_{F_{min}})dW \qquad (3.33)$$

where

W_1 = initial gross weight

$W_2 = W_1$ – *fuel burnoff.*

This integral can be evaluated graphically, since the endurance is equal to the area under the plot of $1/\dot{W}_F$ vs gross weight between W_1 and W_2 (Fig 3.53). If $[1/\dot{W}_F = f(W)]$ is linear between W_1 and W_2, then the endurance is computed bydividing the available loiter fuel by the fuel flow at the average mission weight. Otherwise, the area must be divided into small segments and the Δt of each segment added to get the total endurance.

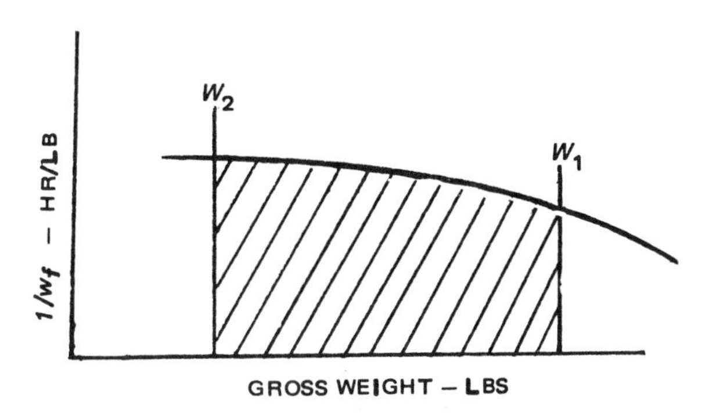

Figure 3.53 Endurance calculation

The procedure for calculating maximum endurance fuel flow is illustrated in Fig 3.54 and outlined below.

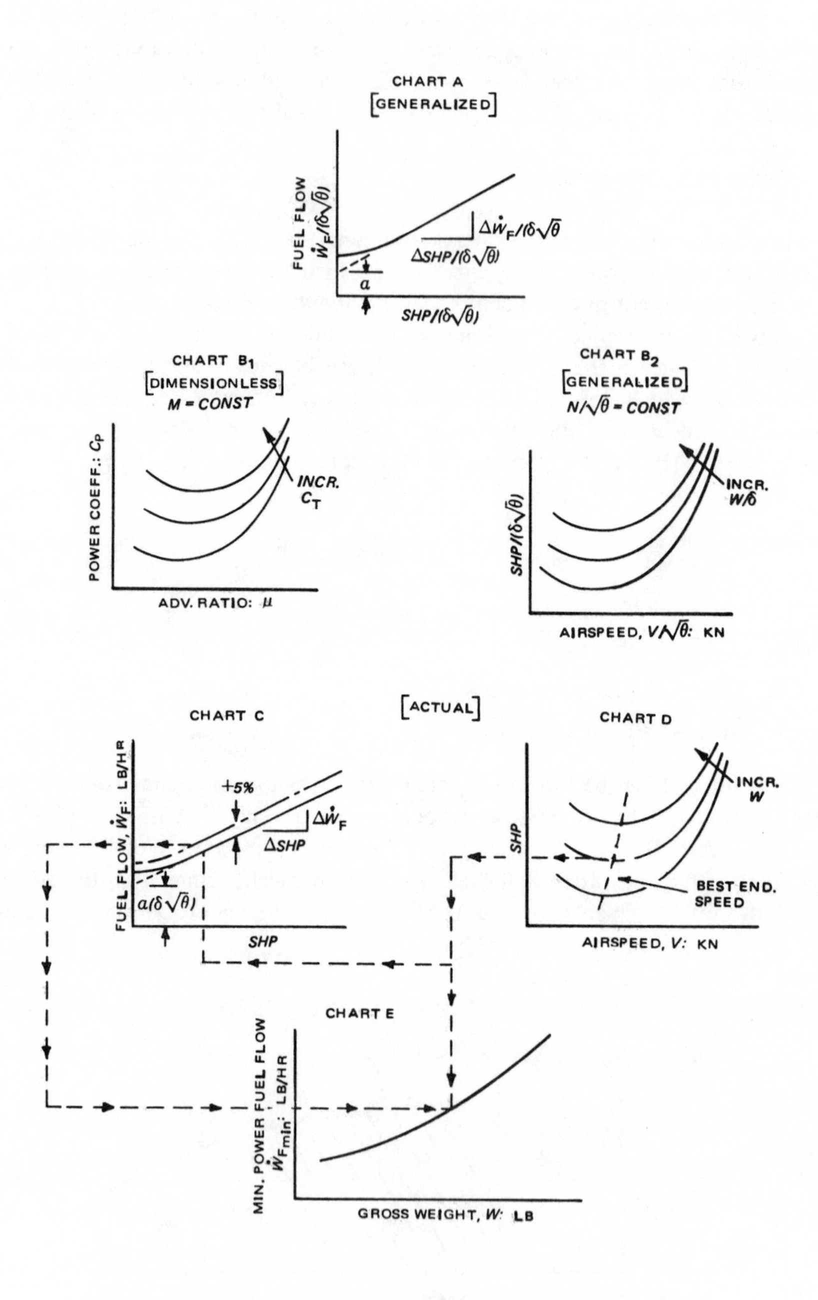

Figure 3.54 Determination of maximum endurance fuel flow

(1) *Determine Power Required* – As noted above, the maximum endurance fuel flow is computed at minimum power required, based on test or theoretical predictions. It is often convenient to plot the power required for weights from W_E to the maximum *TOGW*, in terms of SHP_{min} vs W for calculation purposes.

(2) *Calculate Fuel Flow* – For two-engine aircraft, each engine is assumed to provide 50 percent of the power required. Therefore, the fuel flow can be calculated using the engine manufacturer's fuel flow versus power relationships at $SHP/2$ and operational ambient conditions. These latter variations are usually accounted for by using generalized fuel flow presented in terms of $\dot{W}_F/\delta\sqrt{\theta}$ and $SHP/\delta\sqrt{\theta}$. For performance calculations, the fuel flow specified by engine manufacturers is usually increased by 5 percent.

Maximum endurance fuel flow for the hypothetical helicopter is shown in Fig 3.55. The data is calculated for a 4000 ft/95°F condition and includes a 5-percent SFC increase. A total of 6 points were used to define this line, with gross weights ranging from 9000 to 18,000 lb. Detailed calculations for a gross weight of 15,000 lb are presented below.

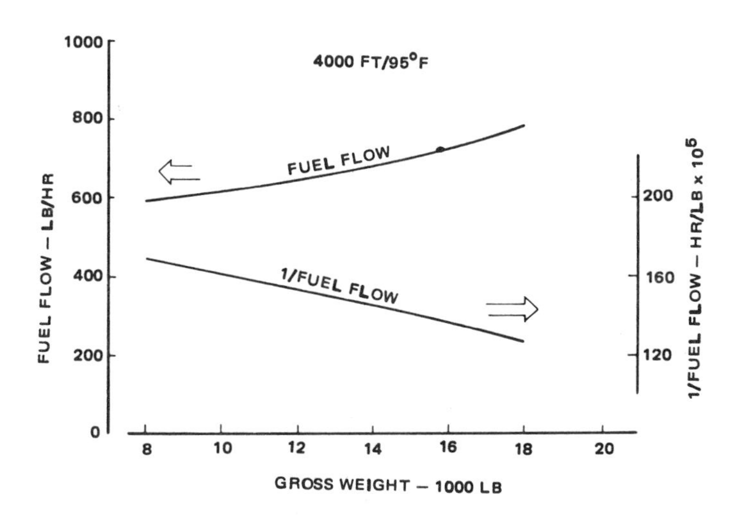

Figure 3.55 Maximum endurance fuel flow

1. Determine the atmospheric constants σ_p and $\delta\sqrt{\theta}$ at 4000 ft/95°F:

$$\sigma_p = 0.8076;$$
$$\delta\sqrt{\theta} = 0.8932.$$

2. Read power required at 15,000 lb from either Fig 3.24 or 3.28 at 95°F.

3. Calculate $SHP/\delta\sqrt{\theta}$ per engine:

$$SHP/\delta\sqrt{\theta} = 909/(2 \times 0.893) = 509\ hp.$$

4. Determine the $\dot{W}_F/\delta\sqrt{\theta}$ per engine from Fig 1.8:
$$\dot{W}_F/\delta\sqrt{\theta} = 379\ lb/hr.$$

5. Calculate total fuel flow:
$$\dot{W}_F = (\dot{W}_F/\delta\sqrt{\theta}) \times 2.1 \times \delta\sqrt{\theta};\ \text{where } 2.1 = 2 \text{ engines} \times 1.05 \text{ SFC increase};$$
$$\dot{W}_F = 711\ lb/hr.$$

The reciprocal of the fuel flow ($1/\dot{W}_F$) for the hypothetical hleicopter is also plotted in Fig 3.55 as a function of gross weight. The $1/\dot{W}_F$ curve is almost linear. Therefore, endurance capability can be computed in one step by dividing the mission fuel by the average fuel flow as described below.

The payload/endurance capability of the hypothetical helicopter is shown in Fig 3.56.

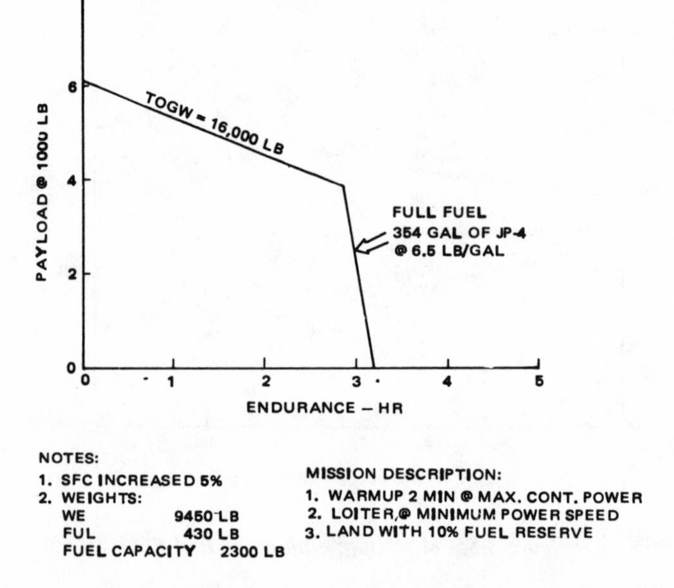

Figure 3.56 Hypothetical helicopter payload/endurance capability

The procedure used to compute this data is very similar to the calculations shown in Table III-5. The only difference in the calculation procedure occurs in steps *17* and *18*, as shown in Table III-6.

PROCEDURE	ZERO RANGE	FULL FUEL CALCULATIONS		
	PL_{max} $(TOGW_{max})$	PL_{max} $(TOGW_{max})$	ZERO PL	$PL = 0.5\,PL_{max}$
17 Fuel Flow @ Avg W (Fig 3.55); lb/hr	–	705	635	668
18 Endurance = *16/17* hr	0	2.89	3.20	3.05

TABLE III-6 PAYLOAD–ENDURANCE CALCULATIONS

7.4 Ferry-Range Capability

Ferry-range capability is the maximum range achievable with zero payload on either internal fuel capacity or with the addition of auxiliary tanks. The delivery of new or refurbished aircraft over extended distances is an example of ferry-range operation. Since there is no payload or cargo, the cabin area can be filled with auxiliary fuel tanks, or external tanks can be added to further increase range capability. These tanks increase the empty weight from 0.3 lb/gal to 1 lb/gal of auxiliary fuel, depending on the type of tank used. Also the drag of external tanks must be accounted for in establishing new power required vs flying-speed relationships. In ferry flights, the aircraft is usually flown in such a way that as the fuel burns off, the cruise altitude is varied in order to retain the maximum mi/lb values.

To secure additional range for a twin-engine aircraft, it may become necessary to shut down one engine in flight, thus forcing the remaining engine to operate at higher power settings with correspondingly lower sfc values. However, this would be done only under emergency conditions where the fuel supply has become critical.

An example of ferry-range performance capability for the hypothetical helicopter is presented as a function of takeoff weight in Fig 3.57. Here, range performance with integral fuel tanks as well as with the addition of internally mounted auxiliary tanks is shown for standard day conditions. The only limitation to the aircraft ferry-range performance is the criteria for establishing the takeoff gross weight. This is determined by the mode of taking off: hover OGE, IGE, or running takeoff (often performed for this type of mission). However, maximum gross weight restrictions must be observed; for example, the hypothetical helicopter at SL/STD has a W_{max} lower than that corresponding to hover OGE, IGE, or running takeoff criteria. The ferry-range capability at maximum weight is 1160 n.mi. Additional calculations at intermediate weights similar to those described below are required to establish the complete ferry-range capability.

(1) *Determination of Optimum Altitude, Speed, and Weight Schedule* – The procedure for optimizing specific range consists of computing (for selected altitudes and gross weights) specific range values corresponding to the best-range-speed of flight and plotting them in the manner shown in Fig 3.58. Although the information presented in this figure is for the hypothetical helicopter, the indicated trends are typical for current helicopter designs. It can be seen that the specific range is significantly better at higher altitudes; due primarily to an increase in turboshaft engine efficiency plus a

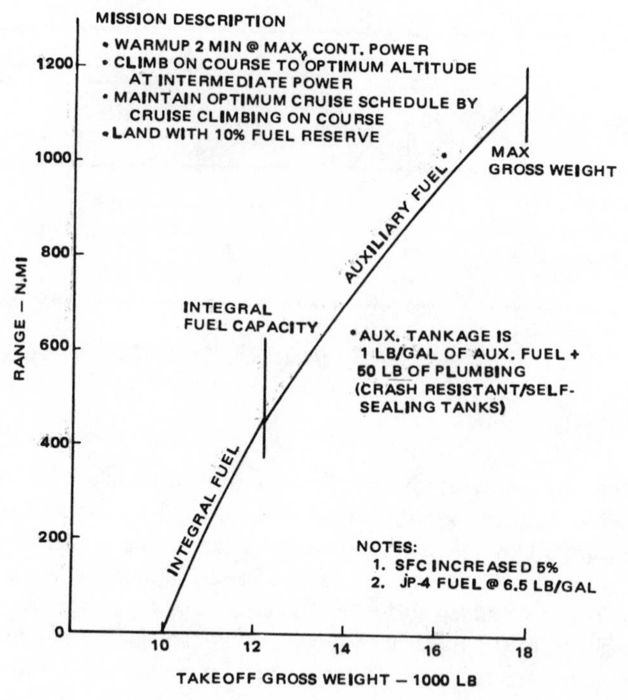

Figure 3.57 Ferry-range capability

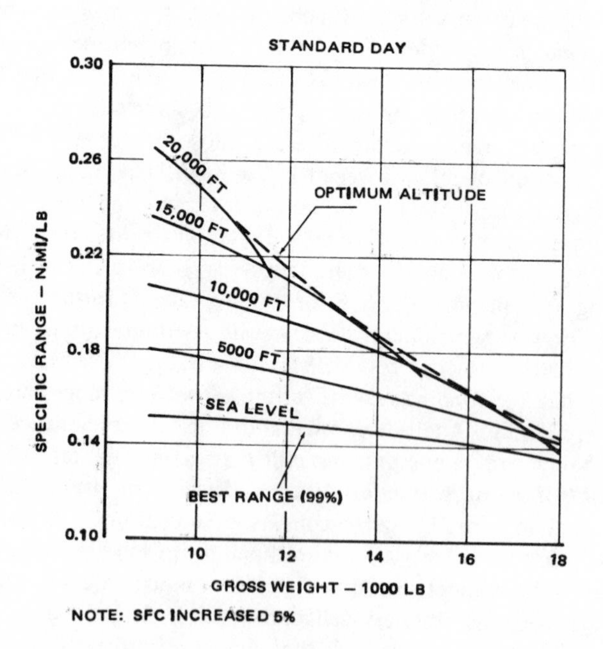

Figure 3.58 Optimum specific range for ferry-range mission

small reduction in power required at low-to-intermediate gross weights. However, at high weights and elevated altitudes, the power required increases rapidly due to stall and compressibility effects, resulting in a large decrease in specific range. The optimum specific range values occur along an envelope tangent to the various constant altitude lines. By plotting the altiutde and airspeed associated with each of the tangent points as a function of gross weight, the optimum cruise-climb schedule is established, as exemplified for the hypothetical helicopter in Fig 3.59. This figure shows that in order to obtain maximum range, the aircraft must increase its cruise altitude(up to 20,000 ft) by 175 ft for every 100 lb of fuel burnoff. The cruise airspeed required is approximately 130 kn.

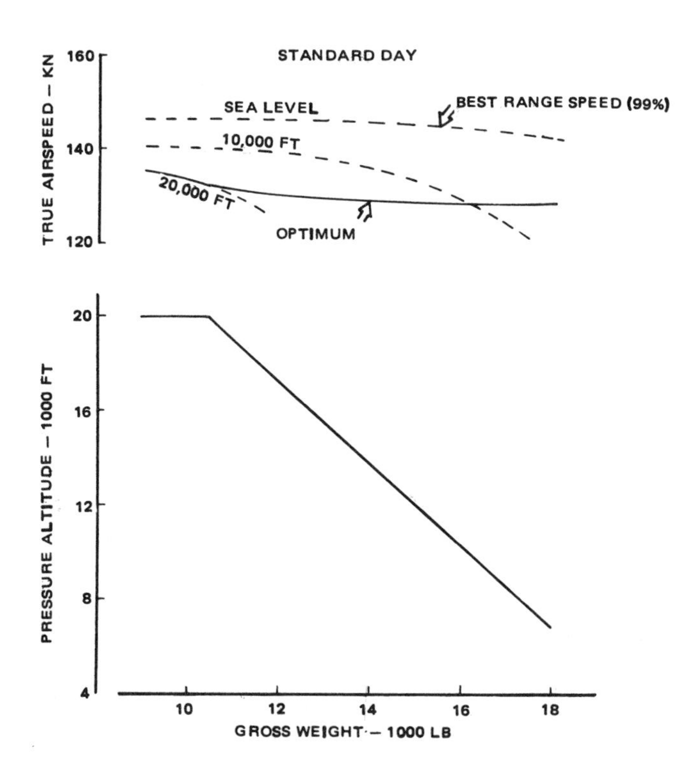

Figure 3.59 Optimum airspeed and altitude for ferry-range mission

(2) *Estimate Initial Climb Fuel and Distance Flown* – Fuel allotments are usually required for climb to the optimum altitude at the beginning of the mission. For example, the hypothetical helicopter operating at a takeoff weight of 18,000 lb must climb to 7000 ft to reach the optimum altitude. At lower takeoff weights, the aircraft must climb to even higher altitudes. To compute the initial climb fuel and distance, it is necessary to know the aircraft rate of climb as a function of both altitude and weight. This performance is usually calculated at either maximum continuous or intermediate power and airspeeds corresponding to minimum power required where the aircraft rate of climb is maximum. This would minimize the time and fuel spent while operating at nonoptimum cruise conditions. The climb fuel and distance calculations are as follows:

$$dt = dh/V_c \tag{3.34}$$

and the initial time to climb (t_{c_i}) from altitude h_1 to h_2

$$t_{c_i} = \int_{h_2}^{h_1} dh/V_c \approx \sum_1^n \left(1/\overline{V}_c\right)_k \Delta h_k \tag{3.35}$$

where

$$\Delta h_k \equiv h_{k+1} - h_k. \tag{3.36}$$

Fuel to climb is

$$W_{F_{c_i}} \approx \sum_1^n \bar{\dot{W}}_{F_k} \Delta t_k \tag{3.37}$$

where

$$\Delta t_k = \Delta h_k / \overline{V}_{c_k}.$$

Ground distance (ℓ) flown during initial climb can be approximated as follows:

$$\ell_{c_i} \approx \sum_1^n \overline{V}_k \Delta t_k; \tag{3.38}$$

where $\overline{V}_k$ is the average speed of flight corresponding to segment k.

An example of climb fuel and distance calculations are given in Table III-7. Data is shown for an intermediate power climb conducted at minimum-power speed (85 kn) from SL to 7000 ft for a gross weight of 18,000 lb. The calculations are divided into two 3500-ft steps. The total climb fuel required is 66 lb and the distance traveled is approximately 4 n.mi. These values will be used in the sample problem ferry-range calculations.

(3) *Climb Fuel Burned During Cruise* – The optimum specific range data presented in Fig 3.58 is based on level-flight power required and fuel consumption, but does not include the additional fuel needed to climb while cruising. As shown in Table III-8, this climb fuel (38 lb) is sufficiently small that it can be computed using the mission mid-point weight (average weight) and the associated specific range ($\overline{SR}$), while the average rate of climb in cruise is obtained from the altitude gained and flight time. The average increase in power required to achieve this rate of climb (ΔSHP) is computed using the potential energy method. The incremental fuel flow needed to climb during cruise $(\Delta \dot{W}_F)_{cr}$ can then be calculated by multiplying the ΔSHP times the average slope of the engine fuel flow vs SHP curve (β):

ALTITUDE – FT	0	3500	7000
FUEL FLOW:			
$\delta\sqrt{\theta}$	1.0	0.875	0.76
SHP_{av} (INSTALLED)	2900*	2900*	2680
$SHP/2(\delta\sqrt{\theta})$	1450	1657	1763
$\dot{W}_F/2(\delta\sqrt{\theta})$: *Generalized Fuel Flow per Fig 1.8;* LB/HR	741	831	880
TOTAL $\dot{W}_F = (\dot{W}_F/2\delta\sqrt{\theta}) \times 2.1 \times \delta\sqrt{\theta}$;LB/HR	1556	1527	1404
TRUE AIRSPEED; KN	85	85	85
RATE OF CLIMB:			
SHP_{req} (Fig 3.28)	1090	1120	1200
$\Delta SHP = SHP_{av} - SHP_{req}$	1810	1780	1480
$V_c = (\Delta SHP \times 33{,}000 \times 0.85)/W$; FPM	2820	2770	2300

AVERAGE ALTITUDE	1750	5250
AVERAGE FUEL FLOW; LB/HR	1542	1466
AVERAGE RATE OF CLIMB; FPM	2795	2535
AVERAGE TRUE AIRSPEED; KN	85	85
Δ ALTITUDE; FT	3500	3500
Δ FUEL; LB	32.2	33.7
Δ TIME; MIN	1.25	1.38
Δ DISTANCE; N.MI	1.77	1.96

SUMMATION	
Σ FUEL; LB	66
Σ TIME; MIN	2.6
Σ DISTANCE; N.MI	3.7

CONDITIONS: 18,000 LB GROSS WEIGHT
INTERMEDIATE POWER/MIN POWER SPEED
STANDARD DAY
SFC INCREASED 5 PERCENT

* TRANSMISSION LIMITED

TABLE III-7 TIME, FUEL, AND DISTANCE TO CLIMB CALCULATIONS (SL TO 7000 FT)

WEIGHTS (LB)	
(1) *Weight Empty*	9,450
(2) *Aux. Tank Wgt (1 lb/gal + 50 lb)*	820
(3) *(2) Crew @ 200 lb ea + Trapped Liquids*	430
(4) *Fuel (Total)*	7,300
(5) *TOGW*	18,000
(6) *2-min Warmup @ Max Cont Power*	47
(7) *W = (5) − (6)*	17,953
(8) *Climb to Opt. Alt. (7000 ft – See Table III-7)*	66
(9) *W = (7) − (8)*	17,887
(10) *1/2 of Cruise Fuel = (26) /2*	3,228
(11) *Average Mission Wgt (9) − (10)*	14,659
(12) *1/2 of Cruise Fuel = (26) /2*	3,229
(13) *Landing Weight = (11) − (12)*	11,430
(14) *10% Fuel Reserve*	730
(15) *W = (13) − (14)*	10,700
(16) *WE + FUL = (1) + (2) + (3)*	10,700

FUEL ANALYSIS (LB)	
(17) *Total Fuel + Aux. Tank Wgt = (5) − [(1) + (3)]*	8,120
(18) *Integral Fuel*	2,300
(19) *Aux Fuel + Aux Tank Wgt = (17) − (18) (Table III-7)*	5,820
(20) *Aux. Fuel**	5,000
(21) *Total Fuel*	7,300
(22) *Warmup 2 min @ Max Cont Power = (6)*	47
(23) *Initial Climb Fuel = (8)*	66
(24) *Reserve (10% of (21))*	730
(25) *= (22) + (23) + (24)*	843
(26) *Total Cruise Climb Fuel = (21) − (25)*	6457

**Assuming a Fuel Weight of 6.5 lb/gal and aux. tank wgt = 1 lb/gal + 50 lb, then (20) + ((20) /6.5) x 1.0 + 50 = 5820; therefore, (20) = 5000 lb.*

CLIMB FUEL (CRUISE)	
(A) *Total Cruise Climb Fuel (lb)*	6,457
(B) *Δ Altitude per Fig 3.59 (ft)*	11,000
(C) *Avg Altitude (ft)*	12,500
(D) *Avg Gross Wgt (lb) = (11)*	14,659
(E) *Avg S.R. per Fig 3.58 (N.Mi)*	0.179
(F) *Avg Cruise Speed per Fig 3.59 (kn)*	129
(G) *Distance = (A) x (E) (N.Mi)*	1,576
(H) *Time = (G) / (F) (hr)*	8.97
(I) *V_c = (B) /((H) x 60) (fpm)*	20.4
(J) *ΔSHP = ((D) x (I))/(33,000 x 0.85)*	10.7
(K) *$\dot{W}_F$ = (F) / (E) (lb/hr)*	721
(L) *$\delta\sqrt{\theta}$ @ Avg Altitude (C)*	0.593
(M) *$\dot{W}_F/\delta\sqrt{\theta}$ x 2.1*	579
(N) *$SHP/\delta\sqrt{\theta}$ (Fig 1.8)*	1,060
(O) *β = Slope of $\dot{W}_F$ curve at (N)* *$\beta = (\Delta\dot{W}_F/\delta\sqrt{\theta})/(\Delta SHP/\delta\sqrt{\theta})$*	0.4
(P) *$\Delta\dot{W}_F$ = (O) x (J) (lb/hr)*	4.3
(Q) *Fuel = (P) x (H) (lb)*	38
(R) *Total Cruise Fuel less Climb Fuel = (A) − (Q) (lb)*	6,419

RANGE CALCULATIONS

W_1 LB	W_2 LB	ΔW LB	AVG W; LB	AVG *SR* Fig 3.60	RANGE N.MI.
17887	15887	2000	16887	0.155	310
15887	13887	2000	14887	0.176	352
13887	11468	2419	12670	0.204	494

INITIAL CLIMB DIST	4 N.MI.
TOTAL RANGE	1160 N.MI.

NOTES: 1. STD DAY CONDITIONS
2. SFC INCREASED 5%

TABLE III-8 DETAILED FERRY RANGE CALCULATIONS

$$(\Delta W_F)_{cr} = \beta \Delta SHP. \tag{3.39}$$

Finally, the climb fuel in cruise is computed by multiplying $(\Delta \dot{W}_F)_{cr}$ by the cruise time (t_{cr}), where $t_{cr} \approx (W_F \times \overline{SR})/\overline{V}$, and $\overline{V}$ is the average cruise speed. This fuel increment is then set aside and is not used for distance calculations.

(4) *Detailed Ferry-Range Calculations* – An example of step-by-step calculations for the following ferry-range mission is presented in Table III-8.

1. Warm up for 2 minutes.
2. Climb to optimum altitude at intermediate engine power with the speed of flight corresponding to minimum power required.
3. Maintain optimum altitude schedule by continuously climbing when in cruise.
4. Land with 10 percent fuel reserve.

Calculations are presented for standard day conditions and $TOGW$ = *18,000 lb*. Auxiliary fuel is added, assuming an auxiliary tankage weight empty of 1 lb/gal (typical of self-sealing, crash-resistant internal tanks) plus 50 lb for plumbing (pumps, lines, etc.).

The tabulations shown in Table III-8 are divided into; a weight and fuel analysis, cruise climb fuel analysis, and range calculations. Subtracting the relatively small 38 lb of cruise climb fuel from the total cruise fuel gives an equivalent level flight quantity of 6,419 lb. This weight is then combined with the level flight specific range data to compute range by integrating the area under the optimum specific range curve between the initial gross weight of 17,887 lb , and the final weight of 11,468 lb after burnoff. The total amount of fuel is so large that it appears advisable to divide the integrated area into two 2000- and one 2419-lb segments, and summing the range increments. The final step in the calculations is to add the ~4 n.mi. initial climb distance from Table III-7 to the cruise distance of 1156 n.mi. for the rotal ferry range of 1160 n.mi.

7.5 Speed Capability

The level flight speed capability of helicopters is determined by matching the power available with the power required while observing transmission and rotor-stall limits. Maximum continuous power available is used to compute maximum normal speeds while intermediate, or 30-min power ratings are used to define dash speed capability. An example of maximum normal speed for the hypothetical helicopter at 4000 ft/95°F and SL/STD ambient conditions as a function of gross weight is shown in Fig 3.60. It should be noted that the aircraft is primarily limited by the maximum continuous power at both ambient conditions.

7.6 Forward-Flight Climb Capability

Forward-flight climb performance is determined from the potential energy relationship discussed in Ch III, Sect 5:

$$V_c = \Delta SHP 33{,}000 k_{p_c} / W \tag{3.40}$$

where

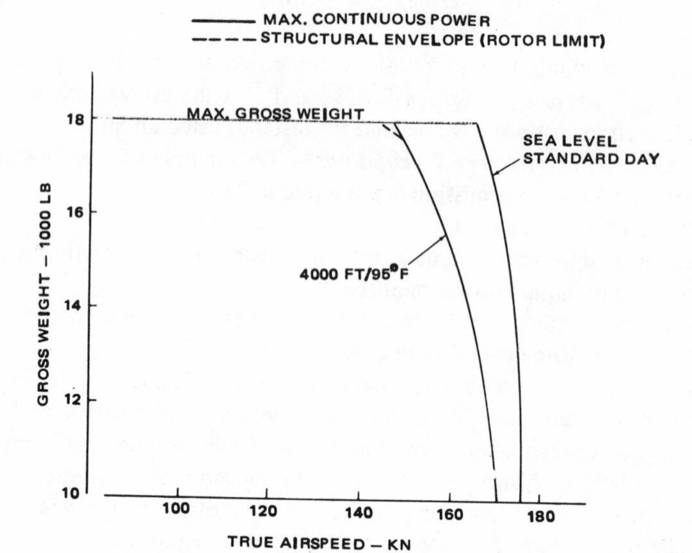

Figure 3.60 Speed capability

V_c = rate of climb; fpm

$\Delta SHP = SHP_{av} - SHP_{req}$

k_{p_c} = climb efficiency factor.

The highest Δ *SHP* values occur at speeds of 70 to 90 kn. Consequently, most climb performance is calculated in this speed range. Maximum continuous power is used for dual engine normal operation, and intermediate power for emergency one-engine-inoperative (*OEI*) conditions. Examples of both dual and single-engine climb performance capabilities of the hypothetical helicopter are presented in Fig 3.61. Using k_{p_c} = *0.85* and the minimum power required given in Fig 3.28, the rate of climb was computed as a function of gross weight at SL/STD and 4000 ft/95°F. In the latter case, the aircraft has a 1100-fpm dual-engine max. cont. power climb capability at a weight of 18,000 lb. Detailed sample calculations for this point are presented below.

1. Determine density ratio at 4000 ft/95°F (see Ch 1, Sect 4):

$$\sigma_\rho = 0.8076.$$

2. Calculate referred weight:

$$W/\sigma_\rho = 22{,}288\ lb.$$

3. Using the data in Fig 3.28, determine referred rotor horsepower required:

$$(RHP/\sigma_\rho)_{req} = 1360\ hp.$$

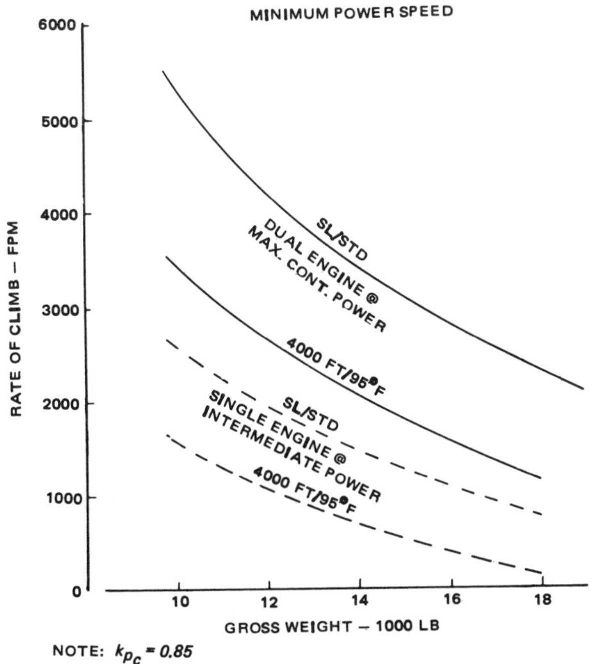

Figure 3.61 Forward flight climb capability

4. Calculate SHP_{req}:

$$SHP_{req} = [(RHP/\sigma_p) \times (\sigma_p/0.98)] + 30 = 1150\,hp.$$

5. From Fig 1.6, determine uninstalled shaft horsepower and then compute installed shaft horsepower available:

$$SHP_{av} = 0.99 SHP_{uninst}.$$

6. Compute excess shaft horsepower:

$$\Delta SHP = SHP_{av} - SHP_{req} = 705\,hp.$$

7. Finally, using Eq (3.40),

$$V_c = 1100\,fpm.$$

7.7 Service Ceiling

Service ceiling (h_{serv}) is the altitude at which the maximum rate of climb is reduced to 100 fpm. Standard day $h_{serv} = f(W)$ of the hypothetical helicopter is shown in Fig 3.62. This figure shows that at $W = 15{,}000\,lb$, $h_{serv} = 19{,}200\,ft$ when both engines are at maximum continuous power, and drops to $h_{serv} = 13{,}700\,ft$ with one engine operating at intermediate power.

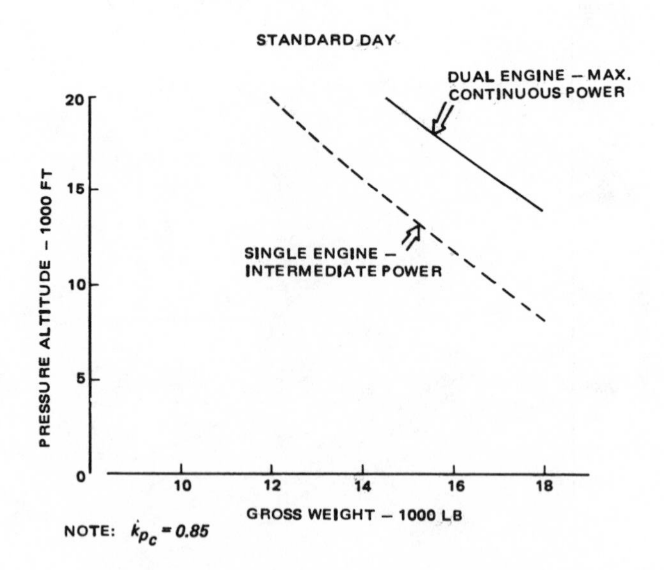

Figure 3.62 Service ceiling vs gross weight

The procedure for calculating service ceiling performance consists of computing the incremental power required to climb at 100 fpm, using Eq (3.23) and adding this increment to the level flight minimum power required plot shown in Fig 3.28. The power available is then compared with power required for various altitudes to obtain the service ceiling gross weight capability at each altitude. In actual operation, altitude restrictions other than power limitations (e.g., structural limits, excessive vibration, and flying qualities) may define the service ceiling. Determination of gross weight vs pressure altitude corresponding to the single-engine service ceiling is shown in Table III-9.

PRESSURE ALTITUDE: FT	0	5,000	10,000	15,000	20,000
TEMPERATURE; °F	59.0	41.2	23.3	5.5	−12.3
σ_ρ	1.0	0.862	0.738	0.629	0.533
SHP_{av} (FIG 1.6)	1584	1405	1236	1072	924
RHP_{av} $0.98(SHP_{av} - 30)$	1521	1349	1181	1021	876
RHP/σ_ρ	1521	1565	1601	1623	1644
W/σ_ρ; LB (FIG 3.28)	22,820	22,900	22,950	22,700	22,600
W; LB	22,820	19,740	16,940	14,280	12,050

NOTE: STANDARD DAY CONDITIONS/INTERMEDIATE POWER
MINIMUM POWER SPEED
ONE-PERCENT ENGINE INSTALLATION LOSSES

TABLE III-9 SINGLE–ENGINE SERVICE CEILING CAPABILITY

7.8 Autorotation

Steady-state autorotational rates of descent (V_d) in forward flight (engines inoperative) are computed using Eq (3.40), where $\Delta SHP = SHP_{req}$:

$$V_d = (SHP \times 33{,}000 \times k_{p_d})/W; \text{ in fpm.} \tag{3.41}$$

Typically, $k_{p_d} \approx 1.0$ in autorotation, as discussed in Sect 5.2 of this chapter.

An example of autorotational rates of descent at $W = 15{,}000\ lb$ and SL/STD conditions is shown in Fig 3.63. Here, V_d is plotted as a function of airspeed, and it can be noted that the $V_d = f(V)$ curve has the shape of a speed power polar with the minimum $V_d = 2000\ fpm$ occurring at minimum-power speed.

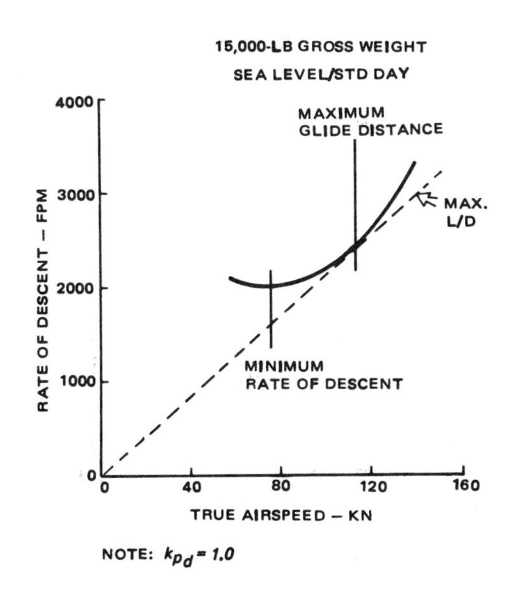

Figure 3.63 Autorotational rate of descent

The glide ratio (horizontal distance flown to altitude lost) is equal to the ratio of the horizontal component of the speed of flight to rate of descent (Fig 3.63). It can be seen that the maximum glide ratio is obtained at a speed of flight representing an abscissa of the point of a tangency of a straight line drawn from the origin of coordinates to the $V_d = f(V)$ curve. For the hypothetical helicopter, the optimum glide speed is 113 kn, or 37 kn above the speed for minimum rate of descent.

The airspeed for maximum glide distance is also the speed for maximum total aircraft L/D_e, where lift is equal to gross weight and equivalent drag, $D_e = 550SHP_{req}/V$. This is shown by solving Eq (3.41) for SHP_{req} and substituting it into the drag expression. The resulting equation is

$$L/D_e = (V/V_d)k_{p_d}$$

where V and V_d are of the same units. It can be seen that L/D_e is proportional to the aircraft glide slope; therefore maximum L/D_e occurs at the speed for maximum glide distance.

References for Ch III

1. Hoerner, S.F. *Fluid Dynamic Drag.* Published by the author, 1965 edition.

2 Gabriel, E. *Drag Estimation V/STOL Aircraft.* Boeing Vertol Report D8-2194-1, September 1968.

3. Keys, C. and Wiesner, R. *Guidelines for Reducing Helicopter Parasite Drag.* Included in AHS Report of AHS Ad Hoc Committee on Rotorcraft Drag. 31st Annual Forum, Washington, D.C. May 1975.

4. Julian, D. *HLH Hub Drag Review.* Boeing HLH Report D301-10100-3, April 1972.

5. Delany, N.K. and Sorensen, N.E. *Low-Speed Drag of Cylinders of Various Shapes.* NACA TN-3038, November 1953.

6. Perkins, C.D. and Hage, R.E. *Airplane Performance Stability and Control.* Published by John Wiley & Sons, Inc., New York. Ninth Printing, May 1963.

7. Biggers, J.L., McCloud, J.L., and Patterakis, P. *Wind-Tunnel Tests of Two Full-Scale Helicopter Fuselages.* NASA TND-1548, October 1962.

8. McDonnell Douglas Corporation. *USAF Stability and Control DATCOM.* AF33/616/6460. Principal Investigator, D.E. Ellison, October 1960.

9. Prouty, R. *Tip Relief for Drag Divergence.* Journal of the Americal Helicopter Society, Vol. 16, No. 4.

10. LeNard, J.M. and Boehler, G.D. *Inclusion of Tip Relief in the Prediction of Compressibility Effects on Helicopter Rotor Performance.* USAAMRDL Tech. Report 73-71, December 1973.

11. LeNard, J.M. *A Theoretical Analysis of the Tip Relief Effect on Helicopter Rotor Performance.* USAAMRDL TR 72-7, August 1972.

12. Davis, S. Jon and Wisniewski, J.S. *User's Manual for HESCOMP, the Helicopter Sizing and Performance Computer Program.* Boeing Vertol Report D210-10699-1 (Contract NAS2-6107), NASA CR-152018, Ames Research Center, Moffett Field, Ca., Sept. 1973.

13. McCormick, B.W., Jr. *Aerodynamics of V/STOL Flight.* Academic Press, 1967.

14. Washizu, Kyuichiro, et al. *Experiments of a Model Helicopter Rotor Operating in the Vortex Ring State.* Journal of Aircraft, Vol 3, No 3, May-June 1966.

15. Dadone, Leo V. and Fukushima, T. *A Review of Design Objectives for Advanced Helicopter Rotor Airfoils.* Presented at the American Helicopter Symposium on Helicopter Aerodynamic Efficiency, Hartford, Conn., March 1975.

16. Benson, R.G., et al. *Influence of Airfoils on Stall Flutter Boundaries of Articulated Helicopter Rotors.* Presented at the 28th Annual National Forum of the American Helicopter Society (Preprint No 621), May 1972.

17. Hoffstedt, D.J., et.al. *Flight Test of Advanced Geometry Boron Blades.* USAAMRDL TR72-65, D210-10486-1, Eustis Directorate, February 1973.

18. Tanner, W.H. *Charts for Estimating Rotary-Wing Performance in Hover and at High Forward Speeds.* NASA CR-114, November 1964.

19. Ormiston, R. A. *Comparison of Several Methods of Predicting Loads on a Hypothetical Helicopter Rotor.* Journal of the American Helicopter Society, Vol 19, No 4, October 1974.

20. Gabel. R. *Current Loads Technology for Helicopter Rotors.* AGARD Conference Proceedings 122, August 1973.

21. Terzanin, F.J., Jr. *Prediction of Control Loads due to Blade Stall.* Journal of the American Helicopter Society, Vol 17, No 2, April 1972.

22. Gormont, R.E. *A Mathematical Model of Unsteady Aerodynamics and Radial Flow for Application to Helicopter Rotors.* USAAMRDL, TR72-67, D210-10492-1, Eustis Directorate, May 1973.

23. Harris, F.D., Tarzanin, F.J., and Fisher, R.K., Jr. *Rotor High-Speed Performance, Theory vs Test.* Journal of the American Helicopter Society, Vol 15, No 3, July 1970.

24. Wells, C.D. and Wood, T.L. *Maneuverability – Theory and Application.* AHS Preprint No 640, May 1972.

25. Brown, E.L. and Schmidt, P.S. *The Effect of Helicopter Pitching Velocity on Lift Capability.* Journal of the American Helicopter Society, Vol 8, No 4, October 1963.

26 Department of the Army. *Engineering Design Handbook – Helicopter, Part One, Preliminary Design.* AMCP 706-201, U.S. Army Material Command, August 1974.

27. Wiesner, Wayne and Kohler, Gary. *Tail-Rotor Design Guide.* USAAMRDL TR-73-99 D210-10687-1, Eustis Directorate, January 1974.

WINGED HELICOPTER PERFORMANCE

A question that both designers and operators often ask themselves is whether any benefits could be derived from adding a lifting wing to a conventional helicopter. To place this problem in a proper perspective, a complete performance envelope is presented in this chapter wherein a fixed wing was added to the hypothetical helicopter described in the preceding chapters; however, the design gross weight remains the same as for the original configuration.

Principle notation for Chapter IV

Symbol	Description	Units
AR	wing aspect ratio	
b	wing span	ft
C_D	wing, or body drag coefficient	
C_f	skin friction drag coefficient	
$C_L = L_w/\frac{1}{2}\rho V^2 S_w$	wing lift coefficient	
$C_T' = L_R/\pi R^2 \rho V_t^2$	rotor lift coefficient	
c	wing chord	ft
c_d	section drag coefficient	
c_ℓ	section lift coefficient	
$c_{\ell a}$	slope of the lift curve	deg^{-1} or rad^{-1}
D	wing, or body drag	lb
d	rotor diameter	ft
f_e	equivalent flat-plate area	ft^2
g	acceleration of gravity	$32.2 fps^2$
h	height, or altitude	ft
IGE	in-ground-effect	
i	angle-of-incidence	deg or rad
k_f	fillet factor	
k_g	ground-effect factor	
k_k	discrete roughness coefficient	
k_ℓ	longitudinal location factor	
k_p	climb efficiency factor	
k_{pd}	descent correlation factor	
k_t	wetted area factor	
$k_{3\text{-}D}$	three-dimensional drag correction factor	
k_v	vertical load factor	
$k_{v\ell}$	vertical location factor	
k_{wf}	wing fuselage interference drag factor	
L	lift	lb
M	moment	ft.lb
OGE	out-of-ground effect	
q	freestream dynamic pressure	psf
R	rotor radius	ft

Performance

R_e	Reynolds number	
S	wing area	ft^2
T	rotor thrust	lb
$TOGW$	takeoff gross weight	lb
V	velocity of aircraft translation	fps or kn
W	weight	lb
w	width	ft
X	propulsive force (+ forward)	lb
α	angle-of-attack	deg or rad
β	rotor downwash angle	deg or rad
Δ	increment	
δ_F	flap deflection	deg
ζ	distance from rotor disc perimeter	ft
κ	induced drag factor	
ρ	air density	slugs/cu.ft
σ	rotor solidity	

Subscripts

c	climb
e	equivalent, or exposed
F	flap
f	fuselage
i	induced
id	ideal
P	planform
R	rotor
r	root
t	tip
t	total
t	horizontal tail
tpp	tip-path-plane
w	wing
we	wetted

Superscript

$\bullet$	derivative with respect to time

1. INTRODUCTORY REMARKS

There are no winged production helicopters at this time except, perhaps, the Mil-6 in the USSR. So-called "wings" on operational aircraft as the CH-46 Sea Knight and AH-1J Cobra are actually stubs used as support structures for stores or as fuel tanks, and contribute very little to lift in forward flight. Development and testing of experimental machines (e.g., the Sikorsky Black Hawk Model S67 and the Boeing Vertol Model 347) should provide an insight into actual and potential gains, as well as problems associated with the addition of lifting wings for pure helicopters.

A wing was added to the Model 347 (Fig 4.1) primarily as a means of improving maneuver capability; however, the overall performance was adversely affected, and efforts were directed toward minimization of the performance penalties, thus reducing the "price" paid for having a much more maneuverable configuration. Although the Model 347 is a tandem, the vast technical documentation – wind tunnel data and flight test results – acquired during the development of this aircraft is of universal value, and is quoted throughout this presentation.

Figure 4.1 Model 347 winged helicopter in flight

Techniques for estimating wing effects on hover and forward flight are given in this chapter; including sample calculations for a 101-ft^2 wing added to the hypothetical helicopter evaluated in Chs II and III. A brief discussion of sizing a wing to meet a given maneuver requirement and methods of optimizing the level flight rotor unloading are provided, and a direct comparison of both wing-off and winf-on performance is obtained with primary emphasis on estimating the incremental wing effects on performance.

It will be shown later in this chapter that the wing installation would increase the weight empty and, unless tilted, would result in increased download in hover as well as in vertical and near-vertical climb. In high-speed regimes of flight (forward flight, climb and V_{max} capability), benefits of installing a wing would accrue only when the rotor operates on the edge of stall inception, where unloading of the rotor would result in a reduction of power required. It should be recognized that performance improvements can be obtained by combining wing unloading with the provision of auxiliary propulsion and slowing down the main rotor. However, these approaches are outside the scope of this text; consequently, in this chapter, efforts will be restricted to the so-called classic winged helicopters.

2 DESCRIPTION OF THE WINGED HELICOPTER

The geometry and primary physical characteristics of the hypothetical winged helicopter are illustrated in Fig 4.2. It is assumed that the design gross and maximum weights of the winged configuration are identical to that of the pure helicopter (see Ch I). However, the weight empty is increased by 350 lb, which includes the weight of the wing (3.5 lb/ft² of wing planform area)[1] and the required fuselage structural modifications.

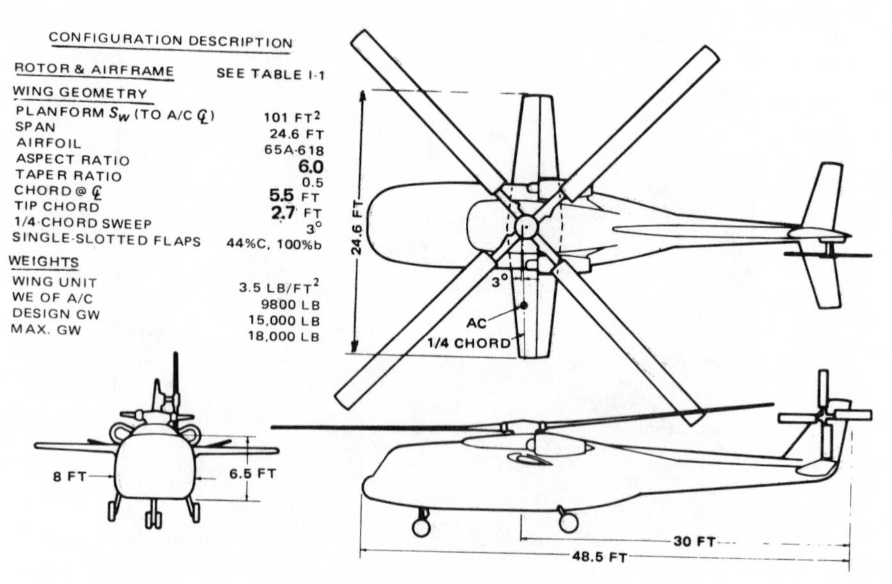

Figure 4.2 Hypothetical winged helicopter

A brief outline of the details involved in defining the hypothetical wing geometry is presented below.

2.1 Planform Area/Flap Geometry

The wing planform area, including the projected area in the fuselage cutout region is $S = 101\ ft^2$. The wing was sized to provide 2-g, or a 60° banked turn maneuver capability at airspeeds from 100 to 170 kn at 4000 ft/95°F (Fig 4.3). Due to rotor stall limits, the pure helicopter has a less than 2-g capability over the entire speed range, and this deficiency increases with increasing airspeed. Installation of the 101 ft² wing meets the 2-g criteria at low speeds and exceeds the requirement at 160 to 180 kn.

To satisfy the 2-g requirement without producing excessive download in hover, full-span, 44-percent chord, single-slotted flaps deflected 30° are used as illustrated at the bottom of Fig 4.3. Without flaps, the wing area needed to satisfy maneuver requirements at 125 to 130 kn would amount to 162 ft² versus 101 ft² with flaps. This reduction in area decreases the hover download by 2.6 percent of the gross weight for a wing having an aspect ratio of 6 (Fig 4.4).

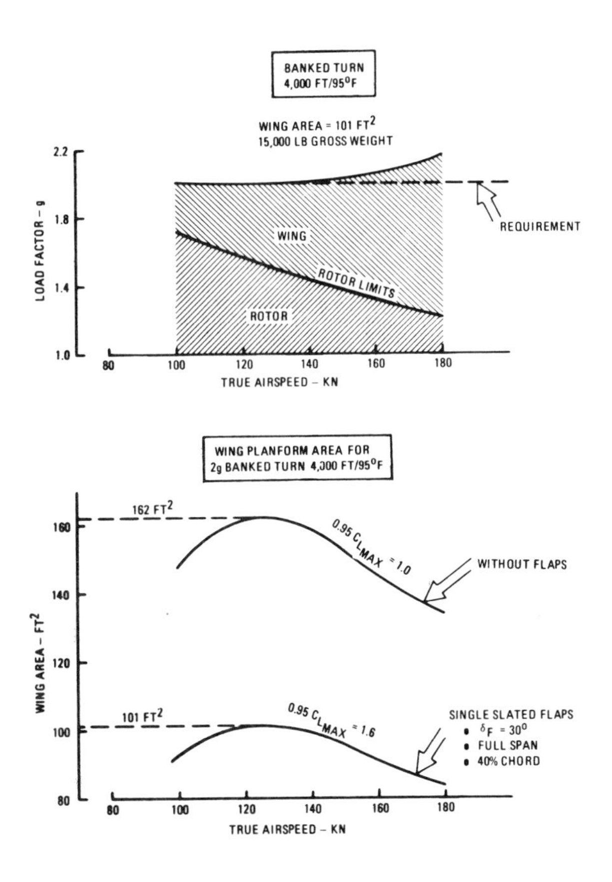

Figure 4.3 Winged helicopter maneuver capability

It can be seen from Fig 4.4 that the flaps, when deflected 80°, provide an additional 0.7 percent reduction in hover download. Other, more complicated, methods of download alleviation are also shown, including a schematic of umbrella installations (later illustrated in Fig 4.11), and wing rotation. However, in order to illustrate the potential magnitude of wing download, flap deflection is assumed as the sole method of reducing this effect.

The flap geometry and aerodynamic characteristics used for the hypothetical helicopter aircraft are based on theoretical and wind-tunnel studies which indicate that flap effectiveness for single-slotted configurations tends to decrease at deflection angles above 30° and for flap chords exceeding 40 to 50 percent of the wing chord. Because of the availability of wind-tunnel data, a 44-percent flap was selected for the hypothetical helicopter. The $C_{L_{max}}$ values used in Fig 4.3 are based on wind-tunnel model test results adjusted for Reynolds number and fuselage cutout effects as shown in Fig 4.5. The Reynolds number correction is based on two-dimensional data[2,3]; and the cutout correction is based on the empirical span calculation described later in this chapter.

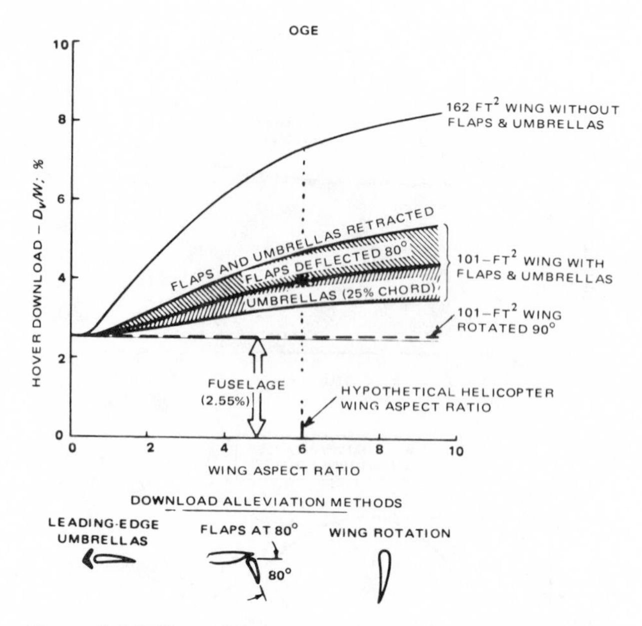

Figure 4.4 Effect of wing geometry on hover download

The design $C_{L_{max}}$ is 95 percent of the true $C_{L_{max}}$ to allow a margin for gusts and to prevent stall buffeting during maneuvers. The flaps can be deflected upwards to decrease wing lift during autorotation, where high wing angles-of-attack ($\alpha \approx 20°$) can compromise autorotation performance and cause roll control problems. Also, flap deflection shifts the wing zero-lift angle-of-attack which, during maneuvers, reduces the fuselage pitch-up attitude to achieve a given wing lift. This provides increased pilot visibility and reduces aircraft deceleration as described later in this chapter.

The method of controlling flap deflections as a function of flight conditions was demonstrated on the Model 347 (Fig 4.6). Vertical acceleration measurements were used to automatically control the flap position to maximize the vehicle g capability. Collective pitch setting was used to position the flaps for autorotational descent. Automatic control of wing incidence and differential flap amgles were employed for download alleviation and roll control. However, as previously stated, wing control of the hypothetical helicopter is limited to flap deflection at a fixed incidence angle.

2.2 Wing Aspect Ratio

An aspect ratio of 6 was selected for the hypothetical helicopter as a reasonable compromise with respect to hover download, induced drag and structural weight. For example, increasing the aspect ratio to 9 would extend the wing further into the higher velocity rotor downwash region and increase the hover download as shown in Fig 4.4. Lowering the aspect ratio below 6 is undesirable because of increased wing induced drag.

Since the hypothetical wing is designed for maximum lift, the question of the effect of aspect ratio on $C_{L_{max}}$ must be answered. Theoretical and test analyses have shown that there is no noticeable variation of $C_{L_{max}}$ with aspect ratio.[4]

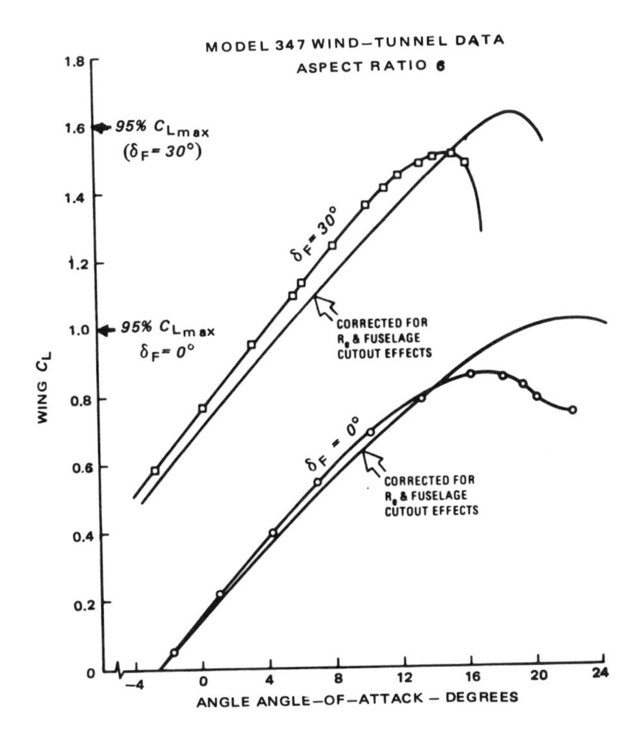

Figure 4.5 Determination of wing $C_{L_{max}}$ from the 1/11-scale Model 347 wind-tunnel data

2.3 Taper Ratio

The taper ratio $c_t/c_r = 0.5$ is based on a root chord (c_r) determined by projecting the wing leading and trailing edges to the fuselage centerline. Wing taper provides considerably more actual wing thickness at the root where bending moments are maximum. $c_t/c_r = 0.5$ also provides a small download benefit by positioning most of the planform inboard. In addition, theoretical induced power calculations[2] indicate that $c_t/c_r = 0.5$ results in the closest approximation to an elliptical lift distribution, required for minimization of induced drag. However, this benefit of taper is relatively small compared to the effect of wing cutout on induced drag as discussed later in this chapter.

In terms of $C_{L_{max}}$, a large taper (low c_t/c_r values) is undesirable because it increases the outboard c_ℓ values at a given total wing C_L, resulting in a small decrease in wing $C_{L_{max}}$[4]. The $c_t/c_r = 0.5$ value therefore, is a compromise between $C_{L_{max}}$ effects, induced drag, and structural requirements.

2.4 1/4–Chord Sweep

The hypothetical wing sweep angle was kept to a minimum (3°) in order to maximize the wing lift capability during maneuvers. Sweep induces spanwise flow which

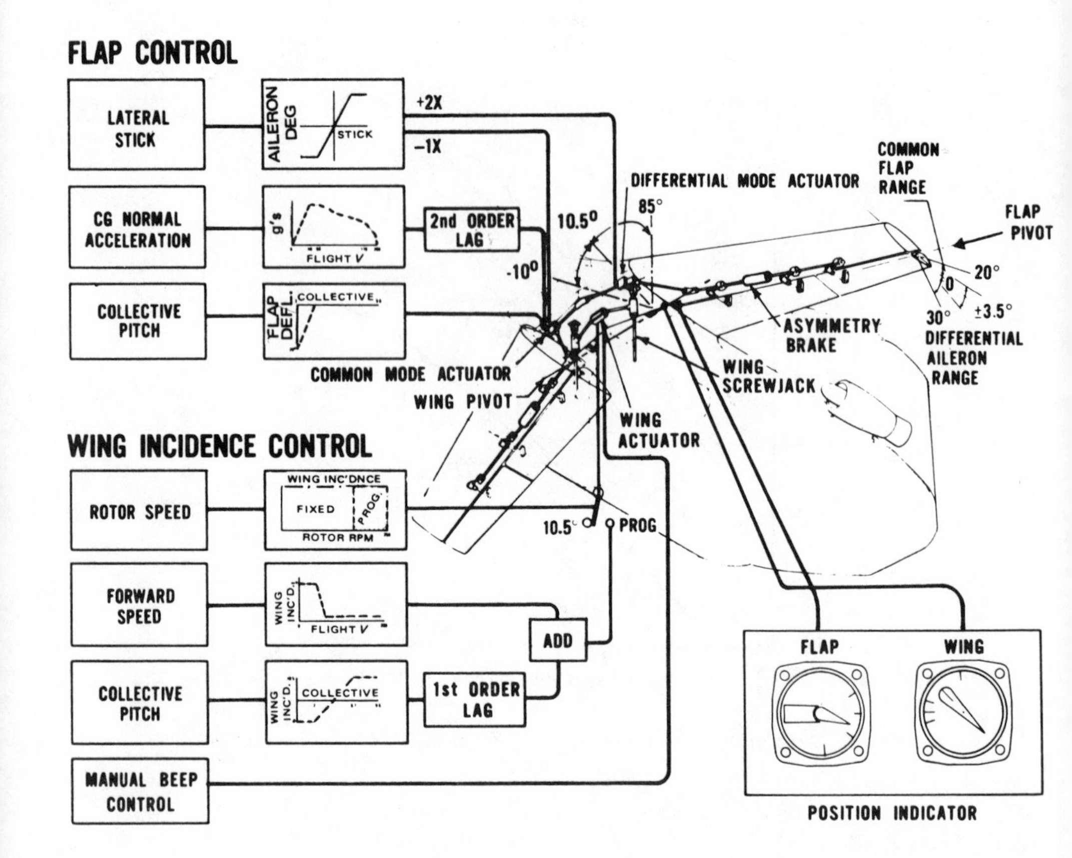

Figure 4.6 Boeing 347 wing flap control systems

causes the boundary layer to build up near the tip, resulting in tip stall. In addition, sweep decreases the wing lift curve slope which is undesirable for rapid maneuvers. The only advantage of sweep for helicopter applications is that it moves the aerodynamic center aft, thus providing a small improvement in static longitudinal stability.

2.5 Wing Location

The wing is positioned on top of the fuselage as shown in Fig 4.2. This is done in order to provide easy access to the cabin and to prevent the wing carry-through-structure and controls from reducing the cabin space. In addition, the high-wing arrangement provides a small reduction in hover download and a small improvement in airframe angle-of-attack stability. However, in this location, the wing is in close proximity to the engine inlet and flow disturbances due to the wing could cause engine performance problems.

In the considered case, the aerodynamic center of the wing is positioned directly beneath the rotor hub. However, if the longitudinal stability of the wing-off configuration is marginal, then the wing aeerodynamic center should be located further aft. This is particularly true for autorotation flight conditions where nose-up pitching must be avoided.

2.6 Airfoil Section

The airfoil for the wing of the hypothetical helicopter is the 65A-618 section, where *A* indicates that the trailing edge cusp, present in the 65_3–618, was removed to simplify the manufacture of the wing[2]. (Fig 4.7). The aerodynamic characteristics leading to the selection of this airfoil are listed below:

Figure 4.7 65A-618 airfoil section

(1) Low profile drag at trim c_ℓ, ($c_d = 0.007$ at $c_\ell = 0.4$)
(2) High $c_{\ell_{max}}$, ($c_{\ell_{max}} = 1.5$)
(3) Gentle trailing edge stall characteristics
(4) High lift-curve slope ($c_{\ell_a} = 0.114$ *per deg*)
(5) Sufficient maximum thickness (18 percent) and favorable chordwise thickness distribution for structural efficiency and low weight.

Low drag is desirable for maximum cruise performance, while the high $c_{\ell_{max}}$, gentle stall characteristics and high lift curve slope are required for good maneuver performance. The two-dimensional characteristics of this airfoil are defined for a Reynolds number of 4.7×10^6, corresponding to a 150-kn, 4000 ft/95°F condition.

3. HOVER AND VERTICAL CLIMB PERFORMANCE

The primary effect of a wing on hover OGE, IGE, and vertical climb performance is an increase in download. The methods used to estimate these effects are identical to the procedures outlined in Ch II. Sample calculations are provided below to illustrate the application of these techniques to the winged aircraft.

3.1 Wing Download OGE

The wing download increment OGE is computed by combining test-measured downwash velocity profiles (Fig 2.19) with the estimated wing drag coefficients (Eqs (2.7) to (2.11). In these equations, the radial stations are measured along the wing span rather than along the fuselage centerline. Detailed sample calculations for the hypothetical helicopter are presented in Table IV-1. As noted, with flaps deflected 80° down, the total download amounts to 3.97 percent of gross weight, or 1.42 percent more than for the baseline aircraft without wings. At 4000 ft/95°F, this penalty is equivalent to a 227-lb reduction in hover OGE gross-weight capability.

STEP	①	②	③	④	⑤	⑥	⑦	⑧	⑨
ITEM	ζ (FT)	ζ/R (%)	$k_{v_{n+1}}$	k_{v_n}	Δk_v	C_{D_v}	w (FT)	C_{D_v} w/4πR	$\Delta D_v/T$ (%)
CALCULATION PROCEDURE		① /25	SEE FIG 2.19		③-④	SEE FIG 4.8	SECTION WIDTH	⑥ × ⑦ /4π × 25	2[⑧ × ⑤]
1. WING TIP	12.5	50	203	134	69	1.06	2.78	0.009385	1.30
2. WING ROOT	20.6	82.4	205	203	2	0.7	3.75	0.008360	.03
3. FUSELAGE CENTERLINE	25.0	100	–	205	–	–	–	–	–

WING D_v/T	1.33
FUSELAGE D_v/T	2.49*
TOTAL D_v/T	3.82
TOTAL D_v/W	3.97

*TABLE II-1

FUSELAGE ℄

BLADE TIP

HUB ℄

R

ζ_2

ζ_1

TABLE IV-1 HOVER OGE DOWNLOAD CALCULATION ($\delta_F = 80°$)

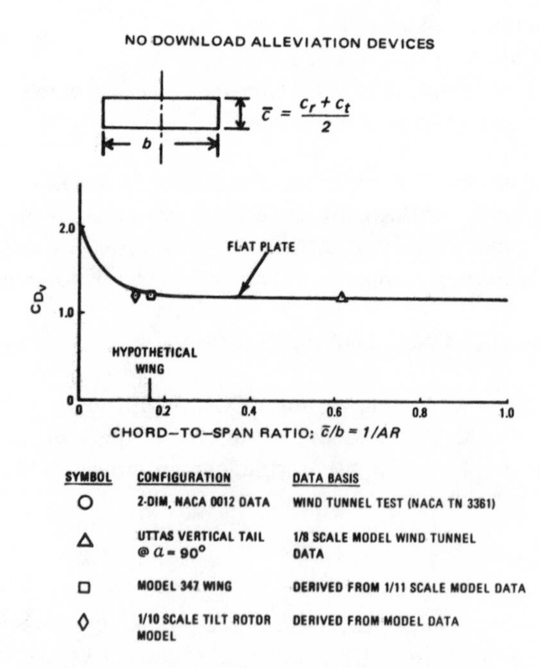

Figure 4.8 Effect of wing aspect ratio on C_{D_v}

The wing vertical drag coefficient used in these calculations depends on the type of alleviation devices employed. In the simplest case of a plain wing, the C_{D_v} is close to the drag of a flat plate normal to the freestream velocity, as shown in Fig 4.8. The drag coefficient of wings with an aspect ratio of $AR \leq 6$ is $C_{D_v} = 1.2$. At $AR \geq 6$, the drag coefficient increases rapidly, becoming 2.0 in the two-dimensional case.

The effect of download alleviation methods on the wing vertical drag coefficient is shown in Fig 4.9. Drag coefficients for flaps deflected 80°, leading edge umbrellas and wing rotation presented in this figure are based on wind-tunnel tests of the 1/11-scale Model 347 and tilt-rotor 1/10-scale models. As noted, the 80° flap deflection reduced the drag coefficient from 1.2 to approximately 1.0. Flap deflections between 70° and 80° represent the minimum download settings, as illustrated by the tilt-rotor test results[5] shown in Fig 4.10. Flap deflections of this magnitude also reduce the projected wing area. For example, deflecting the 44-percent chord flaps 80° reduces the planform area by approximately 30 percent.

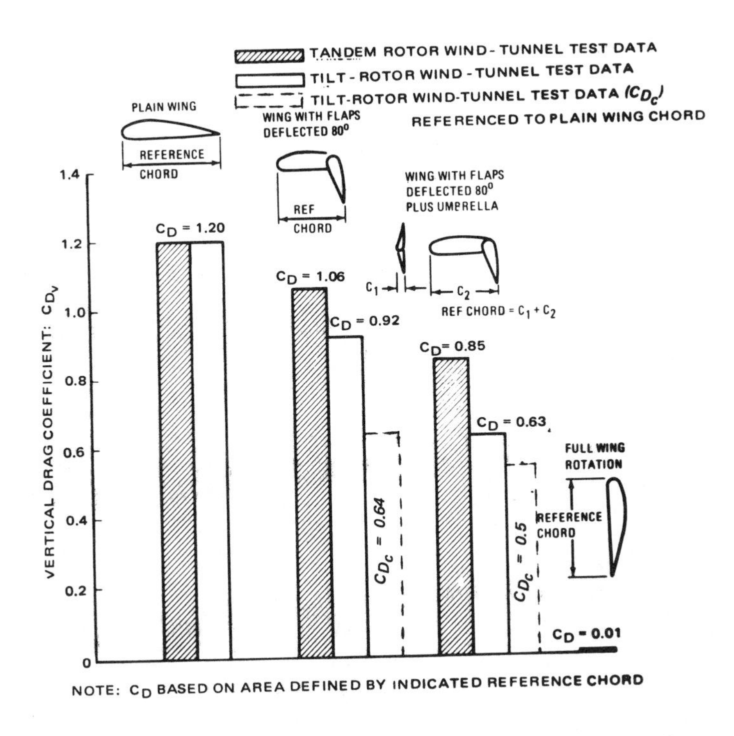

Figure 4.9 Effect of download alleviation methods on C_{D_v}

As indicated in Fig 4.4, the combined effect of flap deflection on C_{D_v} and planform area is to decrease the hypothetical helicopter wing download penalty by 40 to 50 percent. In Table IV-1, C_{D_v} was used to compute the download of the hypothetical wing. In the cutout region, $C_D = 0.7$ was assumed, which is the incremental increase in C_D from the fuselage value (0.4) to the winged configuration ($C_D = 1.1$).

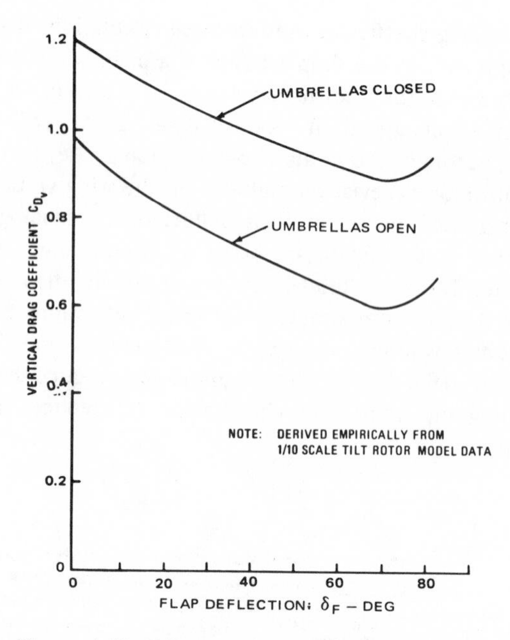

Figure 4.10 Effect of flap deflections on C_{D_V}

By combining the 80° flap deflection with leading-edge umbrellas, the wing drag coefficient can be further reduced. As shown in Fig 4.9, C_{D_V} as low as 0.63 has been measured for this configuration. The improvement in flow conditions below the wing due to the umbrella and flaps is evident in the flow visualization photographs in Fig 4.11. Umbrella installations, when open, typically reduce the exposed wing chord by 15 to 20 percent. The net effect of the combined umbrella and flap deflection is to reduce the wing download penalty by approximately 70 percent, as noted in Fig 4.4.

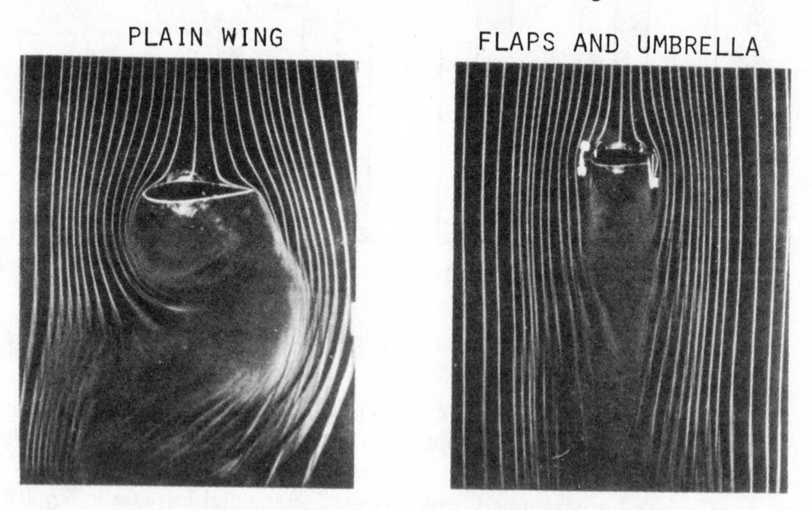

Figure 4.11 Two dimensional smoke studies of a wing

The other download alleviation method (Fig 4.9) is to rotate the wing a full 90°. The C_{D_V} for this wing geometry is $C_D = 0.01$. This technique essentially reduces the wing download to zero (Fig 4.4).

3.2 Wing Download IGE

Due to the increase in local pressure on the lower surface of the wing, the incremental hover download decreases considerably as the aircraft descends from OGE to IGE conditions. Test data for a winged helicopter is not available to define this reduction; however tilt-rotor studies indicate that the fuselage download reductions shown in Fig 2.24 will give a conservative estimate of wing download in ground effect. It can also be seen from this figure that the ratio of IGE to OGE download (k_g) varies with the height of the fuselage or wing above the ground. For example, at a wheel height of 5 ft, the hypothetical wing download correction factor (k_{g_W}) is 0.25 and the corresponding fuselage factor $k_{gf} = 0.09$. The total download factor for the combined fuselage plus wing configuration (k_{g_t}) is

$$k_{g_t} = (\bar{D}_{v_f} + \bar{D}_{v_w})_{IGE} / (\bar{D}_{v_f} + \bar{D}_{v_w})_{OGE} \tag{4.1}$$

where

$\bar{D}_{v_f}$ = relative fuselage vertical drag $\equiv D_{v_f}/W$
$\bar{D}_{v_w}$ = relative wing vertical drag $\equiv D_{v_w}/W$
W = gross weight.

Substituting the OGE download values for the hypothetical aircraft shown in Table IV-1, and the IGE factors described above into Eq (4.1),

$$k_{g_t} = \frac{(0.09 \times 0.0255) + (0.25 \times 0.0142)}{0.0397} = 0.1472.$$

As described in Ch II, IGE performance is computed by correcting the OGE gross weight calculations by the factor W_{IGE}/W_{OGE} determined by rearranging Eq (2.14). The other variables in this equation are k_g, $(D_v/W)_{OGE}$, and thrust ratio T_{IGE}/T_{OGE}, where the thrust ratio is 1.14, as shown in Fig 2.22. Substituting the winged helicopter parameters k_{g_t} and D_v/W in Eq (2.14):

$$\frac{W_{IGE}}{W_{OGE}} = 1.14\left[\frac{1 + 0.0397}{1 + (0.1472 \times 0.0397)}\right] = 1.178.$$

The winged helicopter gross weight ratio, therefore, is 0.8 percent higher than the baseline wingless design ($W_{IGE}/W_{OGE} = 1.17$).

3.3 Hover Ceiling

The OGE and IGE hover ceiling performance at 95°F, with and without the 101-ft² wing installed, is shown in Fig 4.12. These results were obtained using the download corrections defined above and the rotor performance computed in Ch II. The winged

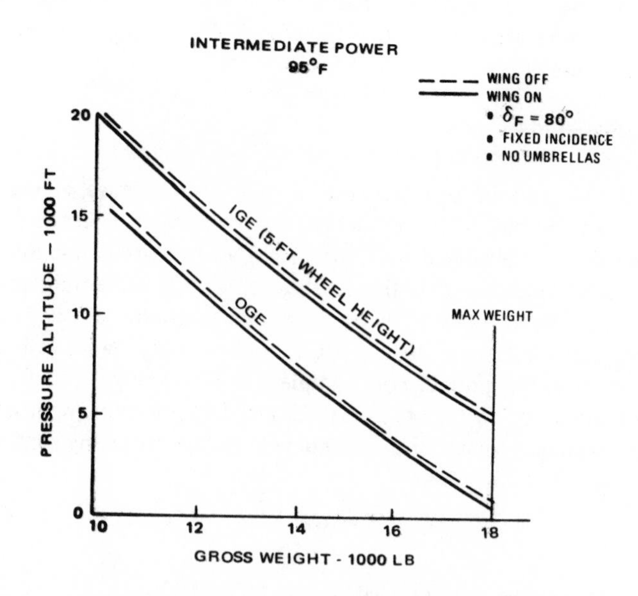

Figure 4.12 Hover ceiling IGE and OGE

configuration includes flaps deflected 80°. It can be seen that the wing causes either a 500-ft loss of altitude or a 200-lb reduction in hover gross weight when hovering OGE. For hover IGE, the losses are 250 ft, or 100 lb.

3.4 Vertical Climb Performance

In vertical climb, the primary effect of installing a wing is to increase the download resulting from the downwash velocity component. At small rates of climb ($V_c < 1000$ *fpm*), the hypothetical winged helicopter will have a 1.42 percent less gross weight capability than shown in Fig 2.26. At 4000 ft/95°F, this is equivalent to a 150-fpm reduction in vertical climb performance.

For a more accurate assessment of wing download effects at higher rates of climb ($V_c > 1000$ *fpm*), the incremental wing download must be added to the fuselage download calculations shown in Table II-4 to determine new factors, k_1, k_2 and k_3. The revised factors are then substituted into Eq (2.27) to compute the climb gross weight, as illustrated in Table II-3.

4. FORWARD FLIGHT PERFORMANCE

A detailed discussion of winged helicopter level flight power required is presented in this section along with an analysis of the wing's effect on climb and descent performance, as well as maneuver capability. Much of the data is based on Vertol winged helicopter wind-tunnel and flight-test programs, and test results summarized by Lynn[6].

4.1 Effect of Wings on Parasite Drag/Power

A summary drag breakdown for the hypothetical wing configuration is presented in Fig 4.13. Wing drag, consisting of wing profile drag, induced drag and rotor/wing interference effects, is shown in this figure as a function of airspeed and wing C_L for a 15,000-lb gross weight aircraft at 4000 ft/95°F. The largest component is the rotor-on-wing interference drag which, at $C_L = 0.4$, accounts for 50 percent of the wing drag at 150 kn. At lower speeds, the interference drag represents an even larger percentage of the total wing drag. As noted in the lower half of this figure, the interference drag at 150 kn does not vary significantly for C_L values above 0.6, while the wing induced drag increases proportionally to $C_L{}^2$. An evaluation of each of these drag components is presented on the following pages.

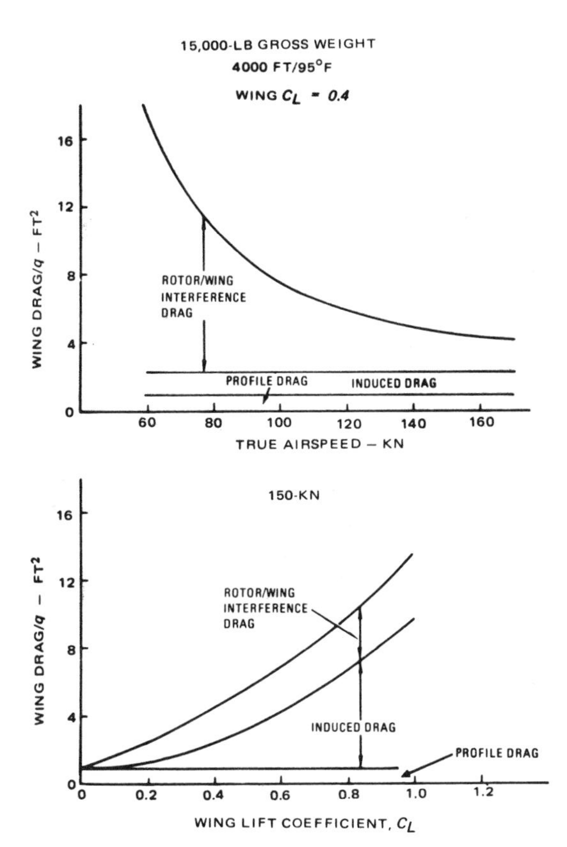

Figure 4.13 Effect of wing C_L and airspeed on wing drag

Wing Profile Drag – The wing profile drag consists of the basic skin friction of the exposed area and pressure drag of the airfoil section, fuselage/wing interference drag, the drag increment due to gaps and tracks associated with the flap installation, and the drag of the wing tip. The equivalent flat plate area corresponding to wing profile drag

(f_{e_w}) can be computed as follows:

$$f_{e_w} = C_f S_{w_e}(1 + k_{3\text{-}D} + k_k) + \Delta f_{e_t} + \Delta f_{e_i} + \Delta f_{e_F} \quad (4.2)$$

where

C_f = skin friction drag coefficient of a flat plate

S_{w_e} = wing exposed wetted area

$k_{3\text{-}D}$ = factor accounting for 3-D effects

k_k = discrete roughness correction

Δf_{e_t} = tip drag

Δf_{e_i} = wing/fuselage interference drag

Δf_{e_F} = drag due to flap tracks and gaps.

The parameter C_f in Eq (4.2) is the turbulent flat plate friction drag defined previously in Fig 3.2 as a function of Reynolds number and surface roughness. It is assumed that turbulent conditions exist over the entire wing on all airfoils including the so-called laminar flow airfoils such as the 65-series employed on the hypothetical helicopter. This is due to the fact that the manufacturing roughness is usually so high that laminar flow cannot be maintained over any significant area. This often results in higher drag values than were measured during two-dimensional wind-tunnel testing on idealized models, which are generally sufficiently smooth to permit the development of some degree of laminar flow near the leading edge. This difference is illustrated by a comparison of the predictions of Eq (4.2) with the two-dimensional tests[2] presented in Table IV-2. This data is determined at $c_\ell = 0$; therefore, the cambered airfoils ($65_3 - 618$) at cruise c_ℓ will show even larger discrepancies between tests of smooth models and predictions, since the minimum c_d for these airfoils occurs at the design c_ℓ and not at $c_\ell = 0$. For example, the $65_3 - 618$ airfoil c_d decreases from 0.0075 at $c_\ell = 0$ to 0.007 at the design c which, for this airfoil, is $c_\ell = 0.6$.

AIRFOIL	PREDICTED c_d	EXPERIMENTAL c_d (REF 2)	$R_e \times 10^6$
65_3–618	0.009	0.0075	4.7
0012	0.00816	0.0059	6
23012	0.00816	0.0072	6

TABLE IV-2 COMPARISON OF PREDICTED AND MEASURED SECTION DRAG COEFFICIENTS AT $c_\ell = 0$

t/c; %	k_t
12	2.042
18	2.077
24	2.12

TABLE IV-3 WETTED AREA FACTOR

The exposed wetted area, S_{we}, is equal to $S_{pe} \times k_t$ where S_{pe} is the exposed planform area outboard of the fuselage cutout (61 ft^2) and k_t is the planform area factor (wing airfoil perimeter divided by the chord). The factor k_t for various airfoil t/c values is given in Table IV-3.

The $k_{3\text{-}D}$ term in Eq (4.2) accounts for the three-dimensional effect of airflow supervelocity on skin friction and pressure drag. As shown in Fig 4.14 (based on empirical data[7,8]), the skin friction drag and pressure drag increase with increasing airfoil thickness-to-chord ratio (t/c). For the 65-series or other laminar flow airfoils where the maximum thickness is located further aft than in conventional ones, there is less skin friction drag, due to lower average supervelocity; however, moving the maximum thickness aft increases the pressure drag. For airfoils with $t/c \leqslant 27$ *percent,* the $k_{3\text{-}D}$ factor decreases as the maximum thickness location moves aft. However, for $t/c > 27$ *percent,* the $k_{3\text{-}D}$ factor increases. For the hypothetical wing design with $t/c = 18$ *percent,* and maximum t/c occurring at approximately 40 percent, $k_{3\text{-}D} = 0.37$.

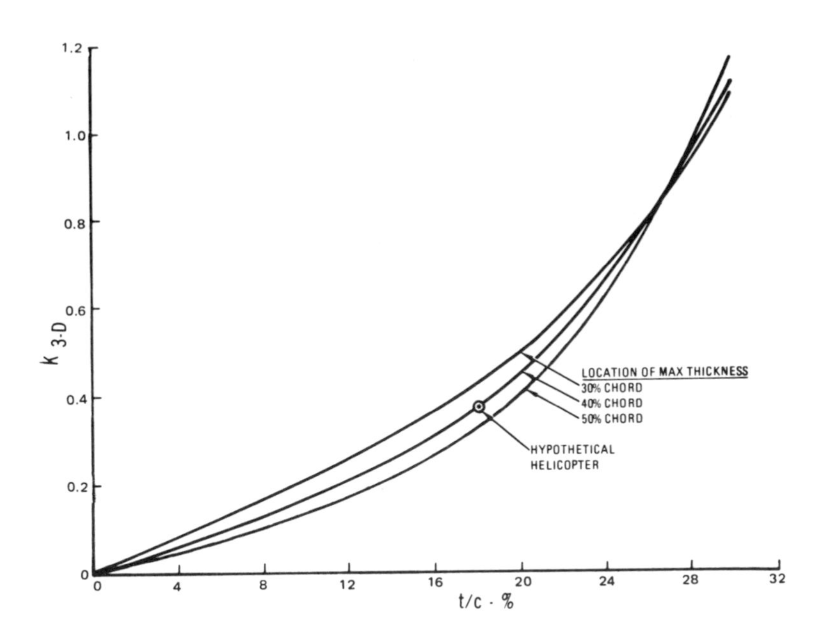

Figure 4.14 Three-dimensional effects for wing and tail surfaces

The terms k_k and Δf_{et} in Eq (4.2) are the roughness correction and wing-tip drag. The three-dimensional flow past the tip has a lower average supervelocity than that of the two-dimensional conditions found inboard; therefore, if rounded tips are employed to prevent local separation, the Δf_{et} term will be zero or negative[7]. The roughness correction for a wing with flush rivets over the first one-third and mushroom rivets over the aft two-thirds, is 7 percent ($k_k = 0.07$). Using round-head rivets over the entire chord increases this correction to 9 percent. The 7-percent value is assumed for the hypothetical helicopter.

The next term in Eq (4.2), Δf_{ei}, is the drag due to wing/fuselage interference effects. This drag component comes from the superposition of supervelocities at the wind-body intersection, resulting in increased pressure drag due to a more adverse trailing edge pressure gradient. The interference drag is a function of the airfoil thickness ratio, wing vertical and longitudinal location, location of airfoil maximum thickness, and fillet size and shape. An empirical expression for the interference drag is as follows[7,8]:

$$\Delta f_{ei} = 1.5(t/c)_{wf}^3 (c_{wf})^2 k_f k_{v\ell} k_\ell \tag{4.3}$$

where

$(t/c)_{wf}$ = wing thickness/chord ratio at the fuselage

c_{wf} = wing chord measured at the fuselage; ft

k_f = fillet factor

$k_{v\ell}$ = vertical location factor, and

k_ℓ = longitudinal location factor.

Eq (4.3) was developed for conventional airfoils; however, Ref 7 indicates that it is valid for laminar flow profiles if optimum size fillets are employed. The optimum fillet extends beyond the trailing edge of the wing, and has a radius-to-chord ratio of approximately 8 percent. It provides an approximate reduction in interference drag of 35 percent ($k_f = 0.65$). The hypothetical helicopter is assumed to have the required fillet radius on the lower surface; however, the upper surface intersects the nacelles and the resulting discontinuity reduces the fillet effectiveness; therefore, $k_f = 1.0$ is used.

The longitudinal location factor in Eq (4.3) is $k_\ell = 1.0$ for wings located approximately at the midpoint (longitudinally) of the fuselage[7]. The vertical location factor varies from $k_{v\ell} = 1.0$ for the mid-wing configuration to $k_{v\ell} = 2.0$ for high or low wing locations.

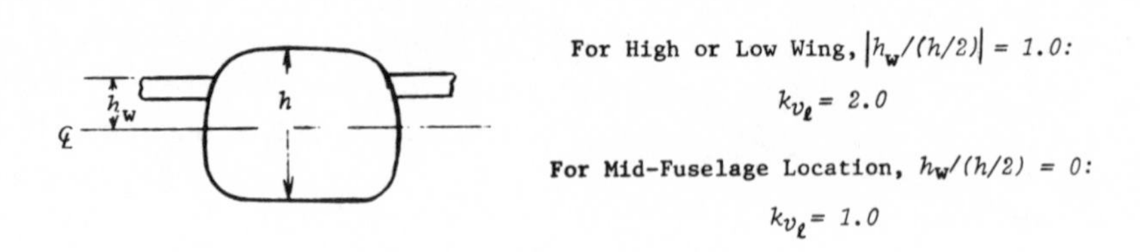

Figure 4.15 Vertical location factors

The interference drag is higher for high- or low-wing installations due to the thicker boundary layer at the acute wing-body intersection. For designs such as the hypothetical helicopter with $|h_w/(h/2)| = 0.7$, the factor $k_{v\ell} = 1.4$. Substituting the hypothetical helicopter location factors in Eq (4.3), and noting thar $t/c = 0.18$ and $c_{wf} = 4.58\ ft$; $\Delta f_{e_i} = 0.257\ ft^2$.

The last term, Δf_{e_F}, in Eq (4.2) is the drag due to exposed flap tracks (or actuators) and gaps remaining after the flaps are retracted. Ref 7 indicates that the gaps due to full-span flaps increase the basic wing drag by approximately 3 percent, while drag contributions of the flap track and exposed actuator vary, depending on their individual locations. The hypothetical wing design is assumed to be relatively clean with a combined gap and track drag equal to 5 percent of the basic wing drag.

The total hypothetical wing profile drag can be calculated using the variables derived above and assuming that $C_f = 0.00357$ and the exposed wetted area $S_{w_e} = 127\ ft^2$. This C_f value corresponds to $R_e = 4.7 \times 10^6$ (150 kn, 4000 ft/95°F); assuming a surface roughness of $k = 1.2 \times 10^{-3}$ inches (Fig 3.2). Substituting these values in Eq (4.2) gives

$$f_{e_w} = [0.00357 \times 127(1 + 0.37 + 0.07) + 0.257]1.05 = 0.955\ ft^2.$$

The total profile C_{D_0} is 0.0094, based on planform area ($S = 101\ ft^2$). A nominal value of 0.009 was used in subsequent performance calculations.

Wing Induced Drag – The wing induced drag coefficient (based on S) as defined from the lifting-line theory[2] is

$$C_{D_i} = C_L^2/\pi AR_e \kappa \tag{4.4}$$

where

AR_e = effective wing aspect ratio

κ = induced drag factor

C_L = wing lift coefficient.

Standard fixed-wing reference texts often refer to κ as the Oswald efficiency factor. It accounts for the increase in induced drag due to deviations of the wing spanwise distribution from the optimum elliptical shape, and is a function of wing taper and aspect ratio[2]. If the wing is twisted, additional factors must be applied.

The effective aspect ratio AR_e in Eq (4.4) is the geometric aspect ratio ($AR = b^2/S$) corrected for fuselage cutout effects, where b is the total span (wing-tip to wing-tip). The effective aspect ratio is lower than the geometric aspect ratio but, because of wing-lift carry-over effects, is higher than that obtained by using the portion of the wing span outboard of the fuselage cutout. Pressure measurements in this region have shown that the lift extends onto the fuselage – resulting in an effective semispan ($b_e/2$) as shown in Fig 4.16. By defining a wing-fuselage induced drag factor, $\kappa_{wf} = (b_e/b)^2$,

$$AR_e = \kappa_{wf} AR. \tag{4.5}$$

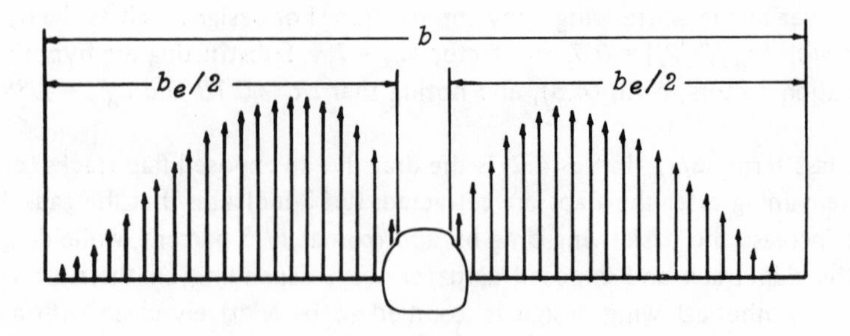

Figure 4.16 Wing-lift carryover

The factor κ_{wf} is determined empirically from wind-tunnel measurements by plotting the incremental wing test results in terms of C_D versus $C_L{}^2$, as shown in Fig 4.17.

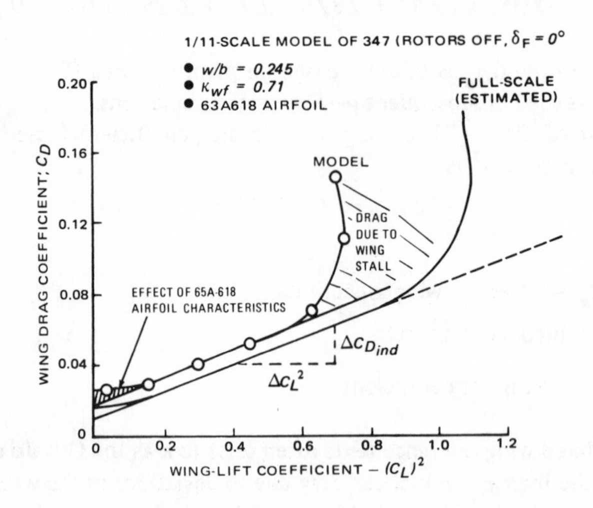

Figure 4.17 Determination of fuselage/wing interference factor κ_{wf}

The nonlinear region at low-lift coefficient values evident in this figure is due to the increase in profile drag with decreasing c_ℓ; characteristic of the 65A-618 airfoil (design $c_\ell = 0.6$). The nonlinear region at high C_L^2 values is due to wing stall. The full-scale wing will have a larger linear $C_L{}^2$ region due to the effect of Reynolds number on $C_{L_{max}}$ and the profile drag will be lower. Using the slope of the linear portion of this curve ($\Delta C_D/\Delta C_L{}^2$) and κ (Ref 2), Eq (4.4), when rearranged, becomes

$$\kappa_{wf} = (\Delta C_L{}^2/\Delta C_D)(1/\pi AR \kappa).$$

Values of κ_{wf} for other wing/fuselage configurations are presented in Fig 4.18 as a function of the fuselage width-to-wing span ratio w/b. As shown, κ_{wf} decreases with increasing amounts of wing cutout, and is independent of flap angle setting and aspect ratio. The factor κ, used to establish κ_{wf}, is also shown as a function of AR and taper ratio c_t/c_r. As κ increases, AR decreases and is maximum at $c_t/c_r \approx 0.5$, indicating that lower AR wings with $c_t/c_r \approx 0.5$ have lift distributions close to being elliptical in shape. For the hypothetical wing, $w/b = 0.325$, $\kappa_{wf} = 0.62$, and $\kappa = 0.996$. The effective aspect ratio of this configuration is $AR_e = 3.72$, or almost 40 percent less than the geometric value.

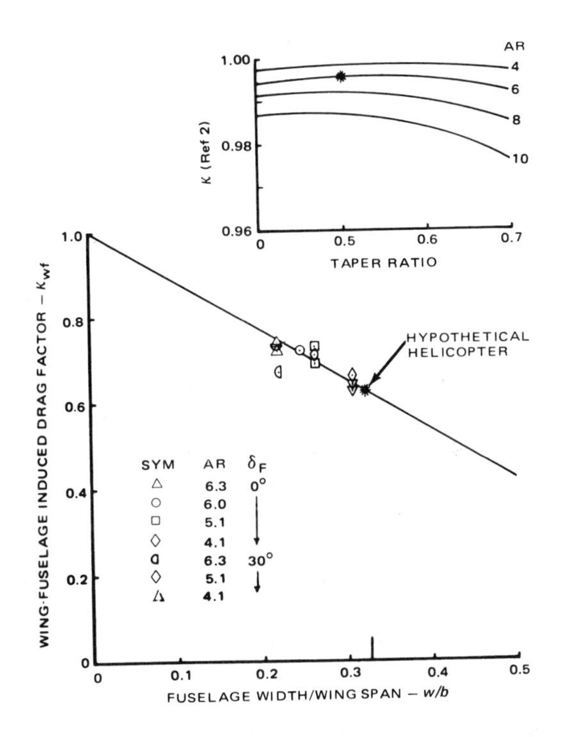

Figure 4.18 Effect of fuselage width on κ_{wf}

Rotor/Wing Interference Drag — Rotor/wing interference drag ($D_{i_{rw}}$) occurs because the wing is located in the downwash of the rotor. As shown in Fig 4.19, the rotor downwash tilts the wing lift vector aft relative to the aircraft forward speed (remote velocity), thus creating a rotor induced wing-drag component $D_{i_{rw}} = L_w \sin \beta$; with β denoting the downwash angle $\beta = \tan^{-1}(v/V_o)$, where v is the rotor downwash at the wing, and V_o is the velocity of the freestream. Using basic momentum theory relationships and assuming fully developed downwash velocities at the wing $[v = (2v_{id})]$, the following relationship for $D_{i_{rw}}$ can be derived:

$$D_{i_{rw}} = L_R L_w / 2\pi R^2 q \tag{4.6}$$

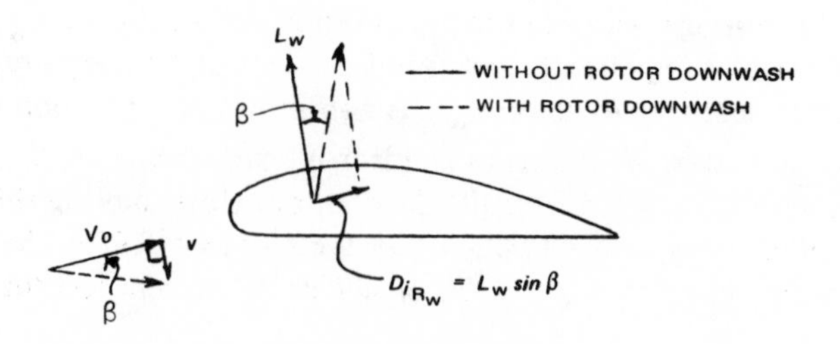

Figure 4.19 Effect of rotor downwash on wing aerodynamic forces

where

L_R = rotor lift

L_w = wing lift

q = freestream dynamic pressure, $\frac{1}{2}V_o^2\rho$.

Therefore, the interference drag decreases rapidly with speed, and increases with aircraft weight.

Although theory predicts rotor downwash amounting to the average induced velocity ($v = v_{id}$) for wing locations close to the rotor disc center (see Vol I, Fig 3.20), the validity of Eq (4.6), based on the $v = 2v_{id}$ assumption, has been confirmed by flight test measurements of local downwash angles obtained on the CH-46 tandem-wing experimental helicopter. In addition, these measurements showed that the rotor downwash on the right wing was higher than on the left, with the fully-developed value at cruise airspeeds being $v = 2v_{id}$. The hypothetical wing is located closer to the rotor than the CH-46 installation; therefore, the assumption of fully-developed flow is conservative. Eq (4.6) can also be developed from the biplane theory[9,10].

Effect of Wing on Fuselage Attitude – The main rotor tip-path plane for the winged aircraft must be tilted further forward than for the pure helicopter trim condition. This is needed to (1) provide the additional propulsive force required to overcome the wing drag, and (2) to compensate for the reduced rotor thrust resulting from the rotor unloading. This is the reason why auxiliary propulsion is required to achieve very high speeds.

The tip-path plane angle required for the hypothetical aircraft with and without the wing installed is shown in the upper half of Fig 4.20. Calculations based on the simplified equation $a_{TPP} = tan^{-1}[(D_f + D_w)/W]$ and on the trim analysis computer program indicate that at $W = 15{,}000\ lb$, wing $C_L = 0.4$, and a cruise speed of $V = 150\ kn$ at $4000\ ft/95°F$ ambient condition, the tip-path plane must be tilted forward about 2° further than for the wing-off case.

In order to achieve trim about the pitch axis ($\Sigma M = 0$), the fuselage attitude must also become more negative in order to alleviate the incremental nose-down moment created by tilting the tip-path plane relative to the plane normal to the shaft. This change in attitude results in an increased drag and download. The change in fuselage angle-of-attack per degree of tip-path plane tilt depends primarily on the size and incidence of the

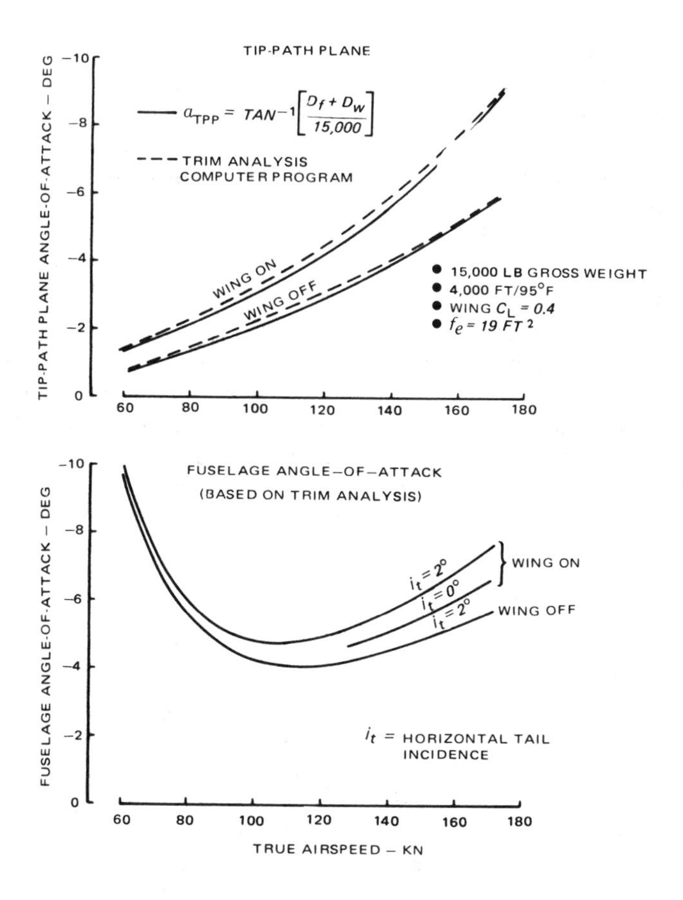

Figure 4.20 Effect of wing on rotor and fuselage angle-of-attack

horizontal tail, the effective hinge offset of the rotor, and c.g. location. Because of the horizontal tail moment contribution, the fuselage angle-of-attack adjustment will be less than the change in tip-path angle as shown at the bottom of Fig 4.20. For the hypothetical helicopter with fixed horizontal tail incidence ($i_t = 2°$), a tip path plane angle inclination change of $-2°$ varies the fuselage cruise angle-of-attack by $-1.3°$. The resulting increase in fuselage drag and download (Fig 3.14) causes a 2 percent rise in cruise power required at 4000 ft/95°F. Two methods of alleviating this penalty are (1) increase the built-in forward inclination of the main rotor shaft to reduce the fuselage nose-down cruise attitude, and (2) reduce the horizontal tail incidence as noted in Fig 4.20, provided the associated tip-path inclination does not lead to an unacceptable excess of structural loads (i.e., shaft or blade flap-bending loads) or excessive longitudinal control travel. Experimental data illustrating the effect of horizontal tail incidence on the fuselage angle-of-attack of a winged helicopter can be found in Ref 11.

4.2 Effect of Wing Unloading on Induced and Profile Power of the Rotor

Unloading the main rotor may, in principle, decrease its induced and profile power. However, in cruise, the induced power represents a relatively small percentage of the total power required, while the profile power is relatively insensitive to thrust changes unless the rotor is operating at high thrust and forward speeds where stall and compressibility effects become significant. Furthermore, it should be remembered that the rotor diameter is almost always at least twice as large as the wing span. Thus, even neglecting the rotor-wing interaction, the transference of the lift from the rotor to the wing is synonymous with shifting the load from a lift generator having a more favorable (lower) span loading to that of a higher one. Consequently, the combined induced drag of the unloaded rotor and the unloading wing becomes higher than for the rotor alone.

The hypothetical helicopter basic wing-off shaft horsepower required developed in Ch III (Fig 3.24) is presented in Fig 4.21 as a function of main rotor lift (thrust component normal to the freestream velocity) for the 4000 ft/95° condition. Here, it is shown that unloading the rotor by 20 percent at 150 kn and 15 000-lb gross weight would reduce the power required by 10 percent if there were no additional power penalties because of the wing. This reduction assumes no change in parasite power.

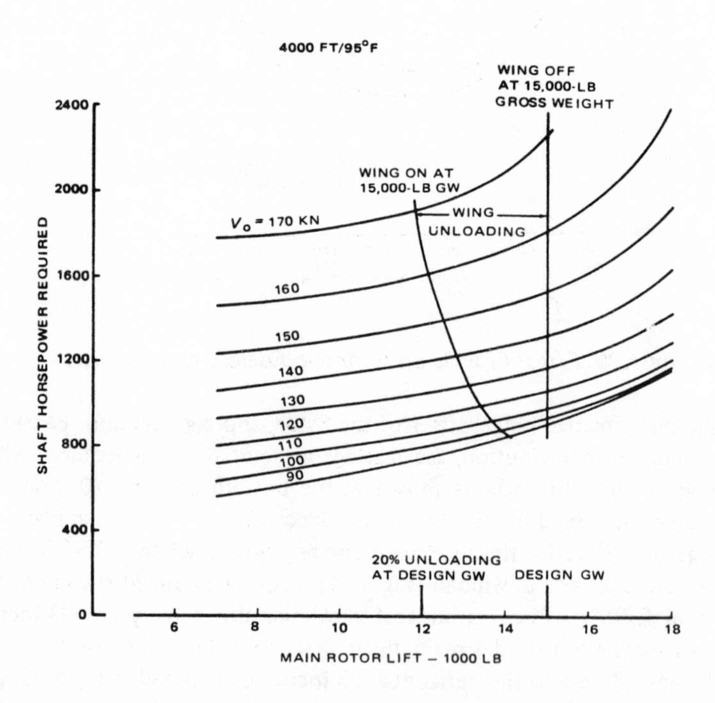

Figure 4.21 Effect of reducing main rotor lift (unloading) on power required

Wind-tunnel testing conducted by Vertol as well as NASA and Bell model tests[6] indicate that the full induced and profile rotor-power required benefits of rotor unloading described above were obtained with no significant unfavorable wing interference effects on the rotor; however these benefits must be combined with wing drag calculations to determine the optimum unloading for minimum total power required as outlined below.

4.3 Determination of Optimum Cruise Unloading

In the considered case, the wing was sized to meet maneuver requirements. Consequently, to minimize level flight performance penalties, the cruise rotor unloading, wing C_L, and incidence angle must be selected in such a way as to make the total power required as low as possible or, in other words, to maximize the total aircraft W/D_e where $D_e = SHP \times 550/V_o$. For example, to make the power required of the winged configuration equal to, or lower than, that of the wingless aircraft, the unloading must be selected in such a way that gains in rotor profile plus induced power must exceed or be equal to the increase in parasite power due to the wing.

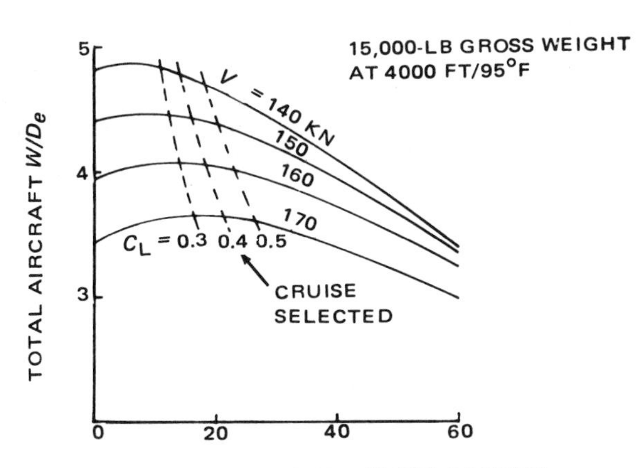

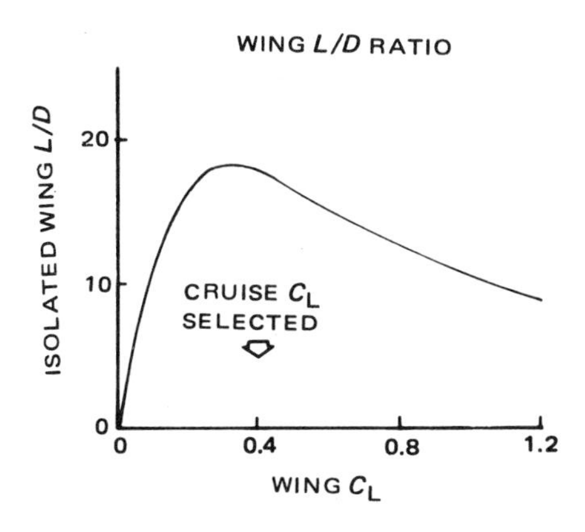

Figure 4.22 Determination of cruise unloading (C_L)

The calculation procedure used to determine wing C_L for the hypothetical helicopter cruising at 4000 ft/95°F, and a gross weight of 15,000 lb is shown in Fig 4.22. In the upper portion of this figure, total aircraft $L/D_e \approx W/D_e$ is presented as a function of wing unloading for cruise speeds of 140 to 170 kn. Isolated wing L/D is presented as a function of wing C_L in the lower half of this figure. The maximum total aircraft W/D_e occurs at an unloading of 10 percent for cruise speeds of 140 to 150 kn and increases to 15 percent at 170 kn. This optimum unloading corresponds to $C_L = 0.3$; however, since the W/D_e curves are relatively flat in this region, a slightly higher C_L of 0.4 was selected in order to reduce the wing download at forward c.g. positions and low airspeeds where the wing α becomes negative. The isolated wing L/D at $C_L = 0.4$ is within 2 percent of the maximum value.

The optimum wing unloading can also be readily determined using rotor maps[12]. Rotor maps are charts of isolated rotor lift versus propulsive force defined for constant power required levels, as illustrated for 170 kn in Fig 4.23 for the hypothetical helicopter. This data, presented in nondimensional form, is based on the Vertol isolated rotor vortex theory computer program (Ch III). The optimum unloading is obtained in one step by simply placing the unloaded rotor thrust versus wing plus fuselage drag trim line on the rotor map and noting where the minimum power required occurs. Rotor maps are particularly useful in cases where the rotor configuration is optimized in conjunction with the wing unloading to achieve improved total aircraft cruise W/D_e. In this case, the geometry of the unloaded rotor can be selected to operate at nondimensional lift levels ($L/q\,d^2\sigma$) corresponding to maximum rotor L/D_e[12]. The hypothetical helicopter maximum rotor L/D_e is also noted in Fig 4.23.

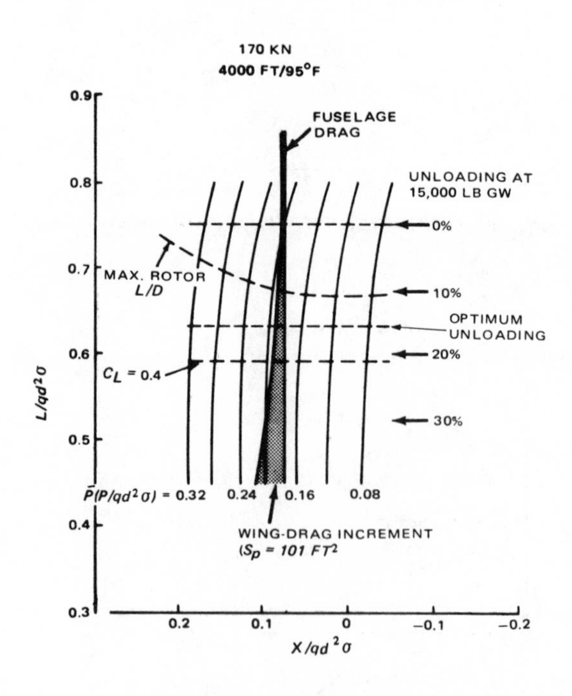

Figure 4.23 Rotor map method of optimizing winged aircraft cruise performance

The total aircraft $W/D_e = f(V)$ relationship associated with operating the wing at $C_L = 0.4$ is shown in Fig 4.24. In spite of the unloading optimization, the W/D_e of the winged helicopter is 2 percent lower than that of the wingless configuration at best range speed. However, at speeds above 160 kn, unloading by the wing provides an improvement due to stall alleviation.

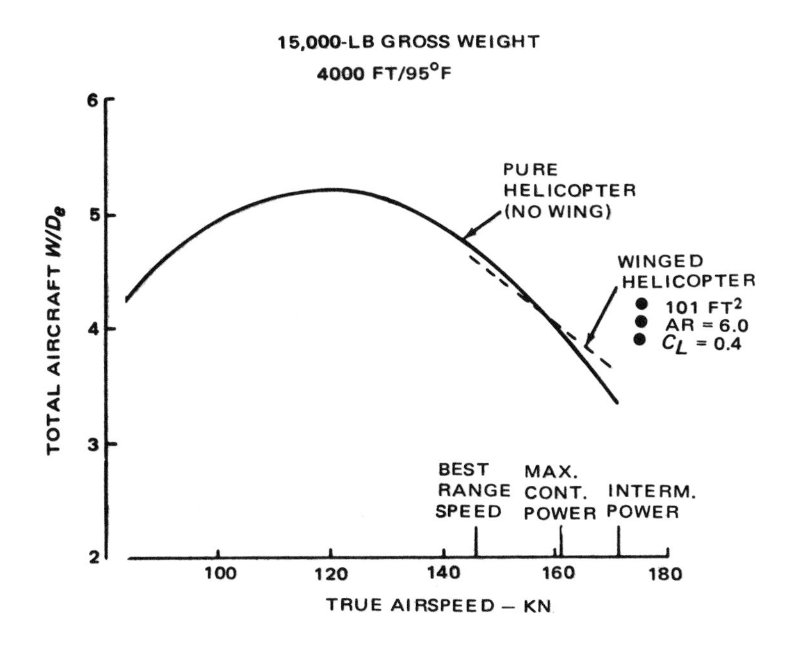

Figure 4.24 Effect of forward speed on total aircraft W/D_e

The wing incidence setting required to achieve the optimum C_L value at the design mission cruise speed is a function of the fuselage angle-of-attack, wing zero-lift angle-of-attack, and wing lift-curve slope. For the hypothetical helicopter, the wing angle-of-attack required for $C_L = 0.4$ is $5°$, as shown in Fig 4.5. The wing lift-curve slope and zero angle-of-attack can also be estimated using basic lifting-line theory relationships, two-dimensional data[2], and the effective aspect ratio described previously. Assuming the fuselage cruise angle-of-attack is $-5°$, the hypothetical wing incidence setting required to achieve $C_L = 0.4°$ is $i_w = 10°$.

The above discussion of wing angle-of-attack refers to the average angle of the right and left-wing panels. However, during the CH-46 tandem-wing program, angle-of-attack measurements showed as much as 4° lateral asymmetry in the downwash field at cruise airspeeds. The right-wing panel (counterclockwise forward rotor rotation) consistently showed lower angles-of-attack and higher downwash angles. Vertol nonuniform downwash analyses and USSR theoretical considerations[13] confirm the existence of a lateral downwash asymmetry. Differential incidence can be used to correct for this effect and prevent asymmetric stall from causing unfavorable roll control problems.

4.4 Level Flight Performance

Sample calculations for the hypothetical helicopter illustrating the effect of the wing on level flight power required and mission performance capability are described in this section.

Total Power Required – The hypothetical single-rotor helicopter level flight power required with and without the wing installed, is presented in Fig 4.25 for 4000 ft/95°F conditions. At low gross weights, the wing increases the power required, while at high gross weights, the power required decreases, due to alleviation of stall and compressibility effects. It should be noted that at $W = 18{,}000\ lb$, the power reducing effect of the wing extends to the minimum power speed.

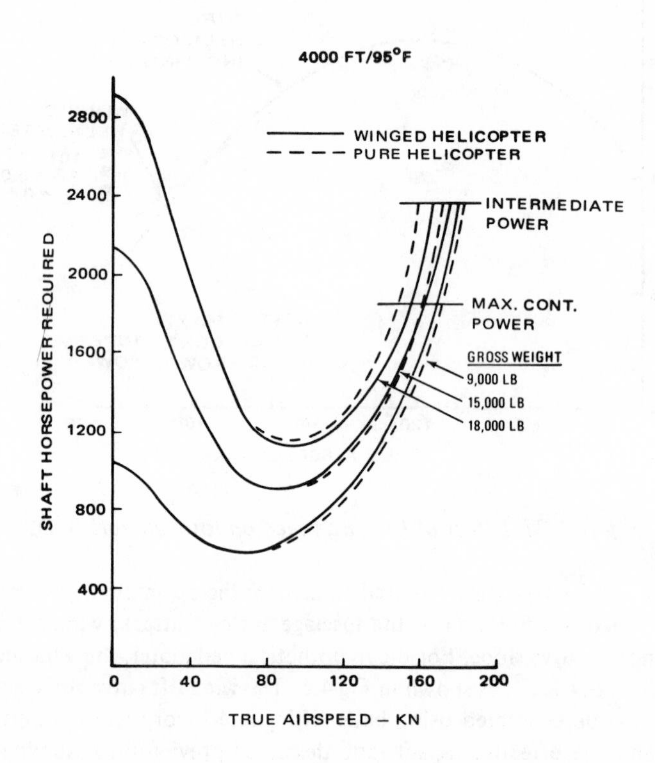

Figure 4.25 Level flight power required at 4000 ft/95°F

The wing-off data shown in this figure was derived in Ch III, and the wing effects were calculated as described above, with the following simplifying assumptions:

(1) Wing $C_L = 0.4$ at all airspeeds
(2) Wing does not affect a_f
(3) Wing profile drag coefficient $C_{D_o} = 0.009$
(4) Rotor downwash is fully developed at the wing location.

These assumptions permit one to calculate power requried for the hypothetical winged helicopter by adjusting the baseline aircraft power required as shown for $W =$

15,000 lb in Table IV-3. The parasite power correction defined in Ch III, Sect 3.3, was applied to all wing calculations.

Payload–Range Capability – The effect of the wing on payload range capability for 4000 ft/95°F conditions is shown in Fig 4.26. The winged helicopter payload capability at zero range is 550 lb lower than that of the baseline aircraft because of (1) reduction of hover OGE gross weight by 200 lb through wing downloading, and (2) a 350-lb increase in weight empty. In terms of range, the wing reduces the full fuel range performance by 10 n.mi, due to increased cruise power required at best range speed as shown in Fig 4.27. The best range speed for a 15,000-lb gross weight aircraft, with and without the wing, is 145 kn.

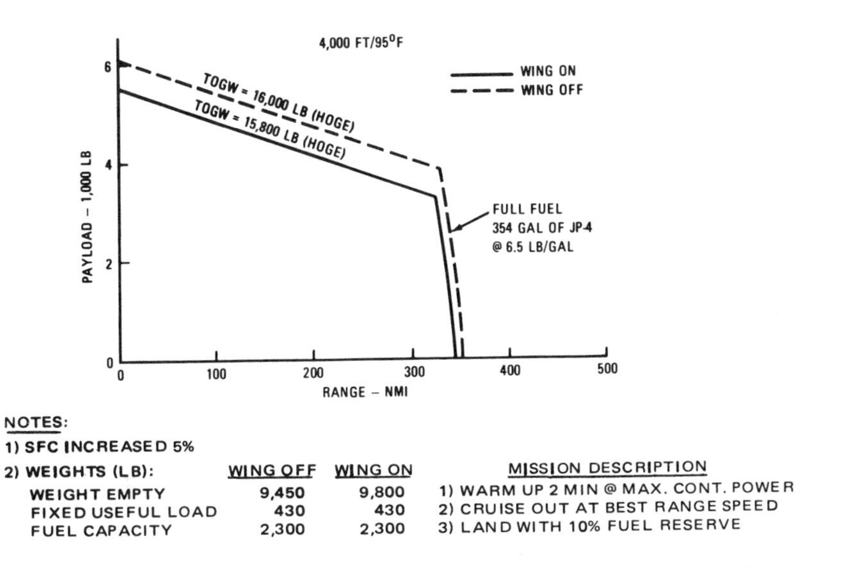

Figure 4.26 Winged helicopter payload–range capability

Speed Capability – The wing has no effect on the hypothetical helicopter maximum continuous power speed capability at 15,000-lb gross weight; however, at heavier weights, its presence is beneficial due to the alleviation of stall and compressibility (Fig 4.27). It can be seen that the hypothetical winged helicopter does not have sufficient power to exceed the structural envelope in level flight if operated within the maximum weight and intermediate power limits. The increase in the level flight (1-g) structural envelope, as defined by the hypothetical rotor control system endurance limit, is presented in Fig 4.28 as a function of density altitude. Additional details concerning the stall boundary can be found in Ch III.

15,000-LB GROSS WEIGHT
4000 FT/95°F

(1)	(2)	(3)	(4)	(5)	(6)	(7)	(8)	(9)	(10)	(11)	(12)	(13)	(14)	(15)	(16)	(17)	(18)
60	9.86	995.9	0.4	0.01375	0.009	0.02275	2.298	398.3	14602	15.23	17.53	0.145	1.0	32.5	−42	1040	1031
80	17.52	1769.5						707.4	14293	8.39	10.69	0.193	0.990	47.4	−50	915	912
100	28.24	2852						1141	13859	5.05	7.35	0.241	0.945	68.8	−55	930	944
120	39.43	3982						1593	13407	3.50	5.80	0.290	0.890	96.6	−65	1060	1092
140	53.67	5421						2168	12832	2.46	4.76	0.338	0.825	135.8	−90	1320	1366
160	70.09	7079						2832	12168	1.79	4.09	0.386	0.735	195.5	−195	1810	1811
170	79.13	7992						3197	11803	1.53	3.83	0.410	0.685	235.7	−370	2260	2126

NOMENCLATURE:

1. SPEED OF FLIGHT, V: KN
2. DYNAMIC PRESSURE, q: PSF
3. $q \times S_W$: (2) × 101
4. WING LIFT COEFFICIENT: C_L
5. WING INDUCED DRAG COEFFICIENT: $C_{D_i} = C_L^2/\pi A \kappa \kappa_{wf}$*
6. WING PROFILE DRAG COEFFICIENT: C_{D_o}
7. WING DRAG COEFFICIENT: C_D = (5) + (6)
8. WING EQUIVALENT DRAG: D_W/q = (7) × 101
9. WING LIFT: L_W = 0.4 × (3)
10. ROTOR LIFT, T_r: LB = 15,000 − (9)
11. Δf_e OF WING DUE TO ROTOR: $D_{i_{RW}}$ = (9) × (10)/$2q^2 \pi R^2$
12. TOTAL f_e OF WING + ROTOR: D/q = (8) + (11)
13. $\mu \equiv V/V_t$
14. PROPULSIVE EFFICIENCY: η_p (Fig 3.21)
15. WING SHP = (12) × (2) × 1.69 × (1) / (550 × 0.98)
16. ΔSHP_r DUE TO UNLOADING: (FIG 4.21)
17. SHP, WING−OFF: (FIG 3.24)
18. SHP, WING−ON: (17) + (16) + (15)

NOTE: *$\pi A \kappa \kappa_{wf}$ = 11.64

TABLE IV-3 WINGED HELICOPTER FORWARD FLIGHT POWER REQUIRED CALCULATIONS

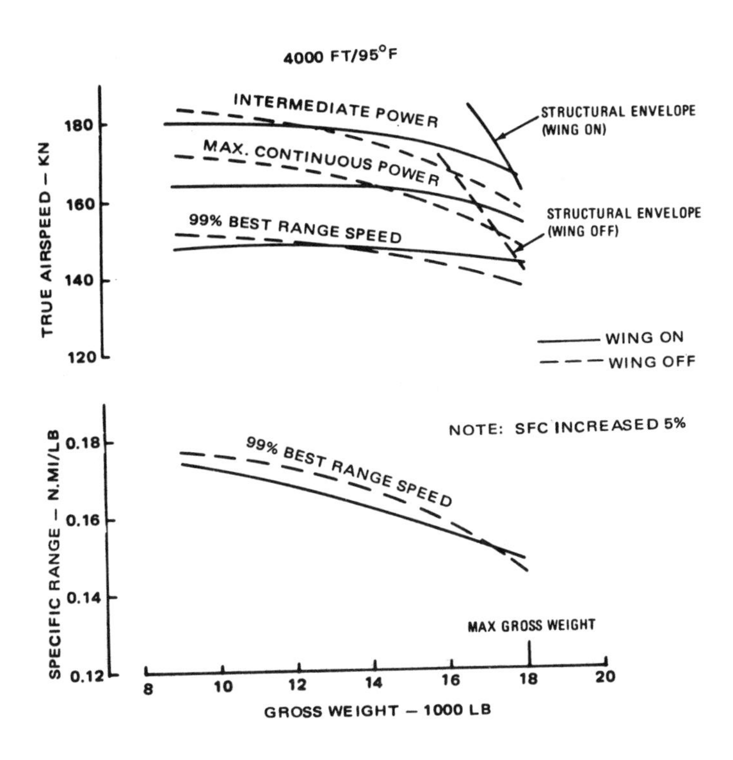

Figure 4.27 Comparison of wing-on and wing-off airspeed and specific range vs gross weight

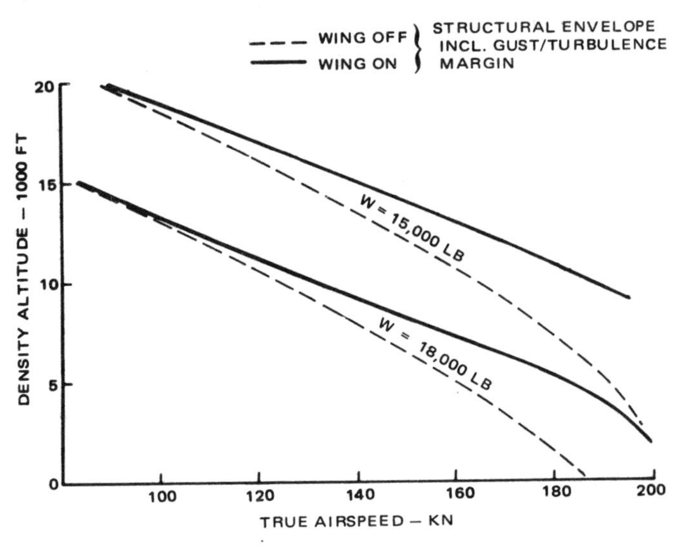

Figure 4.28 Winged helicopter structural envelope

4.5 Climb and Autorotation

The effect of the climb and descent velocity component on wing angle-of-attack and C_L at minimum power speed is shown in Fig 4.29. It can be seen that for rates of climb $V_c > 1000$ *fpm*, the wing with fixed incidence and 0° flap angle produces a download. In descent, the wing angle-of-attack increases and at autorotational rates of descent, the wing angle-of-attack approaches 20° which is well into stall.

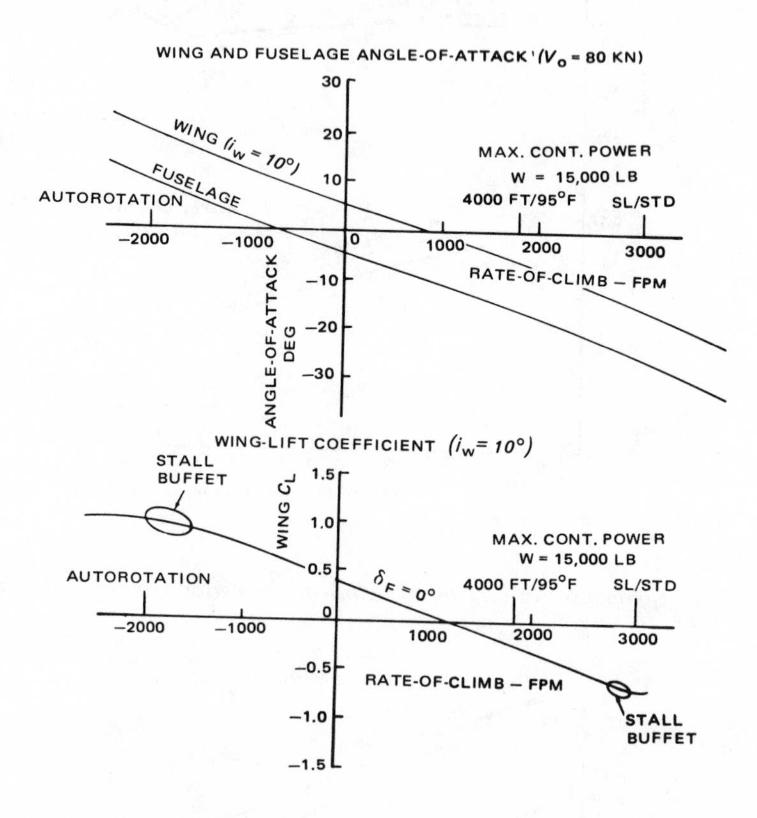

Figure 4.29 Wing C_L and α in climb and descent

These high C_L values and the resulting rotor unloading can cause excessive rotor speed decay during entry into autorotation. Also, control power (moments) is reduced; caused by the reduction in rotor thrust. To alleviate these effects, and the wing download in climb, a variable incidence wing can be used, with the angle-of-attack determined as a function of collective setting and airspeed; however, this design is complex and increases weight empty. Other solutions to the problem are to use flaps or spoilers to control the wing lift. A detailed parametric study is required to optimize the wing-rotor control system.

To illustrate the basic techniques and considerations involved in computing the effect on wings on climb and autorotation performance, sample calculations for the hypothetical helicopter are presented below.

Climb Performance – The effect of installing the wing on dual-engine climb capability at 4000 ft/95°F is shown in Fig 4.30 for the following configurations:

(1) Fixed incidence, flap setting $\delta_F = 0°$
(2) Fixed incidence, variable δ_F $(C_L = 0)$
(3) Variable incidence $(C_L = 0.4)$.

This climb performance was derived by using the simplified climb prediction method described in Ch III (assuming $k_p = 0.85$).

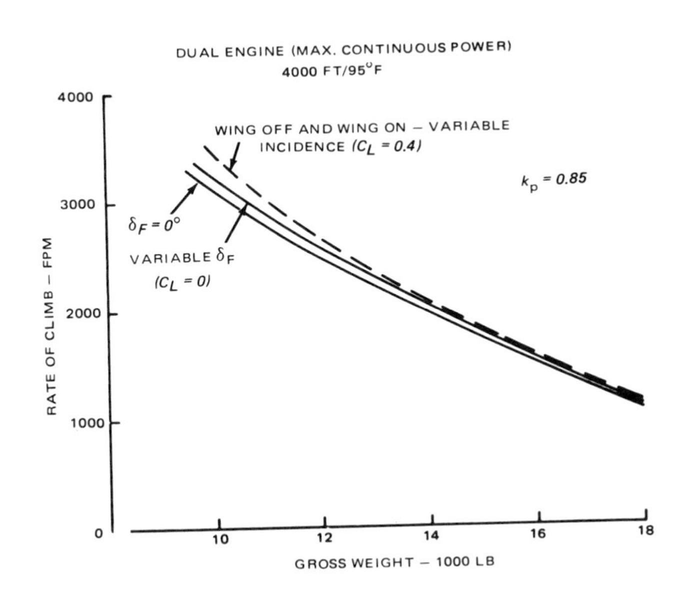

Figure 4.30 Winged helicopter dual-engine climb capability

At higher rates of climb, it can be seen that the fixed incidence, $\delta_F = 0°$, configuration has less capability than the aircraft without wings. This is caused by the wing download and induced drag penalty in this flight condition. If flaps are deflected to eliminate the wing download $(C_L = 0)$, there is an improvement in performance; however, the rate of climb at low gross weights is still less than for the wing-off case because of the incremental wing and flap drag at $C_L = 0$[14]. To eliminate the flap drag, the wing design must be modified to include variable incidence. As noted in Fig 4.30, the variable incidence configuration $(C_L = 0.4)$ has approximately the same climb performance as the aircraft without wings.

The flap angle required to obtain $C_L = 0$ (fixed incidence configuration) is shown in Fig 4.31. It is apparent from this figure that the effectiveness of single-slotted flaps begins to deteriorate at $\delta_F = 30°$, resulting in a maximum practical setting of 45°. At 3000 fpm, the hypothetical helicopter requires $\delta_F = 45°$ to obtain $C_L = 0$. Since wing incidence is fixed, it is not possible to obtain the optimum level flight $C_L = 0.4$ at these rates of climb solely by the use of flaps.

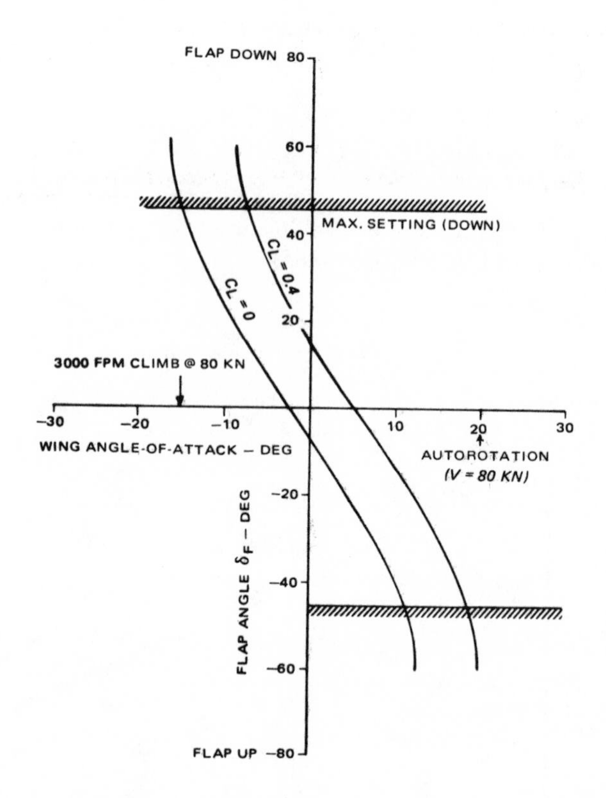

Figure 4.31 Flap settings required to obtain C_L = 0 and 0.4

The effect of wings on single-engine service ceiling capability is shown in Fig 4.32. Here, it is assumed that $\delta_F = 0°$ and $C_L = 0.4$ since, at a 100 fpm rate of climb, the wing angle-of-attack does not significantly vary from that in level flight. By unloading the rotor at high altitudes and high rotor lift coefficients, the wings improve the aircraft ceiling capability through reduced power increments associated with blade stall.

Autorotation – As noted in Fig 4.29, steady-state autorotation results in high wing C_L values. This causes two basic problems.: (1) rotor speed decay during entry, and (2) reduced roll control power due to diminished thrust from unloading of the rotor. In addition, roll disturbances can occur as a result of asymmetric wing stall, thus increasing the roll control problem. As described in Ref 6, the second problem is somewhat alleviated with hingeless rotors since this configuration produces greater control moments at low thrust levels. Articulated rotors with small hinge offsets require more rotor thrust for equivalent roll and pitch control.

Rotor speed decay initially occurs upon autorotation entry, and it will continue unless there is a sufficiently low collective setting available to achieve steady-state autorotation within such various aerodynamic and structural limits as rotor stall, longitudinal cyclic limits, blade flapping, and fuselage attitude restrictions. In addition, if the minimum collective setting is too low, it can cause rotor over-speed at high gross weights.

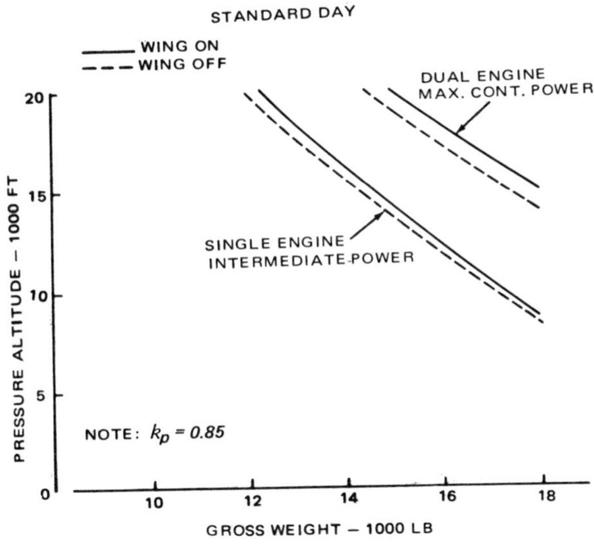

Figure 4.32 Winged helicopter service ceiling

Configuration variables such as adjustable horizontal tail incidence, coupled with wing control devices must be used to achieve autorotation within the various constraints over the full range of operating conditions.

To prevent the difficulties described above, the wing lift must be reduced during autorotation. The complexity of the wing control system or the amount of special devices (spoilers, flaps, incidence controls, etc.) required to reduce the wing lift depends on the wing size and amount of unloading. Ref 6 indicates that small wings with $(L_w/W)_{max} < 0.3$ may be permanently installed. For the hypothetical helicopter at 15,000 lb gross weight and 4000 ft/95°F, $L_w/W = 0.42$ at 150 kn; assuming $C_{L_{max}} = 1.0$ (Fig 4.5). Therefore, because of the size of the wing, special control devices would be necessary.

Flaps were selected as a *special device* to reduce wing lift to illustrate winged helicopter autorotation calculation techniques. As shown in Fig 4.31, the minimum wing C_L that can be obtained for the hypothetical helicopter within the effective range of the flaps ($\delta_F = \pm 45°$) is $C_L \approx 0.4$. The flap schedule and profile drag increment[14] required to keep $C_L = 0.4$ is presented in Fig 4.33 as a function of airspeed. Autorotation capability is computed by taking the wing-on level flight power required (Fig 4.25), and adding the parasite power associated with the flap deflection needed to achieve $C_L = 0.4$. Then the rate of descent is calculated, using the potential energy method described in Ch III, where $k_{pd} = 1.0$.

The resulting performance for 4000 ft/95°F and design gross weight is shown in Fig 4.34. For the wing at a fixed incidence, it can be seen that the flap deflection necessary to maintain $C_L = 0.4$ increases the hypothetical helicopter minimum rate of descent from 2000 fpm to 2300 fpm. This penalty becomes higher with flight velocity, primarily because flap drag also increases with speed. The effect of employing variable incidence rather than flaps to maintain $C_L = 0.4$ is also presented in Fig 3.34. As shown, the variable incidence approach eliminates the flap penalty and reflects only the difference in wing-on and wing-off level flight power required.

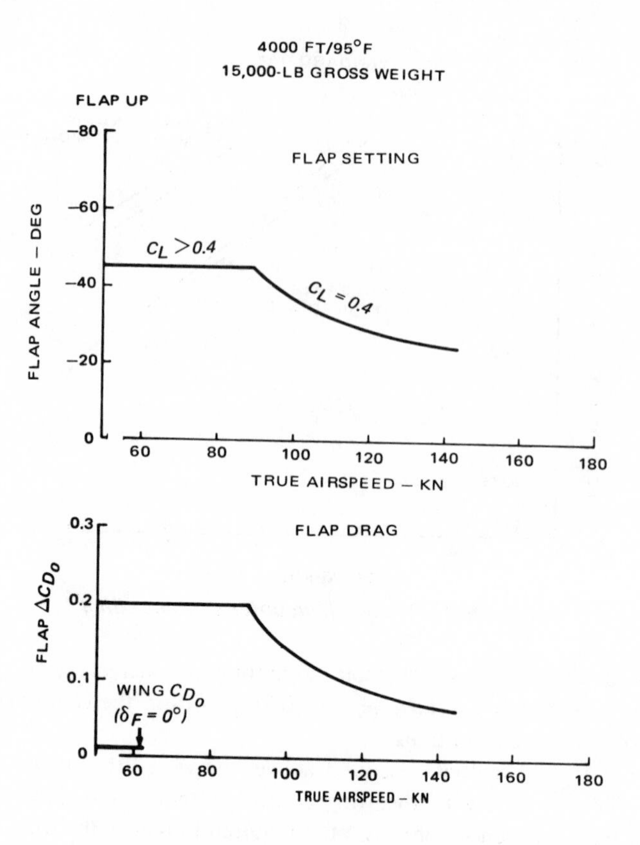

Figure 4.33 Flap angle and flap drag in autorotation ($C_L = 0.4$)

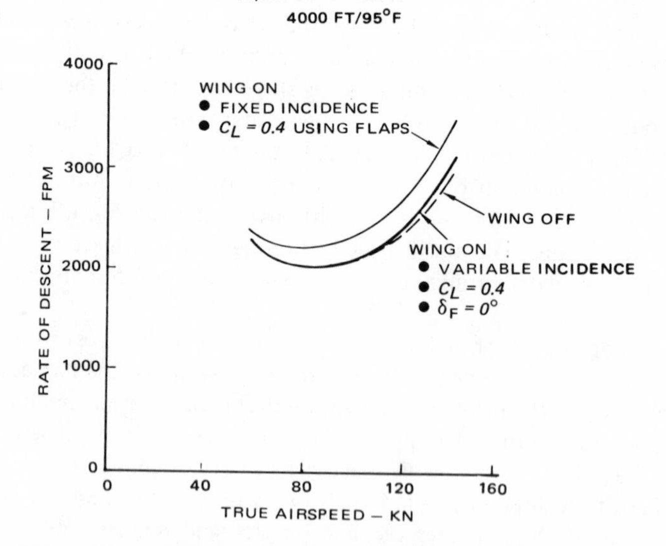

Figure 4.34 Winged helicopter autorotational rate of descent

4.6 Maneuver Capability

A summary of the winged helicopter maneuver capability at 15,000 lb gross weight and 4000 ft/95°F is presented in Fig 4.35 where load factor, deceleration and fuselage attitude required are shown as functions of airspeed for a banked turn. A wing flap setting of $\delta_F = 30°$ is assumed in these calculations, and $0.95\, C_{L_{max}} = 1.6$ (Fig 4.5). This wing-lift coefficient is used to provide a margin for gusts and to reduce wing buffeting during the maneuver. The wing angle-of-attack at $C_L = 1.6$ is $a_w = 17.4°$. To achieve this, the fuselage angle-of-attack (a_f) must increase by approximately 12.5° from the level flight $a_f = -5°$; assuming a nominal wing incidence angle of 10°. The rotor tip-path plane angle-of-attack (a_{tpp}) must also increase in order to initiate the maneuver and obtain a balance of longitudinal rotor, fuselage, and horizontal tail moments during the maneuver. To accomplish this, the hypothetical helicopter will require approximately 1° of a_{tpp} change per $0.6°$ of a_f, as discussed previously in Sect 4.1. The rotor thrust vector is now tilted aft and, when combined with the wing and fuselage drag, results in deceleration ($-\dot{V}$). The deceleration can be expressed in g's as

$$\dot{V} = -(D_w + D_f - X)/W$$

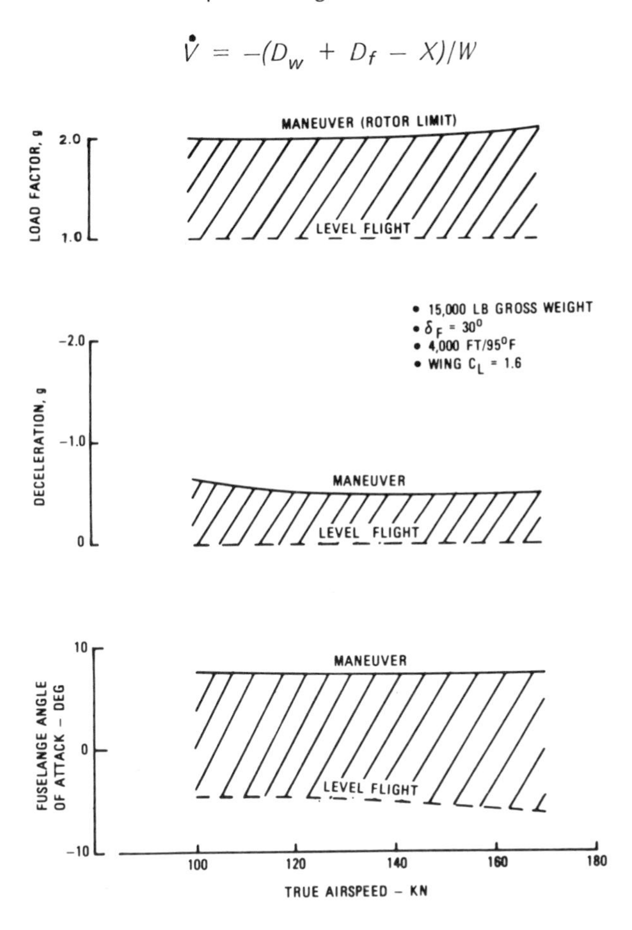

Figure 4.35 Winged helicopter maneuver capability

where

D_w = wing drag; lb
D_f = fuselage drag; lb
X = rotor propulsive force (defined negative if directed aft); lb

As indicated in Fig 4.35, the hypothetical winged aircraft has a $\dot{V} = 0.5g$ deceleration at 150 kn which is equivalent to an airspeed loss of 10 kn per second during the maneuver. No power limits are shown in this figure because the rotor is operating at an equivalent descent angle-of-attack. That is, the rotor propulsive force is sufficiently negative to reduce the power required below the power available. The power margin ($\Delta SHP = SHP_{av} - SHP_{req}$) could have been used to decrease the propulsive force $X = \Delta D_e/\eta_p$ where $\Delta D_e = 550(\Delta SHP)/1.689(V_o)$ and η_p is the propulsive efficiency (the reader is referred to Ch III, Section 3.3). However, this would require changing the horizontal tail incidence during the maneuver in order to keep $\Sigma M = 0$. Deceleration defined by power limits can also be determined using the rotor map method described previously in this chapter (Fig 4.23). This method is also useful in identifying control limits as defined by flying quality or stress considerations.

Decreasing X-force reduces the retreating blade angle-of-attack as well as power required. Therefore, operating at reduced or negative propulsive forces allows the rotor to achieve higher lift levels before stall inception occurs. The stall flutter parameter method or the inplane torque technique (described in Ch III, Sect 6.1) can be used to estimate the propulsive force effects on rotor limits. Model rotor data illustrating the influence of $\bar{X} = X/qd^2\sigma$ on stall flutter type boundaries is presented in Fig 4.36. Additional experimental results can be found in Ref 15. This data indicates that the variation in stall boundaries with $\bar{X}$ is relatively small for μ values up to 0.3. For this reason, the 1-g $(\bar{X} \approx 0.1)$ boundary defined in Ch III was used for the maneuver calculations presented above; however, the level flight limits were adjusted for banked turn pitch rate alleviation effects (Chapter III, Section 6.3).

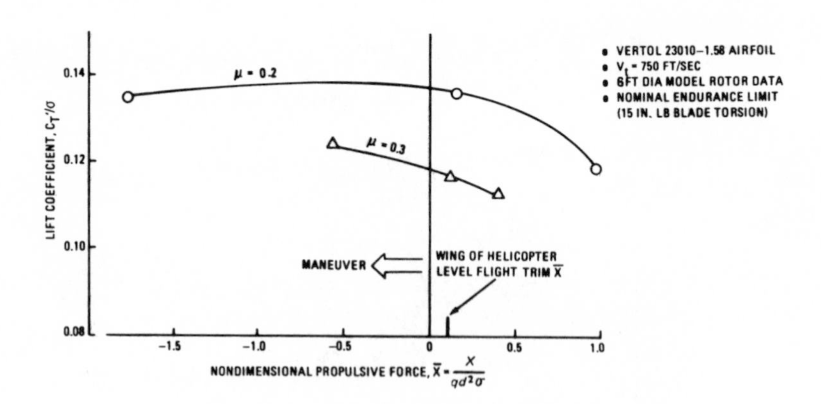

Figure 4.36 Effect of propulsive force on stall boundaries

References for Ch IV

1. Davis, S. Jon and Wisniewski, J. *User's Manual for HESCOMP, the Helicopter Sizing and Performance Computer Program.* Boeing Vertol Report D210-10699-1 (Contract NAS2-6107), NASA CR-152018, Ames Research Center, Moffett Field, Ca. Sept. 1973.

2 Abbott, I.H. and Doenhoff, A.E. *Theory of Wing Sections.* Dover Publications, Inc., 1959 Edition.

3 Loftin, L.K. and Poteat, M.I. *Aerodynamic Characteristics of Several NACA Airfoil Sections at Seven Reynolds Numbers from 0.7×10^6 to 9.0×10^6.* NACA RM No. L8B02, 1948.

4 McCormick, B.W., Jr. *Aerodynamics of V/STOL Flight.* Academic Press, 1967.

5. Fry, B. *Design Studies and Model Tests of the Stowed Tilt-Rotor Concept, Vol. III, Appendices.* AFFDL-TR-71-62, July 1971.

6. Lynn, R.R. *Wing-Rotor Interactions.* Journal of Aircraft, Vol. 3, No. 4, July-August, 1966.

7. Gabriel, E. *Drag Estimation V/STOL Aircraft.* Boeing Vertol Rpt. D8-2194-1, Sept. 1968.

8. Hoerner, S.F. *Fluid Dynamic Drag.* Published by the Author, 1965 Edition.

9. von Mises, Richard. *Theory of Flight.* Dover Publications, Inc., 1959.

10. Durand, W.F. *Aerodynamic Theory,* Vol II. Durand Reprinting Committee, 1943.

11. Dumond, R.C. and Simon, D.R. *Flight Investigation of Design Features of the S-67 Winged Helicopter.* Journal of the American Helicopter Society, Vol. 18, No. 3, July 1973.

12. Kisielowski, E., et.al. *Generalized Rotor Performance.* USAAVLABS TR-66-83, R-390, 1966.

13. Baskin, V.E., et.al. *Theory of the Lifting Airscrew.* Translated from Russian to English by W. Z. Stepniewski. NASA TT F-823, February 1976.

14. Royal Aeronautical Society. *Data Sheets. Vol. 4,* RAeS, London, England.

15. Landgrebe, A.J. and Ballinger, E.C. *A Systematic Study of Helicopter Rotor Stall Using Model Rotors.* Preprint No. 804, 30th Annual National Forum of the American Helicopter Society, May, 1974.

TANDEM–ROTOR HELICOPTER PERFORMANCE

In order to provide an additional comparison of the single-rotor hypothetical helicopter with other conventional helicopter configurations, a complete performance envelope for a tandem is provided in this chapter. As in the case of the winged version, design gross weight and power installed is assumed to be the same as for the hypothetical machine discussed in Chapters I–III.

Principal Notation for Ch V

C_D	body drag coefficient	
$C_P = 550HP/2\pi R^2 \rho V_t^3$	power coefficient	
$C_T = T/2\pi R^2 \rho V_t^2$	helicopter thrust coefficient	
D	drag	lb
d	rotor diameter	ft
d_R	distance between rotor centerlines	ft
d_s	stagger (distance between rotor axes)	ft
g	rotor gap (elevation of one rotor over another)	ft
H	height	ft
h	altitude	ft
IGE	in-ground effect	
k_d	download correction factor	
k_g	download factor IGE	
k_{ind}	induced power correction factor	
k_{ov}	induced power correction factor due to overlap	
k_p	climb efficiency factor	
k_{pd}	descent calculation factor	
k_v	integrated downwash velocity factor	
N	rotational speed	rpm
OGE	out-of-ground effect	
$ov = 1 - d_R/d$	overlap	
P	power	ft-lb/sec or hp
R	rotor radius	ft
r	radial distance from rotor axis	ft
T	total rotor(s) thrust	lb
V	aircraft velocity	fps, or kn
v	induced velocity	fps

W	weight (gross weight in particular)	lb
w	width	ft
y	lateral distance	ft
γ	wake separation angle	deg, or rad
γ_o	aft rotor elevation angle	deg, or rad
Δ	increment	
$\delta = p/p_o$	ambient pressure ratio	
ϵ	downwash angle	deg, or rad
$\theta = T/T_o$	ambient absolute temperature ratio	
θ_f	fuselage pitch attitude	deg, or rad
ρ	air density	slugs/ft^3
$\sigma = bcR/\pi R^2$	rotor solidity	
ψ	yaw angle	deg

Subscripts

c	climb
d	descent
f	fuselage
fr	front rotor
h	hover
id	ideal
ind	induced
iso	isolated
ℓ	local
n	sequence indicator
o	sea level
ov	overlap
rr	rear rotor
t	tip
v	vertical

Superscript

nondimensional

1. INTRODUCTORY REMARKS

The techniques for predicting tandem-rotor helicopter performance presented in this chapter are based on methods developed and refined during the CH-46, CH-47, Model 347, and HLH programs. The basic prediction methods are the same as the techniques described in Chs II and III, combined with empirical interference corrections resulting from the mutual aerodynamic effects of the two rotors. These were obtained from tandem-rotor wind tunnel testing of the Boeing Vertol Universal Helicopter Model (UHM) shown in Fig 2.20. To illustrate the methods, sample performance calculations for a hypothetical tandem-rotor aircraft are presented and the results compared with those of the single-rotor aircraft discussed in Chs I, II, and III.

2. DESCRIPTION OF THE TANDEM–ROTOR HELICOPTER

A three-view drawing of the tandem-rotor helicopter is presented in Fig 5.1, and a detailed configuration definition is given in Table V-1. In this table, the tandem-rotor aircraft is assumed to have the same design gross weight, and the same rotor blade tip speed, airfoil section, and twist as the single-rotor helicopter. A discussion of the rationale behind the tandem-rotor helicopter configuration selected is presented in Sections 2.1 to 2.8.

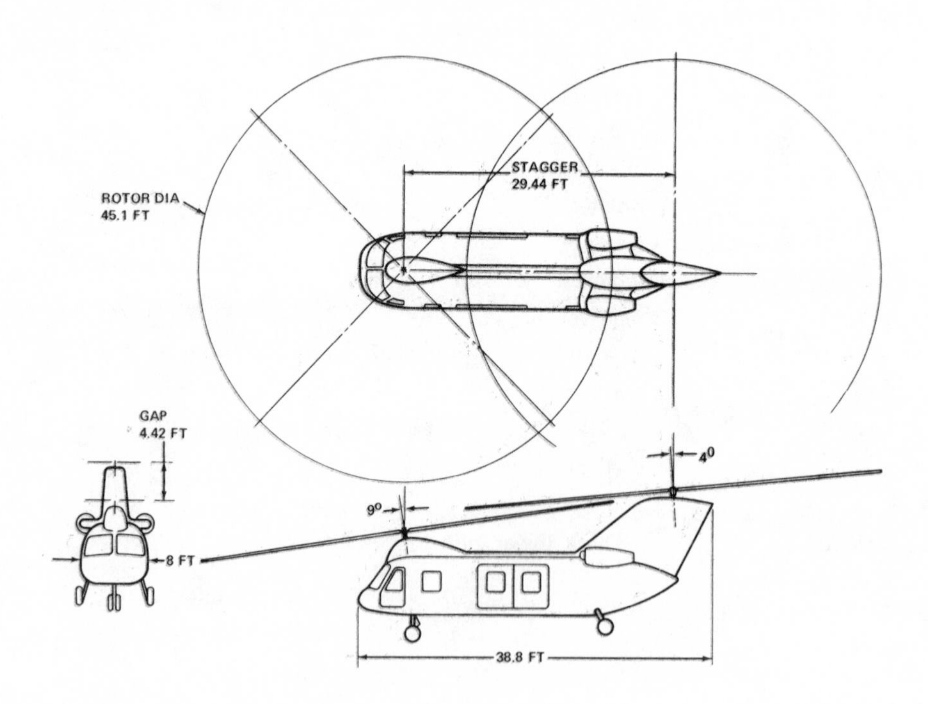

Figure 5.1 Three-view drawing of hypothetical tandem-rotor helicopter

WEIGHTS	
MAXIMUM GROSS WEIGHT	19,450 LBS
DESIGN GROSS WEIGHT	15,000 LBS
DISC LOADING	4.7 LBS/FT^2
WEIGHT EMPTY	9,450 LBS
FIXED USEFUL LOAD	430 LBS
FUEL CAPACITY	2,300 LBS
ROTOR GEOMETRY	
DIAMETER	45.1 FT
CHORD	19.6 IN
SOLIDITY	0.0692
TIP SPEED	700 FT/SEC
NUMBER OF BLADES/ROTOR	3
AIRFOIL	V23010-1.58
TWIST	-10°
CUTOUT (r/R)	20%
RPM	296
OVERLAP/DIAMETER	34%
GAP/STAGGER	0.15
TYPE	ARTICULATED
CYCLIC SCHEDULE	CH-47C
AIRFRAME	
PARASITE DRAG	19.1 FT^2
LANDING GEAR	FIXED
ENGINES (HYPOTHETICAL)	
RATING SL/STD (INTERMEDIATE/MAX. CONT.)	1600/1300
LAPSE RATE	6.0 HP/$^\circ$F
INSTALLATION LOSSES	1%
TRANSMISSION RATINGS	
DUAL ENGINE	2900 SHP
SINGLE ENGINE	1600 SHP

TABLE V-1 CONFIGURATION DESCRIPTION OF THE TANDEM-ROTOR HELICOPTER

2.1 Disc Loading

The historical disc loading trends for existing aircraft presented in Fig. 5.2 show that tandems are designed with lower disc loadings than single-rotor aircraft for the same design gross weight. These lower disc loadings are achieved without increasing aircraft size because of the compact rotor arrangement resulting from overlapped disc areas. Based on these trends at a gross weight of 15,000 lbs, a disc loading of 4.7 lb/ft^2 was selected for the tandem, compared with a disc loading of 7.64 lb/ft^2 for the hypothetical single-rotor helicopter.

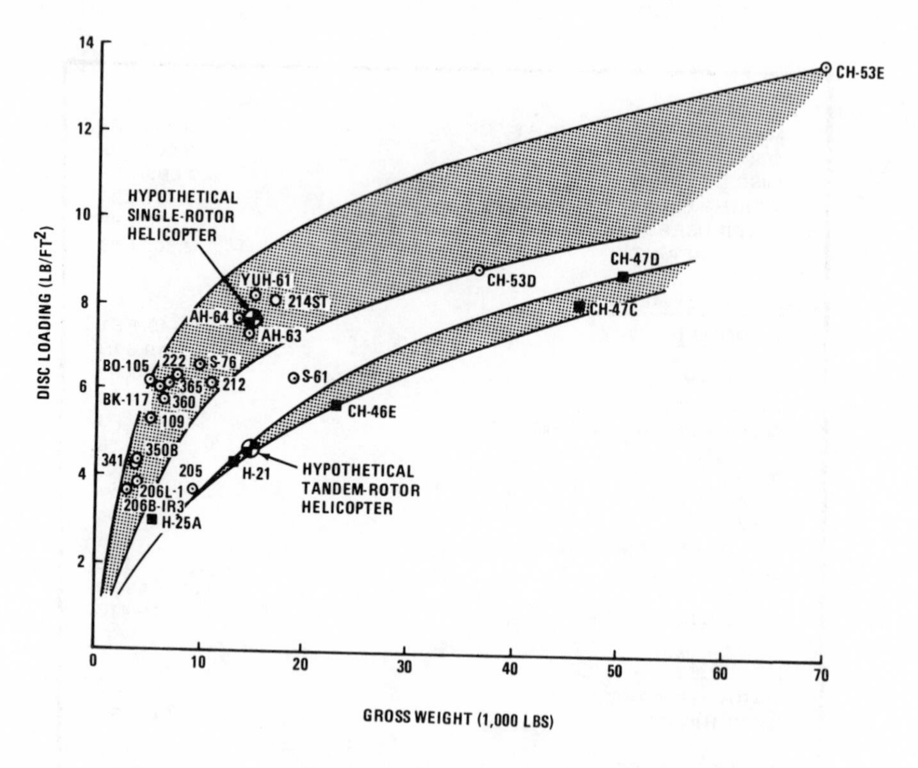

Figure 5.2 Disc loading trends

2.2 Gap/Stagger

The vertical location of the aft rotor relative to the forward rotor is defined as a gap, and the relative horizontal position is defined as stagger (Fig 5.1). This terminology is derived from biplane theory, since the rotor-rotor interference effects in forward flight are similar to the mutual interference between biplane wings. In addition to the rotor-rotor interference, blade-fuselage clearance and the effect on weight empty must be taken into consideration when choosing the gap/stagger ratio. A gap/stagger ratio of 0.15 was selected for this study as representative of existing tandem configurations.

2.3 Overlap/Blade Number

The definition of overlap is $ov = 1 - d_R/d$, where d_R is the slant-line distance between rotors. In terms of gap (g) and stagger (d_s),

$$d_R = \sqrt{g^2 + d_s^2} \tag{5.1}$$

The number of blades, solidity, and blade-clearance considerations define the maximum permissible overlap. A 3-bladed, 34-percent overlap rotor system similar to the CH-46 and CH-47 production aircraft configurations was selected for the tandem.

2.4 Solidity

The tandem-rotor solidity was sized to provide the same maneuver capability and structural envelope as the single-rotor aircraft (Ch III, Sections 6.1 and 6.2). The rotor stall inception limits based on wind-tunnel test data described in Ch III (Fig 3.33) were used; assuming equal thrust on the forward and rear rotors. Based on wind-tunnel test results, an additional $-0.1\,\Delta C_T/\sigma$ increment was applied to account for the effect of operating the rear rotor in the wake of the forward rotor. In practice, this effect can be offset by locating the center of gravity forward to provide a favorable thrust split between the front and rear rotors.

2.5 Weights

The weight empty of the tandem is assumed equal to the single-rotor configuration based on analysis of weight trends for current production aircraft. For the purpose of comparison, the tandem-rotor aircraft design gross weight and fuel capacity are also assumed equal to the single-rotor configuration. The tandem maximum gross weight is 19,450 lb, based on hover capability out-of-ground effect at 4000 ft/95°F.

2.6 Fuselage Configuration

The tandem-rotor aircraft fuselage (Fig 5.1) is assumed to have the same nose and forward cabin geometry as the single-rotor helicopter (Fig 1.1). The aft section of the cabin is lengthened to achieve adequate clearance between the rotors. In addition, the engines are moved aft to drive the combining transmission located in the aft pylon; thus requiring a slight modification to the afterbody.

The geometry of the forward and aft pylons is determined by directional stability considerations. Unlike single-rotor aircraft which use the tail rotor to augment directional stability, the tandem relies on the aft pylon geometry to provide inherent stability. The thickness of the aft pylon is approximately 25 percent of the chord in order to enclose the aft shaft and transmissions. For airfoils of thickness greater than 25 percent, truncation of the trailing edge is often employed to improve pylon effectiveness by reducing the degree of trailing-edge separation[1,2]. The CH-47C, for example, has a truncated aft pylon trailing edge.

The forward pylon produces a directionally unstable yawing moment about the aircraft c.g. The CH-46 and CH-47 have openings near the pylon leading edge (bleed slots) to achieve improved directional stability. The hypothetical helicopter forward pylon has a low profile shape to minimize the destabilizing side force.

2.7 Rotor Hub

The hub of the tandem-rotor aircraft is articulated, with a hinge offset of 2 percent of the radius.

2.8 Engines/Transmissions

The engine performance and transmission limits described in Ch 1 for the single-rotor aircraft are also assumed for the tandem. Consequently, comparisons of tandem and single-rotor aircraft performance will reflect differences only in the power required.

3. HOVER PERFORMANCE

The procedure for computing tandem-rotor helicopter hover performance consists of determining the isolated forward and aft rotor power required (see Ch II) and then applying an empirical overlap correction to the induced power component. In addition, the effect of increased induced velocity on the fuselage download in the overlap region must be taken into account. Details of this calculation procedure are presented in the following sections.

3.1 Isolated Rotor Power Required

Isolated forward and aft rotor power required is determined by using the vortex theory computer analysis or the "shortcut" techniques described in Ch II (Sections 1.2 and 1.3).

3.2 Induced Power Overlap Correction

The induced power of overlapping rotors in hover is higher than that of two isolated rotors combined. By defining k_{ov} as the ratio of total induced power required in hover by both overlapping rotors to the sum of induced power needed (under the same conditions) by the two isolated ones, the total induced power required (P_{ind}) can be expressed as

$$P_{ind} = k_{ov} k_{indh} (T/550)\sqrt{T/4\rho\pi R^2} \tag{5.2}$$

where T = total thrust (forward plus aft rotors), and k_{indh} = isolated rotor induced power correction factor in hover (Figs 2.9 and 2.10).

The above defined factor k_{ov} is a function of the amount of rotor overlap as shown by the model test data in Fig 5.3. The data exhibits considerable scatter, which may be partially explained by the variations in gap or vertical location of the rotors among the various models tested. This results in varying degrees of rotor wake development, or contraction, as the downwash approaches the adjoining rotor; thus leading to a different level of aerodynamic interaction for the same amount of overlap.

The curve faired through the Universal Helicopter Model data in Fig 5.3 has been used successfully to predict CH-46, CH-47, and Model 347 hover power required when combined with the isolated rotor power predictions determined by the vortex theory computer program described in Ch II. Using this faired curve for the hypothetical helicopter results in an overlap induced power correction factor of 1.129 (12.9 percent increase in induced power).

3.3 Fuselage Download, Out-of-Ground Effect (OGE)

Hover download is computed by integrating the local downwash velocity distributions along the fuselage centerline as described in Ch II. The fuselage is divided into segments as shown in Fig 5.4. The vertical drag of each segment D_{v_n} is computed as follows:

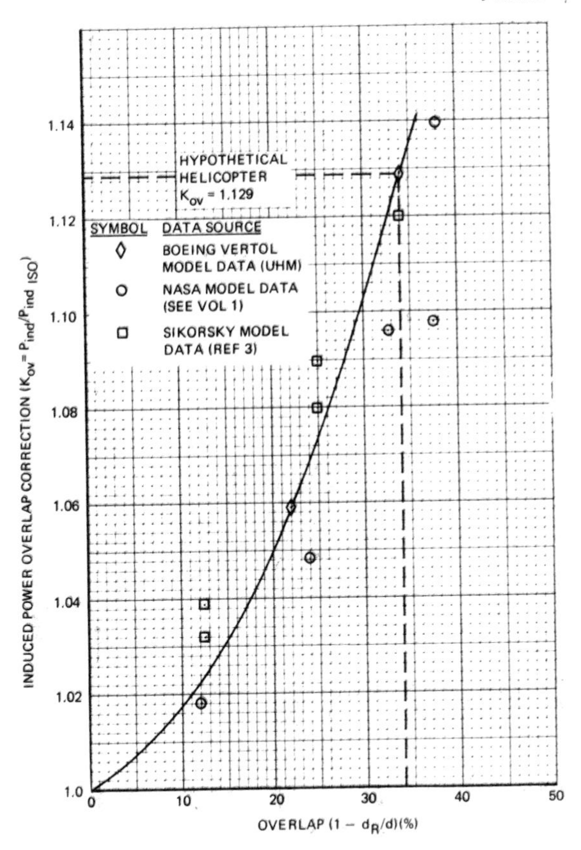

Figure 5.3 Induced power correction for rotor overlap

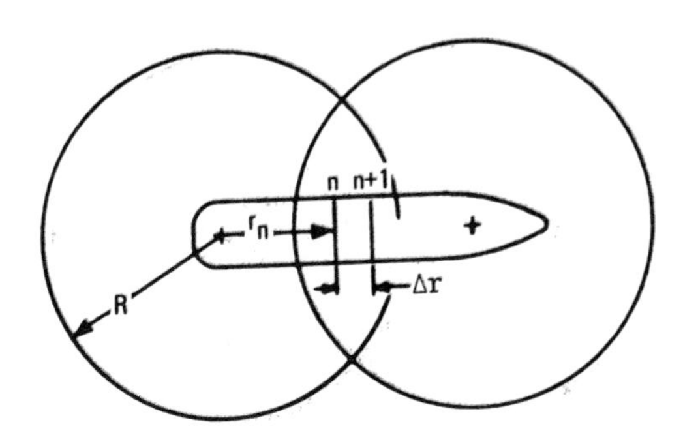

Figure 5.4 Incremental fuselage area for download calculation

$$D_{v_n}/T = (C_{D_{v_n}} w_n / 8\pi R)(k_{v_{n+1}} - k_{v_n}) \tag{5.3}$$

where T = total thrust of both rotors,

and

$$k_{v_n} = \int_0^{(r/R)_n} (v/v_{id})^2 \, d(r/R) \tag{5.4}$$

where r/R = relative distance along fuselage from forward rotor centerline; w_n = fuselage width at station n; and $v_{id} = \sqrt{T/4\rho\pi R^2}$ = ideal induced velocity.

Eq (5.3) is similar to Eq (2.11); however, the constant in the denominator is increased from 4 to 8 to reflect the difference in the number of rotors.

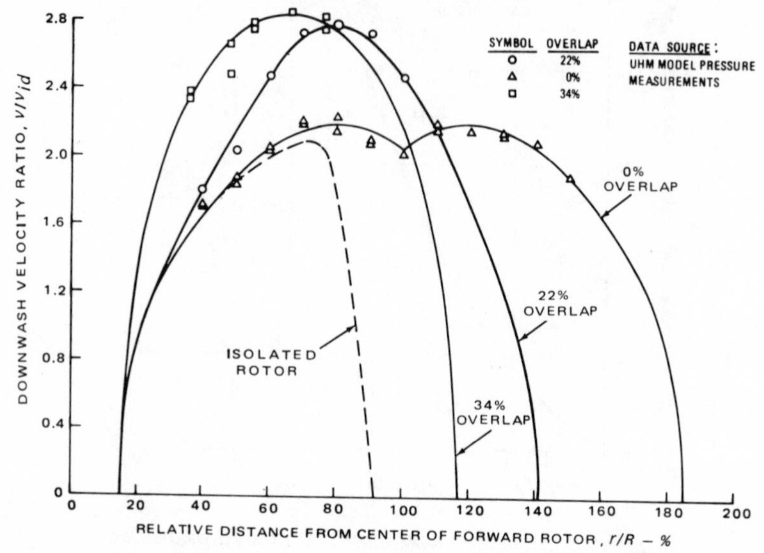

Figure 5.5 Tandem-rotor hover downwash velocity distribution at the fuselage

The wind tunnel test measurements presented in Fig 5.5 were made to define the downwash velocities and to aid in the development of download prediction methods. Pressure tapes located at various stations on the fuselage were employed to record the local downwash velocity in the overlap region. As shown in this figure, the downwash velocity in the overlap area increases as the overlap increases. For example, for 34 percent overlap, the maximum velocity ratio (occurring midway between the rotors) is $v/v_{id} = 2.8$, while for isolated rotors this ratio amounts to 2.1. Integration of the induced velocity profiles of Fig 5.5, designated as the integrated downwash velocity factor k_v, is shown in Fig 5.6.

Sample download calculations are presented in Table V-2. Here, the 34 percent overlap integrated downwash velocity distribution is applied to the helicopter. The fuselage is divided into four segments, with the vertical drag coefficient (C_{D_v}) of each segment equal to the values defined in Ch II. As noted, the total download-to-gross-weight ratio is D_v/W = 4.9 percent. The single-rotor download was 2.5 percent of gross weight (Ch II, Table II-1).

3.4 In-Ground-Effect (IGE) Corrections

The effect of operating IGE is to provide thrust augmentation at a fixed power setting and a reduction in fuselage download. A semi-empirical method of correcting

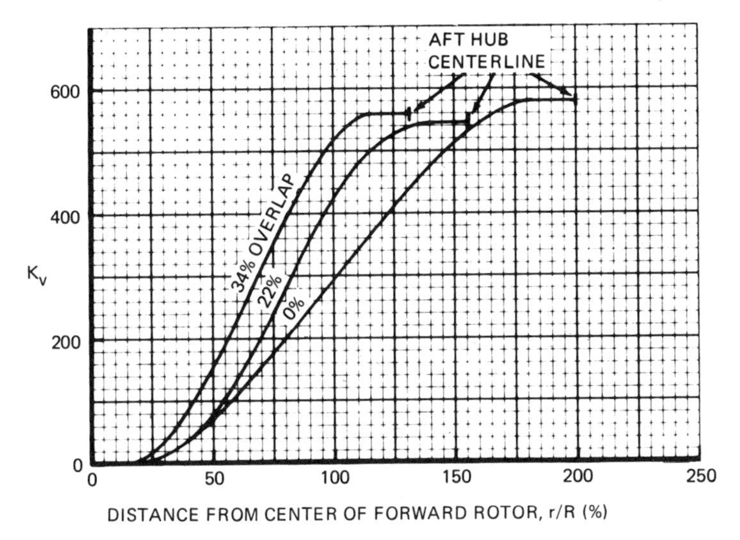

Figure 5.6 Tandem-rotor integrated nondimensional downwash velocity

the OGE gross weight is employed to account for these effects, as described in detail in Ch II. The specific empirical corrections applicable to tandem-rotor helicopters in general, and sample calculations for the study helicopter are presented below.

Thrust Variations, IGE. As shown in Fig 5.7, the ratio of rotor thrust IGE to thrust OGE for the tandems, when presented as a function of rotor height to diameter ratio (*H/d*), agrees well with the single-rotor data fairing in Ch II. It can be seen that the thrust ratio for the hypothetical tandem-rotor aircraft operating at a 5-ft wheel height is 1.077 (*H/d* = 0.43). As shown in Fig 5.8, the hub centerlines are assumed to be at an equal distance from the ground in order to simplify the calculations. The actual trim attitude depends primarily on shaft incidence and longitudinal cyclic input which, for tandems, generally varies automatically as a function of airspeed.

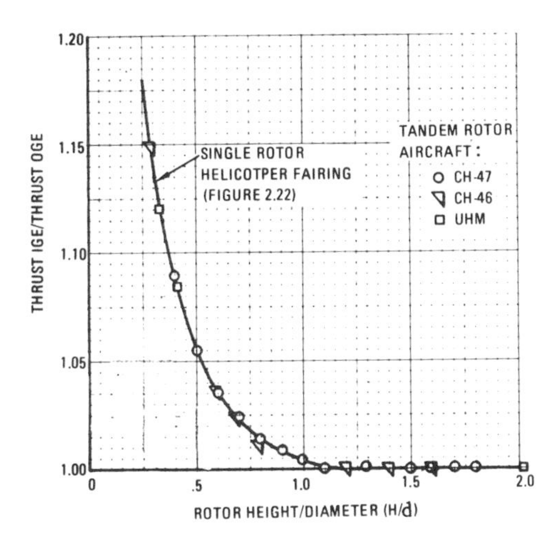

Figure 5.7 Tandem rotor helicopter thrust variation, IGE

STEP	①		②		③	④	⑤	⑥	⑦	⑧	⑨
ITEM	r_n	r_{n+1}	$(r/R)_n$	$(r/R)_{n+1}$	$k_{v_{n+1}}$	k_{v_n}	Δk_v	C_{D_v}	w; ft	$C_{D_v} w/8\pi R$	$\Delta D_v/T$; %
PROCEDURE	DISTANCE FROM FWD ROTOR C_L, ft		① /22.55 (%)		See Figs 5.6 and 2.19		③ − ④	See Fig. 2.17	SECTION WIDTH	$\frac{⑥ \times ⑦}{8\pi \times 22.55}$	⑧ × ⑤
SECTION											
COCKPIT*	4.51	4.9	20.0	21.8	202	200	2	0.5	8.0	0.00706	0.014
CABIN	4.51	6.1	20.0	27.1	24	5	19	0.4	4.9	0.00346	0.066
	4.5	19.5	20.0	86.5	440	5	435	0.4	8.0	0.00565	2.458
NACELLE	19.5	25.7	86.5	114.0	559	440	119	1.2	8.6	0.0182	2.166
AFTERBODY	25.7	26.67	114.0	118.3	560	559	1	0.5	1.0	0.00088	0.001

$\Sigma D_v/T = 4.70$

$\Sigma D_v/W = 4.93$

*NOTE: Use Isolated Rotor Downwash (Fig. 2.19).

TABLE V-2 DETAIL SAMPLE CALCULATION OF THE TANDEM-ROTOR HELICOPTER HOVER DOWNLOAD

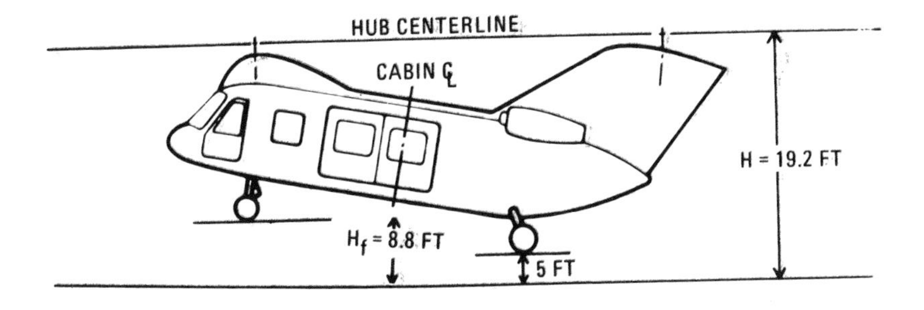

Figure 5.8 Assumed rotor/fuselage attitude in hover IGE

Download Variation, IGE. The variation of tandem and single-rotor aircraft download occurring in ground effect is presented in Fig 5.9 as a function of the fuselage height above the ground (H_f). The download factor k_g, defined as the ratio of download IGE to download OGE, is based on model testing. At a wheel height of 5 ft, the tandem rotor download factor is 0.04 at $H_f/d = 0.195$.

Gross Weight Correction, IGE. Substitution of the hypothetical helicopter thrust ratio, k_g, and OGE download in Eq (2.14) gives the tandem rotor gross-weight ratio as $W_{IGE}/W_{OGE} = 1.128$.

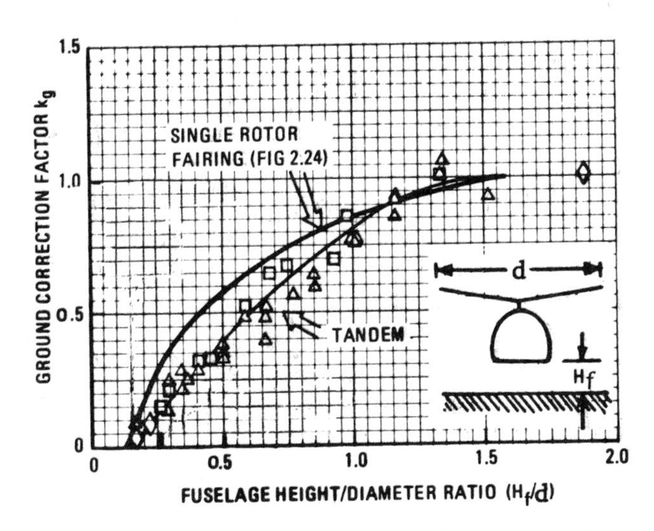

Figure 5.9 Tandem-rotor helicopter IGE download correction

3.5 Total Hover Power Required Sample Calculations

Sample calculations of power required in hover for the hypothetical tandem helicopter at 4000 ft/95°F ambient conditions are presented in tabular form in Table V-3. The isolated rotor performance, rotor overlap corrections, download adjustments, and IGE corrections described previously are used in these calculations, assuming that

① TOTAL T/σ; LBS $(T/\sigma)_{fr}$ + $(T/\sigma)_{rr}$	28482	24765	21092	17335	13620
② Total Isolated Rotor RHP/σ $(RHP/\sigma)_{fr}$ + $(RHP/\sigma)_{rr}$ Vortex Theory (Ch II)	3076	2476	1971	1525	1151
③ $(RHP/\sigma)_{ind}$ $(RHP/\sigma)_{ind_{fr}}$ + $(RHP/\sigma)_{ind_{rr}}$ Vortex Theory (Ch II)	2666.8	2104.8	1617.6	1181.8	806
④ $(\Delta RHP/\sigma)_{ind_{ov}}$ (Overlap Correction) ③ X 0.129	344	272	208	152	104
⑤ Total RHP/σ ② + ④	3420	2748	2179	1677	1255
⑥ SHP (⑤ X σ)/0.98 + 30	2848	2295	1826	1412	1064
⑦ W_{OGE}; lbs ① X σ X (1 – 0.047)	21918	19059	16233	13342	10483
⑧ W_{IGE}; lbs ⑦ X 1.128	24724	21498	18311	15050	11825
NOTE: 4000 ft/95°F					

TABLE V-3 HOVER POWER REQUIRED SAMPLE CALCULATIONS

each rotor carries exactly half of the total thrust. The transmission and accessory losses are assumed to be equal to the single-rotor aircraft values derived in Ch II, Section 1.6.

A comparison of the total hover shaft horsepower required, determined in Table V-3, with the corresponding single-rotor values developed in Ch II, is presented in Fig 5.10. At 4000 ft/95°F and 15,000-lb gross weight, the tandem-rotor helicopter requires 460 hp, or 22 percent less hover OGE power than the single-rotor aircraft, due primarily to the lower disc loading of the tandem and elimination of the tail rotor. The tail rotor requires power to develop trim thrust and increases the main rotor power required because of the interference effects described in Ref 4.

3.6 Hover Ceiling Capability

A comparison of the tandem and single-rotor helicopter hover ceiling capability for 95°F ambient conditions and intermediate power is presented in Fig 5.11. Here, the tandem hover OGE gross weight at 4000 ft/95°F is 19,450 lbs versus 16,000 lbs for the

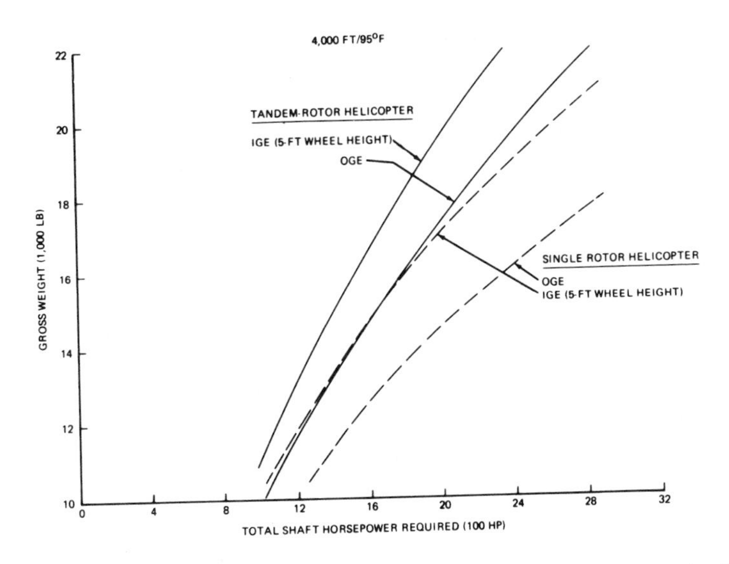

Figure 5.10 Single and tandem-rotor helicopter hover power required

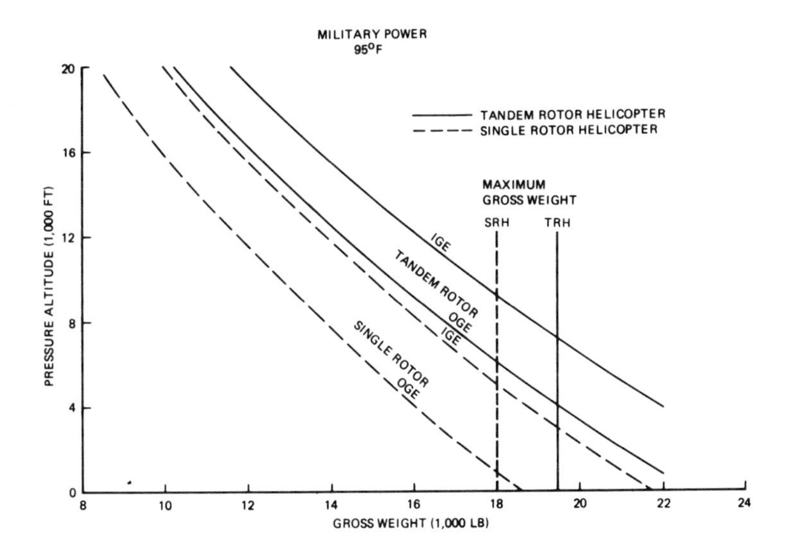

Figure 5.11 Hover ceiling for single and tandem-rotor helicopters

single-rotor configuration. In ground effect, the tendem has a 3,200-lb greater gross-weight capability than the single-rotor configuration.

4. FORWARD FLIGHT PERFORMANCE

Tandem-rotor helicopter level flight performance calculations consist of determining the aircraft drag and then computing the power required; taking into consideration the trim effects and the mutual aerodynamic interference effects of one rotor on the other. This is followed by estimating the power required in climb and descent. Finally, a comparison is made of the performance capability of the tandem helicopter with that of the single-rotor configuration.

4.1 Parasite Drag

The drag trend data presented in Ch 1, Fig 1.4, shows no significant difference between the single and tandem-rotor aircraft. Based on these results, the tandem drag is assumed equal to the single-rotor value of 19.1 ft^2 derived in Ch III.

As described in Ch III, the single-rotor helicopter drag is highest at forward c.g. positions due to the more nose-down fuselage angle of attack required to obtain moment balance and trim the aircraft through utilization of cyclic pitch. The trim fuselage angle of attack for tandems are less sensitive to c.g. variations (see Fig 5.12), because the lift distribution between the rotors required to trim the aircraft can simply be obtained through differential collective pitch. This also provides a larger c.g. range. However because of the nonoptimum load-sharing in the extreme c.g. positions, some penalty in rotor power may be encountered.

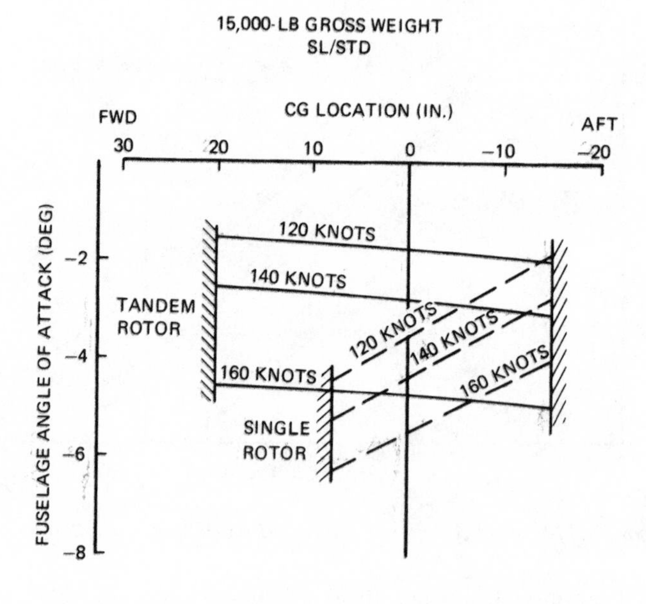

Figure 5.12 Effect of c.g. on fuselage angle of attack

The effect of airspeed on fuselage angle of attack is also evident in Fig 5.12. The tandem angle of attack is less nose-down than the single-rotor aircraft. This can be attributed to the difference in shaft incidence, disc loading, and longitudinal cyclic pitch angles as well as to the differential collective pitch. The tandem longitudinal cyclic angles are assumed to be programmed automatically as a function of forward speed or dynamic pressure, according to a cyclic trim schedule similar to the CH-47 aircraft (Fig 5.13). The automatic longitudinal cyclic control schedule with airspeed is typically selected for minimum noise, vibration, shaft stress levels and maximum performance capability.

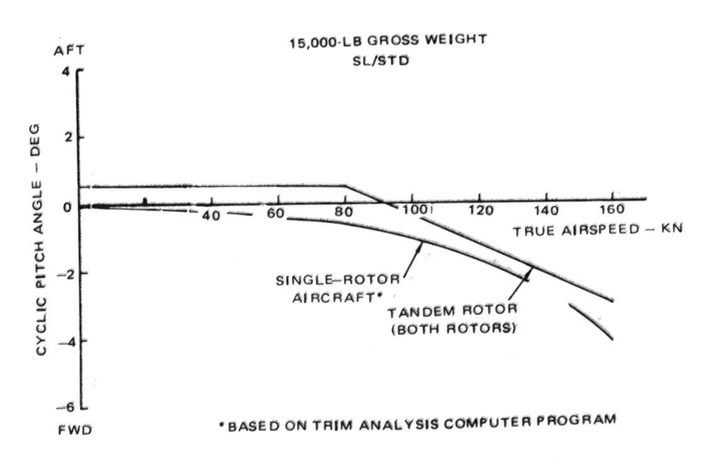

Figure 5.13 Hypothetical tandem-rotor helicopter cyclic trim schedule

4.2 Rotor-Rotor Interference Effects

In forward flight, the aft rotor operates in the wake of the forward rotor. The total axial velocity component of flow through the rear rotor is larger than for the isolated one. Consequently, its induced power is also higher. As described in Vol 1, Ch 11.6, this interference effect can be expressed in terms of the induced power factor k_{ind}, which is defined as $P_{ind}/2P_{id}$, where P_{ind} is the total forward and aft rotor induced power, including rotor-rotor interference effects as well as tip losses and blade root cutout effects. The term P_{id} is the ideal induced power of one rotor. As shown by the wind-tunnel data in Fig 5.14, k_{ind} is a function of the elevation of the rear rotor above the centerline of the forward rotor stream tube, h_{rr} (see Vol 1, Fig 2.25). A summary of this data, which includes a 10 percent correction for the tip losses and blade root cutout effects, was used in the theoretical correlation presented in Vol 1. Fig 5.14 also shows that the factor k_{ind} is maximum ($k_{ind} = 2.11$ at $h_{rr}/R \approx 0$) where the forward rotor wake passes directly through the aft rotor. The magnitude of k_{ind} decreases when the forward rotor stream tube moves above the rotor, as in autorotation; or below the rotor, as in level flight and climb.

The test results shown in Fig 5.14 agree with predictions based on the momentum theory considerations described in Vol 1. In addition, biplane theory[5] has been used successfully to predict rotor-rotor interference effects by assuming an effective wing

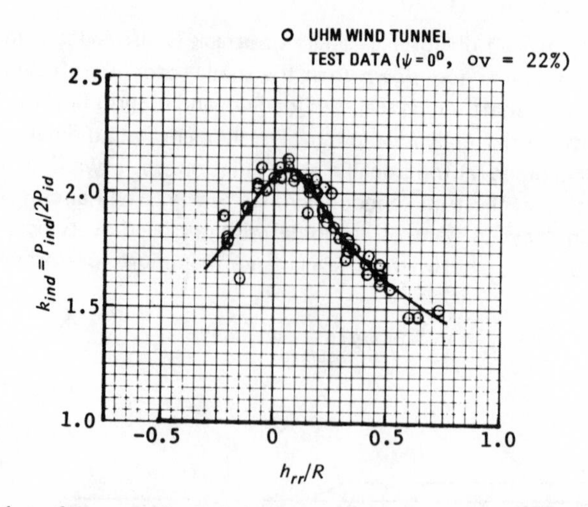

Figure 5.14 Interference effect of the forward rotor on the rear rotor induced power

span-to-diameter ratio of 0.85 in level flight, and 1.0 in autorotation. The theoretical aspects of rotor-rotor interference effects are also discussed in Ref 6.

The test data in Fig 5.14 was obtained for a 22-percent overlap model configuration where rotor wake separation distance h_{rr} was varied by changing the forward rotor thrust, shaft angle, and forward speed. However, the trends agree with 0 and 34-percent overlap test data using the h_{rr}/R technique.

In order to use Fig 5.14 for the hypothetical tandem, the value of parameter h_{rr}/R as a function of the distance between the rotors (d_R) and the forward wake separation angle γ must be determined. The angle γ is a function of aircraft pitch attitude (θ_f), forward rotor downwash angle (ϵ), and aft rotor hub elevation angle (γ_o) as shown in Fig 5.15. The specific relationships are:

$$h_{rr}/R = -2(d_R/d) \sin \gamma \tag{5.5}$$

and

$$\gamma = \theta_f - \epsilon - \gamma_o \tag{5.6}$$

where

$\epsilon = \tan^{-1}(k_d v_{id}/V$; deg

$\gamma_o = \tan^{-1}(g/d_s)$; deg

v_{id} = ideal forward flight induced velocity based on momentum theory

$$v_{id} = v_o\sqrt{-\tfrac{1}{2}(V/v_o)^2 + \sqrt{\tfrac{1}{4}(V/v_o)^4 + 1}} \tag{5.7}$$

$$v_o = \sqrt{T/4\rho\pi R^2}$$

The factor k_d accounts for the variation in local downwash velocity v_ℓ from the ideal value v_{id} where $v_\ell = k_d v_{id}$. The k_d value was determined from test data as follows:

$$k_d = 0.043/(\mu + 0.043). \tag{5.8}$$

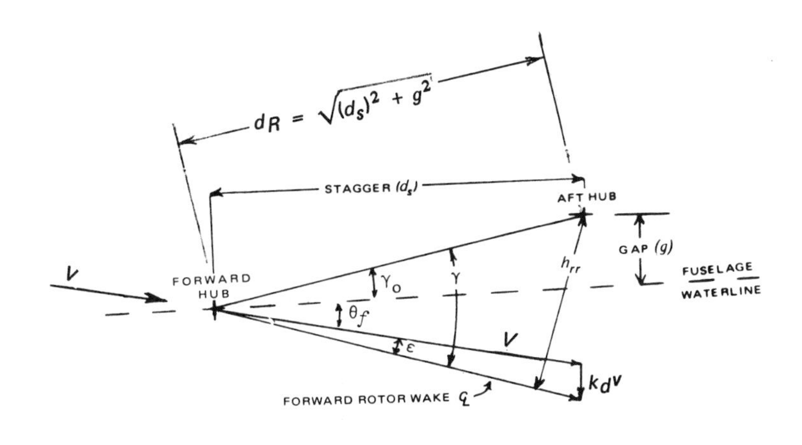

Figure 5.15 Definition of wake separation distance

It should be noted that test data for 0, 22, and 34 percent overlap configurations have shown no significant interference effects of the aft rotor on the forward one. Theoretical predictions[6] have generally confirmed these results.

4.3 Level Flight Power Required Calculations

The tandem rotor forward-flight power required at 4000 ft/95°F, based on a tandem-rotor trim analysis computer program, is presented in Fig 5.16. This program is similar to the single-rotor trim program described in detail in Ch III, and includes the overlap correction shown in Fig 5.14.

A comparison of the single and tandem-rotor helicopter power required at the 15,000-lb design gross weight and 4000 ft/95°F is presented in Fig 5.17. The tandem power required is shown with no yaw ($\psi = 0°$), as well as with optimum yaw angles. Current production aircraft such as the CH-46 utilize yaw to improve single-engine performance as described later in this section. At airspeeds below 60 knots, the tandem, because of its lower disc loading, has lower power required. The tandem power without yaw is 5 percent higher than the single rotor at the minimum power required airspeed and, at optimum yaw angles, the tandem and single-rotor power required are approximately the same. At airspeeds above 150 knots, the tandem and single-rotor aircraft have similar power requirements.

The increase in induced power attributed to rotor interference for the hypothetical tandem having a gross weight of 15,000 lbs at 4000 ft/95°F is presented in Fig 5.18. The largest penalty for interference effects (23 percent of SHP) occurs at the minimum power airspeed, and the increment decreases at both higher and lower airspeeds. Eqs. 5.5 thru 5.8 and Figs 5.14 and 5.15 were used in the detailed sample calcutions for the rotor-rotor interference power presented in Table V-4. The average aircraft pitch attitude of 0° at airspeeds between 80 and 150 kn (see Fig 5.19) used in the sample calculations is based on a trim analysis computer program.

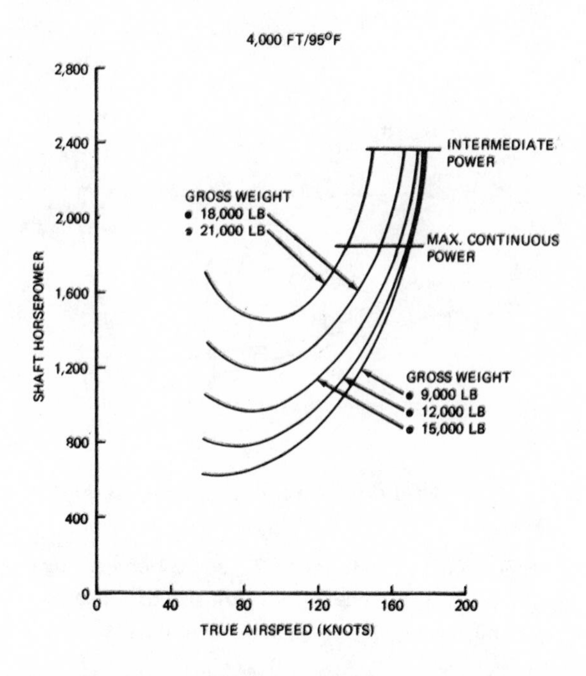

Figure 5.16 Level flight power required for tandem-rotor helicopters

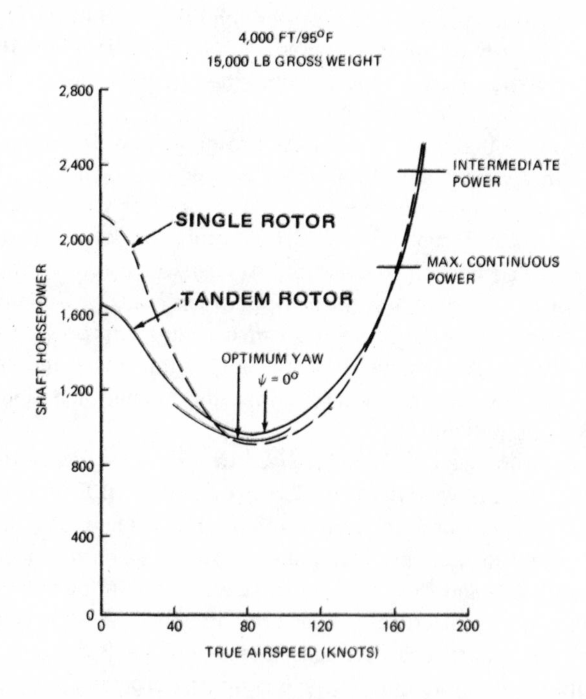

Figure 5.17 Comparison of single and tandem-rotor power required

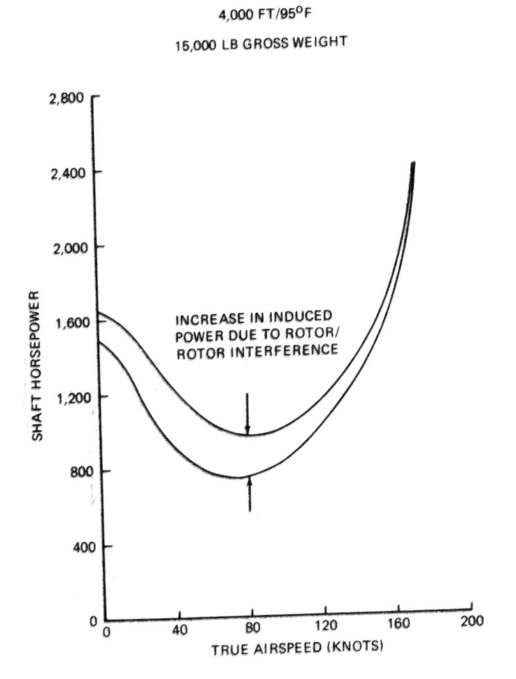

Figure 5.18 Hypothetical tandem rotor-rotor interference effect on power required

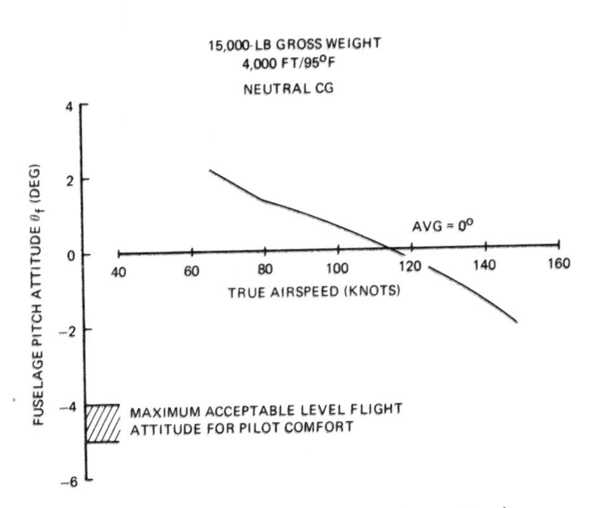

Figure 5.19 Tandem-rotor trim attitude

For emergency one-engine inoperative conditions, the minimum power required can be reduced by decreasing the main-rotor tip speed and flying with nose-right yaw. Decreasing the tip speed reduces the profile power, while employing right yaw tends to reduce the rotor-rotor interference effects by moving the aft rotor out of the wake of

4000 FT/95°F
GROSS WEIGHT (W) = 15,000 LB

STEP NO. ①	②	③	④	⑤	⑥	⑦	⑧	⑨
V (KN)	$\frac{2P_{id}}{W^2}$ / $2200\rho\pi R^2 1.69V$	P_{ind}* ② × 1.1	v_{id} (fps) Eq. 5.7	v_{id}/V ④ / (① × 1.69)	k_d Eq. 5.8	ϵ (deg) $\tan^{-1}$ [⑥ × ⑤]	θ_f (DEG)	γ_o (deg) $\tan^{-1}[g/ds]$
60	329	362	11.98	0.1182	0.229	1.55	0	8.53
80	247	272	9.03	0.0668	0.182	0.70	↓	↓
100	197	217	7.23	0.0428	0.151	0.37		
120	165	182	6.03	0.0298	0.129	0.22		
140	141	155	5.17	0.0219	0.113	0.14		
160	123	135	4.53	0.0168	0.100	0		

①	⑩	⑪	⑫	⑬	⑭	⑮	⑯	⑰
V (KN)	γ (DEG) ⑧ − ⑦ − ⑨ Eq. (5.6)	d_R/R $(1 - ov)$ × ②	h_{rr}/R − ⑪ sin ⑩ Eq. (5.5)	k_{ind} Fig. 5.14	$P_{id} \times k_{ind}$ ② × ⑬	ΔP_{ind} ⑭ − ③	SHP_{TOTAL} Fig. 5.16	SHP_{ISO} ⑯ − ⑮/0.98
60	−10.08	1.32	0.231	1.92	632	270	1048	772
80	− 9.23	↓	0.212	1.95	482	210	967	753
100	− 8.90		0.204	1.96	386	169	1011	839
120	− 8.75		0.201	1.96	323	141	1149	1005
140	− 8.67		0.199	196	276	121	1372	1249
160	− 8.53		0.196	197	242	107	1725	1616

NOTE: *Pind = Pid corrected for tip losses and cutout effects.

TABLE V-4 CALCULATIONS OF FORWARD ON AFT ROTOR INTERFERENCE POWER REQUIRED IN LEVEL FLIGHT

the forward rotor in order to decrease the induced power. The performance benefits of yaw angle have been verified by H-21, CH-46, and CH-47 flight test measurements as well as model tests conducted by both Vertol in the USA and Azuma in Japan[7].

An example of the effect of yaw angle on the CH-47B power required is presented in Figs 5.20 and 5.21. Data is shown for nose-right attitudes only, since less benefit was observed in left yaw. The differences in performance between left and right yaw are attributed to the lateral downwash asymmetry of the forward-rotor wake. For aircraft with the forward rotor turning counterclockwise, it was shown – both theoretically (Vol. 1, Fig 42.8) and experimentally (flight tests and wind tunnel model tests) – that on the advancing side of the rotor, higher downwash velocities are generated than on the retreating side. Therefore, right yaw moves the rear rotor away from the high downwash region of the forward rotor wake, while left yaw produces the opposite effect.

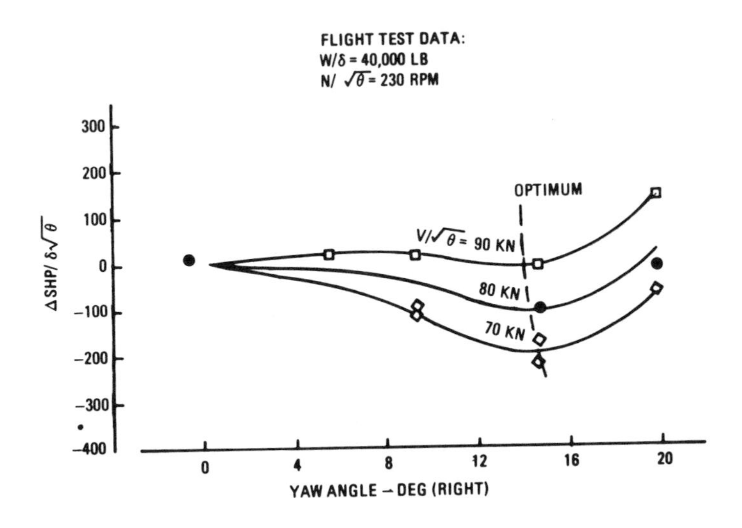

Figure 5.20 Effect of yaw angle on CH-47B power required

It can be seen from Fig 5.20 that the power required decreases as yaw angle increases until an optimum angle is obtained, and then begins to increase as the fuselage parasite drag effects begin to offset the reduction in induced power. The fuselage drag penalty also causes the benefits of yaw angle to diminish with increasing airspeed as shown in Fig 5.21.

The CH-47 flight test data indicates that the optimum yaw angle for minimum power required is between 14° and 15°. Similar testing on the CH-46 helicopter showed 15° to be the optimum angle. The nondimensional power required reduction obtained at these angles for the CH-46 and CH-47 is presented in Fig 5.22. The average reduction in power is $\Delta C_P = -1 \times 10^{-5}$. Since the study helicopter has the same overlap (34 percent) as the CH-47 and CH-46, this increment is applied to predict the minimum power required at 4000 ft/95°F (see Figs 5.23 and 5.17).

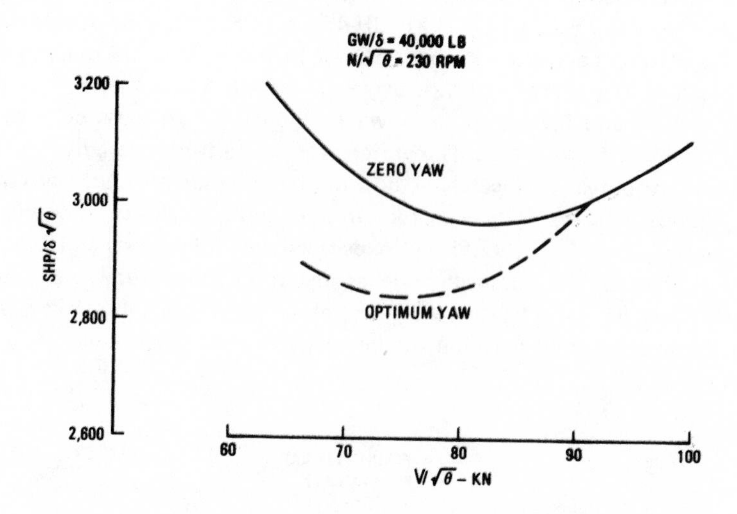

Figure 5.21 Effect of airspeed on yawed performance for the CH-47B

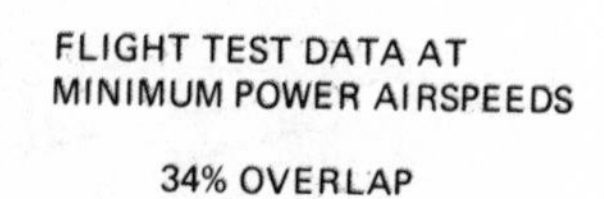

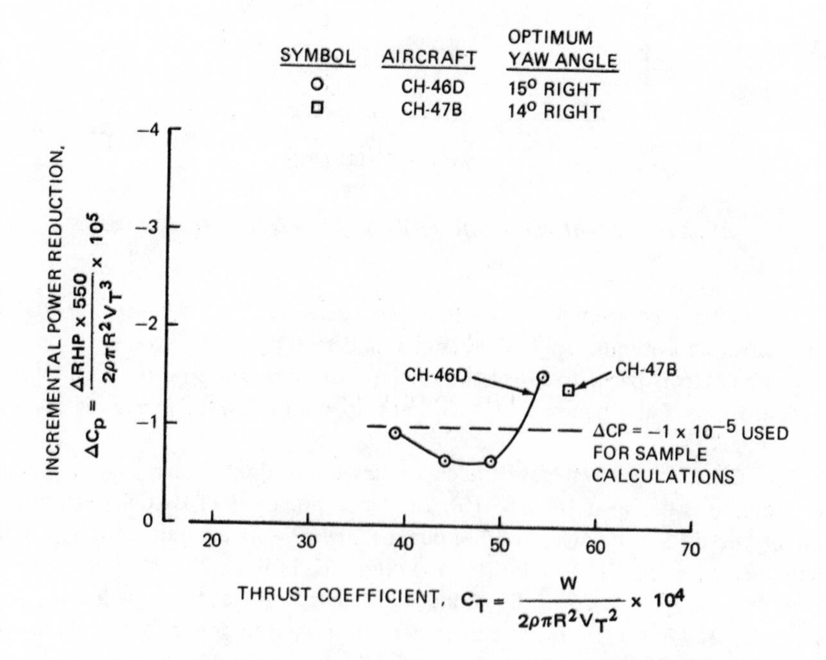

Figure 5.22 Reduction in minimum power required with yaw

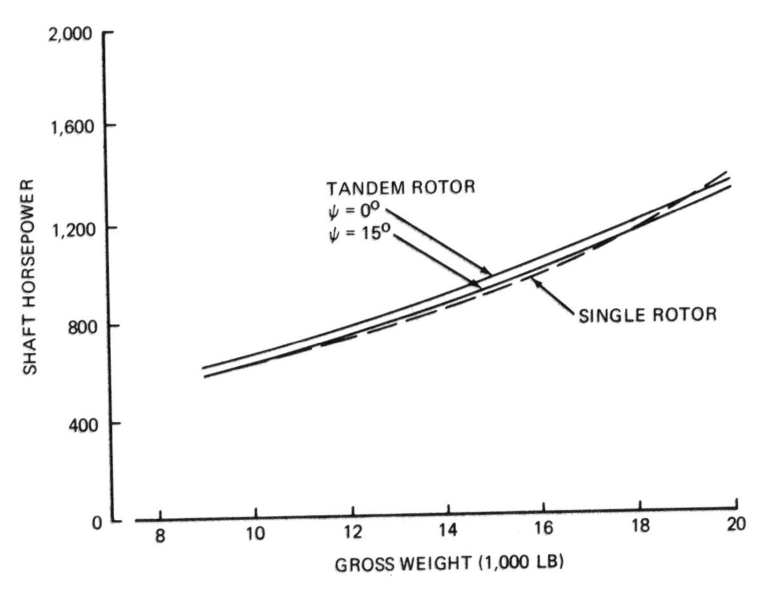

Figure 5.23 Single and tandem-rotor minimum power required at 4000 ft/95° F

4.4 Power Required in Climb and Descent

A comparison of single and tandem-rotor power required in climb and descent is presented in Fig 5.24. This comparison, based on a tandem-rotor trim analysis computer program, is for a gross weight of 15,000 lb, SL/STD conditions, a true airspeed of 80 kn, and includes the rotor-rotor interference correction defined in Fig 5.14. As shown in Fig 5.24, the tandem exhibits a more gradual rise in power required with rate of climb than the single-rotor aircraft. At climb rates between 0 and 1000 fpm, the difference in slope ($\Delta SHP/\Delta V_c$) reflects an increase in the climb efficiency factor (k_p) from 0.85 for the single-rotor helicopter (Ch III) to 0.95 for the tandem. The increased efficiency of the tandem in climb is attributed to (1) the elimination of tail-rotor power requirements, which are sizeable at the higher main-rotor power and torque levels associated with climb, and (2) a decrease in rotor-rotor interference effects with increasing rates of climb. The climb velocity component increases the nose-down fuselage angle of attack which, according to Eqs (5.5) and (5.6), increases the distance between the forward rotor wake and the aft centerline (h_{rr}/R) and thus reduces the induced power factor k_{ind}.

As noted in Ch III, k_p varies with weight and airspeed; however, flight test measurements and theoretical trim analyses for various tandem-rotor aircraft indicate that tandems have average climb k_p values on the order of 0.85 to 0.95 at minimum power speed, while corresponding single-rotor k_p values vary from 0.8 to 0.9. A k_p of 0.95 based on the trim analysis data presented in Fig 5.24 is used for the hypothetical tandem-rotor aircraft.

As indicated in Fig 5.24, the average slopes of the rate of descent vs SHP curves for the two aircraft are almost identical. This implies that the descent correlation factor

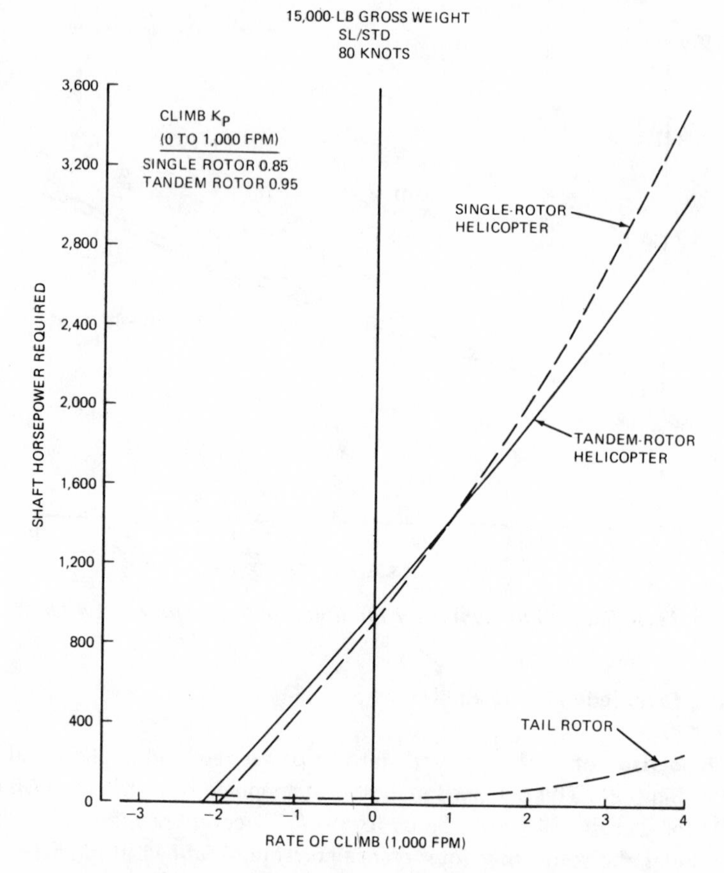

Figure 5.24 Single and tandem-rotor power required in climb and descent

k_{pd} = 1.0 developed in Ch III (Fig 3.29) for the single-rotor aircraft is also applicable to the tandem-rotor helicopter. Test measurements on tandem and single-rotor aircraft show that the use of k_{pd} = 1.0 will usually give a conservative estimate of autorotation rate of descent.

4.5 Performance Capability

Payload-Range Relationship. A comparison of the payload-range performance of the single and tandem-rotor hypothetical helicopters at 4000 ft/95°F is shown in Fig. 5.25. Assuming that the takeoff weight can not exceed the single-engine 100-ft/min (service ceiling) capability, the tandem and single-rotor takeoff weights (17,700 lb) and payload-range capability are approximately the same. The single rotor will require an in-ground-effect takeoff to achieve this weight since its hover OGE weight is only 16,000 lb. The tandem has a hover OGE capability of 19,450 lb. Takeoff weights in excess of the single-engine capability can be used on the tandem for external load

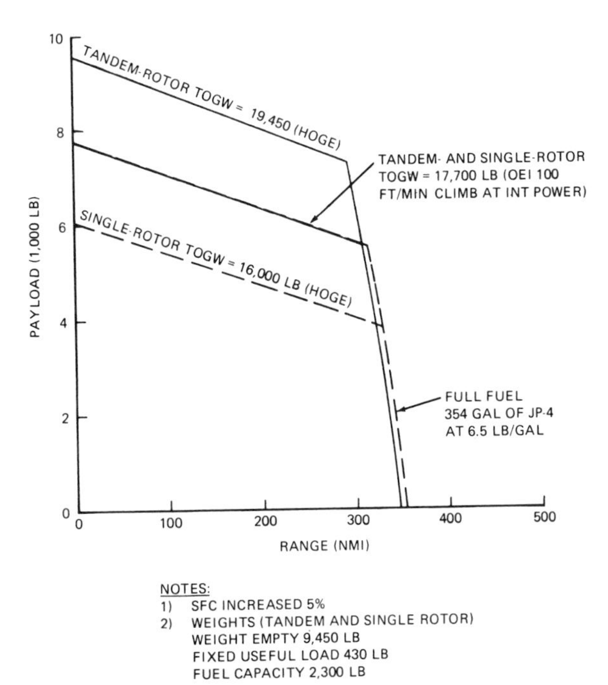

Figure 5.25 Comparison of payload-range capabilities at 4000 ft/95°F

operations where the load can be dropped in the event of an engine failure; however, the range performance would be less than shown in Fig 5.25 because of the increase in drag of the external load.

The cruise performance inputs used in the range calculations for 4000 ft/95°F are presented in Fig 5.26. Specific range data for the tendem and single-rotor aircraft are shown in the lower half of this figure, and speed capability comparisons are given in the upper half. The tandem best range airspeeds are higher than the single-rotor values and the tandem-rotor aircraft specific range is within 2 percent of the single-rotor performance.

Speed Capability. The speed of the tandem and single-rotor helicopters at maximum continuous and intermediate power is also indicated in Fig 5.26. At a gross weight of 15,000 lb, the tandem speed capability is 161 knots for maximum continuous power, versus 160 knots for the single-rotor aircraft.

Climb Cpaability. The tandem and single-rotor helicopter climb capability at 4000 ft/95°F is shown in Fig 5.27. The tandem dual-engine climb performance is defined for 0° yaw and the single-engine climb is at 15° right yaw. At 15,000-lb gross

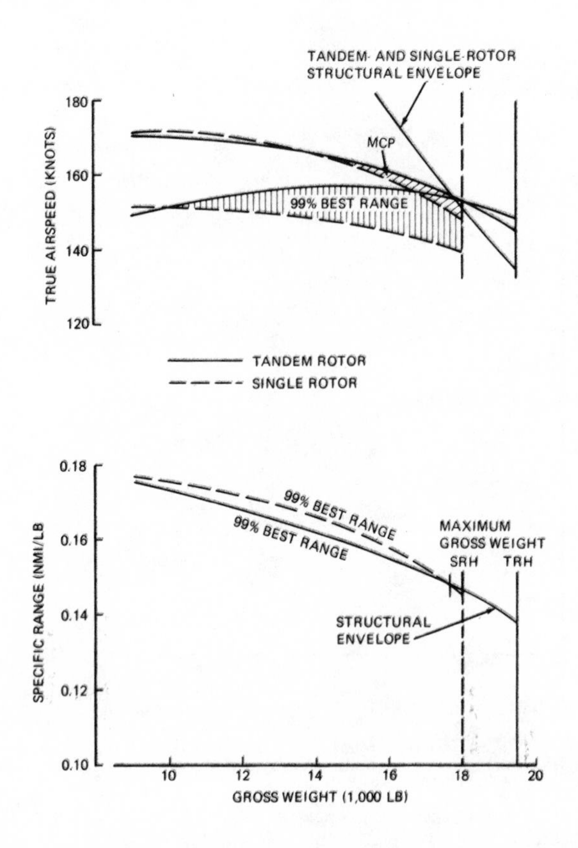

Figure 5.26. Comparison of cruise performance at 4000 ft/95°F

weight, the tandem and single-rotor dual engine climb performance capabilities are within 100 ft/min and their single-engine climb rates are within 40 ft/min.

Autorotation Capability. The tandem-rotor helicopter autorotation rate of descent at 4000 ft/95°F and 15,000-lb gross weight is presented in Fig 5.28 for a descent calculation factor k_{pd} = 1.0. The tandem and single-rotor minimum rate of descent are within 130 ft/min and the maximum glide distance rate of descents are within 300 ft/min.

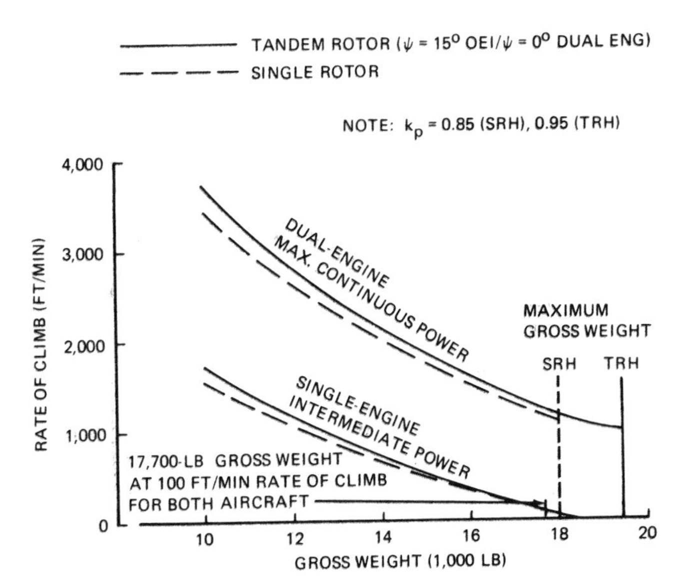

Figure 5.27 Comparison of climb capabilities at 4000 ft/95° F

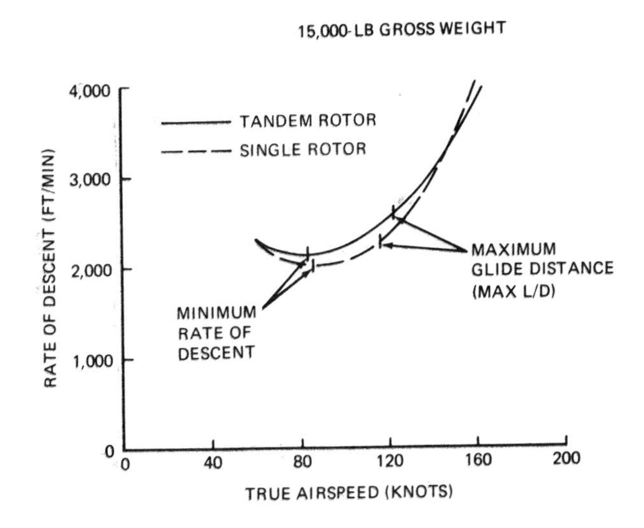

Figure 5.28 Comparison of autorotational rates of descent at 4000 ft/95° F

References for Ch V

1. Hoerner, S.F. *Fluid Dynamic Drag.* Published by the author, 1965 Edition.

2. Hoerner, S.F. *Base Drag and Thick Trailing Edges.* Journal of Aeronautical Sciences. October 1950.

3. Dutton, W.J. *Parametric Analysis and Preliminary Design of a Shaft-Driven Rotor System for a Heavy-Lift Helicopter.* USAAVLABS Technical Report 66-56. Sikorsky Aircraft Div. of United Aircraft Corp. February 1967.

4. Wiesner, Wayne and Kohlar, Gary. *Tail-Rotor Design Guide.* USAAMRDL TR-73-99, D210-10687-1. Eustis Directorate. January 1974.

5. Durand, W.F. *Aerodynamic Theory.* Vol. II, Durand Reprinting Committee, 1943.

6. Baskin, V.E., et al. *Theory of the Lifting Airscrew.* Translated from Russian to English by W.Z. Stepniewski. NASA TT F-823. February 1976.

7. Azuma, A. et al. *Effect of Yaw on Tandem-Rotor Helicopter,* Part III. Institute of Space and Aeronautical Science, University of Tokyo, Japan. February 1969.

PERFORMANCE GUARANTEES

Before initiating performance calculations, the accuracy of the intended theoretical methodology must be established. The level of accuracy required for preliminary design or trade studies is usually lower than that required for final proposals to the government or private customers. An evaluation of the reliability of prediction techniques is especially important prior to presenting guaranteed performance figures, since noncompliance may result in cost penalties or possible cancellation of the program.

The accuracy of performance methodology is best verified by comparing predictions with flight test data. Examples of such comparisons are presented in Fig A.1 for a single-rotor (BO–105), and in Fig A.2 for a tandem-rotor (Model 347) configuration. It can be seen from these figures that the test data agrees reasonably well with the predictions based on the methods described previously, as well as with the material contained in Ref 1 from which the figures were taken. While these comparisons represent a satisfactory substantiation of performance evaluation for preliminary design studies, a more detailed quantitative assessment is required in order to determine specific guaranteed values. With respect to this subject, a statistical approach can be helpful in selecting performance guarantees. The basic method of analysis consists of examining the accuracy

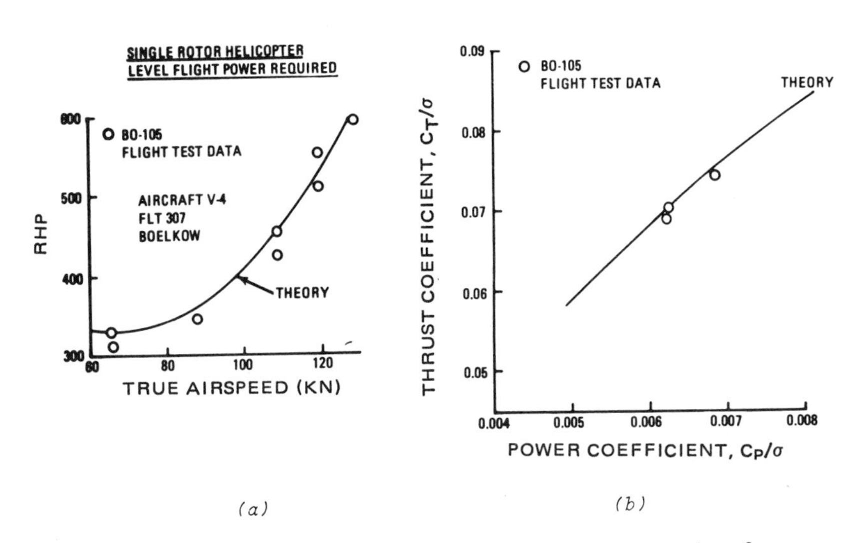

Figure A.1 Comparisons of predicted and flight-test measured performance for a single-rotor helicopter

of the theoretical predictions and evaluating the tolerances associated with the acquisition of flight test data, since final guarantee compliance will ultimately be determined in this manner. Numerical examples for the hypothetical helicopter are included to illustrate the calculation procedure. It should be noted that the theoretical performance tolerances assigned to the sample problems are also hypothetical, and are intended for demonstration purposes only.

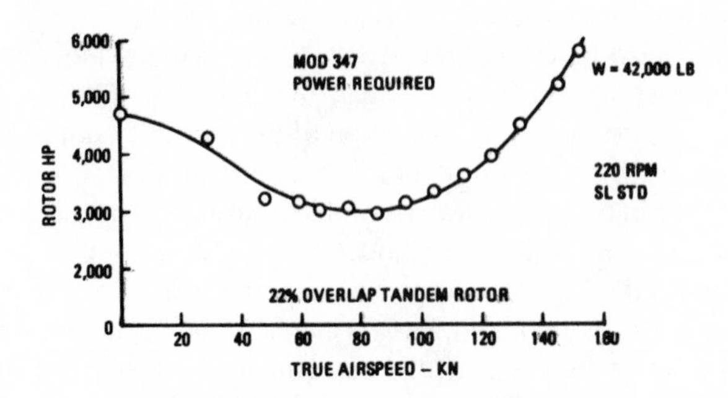

Figure A.2 Comparison of predicted and flight-test measured power required for a tandem

1. Guarantee Items

The performance items to be guaranteed for a new design or for growth versions of existing aircraft are usually specified by the customer and negotiated with the manufacturer. The performance guarantees selected are normally associated with a specific design mission and typically include the following: (a) hover capability, (b) payload, (c) radius, (d) endurance, (e) cruise speed, and (f) one-engine-inoperative service ceiling and speed capability.

When the guarantees are negotiated, the ground rules for determining compliance should be well defined. For instance, the engine performance used to compute the guarantees, the aircraft external configuration (drag), and details of flight testing should be agreed upon and specified in the contract.

2. Method of Analysis

The procedure for establishing performance guarantees consists of determining—as objectively as possible—the probability that a given performance level can be achieved, and then selecting a specific value, depending on the risks and competitive pressures involved. The probability values assigned to a given performance level are determined by evaluating both the accuracy of theoretical predictions and the tolerances associated with flight-test measurements. The most optimistic as well as the most pessimistic incremental values are then assigned to each element or step in the analytical predictions and flight-data reduction system. In this way, for each performance item (say, horizontal speed of flight at given ambient conditions and prescribed power setting) two levels (x'_1 and x'_2) can be established: one – the lowest x'_1, where probability of at least meeting the performance level in the accepted flight measurements will be very high (99.8%); and another – x'_2, where the probability of achieving that level will be very low (0.13%). In other words, it may be stated that the probability (P) of the actual measured value (x), being lower than x'_1, will be $P(x \leqslant x'_1) = 0.13\%$ and conversely,

$P(x \leqslant x_2) = 99.87\%$. Between these two extremes lie intermediate values of that performance item, thus representing various levels of the probability of achieving them.

For those who are not familiar with engineering statistics and thus may not grasp the significance of the above-mentioned 0.13 and 99.87 percentages, some basic techniques of dealing with probabilities are briefly outlined (Ref 2, Ch 7).

Normal Distribution – The so-called normal (Gaussian) distribution represents the most widely used method of determining the probability of a deviation of the random variable (x) from its mean value μ. One states that x has the normal distribution $N(\mu,\sigma^2)$ if its *probability density function (pdf)* is

$$f(x) = \frac{1}{\sigma\sqrt{2\pi}} \exp\left[-\frac{1}{2\sigma^2}(x - \mu)^2\right], \qquad -\infty < x < \infty, \tag{A1}$$

where the symbol σ is the *standard deviation* of x from its *mean value* (μ). σ is determined from the so-called variance (σ^2) which, for continuous functions, is defined as

$$\sigma^2 = \int_{-\infty}^{\infty} (x - \mu)^2 f(x)dx \tag{A2}$$

or for a finite number (n) of known differences between μ and x, σ^2 can be found from the following equation:

$$(n - 1)\sigma^2 = \sum_{j=1}^{n} (x_j - \mu)^2. \tag{A3}$$

Remembering that the arithmetic mean (μ) is $\mu = (1/n) \sum_{j=1}^{n} x_j$,

Eq (A3) can be rewritten in a form more suitable for practical application:

$$(n - 1)\sigma^2 = \sum_{j=1}^{n} x_j^2 - (1/n)\left(\sum_{j=1}^{n} x_j\right)^2 \tag{A3a}$$

The shape of the normal *pdf* curve corresponding to $f(x)$ given by Eq (A1) is shown in Fig A.3.

It can be seen that the normal *pdf* curve is symmetrical about the mean value μ, and points of inflection are located at $\pm\sigma$ from the μ value line.

The probability that some value x will be smaller or equal to a given value x', $P(x \leqslant x')$ can now be expressed from Eq (A1) as follows:

$$P(x \leqslant x') = \int_{-\infty}^{x'} \frac{1}{\sigma\sqrt{2\pi}} \exp\left[-\tfrac{1}{2}\left(\frac{x - \mu}{\sigma}\right)^2\right] dx. \tag{A4}$$

It should be noted that $(x - \mu)/\sigma$ represents the ratio of the difference between x and its mean value (μ) to the standard deviation (σ). This ratio can be treated as a new variable; $z \equiv (x - \mu)/\sigma$, and Eq (A4) can be rewritten as follows:

$$P(x \leqslant x') = \int_{-\infty}^{(x'-\mu)/\sigma} \frac{1}{\sqrt{2\pi}} e^{-z^2/2}\, dz = \Phi\left(\frac{x' - \mu}{\sigma}\right) \tag{A4a}$$

It is interesting to note that for $x = \mu$, $z = 0$; and for $x = \mu \pm 3\sigma, z = \pm 3$. This relationship between the z and x scales can also be seen in Fig A.3. Furthermore it can be shown that Eq (A4a) expresses the statement of probability that $x \leqslant x'$ is the same as the probability that $z \leqslant (x' - \mu)/\sigma \equiv z'$ (shaded area in Fig A.3).

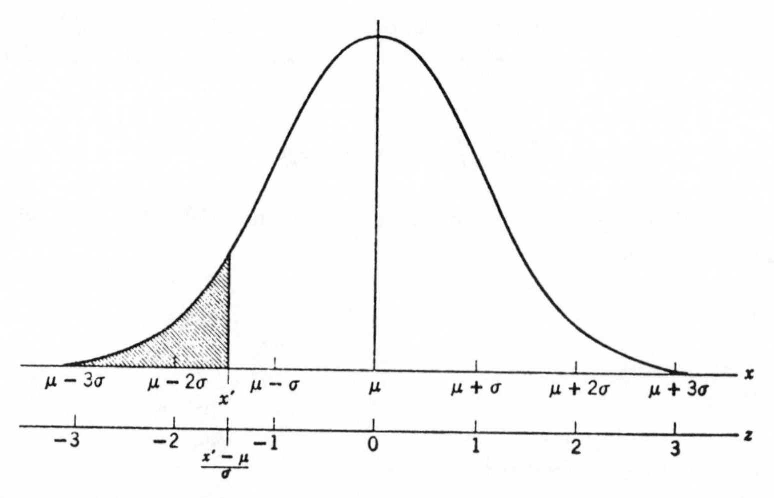

Figure A.3 Graph of the normal probability density function vs x and z scales

When $x' = \pm\infty$, then $z' = \pm\infty$ also; and Eq (A4a) – when integrated from $-\infty$ to ∞ – becomes unity. For $x' = \mu + 3\sigma$ or, in other words, $z' = (\mu + 3\sigma - \mu)/\sigma = 3$ (z-scale in Fig A.3); the limits of integration become $z' = -\infty$ to $z' = 3$, and there is a 99.87 percent probability of finding that $x \leqslant \mu + 3\sigma$; i.e., $P(x \geqslant \mu + 3\sigma) = 0.13$ percent. Similarly, for $x' = \mu - 3\sigma$; $z' = -3$; and $P(x \leqslant \mu - 3\sigma) = 0.13$ percent, or $P(x \geqslant \mu - 3\sigma) = 99.87$ percent.

The character of the complete variation of probabilities within the $x = \mu \pm 3\sigma$ or $z \pm 3$ limits is shown in Fig A.4. However, instead of taking readings from a chart such as this, the so-called probability paper (Fig A.5) is often used for determining probabilities, or checking the distribution of sampling. By plotting the probability expressed by Eq (A4) on logarithmic paper, a straight-line is obtained. Such graphs are used in determining the probability levels when establishing guarantees for various performance items. The procedure described below is an example of guaranteed values for high-speed level-flight velocity.

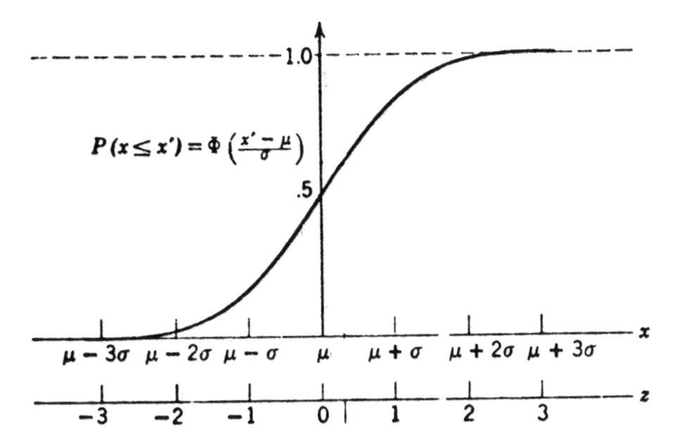

Figure A.4 Character of variation of probabilities corresponding to the normal pdf

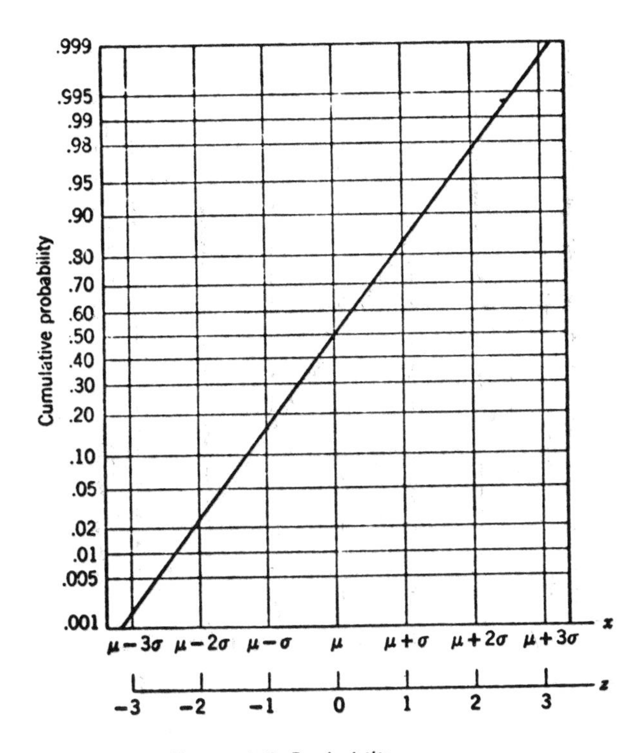

Figure A.5 Probability paper

Example of Performance Guarantee Determination – The procedure of establishing the most pessimistic (V_{pes}) and most optimistic (V_{opt}) true airspeeds achievable under given ambient conditions and power settings are shown in Table A-1.

ITEMS	SENSITIVITY (K)	EXPECTED DEVIATION $(\Delta\ VALUE)$		VARIANCE $\sigma_n^2 = [K(\Delta\ VALUE)/3]^2$			
		MOST PESSIMISTIC	MOST OPTIMISTIC	MOST PESSIMISTIC		MOST OPTIMISTIC	
				$-3\sigma_n$	σ_n^2	$3\sigma_n$	σ_n^2
		PREDICTION TOLERANCES					
ROTOR HP REQUIRED	−0.467 KN/% HP	6% HP	−3% HP	−2.8 KN	0.872	1.4 KN	0.218
ENGINE INSTALLATION LOSSES & RAM RECOVERY	−0.467 KN/%	1%	−1%/HP	−0.467	0.02423	0.467	0.02423
TRANSMISSION & ACCESSORY LOSSES	−0.0182 KN/HP	16 HP	−16 HP	−0.291	0.0094	0.291	0.0094
		FLIGHT TEST TOLERANCES					
POWER REQUIRED	−0.467 KN/% HP	2% HP	−2% HP	−0.934	0.0969	0.934	0.0969
WEIGHT	−0.0015 KN/LB	150 LB	−150 LB	−0.225	0.005625	0.225	0.005625
AIRSPEED	1 KN/KN	2 KN	−2 KN	−2 KN	0.444	2 KN	0.444
			$\sigma^2 = \Sigma\sigma_n^2$	1.452		0.7982	
			3σ, KN	−3.62		2.68	
				$V_{pes} = 173 - 3.61 \approx 169.4$ KN		$V_{opt} = 173 + 2.67 \approx 175.7$ KN	

TABLE A-1 DETERMINATION OF EXTREME V_{max} VALUES

It is assumed that each of the considered items represents the 3σ-type deviation from the calculated value of $V = 173$ *kn*. The total variance (σ^2) is the sum of particular variances (σ_n^2) and it is again assumed that the most pessimistic value will be given by $173 - 3\sigma$ ($V_{pes} = 169.4$ *kn*), and the most optimistic by $173 + 3\sigma$ ($V_{opt} = 175.7$ *kn*).

If V_{opt} represents the $\mu + 3\sigma$ situation, then the probability of reaching or exceeding that value is only 0.13 percent. By contrast, the probability that measured flight velocity values will be equal or better than the most pessimistic ($173 - 3\sigma$) figure will amount to 99.87 percent. Points (0.13%, 175.7 kn) and (99.87%, 169.4 kn) are marked on the probability paper (Fig A.6). It should be noted that in this figure, the coordinate axes are reversed in comparison with Fig A.5.

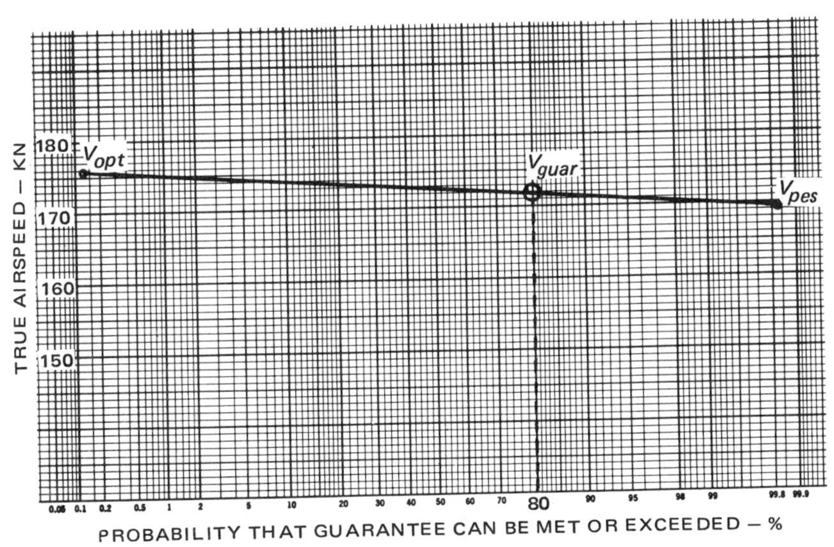

Figure A.6 Determination of probability

The normal probability distribution for the interim velocity values ($169.4 \leqslant V \leqslant 175.7$) is obtained by joining the two points in Fig A.6 by a straight line. It can now be seen that should one desire an 80 percent probability that actual flight-test values will be either better than, or at least equal to, the guaranteed velocity (V_{guar}), then $V_{guar} = 171.5$ *kn* should be quoted.

Prediction Variables – The accuracy of performance predictions depends on both the number and values of the considered variables. For example, if nondimensional data obtained from flight tests of a similar aircraft or prototype is used as the basis for predictions, both the number and tolerance limits of the variables will be low. Conversely, if unsubstantiated theory is used to estimate the performance of advanced aircraft, the number of variables influencing the answer and the magnitude of the tolerance limits will be high. For this reason, it is essential that performance guarantees are based on the best validated performance methodology available.

The major factors affecting the accuracy of hover and forward-flight performance predictions are listed in Table A-II where it is assumed that there is no available prototype test data, and the estimates are based exclusively on the analytical computations presented in Chs II and III.

FLIGHT REGIME	ITEM	PREDICTION VARIABLE
HOVERING	1	MAIN & TAIL ROTOR POWER REQUIRED
	2	DOWNLOAD
	3	ENGINE INSTALLATION LOSSES
	4	TRANSMISSION & ACCESSORY LOSSES
	5	GROUND EFFECT
VERTICAL CLIMB	1–4	SEE HOVER
	5	CLIMB EFFICIENCY FACTOR
HORIZONTAL FLIGHT	1–4	SEE HOVER. IN ITEM 1, ATTENTION SHOULD BE GIVEN TO THE PARASITE DRAG ESTIMATE OF THE AIRFRAME & BLADE PROFILE DRAG CONTRIBUTION

TABLE A-II PREDICTION VARIABLES

Test Tolerances – The tolerances on test data are primarily a function of the type and accuracy of the instrumentation, and the specific prototype or production aircraft selected for the compliance evaluation. A list of the parameters generally recorded during a performance evaluation program and the accuracy of typical instrumentation employed for guarantee compliance testing is presented in Table A-III.

ITEM	MEASUREMENT	TOLERANCE
1	POWER (FROM FUEL FLOW OR ROTOR TORQUE)	±2%
2	TRUE AIRSPEED	±2 KN
3	GROSS WEIGHT	±150 LB*
4	ROTOR RPM	UP TO ±1/4 RPM
5	ALTITUDE	DEPENDS ON ALTIMETER POSITION ERROR & READING ERROR
6	OUTSIDE TEMPERATURE	±1°C
7	WHEEL-HEIGHT, IGE	DEPENDS ON TEST PROCEDURE (TETHERED OR FREE FLIGHT)
NOTE: ITEMS 4, 5, AND 6 INFLUENCE TOLERANCES OF POWER MEASUREMENTS		

*For aircraft of the 15,000-lb gross weight class

TABLE A-III FLIGHT–TEST PERFORMANCE MEASUREMENTS AND THEIR TOLERANCES

References for Appendix A

1. Davis, S. Jon and Wisniewski, J.S. *User's Manual for HESCOMP, the Helicopter Sizing and Performance Computer Program.* Boeing Vertol Report D210-10699-1 (Contract NAS2-6107), NASA CR-152018, Ames Research Center, Moffett Field, Ca. Sept. 1973.

2. Guttman, Irwin; and Wilks, S.S. *Introductory Engineering Statistics.* John Wiley & Sons, N.Y. 1965.

AIRCRAFT GROWTH

The weight empty of the completed prototype aircraft is often higher than originally predicted. This is due to changes in equipment weight, configuration modifications and the accuracy of statistical weight prediction trends employed during early design phases when detailed component drawings were not available. During the service life of an aircraft, its weight empty increases even further with time, as illustrated in Fig B.1 for typical tandem and single-rotor helicopters. **It can be seen from this figure** that even at the first flight of these aircraft, the weight empty was as much as 5 percent higher than the initial predictions. This trend of weight empty increases with time and continues, after the first flight, at a rate of between 0.5 and 1.5 percent per year; due mainly to product improvement programs. For example, four years after the start of production qualification testing, the CH-47 empty weight was approximately 15 percent over the original predicted value.

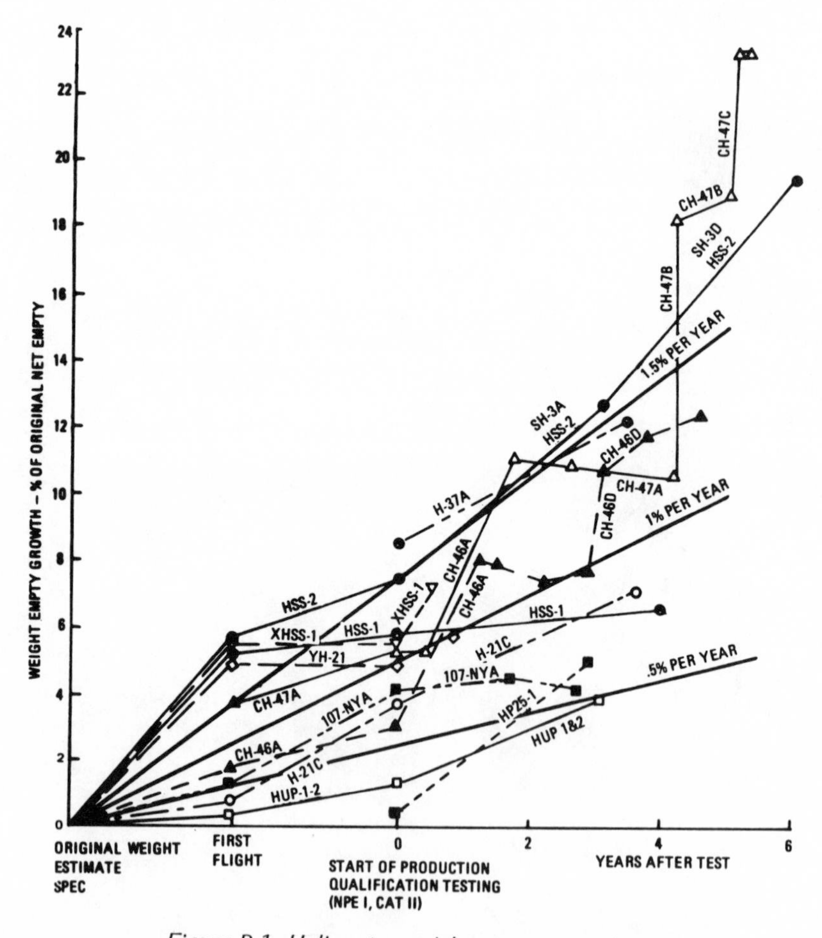

Figure B.1 Helicopter weight-empty growth trends

As a result of this trend, the payload-range capability will decrease, unless the increase in weight empty is counterbalanced by a suitable aircraft growth consisting of various design modifications leading to improvements in performance capability. The effect of increasing the weight empty on the payload/range relationship, assuming no aircraft growth – constant takeoff power available and power required – for the hypothetical single-rotor aircraft is shown in Fig B.2.

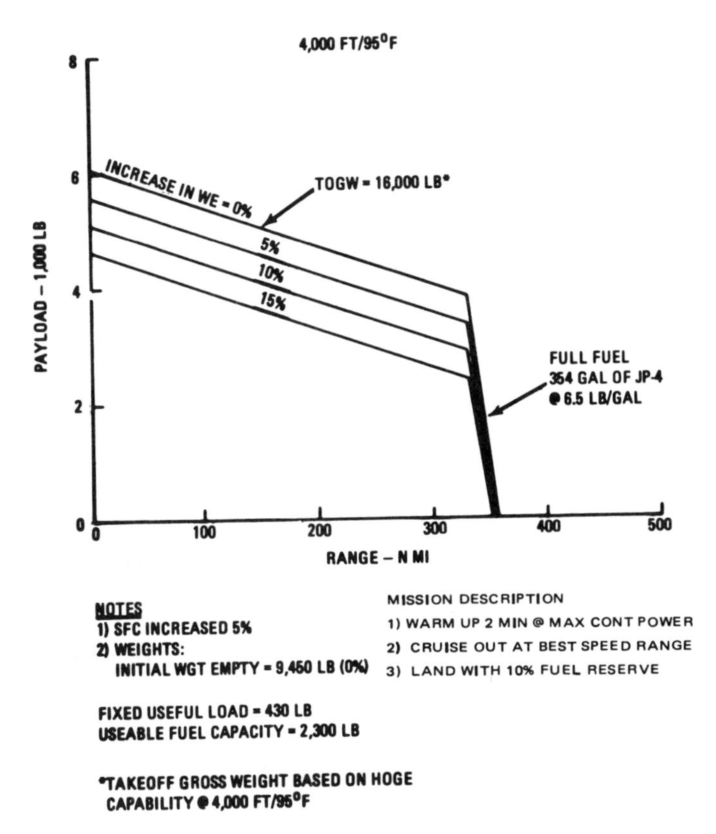

Figure B.2 Hypothetical helicopter payload/range capability

The takeoff gross weight of 16,000 lb noted in this figure is based on a hover OGE takeoff criteria at 4000 ft/95°F and 100 percent intermediate power. Weight empty increases of 5, 10, and 15 percent are indicated. The incremental increase in weight empty results in an identical reduction in payload capability for a fixed takeoff weight (*TOGW*), as defined in the following equation:

$$TOGW = W_E + PL + W_F + FUL$$

where

W_E = weight empty

PL = payload

W_F = weight of fuel

FUL = fixed useful load.

At the full-fuel point, the 5-percent increase in weight-empty causes a 12-percent reduction in maximum payload capability. The variation in range capability at constant payload is relatively small (2 n.mi., or 0.6 percent). This is due to the increase in average mission gross weight associated with increased weight empty.

The payload/range capability can be maintained in spite of increases in weight empty by increasing the takeoff gross weight, or growing the aircraft. This increase in operational gross weight can be achieved by either modification of operational techniques, or changes in design. The latter approach is usually associated with the concept of aircraft growth, as it consists of (1) increasing the power available, (2) modifying the rotor system to decrease the power required, (3) a combination of these items, and (4) selection of alternate takeoff criteria.

The design changes required to achieve higher operational gross weight result in secondary increases in weight empty (ΔW_{E_2}) and mission fuel (ΔW_F) associated with the higher weight. The secondary increases in weight empty are due to the following required design changes:

(1) Fuel tanks must be enlarged to accommodate the additional mission fuel required.

(2) Landing gear must be strengthened to maintain the design touchdown rate of descent at the higher weight.

(3) Rotor control system must be strengthened.

(4) Body structure must be strengthened to maintain the design load factor.

(5) Drive system must be strengthened because of increased power required.

(6) Hub and blades must be modified because of higher loads.

(7) Horizontal tail size must be increased to maintain acceptable stability level.

(8) Weight of engines if uprated.

(9) Tail boom weight increases if the main rotor radius is increased.

The secondary changes in weight empty (ΔW_{E_2}) due to these items can be estimated from trend curves presented as a function of gross weight. Therefore, the total change in gross weight (ΔGW) required to maintain constant payload-radius capability is as follows:

$$\Delta GW = \Delta W_{E_1} + \Delta W_{E_2} + \Delta W_F.$$

Dividing this expression by the initial increase in weight empty (ΔW_{E_1}) gives the growth factor (GF):

$$GF = \Delta GW/\Delta W_{E_1} = 1 + (\Delta W_{E_2}/\Delta W_{E_1}) + (\Delta W_F/\Delta W_{E_1}).$$

Typically, growth factors vary from 1.5 to 2.0. Assuming that $\Delta W_F/\Delta W_{E_1} \approx 0.1$ and $GF = 1.5$, then

$$\Delta W_{E_2}/\Delta W_{E_1} \approx 1/2.5.$$

Therefore, a 2.5-lb initial increase in weight empty results in a total weight empty rise of 3.5 lb ($\Delta W_{E_1} + \Delta W_{E_2}$), and the takeoff gross weight (*TOGW*) required to achieve constant performance is as follows:

$$TOGW = W_{E_0} + FUL + 1.5\Delta W_{E_1} + W_{F_0} + \Delta W_F + PL \tag{B 1}$$

where

W_{E_0} = original weight empty
W_{F_0} = initial fuel
ΔW_F = increase in mission fuel.

The effect of using Eq (B1) to predict the TOGW for 5, 10, and 15 percent increases in initial weight empty is shown in Fig B.3. A fixed mission requirement of 331 n.mi. range and a payload of 3820 lb was assumed for these calculations.

INITIAL INCREASE IN WE – %	TOTAL WEIGHT EMPTY – LB	FIXED USEFUL LOAD – LB	TOGW – LB	PAYLOAD – LB	FUEL* – LB
0	9,450	430	16,000	3,820	2,300
5	10,160	430	16,157	3,820	2,347
10	10,868	430	17,522	3,820	2,404
15	11,577	430	18,307	3,820	2,480

*FUEL FOR FIXED RANGE OF 331 N. MI

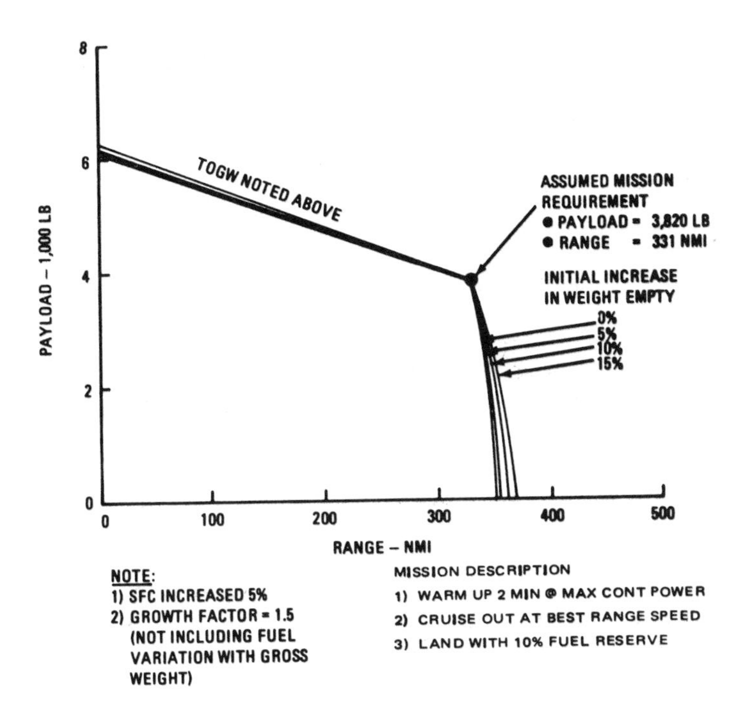

Figure B.3 Gross weight growth required to maintain a fixed payload range capability

Here, an increase of the initial weight empty by 5 percent requires a corresponding 5 percent increase in gross weight. The small variations in payload at zero range and the variation in range approaching zero payload (full-fuel) are due to the nonlinear effect of gross weight on mission fuel. The variation of mission fuel with weight used for sizing the fuel tanks is shown in Fig B.4.

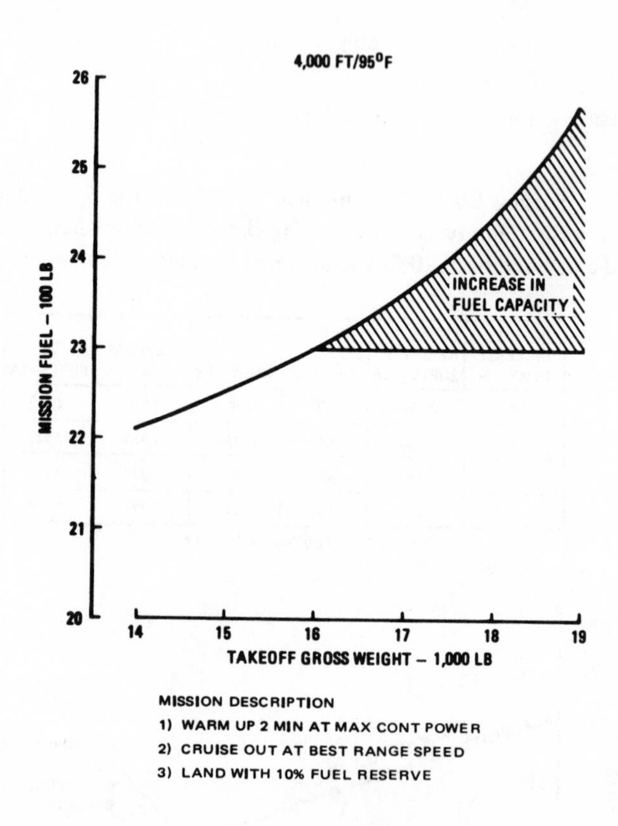

Figure B.4 Fuel required for 331 n.mi. range

In order to take off at the gross weights given in Fig B.3, an increase in engine power is necessary (Fig B.5). It is assumed that the takeoff weight is defined by hover OGE capability at 4000 ft/95°F and 100 percent intermediate power. Power levels required for constant payload as well as constant payload/range are indicated in this figure. It is noted that the percentage increase in power required is almost twice as large as the percentage rise in initial weight empty.

With respect to the powerplant, the question arises whether the required power ratings shown in Fig B.5 can be obtained through growth of the original engine and, if so, what would be the time and cost required to develop these ratings.

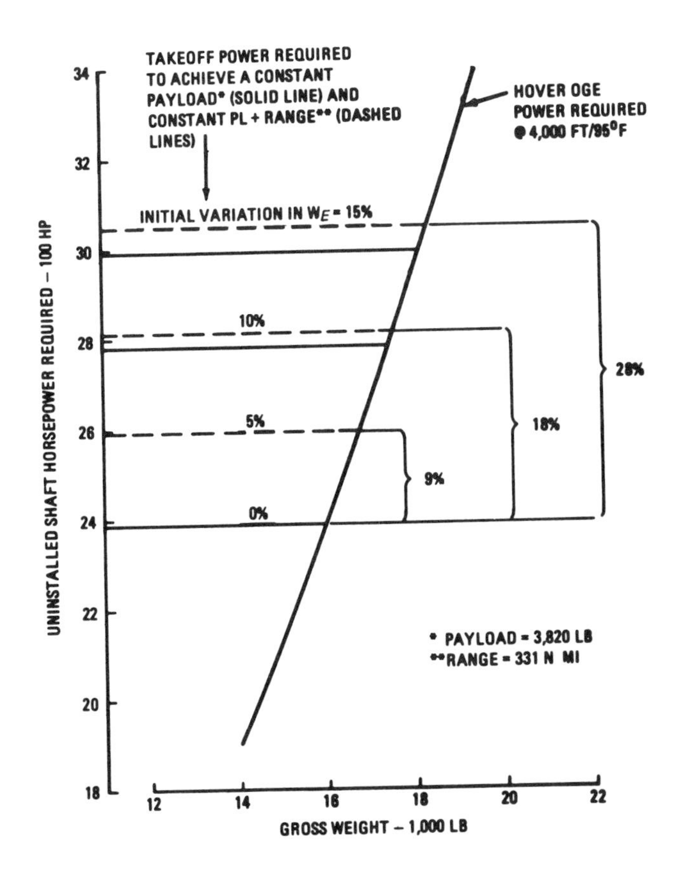

Figure B.5 Takeoff power required to achieve constant payload/range

Some typical production engine growth trends versus years after completion of the initial production testing (MQT) are presented in Fig B.6. Statistics indicate that the first 10 to 15 percent increase in engine power occurs 2 to 4 years after the initial MQT. This growth is generally achieved by simply increasing the turbine inlet temperature (TIT) with no significant engine modification. Extensive modification and redesign of the basic engine is usually required to achieve upratings beyond 10 to 15 percent. For example, to obtain engine growth beyond 10 to 15 percent but less than 50 percent, additional turbine blade cooling airflows or more sophisticated blade cooling concepts are needed, as well as the addition of another compressor stage. The addition of another turbine may be necessary to increase the power available beyond 50 percent.

It is apparent from Figs B.5 and B.6 that the engine modifications required to compensate for initial discrepancies in weight empty of less than 5 percent can be achieved with minimum engine redesign and within approximately 2 to 4 years of the initial MQT. The increase in engine performance required for larger weight-empty variations would call for major engine redesign. Because of the cost and time needed to develop these ratings, the following options are available: (1) redesign the rotor system to decrease

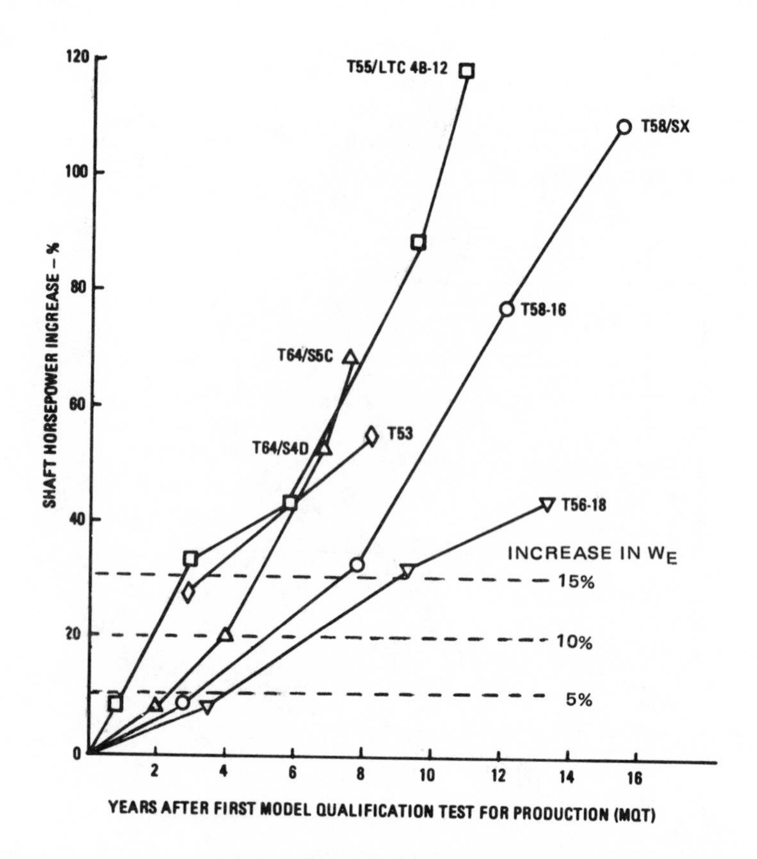

Figure B.6 Engine growth trends

takeoff power required, (2) consider alternate engines with higher ratings, or (3) select an alternate takeoff criteria. These approaches are discussed below.

The takeoff gross weight is usually defined by hover or vertical climb capabilities; therefore, in order to operate at an increased gross weight, the rotor system must be modified to achieve reduced hover power required. Since 70 to 80 percent of the single-rotor aircraft hover power required may be attributed to the main rotor induced power, the largest payoff in takeoff performance can be obtained by reducing this component. As noted in Ch II, the induced power is primarily a function of disc loading and twist. In principle, some gains can be achieved through improved design of the blade tips; but usually the improvements are too small to serve alone as a basis for rotorcraft growth. It should also be noted that the degree of twist is generally limited by forward-flight load considerations. Consequently, rotor radius remains as the most practical design variable.

Dimensionally moderate, but significant performance-wise increases in the main rotor radius can be obtained by adding a section to the tip or by installing a shank extension at the blade root. The tip extension is more desirable from a performance viewpoint since it avoids the small penalty associated with increased blade root cutout[1]. If the chord remains constant, the solidity decreases with increasing radius. The actual blade area, however, increases with radius, resulting in reduced rotor $\bar{c}_\ell$ at a given gross weight. In addition, the rotor rpm must be reduced if the original tip speed and tip Mach number are to be maintained, as assumed in subsequent sample calculations.

When contemplating an extension of the main rotor radius, the side-effects should be considered; for instance, the necessity for, and extent of, fuselage and rotor mast modifications to preserve the necessary clearances and, if the fuselage were extended, what influence would it have on aircraft balance, etc.

The effect of increasing the main rotor radius of the hypothetical single-rotor helicopter on hover OGE gross-weight capability at 4000 ft/95°F is shown in Fig B.7. Takeoff power levels of 100 and 110 percent (uprated engine) are indicated in this figure, while the tip speed and blade chord are assumed constant. The gross-weight capability appears to increase linearly with rotor radius at the rate of 500 lb/ft; however, extrapolation of this data to larger radii shows that thrust (T) developed at a constant power varies as follows: $T \sim R^{2/3}$.

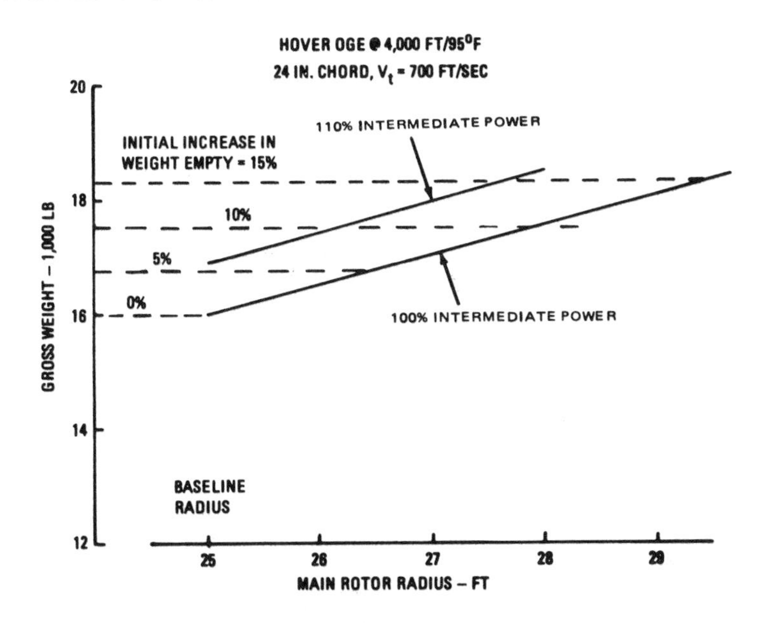

Figure B.7 Effect of radius on HOGE gross weight

The takeoff gross weights required to maintain constant payload and range capability are noted in Fig B.7 for initial empty weight variations of 5, 10, and 15 percent. A growth factor of 1.5 was assumed to account for the effect of radius change on weight empty. The gross weight values also reflect the influence of increasing radius on cruise fuel requirements. The takeoff weight associated with the 5-percent increase in weight empty can be achieved without increasing the radius, if the engines are uprated to 110 percent of initial rating. The larger deviations in empty weight of 10 and 15 percent require the 10 percent power uprating plus radius increases of 1.2 and 2.7 ft, respectively. An additional 2 ft of radius would be required if the engines were not uprated.

The rotor $\bar{c}_\ell$ or C_T/σ values associated with the gross weight variations in Fig B.7 are presented in Fig B.8. This figure shows that the rotor C_T/σ decreases slightly as the radius increases. This occurs because the blade area increases more rapidly than the hover gross weight. Therefore, increasing the radius will not reduce the aircraft airspeed envelope if it is defined by rotor stall considerations.

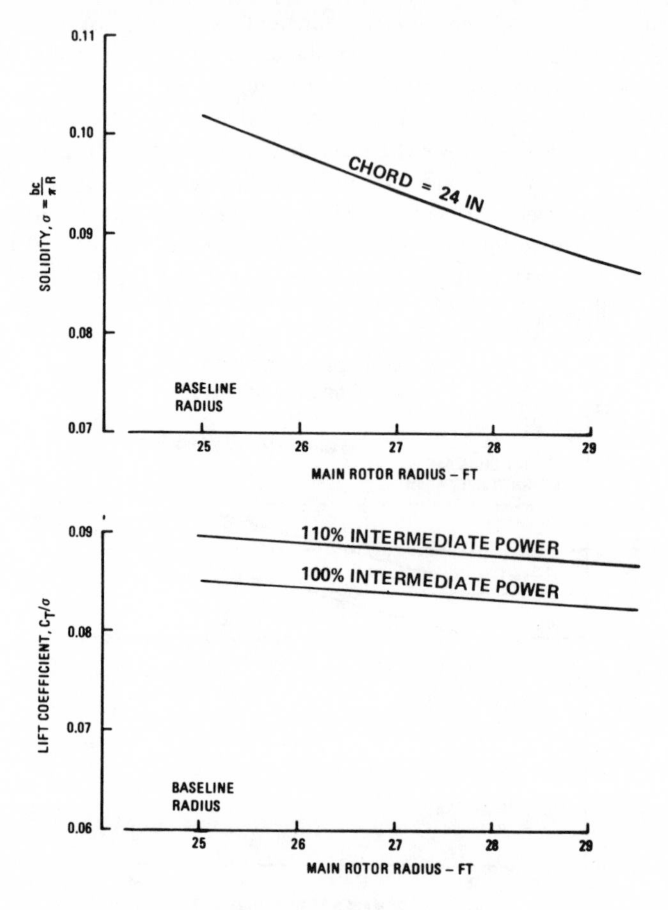

Figure B.8 Effect of rotor radius on solidity and C_T/σ (4000 ft/95°F)

The effect of varying the main rotor radius on hover performance is computed utilizing the nondimensional power required in Fig 2.6. The main rotor profile and induced power components are nondimensionalized separately because the solidity varies with radius (constant chord). The induced power is nondimensionalized in terms of C_P and C_T as it is primarily a function of the disc area, while the profile power must be defined in terms of C_T/σ and C_P/σ as it is a function of blade area. If the solidity rather than the chord is held constant, then the C_P/C_T data can be applied directly to obtain power required. The blade tip Mach number and tip speed are assumed constant in order to avoid the variation of compressibility power increments with radius. In order to simplify the calculations, tail rotor power, transmission and accessory losses, and engine installation losses are assumed to be a fixed percentage of the total SHP available.

Increasing the main rotor radius will require extending the tail boom to maintain clearance between the main and tail rotors. Redesign of the tail rotor may also be needed, due to increased power available and reduced main rotor rotational speed (rpm). The effect of these design changes on weight empty are included in the growth factor as described previously in this appendix. The growth factor of 1.5 however, does not include the variation in the fuel required with that of the rotor radius.

The effect of gross weight on mission fuel required for the 331 n.mi. mission radius is illustrated in Fig B.9. Here, it is noted that a 3-ft increase in the radius resulted in an 83-lb rise in cruise fuel for a 16,000-lb takeoff weight. This penalty is primarily due to the increase in profile power associated with the enlarged blade area. At constant chord and tip speed, the profile power is proportional to the radius, assuming that the average rotor $\bar{c}_d$ remains the same as before.

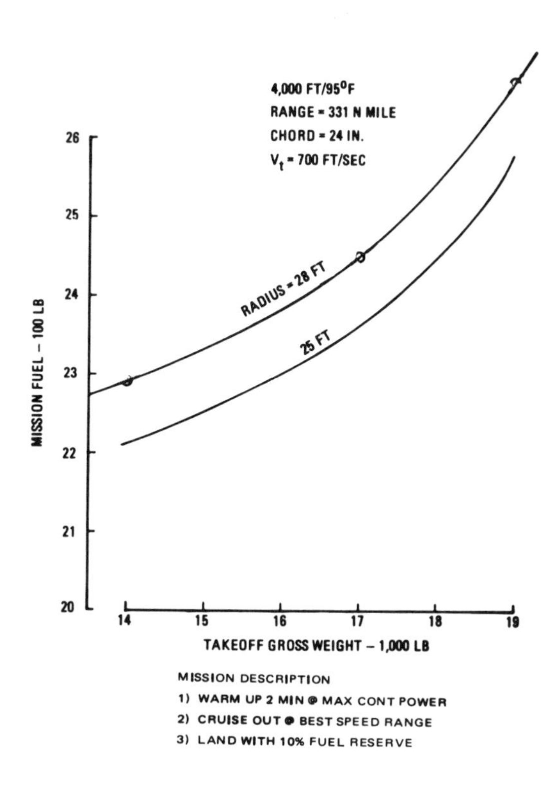

Figure B.9 Effect of rotor radius on mission fuel

Another means of increasing the takeoff gross weight without uprating the engines or modifying the rotor system design is to select an alternate takeoff criteria. For example, if IGE rather than OGE capability were used for the hypothetical helicopter mission, the increase in takeoff weight previously indicated in Fig B.3 could have been met. As noted in this figure, the maximum weight required is 18,307, while the takeoff gross weight capability at 4000 ft/95°F, 100 percent intermediate power IGE (5-ft wheel height) is 18,730 lb. However, the maximum weight limit of 18,000 lb would have to be increased.

The problem with using IGE criteia is that it may not provide a sufficient power margin for climb from a confined area such as a small clearing surrounded by tall trees. The takeoff distance required to clear a given object increases substantially if initiated at full takeoff power from an IGE wheel height because the benefits of ground disappear rapidly during transition. At speeds above 50 kn, there is no significant reduction in power required when operating in ground effect[2]. In fact, more performance capability than provided by hover OGE criteria is often specified to provide a sufficient operational margin. For instance, the current takeoff criteria for Army missions such as the UTTAS mission specify a 500-fpm vertical climb capability at 95 percent of the intermediate power.

References for Appendix B

1. Fradenburgh, E.A. *Aerodynamic Factors Influencing Overall Hover Performance.* Paper No. 7, AGARD—CPP-111, September 1972.

2. Cheeseman, I, and J.J.A. Bennett. *The Effect of the Ground on a Helicopter Rotor in Forward Flight.* Published in British Aerospace and Armament Experimental Establishment Report No. AAEE/2/88, July 1955.

INDEX

A CATALOG OF SELECTED

DOVER BOOKS

IN SCIENCE AND MATHEMATICS

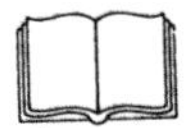